evolution equations, control theory, and biomathematics

PURE AND APPLIED MATHEMATICS

A Program of Monographs, Textbooks, and Lecture Notes

LECTURE NOTES IN PURE AND APPLIED MATHEMATICS

1. *N. Jacobson,* Exceptional Lie Algebras
2. *L.-Å. Lindahl and F. Poulsen,* Thin Sets in Harmonic Analysis
3. *I. Satake,* Classification Theory of Semi-Simple Algebraic Groups
4. *F. Hirzebruch, W. D. Newmann, and S. S. Koh, Differentiable Manifolds and Quadratic Forms*
5. *I. Chavel,* Riemannian Symmetric Spaces of Rank One
6. *R. B. Burckel,* Characterization of C(X) Among Its Subalgebras
7. *B. R. McDonald, A. R. Magid, and K. C. Smith,* Ring Theory: Proceedings of the Oklahoma Conference
8. *Y.-T. Siu,* Techniques of Extension on Analytic Objects
9. *S. R. Caradus, W. E. Pfaffenberger, and B. Yood,* Calkin Algebras and Algebras of Operators on Banach Spaces
10. *E. O. Roxin, P.-T. Liu, and R. L. Sternberg,* Differential Games and Control Theory
11. *M. Orzech and C. Small,* The Brauer Group of Commutative Rings
12. *S. Thomier,* Topology and Its Applications
13. *J. M. Lopez and K. A. Ross,* Sidon Sets
14. *W. W. Comfort and S. Negrepontis,* Continuous Pseudometrics
15. *K. McKennon and J. M. Robertson,* Locally Convex Spaces
16. *M. Carmeli and S. Malin,* Representations of the Rotation and Lorentz Groups: An Introduction
17. *G. B. Seligman,* Rational Methods in Lie Algebras
18. *D. G. de Figueiredo,* Functional Analysis: Proceedings of the Brazilian Mathematical Society Symposium
19. *L. Cesari, R. Kannan, and J. D. Schuur,* Nonlinear Functional Analysis and Differential Equations: Proceedings of the Michigan State University Conference
20. *J. J. Schäffer,* Geometry of Spheres in Normed Spaces
21. *K. Yano and M. Kon,* Anti-Invariant Submanifolds
22. *W. V. Vasconcelos,* The Rings of Dimension Two
23. *R. E. Chandler,* Hausdorff Compactifications
24. *S. P. Franklin and B. V. S. Thomas,* Topology: Proceedings of the Memphis State University Conference
25. *S. K. Jain,* Ring Theory: Proceedings of the Ohio University Conference
26. *B. R. McDonald and R. A. Morris,* Ring Theory II: Proceedings of the Second Oklahoma Conference
27. *R. B. Mura and A. Rhemtulla,* Orderable Groups
28. *J. R. Graef,* Stability of Dynamical Systems: Theory and Applications
29. *H.-C. Wang,* Homogeneous Branch Algebras
30. *E. O. Roxin, P.-T. Liu, and R. L. Sternberg,* Differential Games and Control Theory II
31. *R. D. Porter,* Introduction to Fibre Bundles
32. *M. Altman,* Contractors and Contractor Directions Theory and Applications
33. *J. S. Golan,* Decomposition and Dimension in Module Categories
34. *G. Fairweather,* Finite Element Galerkin Methods for Differential Equations
35. *J. D. Sally,* Numbers of Generators of Ideals in Local Rings
36. *S. S. Miller,* Complex Analysis: Proceedings of the S.U.N.Y. Brockport Conference
37. *R. Gordon,* Representation Theory of Algebras: Proceedings of the Philadelphia Conference
38. *M. Goto and F. D. Grosshans,* Semisimple Lie Algebras
39. *A. I. Arruda, N. C. A. da Costa, and R. Chuaqui,* Mathematical Logic: Proceedings of the First Brazilian Conference
40. *F. Van Oystaeyen,* Ring Theory: Proceedings of the 1977 Antwerp Conference
41. *F. Van Oystaeyen and A. Verschoren,* Reflectors and Localization: Application to Sheaf Theory
42. *M. Satyanarayana,* Positively Ordered Semigroups
43. *D. L Russell,* Mathematics of Finite-Dimensional Control Systems
44. *P.-T. Liu and E. Roxin,* Differential Games and Control Theory III: Proceedings of the Third Kingston Conference, Part A
45. *A. Geramita and J. Seberry,* Orthogonal Designs: Quadratic Forms and Hadamard Matrices
46. *J. Cigler, V. Losert, and P. Michor,* Banach Modules and Functors on Categories of Banach Spaces

47. *P.-T. Liu and J. G. Sutinen,* Control Theory in Mathematical Economics: Proceedings of the Third Kingston Conference, Part B
48. *C. Byrnes,* Partial Differential Equations and Geometry
49. *G. Klambauer,* Problems and Propositions in Analysis
50. *J. Knopfmacher,* Analytic Arithmetic of Algebraic Function Fields
51. *F. Van Oystaeyen,* Ring Theory: Proceedings of the 1978 Antwerp Conference
52. *B. Kadem,* Binary Time Series
53. *J. Barros-Neto and R. A. Artino,* Hypoelliptic Boundary-Value Problems
54. *R. L. Sternberg, A. J. Kalinowski, and J. S. Papadakis,* Nonlinear Partial Differential Equations in Engineering and Applied Science
55. *B. R. McDonald,* Ring Theory and Algebra III: Proceedngs of the Third Oklahoma Conference
56. *J. S. Golan,* Structure Sheaves Over a Noncommutative Ring
57. *T. V. Narayana, J. G. Williams, and R. M. Mathsen,* Combinatorics, Representation Theory and Statistical Methods in Groups: YOUNG DAY Proceedings
58. *T. A. Burton,* Modeling and Differential Equations in Biology
59. *K. H. Kim and F. W. Roush,* Introduction to Mathematical Consensus Theory
60. *J. Banas and K. Goebel,* Measures of Noncompactness in Banach Spaces
61. *O. A. Nielson,* Direct Integral Theory
62. *J. E. Smith, G. O. Kenny, and R. N. Ball,* Ordered Groups: Proceedings of the Boise State Conference
63. *J. Cronin,* Mathematics of Cell Electrophysiology
64. *J. W. Brewer,* Power Series Over Commutative Rings
65. *P. K. Kamthan and M. Gupta,* Sequence Spaces and Series
66. *T. G. McLaughlin,* Regressive Sets and the Theory of Isols
67. *T. L. Herdman, S. M. Rankin III, and H. W. Stech,* Integral and Functional Differential Equations
68. *R. Draper,* Commutative Algebra: Analytic Methods
69. *W. G. McKay and J. Patera,* Tables of Dimensions, Indices, and Branching Rules for Representations of Simple Lie Algebras
70. *R. L. Devaney and Z. H. Nitecki,* Classical Mechanics and Dynamical Systems
71. *J. Van Geel,* Places and Valuations in Noncommutative Ring Theory
72. *C. Faith,* Injective Modules and Injective Quotient Rings
73. *A. Fiacco,* Mathematical Programming with Data Perturbations I
74. *P. Schultz, C. Praeger, and R. Sullivan,* Algebraic Structures and Applications: Proceedings of the First Western Australian Conference on Algebra
75. *L Bican, T. Kepka, and P. Nemec,* Rings, Modules, and Preradicals
76. *D. C. Kay and M. Breen,* Convexity and Related Combinatorial Geometry: Proceedings of the Second University of Oklahoma Conference
77. *P. Fletcher and W. F. Lindgren,* Quasi-Uniform Spaces
78. *C.-C. Yang,* Factorization Theory of Meromorphic Functions
79. *O. Taussky,* Ternary Quadratic Forms and Norms
80. *S. P. Singh and J. H. Burry,* Nonlinear Analysis and Applications
81. *K. B. Hannsgen, T. L. Herdman, H. W. Stech, and R. L. Wheeler,* Volterra and Functional Differential Equations
82. *N. L. Johnson, M. J. Kallaher, and C. T. Long,* Finite Geometries: Proceedings of a Conference in Honor of T. G. Ostrom
83. *G. I. Zapata,* Functional Analysis, Holomorphy, and Approximation Theory
84. *S. Greco and G. Valla,* Commutative Algebra: Proceedings of the Trento Conference
85. *A. V. Fiacco,* Mathematical Programming with Data Perturbations II
86. *J.-B. Hiriart-Urruty, W. Oettli, and J. Stoer,* Optimization: Theory and Algorithms
87. *A. Figa Talamanca and M. A. Picardello,* Harmonic Analysis on Free Groups
88. *M. Harada,* Factor Categories with Applications to Direct Decomposition of Modules
89. *V. I. Istrăţescu,* Strict Convexity and Complex Strict Convexity
90. *V. Lakshmikantham,* Trends in Theory and Practice of Nonlinear Differential Equations
91. *H. L. Manocha and J. B. Srivastava,* Algebra and Its Applications
92. *D. V. Chudnovsky and G. V. Chudnovsky,* Classical and Quantum Models and Arithmetic Problems
93. *J. W. Longley,* Least Squares Computations Using Orthogonalization Methods
94. *L. P. de Alcantara,* Mathematical Logic and Formal Systems
95. *C. E. Aull,* Rings of Continuous Functions
96. *R. Chuaqui,* Analysis, Geometry, and Probability
97. *L. Fuchs and L. Salce,* Modules Over Valuation Domains

98. *P. Fischer and W. R. Smith,* Chaos, Fractals, and Dynamics
99. *W. B. Powell and C. Tsinakis,* Ordered Algebraic Structures
100. *G. M. Rassias and T. M. Rassias,* Differential Geometry, Calculus of Variations, and Their Applications
101. *R.-E. Hoffmann and K. H. Hofmann,* Continuous Lattices and Their Applications
102. *J. H. Lightbourne III and S. M. Rankin III,* Physical Mathematics and Nonlinear Partial Differential Equations
103. *C. A. Baker and L, M. Batten,* Finite Geometrics
104. *J. W. Brewer, J. W. Bunce, and F. S. Van Vleck,* Linear Systems Over Commutative Rings
105. *C. McCrory and T. Shifrin,* Geometry and Topology: Manifolds, Varieties, and Knots
106. *D. W. Kueker, E. G. K. Lopez-Escobar, and C. H. Smith,* Mathematical Logic and Theoretical Computer Science
107. *B.-L. Lin and S. Simons,* Nonlinear and Convex Analysis: Proceedings in Honor of Ky Fan
108. *S. J. Lee,* Operator Methods for Optimal Control Problems
109. *V. Lakshmikantham,* Nonlinear Analysis and Applications
110. *S. F. McCormick,* Multigrid Methods: Theory, Applications, and Supercomputing
111. *M. C. Tangora,* Computers in Algebra
112. *D. V. Chudnovsky and G. V. Chudnovsky,* Search Theory: Some Recent Developments
113. *D. V. Chudnovsky and R. D. Jenks,* Computer Algebra
114. *M. C. Tangora,* Computers in Geometry and Topology
115. *P. Nelson, V. Faber, T. A. Manteuffel, D. L. Seth, and A. B. White, Jr.,* Transport Theory, Invariant Imbedding, and Integral Equations: Proceedings in Honor of G. M. Wing's 65th Birthday
116. *P. Clément, S. Invernizzi, E. Mitidieri, and I. I. Vrabie,* Semigroup Theory and Applications
117. *J. Vinuesa,* Orthogonal Polynomials and Their Applications: Proceedings of the International Congress
118. *C. M. Dafermos, G. Ladas, and G. Papanicolaou,* Differential Equations: Proceedings of the EQUADIFF Conference
119. *E. O. Roxin,* Modern Optimal Control: A Conference in Honor of Solomon Lefschetz and Joseph P. Lasalle
120. *J. C. Díaz,* Mathematics for Large Scale Computing
121. *P. S. Milojević,* Nonlinear Functional Analysis
122. *C. Sadosky,* Analysis and Partial Differential Equations: A Collection of Papers Dedicated to Mischa Cotlar
123. *R. M. Shortt,* General Topology and Applications: Proceedings of the 1988 Northeast Conference
124. *R. Wong,* Asymptotic and Computational Analysis: Conference in Honor of Frank W. J. Olver's 65th Birthday
125. *D. V. Chudnovsky and R. D. Jenks,* Computers in Mathematics
126. *W. D. Wallis, H. Shen, W. Wei, and L. Zhu,* Combinatorial Designs and Applications
127. *S. Elaydi,* Differential Equations: Stability and Control
128. *G. Chen, E. B. Lee, W. Littman, and L. Markus,* Distributed Parameter Control Systems: New Trends and Applications
129. *W. N. Everitt,* Inequalities: Fifty Years On from Hardy, Littlewood and Pólya
130. *H. G. Kaper and M. Garbey,* Asymptotic Analysis and the Numerical Solution of Partial Differential Equations
131. *O. Arino, D. E. Axelrod, and M. Kimmel,* Mathematical Population Dynamics: Proceedings of the Second International Conference
132. *S. Coen,* Geometry and Complex Variables
133. *J. A. Goldstein, F. Kappel, and W. Schappacher,* Differential Equations with Applications in Biology, Physics, and Engineering
134. *S. J. Andima, R. Kopperman, P. R. Misra, J. Z. Reichman, and A. R. Todd,* General Topology and Applications
135. *P Clément, E. Mitidieri, B. de Pagter,* Semigroup Theory and Evolution Equations: The Second International Conference
136. *K. Jarosz,* Function Spaces
137. *J. M. Bayod, N. De Grande-De Kimpe, and J. Martínez-Maurica,* *p*-adic Functional Analysis
138. *G. A. Anastassiou,* Approximation Theory: Proceedings of the Sixth Southeastern Approximation Theorists Annual Conference
139. *R. S. Rees,* Graphs, Matrices, and Designs: Festschrift in Honor of Norman J. Pullman
140. *G. Abrams, J. Haefner, and K. M. Rangaswamy,* Methods in Module Theory

141. *G. L. Mullen and P. J.-S. Shiue*, Finite Fields, Coding Theory, and Advances in Communications and Computing
142. *M. C. Joshi and A. V. Balakrishnan*, Mathematical Theory of Control: Proceedings of the International Conference
143. *G. Komatsu and Y. Sakane*, Complex Geometry: Proceedings of the Osaka International Conference
144. *I. J. Bakelman*, Geometric Analysis and Nonlinear Partial Differential Equations
145. *T. Mabuchi and S. Mukai*, Einstein Metrics and Yang–Mills Connections: Proceedings of the 27th Taniguchi International Symposium
146. *L. Fuchs and R. Göbel*, Abelian Groups: Proceedings of the 1991 Curaçao Conference
147. *A. D. Pollington and W. Moran*, Number Theory with an Emphasis on the Markoff Spectrum
148. *G. Dore, A. Favini, E. Obrecht, and A. Venni*, Differential Equations in Banach Spaces
149. *T. West*, Continuum Theory and Dynamical Systems
150. *K. D. Bierstedt, A. Pietsch, W. Ruess, and D. Vogt*, Functional Analysis
151. *K. G. Fischer, P. Loustaunau, J. Shapiro, E. L. Green, and D. Farkas*, Computational Algebra
152. *K. D. Elworthy, W. N. Everitt, and E. B. Lee*, Differential Equations, Dynamical Systems, and Control Science
153. *P.-J. Cahen, D. L. Costa, M. Fontana, and S.-E. Kabbaj*, Commutative Ring Theory
154. *S. C. Cooper and W. J. Thron*, Continued Fractions and Orthogonal Functions: Theory and Applications
155. *P. Clément and G. Lumer*, Evolution Equations, Control Theory, and Biomathematics

Additional Volumes in Preparation

evolution equations, control theory, and biomathematics

proceedings of the Han-sur-Lesse conference

edited by
Philippe Clément
Delft University of Technology
Delft, The Netherlands

Günter Lumer
University of Mons-Hainaut
Mons, Belgium

Marcel Dekker, Inc. **New York • Basel • Hong Kong**

Library of Congress Cataloging-in-Publication Data

Evolution equations, control theory, and biomathematics: proceedings of the Han-sur-Lesse conference / edited by Philippe Clément, Günter Lumer.
p. cm. -- (Lecture notes in pure and applied mathematics; 155)
Proceedings of the Third International Workshop—Conference on Evolution Equations, Control Theory, and Biomathematics, held Oct. 20–26, 1991, at the Han-sur-Lesse Conference Center of the Belgian Ministry of Education.
Includes bibliographical references.
ISBN 0-8247-8885-0 (acid-free paper)
1. Evolution equations--Congresses. 2. Control theory--Congress 3. Biomathematics--Congresses. I. Clément, Philippe. II. Lumer, Günter. III. International Workshop—Conference on Evolution Equations, Control Theory, and Biomathematics (3rd: 1991: Han-sur-Lesse Conference Center of the Belgian Ministry of Education). IV. Series: Lecture notes in pure and applied mathematics; v. 155.
QA377.E96 1993
515'.353--dc20 93-32127
CIP

The publisher offers discounts on this book when ordered in bulk quantities. For more information, write to Special Sales/Professional Marketing at the address below.

This book is printed on acid-free paper.

MARCEL DEKKER, INC.
270 Madison Avenue, New York, New York 10016

Current printing (last digit):
10 9 8 7 6 5 4 3 2 1

PRINTED IN THE UNITED STATES OF AMERICA

Preface

The Third International Workshop-Conference on Evolution Equations, Control Theory and Biomathematics was held at the Han-sur-Lesse Conference Center of the Belgian Ministry of Education. This conference brought together a particularly distinguished and representative group of researchers, including several French and Russian academicians and a Fields medalist, and indeed many authorities in all the areas covered by this volume. Among the topics discussed were recent developments in evolution equations related to physical, engineering, and biological applications, population models, epidemic models, diffusion models with shocks, control theory, optimal control, the Riccati equation, the Hamilton–Jacobi equation, the Korteweg–de Vries equation, asymptotics of spectral functions, viscosity solutions, the Laplace transform method, perturbation methods, stochastic analysis, and diffusion problems.

This Workshop-Conference was financed to a large extent by the European Community Science Plan Project "Evolutionary Systems." Besides the European Community, important financial support was also provided by

Belgian National Fund for Scientific Research (F.N.R.S.)
Ministère de la Communauté Française de Belgique
University of Mons
University of Antwerp (U.I.A.)
University of Delft
University of Lille
ALPHAGAZ Division of "Air Liquide Belge" (Liège)
The Belgian Caisse Générale d'Epargne et Retraite (C.G.E.R.)

The Organizing Committee (J. van Casteren and the editors) would like to express their warm thanks to the staff of the Conference Center and, in particular, to its manager, Mr. Herman, for their assistance with the arrangements at the Center and for their marvelous hospitality. We are also grateful to the people in charge of the nearby R.T.T. Satellite Communication Center for providing us with a two-hundred-seat lecture room for the general sessions. The editors would further like to thank their collaborators, S. Nicaise, L. Paquet, and J. van Casteren, for their help with many tasks that arose during the meeting, as well as the secretaries in both Mons and Delft, Lyane Bouchez and Tini Nienhuis, for their assistance during the organization of the meeting and the preparation of the Proceedings.

Finally, we thank the contributors, the referees, and Marcel Dekker, Inc., especially Ms. Maria Allegra, for their cooperation in the production of this volume.

Philippe Clément
Günter Lumer

Contents

Contributors

I. ANTONIOU University of Brussels, Brussels, Belgium

J. von BELOW Tübingen University, Tübingen, Germany

IMMANUEL M. BOMZE University of Vienna, Vienna, Austria

M. E. BRADLEY University of Louisville, Louisville, Kentucky

REINHARD BÜRGER University of Vienna, Vienna, Austria

OVIDIU CÂRJĂ University of Iaşi, Romania

J. A. van CASTEREN University of Antwerp, Antwerp, Belgium

IOANA CIORANESCU University of Puerto Rico, Rio Piedras, Puerto Rico

M. G. CRANDALL University of California, Santa Barbara, California

GIUSEPPE DA PRATO Scuola Normale Superiore, Pisa, Italy

RALPH deLAUBENFELS Ohio University, Athens, Ohio

M. DEMUTH University of Potsdam, Potsdam, Germany

O. DIEKMANN CWI Amsterdam, The Netherlands, and Institute of Theoretical Biology, University of Leiden, Leiden, The Netherlands

J. R. DORROH Louisiana State University, Baton Rouge, Louisiana

E. V. EMEL'ANOVA Pedagogical Institute, Belgorod, Russia

JOACHIM ESCHER University of Zürich, Zürich, Switzerland

H. O. FATTORINI University of California, Los Angeles, California

A. FAVINI University of Bologna, Bologna, Italy

MARCO FUHRMAN Polytechnico di Milano, Milan, Italy

GISÈLE RUIZ GOLDSTEIN Louisiana State University, Baton Rouge, Louisiana

JEROME A. GOLDSTEIN Louisiana State University, Baton Rouge, Louisiana

P. GRISVARD University of Nice, Nice, France

MATS GYLLENBERG Luleå University of Technology, Luleå, Sweden

K. P. HADELER Tübingen University, Tübingen, Germany

HERNÁN R. HENRÍQUEZ University of Santiago, Santiago, Chile

MATTHIAS HIEBER University of Zürich, Zürich, Switzerland

THOMAS HILLEN Tübingen University, Tübingen, Germany

ALBRECHT HOLDERRIETH Tübingen University, Tübingen, Germany

SENZHONG HUANG Tübingen University, Tübingen, Germany, and Nankai University, Tianjin, China

MIMMO IANNELLI University of Trento, Trento, Italy

N. KUTEV Bulgarian Academy of Sciences, Sofia, Bulgaria, and University of Heidelberg, Heidelberg, Germany

HORST LANGE University of Cologne, Cologne, Germany

I. LASIECKA University of Virginia, Charlottesville, Virginia

B. M. LEVITAN University of Minnesota, Minneapolis, Minnesota

P. L. LIONS University of Paris-Dauphine, Paris, France

RODICA LUCA Polytechnic Institute of Iaşi, Iaşi, Romania

GÜNTER LUMER University of Mons, Mons, Belgium

V. P. MASLOV Russian Academy of Sciences, Moscow, Russia

MIKLAVŽ MASTINŠEK University of Maribor, Maribor, Slovenia

C. McMILLAN University of Virginia, Charlottesville, Virginia

J. A. J. METZ Leiden University, Leiden, The Netherlands

GHEORGE MOROŞANU University of Iaşi, Iaşi, Romania

B. NAJMAN University of Zagreb, Zagreb, Croatia

P. A. NAUMKIN Moscow State University, Moscow, Russia

FRANK NEUBRANDER Louisiana State University, Baton Rouge, Louisiana

SERGE NICAISE University of Valenciennes, Valenciennes, France

LUC PAQUET University of Mons, Mons, Belgium

S. PISKAREV Luleå University of Technology, Luleå, Sweden

J. POPIOŁEK Warsaw University (Bialystok), Bialystok, Poland

J. PRÜSS University of Paderborn, Paderborn, Germany

F. RÄBIGER Tübingen University, Tübingen, Germany

I. A. SHISHMAREV Moscow State University, Moscow, Russia

GIERI SIMONETT University of Zürich, Zürich, Switzerland

EUGENIO SINESTRARI University of Roma "La Sapienza," Rome, Italy

P. E. SOBOLEVSKII Hebrew University of Jerusalem, Jerusalem, Israel

FRANCISZEK HUGON SZAFRANIEC Jagiellonian University, Krakow, Poland

S. TASAKI International Solvay Institute for Physics and Chemistry, Brussels, Belgium

HORST THIEME Arizona State University, Tempe, Arizona

RENÉ THOM Institut des Hautes Etudes Scientifiques, Bures-sur-Yvette, France

R. TRIGGIANI University of Virginia, Charlottesville, Virginia

CORNELIU URSESCU Romanian Academy, Iaşi, Romania

V. VASIL'EV Pedagogical Institute, Belgorod, Russia

WOLF von WAHL University of Bayreuth, Bayreuth, Germany

A. YAGI Himeji Institute of Technology, Himeji, Hyogo, Japan

Conference Program

- **S. Albeverio** : Dirichletforms, associated semigroups and processes.
- **H. Amann** : Parameter estimation in quasilinear parabolic systems.
- **W. Arendt**: Absorption semigroups and Dirichlet boundary conditions.
- **C. Batty**: Spectral properties of commuting semigroups.
- **J. von Below**: Network diffusion problems with dynamical node conditions.
- **Ph. Bénilan**: BV solutions of nonlinear degenerate parabolic BVP.
- **M. Böhm**: Complex interpolation of nonlinear operators and applications.
- **F. Bucci**: Some results on boundary control problems for damped equations.
- **R. Bürger**: A perturbation problem in the space of finite Borel measures.
- **A. Calsina**: Dynamics of a prey-predator system with mutation.
- **O. Carja**: The Bellman equation for the time optimal control problem.
- **W. Caspers**: Point interactions in L^p.
- **Ph. Clément**: Abstract parabolic quasilinear equations and application to a groundwater flow problem.
- **M.C. Crandall**: Hamilton-Jacobi equations in Hilbert spaces.
- **D. Daners**: Qualitative behaviour of an epidemics model.
- **G. Da Prato**: Some results on operator Riccati equation.
- **R. deLaubenfels**: Simultaneous well-posedness.

- **M. Demuth**: On stochastic spectral analysis for generalized Schrödinger operators.
- **B. de Pagter**: The Banach space valued Hilbert transform.
- **W. Desch**: Asymptotic decomposition of the spectrum in viscoelastic problems.
- **O. Diekmann**: Structured population models, a challenge for semigroup theory.
- **K.J. Engel**: Second order Cauchy problems.
- **J. Escher**: On quasilinear parabolic problems.
- **H.O. Fattorini**: Relaxed controls in infinite demensional systems: Finitely additive Young measures.
- **A. Favini**: On second order implicit differential equations in Banach spaces.
- **J. Goldstein**: Regularity and decay for solutions of nonlinear parabolic problems.
- **J.P. Gossez**: Remarks on unique continuation.
- **P. Grisvard**: Exact controllability of the elastic wave equation.
- **D. Guidetti**: Petrovsky parabolic problems of higher order in time.
- **M. Gyllenberg**: Cumulative output, step response and perturbation of semigroups.
- **K.P. Hadeler**: Differential equations on branched manifolds.
- **H.R. Henriquez**: Regulator problem for linear distributed control systems with delays in outputs.
- **P. Hess**: Generic convergence in discrete monotone dynamical systems.

- **M. Hieber**: Integrated semigroups and systems of partial differential equations.
- **A. Holderrieth**: Multiplicative perturbations of generators of analytic semigroups.
- **N. Huu Duc**: Tichonov's evolution equations.
- **K.M. Hui**: Non negative solutions of the fast diffusion equation with strong reaction term.
- **A. Ivanov**: Global stability and periodic solutions of differential systems with delayed feedback.
- **W. Jäger**: Nonlinear diffusion systems and chemotaxis.
- **N. Kutev**: Blow up of the solutions for a class of nonlinear parabolic equations.
- **H. Lange**: Coupled nonlinear Schrödinger-Heat equations.
- **M. Langlais**: Periodic solutions for epidemic models with age dependance and spatial structure.
- **I. Lasiecka**: Uniform decay rates for the solutions to nonlinear von Karman plate equations.
- **R. Lefever**: Non-linear diffusion models of cellular tissue growth and rejection.
- **G. Lumer**: Diffusion models with abrupt changes of boundary conditions and discontinuous sources. Asymptotic behaviour under periodic heat shocks.
- **W.A.J. Luxemburg**: Generalized initial conditions associated with evolution equations.
- **P. Malliavin**: Equations différentielles stochastiques comme limites d'équations différentielles ordinaires rapidement oscillantes.
- **V. Maslov**: Cauchy problems for Hamilton-Jacobi and Bellman equations via linearization using a new arithmetic.

- **M. Mastinsek**: Dual semigroups and functional differential equations.
- **J.M. Mazon**: A nonlinear version of Hunt theorem.
- **S. Mejza**: Analysis of diallel table experiment carried out in block designs-mixed model.
- **E. Mitidieri**: Blow up results for a class of parabolic systems.
- **M. Mokhtar-Kharroubi**: Limiting absorption principles and wave operators on $L^1(\mu)$ spaces.
- **D. Motreanu**: Critical point methods in nonlinear partial differential equations.
- **R. Nagel**: Operator matrices: the state of the art.
- **B. Najman**: Singularly perturbed evolution equations.
- **P.I. Naumkin**: Large time asymptotic behaviour of solutions to the system of equations for surface waves.
- **F. Neubrander**: The Laplace transform and semigroups.
- **S. Nicaise**: Evolution equations in Hilbert spaces and stable asymptotics for PDE's in cones.
- **S. Oharu**: Semigroup theory for semilinear evolution equations.
- **L. Paquet**: Skeel condition number for operators on $C^o(\Omega)$ and applications.
- **V.Q. Phong**: Almost periodic attractor of abstract dynamical systems.
- **M. Pierre**: Stabilization of second order evolution problems by nonlinear feedback.
- **J. Popiolek**: Parabolic equation of third order.
- **J. Prüss**: Solvability behavior of linear evolutionary integral equations on the line.

- **A. Pugliese**: Multigroup models for the dynamics of HIV/AIDS.
- **F. Raebiger**: Stability and ergodicity of dominated semigroups.
- **G.R. Rieder**: Existence and regularity of solutions of singular quasilinear equations.
- **W. Ruess**: The evolution operator approach to FDE's with delay.
- **S. Sforza**: Some results for a nonlinear integro-differential equation.
- **I.A. Shishmarev**: The decay of the step for the Korteweg - de Vries - Burgers equation.
- **G. Simonett**: On center manifolds for quasilinear parabolic systems.
- **E. Sinestrari**: On the Hille-Yosida operators.
- **P.E. Sobolevskii**: Commutant method and coercitivity of Cauchy problem.
- **W.H Summers**: Operator semigroup for functional differential equations with delay.
- **F. Szafraniec**: On semigroups of unbounded operators.
- **B. Tang**: Effect of mobility on competitive coexistence of microbial communities. A bifurcation analysis.
- **S. Tasaki**: Complex spectrum in evolution equations.
- **H.R. Thieme**: Quasilinear evolution equations modeling structured population dynamics.
- **R. Thom**: From biochemistry to biological meaning.
- **R. Triggiani**: Optimal control problem for boundary control systems and Riccati equations.
- **J. Van Casteren**: On stochastic spectral theory.
- **J. Van Neerven**: The sun-dual of C_0-semigroup is either very large or very small.

- **J.L. Vasquez**: Asymptotic behavior for nonlinear parabolic equations: occurence of anomalous exponents.
- **W. von Wahl**: Regularity of weak solution of some nonlinear systems of PDE.
- **M. Voicu**: Integrated semigroups on locally convex lattices.
- **L. Weis**: The Laplace transform and L^p-solutions of abstract Cauchy problems.
- **S. Zaidman**: The Cauchy problem for some singular operator differential equations.

Conference Participants

E. Alarabiou, Equipe de Mathématiques, Université de Franche-Comté, Route de Gray, F-25030 Besançon Cédex, France.

S. Albeverio, Mathematisches Institut, Ruhr Universität, Universitätstrasse 150, Postfach 102148, D-4630 Bochum, Germany.

H. Amann, Mathematisches Institut, Universität Zürich, Rämistrasse 74, CH-8001 Zürich, Switzerland.

F. Andreu, Departamento Analisis Matematico, Universitad de Valencia, Dr Moliner 50, 46100 Burjassot, Spain.

W. Arendt, Equipe de Mathématiques, Université de Franche-Comté, Route de Gray, F-25030 Besançon Cédex, France.

M. Barcelo Conesa, ETSEIB, Departament de Matematica, Aplicada I, Dagoual 647, 08028 Barcelona, Spain.

L. Barthélemy, Equipe de Mathématiques, Université de Franche-Comté, Route de Gray, F-25030 Besançon Cédex, France.

C.J.K. Batty, St. John's College, Oxford University, Oxford OX1 3JP, United Kingdom.

J. von Below, Lehrstuhl für Biomathematik der Universität Tübingen, Auf der Morgenstelle 10, D-W-7400 Tübingen 1, Germany.

Ph. Bénilan, Equipe de Mathématiques, Université de Franche-Comté, Route de Gray, F-25030 Besançon Cédex, France.

M. Bitteslin, Mathematisches Institut der Universität Zürich, Rämistrasse 74, 8001 Zürich, Switzerland.

M. Böhm, Humboldt Universität zu Berlin, Sektion Mathematik, PSF 1297, Berlin O-1086, Germany.

F. Bucci, Dipartimento di Matematica Pura e Applicata, Via Campi 213/B, I-41100 Modena, Italy.

R. Bürger, Institut für Mathematik, Universität Wien, Strudlhofgasse 4, A-1090 Wien, Austria.

G. Caristi, Universita di Udine, Dipartimento di Matematica e Informatica, Via Zanon 6, I-33100 Udine, Italy.

O. Carja, Faculty of Mathematics, University of Iasi, Iasi-6600, Roumanie.

W. Caspers, Technische Universiteit Delft, Faculteit TWI, Mekelweg 4, 2628 CD Delft, The Netherlands.

I. Cioranescu, Mathematics Department Ciencias Naturales, Rio Piedras, Universidad de Puerto Rico, Porto Rico.

Ph. Clément, Technische Universiteit Delft, Faculteit TWI, Mekelweg 4, 2628 CD Delft, The Netherlands.

M.G. Crandall, Department of Mathematics, University of California, Santa Barbara, CA 93106, U.S.A.

D. Daners, Mathematisches Institut der Universität Zürich, Rämistrasse 74, 8001 Zürich, Switzerland.

G. Da Prato, Scuola Normale Superiore, Piazza dei Cavalieri 7, 56199 Pisa, Italy.

R. deLaubenfels, Ohio University, Athens, Ohio 45701, U.S.A.

M. Demuth, Karl Weierstrass Institut für Mathematik, Mohrenstrasse 39, 1086 Berlin, Germany.

B. de Pagter, Technische Universiteit Delft, Faculteit TWI, Mekelweg 4, 2628 CD Delft, The Netherlands.

W. Desch, Institut für Mathematik, Universität Graz, Heinrichstrasse 36, A-8010 Graz, Austria.

O. Diekmann, C.W.I., Kruislaan 413, 1098 SJ Amsterdam, The Netherlands.

P.G. Dodds, The Flinders University of South Australia, School of Information Science and Technology, GPO Box 2100, Adelaide 5001, Australia.

B. Eberhardt, Mathematisches Institut, Universität Tübingen, Auf der Morgenstelle 10, D-7400 Tübingen, Germany.

P. Egberts, Technische Universiteit Delft, Faculteit TWI, Mekelweg 4, 2628 CD Delft, The Netherlands.

K-J. Engel, Mathematisches Institut, Universität Tübingen, Auf der Morgenstelle 10, D-7400 Tübingen, Germany.

J. Escher, Mathematisches Instiutut der Universität Zürich, Rämistrasse 74, 8001 Zürich, Switzerland.

H. Fattorini, Department of Mathematics, U.C.L.A., Los Angeles, CA 90024, U.S.A.

A. Favini, Dipartimento di Matematica, Universita di Bologna, Piazza Porta S. Donato 5, I-40127 Bologna, Italy.

Q. Fucheng, Department of Mathematics, University of Antwerp, Universiteitsplein 1, 2610 Wilrijk, Antwerpen, Belgium.

M. Fuhrman, Dipartimento di Matematica, Del Politecnico di Milano, Piazza Leonardo da Vinci 32, 20133 Milano, Italy.

J. Goldstein, Department of Mathematics, Louisiana State University, Baton Rouge, LA 70803, U.S.A.

J.P. Gossez, Université Libre de Bruxelles, Département de Mathématique, Campus Plaine, Boulevard du Triomphe, 1050 Bruxelles, Belgium.

A. Grabosch, LB Biomathematik, Biologie II, Auf der Morgenstelle 10, D-7400 Tübingen, Germany.

P. Grisvard, Institut de Mathématiques et Sciences Physique, Université de Nice, Parc Valrose, 06034 Nice Cédex, France.

M. Gyllenberg, Lulea University of Technology, Department of Applied Mathematics, S-95187 Lulea, Sweden.

K. Hadeler, Lehrstuhl für Biomathematik, Universität Tübingen, Auf der Morgenstelle 10, D-7400 Tübingen, Germany.

J. Heikonen, TKK Matematikan Latos, Institute of Mathematics and Systems Analysis, Otakaari 1M, 02150 Espoo, Finland.

H.R. Henriquez, Departamento de Matematica, Universidad de Santiago, Casilla 5659 Correo 2, Santiago, Chili.

P. Hess, Mathematisches Institut, Universität Zürich, Rämistrasse 74, CH-8001 Zürich, Switzerland.

M. Hieber, Mathematisches Institut, Universität Zürich, Rämistrasse 74, CH-8001 Zürich, Switzerland.

F. Hirsch, Ecole Normale Supérieure de Cachan, 22 Rue de Talhouët, 91130-Ris-Orangis, France.

A. Holderrieth, Mathematisches Institut, Universität Tübingen, Auf der Morgenstelle 10, D-7400 Tübingen, Germany.

K. Ming Hui, Institute of Mathematics, Academia Sinica, Nankang, Paipei, Taiwan 11529, Republic of China.

C.B. Huijsmans, Department of Mathematics, Netherlands University of Leiden, Niels Bohrweg 1, P.O. Box 9512, 2300 RA Leiden, The Netherlands.

A. Ivanov, Mathematisches Institut der Universität, Theresienstrasse 39, D-8000 München 2, Germany.

W. Jäger, Institut für Angewandte Mathematik, SFB 123, Universität Heidelberg, Im Neuenheimer Feld 294, 6900 Heidelberg 1, Germany.

V. Keyantuo, Equipe de Mathématiques, Université de Franche-Comté, F-25030 Besançon, France.

P. Koch-Medina, Mathematisches Institut der Universität Zürich, Rämistrasse 74, 8001 Zürich, Switzerland.

N. Kutev, Universität Karlsruhe, Mathematisches Institut I, Englerstrasse 2, Postfach 6380, D-7500 Karlsruhe, Germany
and
Institute of Mathematics, Bulgarian Academy of Sciences, Sofia, Bulgarie.

E. Lami Dozo, Institute Argentino de Matematica, Viamonte 1636, 1055 Buenos Aires, Argentina.

H. Lange, Mathematisches Institut, Universität Köln, Weyertal 86-90, D-5000 Köln 41, Germany.

M. Langlais, U.F.R. Sciences Humaines Appliquées, Université de Bordeaux II, 146 Rue Léo Saignat, F-33076 Bordeaux Cédex, France.

I. Lasiecka, Department of Applied Mathematics, University of Virginia, Thornton Hall, Charlottesville, Virginia 22903-2442, U.S.A.

R. Lefever, Service de Chimie physique, Université Libre de Bruxelles, CP 231, Boulevard du Triomphe, 1050 Bruxelles, Belgium.

B.M. Levitan, 1044 ST SE, Minneapolis MN 54414, U.S.A.

C. Lizama, Universidad de Santiago de Chile, Depto. Matematica y ciencia de la computacion, Casilla 5659, Correo 2, Santiago, Chile.

G. Lumer, Université de Mons-Hainaut, Institut de Mathématique et d'Informatique, Avenue Maistriau 15, 7000 Mons, Belgium.

W.A.J. Luxemburg, Department of Mathematics 253-37, California Institute of Technology, Pasadena, CA 91125, U.S.A.

K.G. Magnusson, Department of Mathematics, University of Iceland, Science Institute, Dunhaga 3, 107 Reykjavik, Iceland.

P. Malliavin, Rue Saint-Louis-en-l'Ile 10, 75004 Paris, France.

J. Martinez Centelles, Departamento de Analisis Matematico, Universitad de Valencia, Dr. Moliner 50, 46100 Burjassot, Spain.

V. Maslov, Institute for Problems in Mechanics, USSR Academy of Sciences, prosp. Vernadskogo 101, 117526 Moscow, URSS.

M. Mastinsek, EPF, Razlagova 14, Univerza v Mariboru, 62000 Maribor, Yougoslavie.

J. Mawhin, Université Catholique de Louvain, Département de Mathématique, Chemin du Cylotron 2, 1348 Louvain-la-Neuve, Belgium.

J.M. Mazon, Departamento Analisis Matematico, Universitad de Valencia, Dr Moliner 50, 46100 Burjassot, Spain.

S. Mejza, Department of Mathematical and Statistical Methods Agricultural University, PL-60-637 Poznan, Wojska Polskiego 28, Akademia Rolnicza, Poland.

S. Merino, Seminar für Angewandte Mathematik ETHZ, Fliederstrasse 23, 8052 Zürich, Switzerland.

D. Merlini, Laboratory of Cellular Pathology, CH-6600 Locarno, Switzerland.

E. Mitidieri, Dipartimento di Matematica e Informatica, Universita di Udine, Via Zanon 6, I-33100 Udine, Italy.

M. Mokhtar-Kharroubi, Equipe de Mathématiques, Université de Franche-Comté, Route de Gray, F-25030 Besançon Cédex, France.

G. Morosanu, Faculty of Mathematics, University of Iasi, Iasi-6600, Roumanie.

D. Motreanu, Universitatea "Al.I.Cuza", Seminarul Matematic "Al.Myller", 6600 Iasi, Roumanie.

R. Nagel, Mathematisches Institut, Universität Tübingen, Auf der Morgenstelle 10, D-7400 Tübingen, Germany.

B. Najman, Department of Mathematics, University of Zagreb, P.O. Box 187, 41001 Zagreb, Yougoslavie.

P.I. Naumkin, Mathematics Faculty, Physics Dept., Moscow State University, Lenin Hills, Moscow 119 899, URSS.

F. Neubrander, Department of Mathematics, L.S.U., Baton Rouge, LA 70803, U.S.A.

S. Nicaise, Université des Sciences et Techniques de Lille Flandres Artois, F-59655 Villeneuve d'Ascq, France.

S. Oharu, Department of Mathematics, Hiroshima University, Naka-ku, Hiroshima 730, Japan.

L. Paquet, Université de Mons-Hainaut, Institut de Mathématique et d'Informatique, Avenue Maistriau 15, 7000 Mons, Belgium.

V. Quoc Phong, Institute of Mathematics, P.O. Box 631, 10000 Hanoi, Vietnam

M. Pierre, Université de Nancy I, Département de Mathématiques, B.P. 239, F-54506 Vandoeuvre, France.

S. Piskarev, Chalmers Tekniska Högskola, University of Göteborg, Matematiscka Institutionen, Sven Hultins gata 6, 41296 Göteborg, Sweden.

J. Popiolek, Institut of Mathematics, Varsaw University (Bialystok), Str. Akademicka 2, 15-267 Bialystok, Poland.

J. Prüss, FB17, Universität Paderborn, Warburger Strasse 100, D-W-4790 Paderborn, Germany.

A. Pugliese, Dipartimento di Matematica, Universita di Trento, I-38050 Povo, Italy.

F. Raebiger, Mathematisches Institut, Universität Tübingen, Auf der Morgenstelle 10, D-7400 Tübingen, Germany.

W. Ruess, F.B. Mathematik, Universität Essen, W-4300 Essen 1, Germany.

G. Ruiz Rieder, Department of Mathematics, Louisiana State University, Baton Rouge, LA 70803, U.S.A.

G. Schätti, Mathematisches Institut der Universität Zürich, Rämistrasse 74, 8001 Zürich, Switzerland.

J. Schmets, Institut de Mathématique, Avenue des Tilleuls 15, 4000 Liege, Belgium.

D. Sforza, Dipartimento di Matematica, Universita di Pisa, Via Buonarroti 2, I-56127 Pisa, Italy.

I.A. Shishmarev, Mathematics Faculty, Physics Dept., Moscow State University, Lenin Hills, Moscow 119 899, URSS.

G. Simonett, Mathematisches Institut der Universität Zürich, Rämistrasse 74, 8001 Zürich, Switzerland.

E. Sinestrari,, Dipartimento di Matematica Universita di Roma I, P. Aldo Moro 2, I-00185 Roma, Italy.

P.E. Sobolevskii, Hebrew University of Jerusalem, P.O. Box 1698, Kyriat Arba 303/19, Jerusalem 90100, Israel.

W.H. Summer, Mathematical Sciences, SCEN 301, University of Arkansas, Fayetteville, AR 72701, U.S.A.

N. Svanstedt, Dept. of Applied Mathematics, University of Technology, Lulea, S-95187 Sweden.

F. Szafraniec, Jagiellonian University, Ul. Raymonta 4, 30059 Krakow, Poland.

B. Tang, Department of Mathematics, Arizona State University, Tempe, AZ 85287, U.S.A.

S. Tasaki, U.L.B., Chimie-Physique, C.P. 231, Campus Plaine, Boulevard du Triomphe, 1050 Bruxelles, Belgium.

H.R. Thieme, Department of Mathematics, Arizona State University, Tempe, AZ 85287, U.S.A.

R. Thom, I.H.E.S., 35 Route de Chartres, 91440 Bures-sur-Yvette, France.

J. Toledo Melero, Departamento de Analisis Matematico, Universitat de Valencia, Dr Moliner 50, 46100 Burjassot, Spain.

R. Triggiani, Department of Applied Mathematics, University of Virginia, Thorton Hall, Charlottesville, Virginia 22903-2442, U.S.A.

K. Tsujioka, Department of Mathematics, Faculty of Science, Saitama University, Shimo-Okuko, Urawa 338, Japan.

J. Van Biesen, Departement Wiskunde, Universitaire Instelling Antwerpen, Universiteitsplein 1, 2610 Wilrijk, Belgium.

J. Van Casteren, Dept. Wiskunde & Informatica, U.I.A., Universiteitsplein 1, 2610 Wilrijk, Belgium.

J. van Neerven, Centre for Mathematics and Computer Science, Kruislaan 413, 1098 SJ Amsterdam, The Netherlands.

V. Vasiljev, Belgorad Pedagogical Institut, Street B Chmelnizkogo 185-22, Belgorod, 308010 URSS.

J. Vazquez, Departamiento de Matematicas, Universidad Autonoma de Madrid, 29049 Madrid, Spain.

P. Vernole, Dipartimento di Matematica, Universita "La Sapienza", Piazzale Aldo Moro 2, 00185 Roma, Italy.

M. Voicu, Inst. de Constructii, Bd. Lacul Tei 124, Sector 2, 79622, Bucuresti, Roumanie.

W. von Wahl, Fachbereich Mathematik und Physik, Universität Bayreuth, Postfach 101251, 8580 Bayreuth, Germany.

P. Weidemaier, Germany.

L. Weis, Department of Mathematics, L.S.U., Baton Rouge, LA 70803, U.S.A.

S. Zaidman, Department of Mathematics and Statistics, Université de Montréal, C.P. 6128, Succ. A., 5620 Darlington Av., H3C3J7, Montréal (Québéc), Canada.

evolution equations, control theory, and biomathematics

Nonlinear and Dynamical Node Transition in Network Diffusion Problems

J. von BELOW Tübingen University, Tübingen, Germany

Suppose a reaction - diffusion process on a ramified network is described by parabolic equations on the edges and by the following transition conditions at the vertices. At the ramification nodes we require continuity. Moreover, we assume the vertices to be partitioned into two classes E_1 and E_K corresponding to inhomogeneous Dirichlet conditions and nonlinear Kirchhoff conditions $\mathcal{V}_i(u,t) = 0$, respectively. Each vertex operator $\mathcal{V}_i$ has the form

$$\mathcal{V}_i(u,t) = \rho_i\big(t, u(e_i,t)\big) - \sum_{j=1}^{N} d_{ij} c_{ij}(t) u_{jx_j}(e_i,t) - \sigma_i\big(t, u(e_i,t), u_t(e_i,t)\big) \quad (e_i \in E_K) \quad (1)$$

with given functions $\rho_i : [0,T] \times \mathbb{R} \to \mathbb{R}$, $c_{ij} : [0,T] \to \mathbb{R}$, and $\sigma_i : [0,T] \times \mathbb{R}^2 \to \mathbb{R}$. If σ_i really depends on u_t, then $\mathcal{V}_i(u,t) = 0$ constitutes a *dynamical* node condition. A special nondynamical case is given by the classical Kirchhoff law

$$\sum_{j=1}^{N} d_{ij} c_{ij} u_{jx_j}(e_i,t) = 0. \quad (2)$$

The conductivities c_{ij} in (1) are always supposed to be positive functions, while the factors d_{ij} indicate the orientation. For constant conductivities and linear second order elliptic differential operators, Lumer [9] has shown that condition (2) at ramification nodes leads to dissipativity iff at each such node all conductivities have the same sign and do not all vanish. In the dynamical case, the σ_i being monotone increasing with respect to u_t constitutes the additional condition ensuring dissipativity, cf. (3b) below and [5].

In the present paper we will first derive some general qualitative properties of classical solutions subject to the transition conditions above. Then we investigate the classical global solvability in some special semilinear cases, where the properties of the nonlinearities in the differential equations and in the transition conditions will be such that they admit an a priori estimate of the time-derivative of a solution. In this way estimates for scalar equations in domains as established in [8] can be applied, and the solvability in the class $C^{2+\alpha,1+\frac{\alpha}{2}}$ is established with a Leray-Schauder argument. More general equations and transition conditions will be treated in the forthcoming paper [7].

As for linear Kirchhoff conditions, we refer to [1 - 3,9] in the nondynamical case and to [4 - 7] in the dynamical one, an example in the latter case arising in neurobiology is mentioned in [4]. Other types of transition replacing the continuity condition in connection with (2), also for hyperbolic equations, can be found in [1,2].

1. NETWORKS AND VERTEX TRANSITION CONDITIONS

Let us recall, e. g. from [3], the notion of a c^ν-network G for $\nu \geq 2$ with finite sets of vertices $E = \{e_i | 1 \leq i \leq n\}$ and edges $K = \{k_j | 1 \leq j \leq N\}$. By definition, G is the union of Jordan curves k_j in $\mathbb{R}^m$ with arc length parametrizations $\pi_j \in C^\nu([0, l_j], \mathbb{R}^m)$. The arc length parameter of an edge k_j is denoted by x_j. The topological graph Γ belonging to G is assumed to be simple and connected. Thus, by definition, $\Gamma = (E, K)$ consists in a collection of N Jordan curves k_j with the following properties: Each k_j has its endpoints in the set E, any two vertices in E can be connected by a path with arcs in K, and any two edges $k_j \neq k_h$ satisfy $k_j \cap k_h \subset E$ and $|k_j \cap k_h| \leq 1$. Endowed with the induced topology G is a connected and compact space in $\mathbb{R}^m$. The valency of each vertex is denoted by $\gamma_i = \gamma(e_i)$. We distinguish the ramification nodes $E_r = \{e_i \in E | \gamma_i > 1\}$ from the boundary vertices $E_b = \{e_i \in E | \gamma_i = 1\}$. The orientation of G is given by the incidence matrix $D = (d_{ij})_{n \times N}$ with

$$d_{ij} = \begin{cases} 1 & \text{if } \pi_j(l_j) = e_i, \\ -1 & \text{if } \pi_j(0) = e_i, \\ 0 & \text{otherwise.} \end{cases}$$

We introduce t as the time variable and for $T > 0$

$$\Omega = G \times [0, T], \quad \Omega_j = [0, l_j] \times [0, T], \quad \Omega_{jp} = (0, l_j) \times (0, T],$$

and for $u : \Omega \to \mathbb{R}$ we define $u_j = u \circ (\pi_j, id) : \Omega_j \to \mathbb{R}$ and use the abbreviations $u_j(e_i, t) := u_j(\pi_j^{-1}(e_i), t)$, $u_{x_j}(e_i, t) := \frac{\partial}{\partial x_j} u_j(\pi_j^{-1}(e_i), t)$ etc. As special subspaces of $C(\Omega)$ we introduce for $\mu \in \mathbb{N}$, $\alpha \in [0, 1), \mu + \alpha \leq \nu$

$$C^{\mu+\alpha, \frac{\mu+\alpha}{2}}(\Omega) = \{u \in C(\Omega) | \forall j \in \{1, ..., N\} : u_j \in C^{\mu+\alpha, \frac{\mu+\alpha}{2}}(\Omega_j)\},$$

where $C^{\mu+\alpha, \frac{\mu+\alpha}{2}}(\Omega_j)$ with a Hölder norm $|\cdot|_{\Omega_j}^{(\mu+\alpha)}$ denotes the Banach space of functions u on Ω_j having continuous derivatives $\frac{\partial^{r+s} u}{\partial t^r \partial x_j^s}$ for $2r + s \leq \mu$ and finite Hölder constants of the indicated exponents in the case $\alpha > 0$, cf. [8.I]. Again, $C^{\mu+\alpha, \frac{\mu+\alpha}{2}}(\Omega)$ is a Banach space endowed with the norm $|u|_\Omega^{(\mu+\alpha)} = \sum_{j=1}^N |u_j|_{\Omega_j}^{(\mu+\alpha)}$.

The basic transition condition at the ramification nodes is given by the *continuity condition*

$$k_j \cap k_s = \{e_i\} \implies u_j(e_i, t) = u_s(e_i, t),$$

that is trivially fulfilled by functions on G or Ω. We decompose the vertex set E into two disjoint parts

$$E = E_1 \uplus E_K$$

with respect to different transition conditions. At the vertices in E_1 we consider Dirichlet conditions of the form

$$u(e_i, t) = \psi(e_i, t) \quad \text{ in } \quad E_1 \times [0, T].$$

Note that E_1 can be empty. At the vertices in E_K we impose nonlinear Kirchhoff conditions for the flows coming from incident edges with $\mathcal{V}_i$ as in (1):

$$\mathcal{V}_i(u, t) = 0 \quad \text{in} \quad E_K \times (0, T].$$

Throughout, c_{ij} and σ_i are assumed to fulfill the following condition (3):

$\forall e_i \in E_K \; \forall t \in [0,T] \; \forall z, p, \tilde{p} \in \mathbb{R} \; \forall j \in \{1, \ldots, N\}$:

$$c_{ij}(t) > 0, \tag{3a}$$

$$p \leq \tilde{p} \quad \Longrightarrow \quad \sigma_i(t,z,p) \leq \sigma_i(t,z,\tilde{p}). \tag{3b}$$

As a Dirichlet condition at a ramification node e_i corresponds to the same Dirichlet condition at γ_i boundary vertices, we may asssume that

$$E_r \subseteq E_K \quad \text{and} \quad E_1 \subseteq E_b.$$

With respect to the decomposition $E = E_1 \uplus E_K$ we define the *parabolic network interior* Ω_p and the *parabolic network boundary* ω_p as

$$\Omega_p = (G \backslash E_1) \times (0,T], \qquad \omega_p = (G \times \{0\}) \cup (E_1 \times (0,T]).$$

Introduce furthermore the notation $\Omega^{\bullet}_{jp} = \Omega_{jp} \cup \left(\{\xi \in [0,l_j] \,|\, \pi_j(\xi) \in E_K\} \times (0,T]\right)$.

2. COMPARISON PRINCIPLES AND BASIC ESTIMATES

Condition (3) guarantees the validity of a basic technique for comparing classical solutions.

LEMMA: *Let $\varphi, \psi \in C(\Omega) \cap C^{2,1}(\Omega_p)$ satisfy $\mathcal{V}_i(\psi,t) \leq \mathcal{V}_i(\varphi,t)$ in $E_K \times (0,T]$ and the test point implication:*

If $\varphi_j = \psi_j$, $\quad \varphi_{jx_j} = \psi_{jx_j}$, $\quad \varphi_{jx_jx_j} \leq \psi_{jx_jx_j}$ at a point in Ω_p, then $\varphi_{jt} < \psi_{jt}$ at this point. (4)

Then $\varphi < \psi$ in ω_p implies $\varphi < \psi$ in Ω_p.

The proof of the linear case given in [5] can be carried over to the present situation. For the details and further applications besides the one below cf. [7.II].

On the edges k_j we first consider general parabolic equations

$$u_{jt} = F_j(x_j, t, u_j, u_{jx_j}, u_{jx_jx_j}) =: F_j[u_j],$$

where each $F_j : \Omega_j \times \mathbb{R}^3 \to \mathbb{R}$ is supposed to be monotone increasing in the variable $q = u_{jx_jx_j}$. For the formulation of the comparison principle with respect to ω_p, we introduce the Lipschitz and monotonicity conditions (5) and (6):

$\exists C > 0 \exists R \geq 0 \forall e_i \in E_K \; \forall j \in \{1, \ldots, N\}, \; \forall t \in [0,T] \; \forall z, \tilde{z}, p, \tilde{p} \in \mathbb{R}$:

$$c_{ij}(t) \geq C, \tag{5a}$$

$$z \geq \tilde{z} \quad \Longrightarrow \quad \rho_i(t,z) - \rho_i(t,\tilde{z}) \leq R(z - \tilde{z}), \tag{5b}$$

$$z \geq \tilde{z} \quad \Longrightarrow \quad \sigma_i(t,z,p) \geq \sigma_i(t,\tilde{z},p). \tag{5c}$$

$\exists L > 0 \forall j \in \{1,..,N\} \forall (x,t) \in \Omega_{jp}^{\bullet} \forall (z,p,q), (\tilde{z},\tilde{p},\tilde{q}) \in \mathbb{R}^3$:

$$z \geq \tilde{z} \Longrightarrow F_j(x,t,z,p,q) - F_j(x,t,\tilde{z},p,q) \leq L(z-\tilde{z}), \tag{6a}$$
$$|F_j(x,t,z,p,q) - F_j(x,t,z,\tilde{p},q)| \leq L|p-\tilde{p}|, \tag{6b}$$
$$q \geq \tilde{q} \Longrightarrow F_j(x,t,z,p,q) - F_j(x,t,z,p,\tilde{q}) \leq L(q-\tilde{q}), \tag{6c}$$

THEOREM 1: *Suppose conditions (5) and (6) are fulfilled. Let $u, v \in C(\Omega) \cap C^{2,1}(\Omega_p)$ satisfy $\mathcal{V}_i(v,t) \leq \mathcal{V}_i(u,t)$ in $E_K \times (0,T]$ and the differential inequalities*

$$u_{jt} - F_j[u_j] \leq v_{jt} - F_j[v_j] \quad \text{in} \quad \Omega_{jp}^{\bullet} \quad \text{for all} \quad j \in \{1,...,N\}. \tag{7}$$

Then $u \leq v$ on ω_p implies $u \leq v$ in Ω_p.

Proof: Define a positive separating function $\varepsilon \in C^{2,1}(\Omega)$ by setting on each edge k_j

$$\varepsilon_j(x_j,t) = \begin{cases} \delta\big((1-\alpha x_j)^4 + 1\big)e^{\lambda t} & \text{for } 0 \leq x_j \leq \xi \\ \delta e^{\lambda t} & \text{for } \xi \leq x_j \leq l_j - \xi \\ \delta\big((1-\alpha(l_j - x_j))^4 + 1\big)e^{\lambda t} & \text{for } l_j - \xi \leq x_j \leq l_j \end{cases}$$

with $\delta > 0$ and $\xi = \alpha^{-1}$ and with fixed positive parameters α and λ satisfying

$$\alpha > \max\left\{\frac{R}{2C\min_{e_i \in E_K}\gamma_i}, \frac{2}{l_1}, ..., \frac{2}{l_N}\right\} \quad \text{and} \quad \lambda > L\big(2 + 4\alpha + 12\alpha^2\big).$$

In connection with $\varepsilon_{jx_j}(e_i,t) = d_{ij}4\alpha\delta\exp(\lambda t)$ and (5) the choice of parameters yields the inequalities:

$$\mathcal{V}_i(v+\varepsilon,t) \leq \mathcal{V}_i(v,t) \quad \text{in} \quad E_K \times (0,T],$$

$$\varepsilon_{jt} > L(\varepsilon_j + |\varepsilon_{jx_j}| + \varepsilon_{jx_jx_j}) \quad \text{in} \quad \Omega_j \quad \text{for all} \quad j \in \{1,...,N\}. \tag{8}$$

Set $\varphi = u$ and $\psi = v + \varepsilon$ such that $\varphi < \psi$ on ω_p. At a point with the hypotheses of (4) we conclude with (6), (7),(8) and with the monotonicity of F_j in q

$$\begin{aligned} 0 &\leq v_{jt} - u_{jt} - F_j[\psi_j] + F_j[u_j] + L(\varepsilon_j + |\varepsilon_{jx_j}| + \varepsilon_{jx_jx_j}) \\ &< v_{jt} - u_{jt} + \varepsilon_{jt} = \psi_{jt} - \varphi_{jt}. \end{aligned}$$

The Lemma shows $\varphi < \psi$ in Ω_p and, since $\delta > 0$ was arbitrary, $u \leq v$ holds in Ω_p. ◇

COROLLARY: *Under the conditions (5) and (6) there exists at most one solution of the initial boundary value problem $(*)$:*

$$(*) \qquad \begin{cases} u \in C(\Omega) \cap C^{2,1}(\Omega_p), \\ u_{jt} = F_j(x_j,t,u_j,u_{jx_j},u_{jx_jx_j}) \quad \text{in } \Omega_{jp}^{\bullet} \text{ for all } j \in \{1,...,N\}, \\ \mathcal{V}_i(u,t) = 0 \quad \text{in} \quad E_K \times (0,T], \\ u\big|_{\omega_p} = \psi \in C(\omega_p). \end{cases}$$

THEOREM 2: *Assume conditions (5) and (6) to hold. Furthermore, suppose there are constants m_0 and m_1 with*

$$F_j(x,t,0,0,0) \le m_0 \quad in \quad \Omega_j \qquad for\ all \quad j \in \{1,...,N\}, \tag{9a}$$
$$\mathcal{V}_i(0,t) \le m_1 \quad in \quad E_K \times (0,T]. \tag{9b}$$

Let $u \in C(\Omega) \cap C^{2,1}(\Omega_p)$ be a solution of the initial boundary value problem (*). Then

$$\max_{\Omega} |u| \le 2e^{\lambda T} \max\Big\{\max_{\omega_p} |u|, m_0, m_1\Big\} =: M_0$$

with

$$\lambda = L\big(2 + 4\alpha + 12\alpha^2\big) + 1 \quad and \quad \alpha = \max\left\{\frac{R+\frac{1}{2}}{2C\min_{e_i \in E_K}\gamma_i}, \frac{2}{l_1}, ..., \frac{2}{l_N}\right\}.$$

Proof. We apply Theorem 1 to u and $\pm\varepsilon \in C^{2,1}(\Omega)$ as defined in the proof of Theorem 1 with the special choice

$$\delta = \max\Big\{\max_{\omega_p} |u|, m_0, m_1\Big\}$$

and α and λ as defined in the assertion. Then the construction of ε yields in connection with (9b) that $\mathcal{V}_i(\varepsilon,t) \le 0$ in $[0,T] \times E_K$, and in connection with (9a)

$$\varepsilon_{jt} \ge L(\varepsilon_j + |\varepsilon_{jx_j}| + \varepsilon_{jx_jx_j}) + m_0 \quad \text{in} \quad \Omega_j \quad \text{for all} \quad j \in \{1,...,N\}.$$

It remains to show $\varepsilon_{jt} - F_j[\varepsilon_j] \ge 0$ in $\Omega^{\bullet}_{jp}$, but this readily follows by (6),(9), and the last differential inequalities:

$$\varepsilon_{jt} \ge L\big(\varepsilon_j + |\varepsilon_{jx_j}| + \varepsilon_{jx_jx_j}\big) + m_0 \ge F_j(\cdot,\cdot,\varepsilon_j,\varepsilon_{jx_j},\varepsilon_{jx_jx_j}) - F_j(\cdot,\cdot,0,0,0) + m_0. \quad \diamondsuit$$

Note that the estimate in Theorem 2 does not depend on bounds for σ_i. Therefore, the results apply especially to the nondynamical case $\rho_i\big(t,u(e_i,t)\big) = \sum_{j=1}^N d_{ij}c_{ij}(t)u_{jx_j}(e_i,t)$. As for ODEs, condition (5b) is essential for the exponential growth with respect to T. When (5b) is violated, blow up in finite time at a node in E_K can occur. On a star graph Γ e. g. with $E_K = \{e_1\}$, $T < T_\infty$ and all $d_{1j} = -1$, $c_{1j} = N^{-1}$, each $u_j(x_j,t) := \big(T_\infty + x_j - t\big)^{-1}$ satisfies the nonlinear equation $u_{jt} = \frac{1}{2}(T_\infty + x_j - t + 1)u_{jx_jx_j} - u_j^3$ in Ω_j. All u_j together constitute a function $u \in C(\Omega)$, that satisfies the nonlinear Kirchhoff condition $u^2(e_1,t) = \frac{1}{N}\sum_{j=1}^N d_{1j}u_{jx_j}(e_1,t)$ in $\{e_1\} \times [0,T]$.

REMARK: The comparison theorem is also valid without condition (6b,c) , when (5) is replaced by the following condition: There exists a constant $b \ge 0$ such that

$$\beta \ge b,\ z \ge \tilde{z} \implies \rho_i(t,z) - \rho_i(t,\tilde{z}) \le \sigma_i\big(t,z,p+\beta(z-\tilde{z})\big) - \sigma_i(t,\tilde{z},p), \tag{5*}$$

while a $|\cdot|_\Omega^{(0)}$ - bound can also be achieved, under (9a),(5*), $\mathcal{V}_i(0,t) \le 0$, and an Osgood condition $zF_j(x,t,z,0,0) \le b_1z^2 + b_2$ instead of (6), cf. [7.II].

3. BOUNDS FOR THE TIME DERIVATIVE AND SOLVABILITY IN $C^{2+\alpha,1+\frac{\alpha}{2}}(\Omega)$

For the sake of simplicity we will investigate the solvability only for special semilinear differential equations and Kirchhoff conditions, to which an easy application of the Leray - Schauder Principle becomes possible. It should be emphasized that similar results hold also in more general situations, especially dependence of the f_j on u_{jx_j} is admitted under usual growth conditions. For $\nu \geq 3$ and for a given function $\psi \in C^{2,1}(\Omega)$ we want to solve the IBVP $(**)$

$$(**)\quad \begin{cases} u \in C^{2,1}(\Omega) & & (10a)\\ u_{jt} = u_{jx_jx_j} + f_j(x_j,t,u_j) \text{ in } \Omega_{jp} & \text{for } 1 \leq j \leq N & (10b)\\ \sum_{j=1}^{N} d_{ij}c_{ij}(t)u_{jx_j}(e_i,t) + \sigma_i(t)u_t(e_i,t) = \rho_i\big(t,u(e_i,t)\big) & \text{in } E_K \times [0,T] & (10c)\\ u|_{\omega_p} = \psi|_{\omega_p} & & (10d) \end{cases}$$

under the conditions (3), (5), (6a), (9), and (11): For all $j \in \{1,...,N\}$ and $e_i \in E_K$

$$f_j \in C(\Omega_j \times [-M_0, M_0]), \tag{11a}$$

$$c_{ij}, \sigma_i \in C([0,T]),\ \rho_i \in C([0,T] \times [-M_0, M_0]) \tag{11b}$$

where $M_0 := M_0(L, C, R, \Omega, m_0, m_1)$ is defined as in Theorem 2. Moreover:
All f_j are differentiable with respect to $z = u_j$ and Lipschitz continuous with respect to t in $\Omega_j \times [-M_0, M_0]$, and there is a constant $K_1 \geq 0$ such that for each f_j in $[0,l_j] \times [0,T] \times [-M_0, M_0]$,

$$\left|\frac{\partial}{\partial z} f_j(x_j,t,z)\right|,\ \sup_{0\leq t+h\leq T, h\neq 0}\left|\frac{f_j(x_j,t+h,z) - f_j(x_j,t,z)}{h}\right| \leq K_1. \tag{11c}$$

By Theorem 2 any solution u of $(**)$ satisfies

$$|u|_\Omega^{(0)} \leq M_0.$$

The conditions (11) allow the application of the interior estimate of $|u_x|$ given in [8.V.3]. For each $d \in (0, \frac{1}{2}\min_j l_j)$ the following estimate holds:

$$\max_{1\leq j\leq N} |u_{jx_j}|^{(0)}_{[d,l_j-d]\times[0,T]} \leq M_1(d) = M_1(d; M_0, L, m_0, \Omega, |\psi|_\Omega^{(1)}). \tag{12}$$

It has been shown in [6] that $|u|_\Omega^{(2)}$ can be estimated from above by known quantities, if there is an upper bound M^* depending only on known quantities such that $|u_t| \leq M^*$ in

$$\mathcal{R} := E_K \times [0,T].$$

Under the conditions imposed at E_K, this estimate over $\mathcal{R}$ is of course the crucial step in showing solvability when using classical domain estimates. Here we demonstrate two estimating techniques.

First, we treat vertex operators not explicitly depending on time

$$\mathcal{V}_i(u,t) = \rho_i\big(u(e_i,t)\big) - \sum_{j=1}^{N} d_{ij}c_{ij}u_{jx_j}(e_i,t) - \sigma_i\, u_t(e_i,t) \tag{13}$$

subject to the additional condition:
All ρ_i are differentiable with respect to u in $[-M_0, M_0]$, and

$$\frac{d}{du}\rho_i(u) \le K_1 \quad \text{in } [-M_0, M_0]. \tag{14}$$

THEOREM 3: *Suppose all $\mathcal{V}_i$ are of the form (13) subject to condition (14). Let u be a solution of $(**)$. Then $|u_t|_{\mathcal{R}}|$ can be estimated from above by a constant M_1^* that depends only on $\Omega, M_0, L, C, R, |\psi|_\Omega^{(2)}$ and K_1.*

Proof: By $(**)$ and (11a) we have $|u_t(\cdot,0)|_G^{(0)} \le |\psi|_\Omega^{(2)}$. By (11c), (12) and the local estimate of $|u_t|$ in [8.V.5] we obtain for each $\varepsilon \in (0, \frac{1}{2}\min_j l_j)$ an estimate

$$\max_j |u_{jt}|^{(0)}_{[\varepsilon/2, l_j-\varepsilon/2]\times[0,T]} \le M_2 = M_2(K_1, |\psi(\cdot,0)|_G^{(1)}, \varepsilon/4, M_1(\varepsilon/4), M_0).$$

The global estimate of $|u_t|$ in [8.V.5] yields the assertion at vertices in E_1. Therefore we may assume without restriction that Γ is a star graph with

$$E_r \subseteq E_K = \{e_i\},\quad N = \gamma_i \ge 1,\quad l_j = \varepsilon > 0,\quad \pi_j(0) = e_i \quad \text{for all} \quad j \in \{1,\dots,N\}.$$

For $h > 0$ sufficiently small set $\Omega_{T-h} = G \times [0, T-h]$, and define $v^h \in C(\Omega_{T-h})$ by the divided differences

$$v_j^h(x_j,t) = \frac{u_j(x_j,t+h) - u_j(x_j,t)}{h}.$$

Furthermore, set $\Xi_i = \big(\tau u_j(e_i,t+h) + (1-\tau)u_j(e_i,t)\big)$,

$$\Xi = \Big(x_j, t+h, \tau u_j(x_j,t+h) + (1-\tau)u_j(x_j,t)\Big),$$

$$r_i^h(t) = \int_0^1 \frac{d}{dz}\rho_i(\Xi_i)d\tau, \qquad C_j^h(x_j,t) = \int_0^1 \frac{\partial}{\partial z} f_j(\Xi)d\tau,$$

$$F_j^h(x_j,t) = \frac{1}{h} f_j\Big(x_j, t+h, u_j(x_j,t)\Big) - \frac{1}{h} f_j\Big(x_j, t, u_j(x_j,t)\Big).$$

Then v^h belongs to $C^{2,1}(\Omega_{T-h})$ and satisfies the linear equations

$$\frac{\partial}{\partial t} v_j^h = a_j v^h_{jx_jx_j} + C_j^h v_j^h + F_j^h \quad \text{in} \quad \Omega_{j,T-h}$$

and at e_i the linear Kirchhoff condition

$$\sum_{j=1}^{N} d_{ij}c_{ij}v^h_{jx_j}(e_i,t) + \sigma_i v_t^h(e_i,t) = r_i(t)v^h(e_i,t).$$

By (11c) and (14) we can apply Theorem 2 to v^h and arrive at an estimate

$$|v^h|^{(0)}_{\Omega_{T-h}} \leq 2e^{\lambda T} \max\Big\{K_1, \max_{t\leq T-h,(x,t)\in\omega_p} |v^h(x,t)|\Big\}$$

with a constant $\lambda = \lambda(R, C, G)$. Letting h tend to zero, the r.h.s. goes to an upper bound $M_1^* = M_1^*(\lambda, T, K_1, |\psi|^{(2)}_\Omega, M_2)$ for $|u_t|^{(0)}_\Omega$. ◊

As the bound M_1^* does not depend on any σ_i, Theorem 3 includes especially the nondynamical case subject to (14), e.g. the condition $\sum_{j=1}^N d_{ij}c_{ij}u_{jx_j}(e_i,t) = -\text{sign}(u(e_i,t))u^2(e_i,t)$.

For the second a priori bound for $|u_t|$ we partition the vertices in E_K into the classes E_2 and E_3 corresponding to the following Kirchhoff laws:

$$\sum_{j=1}^N d_{ij}c_{ij}(t)u_{jx_j}(e_i,t) = 0 \quad \text{in} \quad E_2 \times [0,T] \tag{15a}$$

$$\sigma_i(t)u_t(e_i,t) + \sum_{j=1}^N d_{ij}c_{ij}(t)u_{jx_j}(e_i,t) = \rho_i\big(t, u(e_i,t)\big) \quad \text{in} \quad E_3 \times [0,T] \tag{15b}$$

with

$$c_{ij} \in C^1([0,T]),\ \sigma_i > 0 \quad \text{for all } e_i \in E_K,\ j \in \{1,\ldots,N\}. \tag{16}$$

THEOREM 4: *Suppose all $\mathcal{V}_i$ are of the form (15) subject to condition (16). Let u be a solution of $(**)$. Then $\big|u_t|_{\mathcal{R}}\big|$ can be estimated from above by a constant M_2^* that depends only on $\Omega, L, R, m_0, m_1, M_0, |\psi|^{(2)}_\Omega, K_1$ and on all σ_i and c_{ij}.*

Proof: As in the proof of Theorem 3, it suffices to consider a star graph Γ with $E_r \subseteq E_K = \{e_i\}$, $N = \gamma_i \geq 1$, $l_j = \varepsilon > 0$, $\pi_j(0) = e_i$ for all $j \in \{1,...,N\}$. Furthermore we can assume that $\sum_{j=1}^N c_{ij}(t) \equiv 1$. For $y \in [0,\varepsilon]$ define

$$v(y,t) = \sum_{j=1}^N c_{ij}(t)u_j(y,t), \quad w(y,t) := \begin{cases} v(y,t) & \text{for } y \in [0,\varepsilon], \\ v(-y,t) & \text{for } y \in [-\varepsilon,0]. \end{cases}$$

Then $u(e_i,t) = v(0,t) = w(0,t)$, $u_t(e_i,t) = v_t(0,t) = w_t(0,t)$ and $\mp w_y(0\pm,t) = \rho_i\big(t,w(0,t)\big) - \sigma_i(t)w_t(0,t)$ and w satisfies the linear equation

$$w_t = w_{yy} + f^*(y,t) \qquad \text{in} \quad (-\varepsilon,\varepsilon) \times (0,T]$$

with

$$f^*(y,t) = \sum_{j=1}^N c_{ij}(t)f_j\big(|y|,t,u_j(|y|,t)\big) + \sum_{j=1}^N \dot{c}_{ij}(t)u_j(|y|,t).$$

For $e_i \in E_2$, w lies in $C^{2,1}(Q)$ with $Q := [-\varepsilon,\varepsilon] \times [0,T]$, but, for $e_i \in E_3$, $w \in C^{2,1}([-\varepsilon,0] \times [0,T]) \cap C^{2,1}([0,\varepsilon] \times [0,T])$ is in general not continuously differentiable with respect to y at $y = 0$. Anyhow, in both cases we can apply the interior gradient estimate [8.V.3] and obtain

$$|w_y|^{(0)}_Q \leq M_3 = M_3\big(M_0, K_1, m_0, |c_{ij}|^{(1)}_{[0,T]}, \Omega, |\psi(\cdot,0)|^{(1)}_G\big).$$

In the case $e_i \in E_2$ we conclude with [8.V.5] that $|u_t(e_i,t)| \le M_2^* = M_2^*(M_0, M_3, m_0, K_1, |\psi|_\Omega^{(2)}, |c_{ij}|_{[0,T]}^{(1)})$, while for $e_i \in E_3$ condition (15b) and $\sigma_i > 0$ yield directly $|u_t(e_i,t)| \le M_2^* = M_2^*(M_0, M_3, K_1, |\psi|_\Omega^{(2)}, |\sigma_i^{-1}|_{[0,T]}^{(0)}, R, m_1)$. ◊

Before stating the existence result, we introduce the compatibility condition (17): If $k_j \cap k_s = \{e_i\}$ then

$$\psi_{jx_jx_j}(e_i,0) + f_j\big(e_i,0,\psi_j(e_i,0)\big) = \psi_{sx_sx_s}(e_i,0) + f_s\big(e_i,0,\psi_s(e_i,0)\big), \tag{17a}$$

if $e_i \in E_K$, $d_{is} \neq 0$ then

$$\sigma_i(0)\Big\{\psi_{sx_sx_s}(e_i,0) + f_s\big(e_i,0,\psi_s(e_i,0)\big)\Big\} + \sum_{j=1}^{N} d_{ij}c_{ij}(0)\psi_{jx_j}(e_i,0) = \rho_i\big(0,\psi(e_i,0)\big), \tag{17b}$$

and if $e_i \in E_1$ and $d_{ij} \neq 0$ then $\psi_t(e_i,0) = \psi_{jx_jx_j}(e_i,0) + f_j\big(e_i,0,\psi_j(e_i,0)\big)$. (17c)

THEOREM 5: *Suppose the conditions (3), (5), (6a), (9), (11), and (17) are fulfilled, and let the vertex operators $\mathcal{V}_i$ be either of the form (13) subject to (14) or of the form (15) subject to (16). Suppose there is some $\alpha \in (0,1)$ such that $\psi \in C^{2+\alpha,1+\frac{\alpha}{2}}(\Omega)$ and all $f_j \in C^{\alpha,\frac{\alpha}{2},\alpha}(\Omega_j \times [-M_0,M_0])$, $\rho_i \in C^{1+\frac{\alpha}{2}}([-M_0,M_0])$ in the case (13,$\sigma_i = 0$), $\rho_i \in C^{\frac{\alpha}{2}}([-M_0,M_0])$ in the case (13,$\sigma_i > 0$), $c_i \in C^{1+\frac{\alpha}{2}}([0,T])$ in the case (15,$e_i \in E_2$), and $\sigma_i, c_i \in C^{\frac{\alpha}{2}}([0,T])$ and $\rho_i \in C^{\frac{\alpha}{2},\frac{\alpha}{2}}([0,T]\times[-M_0,M_0])$ in the case (15,$e_i \in E_3$).*
Then problem (**) has a unique solution u belonging to $C^{2+\alpha,1+\frac{\alpha}{2}}(\Omega)$.

Proof: The solvability in the class $C^{2+\alpha,1+\frac{\alpha}{2}}$ for linear parabolic network equations with general inhomogeneous nondynamical and dynamical linear Kirchhoff conditions is established in [3] and [7.III], respectively. A distinctive feature of these results is that the coefficients and inhomogeneities have to belong to $C^{1+\frac{\alpha}{2}}([0,T])$ for vanishing σ_i, while they have to lie only in $C^{\frac{\alpha}{2}}([0,T])$ for $\sigma_i > 0$. Thus, in the case (13) with $\sigma_i > 0$ in E_K or in the case (15) we can proceed as follows. Extend each f_j such that $f_j \in C^{\alpha,\frac{\alpha}{2},\alpha}(\Omega_j \times \mathbb{R}^2)$. Define a nonlinear operator $\theta : u \mapsto \theta(u)$ in the Banach space $B := \big(C^{2,1}(\Omega), |\cdot|_\Omega^{(2)}\big)$ by associating to each $u \in B$ the unique solution $v = \theta(u) \in C^{2+\alpha,1+\frac{\alpha}{2}}(\Omega)$ of the linear problem

$$(***)\qquad \begin{cases} v \in C^{2,1}(\Omega), & \\ v_{jt} = v_{jx_jx_j} + f_j\big(x_j,t,u_j\big) - f_j\big(x_j,0,u_j(x_j,0)\big) & \\ \qquad + f_j\big(x_j,0,\psi_j(x_j,0)\big) & \text{in } \Omega_{jp} \text{ for } 1 \le j \le N, \\ \sigma_i(t)v_t(e_i,t) + \sum_{j=1}^{N} d_{ij}c_{ij}(t)v_{jx_j}(e_i,t) = & \\ \quad \rho_i\big(t,u(e_i,t)\big) - \rho_i\big(0,u(e_i,0)\big) + \rho_i\big(0,\psi(e_i,t)\big) & \text{in } E_K \times [0,T], \\ v\big|_{\omega_p} = \psi\big|_{\omega_p}, & \end{cases}$$

whose unique classical solvability is guaranteed by the more general results in [7.III]. Then, Lemma 1 in [6] and Theorems 3 and 4 yield the a priori estimate for a fixpoint of θ in B. As in the proof of Theorem 5 in [6], θ is shown to be continuous and compact, so that the existence of a solution as asserted follows by the Leray - Schauder Principle.

It remains to handle the nondynamical case for vertex operators (13). Thus, suppose $\sigma_i = 0$ for some $e_i \in E_K$. For these σ_i, ρ_i belongs to $C^{1+\frac{\alpha}{2}}([-M_0,M_0])$ by assumption. For each $k \in \mathbb{N}$ let $u^{(k)} \in C^{2+\alpha,1+\frac{\alpha}{2}}(\Omega)$ denote the unique solution of $(**)$, where $\mathcal{V}_i(u,t) = 0$ for $\sigma_i = 0$, $d_{is} \neq 0$ is replaced by the dynamical condition

$$\frac{1}{k}\Big\{\psi_{jx_sx_s}(e_i,0)+f_s(e_i,0,\psi_s(e_i,0))\Big\}+\rho_i\big(u^{(k)}(e_i,t)\big)-\sum_{j=1}^{N} d_{ij}c_{ij}u^{(k)}_{jx_j}(e_i,t)-\frac{1}{k}u^{(k)}_t(e_i,t) = 0.$$

Note that the first term in this condition is independent of the incident edge k_s by (17a). By Theorem 3, the set $U = \{u^{(k)} \mid k \in \mathbb{N}\}$ is $|\cdot|^{(2)}_{\Omega}$ - bounded, and contains for any fixed $\beta \in (0,1)$ a $|\cdot|^{(1+\beta)}_{\Omega}$ - convergent subsequence that is called $\{u^{(k)} \mid k \in \mathbb{N}\}$ again. Its limit $u \in C^{1+\beta,\frac{1+\beta}{2}}(\Omega)$ satisfies $\rho_i\big(u(e_i,t)\big) - \sum_{j=1}^{N} d_{ij}c_{ij}u_{jx_j}(e_i,t) = 0$ for $\sigma_i = 0$ and $u\big|_{\omega_p} = \psi\big|_{\omega_p}$. Moreover, u is a bounded generalized solution of (10b), and therefore u is a solution of $(**)$ belonging to $C^{2+\alpha,1+\frac{\alpha}{2}}(\Omega)$ by [8.III.12].

Finally, in both cases, the uniqueness assertion follows from the Corollary. ◊

REFERENCES

1. F. Ali Mehmeti and S. Nicaise, Nonlinear interaction problems, *Pub. IRMA, Lille* Vol.23, No.V, 1991.

2. F. Ali Mehmeti and S. Nicaise, Some realizations of interaction problems, in: Ph. Clément, E. Mitidieri, and B. de Pagter (eds.): Semigroup Theory and Evolution Equations, *Lect. Not. Pure and Appl. Math., 135*: 15 -27, M. Dekker Inc. New York (1991).

3. J. v. Below, Classical solvability of linear parabolic equations on networks, *J. Differential Equ., 72*: 316-337 (1988).

4. J. v. Below, Sturm-Liouville eigenvalue problems on networks, *Math. Meth. Applied Sciences, 10*: 383-395 (1988).

5. J. v. Below, A maximum principle for semilinear parabolic network equations, in: J. A. Goldstein, F. Kappel, and W. Schappacher (eds.): Differential equations with applications in biology, physics, and engineering, *Lect. Not. Pure and Appl. Math., 133*: 37-45, M. Dekker Inc. New York (1991).

6. J. v. Below, An existence result for semilinear parabolic network equations with dynamical node conditions, in: C. Bandle, J. Bemelmans, M. Chipot, M. Grüter and J. Saint Jean Paulin (eds.): Progress in partial differential equations: elliptic and parabolic problems, *Pitman Research Notes in Math. Ser., 266*: 274 - 283, Longman Harlow Essex (1991).

7. J. v. Below, Parabolic network equations, *to appear.*

8. O. A. Ladyženskaja, V. A. Solonnikov, and N. N. Ural'ceva, Linear and quasilinear equations of parabolic type, *Amer. Math. Soc. Providence RI*, (1968).

9. G. Lumer, Connecting of local operators and evolution equations on networks, in: Potential Theory Copenhagen 1979, (Proceedings), *Lect. Not. Math., 787*: 219-234, Springer - Verlag Berlin (1980).

On the Balance Between Mutation and Frequency-Dependent Selection in Evolutionary Game Dynamics

IMMANUEL M. BOMZE and REINHARD BÜRGER University of Vienna, Vienna, Austria

Abstract. In this paper we present a dynamic model for the evolution of the type (i.e. strategy) distribution in a population subject to mutation and frequency-dependent selection. Selection is driven by an evolutionary game where individuals may behave according to mixed strategies, i.e. according to a probability distribution of a finite number n of elementary actions. The resulting dynamics is a differential equation on the Banach space of (signed) Borel measures on the standard simplex in n-dimensional Euclidean space, equipped with the variational norm. We prove existence and uniqueness of trajectories following this dynamics, and existence of stationary solutions. Then we characterize the structure of these equilibrium states and investigate the evolution of average population strategies.

1. INTRODUCTION

In this paper we try to combine two approaches to the dynamic theory of evolution at the phenotypic level. One approach is from game dynamics. It focuses on models where individuals may choose from a continuum X of mixed strategies, and goes back to Hines [8,9], Zeeman [15], and Akin [1]. The problem here is to treat differential equations on the infinite-dimensional space of probability distributions on X, which represent the states of the population with respect to the studied conflict. For our purposes, it is convenient to use the methods developed in [3], which do not rely on any absolute continuity properties of the type distributions.

In population genetics, Crow and Kimura [7] have introduced the so-called continuum-of-alleles model, which has the property that new alleles of arbitrary effect may emerge from mutation. This model has been treated by several authors (see [5] for a discussion). While game dynamics ignores mutation, but treats frequency-dependent selection, population genetic models include mutation but deal with frequency-independent selection. A further difference is that in the former case, the types (strategies) are confined to a compact space, while in the latter, types are values of quantitative traits and may vary in an unbounded domain.

Here we present a dynamic model for the evolution of the type distribution in a population subject to mutation and frequency-dependent selection. Selection is driven by an evolutionary game where individuals may behave according to mixed strategies, i.e. according to a probability distribution of a finite number n of elementary actions. The resulting dynamics is a differential equation on the Banach space of (signed) Borel measures on the standard simplex in n-dimensional Euclidean space, equipped with the variational norm.

The paper is organized as follows: in Section 2, we introduce the model and prove existence and uniqueness of trajectories. Section 3 contains an existence theorem for stationary states and a discussion of the structure of these equilibria. The final Section 4 deals with the evolution of the average population strategies.

2. THE MODEL

In the sequel we propose a model which generalizes the classical approach to evolutionary game dynamics (see, e.g. [10] and the references therein) in two aspects: (i) based upon the treatment in [3], it describes the evolution of populations consisting of individuals which may choose from a continuum of mixed strategies, and (ii) it allows for mutation, thus continuing the investigations in [5] and [6].

Let $X = S^n$ be the space of mixed strategies, where

$$S^n = \{x \in \mathbb{R}^n : x_i \geq 0\,, \sum_{i=1}^{n} x_i = 1\}$$

denotes the standard simplex in n-dimensional Euclidean space $\mathbb{R}^n$. The state of the population is described by a Borel probability distribution P on the strategy space X. A monomorphism is described by a point measure $P = \delta_z$ located at the strategy $z \in X$ used by all individuals within the population, while any other distributions P represent polymorphic states.

A mixed strategy $x \in X$ represents an individual's probability distribution over the n pure strategies represented by the standard basis vectors e_i, $1 \leq i \leq n$. Hence the mean payoff $F_P(x)$ to this individual in a population in state P amounts to

$$F_P(x) = x' f(P) = \sum_{i=1}^{n} x_i f_i(P)\,, \tag{1}$$

where x' denotes the transpose of x and $f(P) \in \mathbb{R}^n$ is the response vector consisting of the mean payoffs $f_i(P)$ to the pure strategy e_i in state P. The quantity F_P may – but need not – depend on the state P in a rather complicated way (see, e.g. [4] and the references therein for discussion and examples). Here let us only note that models where F_P depends on P in a non-linear way are frequently termed as "playing-the-field models", e.g. by Maynard Smith [13].

The average mean payoff is given by

$$E_P F_P = \int_X F_P(x)\, P(dx) = (\overline{x}_P)' f(P)\,, \tag{2}$$

where

$$\overline{x}_P = \int_X x\, P(dx) \in X\,, \tag{3}$$

represents the so-called "population strategy".

For ease of presentation we assume that the response vector $f(P)$ depends on P only through its population strategy. This means that there is a function $\overline{f} : \overline{X} \to \mathbb{R}^n$ such that $f(P) = \overline{f}(\overline{x}_P)$ for all $P \in \mathcal{P}$. Here, $\mathcal{P}$ represents the set of all admissible states of the population, and

$$\overline{X} = \{\overline{x}_P : P \in \mathcal{P}\} \subseteq X \tag{3}$$

is the set of all population strategies. The choice of $\mathcal{P}$ depends on the model. In mixed strategy games one usually chooses $\mathcal{P}$ to be the set of all Borel probability measures on X, to obtain the so-called full model. If, however, individuals play pure strategies only, all admissible states are concentrated at the vertices e_i of X:

$$\mathcal{P} = \{\sum_{i=1}^{n} x_i \delta_{e_i} : x \in X\}.$$

In both cases, $\overline{X} = X$. A more general case is the so-called discrete model

$$\mathcal{P} = \{\sum_{j=1}^{m} y_j \delta_{a_j} : y \in S^m\},$$

investigated by Bürger [5] for $X = S^2$, where $a_j \in X$ are arbitrary but fixed, possibly mixed strategies. Here $\overline{X} = \mathrm{co}\,(a_1, \ldots, a_m)$, the convex hull of the a_j's.

Example 1: The special case of pairwise contests emerges if and only if there is some $n \times n$-matrix A such that $\overline{f}(x) = Ax$ for all $x \in \overline{X}$ [4]. Then $F_P(x) = x'A\overline{x}_P$ and the average mean payoff is $E_P F_P = (\overline{x}_P)'A\overline{x}_P$, i.e. a quadratic form in the population strategy.

Taylor and Jonker [14] introduced game dynamics for discrete models where individuals play pure strategies only. This approach has been generalized for important classes of mixed strategy models by Hines [8,9], Zeeman [15], and Akin [1]. For the general setup considered here, it is convenient to use the game (or replicator) dynamics as proposed and investigated in [3]:

$$\dot{P} = \Phi(P) = [F_P - E_P F_P] \cdot P, \tag{4}$$

where $\dot{P}$ represents the time derivative of P, taken with respect to the variational norm. This means that the limit $\dot{P}_t(B) = \lim_{h \to 0} h^{-1}[P_{t+h}(B) - P_t(B)]$ is performed uniformly over all Borel subsets B of X. For a measurable function $\varphi : X \to \mathbb{R}$, the symbol $\varphi \cdot P$ designates the signed Borel measure defined by $\varphi \cdot P(B) = \int_B \varphi(x)\, P(dx)$, where $B \subseteq X$ is a Borel set.

Now we introduce mutation. To this end consider a continuous, strictly positive function $u : X \times X \to (0, +\infty)$ and a probability measure λ on X (e.g. the uniform distribution on X for the full model, or $\frac{1}{m}\sum_{j=1}^{m} \delta_{a_j}$ for the discrete model) such that

$$\mu(x) = \int_X u(y,x)\, \lambda(dy) \leq \|\mu\|_\infty < \infty \quad \text{for all } x \in X, \tag{5}$$

where $\|\mu\|_\infty = \sup_{x \in X} \mu(x)$.

Denote by $\mathcal{L}$ the so-called L-space generated by $\mathcal{P} \cup \{\lambda\}$, which is the smallest Banach space of signed Borel measures on X that contains all measures of the form $\varphi \cdot P + \psi \cdot \lambda$, where $P \in \mathcal{P}$, and φ and ψ are bounded measurable functions from X to $\mathbb{R}$ [2, Proposition 1.1]. In the full model, $\mathcal{L}$ is the space of all signed Borel measures on X. In discrete models, $\mathcal{L} = \{\sum_{j=1}^{m} y_j \delta_{a_j} : y \in \mathbb{R}^m\}$. Put

$$\mathcal{L}_+^1 = \{Q \in \mathcal{L} : Q \geq o \text{ and } Q(X) = 1\}.$$

Then $\mathcal{P} \subseteq \mathcal{L}_+^1$ in general, but with equality in the full and in the discrete models.

The linear operator $U : \mathcal{L} \to \mathcal{L}$ given by

$$UP(B) = \int_B [\int_X u(y,x)\, P(dx)]\lambda(dy) \tag{6}$$

is well defined and bounded by $\|\mu\|_\infty$. Each UP is absolutely continuous with respect to λ, and we denote by

$$u_P(y) = \int_X u(y,x)\, P(dx) > 0\,, \quad y \in X\,, \tag{7}$$

the Radon-Nikodym density of UP with respect to λ.

Assuming that mutation acts independently of selection, we obtain from the replicator dynamics (4)

$$\begin{aligned} \dot{P} = \Psi(P) &= \Phi(P) + UP - \mu \cdot P \\ &= m_P \cdot P + UP\,, \end{aligned} \tag{8}$$

where

$$m_P(x) = F_P(x) - E_P F_P - \mu(x) = (x - \overline{x}_P)' \overline{f}(\overline{x}_P) - \mu(x)\,, \quad x \in X\,. \tag{9}$$

Equation (8) describes the joint action of frequency-dependent selection and mutation. In the discrete case, one obtains the (diploid) selection-mutation model for $\overline{f}(x) = Mx$, where M is a symmetric matrix, cf. the discussion in [5, Section 3]. For mutation and frequency-independent selection, a similar but more general model (X an arbitrary locally compact space) has been proposed by the authors [6].

Theorem 1 (Existence and uniqueness of trajectories). *Assume that $\overline{f}$ is defined in an open neighbourhood of $\overline{X}$, and is continuously differentiable there. Then for any $P \in \mathcal{P}$, there is a unique trajectory $P(t)$ for all $t \geq 0$ satisfying (8) with $P(0) = P$. Furthermore, $P(t)$ is a probability measure on X with $P \in \mathcal{L}_+^1$ for all $t \geq 0$. In particular, if $\mathcal{P} = \mathcal{L}_+^1$ (e.g. in full or discrete models), then $\mathcal{P}$ is positively invariant under (8).*

Proof: The map Ψ as defined in Eq. (8) satisfies a global Lipschitz condition if and only if the map Φ from Eq. (4) does. Hence, similar arguments as in the proofs of Theorem 8 and Lemma 1 in [3] show existence and uniqueness of trajectories $P(t) \in \mathcal{L}$, $t \geq 0$, satisfying Eq. (8). From $[1 - P(t)(X)]^{\bullet} = E_{P(t)} F_{P(t)} [1 - P(t)(X)]$ and $P(0)(X) = 1$ it follows that $P(t)(X) = 1$ for all $t \geq 0$. However, $P(t)$ might be a signed Borel measure on X. It thus remains to show positivity of the solutions. To this end, we proceed as follows: First observe that due to continuity there is a $\delta > 0$ such that $m_{P(t)}(x) = (x - \overline{x}_{P(t)})' \overline{f}(\overline{x}_{P(t)}) - \mu(x)$ is well defined for all $x \in X$ and all $t \in [-\delta, \delta]$. Then define for $x \in X$

$$\varphi_t(x) = \begin{cases} m_{P(t)}(x), & \text{if } |t| \leq \delta, \\ m_{P(\delta)}(x), & \text{otherwise.} \end{cases}$$

Now consider the auxiliary, non-autonomous dynamics

$$\dot{Q} = \varphi_t \cdot Q + UP(t). \tag{10}$$

Then there is a constant K such that $\|\varphi_t\|_\infty \leq K$ for all $t \in \mathbb{R}$ and therefore existence and uniqueness of solutions to (10) are also guaranteed [12]. We now specify an explicit formula for the solution to (10) with $Q(0) = P(0)$: denote by $\psi_0(x) \in [0,1]$ the Radon-Nikodym density of $P(0)$ with respect to the measure $P(0) + \lambda$ and define, for $t \in \mathbb{R}$ and $x \in X$,

$$\psi_t(x) = \exp\Big(\int_0^t \varphi_s(x)\,ds\Big)\Big\{\psi_0(x) + [1 - \psi_0(x)] \int_0^t u_{P(s)}(x) \exp\Big(-\int_0^s \varphi_r(x)\,dr\Big)\,ds\Big\}. \tag{11}$$

Now let $\overline{u} = \min\{u(y,x) : (x,y) \in X \times X\} > 0$. Then (7) yields the inequality $u_{P(0)}(x) \geq \overline{u} > 0$ for all $x \in X$. Since the maps $s \mapsto u_{P(s)}(x)$ are equicontinuous as x ranges over X, we obtain $u_{P(s)}(x) \geq \frac{\overline{u}}{2} > 0$ for all $x \in X$ and all $s \in [0,t]$ whenever $t > 0$ is small enough. Therefore, for these t we have $\psi_t(x) \geq 0$, since also $\psi_0(x) \geq 0$ holds for all $x \in X$. Now put $Q(t) = \psi_t \cdot [P(0) + \lambda] \in \mathcal{L}$ for $t \in \mathbb{R}$. Then $Q(t) \geq o$ if $t > 0$ is small enough, and evidently $\dot{Q}(t) = \dot{\psi}_t \cdot [P(0) + \lambda]$, where

$$\dot{\psi}_t(x) = \frac{d}{dt}\psi_t(x) = \varphi_t(x)\psi_t(x) + [1 - \psi_0(x)] u_{P(t)}(x), \quad x \in X.$$

Hence $\dot{Q}(t) = \varphi_t \cdot (\psi_t \cdot [P(0) + \lambda]) + u_{P(t)} \cdot \lambda = \varphi_t \cdot Q(t) + UP(t)$, so that (10) is satisfied for all $t \in \mathbb{R}$. But of course also $P(t)$ satisfies (10) for $|t| < \delta$ by construction, so that, due to uniqueness of solutions, $P(t) = Q(t) \geq o$ if $t > 0$ is small enough. Since $P(t)(X) = 1$ for all t holds, all $P(t)$ are probability measures for small enough $t > 0$. For the same reason we have $P(t + \varepsilon) \in \mathcal{L}^1_+$ if $\varepsilon > 0$ is small enough, provided that $P(t) \in \mathcal{L}^1_+$. Because the set $\{t \geq 0 : P(t) \in \mathcal{L}^1_+\}$ is closed, this proves $P(t) \in \mathcal{L}^1_+$ for all $t \geq 0$. $\square$

The preceding arguments show that $P(t)$ has the Radon-Nikodym density ψ_t as defined in (11) with respect to the measure $P(0)+\lambda$, and that for any $s>0$ and $t>0$, both $P(s)$ and $P(t)$ have the same null sets.

3. EQUILIBRIUM CONDITIONS

First let us settle the existence of an equilibrium (stationary state) P^* of (8) by an abstract fixed point argument.

Theorem 2 (Existence of stationary states). *Suppose that $\mathcal{P}=\mathcal{L}^1_+$ is weak-star closed (as in the cases of discrete models and the full model) and assume that $\overline{f}$ satisfies the conditions in Theorem 1. Then there is at least one stationary state P^*.*

Proof: Since $\overline{f}$ is continuous on the compact set X, we may pick a constant $c>0$ satisfying

$$m_P(x) \geq -c \quad \text{for all } x \in X \text{ and all } P \in \mathcal{P}\,.$$

Now define $\gamma = \frac{1}{2c} > 0$ and put

$$\Xi(P) = P + \gamma\Psi(P)\,, \qquad P \in \mathcal{P}\,,$$

where $\Psi(P)$ denotes the right-hand side of (8). Then $\Psi(P^*)=o$ if and only if P^* is a fixed point of the transformation Ξ which maps $\mathcal{P}$ into itself. This is a consequence of (i) $\Xi(P)\in\mathcal{L}$ by definition of Ξ and $\mathcal{L}$; (ii) $\Xi(P)(X) = P(X)+\gamma\Psi(P)(X) = 1$; (iii) $\Xi(P) \geq P - \gamma c P + \gamma U P \geq \frac{1}{2}P \geq o$; and (iv) $\mathcal{P}=\mathcal{L}^1_+$.
Now it is easily seen that $m_P(x)$ depends continuously on P uniformly in $x\in X$ and $P\in\mathcal{P}$, if we equip $\mathcal{P}$ with the weak star topology, with respect to which $\mathcal{P}$ is a compact, convex set. Furthermore, from continuity of u, we obtain weak-star continuity of U. Hence we conclude that Ξ is a weak star-continuous map from $\mathcal{P}$ into itself. According to Brouwer's fixed point theorem, Ξ has a fixed point P^*. □

For frequency-independent selection (which means that $\overline{f}$ is a constant) and $X = S^2$ our model reduces to Kingman's mutation-selection model [11], which has been generalized to non-compact X and unbounded, non-linear fitness functions $F_P(x) = m(x)$ in [6]. Here we use ideas from this paper to investigate the structure of stationary states for frequency dependent selection. Note that the examples for non-existence of stationary states in [6] do not apply here, since $X=S^n$ is compact for evolutionary games.

The above Theorem 2 is an abstract existence result and does not reveal the structure of stationary states. To investigate this structure, we decompose P^* into an absolutely continuous and a singular part with respect to λ:

$$P^* = (1-\rho)R + \rho S \,. \tag{12}$$

Here R is a probability measure having a Radon-Nikodym density r with respect to λ, and S is a probability measure satisfying $S(T) = 1$ for some Borel set $T \subset X$ with $\lambda(T) = 0$. Now denote the equilibrium response vector by $f^* = f(P^*)$, the equilibrium population strategy by $x^* = \overline{x}_{P^*}$, and let $m^* = m_{P^*}$ as well as $u^* = u_{P^*}$. Next observe that, due to (7), (8) and (9), the equilibrium condition amounts to the two conditions

$$(1-\rho)[(x^* - y)'f^* + \mu(y)]r(y) = u^*(y) \quad \text{for } \lambda-\text{almost all } y \in X \,, \tag{13a}$$

$$\rho m^* \cdot S = o \,. \tag{13b}$$

From Eq. (12) we get

$$u^*(y) = (1-\rho)\int_X u(y,x)r(x)\,\lambda(dx) + \rho\int_X u(y,x)\,S(dx) \quad \text{for all } y \in X \,. \tag{14}$$

Positivity of u has several important consequences. Before we formulate and prove them in Theorem 3 below, we need some further notation: For $(x,y) \in \overline{X} \times X$ let $g_x(y) = (x-y)'\overline{f}(x) + \mu(y)$ and put

$$X_+ = \{x \in \overline{X} : g_x \geq 0 \text{ and } 1/g_x \in L^1(\lambda)\} \,. \tag{15}$$

Observe that X_+ is not empty provided $\overline{f}$ is continuous on $\overline{X} = X$. Indeed, it is well known that there is an $x \in \overline{X}$ such that

$$(x-y)'\overline{f}(x) \geq 0 \quad \text{for all } y \in X \,.$$

Because $u(y,x) \geq \overline{u} > 0$, we have $g_x \geq \overline{u} > 0$ and hence $x \in X_+$. Now define for all $x \in X_+$ the positive linear operator $A_x : L^1(\lambda) \to L^1(\lambda)$ via

$$A_x r(y) = \frac{1}{(x-y)'\overline{f}(x) + \mu(y)} \int_X u(y,z)r(z)\,\lambda(dz) \,, \quad y \in X \,. \tag{16}$$

The introduction of this operator is motivated by the observation that for $\rho = 0$, i.e. for an equilibrium state P^* which is absolutely continuous with respect to λ, we obtain from (13a) and (14) the integral equation

$$[(x^* - y)'f^* + \mu(y)]r(y) = \int_X u(y,z)r(z)\,\lambda(dz) \quad \text{for } \lambda-\text{almost all } y \in X \,. \tag{17}$$

If we now consider x as an $(n-1)$-dimensional parameter, Eq. (17) can be reformulated as the following problem: For which values of $x = x^*$ does the operator A_x defined in (16) have a positive eigensolution to the eigenvalue 1 ?

Theorem 3 (Structure of equilibria). *At an equilibrium $P^* = (1-\rho)R + \rho S$, we have*

(a) $0 \le \rho < 1$, i.e. a purely singular equilibrium is impossible.

(b) $r(y) > 0$ for $\lambda-$almost all $y \in X$.

(c) $m^*(y) < 0$ for $\lambda-$almost all $y \in X$.

(d) If $0 < \rho < 1$, i.e. if there is a singular part S, then S has to be concentrated on the surface of zeroes of m^*

$$H_{x^*} = \{y \in X : (y - x^*)' f^* = \mu(y)\}\,. \tag{18}$$

(e) If $m^*(y) < 0$ holds even for all $y \in X$, then $\rho = 0$, i.e. P^* is absolutely continuous w.r.t. λ. In this case there is no other equilibrium state P with the same population strategy $\overline{x}_P = x^*$.

Proof: Assume that $\rho = 1$; then from Eqs. (13a) and (14) we got the contradiction

$$0 < \int_X \mu(x)\, S(dx) = \int_X u^*(y)\, \lambda(dy) = 0\,.$$

Hence assertion (a) is established. From positivity of u we deduce positivity of u^*, thus assertions (b) and (c) follow immediately, using again (13a). Similarly, Eq. (13b) implies assertion (d), since H_{x^*} is the set of zeroes of m^*. If this set is empty, then of course $\rho = 0$ has to hold. Finally, the last part of assertion (e) follows from a Perron-Frobenius argument: to this end observe that continuity of $g_{x^*} = -m^*$ on the compact set X yields the existence of a constant $k > 0$ with $g_{x^*}(y) \ge k > 0$ for all $y \in X$, so that $x^* \in X_+$ and that the operator A_{x^*} defined in Eq. (16) is even bounded (and of course positive). Hence there is a unique positive eigensolution of A_{x^*} to the eigenvalue 1. □

Remark: Suppose the assumptions of Theorem 2 are met. If mutation is strong enough, i.e. if the terms $u(y, x)$ are sufficiently large, then we may achieve

$$(x - y)'\overline{f}(x) < \mu(y) \quad \text{for all } (x, y) \in X \times X\,, \tag{19}$$

so that H_{x^*} is empty for all stationary states P^* (we know that there is at least one). Due to Theorem 3(e), this means that there are only absolutely continuous equilibria, which are uniquely determined by their population strategy. Furthermore, we have $X_+ = \overline{X}$, and every A_x is bounded. In this case one can specify sufficient conditions on the behaviour of the response function $\overline{f}$ which guarantee that there is a unique equilibrium (Bomze and Bürger, in preparation). For an important class of models, this unique equilibrium is globally stable in the sense that $P(t) \to P^*$ as $t \to \infty$ for all initial distributions $P(0) \in \mathcal{P}$. See [5] for similar results in the discrete model where

$X = S^2$. It might be interesting to note that the uniqueness conditions are independent of the form of u, provided only that (19) holds.

If there is a singular part S, i.e. if $\rho > 0$, then $m^*(y) = 0$ for S–almost all $y \in X$. Any such state, i.e. any S with $S(H_{x^*}) = 1$ gives rise to a candidate for an equilibrium state, provided that it admits a positive solution $r \in L^1(\lambda)$ to the linear equation

$$(I - A_{x^*})r(y) = \frac{\rho}{1-\rho} \int_X \frac{u(y,x)}{(x^* - y)'f^* + \mu(y)} S(dx) \,. \tag{20}$$

Here $\rho \in (0,1)$ has to be chosen such that $\int_X r(x)\,\lambda(dx) = 1$. Observe that equation (17) can be viewed as the limiting case of (20) for $\rho \searrow 0$. Irrespective of whether $\rho = 0$ or $0 < \rho < 1$, any solution to (20) with $0 \le \rho < 1$ does not automatically yield a fixed point of (8), but has to satisfy in addition the compatibility condition:

Theorem 4 (Compatibility of parameters as population strategies). *Assume that for some $x^* \in \overline{X}$, and some S concentrated on H_{x^*} as defined in (18), where $f^* = \overline{f}(x^*)$, equation (20) has a solution $P = (1-\rho)r \cdot \lambda + \rho S \in \mathcal{P}$ for some ρ with $0 \le \rho < 1$. Then*

(a) the corresponding population strategy $\overline{x}_P = (1-\rho)\int_X x\, r(x)\lambda(dx) + \rho \int_X x\, S(dx)$ satisfies

$$(x^* - \overline{x}_P)'f^* = 0\,;$$

(b) furthermore, P is a stationary solution of (8) if and only if

$$x^* = \overline{x}_P \,.$$

Proof: Integrating (13a) over X with respect to λ, we obtain via (14)

$$(1-\rho)[(x^* - \overline{x}_R)'f^* + \int_X \mu(y)r(y)\,\lambda(dy)] = (1-\rho)\int_X \mu(x)r(x)\,\lambda(dx) + \rho \int_X \mu(x)\,S(dx)$$

and hence $(1-\rho)(x^* - \overline{x}_R)'f^* = \rho \int_X \mu(x)\,S(dx)$. In case of $\rho = 0$ this means $R = P$ and $(x^* - \overline{x}_P)'f^* = 0$. If, on the other hand, $0 < \rho < 1$, we deduce from condition (17) that

$$\int_X \mu(y)\,S(dy) = \int_X (y - x^*)'f^*\,S(dy) = (\overline{x}_S - x^*)'f^* \,.$$

Therefore we arrive at assertion (a):

$$(x^* - \overline{x}_P)'f^* = (1-\rho)(x^* - \overline{x}_R)'f^* + \rho(x^* - \overline{x}_S)'f^* = 0\,.$$

Assertion (b) is evident. □

Example 2: Let $u(y,x) = u_1(y) = 1 = \overline{\mu}$ for all $y \in X$, consider the full model and denote by λ the uniform distribution on X so that $\lambda(X) = 1$. Let $\overline{f}$ be an arbitrary response function which has a zero at some point $x^* \neq \frac{1}{n}e$, the barycentre of X (we put $e = [1,\ldots,1]' \in \mathbb{R}^n$). Then $(x^* - y)'f^* + \overline{\mu} = 1$ and (17) has the solution $r(y) = 1$, i.e. the probability $P = \lambda \in \mathcal{P}$ solves (20) with population strategy $\overline{x}_P = \frac{1}{n}e \neq x^*$. Hence the condition in Theorem 4(b) is not satisfied, and therefore P is not a stationary solution of (8).

For the remainder of this section we consider an important class of models characterized by $u(y,x) = u_1(y)$: the so-called house-of-cards model [11] In this case $\mu(y) = \overline{\mu}$ for all $y \in X$, and Eq. (20) reduces to

$$r(y) = \frac{1}{1-\rho}\,\frac{u_1(y)}{(x^* - y)'f^* + \overline{\mu}}\,,$$

independent of the form of S. Here $x^* \in X_+$ has to be chosen such that $\sigma(x^*) \geq 1$, where

$$\sigma(x) = \int_X \frac{u_1(y)}{(x-y)'\overline{f}(x) + \overline{\mu}}\,\lambda(dy)\,,$$

and $\rho = 1 - \sigma(x^*)^{-1}$. If such an x^* satisfies $\sigma(x^*) > 1$, then $\rho > 0$ and hence the support H_{x^*} of S (which here is a part of a hyperplane) must not be empty. In any case, the compatibility condition in Theorem 4(b) has to be satisfied in order that this x^* gives rise to a stationary solution of (8). Note that such a solution always exists due to Theorem 2.

4. EVOLUTION OF POPULATION STRATEGIES

Let us turn our attention to the time evolution of the population strategies $\overline{x}_P$. A short calculation reveals that

$$\begin{aligned}(\overline{x}_P)^{\cdot} &= \int_X y\,\dot{P}(dy)\\ &= \int_X y[(y-\overline{x}_P)'\overline{f}(\overline{x}_P) - \mu(y)]\,P(dy) + \int_X y[\int_X u(y,x)\,P(dx)]\,\lambda(dy)\\ &= C_P\overline{f}(\overline{x}_P) + k_P\,,\end{aligned} \tag{21}$$

where $C_P = \int_X (x-\overline{x}_P)(x-\overline{x}_P)'\,P(dx)$ denotes the covariance of x over X with respect to the distribution P, while

$$k_P = \int_X [\int_X (y-x)u(y,x)\,\lambda(dy)]P(dx)\,. \tag{22}$$

Hence at equilibrium the quantities x^* and f^* have to satisfy the equation

$$C_{P^*} f^* = -k_{P^*} \,. \tag{23}$$

In the house-of-cards case introduced at the end of the previous section, we have $k_P = \int_X y u_1(y)\, \lambda(dy) - \overline{\mu}\, \overline{x}_P$, so that Eq. (23) can be simplified to

$$C_{P^*} f^* = \overline{\mu}\, x^* - \int_X y u_1(y)\, \lambda(dy)\,. \tag{24}$$

Example 1, continued: Under the house-of-cards assumption, we obtain for pairwise contests the "linear" equation for the population strategy

$$[\overline{\mu} I - C_{P^*} A] x^* = \int_X y u_1(y)\, \lambda(dy)\,. \tag{25}$$

Note that only the first two moments of P^* enter this condition. The following arguments show that fixation of population strategies (in the sense that Eq. (25) holds) does not already imply fixation of the state P of the population. Indeed, it is easy to see that whenever $v = \frac{1}{\overline{\mu}} \int_X y u_1(y)\, \lambda(dy) \in X$ belongs to the relative interior of the standard simplex X, there are several probability distributions P which are (absolutely continuous with respect to λ and) concentrated on X with $\overline{x}_P = x^*$ and a covariance matrix $C_P = C$ such that

$$[I - CA] x^* = v$$

holds. Indeed, there is a $\delta > 0$ and an $\eta > 0$ with the following properties:

(a) whenever an $x \in \mathbb{R}^n$ satisfies $e'x = \sum_{i=1}^n x_i = 1$ and $|x - v| < \delta$, then $x \in X$ (where $|.|$ is an arbitrary norm on $\mathbb{R}^n$);

(b) for any such x and any positive semidefinite matrix C with $Ce = o$ and $\|C\| = \sup\{|Cy| : |y| \leq 1\} < \eta$, there is a probability distribution P concentrated on X with $\overline{x}_P = x$ and $C_P = C$.

Now pick such a matrix C which in addition satisfies $\|CA\| < \frac{\delta}{1+\delta} < 1$, and define

$$x^* = [I - CA]^{-1} v = v + \sum_{k=1}^{\infty} (CA)^k v \in \mathbb{R}^n\,.$$

Then $e'x^* = e'v + 0 = 1$ and $|x^* - v| \leq \frac{\|CA\|}{1 - \|CA\|} < \delta$, so that $x^* \in X$ satisfies (a). Of course not any P fulfilling (b) for $x = x^*$ is fixed under the dynamics (8): in case that the chosen P (is absolutely continuous with respect to λ and) satisfies (17) per coincidence, we may alter it in such a way that the first two moments of P are not affected, but that (17) is now violated.

Acknowledgment. The authors are indebted to B. Pötscher for helpful suggestions.

REFERENCES

[1] AKIN E.: Exponential families and game dynamics. Can. J. Math. **34**, 374–405 (1982).

[2] BOMZE I.M.: A functional analytic approach to statistical experiments. London: Longman 1990.

[3] BOMZE I.M.: Cross entropy maximization in uninvadable states of complex populations. J. math. Biol. **30**, 73—87 (1991).

[4] BOMZE I.M./PÖTSCHER B.M.: Game theoretical foundations of evolutionary stability. Berlin Heidelberg New York: Springer 1989.

[5] BÜRGER R.: Mutation-selection models in population genetics and evolutionary game theory. Acta Applicandae Mathematicae **14**, 75–89 (1989).

[6] BÜRGER R./BOMZE I.M.: Mutation-selection balance and a perturbation theorem on the space of measures. Technical Report **111**, Inst. f. Statistics & Comp. Sci., Univ. Vienna (1991).

[7] CROW, J.F. AND KIMURA, M.: "The theory of genetic loads". Proc. of the XIth Intern. Congr. of Genetics, Pergamon, Oxford, pp. 495-505 (1964).

[8] HINES W.G.S.: Three characterizations of population strategy stability. J. Appl. Prob. **17**, 333–340 (1980a).

[9] HINES W.G.S.: Strategy stability in complex populations. J. Appl. Prob. **17**, 600–610 (1980b).

[10] HOFBAUER J., AND SIGMUND K.: The theory of evolution and dynamical systems. Cambridge: Univ. Press 1988.

[11] KINGMAN J.F.C.: A simple model for the balance between genetic mutation and selection. J. Appl. Prob. **15**, 1–12 (1978).

[12] LANG S.: Differential manifolds. Reading, Mass.: Addison-Wesley 1972.

[13] MAYNARD SMITH J.: Evolution and the theory of games. Cambridge: Univ. Press 1982.

[14] TAYLOR P./JONKER L.: Evolutionarily stable strategies and game dynamics. Math. Biosci. **40**, 145–156 (1978).

[15] ZEEMAN E.C.: Dynamics of the evolution of animal conflicts. J. theor. Biol. **89**, 249–270 (1981).

Uniform Stabilization for the Solutions to a von Kármán Plate

M. E. BRADLEY University of Louisville, Louisville, Kentucky

I. LASIECKA University of Virginia, Charlottesville, Virginia

1 Introduction

1.1 Statement of the Problem

Let Ω be a bounded open domin in R^2 with smooth boundary $\Gamma = \Gamma_0 \cup \Gamma_1$, where Γ_i are relatively open and we admit the possibility that $\Gamma_0 = \emptyset$. We consider the von Kármán system in the variables $w(t,x)$ and $\chi(w(t,x))$:

$$\left.\begin{array}{ll} w_{tt} - \gamma^2 \Delta w_{tt} + \Delta^2 w + b(x) w_t = [w, \chi(w)] & \text{in } Q \\ w(0,\cdot) = w_0 \ ; \ w_t(0,\cdot) = w_1 & \text{in } \Omega \\ w = \frac{\partial}{\partial \nu} w = 0 & \text{on } \Sigma_0 \\ \Delta w + (1-\mu) B_1 w = -\frac{\partial}{\partial \nu} w_t & \text{on } \Sigma_1 \\ \frac{\partial}{\partial \nu} \Delta w + (1-\mu) B_2 w - \gamma^2 \frac{\partial}{\partial \nu} w_{tt} = w_t + w - \frac{\partial^2}{\partial \tau^2} w_t & \text{on } \Sigma_1 \end{array}\right\} \tag{1.1}$$

where $b(x) \in L^\infty(\Omega)$ satisfies $b(x) > 0$ *a.e.* in Ω and where the operators B_1 and B_2 are

$$\left.\begin{array}{lll} B_1 w & = & 2n_1 n_2 w_{xy} - n_1^2 w_{yy} - n_2^2 w_{xx} \\ B_2 w & = & \frac{\partial}{\partial \tau}[(n_1^2 - n_2^2) w_{xy} + n_1 n_2 (w_{yy} - w_{xx})] \end{array}\right\} (1.1)(b).$$

Here, $\chi(w)$ satisfies the system of equations

$$\left.\begin{array}{ll} \Delta^2 \chi = -[w,w] & \\ \chi = \frac{\partial}{\partial \nu}\chi = 0 & \text{on } \Sigma = \Gamma \times (0,\infty) \end{array}\right\} \tag{1.2}$$

where

$$[\phi, \psi] = \frac{\partial^2 \phi}{\partial x^2}\frac{\partial^2 \psi}{\partial y^2} + \frac{\partial^2 \phi}{\partial y^2}\frac{\partial^2 \psi}{\partial x^2} - 2\frac{\partial^2 \phi}{\partial x \partial y}\frac{\partial^2 \psi}{\partial x \partial y}.$$

If we define the bilinear form

$$a(w,v) = \int_\Omega (\Delta w \Delta v + (1-\mu)(2w_{xy}v_{xy} - w_{xx}v_{yy} - w_{yy}v_{xx}))d\Omega \tag{1.3}$$

then the energy of the related uncontrolled von Kármán system (without damping) is

$$E(t) = \frac{1}{2}\int_\Omega \{|w_t|^2 + \gamma^2 |\nabla w_t| + |\Delta \chi|^2\} d\Omega + \frac{1}{2} a(w,w) + \frac{1}{2}\int_{\Gamma_1} |w|^2 d\Gamma_1 \equiv E_1(t) + E_2(t), \tag{1.4}$$

where $E_2(t)$ is defined by
$$E_2(t) = \frac{1}{2}\int_\Omega |\Delta \chi|^2 d\Omega.$$

Our goal is to establish an exponential decay rate for the energy (1.4) by implementing the boundary controls $\frac{\partial}{\partial \nu} w_t$, w_t and $\frac{\partial^2}{\partial \tau^2} w_t$.

1.2 Literature and Orientation

The problem of stabilization and controllability for the von Kármán plate described above has attracted much attention in recent years. Indeed, the results on local controllability and stabilization can be found in works by Bradley and Lasiecka (to appear) and Lagnese

(1989 and 1990) . As for the question of global decay rates (such as we consider here) we refer to Lagnese (1989). There, it was proved that for the unperturbed von Kármán model without rotational inertia or viscous damping (i.e. setting $\gamma = 0$ and $b = 0$) the energy of the resulting system is exponentially stable. This stability was achieved by means of the boundary feedback acting on the whole boundary (i.e. $\Gamma_0 = \emptyset$).

Our main contributions in this paper are the following: (i) We include rotational moments of inertia; (ii) we assume that only a portion of the boundary is available for control actions (i.e. $\Gamma_0 \neq \emptyset$).

It should be noted that the above mentioned features contribute to some major technical difficulties in the problem and require techniques quite different from others found in the literature. Indeed, the basic approach developed in works by Lagnese (1989 and 1991) and Lagnese and Leugering (1991) is based on the construction of an appropriate Lyapunov function which does not appear to be applicable for our problem. This is due to the presence of certain boundary terms (related to the nonlinearity $\chi(w)$) in the estimates for the Lyapunov function. These terms destroy the "good structure" of these estimates, making it impossible to prove a desired differential inequality for the Lyapunov function.

To cope with this difficulty, we propose a technique (see Lasiecka (to appear)) which is based on proving certain functional relations directly for the energy function. This allows us to "build in" an appropriately developed nonlinear compactness-uniqueness argument to "absorb" undesirable lower order terms arising from our energy estimates. Finally, a nonlinear semigroup argument leads to the desired decay rate for the solution to system (1.1). This method is fairly general and allows us to incorporate many other unstructured perturbations and can be applied when the usual Lyapunov technique fails (see Bradley (to appear) where the perturbed Kirchoff plate was considered). On the other hand, the limitation of the method is that we no longer have an exact estimate for the decay rate. (This information is lost in the compactness-uniqueness argument, which is proved by contradiction).

1.3 Statement of Results

We begin by defining the space of finite energy, $\mathcal{H} \equiv H^2_{\Gamma_0}(\Omega) \times H^1_{\Gamma_0}(\Omega)$ where

$$H^2_{\Gamma_0}(\Omega) = \{w \in H^2(\Omega) : w = \tfrac{\partial}{\partial \nu} w = 0 \text{ on } \Gamma_0\}$$

and

$$H^1_{\Gamma_0}(\Omega) = \{w \in H^1(\Omega) : w = 0 \text{ on } \Gamma_0\}.$$

(Observe that in the case $\Gamma_0 = \emptyset$, these reduce to the usual $H^2(\Omega)$ and $H^1(\Omega)$, respectively).

We now state the main result of this paper.

Theorem 1.1 Main Theorem *Assume that the domain $\Omega \subset R^2$ has a sufficiently smooth boundary, $\Gamma = \Gamma_0 \cup \Gamma_1$ and that there exists $x_0 \in R^2$ such that Γ_0 satisfies*

$$\mathbf{h} \cdot \nu \equiv (x - x_0) \cdot \nu \leq 0 \qquad \text{for } x \in \Gamma_0 \tag{1.5}$$

Then for any initial data, $(w_0, w_1) \in \mathcal{H} \equiv H^2_{\Gamma_0}(\Omega) \times H^1_{\Gamma_0}(\Omega)$, there exists a constant C and a constant $\alpha = \alpha(\|w_0\|_{H^2_{\Gamma_0}(\Omega)}, \|w_1\|_{H^1_{\Gamma_0}(\Omega)})$ such that the solution pair for (1.1) satisfies

$$\|(w(t), w_t(t))\|_{\mathcal{H}} \leq Ce^{-\alpha t}\{\|(w_0, w_1)\|_{\mathcal{H}}\}.$$

To prove this theorem we will need several propositions and lemmas, the first of which is the well-posedness of system (1.1) and a related regularity result. These results can be proven by combining the results of Bradley (to appear) with those by Favini and Lasiecka (to appear).

Proposition 1.1 Well-posedness*(i) Given initial data* $(w_0, w_1) \in \mathcal{H} \equiv H^2_{\Gamma_0}(\Omega) \times H^1_{\Gamma_0}(\Omega)$, *there exists a unique solution* (w, w_t) *to (1.1) with*

$$w \in C([0,\infty); H^2_{\Gamma_0}(\Omega)) \quad \text{and} \quad w_t \in C([0,\infty); H^1_{\Gamma_0}(\Omega)). \tag{1.6}$$

(ii) There exists a set $D \subset H^4(\Omega) \times H^3(\Omega)$ *such that* $D \overset{dense}{\subset} \mathcal{H}$ *and such that if* $(w_0, w_1) \in D$ *then the solution pair* (w, w_t) *satisfies*

$$\begin{gathered} w \in C([0,T]; H^4(\Omega) \cap H^2_{\Gamma_0}(\Omega)) \\ w_t \in C([0,T]; H^3(\Omega) \cap H^2_{\Gamma_0}(\Omega)) \text{ and } w_{tt} \in C([0,T]; H^2_{\Gamma_0}(\Omega)) \end{gathered} \tag{1.7}$$

for any $0 < T < \infty$.

Remark: It is interesting to note the role played by the "light" internal damping modeled by $b(x)w_t$. It is obvious that this damping alone (i.e. without the dissipation on the boundary) would not suffice to uniformly stabilize the plate. ("Strong" internal damping which would cause the energy to decay exponentially would be of the form $b(x)\Delta w_t$ where $b(x) \geq b_0 > 0$). On the other hand, the presence of this mild damping seems to be essential in proving the effectivness of the boundary dissipation. Thus, the combination of both the light interior damping and the boundary damping appears to provide the desired energy decay. We only note that the presence of light interior damping is physically motivated, since most vibrating materials possess some degree of interior damping.

The remainder of this paper is organized as follows. We first set out the preliminary energy estimates which will be used in proving the main theorem. These will possess lower order terms which will then be absorbed by a compactness-uniqueness argument. We conclude with an argument from noninear semigroup theory, which will complete the proof of Theorem 1.1.

2 Proof of Theorem 1.1

Our basic strategy in proving Theorem 1.1 is to use the method of multipliers to produce useful energy estimates which may then be used in conjunction with a nonlinear compactness-uniqueness argument and an argument from semigroup theory to obtain the final result. Due to space limitations, the proof of this theorem will be given under the simplifying geometric assumption

$$\mathrm{h} \cdot \nu \geq \beta > 0 \qquad \text{for } x \in \Gamma_1. \tag{2.1}$$

However, the statement of Theorem 1.1 holds true without this additional hypothesis. The details of the more general argument, which are rather lenghthy and based on microlocal analysis will be provided in a forthcoming manuscript.

2.1 Energy Estimates

The first energy estimate we will prove is

Proposition 2.1 *Let* $(w_0, w_1) \in \mathcal{H} \equiv H^2_{\Gamma_0}(\Omega) \times H^1_{\Gamma_0}(\Omega)$, *then the energy of system (1.1) as given by (1.4) satisfies the following estimate*

$$E(T) \leq C_T \left\{ \int_{\Sigma_1} (w_t^2 + |\nabla w_t|^2) d\Sigma_1 + \int_{\Sigma_1} w^2 d\Sigma_1 + \int_Q \tilde{b}(x) w_t^2 dQ + \int_{\Sigma_0} |\Delta \chi| d\Sigma_0 \right\} \tag{2.2}$$

for all T sufficiently large where $\tilde{b}(x) = b(x) + b^2(x)$ and C_T depends on $E(0)$.

Our proof will rely on energy methods which require the regularity of Proposition (1.1)(ii). Since the set D is dense in $\mathcal{H} \equiv H^2_{\Gamma_0}(\Omega) \times H^1_{\Gamma_0}(\Omega)$, we will assume this additional regularity of our initial data for all of our computations. Then the final result will follow by a standard density argument.

Proof: Recalling (1.4) and by using the multiplier w_t, it is straightforward to calculate the identity

$$E(t) + 2\left(\int_0^t \int_{\Gamma_1} \{w_t^2 + |\nabla w_t|^2\} d\Gamma_1\, dt + \int_Q b(x) w_t^2 dQ\right) = E(0). \tag{2.3}$$

This proves that energy is dissipating for the controlled system.

We now consider the multiplier w. We observe that

$$\int_Q [w,\chi] w = \int_Q [w,w]\chi dQ = -\int_Q \Delta^2\chi \cdot \chi dQ = -\int_Q (\Delta\chi)^2 dQ. \tag{2.4}$$

Now, by direct computation, we have

$$\begin{aligned}
&\int_Q \{w_t^2 + \gamma^2 |\nabla w_t|^2\} dQ - \int_0^T a(w,w)dt - \int_Q (\Delta\chi)^2 dQ \\
&= (w_t, w)_\Omega|_0^T + \gamma^2 (\nabla w_t, \nabla w)_\Omega|_0^T + \tfrac{1}{2}(b(x)w, w)_\Omega|_0^T \\
&\quad + (w,w)_{\Gamma_1}|_0^T + (\nabla w, \nabla w)_{\Gamma_1}|_0^T + \int_{\Sigma_1} w^2 d\Sigma_1.
\end{aligned} \tag{2.5}$$

If we now multiply (1.1) by $\mathbf{h} \cdot \nabla w$ and integrate by parts, we obtain

$$\begin{aligned}
\int_Q [w,\chi](\mathbf{h}\cdot\nabla w) dQ &= \int_Q \{w_{tt} - \gamma^2 \Delta w_{tt} + \Delta^2 w + b(x) w_t\} \mathbf{h}\cdot\nabla w dQ \\
&= (w_t, \mathbf{h}\cdot\nabla w)_\Omega|_0^T + \gamma^2(\nabla w_t, \mathbf{h}\cdot\nabla w)_\Omega|_0^T + \int_Q w_t^2 dQ \\
&\quad + \int_0^T a(w,w)dt + \int_Q b(x) w_t (\mathbf{h}\cdot\nabla w) dQ \\
&\quad - \frac{1}{2}\int_{\Sigma_1} (w_t^2 + \gamma^2|\nabla w_t|^2)\mathbf{h}\cdot\nu d\Sigma_1 - \frac{1}{2}\int_{\Sigma_0} \mathbf{h}\cdot\nu(\Delta w)^2 d\Sigma_0 \\
&\quad + \frac{1}{2}\int_{\Sigma_1} \mathbf{h}\cdot\nu\{(\Delta w)^2 + 2(1-\mu)(w_{xy}^2 - w_{xx}w_{yy})\} d\Sigma_1 \\
&\quad + \int_{\Sigma_1}\left\{\left(w_t - \frac{\partial^2 w_t}{\partial\tau^2} + w\right) h\cdot\nabla w + \tfrac{\partial}{\partial\nu}(\mathbf{h}\cdot\nabla w)\tfrac{\partial}{\partial\nu} w_t\right\} d\Sigma_1.
\end{aligned}$$

From Lagnese (1989) (see page 115) we have

$$\int_Q [w,\chi](\mathbf{h}\cdot\nabla w) dQ = -\frac{1}{2}\int_Q (\Delta\chi)^2 dQ - \frac{1}{2}\int_\Sigma \mathbf{h}\cdot\nu(\Delta\chi)^2 d\Sigma.$$

Consequently, we obtain

$$\int_Q w_t^2 + \int_0^T a(w,w)dt + \frac{1}{2}\int_Q (\Delta\chi)^2 dQ$$

$$
\begin{aligned}
&= -\frac{1}{2}\int_{\Sigma} \mathbf{h}\cdot\nu(\Delta\chi)^2 d\Sigma - (w_t, \mathbf{h}\cdot\nabla w)_\Omega\big|_0^T - \gamma^2(\nabla w_t, \mathbf{h}\cdot\nabla w)_\Omega\big|_0^T \\
&\quad +\frac{1}{2}\int_{\Sigma_1} \mathbf{h}\cdot\nu\{w_t^2 + \gamma^2|\nabla w_t|^2\} d\Sigma_1 + \frac{1}{2}\int_{\Sigma_0} \mathbf{h}\cdot\nu(\Delta w)^2 d\Sigma_0 \\
&\quad -\frac{1}{2}\int_{\Sigma_1} \mathbf{h}\cdot\nu\{(\Delta w)^2 - 2(1-\mu)(w_{xy}^2 - w_{xx}w_{yy})\} d\Sigma_1 \\
&\quad -\int_{\Sigma_1} \tfrac{\partial}{\partial\nu}(\mathbf{h}\cdot\nabla w)\tfrac{\partial}{\partial\nu} w_t d\Sigma_1 \\
&\quad -\int_{\Sigma_1} \left\{w_t + w - \frac{\partial^2 w_t}{\partial\tau^2}\right\} \mathbf{h}\cdot\nabla w \, d\Sigma_1 - \int_Q b(x) w_t \mathbf{h}\cdot\nabla w dQ \\
&\leq -\frac{1}{2}\int_{\Sigma_0} \mathbf{h}\cdot\nu(\Delta\chi)^2 d\Sigma_0 - \int_Q b(x) w_t \mathbf{h}\cdot\nabla w \, dQ \\
&\quad + \left|(w_t, \mathbf{h}\cdot\nabla w)_\Omega\big|_0^T\right| + \gamma^2\left|(\nabla w_t, \mathbf{h}\cdot\nabla w)_\Omega\big|_0^T\right| \\
&\quad + C_\varepsilon \int_{\Sigma_1} \{w_t^2 + w^2 + |\nabla w_t|^2\} d\Sigma_1 \\
&\quad + C_1\varepsilon \int_{\Sigma_1} \left\{|\nabla w|^2 + |\mathbf{h}\cdot\nabla w|^2 + \left(\frac{\partial^2 w}{\partial\tau^2}\right)^2 |\mathbf{h}\cdot\tau|^2\right\} d\Sigma_1 \\
&\quad -\frac{1}{2}\int_{\Sigma_1} \mathbf{h}\cdot\nu\{(\Delta w)^2 + 2(1-\mu)(w_{xy}^2 - w_{xx}w_{yy})\} d\Sigma_1. \qquad (2.6)
\end{aligned}
$$

Observing that the last term in our inequality is strictly negative (by virtue of (2.1)) we may choose ε sufficiently small so that the last two terms remain negative. We may also estimate

$$
\begin{aligned}
-\int_Q b w_t(\mathbf{h}\cdot\nabla w) dQ &\leq C_\varepsilon \int_Q b^2 w_t^2 dQ + \varepsilon \int_Q |\mathbf{h}\cdot\nabla w|^2 dQ \\
&\leq C_\varepsilon \int_Q b^2 w_t^2 dQ + C_2\varepsilon \int_Q |\nabla w|^2 dQ \\
&\leq C_\varepsilon \int_Q b^2 w_t^2 dQ + C\cdot\varepsilon \|w\|^2_{H^2_{\Gamma_0}(\Omega)} \\
&\leq C_1 \int_Q b^2 w_t^2 dQ + \frac{1}{2}\int_0^T a(w,w) dt. \qquad (2.7)
\end{aligned}
$$

Here we have used the Sobolev imbedding $H^2_{\Gamma_0}(\Omega) \subset H^1_{\Gamma_0}$. Using (2.6) and (2.7) together, we obtain the estimate

$$
\begin{aligned}
&\int_Q w_t^2 dQ + \frac{1}{2}\int_0^T a(w,w) dt + \frac{1}{2}\int_Q (\Delta\chi)^2 dQ \\
&\leq C_1 \int_Q b^2 w_t^2 dQ + C_2 \int_{\Sigma_0} (\Delta\chi)^2 d\Sigma_0 \\
&\quad + \left|(w_t, \mathbf{h}\cdot\nabla w)_\Omega\big|_0^T\right| + \gamma^2\left|(\nabla w_t, \mathbf{h}\cdot\nabla w)_\Omega\big|_0^T\right| \\
&\quad + C_3 \int_{\Sigma_1} \left\{w_t^2 + w^2 + |\nabla w_t|^2\right\} d\Sigma_1. \qquad (2.8)
\end{aligned}
$$

Multiplying (2.5) by 1/4 and adding to (2.8), we obtain

$$
\frac{1}{4}\int_0^T E(t) dt \leq \frac{5}{4}\int_Q (w_t^2 + \gamma^2|\nabla w_t|^2) dQ + \frac{1}{4}\int_0^T a(w,w) dt + \frac{1}{4}\int_Q (\Delta\chi)^2 dQ + \frac{1}{8}\int_{\Sigma_1} w^2 d\Sigma_1
$$

$$\begin{aligned}
&\leq \frac{1}{4}\left|(w_t,w)_\Omega|_0^T\right| + \frac{\gamma^2}{4}\left|(\nabla w_t,\nabla w)_\Omega|_0^T\right| + \frac{1}{8}\left|(bw,w)_\Omega|_0^T\right| \\
&+\left|(w_t,\mathbf{h}\cdot\nabla w)_\Omega|_0^T\right| + \gamma^2\left|(\nabla w_t,\mathbf{h}\cdot\nabla w)_\Omega|_0^T\right| + \frac{1}{4}\left|(w,w)_{\Gamma_1}|_0^T\right| \\
&+\frac{1}{4}\left|(\nabla w,\nabla w)_{\Gamma_1}|_0^T\right| + C_1\int_{\Sigma_0}(\Delta\chi)^2 d\Sigma_0 + C_2\int_{\Sigma_1} w^2 d\Sigma_1 \\
&+C_3\int_{\Sigma_1}(w_t^2+|\nabla w_t|^2)d\Sigma_1 + C_4\int_Q b^2 w_t^2 dQ. \qquad (2.9)
\end{aligned}$$

By the Sobolev imbeddings and trace theory, we see that

$$\begin{aligned}
C_1E(T)+C_2E(0) \;\geq\; &\left|(w_t,w)_\Omega|_0^T\right| + \left|(\nabla w_t,w)_\Omega|_0^T\right| + \left|(w_t,\mathbf{h}\cdot\nabla w)_\Omega|_0^T\right| \\
&+\gamma^2\left|(\nabla w_t,\mathbf{h}\cdot\nabla w)_\Omega|_0^T\right| + \left|(w,w)_{\Gamma_1}|_0^T\right| + \left|(\nabla w,\nabla w)_{\Gamma_1}|_0^T\right|. \qquad (2.10)
\end{aligned}$$

We now estimate

$$\begin{aligned}
\left|(bw,w)_\Omega|_0^T\right| &\leq \int_\Omega b\left|w^2(T)-w^2(0)\right|d\Omega \\
&\leq \bar{C}\int_\Omega (w^2(T)+w^2(0))d\Omega \qquad (2.11)\\
&\leq C\{E(T)+E(0)\},
\end{aligned}$$

since $b \in L^\infty(\Omega)$.

Using Holder's inequality, we may estimate

$$\begin{aligned}
\int_{\Sigma_0}(\Delta\chi)^2 d\Sigma_0 &\leq \left(\int_{\Sigma_0}|\Delta\chi|^3 d\Sigma_0\right)^{1/2}\left(\int_{\Sigma_0}|\Delta\chi| d\Sigma_0\right)^{1/2} \\
&\leq \varepsilon\|\Delta\chi\|^3_{L^3(\Sigma_0)} + \frac{1}{4\varepsilon}\int_{\Sigma_0}|\Delta\chi|d\Sigma_0. \qquad (2.12)
\end{aligned}$$

By the Sobolev imbeddings (see Adams (1975), theorem 5.22) and by the regularity of χ (see Lions and Magenes (1972) p. 187 and Bradley and Lasiecka (to appear)) we know that for $0<\varepsilon<2/3$

$$\|\Delta\chi(w)\|_{L^3(\Gamma_0)} \leq C_1\|\Delta\chi(w)\|_{H^{1-\epsilon}(\Omega)} \leq C_2\|w\|^2_{H^2(\Omega)}. \qquad (2.13)$$

Consequently,

$$\begin{aligned}
\|\Delta\chi(w)\|^3_{L^3(\Gamma_0)} &\leq \int_0^T \|\Delta\chi(w)\|^3_{H^{1-\epsilon}(\Omega)}dt \\
&\leq \int_0^T \|\Delta\chi(w)\|_{H^{1-\epsilon}(\Omega)}\|\Delta\chi(w)\|^2_{H^{1-\epsilon}(\Omega)}dt \\
&\leq C\cdot E^2(0)\int_0^T \|\Delta\chi(w)\|_{H^{1-\epsilon}(\Omega)}dt \\
&\leq C\cdot E^2(0)\int_0^T \|w(t)\|^2_{H^2_{\Gamma_0}(\Omega)}dt \\
&\leq C\cdot E^2(0)\int_0^T E(t)dt. \qquad (2.14)
\end{aligned}$$

From (2.12)-(2.14), we obtain

$$\int_{\Sigma_0}|\Delta\chi(w)|^2 d\Sigma_0 \leq \varepsilon C\cdot E^2(0)\int_0^T E(t)dt + \frac{1}{4\varepsilon}\int_{\Sigma_0}|\Delta\chi(w)|d\Sigma_0. \qquad (2.15)$$

Selecting $\varepsilon = \frac{1}{2C_1 C \cdot E^2(0)}$ and using (2.10)-(2.14) in (2.9) along with the estimate (2.3), we obtain the estimate

$$\begin{aligned}\int_0^T E(t)dt \;\leq\; & C\left\{E(T) + \int_{\Sigma_1}(w_t^2 + |\nabla w_t|^2)d\Sigma_1\right. \\ & + \int_Q \tilde{b}w_t^2 dQ + \int_{\Sigma_1} w^2 d\Sigma_1 \\ & \left. + C(E(0))\int_{\Sigma_0}|\Delta\chi| d\Sigma_0\right\}, \qquad (2.16)\end{aligned}$$

where $\tilde{b} = b + b^2 > 0$ and $C(E(0))$ is increasing in $E(0)$.

By (2.3), we see that $\frac{d}{dt}E(t) \leq 0$, so that

$$T\,E(T) \leq \int_0^T E(t)dt. \qquad (2.17)$$

Consequently, (2.16) becomes

$$\begin{aligned}\{T - C\}E(T) \;\leq\; & C\left\{\int_{\Sigma_1}(w_t^2 + |\nabla w_t|^2)d\Sigma_1 + \int_Q \tilde{b}w_t^2 dQ + \right. \\ & \left. \int_{\Sigma_1} w^2 d\Sigma_1 + C(E(0))\int_{\Sigma_0}|\Delta\chi| d\Sigma_0\right\}.\end{aligned}$$

Finally, selecting $T > T_0 = C$, we have Proposition 2.1.

2.2 Compactness-Uniqueness Argument

At this point, we have proven that the energy of system (1.1) is bounded by the feedback controls plus lower order and nonlinear terms. To obtain our desired goal, we must prove that the energy is bounded by the feedbacks alone. To accomplish this, we first observe that

$$\int_{\Sigma_0}|\Delta\chi| d\Sigma_0 \leq \int_0^T \|\Delta\chi\|_{L^2(\Gamma)}dt \leq C\int_0^T \|\Delta\chi\|_{H^{1/2+\epsilon}(\Omega)}dt \leq C\int_0^T \|\chi\|_{H^{3-\epsilon}(\Omega)}dt \qquad (2.18)$$

Thus, it suffices to prove

Proposition 2.2 Compactness-Uniqueness *Let (w, w_t) be a solution pair for (1.1). Then for any $\varepsilon > 0$,*

$$\begin{aligned}&\int_{\Sigma_1} w^2 d\Sigma_1 + \int_0^T \|\chi(w)\|_{H^{3-\epsilon}(\Omega)}dt \\ &\leq\; C(E(0))\left\{\int_{\Sigma_1}\{w_t^2 + |\nabla w_t|^2\}d\Sigma_1 + \int_Q \tilde{b}(x)w_t^2 dQ\right\} \qquad (2.19)\end{aligned}$$

where $\tilde{b}(x) = b(x) + b^2(x)$ and where $C(E(0))$ is an increasing function of the initial energy, $E(0)$.

Proof: The proof is by contradiction. Suppose (2.19) does not hold. Then there exists a sequence of functions $\{w_n(t)\}$ in $\mathcal{H}$ which satisfies the system

$$\left.\begin{array}{ll} w_n'' + \gamma^2\Delta w_{tt} + \Delta^2 w_n + bw_n' = [w_n, \chi(w_n)] & \text{in } Q \\ w_n(0,\cdot) = w_{n0}\;;\; w_n'(0,\cdot) = w_{n1} & \text{in } \Omega \\ w_n = \frac{\partial}{\partial\nu}w_n = 0 & \text{on } \Sigma_0 \\ \Delta w_n + (1-\mu)B_1 w_n = -\frac{\partial}{\partial\nu}w_n' & \text{on } \Sigma_1 \\ \frac{\partial}{\partial\nu}\Delta w_n + (1-\mu)B_2 w_n - \gamma^2\frac{\partial}{\partial\nu}w_n'' = w_n' - \frac{\partial^2}{\partial\tau^2}w_n' & \text{on } \Sigma_1 \end{array}\right\} \qquad (2.20)$$

and such that

$$\lim_{n\to\infty} \frac{\int_{\Sigma_1} w_n^2 d\Sigma_1 + \int_0^T \|\chi(w_n)\|_{H^{3-\epsilon}(\Omega)} dt}{\int_{\Sigma_1} \left((w_n')^2 + |\nabla w_n'|^2\right) d\Sigma_1 + \int_Q \tilde{b}(w_n')^2} = \infty \tag{2.21}$$

where the initial energy (as prescribed by initial data (w_{n0}, w_{n1})) are uniformly bounded in n. (Note: for convenience, we denote the time derivatives by $'$.)

Denoting the sequence

$$c_n \equiv \left\{ \int_{\Sigma_1} w_n^2 d\Sigma_1 + \int_0^T \|\chi(w_n)\|_{H^{3-\epsilon}(\Omega)} dt \right\}^{1/2} \tag{2.22}$$

we introduce the new variable

$$v_n \equiv \frac{w_n}{c_n}. \tag{2.23}$$

We observe that v_n satisfies the system

$$\left.\begin{array}{ll} v_n'' - \gamma^2 \Delta v_n'' + \Delta^2 v_n + b v_n' = [v_n, \chi(w_n)] & \text{in } Q \\ v_n(0,\cdot) = v_{n0} \;;\; v_n'(0,\cdot) = v_{n1} & \text{in } \Omega \\ v_n = \frac{\partial}{\partial\nu} v_n = 0 & \text{on } \Sigma_0 \\ \Delta v_n + (1-\mu) B_1 v_n = -\frac{\partial}{\partial\nu} v_n' & \text{on } \Sigma_1 \\ \frac{\partial}{\partial\nu}\Delta v_n + (1-\mu) B_2 v_n - \gamma^2 \frac{\partial}{\partial\nu} v_n'' = v_n' - \frac{\partial^2}{\partial\tau^2} v_n' & \text{on } \Sigma_1. \end{array}\right\} \tag{2.24}$$

By using (2.21), we see that v_n satisfies

$$\lim_{n\to\infty} \int_{\Sigma_1} ((v_n')^2 + |\nabla v_n'|^2) d\Sigma_1 + \int_Q \tilde{b}(v_n')^2 dQ = 0 \tag{2.25}$$

and (by the quadratic dependence of χ on w_n)

$$\begin{aligned} &\int_{\Sigma_1} v_n^2 d\Sigma_1 + \int_0^T \|\chi(v_n)\|_{H^{3-\epsilon}(\Omega)} dt \\ &= \frac{\int_{\Sigma_1} w_n^2 d\Sigma_1 + \int_0^T \|\chi(w_n)\|_{H^{3-\epsilon}(\Omega)} dt}{\int_{\Sigma_1} w_n^2 d\Sigma_1 + \int_0^T \|\chi(w_n)\|_{H^{3-\epsilon}(\Omega)} dt} \equiv 1 \end{aligned} \tag{2.26}$$

By (2.25), we have the following convergence properties:

$$\begin{array}{ll} (i) & v_n' \to 0 \quad \text{in } L^2(Q) \\ (ii) & v_n' \to 0 \quad \text{in } H^1(\Sigma_1). \end{array} \tag{2.27}$$

In order to pass with a limit on (2.24), we need first to determine the convergence properties for v_n and our nonlinear terms.

To determine the convergence properties of v_n, we will use the energy estimates which were derived in the previous section. If we denote the energy of (2.24) by $\widetilde{E}_n(t)$ and the energy of system (1.1) by $E_n(t)$, we have by (2.2) and by (2.3)

$$\begin{aligned} \widetilde{E}_{1n}(0) \leq\; & C_T \left\{ \int_{\Sigma_1} \left((v_n')^2 + |\nabla v_n'|^2\right) d\Sigma_1 + \int_Q \tilde{b} \cdot (v_n')^2 dQ \right. \\ & \left. + \int_{\Sigma_1} v_n^2 d\Sigma_1 + \int_{\Sigma_0} |\Delta\chi(v_n)| d\Sigma_0 \right\} \end{aligned} \tag{2.28}$$

where C_T depends on the initial energy for the system (2.20), $E_n(0)$ (not on $\widetilde{E}_n(0)$) and $\widetilde{E}_{1n}(0)$ represents the linear part of the initial energy (as in (1.4)). By using (2.25), we may assume without loss of generality that

$$\int_{\Sigma_1} \left((v_n')^2 + |\nabla v_n'|^2\right) d\Sigma_1 + \int_Q \tilde{b} \cdot (v_n')^2 dQ \leq 1. \tag{2.29}$$

Consequently, we have by (2.26), (2.29) and (2.18) that $\widetilde{E}_{1n}(0) \leq \widetilde{C}_T$ where $\widetilde{C}_T$ may depend on $E_n(0)$. Since the $E_n(0)$ are uniformly bounded, we conclude that $\widetilde{E}_{1n}(0)$ are uniformly bounded as well.

We may now use the results of our well-posedness theorem to obtain that

$$\begin{aligned} &(i) \quad \|v_n\|_{C([0,T];H^2(\Omega))} \leq M \\ &(ii) \quad \|v_n'\|_{C([0,T];H^1(\Omega))} \leq M \end{aligned} \tag{2.30}$$

which implies, in particular, that $\{v_n\}$ are uniformly bounded in $H^1([0,T] \times \Omega)$. Hence, by the compact Sobolev imbeddings and trace theory, we see that

$$\begin{aligned} & v_n \xrightarrow{w} v \text{ in } L^2([0,T];H^2(\Omega)) \text{ and } v_n' \xrightarrow{w} v' \text{ in } L^2([0,T];H^1(\Omega)) \\ \Longrightarrow \quad & v_n \xrightarrow{w} v \text{ in } H^1([0,T] \times \Omega) \\ \Longrightarrow \quad & v_n \longrightarrow v \text{ in } L^2(\Sigma). \end{aligned} \tag{2.31}$$

Also, since $E_n(0) \leq M$, the well-posedness for (1.1) yields

$$\begin{aligned} & w_n \xrightarrow{w} w \text{ in } L^2([0,T];H^2(\Omega)) \\ \text{and} \quad & w_n' \xrightarrow{w} w' \text{ in } L^2([0,T];H^1(\Omega)), \end{aligned} \tag{2.32}$$

so that $\{w_n\}$ has the same convergence properties as $\{v_n\}$.

We now examine the convergence properties of the von Kármán nonlinearity, $[v_n, \chi(w_n)]$. We wish to prove that

$$[v_n, \chi(w_n)] \longrightarrow [v, \chi(w)] \qquad \text{as } n \to \infty$$

in some meaningful sense on Q. In order to accomplish this, we will first prove

Lemma 2.1 *Let $w_n \xrightarrow{w} w$ in $H^2(\Omega)$. Then $\chi(w_n) \xrightarrow{w} \chi(w)$ in $H_0^2(\Omega)$.*

Proof: Let $\psi \in H_0^2(\Omega)$. Then

$$\begin{aligned} \int_\Omega \Delta\chi(w_n)\Delta\psi d\Omega &= \int_\Omega \Delta^2\chi(w_n)\psi d\Omega \\ &= -\int_\Omega [w_n, w_n]\psi d\Omega = -\int_\Omega [w_n, \psi]w_n d\Omega, \end{aligned}$$

where we have used Lemma 5.2.1 from Lagnese (1989) (see page 111). Now since $w_n \xrightarrow{w} w$ in $H^2(\Omega)$ and dim Ω=2, we have $w_n \to w$ in $L^\infty(\Omega)$. Consequently, we obtain

$$-\int_\Omega [w_n, \psi]w_n d\Omega \longrightarrow -\int_\Omega [w, \psi]w d\Omega.$$

But then

$$-\int_\Omega [w, \psi]w d\Omega = -\int_\Omega [w, w]\psi d\Omega = \int_\Omega \Delta^2\chi(w)\psi = \int_\Omega \Delta\chi(w)\Delta\psi d\Omega,$$

which proves the lemma.

We now seek to obtain the convergence of $\chi(w_n)$ in the space-time cylinder, Q.

Lemma 2.2 *Assume that*

$$\|w_n\|_{C([0,T];H^2(\Omega))} + \|w_n'\|_{C([0,T];H^1(\Omega))} \leq C$$

and

$$\begin{aligned} w_n &\overset{w}{\to} w \quad \textit{in } L^2([0,T];H^2(\Omega)) \\ w_n' &\overset{w}{\to} w' \quad \textit{in } L^2([0,T];H^1(\Omega)). \end{aligned}$$

Then for every $\varepsilon > 0$,

$$\chi(w_n) \to \chi(w) \textit{ in } C([0,T];H^{3-\varepsilon}(\Omega)).$$

Proof: By the results in Bradley and Lasiecka (to appear), we obtain that for any $\varepsilon > 0$,

$$\|\chi(w_n)\|_{L^2([0,T];H^{3-\varepsilon/2}(\Omega))} \leq C_1. \tag{2.33}$$

We shall prove

$$\|\chi(w_n)\|_{H^1([0,T];H^{2-\varepsilon/2}(\Omega))} \leq C. \tag{2.34}$$

To accomplish this, notice first that

$$\frac{d}{dt}\chi(w_n) = -2\mathcal{A}^{-1}[w_n', w_n] \tag{2.35}$$

where $\mathcal{A}$ corresponds to the biharmonic operator with zero boundary conditions as in (1.2).

Now let $\phi \in H_0^{2+\varepsilon/2}(\Omega)$. Then

$$\int_\Omega [w_n', w_n]\phi d\Omega = \int_\Omega [\phi, w_n] w_n' d\Omega. \tag{2.36}$$

By the assumptions imposed on w_n' and by the Sobolev imbeddings, we have

$$\|w_n'\|_{C([0,T];L^p(\Omega))} \leq C \qquad \text{for any } p > 1. \tag{2.37}$$

It can be easily verified that

$$\|[\phi, w_n]\|_{C([0,T];L^{1+\frac{\varepsilon}{4-\varepsilon}}(\Omega))} \leq C. \tag{2.38}$$

From (2.36)-(2.38), taking $p = 4/\varepsilon$ in (2.37), we obtain

$$\|[w_n', w_n]\|_{C([0,T];H^{-2-\varepsilon/2}(\Omega))} \leq C. \tag{2.39}$$

By the results of Grisvard (1985) and by duality, we have

$$H^{-2-\varepsilon/2}(\Omega) \sim \left(\mathcal{D}(\mathcal{A}^{1/2+\varepsilon/8})\right)', \tag{2.40}$$

where this duality is with respect to the L^2 inner product.

Combining (2.35), (2.39) and (2.40), we infer that

$$\left\|\frac{d}{dt}\chi(w_n)\right\|_{C([0,T];\mathcal{D}(\mathcal{A}^{1/2-\varepsilon/8}))} \leq C.$$

This, in particular, implies (2.34).

Now we are in a position to apply the compactness result by Simon (1987) to conclude that (2.33) and (2.34) together imply

$$\chi(w_n) \to z \qquad \text{strongly in } C([0,T];H^{3-\varepsilon}(\Omega)).$$

But by the result of Lemma 2.1, together with our assumptions, $z = \chi(w)$, as desired.

We may now prove

Proposition 2.3 *Suppose that* $v_n \xrightarrow{w} v$ *in* $H^2(\Omega)$ *and* w_n *satisfies the assumptions of Proposition 2.5. Then* $[v_n, \chi(w_n)] \to [v, \chi(w)]$ *in* $\mathcal{D}'(\Omega)$.

Proof: Let $\phi \in \mathcal{D}(\Omega)$. Then

$$\int_\Omega [v_n, \chi(w_n)]\phi d\Omega = \int_\Omega [\phi, \chi(w_n)] v_n d\Omega.$$

Now by the Proposition 2.4, we know that $\chi(w_n) \xrightarrow{w} \chi(w)$ in $H_0^2(\Omega)$, so that

$$\frac{\partial^2}{\partial x_i \partial x_j}\chi(w_n) \xrightarrow{w} \frac{\partial^2}{\partial x_i \partial x_j}\chi(w) \qquad \text{in } L^2(\Omega) \text{ for } i,j \in \{1,2\}.$$

Since $v_n \to v$ in $L^\infty(\Omega)$, we have

$$\int_\Omega [\phi, \chi(w_n)] v_n d\Omega \to \int_\Omega [\phi, \chi(w)] v d\Omega,$$

which gives us the proof. □.

In passing with a limit on (2.24), we will consider two cases.

Case 1: $c_0 = \{\int_{\Sigma_1} w^2 d\Sigma_1 + \int_0^T \|\chi(w)\|_{H^{3-\epsilon}(\Omega)} dt\}^{1/2} \neq 0$. By the result of Lemma 2.2, (2.32) and (2.22), $c_n \to c_0$, hence $v = w/c_0$. Using (2.27), Lemma 2.3 and passing with a limit on (2.24), we obtain the limit system

$$\left.\begin{array}{ll} \Delta^2 v = [v, \chi(w)] = \frac{1}{c_0}[w, \chi(w)] & \text{in } Q \\ v = \frac{\partial}{\partial \nu} v = 0 & \text{on } \Sigma_0 \\ \Delta v + (1-\mu) B_1 v = 0 & \text{on } \Sigma_1 \\ \frac{\partial}{\partial \nu}\Delta v + (1-\mu) B_2 v = 0 & \text{on } \Sigma_1. \end{array}\right\} \tag{2.41}$$

Multiplying (2.41) by v and integrating by parts, we obtain

$$\int_\Omega [v, \chi(w)] v d\Omega = \int_\Omega (\Delta v)^2 + \int_\Gamma \left(\tfrac{\partial}{\partial \nu}\Delta v \cdot v - \Delta v \tfrac{\partial}{\partial \nu} v\right) d\Gamma = a(v,v). \tag{2.42}$$

Next we examine the term

$$\begin{aligned} \int_\Omega [v, \chi(w)] v d\Omega &= \int_\Omega [v, v]\chi(w) d\Omega \\ &= -\int_\Omega \Delta^2 \chi(v)\chi(w) d\Omega \\ &= -\frac{1}{c_0^2}\int_\Omega (\Delta\chi(w))^2 d\Omega. \end{aligned} \tag{2.43}$$

Using (2.43) in (2.42), we obtain

$$0 = a(v,v) + \frac{1}{c_0^2}\int_\Omega (\Delta \chi(w))^2 d\Omega,$$

and conclude, by the positivity of $a(v,v)$ that

$$v \equiv 0 \qquad \text{in } Q. \tag{2.44}$$

In order to complete the contradiction (see (2.26)), we must prove

$$\left(\int_{\Sigma_1} v_n^2 d\Sigma_1 + \int_0^T \|\chi(v_n)\|_{H^{3-\epsilon}(\Omega)} dt\right) \longrightarrow 0.$$

But this follows from Lemma 2.2, (2.30), (2.31) and (2.44). This provides the contradiction, and hence the proof of Proposition (2.2) for Case 1.

Case 2: $c_0 = 0$, (i.e. $\chi(w) \equiv 0$ and $w|_{L^2(\Sigma_1)} = 0$.) In this case, we will again use the result of Lemma 2.2. Here we use the fact that $\chi(w_n) \to 0$ in $C([0,T]; H^{3-\varepsilon}(\Omega))$ in combination with (2.30) and Lemma 2.3 in order to obtain that $[v_n(t), \chi(w_n)(t)] \to 0$ in $\mathcal{D}'(\Omega))$. By using (2.27) and passing with a limit on system (2.41), we obtain

$$\left.\begin{array}{ll} \Delta^2 v = 0 & \text{in } Q \\ v = \frac{\partial}{\partial\nu} v = 0 & \text{on } \Sigma_0 \\ \Delta v + (1-\mu)B_1 v = 0 & \text{on } \Sigma_1 \\ \frac{\partial}{\partial\nu}\Delta v + (1-\mu)B_2 v = 0 & \text{on } \Sigma_1. \end{array}\right\} \tag{2.45}$$

The same argument as in Case 1 yields $v \equiv 0$ and

$$\left(\int_{\Sigma_1} v_n^2 d\Sigma_1 + \int_0^T \|\chi(w_n)\|_{H^{3-\epsilon}(\Omega)} dt\right) \longrightarrow 0,$$

which gives us the contradiction.

2.3 Completion of Proof of Theorem 1.1

By using (2.18) along with Proposition 2.2, we have shown that, for T sufficiently large, the energy for system (1.1) satisfies

$$E(T) \leq C(E(0)) \left(\int_{\Sigma_1} (w_t^2 + |\nabla w_t|^2) d\Sigma_1 + \int_Q \tilde{b} w_t^2 dQ\right), \tag{2.46}$$

where the function $\rho \equiv C(E(0))$ is increasing in $E(0)$. From (2.3) we have that

$$\int_{\Sigma_1} (w_t^2 + |\nabla w_t|^2) d\Sigma_1 + \int_Q \tilde{b} w_t^2 dQ = 1/2(E(0) - E(T))$$

so that by using (2.46) we obtain

$$2E(T) \leq \rho \cdot (E(0) - E(T))$$

or equivalently,

$$E(T) \leq \frac{\rho}{2+2\rho} E(0) \equiv \rho_0(E(0)) \cdot E(0). \tag{2.47}$$

Observe that ρ_0 is an increasing function of $E(0)$ and $\rho_0 < 1$. We are now in a position to complete the proof of Theorem 1.1 by using (2.46) in conjunction with results from nonlinear semigroup theory.

Let $S(t)$ be the semigroup describing the dynamics of (1.1). By Proposition 1.1 we know that such a semigroup of nonlinear operators exists and that

$$\begin{pmatrix} w(t) \\ w_t(t) \end{pmatrix} = S(t) \begin{pmatrix} w_0 \\ w_1 \end{pmatrix}.$$

Now for every $t > 0$ we have $t = nT + \tau$ for some integer n. Then by the semigroup property we can write

$$\begin{pmatrix} w(t) \\ w_t(t) \end{pmatrix} = S(T)\left\{S((n-1)T + \tau) \begin{pmatrix} w_0 \\ w_1 \end{pmatrix}\right\}. \tag{2.48}$$

Viewing

$$S((n-1)T+\tau)\begin{pmatrix} w_0 \\ w_1 \end{pmatrix} \equiv \begin{pmatrix} y_0^n \\ y_1^n \end{pmatrix}$$

as new "initial data", we may apply (2.47) to (2.48) and obtain

$$E(t) \le \rho_0(E_n(0))E_n(0)$$

where

$$E_n(0) = 1/2\|(y_0^n, y_1^n)\|_{\mathcal{H}}^2 + \int_{\Gamma_1} |\Delta\chi(y_0^n)|^2 d\Gamma_1.$$

By applying the semigroup property again, we see that

$$\begin{pmatrix} w(t) \\ w_t(t) \end{pmatrix} = S(T)\left\{S(T)\left\{S((n-2)T+\tau)\begin{pmatrix} w_0 \\ w_1 \end{pmatrix}\right\}\right\}.$$

Again, generating new "initial data"

$$S((n-2)T+\tau)\begin{pmatrix} w_0 \\ w_1 \end{pmatrix} \equiv \begin{pmatrix} y_0^{n-1} \\ y_1^{n-1} \end{pmatrix}$$

and applying (2.47) to this new data, we have

$$E(t) \le \rho_0(E_n(0)) \cdot E_n(0) \le \rho_0(E_n(0)) \cdot \rho_0(E_{n-1}(0)) \cdot E_n(0)$$

where

$$E_{n-1}(0) = 1/2\|(y_0^{n-1}, y_1^{n-1})\|_{\mathcal{H}}^2 + \int_{\Gamma_1} |\Delta\chi(y_0^{n-1})|^2 d\Gamma_1.$$

Continuing in this fashion, we obtain the estimate

$$E(t) \le \rho_{0,n}\rho_{0,n-1}\cdots\rho_{0,1}E(0)$$

where $\rho_{0,i} = \rho_0(E_i)$. Now since $E_i(0) = E(T(i-1)+\tau)$ and since $E(t) \le E(0)$, we see that

$$E_1(0) \ge E_2(0) \ge \cdots \ge E_{i+1}(0).$$

Also, since ρ_0 is an increasing function of $E(0)$, we have that

$$\rho_{0,1} \ge \rho_{0,2} \ge \cdots \ge \rho_{0,n}.$$

Consequently, we have

$$E(t) \le \rho_{0,1}^n E(0). \tag{2.49}$$

Since $\rho_0 < 1$, we know that $\rho_{0,i}(E(0)) < 1$ for all i and for all values of $E(0)$. Setting $w = w(E(0)) = \frac{-\ln \rho_{0,1}}{T}$, we have that

$$E(t) \le Ce^{-wt}E(0)$$

which was to be proved. æ

References

[1] R. Adams. *Sobolev Spaces.* Academic Press, New York, 1975.

[2] M. E. Bradley. Global exponential decay rates for a Kirchoff plate with boundary nonlinearities. To appear.

[3] M. E. Bradley and I. Lasiecka. Local exponential stabilization of a nonlinearly perturbed von Kármán plate. *Journal of Nonlinear Analysis: Techniques, Methods and Applications.* To appear.

[4] A. Favini and I. Lasiecka. Well-posedness and regularity of second order abstract nonlinear evolution equations. To appear.

[5] P. Grisvard. *Elliptic Problems in Nonsmooth Domains.* Pitman, London, 1985.

[6] J. Lagnese. *Boundary Stabilization of Thin Plates.* Society for Industrial and Applied Mathematics, Philadelphia, 1989.

[7] J. Lagnese. Local controllability of dynamic von Karman plates. *Control and Cybernetics*, 1990.

[8] J. Lagnese. Uniform asymptotic energy estimates for solutions of the equations of dynamical plane elasticity with nonlinear dissipation at the boundary. *Nonlinear Analysis*, 16:35–54, 1991.

[9] J. Lagnese and G. Leugering. Uniform stabilization of a nonlinear beam by nonlinear feedback. *Journal of Differential Equations*, 91:355–388, 1991.

[10] I. Lasiecka. Global uniform decay rates for the solutions to wave equations with nonlinear boundary conditions. To appear. *Applicable Analysis.*

[11] J. L. Lions and E. Magenes. *Non-Homogeneous Boundary Value Problems and Applications*, volume 1. Springer-Verlag, New York, 1972.

[12] J. Simon. Compact sets in the space $L^p(0,T;B)$. *Annali di Mathematica Pura et Applicata*, 4:65–96, 1987.

Viscosity Solutions and Partial Differential Inequalities

OVIDIU CÂRJĂ University of Iaşi, Romania

CORNELIU URSESCU Romanian Academy, Iaşi, Romania

1 INTRODUCTION

Let A be a nonempty, open subset of a finite dimensional space V, let $H : A \times R \to V \times R$ be a multifunction with nonempty values, and consider the first-order partial differential equation

$$\inf_{(u,v)\in H(x,w(x))} (D_w(x)(u) - v) = 0 \tag{1}$$

where D denotes the differentiability concept of Severi (1935, p.10) (see also Ursescu, 1973, p. 200), i.e.,

$$D_w(x)(u) = \lim_{\substack{s\downarrow 0 \\ p\to 0}} (l/s)(w(x + s(u + p)) - w(x)).$$

Particular cases of equation (1) are well known. These are the quasilinear first-order partial differential equations and the Hamilton Jacobi Bellman equations for variational problems. In Cârjă and Ursescu (1991), in a slightly more general setting where the set A is locally closed, the characteristics method for equation (1) has been developed and it has been shown that the solutions w to (1) can be characterized by means of the behavior of the differentiable functions w along solutions (X, Y) to the differential inclusion

$$(X'(s), Y'(s)) \in H(X(s), Y(s)). \tag{2}$$

The characterization consists of the two conditions:

$$\begin{aligned}&\text{for every } x \in A, \text{ there exists a solution}\\&(X,Y):[0,\sigma)\to V\times R \text{ to the inclusion (2)}\\&\text{with } (X(0),Y(0))=(x,w(x)) \text{ such that, for}\\&\text{every } s\in(0,\sigma), w(X(s))\le Y(s);\end{aligned}\tag{3}$$

$$\begin{aligned}&\text{for every } x \in A, \text{ for every solution}\\&(X,Y):[0,\sigma)\to V\times R \text{ to the inclusion (2)}\\&\text{with } (X(0),Y(0))=(x,w(x)), \text{ and for}\\&\text{every } s\in(0,\sigma), Y(s)\le w(X(s)).\end{aligned}\tag{4}$$

In fact, in the absence of the differentiability condition, the functions w satisfying (3) and (4) are just the solutions w to the inequations

$$\inf_{(u,v)\in H(x,w(x))}(\underline{D}_w(x)(u)-v)\le 0,\tag{5}$$

$$0\le \inf_{(u,v)\in H(x,w(x))}(\bar{D}_w(x)(u)-v),\tag{6}$$

respectively. Here

$$\underline{D}_w(x)(u)=\varliminf_{\substack{s\downarrow 0\\ p\to 0}}(l/s)(w(x+s(u_p))-w(x)),$$

$$\bar{D}_w(x)(u)=\varlimsup_{\substack{s\downarrow 0\\ p\to 0}}(l/s)(w(x+s(u+p))-w(x))$$

(see Penot, 1974, p. 1553). In this paper we show that the couple of inequations (5) and (6), which are more general than equation (1), can be identified with equation (1) from the viscosity point of view. Consequently, viscosity solutions to (1) can also be characterized by (3) and (4).

2 MAIN RESULTS

Consider the equation

$$\sup_{(u,v)\in H(x,w(x))}(v-D_w(x)(u))=0\tag{7}$$

as well as the two inequations

$$0\le \sup_{(u,v)\in H(x,w(x))}(v-\underline{D}_w(x)(u))\tag{8}$$

$$\sup_{(u,v)\in H(x,w(x))}(v-\bar{D}_w(x)(u))\le 0,\tag{9}$$

which are equivalent to equation (1) as well as to the inequations (5) and (6), respectively. Our first main result states that the viscosity solutions to equation (7) (see Crandall et al., 1984, p. 488) are just the solutions to the couple of inequations (8) and (9).

THEOREM 1 Let the multifunction H be continuous and let its values be compact and convex. A continuous function $w : A \to$ is a viscosity solution to equation (7) if and only if it is a solution to inequations (8) and (9).

Theorem 1 above is an immediate corollary of Theorem 2 below, where viscosity sub and super solutions are involved.

THEOREM 2 (a) Let the multifunction H be upper semicontinuous and let its values be compact and convex. A continuous function $w : A \to R$ is a viscosity supersolution to equation (7) if and only if it is a solution to inequation (8).

(b) Let the multifunction H be lower semicontinuous and let its values be closed and convex. A continuous function $w : A \to R$ is a viscosity subsolution to equation (7) if and only if it is a solution to inequation (9).

The continuity hypotheses of the function w can be slightly weakened in case the multifunction H satisfies the condition

$$\begin{aligned}&\text{for every } x \in A, \text{ for every } y_1 \in R, \text{ and for}\\&\text{every } y_2 \in R \text{ with } y_1 \le y_2, H(x,y_1) \subseteq H(x,y_2).\end{aligned} \tag{10}$$

THEOREM 3 (a) Let the multifunction H be upper semicontinuous and let its values be compact and convex. In addition, let the multifunction H satisfy condition (10). A lower semicontinuous function $w : A \to R$ is a viscosity supersolution to equation (7) if and only if it is a solution to inequation (8).

(b) Let the multifunction H be lower semicontinuous and let its values be closed and convex. In addition, let the multifunction H satisfy condition (10). An upper semicontinuous function $w : A \to R$ is a viscosity subsolution to equation (7) if and only if it is a solution to inequation (9).

3 VISCOSITY SOLUTIONS AND INEQUATIONS

Consider the equation

$$\mathcal{H}(x, w(x), D_w(x)) = 0 \tag{11}$$

where the real function $\mathcal{H}$ is defined at least for all $x \in A$, for all $y \in R$, and for all real linear functions $z : V \to R$, i.e., for all $z \in V^*$. A function $w : A \to R$ is said to be a solution to equation (11) if, for every $x \in A$, w is differentiable at x, in the sense of Severi, $\mathcal{H}$ is defined at the point $(x, w(x), D_w(x))$, and (11) is satisfied. One way to get rid of the differentiability condition in (11) is to consider the viscosity solutions to (11). Here we show that viscosity solutions to (11) are just the solutions to a certain couple of inequations involving the points $(x, w(x), \underline{D}_w(x))$ and $(x, w(x), \bar{D}_w(x))$. For this purpose we have to extend the function $\mathcal{H}$ to all extended real functions z.

Define $\underline{\delta}$ and $\bar{\delta}$ for every extended real function $z : V \to \bar{R}$ by

$$\underline{\delta}(z) = \{q \in V^*; \forall u \in V, \quad q(u) \le z(u)\},$$
$$\bar{\delta}(z) = \{q \in V^*; \forall u \in V, \quad z(u) \le q(u)\},$$

define $\underline{\mathcal{H}}$ and $\bar{\mathcal{H}}$ for every $x \in A$, for every $y \in R$, and for every extended real function $z : V \to \bar{R}$ by

$$\underline{\mathcal{H}}(x,y,z) = \inf_{q \in \underline{\delta}(z)} \mathcal{H}(x,y,q),$$

$$\bar{\mathcal{H}}(x,y,z) = \sup_{q \in \bar{\delta}(z)} \mathcal{H}(x,y,q),$$

and consider the inequations

$$0 \leq \underline{\mathcal{H}}(x, w(x), \underline{D}_w(x)), \tag{12}$$

$$\bar{\mathcal{H}}(x, w(x), \bar{D}_w(x)) \leq 0. \tag{13}$$

LEMMA 1 A function $w : A \to R$ is a viscosity solution to equation (11) if and only if it is a solution to inequations (12) and (13).

In fact, there holds true a stronger result involving viscosity sub and super solutions.

LEMMA 2 (a) A function $w : A \to R$ is a viscosity supersolution to equation (11) if and only if it is a solution to inequation (12).

(b) A function $w : A \to R$ is a viscosity subsolution to equation (11) if and only if it is a solution to inequation (13).

Proof. (a) Recall that a function $w : A \to R$ is said to be a viscosity supersolution to equation (11) if

$$0 \leq \mathcal{H}(x, w(x), q)$$

for every $x \in A$ and for every $q \in \underline{\mathcal{D}}_w(x)$ where

$$\underline{\mathcal{D}}_w(x) = \left\{ q \in V^*; 0 \leq \varliminf_{u \to x} (w(u) - w(x) - q(u-x))/\|u - x\| \right\}.$$

Since $\underline{\mathcal{D}}_w(x) = \underline{\delta}(\underline{D}_w(x))$ (see Frankowska, 1989, p. 300), the conclusion follows at once.

(b) Recall that a function $w : A \to R$ is said to be a viscosity subsolution to equation (11) if

$$\mathcal{H}(x, w(x), q) \leq 0$$

for every $x \in A$ and for every $q \in \bar{\mathcal{D}}_w(x)$ where

$$\bar{\mathcal{D}}_w(x) = \left\{ q \in V^*; \varlimsup_{u \to x} (w(u) - w(x) - q(u-x))/\|u - x\| \leq 0 \right\}.$$

Since $\bar{\mathcal{D}}_w(x) = \bar{\delta}(\bar{D}_w(x))$ (see Frankowska, 1989, p. 300), the conclusion follows at once.

4 PROOFS OF THE MAIN RESULTS

Given any extended real function $z : V \to \bar{R}$, denote by $\underline{\gamma}(z) : V \to \bar{R}$ the extended real function whose epigraph equals the closed convex hull of the epigraph of z and denote by $\bar{\gamma}(z) : V \to \bar{R}$ the extended real function whose hypograph equals the closed convex hull of the hypograph of z. Consider now the inequations

$$0 \leq \sup_{(u,v) \in H(x,w(x))} (v - \underline{\gamma}(\underline{D}_w(x))(u)), \tag{14}$$

$$\sup_{(u,v) \in H(x,w(x))} (v - \bar{\gamma}(\bar{D}_w(x))(u)) \leq 0. \tag{15}$$

LEMMA 3 (a) Let the values of the multifunction H be compact and convex. A function $w : A \to R$ is a viscosity supersolution to equation (7) if and only if it is a solution to inequality (14).

(b) A function $w : A \to R$ is a viscosity subsolution to equation (7) if and only if it is a solution to inequality (15).

Proof. Denote

$$\mathcal{H}(x,y,z) = \sup_{(u,v)\in H(x,y)} (v - z(u)).$$

In the first case, (a), it follows from a result of Kneser (1952, pp. 2418, 2419) that

$$\underline{\mathcal{H}}(x,y,z) = \sup_{(u,v)\in H(x,y)} (v - \sup_{q\in\underline{\delta}(z)} q(u)).$$

Since

$$\sup_{q\in\underline{\delta}(z)} q(u) = \underline{\gamma}(z)(u)$$

whenever the graph of z is a cone, the conclusion follows from Lemma 2, part (a). In the second case, (b),

$$\bar{\mathcal{H}}(x,y,z) = \sup_{(u,v)\in H(x,y)} (v - \inf_{q\in\bar{\delta}(z)} q(u)).$$

Since

$$\inf_{q\in\bar{\delta}(z)} q(u) = \bar{\gamma}(z)(u)$$

whenever the graph of z is a cone, the conclusion follows from Lemma 2, part (b).

We can now proceed to the proof of Theorems 2 and 3.

(a) It has been shown in Cârjă and Ursescu (1991, pp.15,17) that w is a solution to (8) if and only if

$$0 \le \sup_{(u,v)\in H(x,y)} (v - \underline{D}_w(x,y)(u))$$

for every $(x,y) \in \text{epigraph}(w)$, where

$$\underline{D}_w(x,y)(u) = \varliminf_{\substack{s\downarrow 0\\ p\to 0}} (1/s)(w(x+s(u+p)) - y).$$

In a similar way, it can be shown that w is a solution to (14) if and only if

$$0 \le \sup_{(u,v)\in H(x,y)} (v - \underline{\gamma}(\underline{D}_w(x,y))(u))$$

for every $(x,y) \in \text{epigraph}(w)$. The preceding inequalities state that

$$\emptyset \ne H(x,y) \cap K_{\text{epigraph}(w)}(x,y))$$

respectively. Here K denotes the tangency concept of Bouligand (1930, p. 32) and Severi (1930, p. 99). The conclusion follows now from a result in Aubin and Frankowska (1991, p. 18).

(b) It has been shown in Câjă and Ursescu (1991, pp. 15, 17) that w is a solution to (9) if and only if

$$\sup_{(u,v)\in H(x,y)} (v - \bar{D}_w(x,y)(u)) \le 0$$

for every $(x, y) \in \text{hypograph}(w)$, where

$$\bar{D}_w(x,y)(u) = \overline{\lim_{\substack{s \downarrow 0 \\ p \to 0}}} (l/s)(w(x + s(u+p)) - y).$$

In a similar way, it can be shown that w is a solution to (15) if and only if

$$\sup_{(u,v) \in H(x,y)} (v - \bar{\gamma}(\bar{D}_w(x,y))(u)) \leq 0$$

for every $(x, y) \in \text{hypograph}(w)$. The preceding inequalities state that

$$H(x,y) \subseteq K_{\text{hypograph}(w)}(x,y),$$
$$H(x,y) \subseteq \text{cl co}(K_{\text{hypograph}(w)}(x,y))$$

respectively. The conclusion follows now from a result in Aubin and Frankowska (1990, p. 130).

REFERENCES

Aubin, J. P., and Frankowska, H. (1990). *Set Valued Analysis*. Boston: Birkhäuser.

Aubin, J. P., and Frankowska, H. (1991). Feedback controls for uncertain systems. In *Modeling, Estimation and Control of Systems with Uncertainty*, edited by G. B. Di Masi, A. Gombani, and A. B. Kurzhansky. Proceedings of a conference held in Sopron, Hungary, September 1990. Boston: Birkhäuser, pp. 1–21.

Bouligand, G. (1930). Sur les surfaces dépourvues de points hyperlimites (ou: un théorème d'existence du plan tangent). *Ann. Soc. Polon. Math.* *9*:32–41.

Cârjă, O., and Ursescu, C. (1991). The characteristics method for a first order partial differential equation. *Preprint Series in Mathematics of "A. Myller" Mathematical Seminary*, No. 1, "Al. I. Cuza" University, Iaşi.

Crandall, M., Evans, L. C., and Lions, P. L. (1984). Some properties of viscosity solutions of Hamilton Jacobi equations. *Trans. Am. Math. Soc.* *282*:487–502.

Frankowska, H. (1989). Optimal trajectories associated with a solution of the contingent Hamilton Jacobi equation. *Appl. Math. Optim.* *19*:291–311.

Kneser, H. (1952). Sur un théorème fondamental de la théorie des jeux. *C. R. Acad. Sci. Paris* *234*:2418–2420.

Penot, J. P. (1974). Sous-différentiels de fonctions numériques non convexes. *C. R. Acad. Sci. Paris* *278*:1553–1555.

Severi, F. (1930). Su alcune questioni di topologia infinitesimale. *Ann. Soc. Polon. Math.* *9*:97–108.

Severi, F. (1935). Sulla differenziabilità totale delle funzioni di piu variabili reali. *Ann. Mat. Pura Appl.* *13*:1–35.

Ursescu, C. (1973). Sur une généralisation de la notion de différentiabilité. *Atti Accad. Naz. Lincei Rend. Cl. Sci. Fis. Mat. Natur.* *54*(7):199–204.

On a Class of C-Regularized Semigroups

IOANA CIORANESCU University of Puerto Rico, Rio Piedras, Puerto Rico

Abstract: This work is concerned with characterizations of closed densely defined operators A on a Banach space X with nonempty resolvent set for which the abstract Cauchy Problem $u'(t, x) = Au(t, x)$, $u(0, x) = x$ has a unique solution for every initial value $x \in D(Ak)$ which is $0\left(e^{t^w}\right)$, $w > 1$.

Introduction. Let X be a Banach space and $A:D(A) \to X$ a linear operator; consider the following abstract Cauchy Problem:

$$u'(t, x) = Au(t, x), \; u(0, x) = x, \; t \geq 0 \qquad (*)$$

where we are looking for D(A)-valued continuously differentiable solutions.

Motivated by the equation (*) three generalizations of the semigroups of class c_0 appeared in the last decades: the distribution semigroups, the C-regularized semigroups and the n-times integrated semigroups.

In this section we shall review shortly the connection between these classes of operators and the abstract Cauchy Problem (*) for $x \in D\left(A^{k+1}\right)$, $k \in N$. The results can be devided in two parts: one concerns exponentially bounded solution of (*) and the other, general solutions.

In the next section we shall consider an intermediate case to the two above, namely solutions satisfying $u(t,x) = 0\left(e^{\tau t^w}\right)$, $w > 1$, $\tau > 0$, $x \in D\left(A^{k+1}\right)$. This situation was previously studied by Foias (1960) and the author (1971, 1977) from the distributional point of view. We are now revisiting the distribution semigroups of growth order $\leq w$, interpreting their generators as generators of C-regularized semigroups and giving a Hille-Yosida type result for them.

Throughout this work A is linear and densely defined with domain $D(A) \subset X$, $\rho(A)$ is the resolvent set of A, $R(\lambda, A)$ the resolvent function at

$\lambda \in \rho(A)$, $L(X)$ the space of all linear and bounded operators in X. $D(A^k)$ considered as a Banach space with respect to the norm $\|x\|_k = \sum_{j=0}^{n} \|A^j x\|$, will be denoted by $[D(A^k)]$.

In the first two theorems several characterizations of densely defined operators for which the equation (*) is exponentially well-posed on $D(A^{k+1})$ are given.

THEOREM I. Suppose $\omega \geq 0$ and that $(\omega,\infty) \subset \rho(A)$; then the following statements are equivalent:

i) the equation (*) has a unique solution $u(.,x)$ for all $x \in D(A^{k+1})$, s.t. $\|u(t,x)\| + \|u'(t,x)\| = 0(e^{\omega t})$

ii) there exists M > 0 s.t.

$$\|(\lambda - \omega)^n R(\lambda; A)^n x\| \leq M\|x\|_k, \text{ for all } \lambda > \omega,\ n \in N,\ x \in D(A^k)$$

iii) A generates for all $\lambda > \omega$ an $(A - \lambda)^{-k}$-regularizaed semigroup which is $\omega(e^{wt})$.

iv) A generates a k-times integrated semigroup which is $0(e^{\omega t})$.

(see Sanekata (1975) for i)$\Rightarrow$ii), Oharu (1971-72), for ii)$\Rightarrow$i), de Laubenfels (1990) for iii)$\Leftrightarrow$iv) and Neubrander (1988) for ii)$\Leftrightarrow$iii)).

NOTATION 1. For $k \in N$ and $\omega \geq 0$ we denote by $G_\omega(k)$ the class of operators A s.t. $(\omega,\infty) \subset \rho(A)$ and A satisfies one of the equivalent conditions in the above theorem.

THEOREM II. The following statements for A are equivalent:

a) $A \in \bigcup_{\omega,k} G_\omega(k)$

b) A generates an exponentially bounded distribution semigroup.

c) there exist ω, $M > 0$ and $m \in N$ s.t.

$$\|R(\lambda; A)\| \leq M(1 + |\lambda|)^m, \text{ for } \mathrm{Re}\lambda > \omega$$

(see Oharu (1971-72) for a)$\Leftrightarrow$b) and Lions (1960) for b)$\Leftrightarrow$c)).

THEOREM III. Suppose $\omega \geq 0$ and that $(\omega,\infty) \subset \rho(A)$, then the following statements are equivalent:

i) the equation (*) has a unique solution $u(.,x)$ for all $x \in D(A^{k+1})$.

ii) for every $t > 0$, there exists $M(t) > 0$ s.t.

$$\left\|\lambda^n R(\lambda;A)^n x\right\| \le M(t)\|x\|_k, \text{ for } \lambda > \omega\ ,\ 0 \le \frac{n}{\lambda} \le t,\ \ x \in D\left(A^k\right)$$

iii) A generates for all $\lambda > \omega$ an $(A-\lambda)^{-k}$-regularized semigroup.

(see Sanekata (1975) for i$\Rightarrow$ii), Oharu (1971-72) for ii)$\Rightarrow$i), and de Laubenfels (1991) for i$\Leftrightarrow$iii)).

NOTATION 2. For $k \in N$ and $\omega \ge 0$ we denote by $\mathfrak{G}(\omega,k)$ the class of operators A s.t.$(\omega,\infty) \subset \rho(A)$ and A satisfies one of the equivalent conditions in Theorem III.

REMARK 1. S. Oharu (1971-72) proved that if $A \in \mathfrak{G}(\omega,k)$, then the solution of (*) satisfies

$$\|u(t,x)\| + \|u'(t,x)\| = o(M(t)),\ \ x \in D\left(A^{k+1}\right),\ \ t > 0$$

where M(.) is the function of ii).

THEOREM IV. The following statements for A are equivalent:

a) $A \in \bigcup_{\omega,k} \mathfrak{G}(\omega,k)$

b) A generates a distribution semigroup of finite order.

One can prove that a)$\Rightarrow$b) exactly as in Theorem 4.2 of Ushijima (1972) For b)$\Rightarrow$a) we note that if A generates the distribution semigroup $\mathfrak{T}$ s.t. $\mathfrak{T} = F^{(k)}$ where $F{:}R \to L(X)$ is continuous, then by results of Ushijima (1972), $U_t x = F(t)A^k x,\ x \in D\left(A^{k+1}\right)$ is the solution to (*).

REMARK 2. Each of the conditions in Theorem IV yield (Chazarain, 1971)
there exist $\alpha,\ \beta,\ M,\ \omega > 0$ and $m \in N$ s.t.

$\Lambda = \left\{\lambda \in C; \mathrm{Re}\lambda \ge \alpha \ln|\lambda| + \beta,\ \mathrm{Re}\lambda > \omega\right\} \subset \rho(A)$ and $\|R(\lambda;A)\| \le M(1+|\lambda|)^m, \lambda \in \Lambda$.

Main Result. We present next our main result which is:

THEOREM V. Suppose $\tau > 0,\ \omega \ge 0,\ w > 1$ and that $(\omega,\infty) \subset \rho(A)$; then the following statements are equivalent:

i) the equation (*) has a unique solution for all $x \in D\left(A^{k+1}\right)$ s.t. $\|u(t,x)\| + \|u'(t,x)\| = o\left(e^{\tau t^w}\right),\ \ t \ge 0.$

ii) there is $M > 0$ s.t. for every $t > 0$

$$\left\|\lambda^n R(\lambda;A)^n x\right\| \le M e^{\tau t^w} \|x\|_k, \text{ for } \lambda > \omega,\ 0 \le \frac{n}{\lambda} \le t \text{ and } x \in D\left(A^k\right)$$

iii) A generates for all $\lambda > \omega$ an $(A-\lambda)^{-k}$-regularized semigroup which is $O\left(e^{\tau t^w}\right)$.

Proof: $i \Rightarrow ii)$. For $x \in D\left(A^{k+1}\right)$ and $t \geq 0$ let $U_t x = u(t,x)$. Also we denote by $Z_{\tau,w}$ the Banach space which consists of all functions $u \in C([0,\infty); [D(A)])$ s.t. $\sup_{t\geq 0} e^{-\tau t^w}\|u(t)\|_1 < \infty$. Then $x \to U_t x$ defined on $\left[D\left(A^{k+1}\right)\right]$ with range $Z_{\tau,w}$ is a closed operator for every $t \geq 0$; therefore it is continuous. Let $\lambda \in \rho(A)$, $x \in D\left(A^{k+1}\right)$ and put $y = R(\lambda;A)x$; then $\|U_t x\| = \|(\lambda - A)U_t y\| \leq$ const. $\|U_t y\|_1 \leq$ const $e^{\tau t^w}\|y\|_{k+1} \leq$ const. $e^{\tau t^w}\|x\|_k$.
Thus, we have:

$$\|U_t x\| \leq \text{const. } e^{\tau t^w}\|x\|_k, \; x \in D\left(A^{k+1}\right), \; t \geq 0 \tag{1}$$

We follow now the proof of the main Theorem in Sanekata (1975), to obtain the formula

$$\begin{aligned}\lambda^n R(\lambda;A)^n &= \frac{\lambda^n}{(n-1)!}\int_0^t e^{-\lambda s}s^{n-1}U_s x \; ds + \frac{1}{2\pi i}\int_\Gamma e^{-zt}\left(1-\frac{t}{\lambda}\right)^{-n} R(z;A)U_t x \; dz \\ &= I_1, + I_2, t > 0, \lambda > \omega\end{aligned}$$

where Γ is the boundary of the domain Λ in Remark 2.
Using (1) we obtain at once:

$$\|I_1\| \leq \text{const. } e^{\tau t^w}\|x\|_k \text{ for } x \in D\left(A^k\right), \; n \in N. \tag{2}$$

In order to estimate I_2, we change Γ conveniently and use Lemma 5.2 of Sanekata (1975), we obtain:

$$\|I_2\| \leq \text{const. } e^{\tau t^w}\|x\|_k \text{ for } x \in D\left(A^k\right), \; \lambda > \omega \text{ and } 0 \leq \frac{n}{\lambda} \leq t. \tag{3}$$

From (2) and (3) we obtain ii)
$ii) \Rightarrow i)$ is a consequence of Remark 1.
$i) \Rightarrow iii)$ By Theorem III iii), A generates an $(A-\lambda)^{-k}$-regularized semigroup $\{S_t\}_{t\geq 0}$, $\forall\lambda > \omega$ and by Corollary 4.5. of de Laubenfels (1991) we have

$$S_t x = (A-\lambda)u\left(t,(A-\lambda)^{-(k+1)}x\right), \quad x \in X.$$

Then by hypothesis, for $x \in X$,

$$\sup_{t\geq 0}\left\|e^{-\tau t^w}S_t x\right\| = \sup_{t\geq 0}\left\|e^{-\tau t^w}\left[u'\left(t,(A-\lambda)^{-(k+1)}x\right) - \lambda u(t,(A-\lambda)^{-(k+1)}x\right\| < \infty.$$

Using the Banach-Steinhaus theorem we obtain that $\|S_t\| = O\left(e^{-\tau t^w}\right)$.

iii) $\Rightarrow$ i) By Corollary 4.1 of de Laubenfels (1991) the equation (*) has a unique solution for all $x \in D\left(A^{k+1}\right)$ given by

$$u(t,x) = S_t(A-\lambda)^{-(k+1)}x, \ t \geq 0$$

where $\{S_t\}_{t\geq 0}$ is the $(A-\lambda)^{-k}$-regularized semigroup generated by A. Since $\|S_t\| = O\left(e^{-\tau t^w}\right)$ if follows that $\|u(t,x)\|$ and $\|u'(t,x)\| = \left\|S_t A(A-\lambda)^{(k+1)}x\right\|$ are also $O\left(e^{\tau t^w}\right)$.

NOTATION 3. For $k \in N, \omega, \ \tau > 0, \ w > 1$ denote by $\mathfrak{G}^w(\omega,k,\tau)$ the class of operators A s.t. $(\omega,\infty) \subset \rho(A)$ and A satisfies one of the quivalent conditions in the Theorem V.

We can finally connect these new results with previous ones by the:

COROLLARY. The following statements for A are equivalent:

a) $A \in \bigcup\limits_{\omega,k,\tau} \mathfrak{G}^w(\omega,k,\tau)$

b) A generates a distribution semigroup $\mathfrak{E}$ of growth order $\leq w$, i.e. there exists $\tau > 0$ s.t. $e^{-\tau\left(1+t^2\right)^{w/2}}$ $\mathfrak{E}$ is tempered.

c) there are $\alpha, \beta, M, \omega, \ \tau > 0$ and $m \in N$ s.t.

$\Lambda_\ell = \left\{\lambda; \mathrm{Re}\lambda \geq (\alpha \ \ln|\lambda| + \beta)^\ell, \ \mathrm{Re}\lambda > \omega\right\} \subset \rho(A)$ and $\|R(\lambda;A)\| \leq M(1+|\lambda|)^n$, $\lambda \in \Lambda_l$

where $\ell = \dfrac{w-1}{w}$.

PROOF. For $a \Leftrightarrow b$ see Cioranescu (1977) and for $b \Leftrightarrow c$ see Cioranescu (1971).

REMARK 3. By results of Foias (1960) each normal operator on a Hilbert space with $\rho(A) \supset \Lambda_\ell$ generates a distribution semigroup with a growth order $\leq w$.

References

1. Chazarain, J. (1971), Problemes de Cauchy abstraits et application a quelques problemes mixtes, J. Funct. Anal., 7, p. 386.

2. Cioranescu, I. (1971), La caracterisation spectrale d' opérateurs generateurs de semigroupes distribution d'ordre fini de croissance, J. Math. Anal. Appl. 34, p. 227.

3. Cioranescu, I. (1977), A characterization of distribution semigroups with finite growth order, Rev. Roum. Math. pures et. Appl. XXII, No. 8, p. 1053.

4. de Laubenfels, R. (1990), Integrated semigroups, C-semigroups and the abstract Cauchy Problem, Semigroup Forum, Vol. 41, p. 83.

5. de Laubenfels, R. (1991), C-semigroups and the Cauchy Problem, J. Funct. Anal. (to appear)

6. Foias, C., (1960), Remarques sur les semigroupes distribution d'opératerus normeaux, Portugalice Math, 19, 4, p. 227.

7. Lions, J. L. (1960), Les semigroupes distributions, Portugaliae Math, 19, 3. p. 141.

8. Neubrander, F. (1988) Integrated semigroups and their applications, Pacific J. of Math., Vol. 135, No. 1, p. 111.

9. Oharu, S. (1971-72) Semigroups of linear operators in a Banach space, Publ. R.I.M.S., Kyoto Univ. J., p. 205.

10. Sanekata, N. (1975), Some remarks on the ACP, Publ. RIMS, Kyoto Univ. 11, p. 51.

11. Ushijima, T. (1972), On the generation and the smoothness of semigroups of linear operators, J. Fac. Sci. Univ. Tokyo, 19, 1, p. 65.

Hamilton–Jacobi Equations in Infinite Dimensions, Part VI: Nonlinear A and Tataru's Method Refined

M. G. CRANDALL University of California, Santa Barbara, California

P. L. LIONS University of Paris-Dauphine, Paris, France

0 INTRODUCTION

Our concern in this work is the theory of "viscosity solutions" of the stationary Hamilton-Jacobi equation

$$u(x) + \langle Ax, \nabla u(x)\rangle + F(x, u(x), \nabla u(x)) = 0 \quad \text{for} \quad x \in D(A) \tag{1}$$

where H is a Hilbert space with the inner-product $\langle \cdot, \cdot \rangle$, A is an m-accretive (equivalently, maximal monotone) operator in H, u and F are real-valued and ∇u corresponds to the gradient of u with respect to $x \in H$, as it has been developed in a series of papers by the authors [7, Parts I - VI] and D. Tataru [15], [16]. We are likewise concerned with the Cauchy problem analogue of (1):

$$\begin{cases} \text{(E)}\ u_t(t,x) + \langle Ax, \nabla u(t,x)\rangle + F(t, x, u(t,x), \nabla u(t,x)) = 0 & \text{for} \quad t > 0, x \in D(A), \\ \text{(IC)}\ u(0,x) = u_0(x) & \text{for} \quad x \in \overline{D(A)}. \end{cases} \tag{2}$$

In these equations the nonlinearity $F(t, x, r, p)$ is assumed to be relatively well-behaved while the expression "$\langle Ax, p\rangle$" is not. The primary contribution of the current work is a significant

(†)Supported in part by Army Research Office Contract DAAL03-90-G-0102 and National Science Foundation Grant DMS90-02331

amelioration and the attendant simplification of some of the major contributions made in [15], [16], as explained below.

We recount some of the relevant developments in an informal way, after alerting the reader that the main text is *self-contained* even if this introduction is not.

A primary problem encountered in the theory of these equations is that of interpreting them in the "viscosity sense". Two difficulties arise: the lack of local compactness in infinite dimensions — which prohibits the immediate transfer of the now highly developed finite dimensional theory even if $A \equiv 0$ — and the treatment of the "unbounded term" $\langle Ax, \nabla u(x)\rangle$, which is difficult to interpret. (We refer the reader to [5] for the history of the finite dimensional theory and many references.)

In [7, Parts I-III] the case $A = 0$ was discussed and it was shown that the lack of local compactness could be overcome via use of perturbed optimization. This was done in two versions. One, developed in [7, Parts I-II], took care of the Hilbert space case (or, more generally, the case in which H is replaced by suitable spaces with the Radon – Nikodym property). In these works perturbed optimization was used in the guise of theorems stating that one may perturb functions by arbitrarily small linear functionals to produce extrema. A second and different use of perturbed optimization was required to treat spaces without the Radon–Nikodym property in [7, Part III]. This time, Ekeland's principle was combined with a notion of "strict viscosity solutions" to obtain a good theory. Recently, Deville, Godefroy and Zizler [8] have eliminated the need for this notion in the case $A = 0$.

Turning to the "unbounded case", in which the term $\langle Ax, \nabla u(x)\rangle$ is present, in [7, Parts IV - V] the case of linear operators A was shown to be accessible provided that A enjoyed a certain natural structure present in important examples and delicate continuity properties then enjoyed by solutions were taken into account. Weak continuity was used to handle the lack of local compactness in IV and perturbed optimization played no role, restricting the theory. This was remedied by more subtle considerations in Part V, which again used perturbed optimization. In [7, Part VII] another case is taken up; it corresponds to the limiting possibility for the choice of the dependence of $F(t,x,p)$ on (x,p) – in this setting, if A is self-adjoint, cases are treated in which $F(t,x,p)$ has continuity properties in (x,p) in the product of the graph topology on $D((I+A)^{1/2})$ in each slot. Part VIII will concern "intermediate cases" with more regularity than this limiting case – e.g., (x,p) will be measured in $D((I+A)^{r/2}) \times D((I+A)^{s/2})$ where $r+s<2$, $r,s \leq 1$.

The general case of a nonlinear m-accretive operator A, without further restrictions, and well-behaved F, was treated in [15], [16]. In [15] a new interpretation of the term $\langle Ax, \nabla u(x)\rangle$ was introduced and a strict viscosity interpretation of the nonlinear term $F(x, u(x), \nabla u(x))$ was used. Comparison results were proved via the introduction of new test functions and a perturbed optimization lemma involving a new nonlinear "distance" which measures continuity along trajectories of the problem $w' + Aw = 0$. Given these results, existence could be established by the program initiated in [7, Part III] and used again in [7, Part V]. In [16] an important observation extended the scope of [15] and permitted the use of Perron's method, previously introduced into the general theory of viscosity solutions by H. Ishii [13], for existence theory. (See also [14] for another context.)

The notions used in [15], [16] are complex and not transparent, although they are simplified in [16]. In the current work, we introduce related but pleasantly simplified notions, rendering the theory more elegant, transparent and accessible. We present proofs of uniqueness which use the tools developed in [15]. However, we use different definitions and our proof is organized differently. The original research for this manuscript, as announced at the conference, preceded the appearance of [16]; the current version takes into account the important observation in Lemma 1 of [16]. Thus we may use Perron's method to treat existence questions.

The text is organized as follows: following a prologue in Section 0, Section 1 is devoted to the definitions and some results and examples which build familiarity with them. Some of this material is not essential for what follows (although we think it is pleasantly written and illuminating), and we indicate to the reader what may be skipped for a quick run through. Similar remarks apply to later sections. Section 2 introduces a main ingredient for use in perturbed optimizations, namely, "Tataru's distance". It also contains calculations needed later and explains Tataru's perturbed optimization results using his distance.

Section 3 gives sample comparison proofs in the stationary case of (1). Here we adopt a strategy for the reader's convenience – we are not concerned about technical generality. Indeed, there is a large lore concerning the restrictions on F under which one can treat problems like (1) and our point here is that all these results generalize to the current setting. We simply show how this is done in some cases. However, we explain enough so that the reader new to the subject can obtain a feeling for some of the techniques and we give complete proofs in a particular case. Perusing conditions and examples in [7] and referring to the references therein (as well as in [5]) gives a complete picture. By the way, we have similarly restricted ourselves to the case of a Hilbert space H, but all that we do here generalizes to Banach spaces X which are, together with their duals, uniformly convex (the framework used in [15], [16]). Section 4 contains the basics of Perron's method, applies it to the stationary problem and discusses the "consistency property", which roughly asserts that limits of solutions are solutions of limit equations. Section 5 discusses the Cauchy problem (2). Here the situation is less familiar, due to relatively delicate problems with time dependence, and we spend more time on a sample problem, resolving it in detail. The issue of constructing sub and supersolutions for use with Perron's method here is not standard (as opposed to the stationary case), and we show how this is done. We conclude with an appendix containing the proof of a keystone result, the Doubling Theorem, which is used without proof in Sections 3 and 5. In the Appendix we also, for the reader's convenience, briefly comment on Tataru's notions and results and their relationship with ours.

One topic we omit here is a discussion of the fact that value functions for optimal control problems and differential game problems in infinite dimensions are solutions of the associated partial differential equations of dynamic programming in our sense. The relevance of Hamilton-Jacobi equations in this regard may be seen in [2]; the viscosity approach is developed in [7, Parts IV - VI]. This issue has been treated by Tataru [15], [17] under his presentation; it is more straightforward to discuss in our setting and will be addressed elsewhere.

0 PROLOGUE

We foreshadow what is going to happen below. Suppose that $A: H \to H$ is everywhere defined, single-valued, monotone, $0 = A0$ and that we have defined viscosity subsolutions and supersolutions of

$$u(x) + \langle Ax, \nabla u(x)\rangle + F(x, u(x), \nabla u(x)) = 0 \quad \text{for} \quad x \in H \tag{3}$$

so that if u (respectively, v) is a subsolution (respectively, supersolution), then at any local maximum $\hat{x}$ (respectively, local minimum $\hat{y}$) of $u - \varphi$ (respectively, $v - \varphi$) where $\varphi \in C^1(H)$ we have

$$u(\hat{x}) + \langle A\hat{x}, \nabla\varphi(\hat{x})\rangle + F(\hat{x}, \varphi(\hat{x}), \nabla\varphi(\hat{x})) \leq 0$$

(respectively,

$$v(\hat{y}) + \langle A\hat{y}, \nabla\varphi(\hat{y})\rangle + F(\hat{y}, \varphi(\hat{y}), \nabla\varphi(\hat{y})) \geq 0).$$

Assume, moreover, that if φ is perturbed by functions with small Lipschitz constants, then this remain approximately true. Then to prove *comparison*, that is, $u \leq v$ whenever u is a subsolution and v is a supersolution satisfying some restrictions on their growth, we produce a maximum point $(\hat{x}, \hat{y}) \in H \times H$ of something like

$$\Phi(x, y) = u(x) - v(y) - \frac{\alpha}{2}\|x - y\|^2 - \frac{\delta}{2}\left(\|x\|^2 + \|y\|^2\right) - \lambda \mathrm{Pert}(x, y)$$

where $\alpha, \delta, \lambda > 0$ and "Pert" is a perturbation function corresponding to using a perturbed optimization principle to produce the maximum point. Then the fact that $x \to \Phi(x, \hat{y})$ (respectively, $y \to \Phi(\hat{x}, y)$) is maximized at $\hat{x}$ (respectively, minimized at $\hat{y}$) is plugged into the subsolution (respectively, supersolution) inequalities, the results are subtracted one from the other and estimates deduced. Then, in succession, the limits $\lambda \downarrow 0$ (to remove the perturbation), $\delta \downarrow 0$ (to remove the "control at infinity"), $\alpha \to \infty$ (to penalize the doubling of variables) are taken; assumptions are made on F so that this succeeds and the conclusion $u \leq v$ is reached.

A quick computation shows that differencing the terms arising from the presence of $\langle Ax, \nabla u(x)\rangle$ in (3) in the above process leads to

$$\langle A\hat{x} - A\hat{y}, \hat{x} - \hat{y}\rangle + \delta\langle A\hat{x}, \hat{x}\rangle + \delta\langle A\hat{y}, \hat{y}\rangle + \text{ (stuff from Pert)} .$$

Since A is monotone and $A0 = 0$, the first three expressions are nonnegative. In [7, Parts IV & V], the "nonnegativity" of the middle two terms was exploited in contexts where they were not explicitly defined, while a regularized version of the nonnegativity of the first term was used which, however, required strong additional assumptions on A.

The presentation below makes this outline work in the case of general maximal monotone operators A by defining the notions of sub and supersolutions of (3) in such a way that $w(x, y) = u(x) - v(y)$ becomes a subsolution of the "doubled equation"

$$w(x, y) + \langle \mathcal{A}(x, y), \nabla w(x, y)\rangle + F(x, u(x), \nabla_x w(x, y)) - F(y, v(y), -\nabla_y w(x, y)) \leq 0$$

where $\mathcal{A}(x, y) = Ax \times Ay$ defines the maximal monotone operator $\mathcal{A}$ in $H \times H$. Moreover, this implies that "$\langle A\hat{x} - A\hat{y}, \hat{x} - \hat{y}\rangle$" ≥ 0 has a proper interpretation in the sketch above. Tataru's choices of "Pert" are involved throughout the process. Roughly speaking, A then "disappears" and we may immediately adapt the proofs of standard finite dimensional theorems to the infinite dimensional case in a routine way. (However, as mentioned before, there are subtleties in the case of the Cauchy problem corresponding to the the delicate continuity of semigroups in time.) Presented this way, the Doubling Theorem 3.1 in Section 3, which asserts that w is a subsolution, becomes a keystone.

The entire program succeeds in the general case due to the appropriate viscosity interpretation of the term $\langle Ax, \nabla u\rangle$, where we simplify Tataru's ideas significantly, and the choice of "Pert" due to Tataru and explained below.

1 DEFINITION OF VISCOSITY SOLUTIONS

We always assume that

$$A \quad \text{is an } m-\text{accretive operator in} \quad H. \tag{4}$$

We review some basic facts concerning nonlinear semigroup theory. A general m-accretive operator A in H is typically multivalued and may be regarded as a mapping of H to its subsets. The notation "$y \in Ax$" means that $x \in D(A)$ and y is an element of Ax. For $x \in D(A)$, Ax is a closed and convex set, therefore containing a unique element $A^\circ x$ of minimal norm. The (single-valued) operator A° is called the minimal section of A. Let $S(t) = e^{-tA}$ be the semigroup generated by $-A$ on $\overline{D(A)}$, that is $u(t) = S(t)x$ is the unique mild solution of $u' + Au \ni 0$, $u(0) = x$ on $[0, \infty)$. (Brezis [3] and Barbu [1] are references for this theory.) If $x \in D(A)$, then $t \mapsto S(t)x$ is Lipschitz continuous and

$$d_r u(t) + A^\circ u(t) = 0 \quad \text{for} \quad t \geq 0$$

where d_r denotes differentiation from the right.

Hence if $\varphi \in C^1$ we have

$$d_r \varphi(S(t)x) = -\langle A^\circ S(t)x, \nabla\varphi(S(t)x)\rangle \tag{5}$$

for $x \in D(A)$ and $t \geq 0$. With these remarks in mind, it is natural to interpret the term $\langle Ax, \nabla u(x)\rangle$ in (1) in terms of limits of the expressions $(u(x) - u(S(h)x))/h$ as $h \downarrow 0$. This was successfully done, in a manner too complex to reproduce here, in [15] (see also [16] and our appendix). One of our main points is to simplify this aspect.

We will understand the term $\langle Ax, \nabla u(x)\rangle$ in (1) in the above spirit in a manner made precise below. In addition, we will understand (1) in a "viscosity sense", the definition of which, as usual, requires notions of subsolutions and supersolutions and that a solution be both a subsolution and a supersolution. Subsolutions and supersolutions are defined by placing derivatives on test functions (in one sense or another) at extrema and asking that natural inequalities be satisfied. For later convenience, we carry out the discussion for a general equation

$$\langle Ax, \nabla u(x)\rangle + G(x, u(x), \nabla u(x)) = 0 \tag{6}$$

where $G: \overline{D(A)} \times \mathbb{R} \times H \to \mathbb{R}$ is continuous and $r \to G(x, r, p)$ is nondecreasing for $x \in \overline{D(A)}$, $p \in H$.

We ask the reader's indulgence as we leisurely peel off the layers of the precise definitions below, as undefined expressions will occur at every stage until the end. We first define the notion of a subsolution of (6) on a set $\Omega \subset H$. A subsolution u on Ω is a solution of the "inequality"

$$\liminf_{\substack{h \downarrow 0 \\ y \to x}} \frac{u(y) - u(S(h)y)}{h} + G(x, u(x), \nabla u(x)) \leq 0 \text{ on } \Omega. \tag{7}$$

It will not be apparent why the limit $y \to x$ is inserted in the first term of (7) for quite a while.

Of course, we still have to interpret the various ingredients of these expressions. $S(h)y$ makes sense only for $y \in \overline{D(A)}$, so hereafter we ask that $\Omega \subset \overline{D(A)}$ and (for technical reasons) that Ω be relatively open in $\overline{D(A)}$. Interpretation of (for example) (7) in the "viscosity sense" requires that if Φ is a "subtest function" and $\hat{x} \in \Omega$ is a local maximum of $u - \Phi$, that is

$$u(x) - \Phi(x) \leq u(\hat{x}) - \Phi(\hat{x}) \quad \text{for} \quad x \in \Omega, \ \|x - \hat{x}\| \leq r \tag{8}$$

for some $r > 0$, then

$$\liminf_{\substack{h \downarrow 0 \\ y \to \hat{x}}} \frac{\Phi(y) - \Phi(S(h)y)}{h} + G(\hat{x}, u(\hat{x}), \nabla\Phi(\hat{x})) \leq 0; \tag{9}$$

that is, we replace u by Φ where differentiations are involved. We are still not done, for we have to explain the class of subtest functions from which Φ may be chosen and to explain how $G(\hat{x}, u(\hat{x}), \nabla\Phi(\hat{x}))$ is interpreted above if Φ is not itself differentiable, as may be the case.

For $x \in \overline{D(A)}$

$$\liminf_{\substack{h \downarrow 0 \\ y \to x}} \frac{\Phi(y) - \Phi(S(h)y)}{h} = \liminf_{r \downarrow 0} \left\{ \frac{\Phi(y) - \Phi(S(h)y)}{h} : 0 < h \leq r, y \in \overline{D(A)}, \|y - x\| \leq r \right\}$$

can also be defined as the least extended real number $L \in [-\infty, \infty]$ obtainable as a limit of a sequence $(\Phi(y_n) - \Phi(S(h_n)y_n))/h_n$ where $(y_n, h_n) \in \overline{D(A)} \times (0, \infty)$ converges to $(x, 0)$. The restrictions y, $y_n \in \overline{D(A)}$ above are not otherwise recorded in the notation "$\liminf_{\substack{h \downarrow 0 \\ y \to x}}$", but arise naturally from the fact that $S(h)y$ in only defined for $y \in \overline{D(A)}$.

In order to interpret "$G(\hat{x}, u(\hat{x}), \nabla\Phi(\hat{x}))$" for subsolutions of (6) when Φ is not differentiable at $\hat{x}$, we assume that $\Phi = \varphi + \psi \in C^1(H) + \text{Lip}(H)$ where $\varphi \in C^1(H)$ and $\psi \in \text{Lip}(H)$; that is $\psi: H \to \mathbb{R}$ is Lipschitz continuous. It will hereafter be assumed that if we write $\Phi = \varphi + \psi \in C^1(H) + \text{Lip}(H)$, then $\varphi \in C^1(H)$ and $\psi \in \text{Lip}(H)$. We need some more notation. $L(\psi)$ will denote the least Lipschitz constant of $\psi \in \text{Lip}(H)$, and for $\lambda > 0$ and $p \in H$ we set

$$G_\lambda(x, r, p) = \inf_{\|q\| \leq \lambda} G(x, r, p + q) \quad \text{and} \quad G^\lambda(x, r, p) = \sup_{\|q\| \leq \lambda} G(x, r, p + q).$$

REMARK 1.1. It is obvious that if G is continuous, then $G_\lambda(x, r, p)$ is upper semicontinuous in (x, r, p) and nonincreasing in λ while G^λ is lower semicontinuous in (x, r, p) and nondecreasing in λ. Moreover, if G is uniformly continuous on bounded sets, then both functions are uniformly continuous in all variables (λ, x, r, p) on bounded sets and if G is uniformly continuous in p on bounded sets, then both functions are uniformly continuous in λ, p on bounded sets. Sometimes below we assume functions to be uniformly continuous on bounded sets for simplicity, but significantly weaker assumptions suffice.

The precise interpretation of (7) is that

$$\liminf_{\substack{h \downarrow 0 \\ y \to \hat{x}}} \frac{\Phi(y) - \Phi(S(h)y)}{h} + G_{L(\psi)}(\hat{x}, u(\hat{x}), \nabla\varphi(\hat{x})) \leq 0 \tag{10}$$

whenever $\hat{x}$ is a local maximum of $u - \Phi$ and $\Phi = \varphi + \psi \in C^1(H) + \text{Lip}(H)$ and Φ is a *subtest function*; the subtest functions are those $\Phi \in C^1(H) + \text{Lip}(H)$ which satisfy one further condition, namely (12) below .

Since A is m-accretive, $\overline{D(A)}$ is closed and convex and the projection P of H on $\overline{D(A)}$ defined by

$$Pz \in \overline{D(A)} \quad \text{and} \quad \|z - Pz\| \leq \|z - x\| \quad \text{for} \quad x \in \overline{D(A)} \tag{11}$$

is nonexpansive. In order that $\Phi = \varphi + \psi \in C^1(H) + \text{Lip}(H)$ be a subtest function, we further require that

$$\varphi(Px) \leq \varphi(x) \quad \text{and} \quad \psi(Px) \leq \psi(x) \quad \text{for} \quad x \in H. \tag{12}$$

One proceeds similarly for supersolutions; (10) will be replaced by

$$\limsup_{\substack{h \downarrow 0 \\ y \to \hat{x}}} \frac{\Phi(y) - \Phi(S(h)y)}{h} + G^{L(\psi)}(\hat{x}, u(\hat{x}), \nabla\varphi(\hat{x})) \geq 0 \tag{13}$$

and (12) will be replaced by

$$\varphi(Px) \geq \varphi(x) \quad \text{and} \quad \psi(Px) \geq \psi(x) \quad \text{for} \quad x \in H. \tag{14}$$

REMARK 1.2. The restrictions on ψ above are made only for convenience in the presentation. Indeed, if $\psi \in \text{Lip}(\overline{D(A)})$, then the extension $\tilde{\psi}(x) = \psi(Px)$ for $x \in H$ satisfies $\tilde{\psi}(Px) = \tilde{\psi}(x)$ and $L(\tilde{\psi}) = L(\psi)$. It is only the behavior of ψ on $\overline{D(A)}$ that is significant. On the other hand, the restrictions on φ are material, since there is an interaction with the condition that φ be C^1. In the proofs of comparison, we will make the most use of the simple choice $\varphi(x) = \|x - y\|^2$ for some $y \in \overline{D(A)}$ as the C^1 part of a subtest function, in which case (12) evidently holds. In the proofs of existence, this condition is crucial.

If $\mathcal{M}$ is a metric space, we put

$$USC(\mathcal{M}) = \{\text{upper-semicontinuous maps } u\colon \mathcal{M} \to \mathbb{R}\},$$
$$LSC(\mathcal{M}) = \{\text{lower-semicontinuous maps } u\colon \mathcal{M} \to \mathbb{R}\}.$$

DEFINITION 1.3. *Let A be an m-accretive operator in H, $S(t)$ be the semigroup on $\overline{D(A)}$ generated by $-A$ and P be the projection on $\overline{D(A)}$ defined by (11). Let Ω be a relatively open subset of $\overline{D(A)}$. Then $u \in USC(\Omega)$ (respectively, $u \in LSC(\Omega)$) is a subsolution (respectively, supersolution) of (6) on Ω if for every $\Phi = \varphi + \psi \in C^1(H) + \text{Lip}(H)$ satisfying (12) (respectively, (14)) and local maximum (respectively, minimum) $\hat{x}$ of $u - \Phi$ (10) (respectively, (13)) holds. Finally, $u \in C(\mathcal{M})$ is a solution of (1) on Ω if it is simultaneously a subsolution and a supersolution on Ω.*

We will also indicate that u is a subsolution of (1) on Ω by saying "u satisfies"

$$\langle Ax, \nabla u(x)\rangle + G(x, u(x), \nabla u(x)) \leq 0 \quad \text{on} \quad \Omega;$$

analogous conventions will be used for supersolutions and solutions.

It will be convenient to use the simpler notations

$$D_A^-\Phi(x) = \liminf_{\substack{h\downarrow 0\\ y\to x}} \frac{\Phi(y)-\Phi(S(h)y)}{h} \quad \text{and} \quad D_A^+\Phi(x) = \limsup_{\substack{h\downarrow 0\\ y\to x}} \frac{\Phi(y)-\Phi(S(h)y)}{h} \tag{15}$$

hereafter.

At this point the reader may pass to Remark 1.6 below, forgoing for the moment the intervening material.

In order to obtain some feeling for these definitions and for later use, we prove the following result concerning construction of viscosity subsolutions.

LEMMA 1.4. *Let $\Phi = \varphi + \psi \in C^1(H) + \mathrm{Lip}(H)$ be a supertest function and $\Omega \subset \overline{D(A)}$ be relatively open. If the pointwise inequality*

$$D_A^+\Phi(x) + G^{L(\psi)}(x, \Phi(x), \nabla\varphi(x)) \leq 0 \quad \textit{for} \quad x \in \Omega \tag{16}$$

holds, then

$$\langle Ax, \nabla\Phi(x)\rangle + G(x, \Phi(x), \nabla\Phi(x)) \leq 0 \quad \textit{in} \quad \Omega$$

in the viscosity sense. Similarly, if Φ is a subtest function and

$$D_A^-\Phi(x) + G_{L(\psi)}(x, \Phi(x), \nabla\varphi(x)) \geq 0 \quad \textit{for} \quad x \in \Omega,$$

then Φ is a viscosity supersolution.

Proof. Let $\tilde\Phi = \tilde\varphi + \tilde\psi \in C^1(H) + \mathrm{Lip}(H)$ be a subtest function, $\hat x \in \Omega$ and $\Phi - \tilde\Phi$ have a local (relative to $\overline{D(A)}$) maximum at $\hat x$. Then

$$\Phi(x) - \tilde\Phi(x) \leq \Phi(\hat x) - \tilde\Phi(\hat x) \tag{17}$$

for $x \in \overline{D(A)}$ near $\hat x$ implies

$$\tilde\Phi(\hat x) - \tilde\Phi(S(h)\hat x) \leq \Phi(\hat x) - \Phi(S(h)\hat x)$$

for small h; it follows that

$$D_A^-\tilde\Phi(\hat x) \leq D_A^+\Phi(\hat x). \tag{18}$$

Next, since Φ is a supertest function and $\tilde\Phi$ is a subtest function, we claim that (17) holds in a full neighborhood of $\hat x$. Indeed, if P is the projection on $\overline{D(A)}$, for $x \in H$ near $\hat x$,

$$\Phi(x) - \tilde\Phi(x) \leq \Phi(Px) - \tilde\Phi(Px) \leq \Phi(\hat x) - \tilde\Phi(\hat x). \tag{19}$$

But (19) implies

$$\varphi(x) - \tilde\varphi(x) \leq \varphi(\hat x) - \tilde\varphi(\hat x) + L\left(\psi - \tilde\psi\right)\|x - \hat x\|$$

and then an easy calculus exercise shows that

$$\|\nabla\varphi(\hat{x}) - \nabla\tilde{\varphi}(\hat{x})\| \leq L\left(\psi - \tilde{\psi}\right)$$

Since $L\left(\psi - \tilde{\psi}\right) \leq L(\psi) + L\left(\tilde{\psi}\right)$ we thus have

$$G^{L(\psi)}(\hat{x}, \Phi(\hat{x}), \nabla\varphi(\hat{x})) \geq G_{L(\tilde{\psi})}(\hat{x}, \Phi(\hat{x}), \nabla\tilde{\varphi}(\hat{x})). \tag{20}$$

In view of (17), (18), and (20), Φ is a subsolution. □

EXAMPLE 1.5. In order to further illustrate the nature of the subtest condition in the definition of a subtest function, we take a simple example. Let $H = \mathbb{R}$ and A be the subdifferential of the indicator function of the unit interval, $A = \partial I_{[0,1]}$. Here the indicator function I_K of $K \subset H$ is given by

$$I_K = \begin{cases} \infty & \text{if } x \notin K \\ 0 & \text{if } x \in K. \end{cases}$$

Then ∂I_K assigns to $x \in K$ the normal cone to K at x; this is maximal monotone if K is closed and convex. In the case $K = [0,1]$, $A = \partial I_{[0,1]}$ we have

$$Ax = \begin{cases} [0,\infty) & \text{if } x = 1 \\ \{0\} & \text{if } 0 < x < 1 \\ (-\infty, 0] & \text{if } x = 0 \\ \emptyset & \text{otherwise,} \end{cases}$$

$D(A) = [0,1] = \overline{D(A)}$ and $e^{-tA} = S(t)$ is the identity semigroup on $[0,1]$. Thus

$$D_A^-\Phi(x) = D_A^+\Phi(x) = 0$$

for every Φ, $x \in [0,1]$. We seek a solution of the equation

$$u + \langle Ax, \nabla u\rangle + u' = 1 \quad \text{on} \quad [0,1]$$

of the form

$$u = 1 + ae^{-x}$$

for some $a \in \mathbb{R}$. Since

$$u + D_A^+ u + u' = u + u' = 1 \quad \text{on} \quad [0,1],$$

by Lemma 1.4 u is a subsolution if it is a supertest function. On the other hand, u is a supertest function if $u(0) \leq u(x)$ for $x \leq 0$ and $u(1) \leq u(x)$ for $x \geq 1$. The only choice of a which achieves this is $a = 0$ and then, by Lemma 1.4, u is also a supersolution and hence a solution. In this example, the unique solution is $u \equiv 1$ and if (for example) we test the subsolution property with test functions Φ which are not subtest functions, we will not find the defining inequalities to be satisfied. The subtest condition on subtest functions is *necessary* to have existence.

The above is a special case of the linear equation

$$u(x) + \langle Ax, \nabla u(x)\rangle = f(x) \quad \text{for} \quad x \in \overline{D(A)}.$$

The solution of this equation should be

$$u(x) = \int_0^\infty f(S(t)x)e^{-t}\,dt.$$

Let us verify that this is so. Suppose $\Phi = \varphi + \psi$ is a subtest function and

$$u(x) - \Phi(x) \leq u(\hat{x}) - \Phi(\hat{x}) \quad \text{for} \quad x \in \overline{D(A)} \cap B_r(\hat{x}).$$

We need to verify that

$$u(\hat{x}) + D_A^- \Phi(\hat{x}) \leq f(\hat{x}).$$

However, $\Phi(\hat{x}) - \Phi(S(h)\hat{x}) \leq u(\hat{x}) - u(S(h)\hat{x})$ for small h, so it suffices to show that

$$u(\hat{x}) + \liminf_{h \downarrow 0} \frac{u(\hat{x}) - u(S(h)\hat{x})}{h} \leq f(\hat{x})$$

or

$$\int_0^\infty f(S(t)\hat{x})e^{-t}\,dt + \liminf_{h \downarrow 0} \int_0^\infty \frac{f(S(t)\hat{x}) - f(S(t+h)\hat{x})}{h} e^{-t}\,dt \leq f(\hat{x})$$

however

$$\int_0^\infty \frac{f(S(t)\hat{x}) - f(S(t+h)\hat{x})}{h} e^{-t}\,dt = \int_0^\infty f(S(t)\hat{x}) \frac{e^{-t} - e^{-(t+h)}}{h}\,dt + \frac{1}{h} e^h \int_0^h e^{-t} f(S(t)\hat{x})\,dt$$

so in fact

$$\int_0^\infty f(S(t)\hat{x})e^{-t}\,dt + \lim_{h \downarrow 0} \int_0^\infty \frac{f(S(t)\hat{x}) - f(S(t+h)\hat{x})}{h} e^{-t}\,dt = f(\hat{x}).$$

REMARK 1.6. If the reader reviews all the above, it is clear that in fact "A" is everywhere irrelevant and the essential data are the semigroup $S(t)$ and the set C (written "$\overline{D(A)}$" above) on which it acts. For example, the elaborate description of A in Example 1.5 played no role; only the identity semigroup on [0,1] entered. The projection P on C should be defined, so we need C to be closed and convex. Convexity will be important in calculations done later. In this spirit, we could (but won't) write $D_S^{\pm}$ in place of $D_A^{\pm}$.

2 SUBTEST FUNCTIONS AND TATARU'S LEMMAS

In this section, we estimate $D_A^- \Phi(x)$ for some choices of subtest functions Φ and note simple general properties of this "limit" which will be used later. One of these subtest functions is "Tataru's distance", which is introduced. The section ends with Tataru's perturbed optimization results.

Hereafter, for simplicity in the presentation, we will always assume that

$$0 \in A0. \tag{21}$$

In consequence,

$$S(t)0 \equiv 0 \tag{22}$$

and then, since $S(t)$ is nonexpansive,

$$\|S(t)y\| = \|S(t)y - S(t)0\| \leq \|y\| \quad \text{for} \quad y \in \overline{D(A)},\ t \geq 0. \tag{23}$$

If $g\colon [0,\infty) \to \mathbb{R}$, we abuse notation and write $g(y)$ for the associated "radial function" $g(\|y\|)$ and we call this radial function nondecreasing if the scalar function g is nondecreasing. Since $g(y) - g(S(t)y) \geq 0$ if g is radial and nondecreasing by (23), we have

$$D_A^- g(x) \geq 0 \quad \text{for} \quad x \in \overline{D(A)} \tag{24}$$

if g is radial and nondecreasing. This is our first estimate of an expression $D_A^-\Phi(x)$. Moreover, if P is the projection on $\overline{D(A)}$, we also have

$$\|Px\| = \|Px - P0\| \leq \|x - 0\| = \|x\| \quad \text{for} \quad x \in \overline{D(A)}$$

since P is nonexpansive. In consequence, the $g(x)$ above is a subtest function.

For a second example, let $\xi \in D(A)$ and $\Phi(x) = \|x - \xi\|^2$. For $x \in \overline{D(A)}$, if $y \in \overline{D(A)}$ we have

$$\begin{aligned}
\Phi(y) - \Phi(S(h)y) &= \|y - \xi\|^2 - \|S(h)y - \xi\|^2 \\
&= \|y - \xi\|^2 - \|S(h)y - S(h)\xi + S(h)\xi - \xi\|^2 \\
&= \|y - \xi\|^2 - \|S(h)y - S(h)\xi\|^2 - 2\langle S(h)y - S(h)\xi, S(h)\xi - \xi\rangle - \|S(h)\xi - \xi\|^2 \\
&\geq -2\langle S(h)\xi - \xi, S(h)y - \xi\rangle + \|S(h)\xi - \xi\|^2.
\end{aligned}$$

We divide this relation by $h > 0$ and use $d_r S(t)\xi = -A^\circ S(t)\xi$ while taking the limit inferior to conclude that

$$D_A^-\Phi(x) \geq 2\langle A^\circ\xi, x - \xi\rangle. \tag{25}$$

Moreover,

$$\begin{aligned}
\|x - \xi\|^2 &= \|x - Px + Px - \xi\|^2 \\
&= \|x - Px\|^2 + 2\langle x - Px, Px - \xi\rangle + \|Px - \xi\|^2 \\
&\geq \|Px - \xi\|^2
\end{aligned} \tag{26}$$

since the middle term in the second quantity is nonnegative by the variational characterization of P. Thus Φ is a subtest function for $\xi \in D(A)$.

REMARK 2.1. In Remark 1.6, we noted that A had not yet played a role; it appears explicitly now in (25). However, the object A° in use here is actually -(the infinitesimal generator of S). Moreover, we note one way to lift the assumption (22) in what follows. This assumption is used to verify (24) and the case $g(x) = (\delta/2)\|x\|^2$ will occur many times below. Instead, one may fix $\xi \in D(A)$ and use $(\delta/2)\|x - \xi\|^2$ with appropriate changes in the assumptions on the nonlinearities and the arguments given. This adds terms to the considerations.

For a third example, we introduce the "distance" (a term we use to emphasize that it is not a metric) from [15] and given by

$$d(x,y) = \inf_{t\geq 0}(t + \|x - S(t)y\|); \tag{27}$$

this prescription makes sense for $x \in H, y \in \overline{D(A)}$. When necessary, we indicate the dependence of d on A by writing $d = d_A$. It is d from which the "Pert" of the prologue will be built. The next computation shows one of its virtues established in [15]; others appear below. In view of the definition, for $x \in H, y \in \overline{D(A)}$, there is a $t \geq 0$ such that $d(x,y) = t + \|x - S(t)y\|$.

Consider the choice $\Phi(x) = d(x,y)$ where d is given by (27) and $y \in \overline{D(A)}$. Choosing a $t > 0$ for which $t + \|x - S(t)y\| = d(x,y)$ we have

$$t + h + \|x - S(t)y\| \geq t + h + \|S(h)x - S(t+h)y\| \geq d(S(h)x, y) = \Phi(S(h)x)$$

and therefore

$$\frac{\Phi(x) - \Phi(S(h)x)}{h} \geq \frac{t + \|x - S(t)y\| - (t + h + \|x - S(t)y\|)}{h} = -1$$

which yields,

$$\Phi(x) = d(x,y) \quad \text{satisfies} \quad D_A^- \Phi(x) \geq -1 \quad \text{for} \quad x, y \in \overline{D(A)}.$$

Tataru's distance (27) obviously has the property

$$|d(x,y) - d(\hat{x},\hat{y})| \leq |x - \hat{x}| + |y - \hat{y}| \quad \text{for} \quad x, y, \hat{x}, \hat{y} \in \overline{D(A)}; \tag{28}$$

in particular, it is a candidate for the Lipschitz part of a subtest function. As a Lipschitz perturbation, in accordance with Remark 1.2, we need not check the subtest condition. But $d(x,y)$ is defined for $x \in H$ and $\|Px - S(t)y\| \leq \|x - S(t)y\|$ for $y \in \overline{D(A)}$ by (26), so indeed $d(Px,y) \leq d(x,y)$ and Φ is a subtest function.

We record the results above in a lemma.

LEMMA 2.2. *Let (23) hold* $g\colon [0,\infty) \to \mathbb{R}$ *be nondecreasing,* $y \in \overline{D(A)}$, $\alpha, \lambda, \delta > 0$, $\xi \in D(A)$ *and*

$$\Phi(x) = \frac{\alpha}{2}\|x - \xi\|^2 + \delta g(x) + \lambda d(x,y)$$

where d *is given by (27). Then* Φ *is a subtest function and*

$$D_A^- \Phi(x) \geq \alpha \langle A^\circ \xi, x - \xi \rangle - \lambda \quad \textit{for} \quad x \in \overline{D(A)}.$$

One more example and we will have treated all of the expressions discussed in the prologue. We form the operator $\mathcal{A} = A \times A$ on $H \times H$ by

$$\mathcal{A}(x,y) = Ax \times Ay \quad \text{and} \quad D(\mathcal{A}) = D(A) \times D(A). \tag{29}$$

Here $H \times H$ is equipped with the inner product and norm

$$\langle (x,y), (\hat{x},\hat{y}) \rangle_{H\times H} = \langle x, \hat{x} \rangle + \langle y, \hat{y} \rangle, \qquad \|(x,y)\|_{H\times H}^2 = \|x\|^2 + \|y\|^2.$$

$\mathcal{A}$ is m-accretive and $-\mathcal{A}$ generates the semigroup $\mathcal{S}(t) = e^{-t\mathcal{A}}$ on $\overline{\mathcal{A}} = \overline{D(A)} \times \overline{D(A)}$ given by

$$\mathcal{S}(t)(x,y) = (S(t)x, S(t)y) \quad \text{for} \quad x, y \in \overline{D(A)}.$$

Let $\Phi(x,y) = \|x - y\|^2$; then

$$\Phi((x,y)) - \Phi(\mathcal{S}(h)(x,y)) = \|x-y\|^2 - \|S(h)x - S(h)y\|^2 \geq 0.$$

It follows that if $g\colon \mathbb{R} \to \mathbb{R}$ is nondecreasing, we have

$$\text{if} \quad \Phi((x,y)) = g(\|x-y\|) \quad \text{then} \quad D_{\mathcal{A}}^{-}\Phi(x,y) \geq 0 \quad \text{for} \quad x, y \in \overline{D(A)}. \tag{30}$$

It is obvious that Φ is a subtest function (for $\mathcal{A}$).

Here us a lemma which may be skipped until needed later.

LEMMA 2.3. *If $\Phi\colon \overline{D(A)} \to \mathbb{R}$, then $D_A^{-}\Phi\colon \overline{D(A)} \to [-\infty, \infty]$ is lower-semicontinous and $D_A^{+}\Phi\colon \overline{D(A)} \to [-\infty, \infty]$ is upper-semicontinuous.*

Proof. We treat $D_A^{-}\Phi$. Let $x_n \in \overline{D(A)}$ and $x_n \to x$. Choose $h_n \downarrow 0$, $z_n \to \hat{x}$ such that

$$\frac{\Phi(z_n) - \Phi(S(h_n)z_n)}{h_n} \leq \max\left\{-n, D_A^{-}\Phi(x_n) + \frac{1}{n}\right\};$$

this is always possible. Then

$$\begin{aligned} D_A^{-}\Phi(x) &\leq \liminf_{n\to\infty} \frac{\Phi(z_n) - \Phi(S(h_n)z_n)}{h_n} \\ &\leq \liminf_{n\to\infty} \max\left\{-n, D_A^{-}\Phi(x_n) + \frac{1}{n}\right\} = \liminf_{n\to\infty} D_A^{-}\Phi(x_n). \end{aligned}$$

□

We have one more story to tell, that of perturbed optimization and Tataru's distance, about the interesting ingredients in this saga – has it ever taken so long to get started? – before turning to some proofs of uniqueness and existence.

If $x \in H$, $z, y \in \overline{D(A)}$, and $a, b \geq 0$ we have

$$\|x - S(a+b)y\| \leq \|x - S(a)z\| + \|S(a)z - S(a+b)y\| \leq \|x - S(a)z\| + \|z - S(b)y\| \tag{31}$$

and then $d(x,z) = a + \|x - S(a)z\|$, $d(z,y) = b + \|z - S(b)y\|$ implies

$$d(x,y) \leq a + b + \|x - S(a+b)y\| \leq a + \|x - S(a)z\| + b + \|z - S(b)y\| = d(x,z) + d(z,y)$$

so we have

$$d(x,y) \leq d(x,z) + d(z,y) \quad \text{for} \quad x, y, z \in \overline{D(A)} \quad \text{and} \quad d(x,y) = 0 \text{ iff } x = y \tag{32}$$

as noted in [15]. Thus d is "almost" a metric, but it is not symmetric (and hence not a metric).

The lack of symmetry is emphasized by noting that if x_n, y_n are sequences in H and $x, y \in H$, then

$$d(x_n, y) \to 0 \quad \text{if and only if} \quad \|x_n - y\| \to 0 \tag{33}$$

while

$$d(x, y_n) \to 0 \quad \text{if and only if} \quad \exists t_n \downarrow 0 \quad \text{such that} \quad \|x - S(t_n)y_n\| \to 0 \tag{34}$$

and these are different types of convergence (see Examples 2.6 below). Both assertions are obvious; the first may obtained from a more explicit estimate. If $d(x,y) = t + \|x - S(t)y\|$, then $t \le d(x,y)$ and so

$$\begin{aligned} d(x,y) &= t + \|x - S(t)y\| \\ &\ge (t + \|x - y\| - \|S(t)y - y\|) \\ &\ge \|x - y\| - g_y(d(x,y)) \end{aligned}$$

where

$$g_y(r) = \sup_{0 \le t \le r} (\|S(t)y - y\| - t);$$

we conclude that

$$d(x,y) + g_y(d(x,y)) \ge \|x - y\|. \tag{35}$$

Whenever $\mathcal{M}$ is a set and $d: \mathcal{M} \times \mathcal{M} \to [0, \infty)$ satisfies (32) with $\mathcal{M}$ in place of $\overline{D(A)}$, we will say that $(\mathcal{M}, d)$ is a *distance space*. We say that a distance space $(\mathcal{M}, d)$ is *complete* if whenever $\{x_n\}$ is a sequence in $\mathcal{M}$ such that

$$\sum_{j=1}^{\infty} d(x_{n+1}, x_n) < \infty \tag{36}$$

then there exists $x \in \mathcal{M}$ such that $d(x, x_n) \to 0$. If $x_n, x \in \mathcal{M}$ for $n = 1, 2, \ldots$, we define:

$$x_n \to_d x \quad \text{if (36) holds and} \quad d(x, x_n) \to 0. \tag{37}$$

Notice that an ordinary metric space $(\mathcal{M}, d)$ is complete if and only if it is complete as a distance space. This is due to the fact that Cauchy sequences are exactly those sequences such that from any subsequence a further subsequence can be extracted such that (36) holds along it (a fact often used in analysis).

In this language, Tataru observed in [15] that Ekeland's lemma holds in complete distance spaces with the usual proof. We need only define "semicontinuous" in the obvious way: we say that $u: \mathcal{M} \to [-\infty, \infty]$ is *d-upper semicontinuous* if $x_n \to_d x$ implies $\limsup_{n\to\infty} u(x_n) \le u(x)$. Thus we have:

LEMMA 2.4. *Let $(\mathcal{M}, d)$ be a complete distance space and $u: \mathcal{M} \to [-\infty, \infty)$ be bounded above and d-upper semicontinuous. Then for any $x_0 \in \mathcal{M}$ and $\lambda > 0$ there exists $x_1 \in \mathcal{M}$ such that*

$$\begin{cases} \text{(i) } u(x_0) + \lambda d(x_0, x_1) \le u(x_1), \\ \text{(ii) } u(x) - \lambda d(x, x_1) \le u(x_1) \quad \textit{for} \quad x \in \mathcal{M}. \end{cases}$$

Tataru's second observation ([16, Lemma 1]) in this context is more surprising, especially in view of the Examples 2.6 below. Perhaps we package this beautiful observation more clearly.

LEMMA 2.5. *If d is given by (27), then $(\overline{D(A)}, d)$ is a complete distance space and $x_n \to_d x$ implies $x_n \to x$. In particular, any function $u\colon \overline{D(A)} \to [-\infty, \infty]$ which is upper-semicontinuous in the norm topology is also d-upper semicontinuous.*

Proof. Let $x_n \in \overline{D(A)}$ and (36) hold with

$$d(x_{n+1}, x_n) = s_n + \|x_{n+1} - S(s_n)x_n\|.$$

Using (31), for any $z \in H$, $\tau \geq 0$, $m \leq n$,

$$\|z - S(\tau + \sum_{j=m}^{n} s_j)x_m\| \leq \|z - S(\tau + s_n)x_n\| + \sum_{j=m}^{n-1} d(x_{j+1}, x_j); \tag{38}$$

putting $\tau = \sum_{j=n+1}^{\infty} s_j$, $z = S(\tau)x_{n+1}$, $y_k = S(\sum_{j=k}^{\infty} s_j)x_k$ yields

$$\|y_{n+1} - y_m\| \leq \sum_{j=m}^{n} d(x_{j+1}, x_j) \leq \sum_{j=m}^{\infty} d(x_{j+1}, x_j).$$

This shows that $x = \lim_{n\to\infty} y_n$ exists so (by (34) with $t_n = \sum_{j=n}^{\infty} s_j$) $x_n \to_d x$. Putting $z = x_{n+1}$ and $\tau = 0$ we find

$$\|x_{n+1} - S(\sum_{j=m}^{n} s_j)x_m\| \leq \sum_{j=m}^{n} d(x_{j+1}, x_j) \leq \sum_{j=m}^{\infty} d(x_{j+1}, x_j);$$

since $S(\sum_{j=m}^{n} s_j)x_m \to y_m$ as $n \to \infty$, we then have

$$\limsup_{n\to\infty} \|x_{n+1} - y_m\| \leq \sum_{j=m}^{\infty} d(x_{j+1}, x_j).$$

Since $y_m \to x$, we also have $x_n \to x$. □

EXAMPLE 2.6. The reader may skip this, but it's fun. As noted above, the convergences (33) and (34) are different in general, but not always, even if A is an unbounded linear operator. Let us give some examples. If $S(t)$ is in fact a group of isometries on $H = \overline{D(A)}$, then

$$\begin{aligned} t + \|x - S(t)y\| &= t + \|x - S(t)x + S(t)x - S(t)y\| \\ &\geq t + \|S(t)x - S(t)y\| - \|x - S(t)x\| = t + \|x - y\| - \|x - S(t)x\| \end{aligned}$$

shows that $d(x, y_n) \to 0$ if and only if $\|x - y_n\| \to 0$. In the opposite extreme, consider the example of translations on $L^2(0,1)$; that is

$$S(t)f(s) = \begin{cases} f(s+t) & \text{if } s + t < 1 \\ 0 & \text{if } s + t > 1. \end{cases}$$

Then any sequence f_n supported in intervals $[1 - t_n, 1]$ where $t_n \downarrow 0$ satisfies $S(t_n)f_n = 0$ and so $d(0, f_n) \to 0$. We may therefore have $\|f_n\| \to \infty$ and $d(0, f_n) \to 0$.

3 PROOFS OF COMPARISON IN THE STATIONARY CASE

We will, under appropriate assumptions on F, show how to prove that if $u, -v \in USC(\overline{D(A)})$ satisfy

$$\begin{cases} u(x) + \langle Ax, \nabla u(x)\rangle + F(x, u(x), \nabla u(x)) \leq 0 & \text{on} \quad \overline{D(A)} \\ v(x) + \langle Ax, \nabla v(x)\rangle + F(x, v(x), \nabla v(x)) \geq 0 & \text{on} \quad \overline{D(A)}, \end{cases} \tag{39}$$

and the linear growth estimates

$$u(x) \leq B\|x\| + C, \quad -v(x) \leq B\|x\| + C \quad \text{for} \quad x \in \overline{D(A)} \tag{40}$$

for some $B, C \geq 0$, then $u \leq v$ on $\overline{D(A)}$. For simplicity in writing, we will assume here after that $F(x, r, p) = F(x, p)$, i.e., F is independent of r. (In general, F nondecreasing in r works.)

We begin attempting to show that $u \leq v$, recalling for the reader the way this game is played. To have a simple case, let us assume that $B = 0$ in (40), so that $u, -v$ are bounded above. Now we let $\alpha, \delta > 0$ and put

$$\Phi(x, y) = u(x) - v(y) - \frac{\alpha}{2}\|x - y\|^2 - \frac{\delta}{2}\left(\|x\|^2 + \|y\|^2\right).$$

We assume that $u(z) - v(z) = 2\gamma > 0$ for some $z \in \overline{D(A)}$ and seek to contradict this assumption; then for all $\alpha > 0$ and all sufficiently small $\delta > 0$

$$0 < \gamma \leq u(z) - v(z) - \delta\|z\|^2 = \Phi(z, z) \leq \sup_{\overline{D(A)} \times \overline{D(A)}} \Phi(x, y). \tag{41}$$

Moreover, $\Phi \in USC(\overline{D(A)} \times \overline{D(A)})$ and Φ is bounded above (since $B = 0$ in (40)). Hence, by Lemmas 2.4 and 2.5, for $\lambda > 0$ there exists $(\hat{x}, \hat{y}) \in \overline{D(A)} \times \overline{D(A)}$ such that

$$\begin{cases} \sup_{\overline{D(A)} \times \overline{D(A)}} \Phi(x, y) \leq \Phi(\hat{x}, \hat{y}) + \lambda \\ \Phi(x, y) - \lambda\left(d(x, \hat{x}) + d(y, \hat{y})\right) \leq \Phi(\hat{x}, \hat{y}). \end{cases} \tag{42}$$

In particular,

$$\begin{cases} u(x) - v(y) - \frac{\alpha}{2}\|x - y\|^2 - \frac{\delta}{2}\left(\|x\|^2 + \|y\|^2\right) - \lambda\left(d(x, \hat{x}) + d(y, \hat{y})\right) \\ \text{is maximized at} \quad (\hat{x}, \hat{y}). \end{cases} \tag{43}$$

We would like to use this with the assumptions that u, v are sub and supersolutions. What allows us to do so is the doubling theorem:

DOUBLING THEOREM 3.1. *Let Ω be a relatively open subset of $\overline{D(A)}$, $F, G: \overline{\Omega} \times \mathbb{R} \times H \to \mathbb{R}$ be uniformly continuous on bounded sets and nondecreasing in r for $(x, r, p) \in \overline{\Omega} \times \mathbb{R} \times H$. Let, $u, -v \in USC(\Omega)$ satisfy*

$$\langle Ax, \nabla u(x)\rangle + F(x, u(x), \nabla u(x)) \leq 0 \quad \textit{or} \quad \Omega \tag{44}$$

and

$$\langle Ax,\nabla v(x)\rangle + G(x,v(x),\nabla v(x)) \geq 0 \quad on \quad \Omega. \tag{45}$$

Then $w(x,y) = u(x) - v(y)$ *is a solution of*

$$\langle \mathcal{A}(x,y),\nabla w(x,y)\rangle_{H\times H} + H(x,y,\nabla w(x,y)) \leq 0 \quad on \quad \Omega\times\Omega \tag{46}$$

where

$$H(x,y,p,q) = F(x,u(x),p) - G(y,v(y),-q) \quad for \quad (x,y)\in\Omega\times\Omega,(p,q)\in H\times H \tag{47}$$

and $\mathcal{A}$ *is given by (29).*

Once we have the doubling theorem in hand, life is sweet. Using it in conjunction with (41), (43) and the computations of Section 2 (which show also that $(\alpha/2)\|x-y\|^2+\lambda(d(x,\hat{x})+d(y,\hat{y}))$ is a subtest function for the doubled equation), we deduce that

$$u(\hat{x}) - v(\hat{y}) \leq \sup_{\{\|p\|^2+\|q\|^2\leq 2\lambda^2\}} \left(F(\hat{y},\alpha(\hat{x}-\hat{y}) - \delta\hat{y} + q) - F(\hat{x},\alpha(\hat{x}-\hat{y}) + \delta\hat{x} + p)\right).$$

We are now on familiar ground. We will have to struggle with the three parameters α,δ,λ. A version of this old story goes like this: in view of (41) and (42) and the upper bound on $u,-v$

$$\frac{\gamma}{2} + \frac{\alpha}{2}\|\hat{x}-\hat{y}\|^2 + \frac{\delta}{2}\left(\|\hat{x}\|^2 + \|\hat{y}\|^2\right) \leq u(\hat{x}) - v(\hat{y}) \leq 2C \quad \text{if} \quad 2\lambda\leq\gamma. \tag{48}$$

In view of this estimate, $\hat{x},\hat{y}$ are bounded independently of α,λ if δ is fixed, $\delta\hat{x},\delta\hat{y}\to 0$ (uniformly in α,λ) as $\delta\downarrow 0$, and $\alpha(\hat{x}-\hat{y})$ remains bounded independently of δ,λ if α is fixed. We exhibit simple assumptions which will allow us to conclude that $u\leq v$.

For convenience in the statements, we use continuous functions $m:[0,\infty)\to[0,\infty)$ and $\sigma:[0,\infty)\times[0,\infty)\to[0,\infty)$ to measure continuity properties. If these functions are nondecreasing in their arguments, $m(0)=\sigma(0,R)=0$ for $R\geq 0$, and $m(r),\sigma(r,R)$ are subadditive in r, we call m a "modulus" and σ a "local modulus".

We assume that there is a modulus ω and a local modulus σ_F such that

$$\begin{cases} \text{(i) } |F(x,p) - F(x,q)| \leq \sigma_F(\|p-q\|,\|p\|+\|q\|) \quad \text{for} \quad p,q\in H, x\in\overline{D(A)} \\ \text{(ii) } F(x,\alpha(x-y)) - F(y,\alpha(x-y)) \leq \omega(\|x-y\| + \alpha\|x-y\|^2) \\ \qquad \text{for } x,y\in\overline{D(A)}, 0\leq\alpha. \end{cases} \tag{49}$$

For example, (49) trivially holds if $F(x,p) = G(p) - f(x)$ where G is uniformly continuous on bounded sets and $f\in UC(H)$. The first condition is uniform continuity of $F(x,p)$ in p when p is bounded uniform in $x\in\overline{D(A)}$.

The above remarks show $\delta\hat{x},\delta\hat{y}\to 0$ as $\delta\downarrow 0$ and then (49) and the bound on $\alpha(\hat{x}-\hat{y})$ imply

$$u(\hat{x}) - v(\hat{y}) \leq \omega(\|\hat{x}-\hat{y}\| + \alpha\|\hat{x}-\hat{y}\|^2) + \epsilon_\alpha(\delta,\lambda)$$

where $\epsilon_\alpha(\delta,\lambda)\to 0$ as $\delta,\lambda\downarrow 0$ for fixed α. Coupled with (48) we have

$$\frac{\gamma}{2} \leq \omega(\|\hat{x}-\hat{y}\| + \alpha\|\hat{x}-\hat{y}\|^2) + \epsilon_\alpha(\delta,\lambda) \tag{50}$$

and we wish to take the limit $\lambda, \delta \downarrow 0$ and then $\alpha \to \infty$ to reach a contradiction. We need $\alpha\|\hat{x} - \hat{y}\|^2 \to 0$ to accomplish this.

Put

$$\begin{cases} M_0 = \sup\limits_{\overline{D(A)}} (u(x) - v(x)) \quad \text{and} \\ M_\alpha = \sup\limits_{\overline{D(A)}\times\overline{D(A)}} (u(x) - v(y) - \frac{\alpha}{2}\|x - y\|^2) \quad \text{for} \quad \alpha > 0. \end{cases} \tag{51}$$

Obviously $\alpha \to M_\alpha$ is nonincreasing and $M_\alpha \geq M_0$. A simple standard lemma turns out to be useful.

LEMMA 3.2. *If* $u, v: \overline{D(A)} \to \mathbb{R}$, $\alpha, \epsilon > 0$, $M_{\alpha/2} < \infty$ *and*

$$u(x) - v(y) - \frac{\alpha}{2}\|x - y\|^2 \geq M_\alpha - \epsilon \tag{52}$$

then

$$4(M_{\alpha/2} - M_\alpha + \epsilon) \geq \alpha\|x - y\|^2. \tag{53}$$

In particular, if $x_\alpha, y_\alpha \in \overline{D(A)}$ *for* $\alpha > 0$, $\lim_{\alpha\to\infty} M_\alpha < \infty$, *and*

$$M_\alpha - \left(u(x_\alpha) - v(y_\alpha) - \frac{\alpha}{2}\|x_\alpha - y_\alpha\|^2\right) \to 0 \quad as \quad \alpha \to \infty \tag{54}$$

then $\alpha\|x_\alpha - y_\alpha\|^2 \to 0$.

Proof. By (52)

$$\begin{aligned} M_{\alpha/2} &\geq u(x) - v(y) - \frac{\alpha}{4}\|x - y\|^2 \\ &= u(x) - v(y) - \frac{\alpha}{2}\|x - y\|^2 + \frac{\alpha}{4}\|x - y\|^2 \\ &\geq M_\alpha - \epsilon + \frac{\alpha}{4}\|x - y\|^2, \end{aligned}$$

establishing (53). The last assertion follows from the fact that $M_{\alpha/2} - M_\alpha \to 0$ as $\alpha \to \infty$, since M_α is nonincreasing and finite for large α. □

The final fact to observe is that (42) implies that

$$\lim_{\alpha\to\infty} \lim_{(\delta,\lambda)\to 0} \left(M_\alpha - (u(\hat{x}) - v(\hat{y}) - \frac{\alpha}{2}\|\hat{x} - \hat{y}\|^2)\right) = 0;$$

in view of this, (50), Lemma 3.2 and $\gamma > 0$, we have a contradiction.

Other, more general assumptions, may be used as well and the reader may find them in the literature. Since the above example illustrates all the new arguments (i.e., the use of the doubling theorem and Tataru's distance) needed to transform them to the current case, we just illustrate their nature. Instead of taking the double limit $\delta, \lambda \downarrow 0$ above, one can make more general assumptions for which the iterated limit $\lambda \downarrow 0$ and then $\delta \downarrow 0$ succeeds (using that $\hat{x}$ remains bounded as $\lambda \downarrow 0$). For example, this works if (49) (i) is replaced by uniform continuity of F on bounded sets and

$$F(x, p) - F(x, p + \delta x) \leq \sigma(\delta\|x\|, \|p\| + \delta\|x\|) \tag{55}$$

for a local modulus σ.

Moreover, the general case of (40) can be treated by the device of first showing that $u(x) - v(y) - K\|x - y\|$ is bounded for some K (then $M_{2K} < \infty$); the arguments above need no essential change after this is done. One way to accomplish this is to choose a smooth nondecreasing $\beta: [0, \infty) \to [0, \infty)$ with the properties

$$\begin{cases} \beta(r) = 0 \quad \text{for} \quad 0 \le r \le 1 \\ \lim_{r\to\infty} \dfrac{\beta(r)}{r} = 1 \\ 0 \le \beta'(r) \le 2 \quad \text{for} \quad 0 \le r \end{cases} \tag{56}$$

and then one assumes that

$$u(x) - v(y) - K\left(\|x - y\|^2 + 1\right)^{1/2} - 2B\left(\beta(\|x\| - M) + \beta(\|y\| - M)\right), \tag{57}$$

where $M > 0$ is a parameter, is positive somewhere and (approximately) maximizes it (it is bounded above by (40) and (56). Plugging this into the doubling theorem and analyzing the result leads to

$$\begin{aligned} K\left(\|\hat{x} - \hat{y}\|^2 + 1\right)^{1/2} &\le u(\hat{x}) - v(\hat{y}) \\ &\le \sup_{\{\|p\|^2 + \|q\|^2 \le 2\lambda^2\}} \left\{ F\left(\hat{y}, K\frac{(\hat{x} - \hat{y})}{(\|\hat{x} - \hat{y}\|^2 + 1)^{1/2}} - 2B\beta'(\|\hat{y}\| - M)\frac{\hat{y}}{\|\hat{y}\|} + q\right)\right. \\ &\qquad \left. - F\left(\hat{x}, K\frac{(\hat{x} - \hat{y})}{(\|\hat{x} - \hat{y}\|^2 + 1)^{1/2}} + 2B\beta'(\|\hat{x}\| - M)\frac{\hat{x}}{\|\hat{x}\|} + p\right)\right\} \end{aligned} \tag{58}$$

Thus we will succeed in bounding (57) uniformly in M if $F(x, p)$ is uniformly continuous in p when (x, p) is bounded and

$$F(x, K\alpha(x - y) + 2B\mu x) - F(y, K\alpha(x - y) - 2B\mu y)$$

is bounded when

$$K\left(\|x - y\|^2 + 1\right)^{1/2} \le F(x, K\alpha(x - y) + 2B\mu x) - F(y, K\alpha(x - y) - 2B\mu y) \tag{59}$$

$0 \le \alpha, \mu$ and $\alpha\|x - y\| \le 1, \mu\|x\|, \mu\|y\| \le 2$. If $F(x, p) = G(p) - f(x)$ and G is bounded on bounded sets, this condition amounts to producing a bound on $x - y$ from an estimate

$$K\|x - y\| \le |f(x) - f(y)| + C$$

which works if $f \in UC(\overline{D(A)})$ and K is large. If we assume that (49) (ii), (55) hold, then (59) implies

$$K\left(\|x - y\|^2 + 1\right)^{1/2} \le \omega(\|x - y\|(1 + K)) + C_K$$

and we may bound $\|x - y\|$ if K is large depending on the nature of ω. For example,

$$\lim_{r\to\infty} \frac{\omega(r)}{r} < 1 \tag{60}$$

suffices. The constant 1 appears here due to the way u enters the equation; if u is replaced by ηu, then 1 is replaced by η. Other devices avoid this type of condition if one of u or v is uniformly continuous; however, as we see in the next section, comparison of semicontinuous sub and supersolutions correspond to existence, and this is not the case for comparison of uniformly continuous sub and supersolutions. See Section 5 of [7, Part II] and Remark 3.3 below for expansion on this point. Other growth and structure conditions can be considered, the role of (55) divided into two parts, etc., and there are other techniques; the reader is referred to [4], [6], and [9] – [12] for this lore and its origins.

When (57) is bounded uniformly in M, letting $M \to \infty$ produces a bound on $u(x) - v(y) - K\|x-y\|$ and hence $u(x) - v(y) - K\|x-y\|^2$.

Finally, we note that when this line of proof succeeds, it produces a modulus of continuity of any continuous solution u. Indeed, the proof shows that if u is a solution and $\gamma > 0$, then

$$u(x) - u(y) \le \gamma + \frac{\alpha}{2}\|x-y\|^2 + \frac{\delta}{2}\left(\|x\|^2 + \|y\|^2\right)$$

when δ is small and α is large. Letting $\delta \downarrow 0$, we see that u is uniformly continuous.

REMARK 3.3. Since we have been informal above, let us summarize and augment the discussion here. Let $u, -v \in USC(\overline{D(A)})$ be a subsolution and a supersolution of the equation $w + \langle Ax, \nabla w\rangle + F(x, \nabla w) = 0$ where F is uniformly continuous on bounded sets. Assume that (49) (ii) and (55) hold. If $u, -v$ are bounded above, then we (essentially) proved that $u \le v$ and that any bounded solution w is uniformly continuous. In the next section, we will see that there is then a bounded continuous solution if $F(x, 0)$ is bounded. If we allow $u, -v$ to grow linearly, then (49) (ii), (60), (55) imply that $u \le v$ and that any continuous solution which grows at most linearly is uniformly continuous. In the next section, we will see that under these assumptions there is a uniformly continuous solution. If one of $u, -v$ is uniformly continuous, it can be shown that $u \le v$ holds even if (60) is not assumed. The equation $u - xu' = 0$ in $\mathbb{R}$ provides an example in which $\omega(r) = r$ may be used in (49) (ii) and there are the two distinct uniformly continuous solutions $u \equiv 0$ and $u \equiv x$; both (60) and (55) fail in this example. In Section 5 of [7, Part II], an example of the form $u + b(x)u' - |x| = 0$ in $\mathbb{R}$ is provided for which (49) (ii) holds with $\omega(r) = r$, but there is no uniformly continuous solution. As mentioned above, [7, Part II] contains further dissections of the various conditions.

4 PERRON'S METHOD AND CONSISTENCY

Existence proofs can be given by Perron's method; we show how the method of Ishii [13] adapts to our situation. To employ this method with a given notion of solution we need to check two things. To state them, we define the upper and lower semicontinuous envelopes u^* and u_* of $u: \Omega \to \mathbb{R}$:

$$u^*(x) = \limsup_{r \downarrow 0} \{u(y) : y \in \Omega, \|y-x\| \le r\},$$

$$u_*(x) = \liminf_{r \downarrow 0} \{u(y) : y \in \Omega, \|y-x\| \le r\}.$$

The first thing we will check is formulated in the next lemma in terms of a general equation

$$\langle Ax, \nabla u\rangle + G(x, u, \nabla u) = 0. \tag{61}$$

LEMMA 4.1. *Let G be locally uniformly continuous, Ω be relatively open in $\overline{D(A)}$ and $\mathcal{F}$ be a nonempty family of upper-semicontinuous subsolutions of (61) on Ω. Let*

$$U(x) = \sup_{u\in\mathcal{F}} u(x) \quad \textit{for} \quad x \in \Omega. \tag{62}$$

If $U^* < \infty$ on Ω, then U^* is a subsolution of (61) on Ω.

Proof. The arguments being local, we may as well assume that $\Omega = \overline{D(A)}$. Let $\Phi = \varphi + \psi \in C^1(H) + \mathrm{Lip}(H)$ be a subtest function for (61) and $U^* - \Phi$ have a local maximum at $\hat{x} \in \overline{D(A)}$. We seek to show that

$$D_A^- \Phi(\hat{x}) + G_{L(\psi)}(\hat{x}, U^*(\hat{x}), \nabla\varphi(\hat{x})) \le 0. \tag{63}$$

Replacing Φ by $\Phi(x) - \lambda d(x, \hat{x})$ with $\lambda > 0$ we may assume that there is a closed neighborhood $N = B_r(\hat{x}) \cap \overline{D(A)}$ of $\hat{x}$ such that if $x_n \in N$ and $U^*(x_n) - \Phi(x_n) \to U^*(\hat{x}) - \Phi(\hat{x})$, then $x_n \to x$ and $U^*(x_n) \to U^*(\hat{x})$. (At the end, one lets $\lambda \downarrow 0$.)

Choose $u_n \in \mathcal{F}$, $x_n \in \overline{D(A)}$ such that

$$x_n \to \hat{x} \quad \text{and} \quad u_n(x_n) \to U^*(\hat{x}). \tag{64}$$

Choose, using Lemma 2.5, $\hat{x}_n \in N$ such that

$$u_n(x) - \Phi(x) - \frac{1}{n} d(x, \hat{x}_n) \le u_n(\hat{x}_n) - \Phi(\hat{x}_n) \quad \text{for} \quad x \in N. \tag{65}$$

Putting $x = x_n$ above, we find

$$\begin{aligned} u_n(x_n) - \Phi(x_n) - \frac{1}{n} d(x_n, \hat{x}_n) &\le u_n(\hat{x}_n) - \Phi(\hat{x}_n) \\ &\le U^*(\hat{x}_n) - \Phi(\hat{x}_n) \\ &\le U^*(\hat{x}) - \Phi(\hat{x}); \end{aligned}$$

since

$$u_n(x_n) - \Phi(x_n) - \frac{1}{n} d(x_n, \hat{x}_n) \to U^*(\hat{x}) - \Phi(\hat{x})$$

by (64), $U^*(\hat{x}_n) - \Phi(\hat{x}_n) \to U^*(\hat{x}) - \Phi(\hat{x})$ and then

$$\hat{x}_n \to \hat{x} \quad \text{and} \quad u_n(\hat{x}_n) \to U^*(\hat{x}).$$

Since u_n is a subsolution, we have

$$D_A^- \Phi(\hat{x}_n) + G_{L(\psi)+1/n}(\hat{x}_n, u_n(\hat{x}_n), \nabla\varphi(\hat{x}_n)) \le 0$$

and we may pass to the limit using the above and the lower semicontinuity of $D_A^- \Phi$ (Lemma 2.3) to find (63). □

The next thing to establish is:

LEMMA 4.2. *Let $v \in LSC(\Omega)$ be a supersolution of (61) and*

$$\mathcal{F} = \{w : w^* \text{ is a subsolution of (61) on } \Omega \text{ and } w \le v\}, \tag{66}$$

be nonempty. If

$$u(x) = \sup_{w \in \mathcal{F}} w(x), \tag{67}$$

then u_* is a supersolution of (61).

Proof. Assume $\Omega = \overline{D(A)}$ and that u_* is not a subsolution. Then there exists a supertest function $\Phi = \varphi + \psi \in C^1(H) + \mathrm{Lip}(H)$ and a local minimum point $\hat{x}$ of $u_* - \Phi$ such that

$$D_A^+\Phi(\hat{x}) + G^{L(\psi)}(\hat{x}, u_*(\hat{x}), \nabla\varphi(\hat{x})) < 0. \tag{68}$$

Since $u_* \le v$ (because $v \in LSC(\overline{D(A)})$), if $v(\hat{x}) = u_*(\hat{x})$ then $\hat{x}$ is a minimum point of $v - \Phi$, and (68) contradicts the assumption that v is a supersolution. Hence

$$u_*(\hat{x}) < v(\hat{x}). \tag{69}$$

In view of the lower-semicontinuity of $D_A^+\Phi$ (Lemma 2.3) and $G^{L(\psi)}$, for sufficiently small $\gamma, r_0, \delta > 0$ we have

$$\begin{cases} D_A^+\Phi(x) + G^{L(\psi)+r_0}(x, u_*(\hat{x}) + \delta + \Phi(x) - \Phi(\hat{x}), \nabla\varphi(x)) < -\gamma \\ \text{for} \quad x \in \overline{D(A)},\ 0 \le \delta, r, \|x - \hat{x}\| \le 2r_0. \end{cases} \tag{70}$$

Now put

$$N_r = \left\{x \in \overline{D(A)} : d(x, \hat{x}) \le 2r\right\} \quad \text{and} \quad C_r = \left\{x \in \overline{D(A)} : r \le d(x, \hat{x}) \le 2r\right\}$$

and

$$U(x) = u_*(\hat{x}) + \delta + \Phi(x) - \Phi(\hat{x}) - Md(x, \hat{x}). \tag{71}$$

Putting

$$\tilde{\varphi}(x) = u_*(\hat{x}) + \delta + \varphi(x) - \Phi(\hat{x}), \ \tilde{\psi}(x) = \psi(x) - Md(x, \hat{x}) \tag{72}$$

we have $U = \tilde{\varphi} + \tilde{\psi} \in C^1(H) + \mathrm{Lip}(H)$, and, by Section 2, U is a supertest function satisfying

$$D_A^+U(x) \le D_A^+\Phi(x) + M, \ L\left(\tilde{\psi}\right) \le L(\psi) + M. \tag{73}$$

We seek to choose the parameters $r, M, \delta, \epsilon > 0$ so that

$$\begin{cases} \text{(i) } D_A^+U(x) + G^{L(\tilde{\psi})}(x, U(x), \nabla\tilde{\varphi}(x)) \le 0 \quad \text{for} \quad x \in N_r \\ \text{(ii) } U(x) \le u(x) - \epsilon \quad \text{for} \quad x \in C_r, \\ \text{(iii) } U(x) \le v(x) \quad \text{for} \quad x \in N_r. \end{cases} \tag{74}$$

In view of (70), (72) and (73), (i) holds if r, M and δ are sufficiently small. Part (ii) reads

$$u_*(\hat{x}) + \delta + \Phi(x) - \Phi(\hat{x}) - Md(x, \hat{x}) \le u(x) - \epsilon \quad \text{for} \quad x \in C_r$$

or

$$\delta - Md(x,\hat{x}) + \epsilon \le u(x) - \Phi(x) - (u_*(\hat{x}) - \Phi(\hat{x})) \quad \text{for} \quad x \in C_r$$

which is satisfied if r is small and

$$\delta - Mr + \epsilon \le 0$$

since $u_* \le u$ and $u_* - \Phi$ has a minimum at $\hat{x}$, while $d \ge r$ on C_r. Hence (i) and (ii) hold with suitable choices. The lower semicontinuity of v and (69) implies that we may satisfy (iii) as well.

By Lemma 1.4, U is a subsolution on $d < 2r$. It is now immediate from a check of definitions and the construction that if

$$W(x) = \begin{cases} U(x) \wedge u(x) & \text{for} \quad x \in N_r \\ u(x) & \text{for} \quad x \in \overline{D(A)} \setminus N_r \end{cases}, \tag{75}$$

then

$$W^*(x) = \begin{cases} U(x) \wedge u^*(x) & \text{for} \quad x \in N_r \\ u^*(x) & \text{for} \quad x \in \overline{D(A)} \setminus N_r \end{cases}$$

is a subsolution since U and $u^* = (\sup_{\mathcal{F}} w^*)^*$ are subsolutions by Lemma 4.1. Moreover, $u \le W \le v$ and $W(x) > u(x)$ holds at some points since there exists $x_n \to \hat{x}$ such that $u(x_n) \to u_*(\hat{x}) < W(\hat{x}) = u_*(\hat{x}) + \delta$. This is a contradiction to definition of u; we conclude that u_* is a supersolution. □

Here is our version of the punch line:

THEOREM 4.3. *Assume that $u, -v \in USC(\Omega)$ are a subsolution and a supersolution of (61) on Ω, $v^* < \infty$ on Ω, and given any subsolution $U \in USC(\Omega)$ and supersolution $V \in LSC(\Omega)$ with $u_* \le U, V \le v^*$ we have $U \le V$. If*

$$\mathcal{F} = \{w \in USC(\Omega) : w \textit{ is a subsolution of } (61) \textit{ on } \Omega \textit{ and } w \le v\}, \tag{76}$$

and

$$Z(x) = \sup_{w \in \mathcal{F}} w(x),$$

then $Z \in C(\Omega)$ and Z is the one and only continuous solution of (61) on Ω satisfying $U_* \le Z \le V^*$.

Proof. Z^* is a subsolution by Lemma 4.1. Suppose $w \le v$ and w^* is a subsolution. Then $w \le u \vee w$ and $(u \vee w)^* = u \vee w^*$ is a subsolution by Lemma 4.1. Since $u_* \le u \le u \vee w^* \le v_*$, we have $w^* \le u \vee w^* \le v$ by the assumed comparison applied to $U = u \vee w^*$ and $V = v$. It follows that the sups over (66) and (76) coincide and then from Lemma 4.2 that Z_* is a supersolution. Since $u_* \le Z_*, Z^* \le v^*$, we have $Z^* \le Z_*$ by assumption, so $u \le Z^* = Z = Z_* \le v$ and Z is a continuous solution. Z is unique by the comparison assumption. □

EXAMPLE 4.4. We recall for the reader how to construct sub and supersolutions for the case

$$u + \langle Ax, \nabla u(x) \rangle + F(x, \nabla u) = 0$$

where F is uniformly continuous on bounded sets and satisfies (49) (ii) and (60). (The reader will enjoy generalizing this construction.) Put

$$V(x) = C + D\left(\|x\|^2 + 1\right)^{1/2}.$$

By Lemma 1.4, we need only choose C, D so that

$$C + D\left(\|x\|^2 + 1\right)^{1/2} + D_A^- V(x) + F\left(x, D\frac{x}{(\|x\|^2+1)^{1/2}}\right) \geq 0.$$

Since $D_A^- V(x) \geq 0$ and, by (49) (ii),

$$F\left(x, D\frac{x}{(\|x\|^2+1)^{1/2}}\right) \geq F\left(0, D\frac{x}{(\|x\|^2+1)^{1/2}}\right) - \omega(D\|x\|) \geq F_D(0,0) - \omega((D+1)\|x\|)$$

while, by (60), there exists $E < 1$ such that

$$\omega(r) \leq 1 + Er \quad \text{for} \quad r \geq 0.$$

Thus if $D > 1/(1-E)$ is fixed and C is large enough, we succeed. Similarly, $-V$ is a subsolution. Any assumptions that imply the comparison of sub and supersolutions which grow at most linearly and Theorem 4.3 then guarantee the unique existence of a solution of at most linear growth. For example, (49) (ii), (55), (60) suffice. A trivial case arises if $F(x,0)$ is bounded, for then we may use constant sub and supersolutions. In accordance with Remark 3.3, we then have the unique existence of a bounded uniformly continuous solution provided (49) (ii) and (55) hold.

REMARK 4.5. Note that, for example, the trivial existence of constant sub and supersolutions in the case in which $F(x,0)$ is bounded and the lemmas above provide a function whose upper envelope is a subsolution and whose lower envelope is a supersolution. However, we do not necessarily obtain a continuous solution in the absence of a companion comparison result. Translating the terminology from [13] to this situation, this perhaps discontinuous function would be called a "solution". According to this notion, the characteristic function of the rationals is a solution of $u' = 0$ on $\mathbb{R}$.

For the final topic in this section, we show "consistency".

THEOREM 4.6. *Let $G_n: \overline{D(A)} \times \mathbb{R} \times H \to \mathbb{R}$ for $n = 1, 2, \ldots$ and let Ω be a relatively open subset of $\overline{D(A)}$. Let $u_n \in USC(\Omega)$ (respectively, $u_n \in LSC(\Omega)$) be a subsolution (respectively, supersolution) of*

$$\langle Ax, \nabla u_n \rangle + G_n(x, u_n(x), \nabla(x)) = 0 \tag{77}$$

in Ω for $n = 1, 2, \ldots$. Assume, moreover, that the limits

$$\lim_{n\to\infty} u_n = u, \qquad \lim_{n\to\infty} G_n = G$$

exist locally uniformly and that G is uniformly continuous on bounded sets. Then $\langle Ax, \nabla u(x)\rangle + G(x, u(x), \nabla u(x)) \leq 0$ (respectively, ≥ 0) in Ω.

Proof. It suffices to treat the subsolution case. Suppose that $\Phi = \varphi + \psi$ is a subtest function and $\hat{x}$ is a local maximum of $u - \Phi$; that is

$$u(x) - \Phi(x) \leq u(\hat{x}) - \Phi(\hat{x}) \quad \text{for} \quad x \in B_r(\hat{x}) \cap \overline{D(A)}. \tag{78}$$

Choosing r sufficiently small, we may assume that $u_n \to u$ uniformly on $B_r(\hat{x}) \cap \overline{D(A)}$. Choose $\gamma > 0$. By perturbed optimization we can find $\hat{x}_n \in B_r(\hat{x}) \cap \overline{D(A)}$ such that

$$u_n(\hat{x}) - \Phi(\hat{x}) = u_n(\hat{x}) - \Phi(\hat{x}) - \gamma d(\hat{x}, \hat{x}) \leq u_n(\hat{x}_n) - \Phi(\hat{x}_n) - \gamma d(\hat{x}_n, \hat{x}) \tag{79}$$

and

$$u_n(x) - \Phi(x) - \gamma d(x, \hat{x}) - \frac{1}{n} d(x, \hat{x}_n) \leq u_n(\hat{x}_n) - \Phi(\hat{x}_n) - \gamma d(\hat{x}_n, \hat{x})$$

for $x \in B_r(\hat{x}) \cap \overline{D(A)}$. From (79), (78) and the uniform convergence of u_n to u, it follows that

$$\limsup_{n\to\infty} \gamma d(\hat{x}_n, \hat{x}) \leq \limsup_{n\to\infty} \sup_{B_r(\hat{x})\cap\overline{D(A)}} (u_n(x) - \Phi(x)) - \lim_{n\to\infty} (u_n(\hat{x}) - \Phi(\hat{x})) = 0$$

so $d(\hat{x}_n, \hat{x}) \to 0$, which implies $\hat{x}_n \to \hat{x}$. We also have $u_n(x_n) \to u(\hat{x})$.

To conclude, we use that u_n is a subsolution of (77), so

$$u_n(\hat{x}_n) + \liminf_{\substack{h\downarrow 0 \\ y\to\hat{x}_n}} \frac{\Phi(y) - \Phi(S(h)y)}{h} + (G_n)_{L(\psi)+\gamma+\frac{1}{n}}(\hat{x}_n, \nabla\varphi(\hat{x}_n)) \leq 0;$$

the result follows upon first letting $n \to \infty$ and then $\gamma \downarrow 0$. □

5 THE CAUCHY PROBLEM

The stationary problem (1) is the special case of (6) in which $G(x, r, p) = r + F(x, r, p)$. In order to include the time-dependent equation in the Cauchy problem

$$\begin{cases} \text{(E)}\ u_t(t,x) + \langle Ax, \nabla u(t,x)\rangle + F(t, x, u(t,x), \nabla u(t,x)) = 0 & \text{for} \quad t > 0, x \in D(A), \\ \text{(IC)}\ u(0,x) = \eta(x) & \text{for} \quad x \in \overline{D(A)}. \end{cases} \tag{80}$$

we use a device. Observe that the operator defined by

$$\mathcal{A}(t, x) = \{(1, y) : y \in Ax\} \quad \text{for} \quad t \in \mathbb{R}, x \in H \tag{81}$$

is m-accretive in the Hilbert space $\mathbb{R} \times H$ and, formally,

$$\nabla_{(t,x)} u(t,x) = (u_t(t,x), \nabla u(t,x));$$

"∇" indicating differentiation with respect to $x \in H$. Thus, formally,

$$u_t(t,x) + \langle Ax, \nabla u(t,x)\rangle = \langle \mathcal{A}(t,x), \nabla_{(t,x)} u(t,x)\rangle_{\mathbb{R}\times H}.$$

If $\mathcal{S}(t) = e^{-t\mathcal{A}}$ is the semigroup on $R \times \overline{D(A)}$ corresponding to $\mathcal{A}$, then

$$\mathcal{S}(h)(t,x) = (t-h, S(h)x) \quad \text{for} \quad 0 \le h, t \in \mathbb{R} \quad \text{and} \quad x \in \overline{D(A)}.$$

We will be interested in fixing a closed upper-bound for t, $t \le T$. This fits formally in our framework either by changing $\mathcal{A}$ according to

$$\mathcal{A} = \partial I_{(-\infty,T]} \times A$$

(see Example 1.5) or observing that $t-h \le T$ if $t \le T$ and recalling Remark 1.6.

By these means, we can interpret sub and supersolutions of

$$u_t(t,x) + \langle Ax, \nabla u(t,x)\rangle + F(t,x,u(t,x),\nabla u(t,x)) = 0 \tag{82}$$

on relatively open subsets of $(0,T] \times \overline{D(A)}$ via:

DEFINITION 5.1. *Let A be an m-accretive operator in H, $F : [0,T] \times H \times \mathbb{R} \times H \to \mathbb{R}$ be continuous and $\mathcal{A}$ be given by (81). Let Ω be a relatively open subset of $(0,T] \times \overline{D(A)}$. Then $u \in USC(\Omega)$ (respectively, $u \in LSC(\Omega)$) is a subsolution (respectively, supersolution) of (82) on Ω if u is a subsolution (respectively, supersolution) of the equation*

$$\langle \mathcal{A}(t,x), \nabla_{(t,x)} u(t,x)\rangle_{\mathbb{R}\times H} + F(t,x,u(t,x),\nabla u(t,x)) = 0$$

on Ω. A solution $u \in C(\Omega)$ of (2) on Ω is a function which is simultaneously a subsolution and a supersolution on Ω.

In using this definition, we have

$$D_{\mathcal{A}}^{-}\Phi(t,x) = \liminf_{\substack{h\downarrow 0 \\ (s,y)\to(t,x)}} \frac{\Phi(s,y) - \Phi(s-h, S(h)y)}{h} \tag{83}$$

and

$$D_{\mathcal{A}}^{+}\Phi(t,x) = \limsup_{\substack{h\downarrow 0 \\ (s,y)\to(t,x)}} \frac{\Phi(s,y) - \Phi(s-h, S(h)y)}{h}. \tag{84}$$

The homogeneous linear case of (80) (i.e., $F \equiv 0$) is formally solved by

$$u(t,x) = \eta(S(t)x).$$

The corresponding solution for the inhomogeneous linear problem

$$u_t(t,x) + \langle Ax, \nabla u(t,x)\rangle - f(t,x) = 0 \tag{85}$$

is given, via variation of parameters, by

$$u(t,x) = \eta(S(t)x) + \int_0^t f(s, S(t-s)x)\, ds. \tag{86}$$

A quick computation shows that if f is continuous, then $D_{\mathcal{A}}^{-}u(t,x) = D_{\mathcal{A}}^{+}u(t,x) = 0$, and thus Definition 5.1 is perfect for checking that (86) is indeed a solution of (85). These formulas also show that in defining a solution of the (80) the strongest assumption we can reasonably make concerning the assumption of initial values is

$$\lim_{t\downarrow 0}(u(t,x)-\eta(S(t)x))=0 \quad \text{uniformly on bounded subsets of} \quad \overline{D(A)}. \tag{87}$$

The theory of the Cauchy problem is not yet fully understood; there are subtleties involving continuity in t and the assumption of the initial condition hinted at above which complicate the situation. We will make strong assumptions concerning the interaction of the initial condition and $S(t)x$, state a general theorem and and prove it, for simplicity, in the model "separated" case

$$F(t,x,r,p)=G(p)-f(x) \quad \text{for} \quad t,r\in \mathrm{I\!R}, x\in\overline{D(A)}, p\in H. \tag{88}$$

We could include t,u dependence as well); we merely want to keep the exposition clear. (Well, clearer, anyway.)

THEOREM 5.2. *Let (49) (ii), (55) hold. Let $\eta\in UC(\overline{D(A)})$ have the property*

$$\eta(S(t)x) \quad \textit{is uniformly continuous in} \quad t \quad \textit{uniformly for bounded} \quad x\in\overline{D(A)}. \tag{89}$$

Let $T>0$ and $u,-v\in USC([0,T]\times\overline{D(A)})$ be, respectively, a subsolution and a supersolution of (82) on $(0,T]\times\overline{D(A)}$ which satisfy

$$u(0,x)\le\eta(x)\le v(0,x) \quad \textit{for} \quad x\in\overline{D(A)} \tag{90}$$

and either (i) or (ii) of

$$\begin{cases} \text{(i)} \ \lim_{t\downarrow 0}\max(u(t,x)-\eta(S(t)x),0)=0 \\ \text{(ii)} \ \lim_{t\downarrow 0}\max(\eta(S(t)x)-v(t,x),0)=0 \end{cases} \tag{91}$$

holds uniformly for bounded $x\in\overline{D(A)}$. Let $C,D\in\mathrm{I\!R}$ and

$$u(t,x),-v(t,x)\le C+D\|x\| \quad \textit{for} \quad 0\le t\le T, x\in\overline{D(A)}. \tag{92}$$

Then $u\le v$ on $(0,T]\times\overline{D(A)}$. Moreover, there exists a unique $u\in C([0,T]\times\overline{D(A)})$ which is uniformly continuous in x uniformly in t, uniformly continuous in $t\in[0,T]$ for bounded x, and which is a solution of (82) on $(0,T]\times\overline{D(A)}$ and satisfies (87).

REMARK 5.3. As regards the disturbing condition (89), the reader will see how it enters the proofs. It does hold in many situations. For example, suppose that $S(\epsilon)$ maps bounded sets in $\overline{D(A)}$ into compact sets and $g\in UC(\overline{D(A)})$. Then $\eta(x)=g(S(\epsilon)x)$ satisfies (89). A different sort of example is provided by the linear case. Suppose A is linear and densely defined. Then $\eta(x)=\langle p,x\rangle$ satisfies (91) for $p\in H$; this follows from $\eta(S(t)x)=\langle S(t)^*p,x\rangle$ where $S(t)^*$ is the adjoint of $S(t)$. One could also take nonlinear functions of combinations of this sort of initial condition. See also Remark 5.6 below. Perhaps it is possible to avoid this condition.

REMARK 5.4. The reader may notice that we did not assume any analogue of (60). This may be understood via rescaling in time, which essentially allows us to replace F by κF for any $\kappa > 0$.

REMARK 5.5. The significance of the "either (i) or (ii)" of (91) is illustrated by the solution formula (86); it provides a solution with the properties (i) and (ii) of (91) and is therefore unique in the class of linearly growing continuous solutions which assume the same initial value satisfying (89).

Proof of Theorem 5.2. We write the proof in the case (88); the reader can supply the modifications necessary for the general case. We begin the proof by verifing the comparison assertion. We will assume that $D = 0$ in (92) for simplicity – the general case is handled as explained in Section 4. We assume that

$$0 < \gamma \leq u(t,x) - v(t,x) \quad \text{for some} \quad 0 < t \leq T, x \in \overline{D(A)} \tag{93}$$

and choose $c_0 > 0$ such that

$$2c_0 \leq \frac{\gamma}{4}. \tag{94}$$

Choose $\alpha, \beta, \delta > 0$ and put

$$\Gamma_{\beta,\alpha,\delta}(t,s,x,y) = u(t,x) - v(s,y) - \left(\frac{\alpha}{2}\|x-y\|^2 + \frac{\beta}{2}(t-s)^2 + \frac{\delta}{2}\left(\|x\|^2 + \|y\|^2\right) + c_0(t+s)\right);$$

in view of (93) and (94), as soon as $\delta > 0$ is sufficiently small, there are $(\hat{t},\hat{x}), (\hat{s},\hat{y}) \in [0,T] \times \overline{D(A)}$ such that

$$\frac{\gamma}{2} < \Gamma_{\beta,\alpha,\delta}(\hat{t},\hat{s},\hat{x},\hat{y}) \tag{95}$$

and for $\lambda > 0$, by perturbed optimization, we may also assume that

$$\Gamma_{\beta,\alpha,\delta}(t,s,x,y) - \lambda\left(|t-\hat{t}| + |s-\hat{s}| + d_A(x,\hat{x}) + d_A(y,\hat{y})\right) \leq \Gamma_{\beta,\alpha,\delta}(\hat{t},\hat{s},\hat{x},\hat{y}) \tag{96}$$

for $(t,x),(s,y) \in [0,T] \times \overline{D(A)}$.

We first discuss the possibility that $\hat{t} = 0$ and, e.g., it is (ii) of (91) that holds. By (95) and the assumption that $u, -v$ are bounded above, we have

$$\frac{\gamma}{4} + \left(\frac{\alpha}{2}\|\hat{x}-\hat{y}\|^2 + \frac{\beta}{2}(\hat{t}-\hat{s})^2 + \frac{\delta}{2}\left(\|\hat{x}\|^2 + \|\hat{y}\|^2\right)\right) \leq u(\hat{t},\hat{x}) - v(\hat{s},\hat{y}) \leq 2C. \tag{97}$$

By (91) (ii), there is a function $g\colon [0,\infty) \times [0,\infty) \to [0,\infty]$ such that $g(r,R)$ is nondecreasing in r, R, $g(0+,R) = 0$ for $R > 0$ and

$$\eta(S(s)y) \leq v(s,y) + g(s,\|y\|). \tag{98}$$

Thus, if $\hat{t} = 0$ in (97), (91), (98) and the middle inequality of (97) imply

$$\begin{aligned}
\frac{\gamma}{4} &\leq \eta(\hat{x}) - \eta(S(\hat{s})\hat{y}) + g(\hat{s},\|\hat{y}\|) \\
&\leq \eta(\hat{x}) - \eta(\hat{y}) + \eta(\hat{y}) - \eta(S(\hat{s})\hat{y}) + g(\hat{s},\|\hat{y}\|) \\
&\leq \rho_\eta(\|\hat{x}-\hat{y}\|) + \eta(\hat{y}) - \eta(S(\hat{s})\hat{y}) + g(\hat{s},\|\hat{y}\|),
\end{aligned}$$

where ρ_η is the modulus of continuity of η. The extreme inequality of (97) yields

$$\|\hat{x}\|, \|\hat{y}\| \le (2C/\delta)^{1/2}, \quad \hat{s} \le (2C/\beta)^{1/2}, \quad \|\hat{x}-\hat{y}\| \le (2C/\alpha)^{1/2}, \tag{99}$$

and then

$$\frac{\gamma}{4} \le \rho_\eta\left((2C/\alpha)^{1/2}\right) + \eta(\hat{y}) - \eta(S(\hat{s})\hat{y}) + g\left((2C/\beta)^{1/2}, (2C/\delta)^{1/2}\right).$$

Choose α_0 such that $\rho_\eta((2C/\alpha_0)^{1/2}) < \gamma/8$ and then

$$\frac{\gamma}{8} \le \eta(\hat{y}) - \eta(S(\hat{s})\hat{y}) + g\left((2C/\beta)^{1/2}, (2C/\delta)^{1/2}\right)$$

if $\alpha > \alpha_0$. Hereafter we only consider $\alpha > \alpha_0$. If for some fixed $\delta > 0$ we can send $\beta \to \infty$, we have a bound on $\hat{y}$ and $\hat{s} \to 0$ by (99) and then a contradiction via (89), since then both terms on the right of the inequality above tend to zero as $\beta \to \infty$. The conclusion is that if $\alpha > \alpha_0$, then for fixed $\delta > 0$, $0 < \hat{t}$ as soon as β is sufficiently large. Likewise, $\hat{s} > 0$. In this way, we reduce to the case $0 < \hat{t}, \hat{s}$.

Thus our maximum point of (96) may be used in the Doubling Theorem together with the computations of Section 2 to assert that

$$\begin{aligned} 2c_0 - 2\lambda &\le F^\lambda(\hat{y}, \alpha(\hat{x}-\hat{y}) - \delta\hat{y}) - F_\lambda(\hat{x}, \alpha(\hat{x}-\hat{y}) + \delta\hat{x}) \\ &= G^\lambda(\alpha(\hat{x}-\hat{y}) - \delta\hat{y}) - G_\lambda(\alpha(\hat{x}-\hat{y}) + \delta\hat{x}) + f(\hat{x}) - f(\hat{y}) \\ &\le G^\lambda(\alpha(\hat{x}-\hat{y}) - \delta\hat{y}) - G_\lambda(\alpha(\hat{x}-\hat{y}) + \delta\hat{x}) + \rho_f(\|\hat{x}-\hat{y}\|) \end{aligned}$$

when $\alpha > \alpha_0$ and β is sufficiently large (depending on δ). Letting $\lambda \downarrow 0$, $\delta \downarrow 0$ and $\alpha \to \infty$ while using the estimates (99), we obtain a contradiction. (We remark that the role of β will be more clear if $f(x)$ is replaced by $f(t,x)$ above.)

The comparison assertion being proved, we turn to the question of existence. This is a bit tricky. Suppose first that η is Lipschitz continuous with constant L. We claim that with a proper selection of the constants $a, C > 0$ for each $x_0 \in \overline{D(A)}$

$$\begin{cases} \overline{U}(t,x) = \eta(x_0) + \displaystyle\int_0^t ae^{as}\|S(t-s)x - x_0\|\,ds + L\|S(t)x - x_0\| + Ct, \\ \underline{U}(t,x) = \eta(x_0) - \displaystyle\int_0^t ae^{as}\|S(t-s)x - x_0\|\,ds - L\|S(t)x - x_0\| - Ct, \end{cases} \tag{100}$$

defines a supersolution $\overline{U}$ and a subsolution $\underline{U}$ of (80) (E) (with $F = G(p) - f(x)$).

From the formula (86) for the solution of the linear problem, in fact $\overline{U}$ solves

$$\begin{cases} \overline{U}_t + \langle Ax, \nabla\overline{U}\rangle = ae^{at}\|x - x_0\| + C \\ \overline{U}(0,x) = \eta(x_0) + L\|x - x_0\|. \end{cases}$$

On the other hand, $\overline{U}(t,x)$ is Lipschitz in x with constant

$$\int_0^t ae^{as}\,ds + L = (e^{at} - 1) + L \le (e^{aT} - 1) + L \quad \text{for} \quad 0 \le t \le T.$$

Thus, in the viscosity sense,

$$\text{“}G(\nabla\overline{U})\text{”} \geq G_{(e^{aT}+L-1)}(0)$$

and $\overline{U}$ is a supersolution of (80) provided that

$$ae^{at}\|x - x_0\| + C \geq -G_{(e^{aT}+L-1)}(0) - f(x).$$

Since

$$f(x) \geq f(x_0) - \rho_f(\|x - x_0\|) \geq f(x_0) - (1 + K\|x - x_0\|)$$

for a suitable constant K (a modulus of continuity has at most linear growth) we need only choose a, C so that

$$a > K \quad \text{and then} \quad C > -G_{(e^{aT}+L-1)}(0) - f(x_0) + 1.$$

Note that a is independent of x_0 and bounded for bounded L, while C is bounded for bounded L, x_0. Via a similar analysis, we make $\underline{U}$ to be subsolution. In this analysis, a depends on L and C on L, x_0; let us write $C_{(L,x_0)}$ and also choose $C_{(L,x_0)} \geq 0$.

We now index $\overline{U}$, $\underline{U}$ by x_0 to obtain $\overline{U}_{x_0}$, $\underline{U}_{x_0}$. The next thing to notice is that $\overline{U}_y \geq \underline{U}_z$ for $y, z \in \overline{D(A)}$ (since L is a Lipschitz constant for η). Let

$$\begin{cases} \overline{u}(t,x) = \inf\limits_{x_0 \in \overline{D(A)}} \overline{U}_{x_0}(t,x), \\ \underline{u}(t,x) = \sup\limits_{x_0 \in \overline{D(A)}} \underline{U}_{x_0}(t,x); \end{cases}$$

We then have

$$\begin{aligned} \underline{U}_{S(t)x}(t,x) &= \eta(S(t)x) - \int_0^t ae^{as}\|S(t-s)x - S(s)x\|\,ds - C_{(L,S(t)x)}t \\ &\leq \underline{u}(t,x) \leq \overline{u}(t,x) \leq \overline{U}_{S(t)x}(t,x) \\ &= \eta(S(t)x) + \int_0^t ae^{as}\|S(t-s)x - S(s)x\|\,ds + C_{(L,S(t)x)}t. \end{aligned} \tag{101}$$

By Lemma 4.1, $\underline{u}^*$ is a supersolution and $\overline{u}_*$ is a subsolution of the equation. By (101), both these functions assume the initial value η in the sense (87). Theorem 4.3 then shows that $\sup \mathcal{F}$, where

$$\mathcal{F} = \{w : w \quad \text{is a upper semicontinuous subsolution of} \quad (E) \quad \text{and} \quad \underline{u} \leq w \leq \overline{u}\}$$

is a continuous solution of (80) which, by (101), assumes the initial value η in the sense (87).

The general case in which $\eta \in UC(\overline{D(A)})$ is not necessarily Lipschitz continuous can be handled by a slight generalization. For $\gamma > 0$, choose L_γ such that

$$|\eta(x) - \eta(y)| \leq \gamma + L_\gamma|x - y|;$$

then construct supersolutions

$$\overline{U}(t,x) = \eta(x_0) + \int_0^t ae^{as}\|S(t-s)x - x_0\|\,ds + \gamma + L_\gamma\|S(t)x - x_0\| + Ct$$

as above by suitable choices of a, C. Now $a = a_\gamma$ will depend on γ and $C = C_{(L_\gamma, x_0)}$. For example, one now checks that

$$\begin{aligned}\overline{u}(t,x) &= \inf_{x_0,\gamma>0} \overline{U}(t,x)\\ &\le \eta(S(t)x) + \int_0^t a_\gamma e^{a_\gamma s}\|S(t-s)x - S(t)x\|\, ds + \gamma + C_{(L_\gamma, S(t)x)}t,\end{aligned}$$

etc. Since we may choose γ small and then t small, the analysis above may be adapted to this case as well. Example 4.4 indicates how to generalize the construction for general F.

To establish the continuity properties of the solution, we can take the following tack. Let u be a solution with all the desired properties and look back into the proof of comparison with $v(s,y)$ replaced by $u(s,y)$. The contradiction which was reached shows that if $\gamma > 0$ is fixed, then for

$$\begin{cases} \alpha(\rho_\eta, \gamma) \le \alpha, \\ \delta \le \delta(\alpha,\gamma), \\ \beta(\delta,\gamma,(89),(91)) \le \beta \end{cases}, \tag{102}$$

we have

$$u(t,x) - u(s,y) \le \gamma + \frac{\alpha}{2}\|x-y\|^2 + \frac{\delta}{2}\|x\|^2 + \frac{\delta}{2}\|y\|^2 + \frac{\beta}{2}(t-s)^2. \tag{103}$$

In (102), we mean that α should be chosen sufficiently large depending on γ and the modulus of continuity of η, then δ sufficiently small according to the choice of α and γ and then β sufficiently large according to the choice of δ, γ and the manner in which (89) and (87) hold. (We are regarding G, f as fixed). If $t = s$ in (103), we may send $\beta \to \infty$ and then $\delta \downarrow 0$ to conclude that $u(t,x)$ is uniformly continuous in x uniformly in t. If $x = y$, we conclude that $u(t,x)$ is uniformly continuous in t on bounded sets. The continuity property in t so exhibited is implicit in that it depends on the manner in which (91) holds, which depends on u. However, (101) gives explicit estimates of this, depending on G, f, η. □

REMARK 5.6. The reader can verify that the comparison proof and then the existence to the case in which (89) is generalized to: for $\epsilon > 0$

$$\eta(S(t)x) \quad \text{is uniformly continuous in} \quad t \in [\epsilon, T] \quad \text{uniformly for bounded} \quad x \in \overline{D(A)}. \tag{104}$$

The proof introduces an additional parameter $\epsilon > 0$ and proceeds by (perturbed) optimizing on $[\epsilon, T]$, etc. Everything goes much as before, first choosing $\epsilon > 0$ sufficiently small and then proceeding to show that $0 < \hat{t}, \hat{s}$, etc. Points like this occur in [7, Part IV]. The continuity in t of the solution will no longer be uniform on bounded sets, as simple examples (e.g., the linear problem) show. However, $u(t,x) - u(t-h, S(h)x)$ tends to zero with h uniformly on bounded sets, in view of (101), which "propagates".

REMARK 5.7. Suppose that u, v are a two solutions of (E) which assume intial values $u(0,x) \le v(0,x)$ in the desired manner and these initial values satisfy (104) in place of η. Then

$$u(t,x) - v(t,x) \le \sup_{z \in \overline{D(A)}} u(0,z) - v(0,z) \equiv K;$$

we prove this by comparing u and $v+K$, using $\eta(x)=u(0,x)$ in the theorem. That is, we finally remind the reader, that continuity estimates (of which this is only one of many true for the stationary and Cauchy problems) follow from comparison in standard ways.

APPENDIX: PROOF OF THE DOUBLING THEOREM 3.1

The proof which follows involves some clever arguments we extracted from [15].

Adopting the notation of the theorem, let us review what must be shown. Let $\Phi = \varphi+\psi \in C^1(H\times H)+\mathrm{Lip}(H\times H)$ and $(\hat{x},\hat{y})$ be a local maximum of $u(x)-v(y)-\Phi(x,y)$; say

$$\begin{cases} u(x)-v(y)-\Phi(x,y)\le u(\hat{x})-v(\hat{y})-\Phi(\hat{x},\hat{y}) \\ \text{for } (x,y)\in B_{2r}((\hat{x},\hat{y}))\cap\left(\overline{D(A)}\times\overline{D(A)}\right) \end{cases} \tag{105}$$

where $r>0$ is sufficiently small to guarantee

$$B_{2r}((\hat{x},\hat{y}))\cap\left(\overline{D(A)}\times\overline{D(A)}\right)=B_{2r}((\hat{x},\hat{y}))\cap(\Omega\times\Omega).$$

If Φ is a subtest function, which now means that

$$\varphi(Px,Py)\le\varphi(x,y) \quad\text{and}\quad \psi(Px,Py)\le\psi(x,y) \tag{106}$$

then we want to prove that

$$\liminf_{\substack{h\downarrow 0\\(x,y)\to(\hat{x},\hat{y})}}\frac{\Phi(x,y)-\Phi(S(h)x,S(h)y)}{h}+\inf_{\|p\|^2+\|q\|^2\le L(\psi)^2}H(\hat{x},\hat{y},\nabla\varphi(\hat{x},\hat{y})+(p,q))\le 0. \tag{107}$$

Step 1. Reduction to a Strict Maximum

Let $\gamma>0$ and set $u_\gamma(x)=u(x)-\gamma d(x,\hat{x})$, $v_\gamma(y)=v(y)+\gamma d(y,\hat{y})$. One checks that u_γ satisfies

$$\langle Ax,\nabla u_\gamma\rangle+F_\gamma(x,u_\gamma(x),\nabla u_\gamma(x))\le\gamma$$

and v_γ satisfies

$$\langle Ax,\nabla v_\gamma\rangle+G_\gamma(x,v_\gamma(x),\nabla v_\gamma(x))\ge-\gamma.$$

We thus expect $w_\gamma(x,y)=u_\gamma(x)-v_\gamma(y)$ to solve

$$\langle\mathcal{A}(x,y),\nabla w_\gamma(x,y)\rangle+F_\gamma(x,u_\gamma(x),\nabla_x w_\gamma(x,y))-G^\gamma(y,v_\gamma(y),-\nabla_y w_\gamma(x,y))\le 2\gamma.$$

If this is verified, since $u_\gamma(x)-v_\gamma(y)-\Phi(x,y)$ has a maximum at $(\hat{x},\hat{y})$, we have

$$D_{\mathcal{A}}^{-}\Phi(\hat{x},\hat{y})+H_{L(\psi)+2\gamma}(\hat{x},\hat{y},\nabla\varphi(\hat{x},\hat{y}))\le 2\gamma$$

and we may let $\gamma\downarrow 0$ to obtain the desired result.

Not only does $u_\gamma(x)-v_\gamma(y)-\Phi(x,y)$ have a maximum at $(\hat{x},\hat{y})$, any maximizing sequence $(x_n,y_n)\in B_{2r}((\hat{x},\hat{y}))\cap\left(\overline{D(A)}\times\overline{D(A)}\right)$ of $u_\gamma(x)-v_\gamma(y)-\Phi(x,y)$ converges to $(\hat{x},\hat{y})$ by

(33) (since then $d(x_n, \hat{x}) \to 0$, etc.). We reduce to the case in which this holds; i.e., we may assume that in addition to (105) we have

$$\begin{cases} (x_n, y_n) \in B_{2r}((\hat{x}, \hat{y})) \cap \left(\overline{D(A)} \times \overline{D(A)}\right) \quad \text{and} \\ u(x_n) - v(y_n) - \Phi(x_n, y_n) \to u(\hat{x}) - v(\hat{y}) - \Phi(\hat{x}, \hat{y}) \\ \text{implies } (x_n, y_n) \to (\hat{x}, \hat{y}) \text{ and } u(x_n) \to u(\hat{x}), v(x_n) \to v(\hat{y}). \end{cases}$$

Step 2. Basic Estimates and Perturbed Optimization

Given $\epsilon, \delta > 0$, we define the three functions

$$\Psi(x, y) = u(x) - v(y) - \Phi(x, y),$$

$$\Psi_\epsilon(x, y, \xi, \eta) = u(x) - v(y) - \Phi(\xi, \eta) - \frac{1}{2\epsilon}\left(\|x - \xi\|^2 + \|y - \eta\|^2\right)$$

and

$$\Psi_{\epsilon,\delta}(x, y, \xi, \eta) = u(x) - v(y) - \Phi(\xi, \eta) - \frac{1}{2\epsilon}\left(\|x - \xi\|^2 + \|y - \eta\|^2\right) - \delta\left(\|A^\circ\xi\| + \|A^\circ\eta\|\right)$$

for $(x, y) \in B_{2r}((\hat{x}, \hat{y})) \cap \left(\overline{D(A)} \times \overline{D(A)}\right)$, $(\xi, \eta) \in B_{2r}((\hat{x}, \hat{y})) \cap \left(\overline{D(A)} \times \overline{D(A)}\right)$. Here we use the convention

$$\|A^\circ\xi\| = \infty \quad \text{for} \quad \xi \notin D(A).$$

Put

$$N = B_{2r}((\hat{x}, \hat{y})) \cap \left(\overline{D(A)} \times \overline{D(A)}\right), \quad N^2 = N \times N.$$

The supremums of these functions we denote as follows:

$$M = \sup_N \Psi = u(\hat{x}) - v(\hat{y}) - \Phi(\hat{x}, \hat{y}), \quad M_\epsilon = \sup_{N^2} \Psi_\epsilon, \ M_{\epsilon,\delta} = \sup_{N^2} \Psi_{\epsilon,\delta}.$$

We claim that

$$M_\epsilon \downarrow M \quad \text{as} \quad \epsilon \downarrow 0 \quad \text{and} \quad M_{\epsilon,\delta} \uparrow M_\epsilon \quad \text{as} \quad \delta \downarrow 0. \tag{108}$$

It is clear that $M \le M_\epsilon$ and $M_\epsilon \ge M_{\epsilon,\delta}$ and a little thought shows that $M_{\epsilon,\delta} \uparrow M_\epsilon$ as $\delta \downarrow 0$. To establish the first convergence of (108), let L be a Lipschitz constant for Φ on $B_{2r}((\hat{x}, \hat{y})) \cap \left(\overline{D(A)} \times \overline{D(A)}\right)$ so that for $(x, y), (\xi, \eta) \in B_{2r}((\hat{x}, \hat{y})) \cap \left(\overline{D(A)} \times \overline{D(A)}\right)$ we have

$$\begin{aligned} &\Phi(\xi, \eta) + \frac{1}{2\epsilon}\left(\|x - \xi\|^2 + \|y - \eta\|^2\right) \\ &\quad \ge \Phi(x, y) + \frac{1}{2\epsilon}\left(\|x - \xi\|^2 + \|y - \eta\|^2\right) - L\left(\|x - \xi\|^2 + \|y - \eta\|^2\right)^{1/2} \\ &\quad \ge \Phi(x, y) - \frac{L^2\epsilon}{2}; \end{aligned}$$

here we used that $s^2/2\epsilon - Ls \ge -(L^2/2)\epsilon$ for $s \ge 0$. Thus $M_\epsilon \le M + (1/2)L^2\epsilon$ and (108) holds.

In what follows all points are taken from and inequalities hold on N or N^2 as dictated by the context.

It follows from (108) that for $\theta > 0$ we can choose $\delta = \delta(\epsilon, \theta)$ and $(x_{\epsilon,\theta}, y_{\epsilon,\theta}, \xi_{\epsilon,\theta}, \eta_{\epsilon,\theta})$ such that $\delta(\epsilon,\theta) \downarrow 0$ as $\epsilon, \theta \downarrow 0$ and

$$M_\epsilon - \theta \le \Psi_{\epsilon,\delta(\epsilon,\theta)}(x_{\epsilon,\theta}, y_{\epsilon,\theta}, \xi_{\epsilon,\theta}, \eta_{\epsilon,\theta}). \tag{109}$$

Moreover, by Lemmas 2.4 and 2.5 we can also assume that

$$\begin{aligned}\Psi_{\epsilon,\delta(\epsilon,\theta)}(x,y,\xi,\eta) - \epsilon\left(d(x,x_{\epsilon,\theta}) + d(y,y_{\epsilon,\theta}) + d(\xi,\xi_{\epsilon,\theta}) + d(\eta,\eta_{\epsilon,\theta})\right)& \\ \le \Psi_{\epsilon,\delta(\epsilon,\theta)}(x_{\epsilon,\theta}, y_{\epsilon,\theta}, \xi_{\epsilon,\theta}, \eta_{\epsilon,\theta}).&\end{aligned} \tag{110}$$

We note that putting $x = x_{\epsilon,\theta}, y = y_{\epsilon,\theta}$ above leads to the relation

$$\begin{aligned}\Phi(\xi,\eta)+&\frac{1}{2\epsilon}\left(\|x_{\epsilon,\theta}-\xi\|^2 + \|y_{\epsilon,\theta}-\eta\|^2\right)\\ &+\delta(\epsilon,\theta)\left(\|A^\circ\xi\| + \|A^\circ\eta\|\right) + \epsilon\left(d(\xi,\xi_{\epsilon,\theta}) + d(\eta,\eta_{\epsilon,\theta})\right)\\ &\ge \Phi(\xi_{\epsilon,\theta},\eta_{\epsilon,\theta}) + \frac{1}{2\epsilon}\left(\|x_{\epsilon,\theta}-\xi_{\epsilon,\theta}\|^2 + \|y_{\epsilon,\theta}-\eta_{\epsilon,\theta}\|^2\right)\\ &\qquad + \delta(\epsilon,\theta)\left(\|A^\circ\xi_{\epsilon,\theta}\| + \|A^\circ\eta_{\epsilon,\theta}\|\right).\end{aligned} \tag{111}$$

It is a consequence of (109) and $\Psi_\epsilon \ge \Psi_{\epsilon,\delta(\epsilon,\theta)}$ that

$$\Psi_\epsilon(x,y,\xi,\eta) \le \Psi_\epsilon(x_{\epsilon,\theta}, y_{\epsilon,\theta}, \xi_{\epsilon,\theta}, \eta_{\epsilon,\theta}) + \theta$$

and this with $x = x_{\epsilon,\theta}$, $y = y_{\epsilon,\theta}$, in turn, implies that

$$\begin{aligned}\theta + \Phi(\xi,\eta)+&\frac{1}{2\epsilon}\left(\|x_{\epsilon,\theta}-\xi\|^2 + \|y_{\epsilon,\theta}-\eta\|^2\right)\\ &\ge \Phi(\xi_{\epsilon,\theta},\eta_{\epsilon,\theta}) + \frac{1}{2\epsilon}\left(\|x_{\epsilon,\theta}-\xi_{\epsilon,\theta}\|^2 + \|y_{\epsilon,\theta}-\eta_{\epsilon,\theta}\|^2\right).\end{aligned} \tag{112}$$

Since Φ is a subtest function and $(x_{\epsilon,\theta}, y_{\epsilon,\theta}) \in \overline{D(A)} \times \overline{D(A)}$, the left-hand side above is not increased by replacing (ξ,η) by $(P\xi, P\eta)$ — we conclude that this relation in fact holds for $(\xi,\eta) \in B_{2r}((\hat{x},\hat{y}))$.

Step 3. Convergence of Approximate Maxs

From (109) we deduce that

$$M_\epsilon \le \Psi_{\epsilon,\delta(\epsilon,\theta)}(x_{\epsilon,\theta}, y_{\epsilon,\theta}, \xi_{\epsilon,\theta}, \eta_{\epsilon,\theta}) + \theta \le \Psi_\epsilon(x_{\epsilon,\theta}, y_{\epsilon,\theta}, \xi_{\epsilon,\theta}, \eta_{\epsilon,\theta}) + \theta \le M_\epsilon + \theta$$

so that $|\Psi_\epsilon(x_{\epsilon,\theta}, y_{\epsilon,\theta}, \xi_{\epsilon,\theta}, \eta_{\epsilon,\theta}) - M_\epsilon| \to 0$ as $\theta, \epsilon \downarrow 0$. It follows, using Lemma 3.2, that $M - \Psi(x_{\epsilon,\theta}, y_{\epsilon,\theta}) \to 0$ as well, and thus

$$(x_{\epsilon,\theta}, y_{\epsilon,\theta}), (\xi_{\epsilon,\theta}, \eta_{\epsilon,\theta}) \to (\hat{x},\hat{y}),\ u(x_{\epsilon,\theta}) \to u(\hat{x}),\ v(y_{\epsilon,\theta}) \to v(\hat{y}) \tag{113}$$

as $\epsilon, \theta \downarrow 0$ by Step 1.

Hence we may assume that

$$(x_{\epsilon,\theta}, y_{\epsilon,\theta}), (\xi_{\epsilon,\theta}, \eta_{\epsilon,\theta}) \in B_r(\hat{x}, \hat{y}) \cap \left(\overline{D(A)} \times \overline{D(A)}\right) \tag{114}$$

by choosing ϵ, θ sufficiently small.

Step 4. The Consequence of (110)

It follows from putting $\xi = \xi_{\epsilon,\theta}$, $\eta = \eta_{\epsilon,\theta}$, $y = y_{\epsilon,\theta}$ in (110) that $x_{\epsilon,\theta}$ is a local maximum of

$$x \mapsto u(x) - \frac{1}{2\epsilon}\|x - \xi_{\epsilon,\theta}\|^2 - \epsilon d(x, x_{\epsilon,\theta})$$

and then, from the assumption (44), Lemma 2.2, and $L\left(d(\cdot, \xi_{\epsilon,\theta})\right) \leq 1$ that

$$\frac{1}{\epsilon}\langle A^\circ \xi_{\epsilon,\theta}, x_{\epsilon,\theta} - \xi_{\epsilon,\theta}\rangle + F_\epsilon\left(x_{\epsilon,\theta}, u(x_{\epsilon,\theta}), \frac{x_{\epsilon,\theta} - \xi_{\epsilon,\theta}}{\epsilon}\right) \leq \epsilon.$$

Similarly,

$$\frac{1}{\epsilon}\langle A^\circ \eta_{\epsilon,\theta}, \eta_{\epsilon,\theta} - y_{\epsilon,\theta}\rangle + G^\epsilon\left(y_{\epsilon,\theta}, v(y_{\epsilon,\theta}), \frac{\eta_{\epsilon,\theta} - y_{\epsilon,\theta}}{\epsilon}\right) \geq -\epsilon.$$

Subtracting these and recalling $w(x,y) = u(x) - v(y)$ and (47) leads to

$$\begin{aligned}&\frac{1}{\epsilon}\left(\langle A^\circ \xi_{\epsilon,\theta}, x_{\epsilon,\theta} - \xi_{\epsilon,\theta}\rangle + \langle A^\circ \eta_{\epsilon,\theta}, y_{\epsilon,\theta} - \eta_{\epsilon,\theta}\rangle\right)\\ &\quad + H_{2\epsilon}\left(x_{\epsilon,\theta}, y_{\epsilon,\theta}, \frac{x_{\epsilon,\theta} - \xi_{\epsilon,\theta}}{\epsilon}, \frac{y_{\epsilon,\theta} - \eta_{\epsilon,\theta}}{\epsilon}\right) \leq 2\epsilon.\end{aligned} \tag{115}$$

Next we use (109) to relate $(x_{\epsilon,\theta} - \xi_{\epsilon,\theta}, y_{\epsilon,\theta} - \eta_{\epsilon,\theta})/\epsilon$ to $\nabla\varphi(\xi_{\epsilon,\theta}, \eta_{\epsilon,\theta})$. For this we use a standard lemma; in the lemma φ and ψ are taken to be defined on a Hilbert space Z and do not have the above meanings.

LEMMA A.8. *Let Z be a Hilbert space and $\varphi + \psi \in C^1(Z) + \mathrm{Lip}(Z)$. Suppose $z \in Z$, $\kappa, \epsilon, r > 0$ and*

$$\kappa + \varphi(\zeta) + \psi(\zeta) + \frac{1}{2\epsilon}\|z - \zeta\|^2 \geq \varphi(\hat{\zeta}) + \psi(\hat{\zeta}) + \frac{1}{2\epsilon}\|z - \hat{\zeta}\|^2 \quad \text{for} \quad \zeta \in B_r(\hat{\zeta}). \tag{116}$$

Let

$$g(\lambda) = \sup_{\|p\| \leq \lambda}\left\{\|\nabla\varphi(\hat{\zeta} + p) - \nabla\varphi(\hat{\zeta})\|\right\} \tag{117}$$

Then

$$L(\psi) + \frac{\kappa}{\lambda} + g(\lambda) + \frac{\lambda}{2\epsilon} \geq \|\frac{z - \hat{\zeta}}{\epsilon} - \nabla\varphi(\hat{\zeta})\| \quad \text{for} \quad 0 \leq \lambda \leq r. \tag{118}$$

Proof. Without loss of generality, let $\hat{\zeta} = 0$. Let $\omega \in Z$, $\|\omega\| = 1$, $0 \leq \lambda \leq r$ and put $\zeta = \lambda\omega$ in (116). Note that

$$\begin{gathered}|\varphi(\lambda\omega) - \varphi(0) - \lambda\langle\nabla\varphi(0), \omega\rangle| = \left|\int_0^\lambda \langle\nabla\varphi(s\omega) - \nabla\varphi(0), \omega\rangle\, ds\right| \leq \lambda g(\lambda)\\ |\psi(\lambda\omega) - \psi(0)| \leq \lambda L(\psi)\\ \|z\|^2 - \|z - \lambda\omega\|^2 = 2\lambda\langle z, \omega\rangle - \lambda^2.\end{gathered}$$

In conjunction with (116), this implies

$$\frac{\kappa}{\lambda} + g(\lambda) + \frac{\lambda}{2\epsilon} + L(\psi) \geq \langle \frac{1}{\epsilon} z - \nabla\varphi(0), \omega\rangle \quad \text{for} \quad \|\omega\| = 1$$

or

$$\frac{\kappa}{\lambda} + g(\lambda) + \frac{\lambda}{2\epsilon} + L(\psi) \geq \|\frac{1}{\epsilon} z - \nabla\varphi(0)\|$$

which amounts to (118). □

Applying Lemma A.8 to (112) in a straightforward way leads to

$$\|\frac{1}{\epsilon}(x_{\epsilon,\theta} - \xi_{\epsilon,\theta}, y_{\epsilon,\theta} - \eta_{\epsilon,\theta}) - \nabla\varphi(\xi_{\epsilon,\theta}, \eta_{\epsilon,\theta})\|_{H\times H} \leq \frac{\theta}{\lambda} + g_{\epsilon,\theta}(\lambda) + \frac{\lambda}{2\epsilon} + L(\psi)$$

where

$$g_{\epsilon,\theta}(\lambda) = \sup_{\|(\xi - \xi_{\epsilon,\theta}, \eta - \eta_{\epsilon,\theta})\|_{H\times H} \leq \lambda} \left\{\|\nabla\varphi(\xi,\eta) - \nabla\varphi(\xi_{\epsilon,\theta}, \eta_{\epsilon,\theta})\|\right\}.$$

Put $\lambda = \epsilon^2$, $\theta = \epsilon^3$ and use (113) and the continuity of $\nabla\varphi$ at $(\hat{x}, \hat{y})$ to conclude that $\lim_{\epsilon\downarrow 0} g_{\epsilon,\epsilon^3}(\epsilon^2) = 0$. Thus, in view of (113), if

$$x_\epsilon = x_{\epsilon,\epsilon^3}, y_\epsilon = y_{\epsilon,\epsilon^3},\ \xi_\epsilon = \xi_{\epsilon,\epsilon^3},\ \eta_\epsilon = \eta_{\epsilon,\epsilon^3}, \tag{119}$$

we have

$$\limsup_{\epsilon\downarrow 0} \|\frac{1}{\epsilon}(x_\epsilon - \xi_\epsilon, y_\epsilon - \eta_\epsilon) - \nabla\varphi(\hat{x}, \hat{y})\|_{H\times H} \leq L(\psi). \tag{120}$$

This estimate will allow us to substitute $\nabla\varphi(\hat{x}, \hat{y})$ for $(x_\epsilon - \xi_\epsilon, y_\epsilon - \eta_\epsilon)/\epsilon$ in the argument of H in (115) and then pass to the limit. However, before this, we obtain the remaining element of the proof.

Step 5. The Consequence of (111)

We use the notation (119). Note that by (114) and, for example,

$$\begin{aligned}\|S(h)\xi_\epsilon - \hat{x}\| &\leq \|S(h)\xi_\epsilon - S(h)\hat{x}\| + \|S(h)\hat{x} - \hat{x}\| \\ &\leq \|\xi_\epsilon - \hat{x}\| + \|S(h)\hat{x} - \hat{x}\| \\ &\leq r + \|S(h)\hat{x} - \hat{x}\|\end{aligned}$$

we have $(S(h)\xi_\epsilon, S(h)\eta_\epsilon) \in B_{2r}((\hat{x}, \hat{y})) \cap \left(\overline{D(A)} \times \overline{D(A)}\right)$ for h small. Thus we may put $\theta = \epsilon^3$, $\xi = S(h)\xi_\epsilon, \eta = S(h)\eta_\epsilon$ in (111), use $\|A^\circ S(h)\xi_\epsilon\| \leq \|A^\circ\xi_\epsilon\|$, $\|A^\circ S(h)\eta_\epsilon\| \leq \|A^\circ\eta_\epsilon\|$ and rearrange to find

$$\begin{aligned}\frac{\Phi(\xi_\epsilon, \eta_\epsilon) - \Phi(S(h)\xi_\epsilon, S(h)\eta_\epsilon))}{h} \leq \frac{1}{2\epsilon} &\frac{\|(x_\epsilon, y_\epsilon) - S(h)(\xi_\epsilon, \eta_\epsilon)\|^2_{H\times H} - \|(x_\epsilon, y_\epsilon) - (\xi_\epsilon, \eta_\epsilon)\|^2_{H\times H}}{h} \\ &+ \epsilon\frac{d(S(h)\xi_\epsilon, \xi_\epsilon) + d(S(h)\eta_\epsilon, \eta_\epsilon)}{h}.\end{aligned}$$

Letting $h \downarrow 0$ we conclude, via Lemma 2.2, that

$$\begin{aligned}\liminf_{\substack{h\downarrow 0\\(x,y)\to(\xi_\epsilon,\eta_\epsilon)}} & \frac{\Phi(x,y)-\Phi(S(h)x,S(h)y)}{h}\\ &\leq \limsup_{h\downarrow 0}\frac{1}{2\epsilon}\frac{\|(x_\epsilon,y_\epsilon)-\mathcal{S}(h)(\xi_\epsilon,\eta_\epsilon)\|^2_{H\times H}-\|(x_\epsilon,y_\epsilon)-(\xi_\epsilon,\eta_\epsilon)\|^2_{H\times H}}{h}\\ &\qquad + \liminf_{h\downarrow 0}\epsilon\frac{d(S(h)\xi_\epsilon,\xi_\epsilon)+d(S(h)\eta_\epsilon,\eta_\epsilon)}{h}\\ &\leq \frac{1}{\epsilon}\left(\langle A^\circ\xi_\epsilon, x_\epsilon-\xi_\epsilon\rangle+\langle A^\circ\eta_\epsilon, y_\epsilon-\eta_\epsilon\rangle\right)+2\epsilon\end{aligned} \tag{121}$$

Step 6. Putting it All Together

Using (115) and (121) we have

$$\liminf_{\substack{h\downarrow 0\\(x,y)\to(\xi_\epsilon,\eta_\epsilon)}} \frac{\Phi(x,y)-\Phi(S(h)x,S(h)y)}{h}+H_{2\epsilon}\left(x_\epsilon,y_\epsilon,\frac{x_\epsilon-\xi_\epsilon}{\epsilon},\frac{y_\epsilon-\eta_\epsilon}{\epsilon}\right)\leq 4\epsilon.$$

Now using (120) as well, we conclude that

$$\liminf_{\substack{h\downarrow 0\\(x,y)\to(\xi_\epsilon,\eta_\epsilon)}} \frac{\Phi(x,y)-\Phi(S(h)x,S(h)y)}{h}+H_{L(\psi)+k(\epsilon)}(\hat{x},\hat{y},\nabla\varphi(\hat{x},\hat{y}))\leq 4\epsilon$$

where $k(\epsilon)\to 0$ as $\epsilon\downarrow 0$. In view of (113) and Lemma 2.3, we find (107) in the limit as $\epsilon\downarrow 0$. The proof of the Doubling Theorem is complete.

We conclude this appendix by remarking on the relationships between Tataru's notions and ours. We do not attempt to study in detail the relations between these notions (certainly, they identify the same solutions whenever existence and uniqueness hold), since we expect our presentation (or some variant thereof) to be adopted in place of Tataru's more complex one. In our opinion, it is easier and preferable to prove things directly in our framework than to transfer them from Tataru's via any equivalence.

Tataru's "simplified" notion, presented in [16], of an upper semicontinuous solution u of

$$\langle Ax,\nabla u(x)\rangle+G(x,u(x),\nabla u(x))\leq 0,$$

clarified still further, reads roughly as follows:

Stage 1: Redefine D_A^+, D_A^- by

$$D_A^+\Phi(x)=\limsup_{h\downarrow 0}\frac{\Phi(x)-\Phi(S(h)x)}{h},\qquad D_A^-\Phi(x)=\liminf_{h\downarrow 0}\frac{\Phi(x)-\Phi(S(h)x)}{h}.$$

For $z\in H$, introduce the semigroups $S_z(t)$ generated by $A_z=A-z$ so that $D_{A_z}^+, D_{A_z}^-$ are defined.

Stage 2: Define Φ to be a subtest function at $\hat{x}$ if there are functions $f:\overline{D(A)} \to \mathbb{R}$, $a:\overline{D(A)} \to [0,\infty)$ which are continuous at $\hat{x}$ and $p \in H$ such that

$$\limsup_{w\in\overline{D(A)},w\to y} \frac{G(w)-G(y)-\langle p, w-y\rangle}{\|w-y\|} \le a(y)$$

and

$$D^-_{A_z}\Phi(y) \ge f(y) - \langle p,z\rangle - a(y)\|z\|$$

for $z \in H$ and $y \in \overline{D(A)}$ near $\hat{x}$; we regard Φ as carrying the data $f(\hat{x}), \langle p,z\rangle, a(\hat{x})$.

Stage 3: Require that if Φ is a subtest function at $\hat{x}$, with data as above, and $u-\Phi$ has a maximum at $\hat{x}$, then

$$f(\hat{x}) + G_{a(\hat{x})}(\hat{x}, u(\hat{x}), p) \le 0.$$

A main result of [16] is that this notion is equivalent to a much more complex one given in [15]. The more complex notion is a sort of doubled version of the simpler one, and this result is related to our Doubling Theorem. Uniqueness is proved in [15] using the more complex notion and no ameliorations are introduced by the simplified notion. The spirit of the Doubling Theorem is also embedded in the uniqueness proof. Note that, as opposed to our notion, Stage 2 requires the full m-accretive operator A, while we need only the semigroup $S(t)$ itself (Remark 1.6)). (See, however, the discussion in [16].) Our subtest condition somehow replaces this complexity in Stage 2; likewise our D^+_A, D^-_A take care of other complexities.

REFERENCES

[1] Barbu, V., **Nonlinear Semigroups and Differential Equations in Banach Spaces**, Nordhoff, Leyden, 1976.

[2] Barbu, V. and G. Da Prato, **Hamilton-Jacobi equations in Hilbert spaces**, *Research Notes in Mathematics* 86, Pitman, Boston (1983)

[3] Brezis, H., **Opérateurs Maximaux Monotones et Semi-groups de Contractions dan les Espaces de Hilbert**, *North Holland Mathematics Studies* 5, North Holland Pub. Co., Amsterdam, 1973.

[4] M. G. Crandall, H. Ishii and P. L. Lions, Uniqueness of viscosity solutions revisited, *J. Math. Soc. Japan* 39 (1987), 581–596.

[5] Crandall, M. G., H. Ishii and P. L. Lions, User's Guide to viscosity solutions of second order partial differential equations, *Bull. A. M. S.* to appear.

[6] M. G. Crandall, and P. L. Lions, Unbounded viscosity solutions of Hamilton-Jacobi equations, *Illinois J. Math.* 31 (1987), 665–688.

[7] Crandall, M. G., and P. L. Lions, Hamilton-Jacobi equations in infinite dimensions: *J. Func. Anal.* Part I. Uniqueness of viscosity solutions, vol. 62 (1985), 379 - 396. Part II. Existence of viscosity solutions, vol. 65 (1986), 368 - 405. Part III. vol. 68 (1986),

214 - 247. Part IV. Unbounded linear terms, vol. 90 (1990), 237–283. Part V. B-continuous solutions 97 (1991), 417-465. Part VII. The HJB equations is not always satisfied, submitted.

[8] Deville, R., G. Godefroy and V. Zizler, A smooth variational principle with applications to Hamilton-Jacobi equations in infinite dimensions, preprint.

[9] Ishii, H., Remarks on existence and uniqueness of viscosity solutions of Hamilton-Jacobi equations , *Bull. Fac. Sci. Eng. Chuo Univ.* 26 (1983), 5–24.

[10] Ishii, H., Uniqueness of unbounded viscosity solutions of Hamilton-Jacobi equations, *Ind. Univ. Math. J.* 33 (1984), 721–748.

[11] Ishii, H., Hamilton-Jacobi equations with discontinuous Hamiltonians on arbitrary open sets, *Bull. Fac. Sci. Eng. Chuo Univ.* 28 (1985), 33–77.

[12] Ishii, H., Existence and uniqueness of solutions of Hamilton-Jacobi equations, *Funkcial. Ekvac.* 29 (1986), 167–188.

[13] Ishii, H., Perron's method for Hamilton-Jacobi equations , *Duke Math. J.* 55 (1987), 369–384.

[14] Ishii, H., Viscosity solutions for a class of Hamilton-Jacobi equations in Hilbert spaces, to appear.

[15] Tataru, D., Viscosity solutions for Hamilton - Jacobi equations with unbounded nonlinear term, *J. Math. Anal. Appl.*, to appear.

[16] Tataru, D., Viscosity solutions for Hamilton - Jacobi equations with unbounded nonlinear term - a simplified approach, preprint.

[17] Tataru, D., On the equivalence between the dynamic programming principle and the dynamic programming equation, preprint.

Boundary Control Problem for Age-Dependent Equations

GIUSEPPE DA PRATO Scuola Normale Superiore, Pisa, Italy

MIMMO IANNELLI University of Trento, Trento, Italy

1 Introduction and setting of the problem

We are here concerned with a dynamical system, describing the evolution of a certain population, governed by the equations:

$$\begin{cases} p_t(t,a) + p_a(t,a) + \mu(a)p(t,a) = 0, & a \in [0,a_\dagger], \quad t \geq 0 \\ p(t,0) = \int_0^{a_\dagger} \beta(\sigma)p(t,\sigma)d\sigma + u(t), & t \geq 0 \\ p(0,a) = p_0(a), & a \in [0,a_\dagger] \end{cases} \tag{1.1}$$

Here $p(t,a)$ is the density of the population of age a at time t, $\mu(a)$ is the age specific mortality rate, $\beta(a)$ the age specific birth rate, $a_\dagger < +\infty$ the maximal age and $u(t)$ the control. We shall assume on (1.1) the following usual assumptions:

$$\beta \quad \textit{is non-negative and belongs to} \quad L^\infty(0,a_\dagger) \tag{1.2}$$

$$\mu \quad \textit{is non-negative and belongs to} \quad L^1_{loc}([0,a_\dagger)) \tag{1.3}$$

$$\int_0^{a_\dagger} \mu(\sigma)d\sigma = +\infty \tag{1.4}$$

moreover we will suppose:

$$\int_0^{a_\dagger} \beta(a)e^{-\int_0^a \mu(b)db}da \neq 1 \tag{1.5}$$

that is, we assume that the population has not an intrinsic net rate equal to 1 .

We want to minimize a quadratic cost function such as

$$J(u) = \int_0^T \left[\int_0^{a_\dagger} p^2(t,a)da + u^2(t)\right] dt \ , \tag{1.6}$$

over all controls $u \in L^2(0,T)$ subject to the state equation (1.1). T can be taken finite or infinite.

[1]Partially supported by the Italian National Project M.P.I. "Equazioni di Evoluzione e Applicazioni Fisico-Matematiche"

In §2 we write problem (1.1) in the abstract form

$$p(t) = S(t)p_0 - \alpha A \int_0^t S(t-s)Du(s)ds\ , \qquad t \in [0,T], \tag{1.7}$$

where $S(\cdot)$ is a C_0 semigroup on $H = L^2(0, a_\dagger)$, with infinitesimal generator A, and D is a mapping from $\mathbf{R}$ into H.

The special feature of problem (1.7), typical of boundary control problems, is that Im $D \cap D(A) = \{0\}$ and so the meaning of (1.7) is not clear. However it is possible to show that $\int_0^t S(t-s)Du(s)ds \in D(A)$ for all $u \in L^2(0,T)$ and so (1.7) is meaningful. This will be shown in §2 : see Theorem 2.3.

§3 is devoted to study the control problem (1.6) when T is finite. We use here the Dynamic Programming approach based on the solution to the operator Riccati equation, which is solved, in a generalized sense, by using a recent result in Barbu et al. (1991).

Finally in §4 we consider the infinite horizon problem. Here the main result is to prove stabilizability: see Proposition 4.1 .

2 Abstract setting

Let us first introduce some notation. We set

$$\Pi(a) = e^{-\int_0^a \mu(\sigma)d\sigma}, \quad a \in [0, a_\dagger] \tag{2.1}$$

this function $\Pi(\cdot)$ belongs to $C([0,a_\dagger])$.

Besides, for any $\varphi \in L^2([0,a_\dagger])$ we denote by B_φ the solution of the following integral equation

$$B(t) = F_\varphi(t) + \int_0^t K(t-s)B(s)ds\ , \quad t \geq 0 \tag{2.2}$$

where

$$K(t) = \begin{cases} \beta(t)\Pi(t) & \text{if } t \in [0,a_\dagger] \\ 0 & \text{if } t > a_\dagger \end{cases} \tag{2.3}$$

and

$$F_\varphi(t) = \begin{cases} \displaystyle\int_t^{a_\dagger} \varphi(s-t)\beta(s)\frac{\Pi(s)}{\Pi(s-t)}ds & \text{if } t \in [0,a_\dagger] \\ 0 & \text{if } t > a_\dagger \end{cases} \tag{2.4}$$

By the previous assumptions (1.2)-(1.4), equation (2.2) has a unique solution $B_\varphi \in C([0,+\infty))$.

Next we want to transform (1.1) into a problem with homogeneous boundary conditions. To this purpose we introduce the following change of variables.

$$q(t,a) = p(t,a) - \gamma(t,a), \qquad t \geq 0,\ a \in [0,a_\dagger], \tag{2.5}$$

where γ is chosen such that it solves the problem:

$$\begin{cases} \gamma_a(t,a) + \mu(a)\gamma(t,a) = 0, & t \geq 0,\ a \in [0, a_\dagger] \\ \gamma(t,0) = u(t), & t \geq 0 \end{cases}$$

That is:

$$\gamma(t,a) = \alpha\Pi(a)u(t), \qquad a \in [0, a_\dagger],\ t \geq 0 \tag{2.6}$$

with

$$\alpha = \frac{1}{1 - \int_0^{a_\dagger} K(s)ds}, \tag{2.7}$$

where the denominator does not vanish by 1.5.

Assume for the moment that u is differentiable, then after the transformation (2.5), problem (1.1) reduces to:

$$\begin{cases} q_t(t,a) + q_a(t,a) + \mu(a)q(t,a) = -\gamma_t(t,a), & a \in [0, a_\dagger],\ t \geq 0 \\ q(t,0) = \int_0^{a_\dagger} \beta(\sigma)q(t,\sigma)d\sigma, & t \geq 0 \\ q(0,a) = p_0(a) - \alpha u(0), & a \in [0, a_\dagger] \end{cases} \tag{2.8}$$

Now we are ready to introduce the abstract setting.

We set $H = L^2(0, a_\dagger)$, with the inner product $< \cdot,\cdot >$ and norm $\|\cdot\|$, and introduce the linear operator in H

$$\begin{cases} D(A) = \{\varphi \in H :\ \varphi'(\cdot) + \mu(\cdot)\varphi(\cdot) \in H,\ \varphi(0) = \int_0^{a_\dagger} \beta(\sigma)\varphi(\sigma)d\sigma\} \\ (A\varphi)(a) = -\varphi'(a) - \mu(a)\varphi(a). \end{cases} \tag{2.9}$$

It is known (see Webb (1985)) that A is the infinitesimal generator of a strongly continuous semigroup $S(\cdot)$ in H given by

$$(S(t)\varphi)(a) = \begin{cases} B_\varphi(t-a)\Pi(a) & \textit{if } 0 \leq a \leq t \\ \varphi(a-t)\dfrac{\Pi(a)}{\Pi(a-t)} & \textit{if } a > t \end{cases} \tag{2.10}$$

where $t \geq 0$, $\varphi \in L^2(0, a_\dagger)$.

We recall that $S(t)$ is compact for $t > a_\dagger$.

Lemma 2.1 *Let A be the linear operator defined by* (2.9). *The spectrum $\sigma(A)$ consists of simple eigenvalues λ that are exactly all the solutions to the equation*

$$1 = \int_0^{a_\dagger} e^{-\lambda\sigma}K(\sigma)d\sigma. \tag{2.11}$$

If λ is an eigenvalue of A then a corresponding eigenvector is given by

$$\varphi_\lambda(a) = e^{-\lambda a}\Pi(a), \quad a \in [0, a_\dagger]. \tag{2.12}$$

Moreover the adjoint A^* of A is given by

$$\begin{cases} D(A^*) = \{\psi \in H,\ \psi'(\cdot) - \mu(\cdot)\psi(\cdot) \in H,\ \psi(T) = 0\}, \\ (A^*\psi)(a) = \psi'(a) - \mu(a)\psi(a) + \beta(a)\psi(0). \end{cases} \tag{2.13}$$

Finally $\sigma(A^*) = \sigma(A)$ and if $\lambda \in \sigma(A^*)$ then a corresponding eigenvector is given by

$$\psi_\lambda(a) = \frac{1}{\Pi(a)} \int_a^{a_\dagger} e^{\lambda(a-\sigma)} K(\sigma)d\sigma, \quad a \in [0, a_\dagger]. \tag{2.14}$$

Proof —

The proof is straightforward, let us check for instance formula (2.13). Let Λ be the linear operator

$$\begin{cases} D(\Lambda) = \{\psi \in H,\ \psi'(\cdot) - \mu(\cdot)\psi(\cdot) \in H,\ \psi(T) = 0\ \}, \\ (\Lambda\psi)(a) = \psi'(a) - \mu(a)\psi(a) + \beta(a)\psi(0). \end{cases} \tag{2.15}$$

For $\varphi \in D(A), \psi \in D(\Lambda)$ one can check easily that

$$< A\varphi, \psi >=< \varphi, \psi'(\cdot) - \mu(\cdot)\psi + \beta(\cdot)\psi(0) > .$$

This implies $\psi \in D(A^*)$ and $A^*\psi = \Lambda\psi$ that is $\Lambda \subset A^*$. Since, as easily checked, the resolvent set of Λ, as well as the one of A^*, is not empty, we have necessarily $\Lambda = A^*$. ∎

Now we define $D : \mathbf{R} \mapsto H$ as follows:

$$x \to (Dx)(a) = \Pi(a)x. \tag{2.16}$$

and

$$q_0(a) = p_0(a) - \alpha u(0) \tag{2.17}$$

Then, setting $q(t,\cdot) = q(t),\ p(t,\cdot) = p(t),\ p_0(\cdot) = p_0,\ q_0(\cdot) = q_0,$ problem (2.8) can be written as the following abstract problem in H

$$\begin{cases} q'(t) = Aq(t) - \alpha Du'(t) \\ q(0) = q_0, \end{cases}$$

and also in its mild form:

$$q(t) = S(t)q_0 - \alpha \int_0^t S(t-s)Du'(s)ds.$$

Integrating by parts in (2) , we get finally that the state equation can be written as

$$p(t) = S(t)p_0 - \alpha A \int_0^t S(t-s)Du(s)ds, \tag{2.18}$$

Remark that in (2.18) it does not appear the derivative of u; from now on we shall consider this equation as the state equation. We have now to show that equation (2) is meaningful for any $u \in L^2(0,T)$. To this purpose we must analize the term

$$z(t) = \int_0^t S(t-s)Du(s)ds, \quad t \in [0,T]. \tag{2.19}$$

In fact we have:

Theorem 2.2 *Let $z(t)$ be defined as in (2.19) and let $u \in L^2(0,T)$. Then $z(t) \in D(A)$ for all $t \in [0,T]$ and $Az(\cdot) \in C([0,T];H)$. Moreover there exists a constant $L > 0$ such that*

$$\|Az(t)\| \leq L|u|_{L^2(0,T)}, \quad t \in [0,T]. \tag{2.20}$$

Proof —

By (2.10) and (2.16) we have:

$$z(t)(a) = \Pi(a)\left(\int_{(t-a)\vee 0}^t u(\sigma)d\sigma + \int_0^{(t-a)\vee 0} B_\Pi(t-a-\sigma)u(\sigma)d\sigma\right) \tag{2.21}$$

where we note that $B_\Pi(\cdot) \in H^1(0,T)$, in fact we have

$$F_\Pi(t) = \int_t^{+\infty} K(\sigma)d\sigma \tag{2.22}$$

that is $F_\Pi(\cdot) \in H^1(0,T)$.

Now it is easy to check that:

$$z(t)'(a) + \mu(a)z(t)(a) =$$

$$= \begin{cases} \Pi(a)\left[(1-B_\Pi(0))\,u(t-a) - \int_a^t B'_\Pi(s-a)u(t-s)ds\right] & \textit{if } a \leq t \\ 0 & \textit{if } a > t, \end{cases} \tag{2.23}$$

which implies:

$$z(t)'(\cdot) + \mu(\cdot)z(t)(\cdot) \in L^2(0,a_\dagger) \tag{2.24}$$

Besides we check the equality:

$$z(t)(0) = \int_0^{a_\dagger} \beta(\sigma)z(t)(\sigma)d\sigma \tag{2.25}$$

Actually from (2.21) we have:

$$z(t)(0) = \int_0^t B_\Pi(t-\sigma)u(\sigma)d\sigma \tag{2.26}$$

and also:

$$\int_0^{a_\dagger} \beta(\sigma)z(t)(\sigma)d\sigma =$$

$$= \int_0^t K(a)\int_{t-a}^t u(\sigma)d\sigma da + \int_0^t K(a)\int_0^{t-a} B_\Pi(t-a-\sigma)u(\sigma)d\sigma da +$$
$$+ \int_t^{+\infty} K(a)da\int_0^t u(\sigma)d\sigma =$$
$$= \int_0^t u(\sigma)\int_{t-\sigma}^t K(a)dad\sigma + \int_0^t u(\sigma)\int_0^{t-\sigma} K(a)B_\Pi(t-a-\sigma)dad\sigma + \tag{2.27}$$
$$+ \int_t^{+\infty} K(a)da\int_0^t u(\sigma)d\sigma =$$
$$= \int_0^t u(\sigma)\left[F_\Pi(t-\sigma) + \int_0^{t-\sigma} K(a)B_\Pi(t-a-\sigma)da\right]d\sigma =$$
$$= \int_0^t B_\Pi(t-\sigma)u(\sigma)d\sigma$$

so that (2.25) is true and have proven that $z(t) \in D(A)$ for all $t \in [0,T]$; moreover we note that by (2.21) it is also easy to show the inequality (2.20). To finish the proof we have to prove that $Az(\cdot) \in C([0,T];H)$.

Let $\{u_n\}$ be a sequence in $C^1(0,T)$ such that $\lim_{n\to+\infty} u_n = u$ in $L^2(0,T)$. Then Du_n belongs to $C^1([0,T];H)$ so that, setting

$$z_n(t) = \int_0^t S(t-s)Du_n(s)ds,$$

it follows that $z_n(\cdot) \in C([0,T];D(A))$, and, since by (2.20)

$$\|Az(t) - Az_n(t))\| \le L|u-u_n|_{L^2(0,T)}$$

we have that

$$\lim_{n\to+\infty} Az_n(\cdot) = Az(\cdot) \quad in \quad C([0,T];H)$$

and the proof of the theorem is complete. ∎

As a consequence of the previous result, we can define the following continuous mapping $G : L^2(0,T) \to C([0,T];H)$ by defining:

$$(G(u))(t) = A\int_0^t S(t-s)Du(s)ds \tag{2.28}$$

then formula (2.18) defines a function $p \in C([0,T];H)$ that can be written as

$$p(t) = S(t)p_0 - \alpha(G(u))(t) \tag{2.29}$$

3 The finite horizon control

Let T be a fixed positive number. Here we assume the assumptions (1.2) – (1.5) and consider the optimal control problem of minimizing:

$$J_T(u) = \int_0^T \left[|p(t, p_0; u)|_H^2 + |u(t)|^2 \right] dt \tag{3.1}$$

over all controls $u \in L^2(0,T)$, where $p(t, p_0; u)$ is given by (2.29), that is

$$p(t, p_0; u) = S(t)p_0 - \alpha G(u)(t) \tag{3.2}$$

Since J_T is a quadratic coercive form, it has a unique minimum u^*, called the *optimal control*. The corresponding function $p^* = p(\cdot, p_0; u^*)$ is called the *optimal state* and the pair (u^*, p^*) the *optimal pair*. As it is well known, the goal of the Dynamic Programming is to express $u^*(t)$ as a linear function of $p^*(t)$ for any $t \in [0,T]$, or equivalently, to find a feedback optimal control, see for instance Bensoussan et al. (to appear). This is difficult on the present case because the state equation (2) is very irregular, in fact, if it is written formally as an evolution equation, it looks like

$$p'(t) = Ap(t) - \alpha ADu(t)\ , \quad p(0) = p_0 \tag{3.3}$$

and this is not meaningful because $Du(t)$ does not belong to $D(A)$ unless $u(t) \equiv 0$. To overcome this difficulty we introduce, for any positive integer n, the approximating problem:

$$\begin{cases} p_n'(t) = Ap_n(t) - \alpha A_n Du(t) \\ \\ p_n(0) = p_{0,n} \end{cases} \tag{3.4}$$

where $A_n = nA(nI - A)^{-1}$ are the Yosida approximations of A and $p_{0,n} = n(nI - A)^{-1}p_0$. Then we consider the problem of minimizing:

$$J_{T,n}(u) = \int_0^T \left[\|p_n(t, p_0; u)\|^2 + |u(t)|^2 \right] dt \tag{3.5}$$

over all controls $u \in L^2(0,T)$, where $p_n(t, p_0; u)$ is the solution to the problem (3.4) .

We first note that the mild solution to problem (3.4) has the form:

$$p_n(t) = S(t)p_{0,n} - \alpha A_n \int_0^t S(t-s)Du(s)ds = n(nI - A)^{-1}p(t, p_0; u), \tag{3.6}$$

so that:

$$\lim_{n \to +\infty} p_n(\cdot) = p(\cdot, p_0; u) \quad in\ \ C([0,T]; H) \tag{3.7}$$

In addition we recall that by a well known result concerning problem (3.5), (see for instance Bensoussan et al. (to appear)), if (u_n^*, p_n^*) is an optimal pair for problem (3.5), the following feedback formula holds

$$u_n^*(t) = \alpha D^* A_n^* P_n(T-t)p_n^*(t), \quad t \in [0,T] \tag{3.8}$$

where $P_n(\cdot)$ is the solution to the operator Riccati equation

$$\begin{cases} P_n' = A^*P_n + P_nA - \alpha^2 P_n A_n DD^* A_n^* P_n + I, \; t \in [0,T] \\ P_n(0) = 0. \end{cases} \tag{3.9}$$

and is strongly continuous on $[0,T]$ .We are going to prove:

Theorem 3.1 *Assume* (1.2) – (1.5) *and let* (u^*, p^*) *and* (u_n^*, p_n^*) *be the optimal pairs of problem* (3.1) *and* (3.5) *respectively. Then we have:*

$$\lim_{n\to\infty} u_n^* = u^* \quad in \; L^2(0,T) \quad , \quad \lim_{n\to\infty} p_n^* = p^* \quad in \quad L^2(0,T;H) \tag{3.10}$$

Proof —

First of all we notice that, in virtue of the estimate (2.20), by a result in Barbu et al. (1991) , there exists the function $t \to P(t)$ strongly continuous on $[0,T]$ such that

$$\lim_{n\to\infty} P_n(t)x = P(t)x, \tag{3.11}$$

for all $x \in H$ uniformly in $t \in [0,T]$.

Let now $u \in L^2(0,T)$, and let p_n be the solution of (3.4); we have already seen that

$$\lim_{n\to+\infty} p_n = p(t, u_0; u) \quad in \; C([0,T];H)$$

so that we also have

$$\lim_{n\to\infty} J_{T,n}(u) = J_T(u) \quad \forall u \in L^2(0,T) \tag{3.12}$$

Now, computing $\frac{d}{ds} < P_n(T-s)p_n(s), p_n(s) >$, taking into account equations (3.4)–(3.9), and finally, integrating with respect to s between 0 and T, we arrive at the identity

$$\begin{aligned} &< P_n(T)p_{0,n}, p_{0,n} > + \int_0^T |u(s) - \alpha D^* A_n^* P_n(T-s)p_n(s)|^2 ds \\ &= \int_0^T \left[\|p_n(s)\|^2 + |u(s)|^2 \right] ds = J_{T,n}(u) \end{aligned} \tag{3.13}$$

Thus, by (3.11) and (3.12) we get:

$$< P(T)p_0, p_0 > \le \; J_T(u), \quad \forall \, u \in L^2(0,T) \tag{3.14}$$

and moreover, setting $u = u_n^*$ and, taking into account (3.8) we find

$$< P_n(T)p_{0,n}, p_{0,n} >= \int_0^T \left[\|p_n^*(s)\|^2 + |u_n^*(s)|^2 \right] ds \tag{3.15}$$

Now, (3.15) and (3.14) imply that the sequence $\{u_n^*\}$ is bounded in $L^2(0,T)$ and consequently there exists a subsequence $\{u_{n_k}^*\}$ weakly convergent to an element $\widetilde{u}$ of $L^2(0,T)$. Since

$$p_{n_k}^*(t) = n_k(n_k I - A)^{-1}(S(t)p_0 - \alpha G(u_{n_k}^*))$$

and

$$G(u_{n_k}^*) \rightharpoonup G(\widetilde{u}) \quad in \;\; L^2(0,T;H),$$

we have

$$p_{n_k}^* \rightharpoonup \widetilde{p} \text{ in } L^2(0,T;H),$$

where

$$\widetilde{p} = S(\cdot)p_0 - \alpha G(\widetilde{u}) = p(\cdot, u_0; \widetilde{u}),$$

and the symbol $\rightharpoonup$ denotes weak convergence.

By letting n tend to infinity in (3.15) we have

$$< P(T)p_0, p_0 > \geq \int_0^T \left[\|\widetilde{p}(s)\|^2 + |\widetilde{u}(s)|^2\right] ds = J_T(\widetilde{u}),$$

which, along with (3.14), yields $\widetilde{u} = u^*$, $\widetilde{p} = p^*$ and

$$< P(T)p_0, p_0 >= J_T(u^*) \tag{3.16}$$

Thus the sequences $\{u_n^*\}$ and $\{p_n^*\}$ converge weakly to u^* and p^* respectively; finally , by (3.15) and (3.16) we have

$$|u^*|^2_{L^2(0,T)} + |p^*|^2_{L^2(0,T;H)}$$

$$\leq \lim_{n\to\infty} \left[|u_n^*|^2_{L^2(0,T)} + |p_n^*|^2_{L^2(0,T;H)}\right] =$$

$$= J_T(u^*) = |u^*|^2_{L^2(0,T)} + |p^*|^2_{L^2(0,T;H)}$$

so that $\{u_n^*\}$ and $\{p_n^*\}$ converge strongly and the proof is complete. ∎

4 The infinite horizon problem

In this section we consider the optimal control problem of minimizing:

$$J_\infty(u) = \int_0^\infty \left[\|p(t,p_0;u)\|^2 + |u(t)|^2\right] dt \tag{4.1}$$

over all controls $u \in L^2(0,\infty)$, where $p(t) = p(t,p_0;u)$ is given by

$$p(t) = S(t)p_0 - \alpha G(u)(t), \quad t \geq 0. \tag{4.2}$$

We say that the system (4.2) is *stabilizable* if, for any $p_0 \in H$, there exists a control $u \in L^2(0,\infty)$ such that $p(t,p_0;u) \in L^2(0,\infty;H)$, or, equivalently, such that $J_\infty(u) < \infty$.

Proposition 4.1 *Assume* (1.2) – (1.5) *, then the system* (4.2) *is stabilizable.*

Proof —

Set

$$\sigma_1(A) = \{\lambda \in \sigma(A) : \ \text{Re}\ \lambda < 0\}, \quad \sigma_2(A) = \{\lambda \in \sigma(A) : \ \text{Re}\ \lambda \geq 0\},$$

and denote by Q_1 and Q_2 the spectral projectors on $\sigma_1(A)$ and $\sigma_2(A)$ respectively. We note that the subspaces $Q_1(H)$ and $Q_2(H)$ are left invaried by the semigroup $S(t)$ and that, by Lemma 2.1, Q_2 is a finite dimensional projector. Moreover, since $S(t)$ is compact for $t > a_\dagger$ the spectral determining condition holds true (see Pazy (1983)) and there exists $M > 0$ and $r > 0$ such that

$$\|S(t)Q_1\| \leq Me^{-rt}, \quad t \geq 0. \tag{4.3}$$

For any $t \geq 0$, we set $p_1(t) = Q_1 p(t)$ and $p_2(t) = Q_2 p(t)$, so that $p(t) = p_1(t) + p_2(t)$.

By (4.3) it follows easily that $p_1 \in L^2(0, \infty; H)$ for any $u \in L^2(0, \infty)$. , thus we only have to check that, for any $p_0 \in H$ there exists $u \in L^2(0, \infty)$ such that $p_2 \in L^2(0, \infty; H)$.

Now we denote by A_2 the restriction of A to $Q_2(H)$; A_2 is a bounded operator of $Q_2(H)$ into itself and p_2 is the solution of the finite dimensional Cauchy problem

$$\begin{cases} p_2'(t) = A_2 p_2(t) - \alpha A_2 Q_2 D u(t) \\ \\ p_2(0) = Q_2 p_0. \end{cases} \tag{4.4}$$

By a well known Hautus result (see Hautus (1970)), the system (4.4) is stabilizable if and only if

$$\text{Ker}\ (\lambda - A_2^*) \cap \ \text{Ker}\ (D^* Q_2^* A_2^*) = \{0\}, \quad \forall\, \lambda \in \sigma_2(A). \tag{4.5}$$

To check this condition let $\lambda \in \sigma_2(A)$ and $\psi^* \in \text{Ker}\ (\lambda - A_2^*)$. It is easy to see that setting $\psi = Q_2^* \psi^*$ then $\psi \in \text{Ker}(\lambda - A^*)$, that is, ψ is an eigenvector of A^* relative to λ ; then by Lemma 2.1 we have

$$1 = \int_0^{a_\dagger} e^{-\lambda\sigma} K(\sigma) d\sigma. \tag{4.6}$$

and

$$\psi(a) = \frac{c}{\Pi(a)} \int_a^{a_\dagger} e^{\lambda(a-\sigma)} K(\sigma) d\sigma, \quad a \in [0, a_\dagger]. \tag{4.7}$$

where c is a suitable constant. It follows (see 1.5)

$$D^* Q_2^* A_2^* \psi^* = \lambda D^* Q_2^* \psi^* = \lambda D^* \psi = \lambda < \psi, \Pi(\cdot) >= c \left[\int_0^{a_\dagger} K(\sigma) d\sigma - 1 \right] \neq 0. \tag{4.8}$$

and 4.5 is verified. ∎

Simultaneous Well-Posedness

RALPH deLAUBENFELS Ohio University, Athens, Ohio

I. INTRODUCTION. We would like to consider the following problem. Given a family of closed operators, $\mathcal{A} \equiv \{A_\alpha\}_{\alpha \in I}$, on a Banach space X, find a maximal Banach space Z, continuously embedded in X, such that $A_\alpha|_Z$ generates a strongly continuous semigroup, for all $\alpha \in I$.

For a single operator, this was first considered, independently, by Krein, Laptev and Cretkova[12] and Kantorovitz[11]; see also [8] for another approach. In [11], the maximal subspace on which an operator A generates a strongly continuous semigroup of contractions is called the *Hille-Yosida space.* Recently, such spaces have been considered in connection with integrated semigroups and regularized semigroups; see [1], [5] and [13].

Here are some examples of where our problem might be of interest. We will(after showing it exists) write $Z_0(\mathcal{A})$ for the maximal Banach subspace on which all members of $\mathcal{A}$ generate a strongly continuous semigroup of contractions.

Example 1.1: Time-dependent evolution equations. In considering

$$\frac{d}{dt}u(t) = A(t)(u(t))\,(0 \leq s \leq t \leq T),\ u(s) = x, \tag{1.2}$$

a common hypothesis is to have $A(t)$ generate a strongly continuous semigroup of contractions, for each $t \geq 0$.

We will give an example in Section VI of how we may use $Z_0(\{A(t)\}_{t \geq 0})$ to consider (1.2).

Example 1.3: Semi-simplicity manifold. If X is reflexive and $\mathcal{A} \equiv \{\pm iA^n\}_{n \in \mathbf{N}}$, for some closed operator A, then it may be shown that $Z_0(\mathcal{A})$ is the *semi-simplicity manifold* of Kantorovitz([10, p. 157]); this is a maximal Banach subspace such that, for every $x \in Z_0(\mathcal{A})$, there exists a countably additive vector-valued measure E_x such that

$$e^{itA_{Z_0(\mathcal{A})}}x = \int_{\mathbf{R}} e^{-ist}\, dE_x(s)\, (t \in \mathbf{R}).$$

This space may also be shown to be the maximal subspace on which A is a scalar type spectral operator, with real spectrum.

Example 1.4: Semi-simplicity manifold with positive spectrum. If X is reflexive and $\mathcal{A} \equiv \{-A, \pm iA^n\}_{n \in \mathbf{N}}$, then we obtain a maximal Banach subspace on which $-A$ generates a strongly continuous semigroup that is a Laplace transform,

$$e^{-tA|_{Z_0(\mathcal{A})}}x = \int_0^\infty e^{-st}\, dE_x(s)\, (t \geq 0),$$

where E_x is as in Example 1.3.

All operators are linear, on a Banach space, X. Throughout this paper, $\mathcal{A}$ will be a collection of closed operators $\{A_\alpha\}_{\alpha \in I}$. The empty product of operators $\prod_{k=1}^{0} G_k$ will mean the identity operator. Basic information on strongly continuous semigroups may be found in [3], [9], [15], [16], and [19].

II. SIMULTANEOUS SOLUTION SPACE.

For a single operator, we introduced the following in [6] and gave it a Frechet space topology, with respect to which the operator generated a strongly continuous locally equicontinuous semigroup.

Definition 2.1. If A is a closed operator, by the *solution space*, $Z(A)$, for A, we will mean the set of all x in X for which the abstract Cauchy problem

$$\frac{d}{dt}u(t,x) = A(u(t,x))\,(t \geq 0),\ u(0,x) = x \tag{2.2}$$

has a mild solution. By a mild solution, we mean a function $t \mapsto u(t,x) \in C([0,\infty), X)$ such that, $\forall t \geq 0$, $\int_0^t u(s,x)\, ds \in \mathcal{D}(A)$ and

$$A\left(\int_0^t u(s,x)\, ds\right) = u(t,x) - x.$$

To construct a Banach space on which A generates a strongly continuous semigroup, we need to restrict ourselves to exponentially bounded solutions. We will denote by $Z_{exp}(A)$ the *exponentially bounded solution space* for A, the set of all x in $Z(A)$ for which $t \mapsto u(t,x)$, from $[0,\infty)$ into X, is exponentially bounded.

The construction that follows may be done with Z_{exp} replaced by Z, to yield a space with a locally convex topology, on which A_α will generate a strongly continuous semigroup, for all α. We will do this in a future paper.

Definition 2.3. Suppose A is a closed operator such that $(r - A)$ is injective, for r large. Then a Laplace transform argument shows that, for any $x \in Z_{exp}(A)$, the exponentially bounded mild solution of (2.2) is unique. We then define, for $t \geq 0$, the operator e^{tA} by

$$e^{tA}x \equiv u(t,x),\ \mathcal{D}(e^{tA}) \equiv Z_{exp}(A).$$

There is no reason to believe that e^{tA} will be closed, or even closable, in general.

Definition 2.4. Suppose that, $\forall \alpha \in I, A_\alpha$ is a closed operator such that $(r - A_\alpha)$ is injective, for r large. We will write $Z_{exp}(\mathcal{A})$ for the *exponentially bounded simultaneous solution space*

$$\bigcap \mathcal{D}(\prod_{k=1}^{n} e^{t_k A_{\alpha_k}}),$$

over all finite sequences $< t_k >, < \alpha_k >$.

Definition 2.5. We will write $\vec{t}, \vec{\alpha}$ for $< t_k >_{k=1}^{m}, < \alpha_k >_{k=1}^{m}, |\vec{t}| \equiv \sum_{k=1}^{m} t_k$.

For $x \in Z_{exp}(\mathcal{A}), \omega \in \mathbf{R}$, define

$$\|x\|_{\mathcal{A},\omega} \equiv \sup\{\|e^{-\omega|\vec{t}|}(\prod_{k=1}^{n} e^{t_k A_{\alpha_k}})x\| \mid t_k \geq 0, \alpha_k \in I\}.$$

$Z_\omega(\mathcal{A})$ is defined to be the set of all $x \in Z(\mathcal{A})$ such that

(1) $\|x\|_{\mathcal{A},\omega} < \infty$.

(2) The map $s \mapsto e^{sA_\alpha}x$, from $[0, \infty)$ into $(Z_{exp}(\mathcal{A}), \| \ \|_{\mathcal{A},\omega})$ is continuous, $\forall \alpha \in I$.

We will write $Z_\omega(\mathcal{A})$ for the normed vector space $(Z_\omega(\mathcal{A}), \| \ \|_{\mathcal{A},\omega})$

Theorem 2.6. *Suppose that,* $\forall \alpha \in I, \exists r_\alpha$ *such that* $(r - A_\alpha)$ *is injective,* $\forall r > r_\alpha$. *Then, for any real* ω,

(1) $Z_\omega(\mathcal{A})$ *is a Banach space.*

(2) $Z_\omega(\mathcal{A}) \hookrightarrow X$.

(3) $\forall \alpha \in I, A_\alpha|_{Z_\omega(\mathcal{A})}$ *generates a strongly continuous semigroup, with* $\|e^{tA_\alpha|_{Z_\omega}}\| \leq e^{\omega t}, \forall t \geq 0$.

(4) Z *is maximal-unique, that is, if* Y *is a normed vector space satisfying (2) and (3), then* $Y \hookrightarrow Z$.

Example 2.7. It is clear that $Z_0(\{A_1, A_2\}) \subseteq Z_0(A_1) \bigcap Z_0(A_2)$, for any pair of closed operators A_1, A_2.

We give here an example of commuting operators A_1, A_2 such that $Z_0(\{A_1, A_2\}) \neq Z_0(A_1) \bigcap Z_0(A_2)$.

For Ω a region in $\mathbf{R}^2$, write $BUC(\Omega)$ for the set of all bounded uniformly continuous complex-valued functions on $\Omega, BUC^1(\Omega)$ for the set of all $f \in BUC(\Omega)$ such that $\frac{\partial f}{\partial x}$ and $\frac{\partial f}{\partial y}$ are in $BUC(\Omega)$.

Let $A_1 \equiv \frac{\partial}{\partial x}$, the generator of left translation,

$$e^{tA_1} f(x, y) \equiv f(x + t, y),$$

$A_2 \equiv \frac{\partial}{\partial y}$, the generator of downward translation,

$$e^{tA_2} f(x, y) \equiv f(x, y + t).$$

We let both these operators act on a space where translation is not strongly continuous, $X \equiv BUC(\mathbf{R}^2) \bigcap BUC^1(\{(x,y) \in \mathbf{R}^2 \,|\, x,y \leq 0\})$.

BUC	BUC
BUC^1	BUC

Then it is not hard to see that $Z_0(\{A_1, A_2\}) = BUC^1(\mathbf{R}^2)$,

BUC^1	BUC^1
BUC^1	BUC^1

while $Z_0(A_1) \bigcap Z_0(A_2) = BUC(\mathbf{R}^2) \bigcap BUC^1\{(x,y) \in \mathbf{R}^2 \,|\, x \leq 0 \text{ or } y \leq 0\})$.

BUC^1	BUC
BUC^1	BUC^1

Example 2.8. Even for a single operator, the growth condition of (3) is necessary for maximality, in Theorem 2.6. That is, in general, given a closed operator A, there exists no maximal Banach space such that the restriction of A to that space generates a strongly continuous semigroup; given any Banach space Y, such that $A|_Y$ generates a strongly continuous semigroup, we can find a bigger Banach space on which A generates a strongly continuous semigroup. Here is an example.

Let $X \equiv C_0(\mathbf{R}) \cap C_0^1((-\infty, 0])$, where $C_0^1((-\infty,0]) \equiv \{f \in C^1((-\infty,0]) \cap C_0((-\infty,0]) \,|\, \lim_{x\to-\infty} f'(x) = 0\}$. Let $A \equiv \frac{d}{dx}$, with maximal domain, so that

$$e^{tA}f(x) = f(x+t),$$

left translation, for $f \in Z(A)$.

Note first that, if Y is any Banach space such that $A|_Y$ generates a strongly continuous semigroup and $Y \hookrightarrow X$, then $\exists \omega \in \mathbf{R}$ such that $Y \hookrightarrow Z_\omega(A)$. Mainly, choose any ω such that $e^{-\omega t}\|e^{tA|_Y}\|$ is bounded (this is true for any A).

It is clear from (4) of Theorem 2.6 that $Z_{\omega_1} \hookrightarrow Z_{\omega_2}$, whenever $\omega_1 < \omega_2$. To show that we have the desired example, it is sufficient to show that, when $\omega_1 < \omega_2$, then $Z_{\omega_1} \neq Z_{\omega_2}$.

Define, for $\omega_1 < \omega < \omega_2$,

$$f(x) \equiv g(x)1_{(-\infty,0)} + \sin(e^{\omega x})1_{[0,\infty)},$$

where g is chosen so that $f \in X$.

Then, for $t \geq 0, x \in \mathbf{R}$

$$(e^{tA}f)'(x) = g'(x+t)1_{(-\infty,-t)} + \omega e^{\omega(x+t)}\cos(e^{\omega(x+t)})1_{[-t,\infty)};$$

in particular,

$$(e^{tA}f)'(0) = \omega e^{\omega t}\cos(e^{\omega t}).$$

Thus $f \in Z_{\omega_2}$, but is not in Z_{ω_1}.

Proof of Theorem 2.6: Without loss of generality, we may assume $\omega = 0$(otherwise work with $(A_\alpha - \omega)$).

Properties (2) and (3) are clear from the definition of $\| \quad \|_{\mathcal{A},0}$ and (2) of Definition 2.5.

To prove (1), suppose $< x_n >$ is Cauchy in $Z_0(\mathcal{A})$. Then $\exists x \in X$ such that $x_n \to x$ in X. We will show, by induction on m, the number of coordinates in $\vec{t}$ and $\vec{\alpha}$, that for all sequences $\vec{t}$ and $\vec{\alpha}$,

$$x \in \mathcal{D}\left(\prod_{k=1}^{m} e^{t_k A_{\alpha_k}}\right),$$

and

$$\lim_{j\to\infty} \|(\prod_{k=1}^{m} e^{t_k A_{\alpha_k}})(x_j - x)\| = 0. \qquad (*)$$

For m equal to zero, (*) clearly holds. Suppose (*) is valid for all sequences of length m. Given $t_1, \ldots, t_{m+1} \geq 0, \alpha_1, \ldots, \alpha_{m+1} \in I$, let

$$B \equiv A_{\alpha_{m+1}},\ y_j \equiv (\prod_{k=1}^{m} e^{t_k A_{\alpha_k}})x_j,\ y \equiv (\prod_{k=1}^{m} e^{t_k A_{\alpha_k}})x.$$

Since $< x_n >$ is Cauchy in $Z_0(\mathcal{A})$, the functions $t \mapsto e^{tB}y_j$, from $[0,\infty)$ into X, are uniformly Cauchy, thus converge uniformly to a continuous $\phi : [0,\infty) \to X$. Note that, by the induction hypothesis, $\phi(0) = y$. For any $j \in \mathbf{N}, t \geq 0$,

$$B\left(\int_0^t e^{sB}y_j\,ds\right) = e^{tB}y_j - y_j,$$

thus, since B is closed and the convergence of $e^{sB}y_j$ is uniform,

$$B\left(\int_0^t \phi(s)\,ds\right) = \phi(t) - y.$$

This is saying that $y \in \mathcal{D}(e^{tB})$, and $e^{tB}y = \phi(t)$, so that $e^{tB}(y_j - y) \to 0, \forall t \geq 0$. This concludes the induction, proving (*).

To see that $\|x\|_{\mathcal{A},0}$ is finite, first note that, since $< x_n >$ is Cauchy, $\exists$ a constant K such that $\|x_n\|_{\mathcal{A},0} \leq K, \forall n \in \mathbf{N}$. Thus, by (*), for any $\vec{t}, \vec{\alpha}$,

$$\|(\prod_{k=1}^m e^{t_k A_{\alpha_k}})x\| = \lim_{j\to\infty} \|(\prod_{k=1}^m e^{t_k A_{\alpha_k}})x_j\| \leq K,$$

so that $\|x\|_{\mathcal{A},0} \leq K < \infty$.

Next, we will show that $\|(x_n - x)\|_{\mathcal{A},0}$ converges to 0, as $n \to \infty$. Fix $\epsilon > 0$ and choose N so that $\|(x_n - x_j)\|_{\mathcal{A},0} < \epsilon, \forall n, j > N$. For $n > N$,

$$\|(\prod_{k=1}^m e^{t_k A_{\alpha_k}})(x_n - x)\| = \lim_{j\to\infty} \|(\prod_{k=1}^m e^{t_k A_{\alpha_k}})(x_n - x_j)\|,$$

for any $\vec{t}, \vec{\alpha}$, by (*), thus is less than ϵ. Taking suprema, we conclude that $\|(x_n - x)\|_{\mathcal{A},0} \leq \epsilon, \forall n > N$.

All that remains is to show that x satisfies (2) of Definition 2.5. It is clear from the definition of the norm that $\|e^{sA_\alpha}(x_n - x)\|_{\mathcal{A},0} \leq \|(x_n - x)\|_{\mathcal{A},0}, \forall s \geq 0, n \in \mathbf{N}, \alpha \in I$. Thus the functions $s \mapsto e^{sA_\alpha}x_n$, from $[0,\infty) \to Z_0(\mathcal{A})$, converge uniformly to $e^{sA_\alpha}x$. This implies that $s \mapsto e^{sA_\alpha}x$, from $[0,\infty)$ into $(Z_{exp}(\mathcal{A}), \|\ \|_{\mathcal{A},0})$, is continuous, as desired.

For (4), suppose Y is a normed vector space satisfying (2) and (3). There exists a constant M such that $\|x\| \leq M\|x\|_Y, \forall x \in Y$. For any $\vec{t}, \vec{\alpha}$, $x \in Y$, by the Hille-Yosida theorem,

$$\|(\prod_{k=1}^m e^{t_k A_{\alpha_k}})x\| \leq M\|(\prod_{k=1}^m e^{t_k A_{\alpha_k}|_Y})x\|_Y \leq M\|x\|_Y.$$

Thus, $\|x\|_{\mathcal{A},0} \leq M\|x\|_Y$. To show that $Y \hookrightarrow Z_0(\mathcal{A})$, all that remains is to show that x satisfies (2) of Definition 2.5. The map $s \mapsto e^{sA_\alpha}x$, from $[0,\infty) \to Y$, is continuous, thus since $\|x\|_{\mathcal{A},0} \leq M\|x\|_Y$, the same is true as a map from $[0,\infty)$ into $(Z_{exp}(\mathcal{A}), \|\ \|_{\mathcal{A},0})$. ∎

III. ANOTHER CONSTRUCTION, WHEN RESOLVENT SETS ARE LARGE In this section, we suppose that $(\omega, \infty) \subseteq \rho(A_\alpha)$, for all $\alpha \in I$, and show how we may then use the resolvents to construct $Z_\omega(\mathcal{A})$. Without loss of generality, suppose $\omega = 0$. Then we may mimic the construction and proof in [11] as follows.

Let $Y(\mathcal{A})$ be the set of all $x \in X$ such that

$$\|x\|_{Y(\mathcal{A})} \equiv \sup\{\|(\prod_k \lambda_k(\lambda_k - A_{\alpha_k})^{-1})x\| \mid \lambda_k > 0, \alpha_k \in I\}$$

is finite.

Then, using the maximality of our construction in Section II and (an imitation of) the construction in [11], we have the following.

Theorem 3.1.

(1) *$Z_0(\mathcal{A})$ equals the closure, in $Y(\mathcal{A})$, of* $\bigcap\{\mathcal{D}(\prod_k A_{\alpha_k}|_{Y(\mathcal{A})}) \mid \alpha_k \in I\}$.

(2) $\|x\|_{Z_0(\mathcal{A})} = \|x\|_{Y(\mathcal{A})}, \forall x \in Z_0(\mathcal{A})$.

We should mention that, for many interesting examples, the resolvent condition of this section will not be satisfied; for example, the families of matrices of Section V.

IV. SIMULTANEOUS EXISTENCE FAMILIES. The disadvantage of these constructions is that they are too mysterious. The desired space exists, and is maximal, but how big is it? When is it even nontrivial? This is not clear from the construction. We need simple criteria for determining what $Z_\omega(\mathcal{A})$ contains. In this section, we introduce a multivariable version of C-existence families(see [4]), that is equivalent to $Z_\omega(\mathcal{A})$ containing the image of a bounded operator C.

When C commutes with $A_\alpha, \forall \alpha \in I$, this is a multivariable version of a C-regularized semigroup, introduced by Da Prato[2].

Throughout this section, C will be a bounded operator.

Definition 4.1. If $Im(C) \subseteq Z_{exp}(\mathcal{A})$ and

$$W(\vec{t}, \vec{\alpha}) \equiv \left(\prod_k e^{t_k A_{\alpha_k}}\right) C$$

is a bounded operator from X to itself, for all $\vec{t}, \vec{\alpha}$, then $\{W(\vec{t}, \vec{\alpha})\}_{t_k \geq 0, \alpha_k \in I}$ is a *simultaneous C-existence family* for $\mathcal{A}$.

Note that strong continuity, in each t variable, follows automatically.

We need to define exponential boundedness. We will say that this family of operators is $O(e^{\omega|\vec{t}|})$ if $\exists M$ such that

$$\|W(\vec{t}, \vec{\alpha})\| \leq M e^{\omega|\vec{t}|},$$

$\forall \vec{t}, \vec{\alpha}$.

In fact, Theorem 4.2 shows that the boundedness of $W(\vec{t}, \vec{\alpha})$ follows automatically.

In the following theorem, we will write $[Im(C)]$ for the Banach space with norm $\|y\|_{[Im(C)]} \equiv \inf\{\|x\| \mid Cx = y\}$.

Theorem 4.2. *Suppose that $(\lambda - A_\alpha)$ is injective, $\forall \alpha \in I, \lambda > \omega$. Then the following are equivalent.*

(a) $Im(C) \subseteq Z_\omega(\mathcal{A})$.

(b) $[Im(C)] \hookrightarrow Z_\omega(\mathcal{A})$.

(c) *$\exists$ a simultaneous C-existence family for $\mathcal{A}$ that is $O(e^{\omega|\vec{t}|})$.*

If $\bigcap_{\alpha \in I} \mathcal{D}(A_\alpha)$ is dense in X, then these are equivalent to

(d) $Im(C) \subseteq Im(\prod_k (\lambda_k - A_{\alpha_k})), \forall \alpha_k \in I, \lambda_k > \omega$ and

$$\sup\{\|\left(\prod_k (\lambda_k - \omega)(\lambda_k - A_{\alpha_k})^{-1}\right) C\| \mid \lambda_k > \omega, \alpha_k \in I\}$$

is finite.

Proof: By translating, we may assume that $\omega = 0$.
(c) $\to$ (b). There exists M such that

$$\|(\prod_{k=1}^{n} e^{t_k A_{\alpha_k}})C\| \leq M,$$

for any $\vec{t}, \vec{\alpha}$. Thus, for $y \in Im(C), Cx = y$,

$$\| \left(\prod_{k=1}^{n} e^{t_k A_{\alpha_k}} \right) y\| \leq M\|x\|,$$

so that taking suprema on the left and infima on the right implies that $\|y\|_{\mathcal{A},0} \leq M\|y\|_{[Im(C)]}$, as desired.
(a) $\to$ (c). We must show that

$$W(\vec{t}, \vec{\alpha})(x) \equiv \left(\prod_{k=1}^{m} e^{t_k A_{\alpha_k}} \right) Cx$$

defines a bounded operator on X, for all $\vec{t}, \vec{\alpha}$.

We will show this by induction on m, the number of coordinates in $\vec{t}$ and $\vec{\alpha}$, where $\prod_{k=1}^{0} \equiv I$. For $m = 0$ the assertion is clear. Suppose $W(\vec{t}, \vec{\alpha})$ is bounded, for vectors with $m - 1$ coordinates($m \in \mathbf{N}$). Given vectors with m coordinates, define $W : X \to BUC([0, \infty), X)$ by

$$(Wx)(t_m) \equiv \left(\prod_{k=1}^{m} e^{t_k A_{\alpha_k}} \right) Cx \ \ (t_m \geq 0),$$

for fixed $t_1, ...t_{m-1}, \alpha_1, ...\alpha_m$. Let $G \equiv \left(\prod_{k=1}^{m-1} e^{t_k A_{\alpha_k}} \right) C, B \equiv A_{\alpha_m}$.

Suppose $x_n \to x, Wx_n \to \phi$. Then, by the induction hypothesis, $Gx_n \to Gx$. For any $t \geq 0$, since B is closed, and $\forall n$,

$$B\left(\int_0^t Wx_n(s)\, ds \right) = Wx_n(t) - Gx_n,$$

it follows that $\int_0^t \phi(s)\, ds \in \mathcal{D}(B)$, with

$$B\left(\int_0^t \phi(s)\, ds \right) = \phi(t) - Gx. \tag{*}$$

Also note that $\phi(0) = \lim_{n\to\infty} Wx_n(0) = \lim_{n\to\infty} Gx_n = Gx$. By (*), this implies that $\phi(t) = e^{tB}Gx = Wx(t)$.

Thus W is closed, hence, by the closed graph theorem, bounded, which clearly implies that $W(\vec{t}, \vec{\alpha})$ is bounded, for vectors $\vec{t}, \vec{\alpha}$ with m coordinates. This completes the induction. The uniform boundedness of $W(\vec{t}, \vec{\alpha})$ follows from the Banach-Steinhaus theorem.

Now suppose that $\mathcal{D} \equiv \cap_{\alpha in I}\mathcal{D}(A_\alpha)$ is dense.

Let $Z \equiv Z_0(\mathcal{A})$.

(b) $\rightarrow$ (d). This follows from the Hille-Yosida theorem, applied to $A_\alpha|_Z$. First, for $\lambda_k > 0, \alpha_k \in I$, since $(\lambda_k - A_{\alpha_k}|_Z)$ is surjective, $Im(C) \subseteq Im(\prod_k(\lambda_k - A_{\alpha_k}))$.

If $\|y\|_Z \leq M\|y\|_{[Im(C)]}, \forall y \in Im(C)$, then for any $x \in X$,

$$\| \left(\prod_{k=1}^{m} \lambda_k(\lambda_k - A_{\alpha_k})^{-1} \right) Cx\| \leq \| \left(\prod_{k=1}^{m} \lambda_k(\lambda_k - A_{\alpha_k})^{-1} \right) Cx\|_Z$$
$$\leq \|Cx\|_Z \leq M\|Cx\|_{[Im(C)]} \leq M\|x\|.$$

(d) $\rightarrow$ (a). Define, for $x \in \bigcap\{Im(\prod_k(\lambda_k - A_{\alpha_k})) \mid \lambda_k > 0, \alpha_k \in I\}$,

$$\|x\|_W \equiv \sup\{\| \left(\prod_k \lambda_k(\lambda_k - A_{\alpha_k})^{-1} \right) x\| \mid \lambda_k > 0, \alpha_k \in I\},$$

and let W be the normed vector space of all x such that $\|x\|_W < \infty$. Let $\tilde{W}$ be the completion, with respect to $\| \ \|_W$, of W. Note that $\tilde{W} \hookrightarrow X$. Let Y be the closure, in $\tilde{W}$, of $\cap_{\alpha \in I}\mathcal{D}(A_\alpha|_{\tilde{W}})$.

Then $\forall \alpha \in I$, by the Hille-Yosida theorem, $A_\alpha|_Y$ generates a strongly continuous semigroup of contractions.

By Theorem 2.6(4), $Y \hookrightarrow Z$. Clearly $[Im(C)] \hookrightarrow W$, hence $[Im(C)] \hookrightarrow \tilde{W}$. Thus $[C(\mathcal{D})] \hookrightarrow Y$. For $x \in X$, choose $x_n \in \mathcal{D}$ such that $x_n \rightarrow x$. Then $Cx_n \rightarrow Cx$, in $[Im(C)]$, hence in $\tilde{W}$. Thus $[Im(C)] \hookrightarrow Y$.

V. MATRICES OF OPERATORS. In this section, we will show that certain families of matrices of constant coefficient differential operators have an exponentially bounded simultaneous $(1 + \Delta)^{-r}$-existence family, for appropriate r. More generally, we will consider matrices of polynomials of commuting generators of bounded strongly continuous groups.

Throughout this section, m and n are fixed natural numbers.

We will need some standard multivariable terminology. We will write $x = (x_1, \cdots x_n)$, for vectors in $\mathbf{R}^n, \alpha = (\alpha_1, \cdots \alpha_n)$ for vectors in $\mathbf{N}^n$. We will write $x^\alpha \equiv x_1^{\alpha_1} \cdots x_n^{\alpha_n}, |\alpha| \equiv \sum_{j=1}^n \alpha_j, \|x\| \equiv \left(\sum_{j=1}^n x_j^2\right)^{\frac{1}{2}}$.

Operator Terminology 5.1. Throughout this section, $iB_1, \cdots, iB_n$ will be commuting generators of bounded strongly continuous groups of operators, $\{e^{itB_k}\}_{t \in \mathbf{R}}$. Without loss of generality (after an equivalent renorming) we may assume that $\|e^{itB_k}\| \leq 1, \forall t \in \mathbf{R}, 1 \leq k \leq n$.

We will write $e^{i(x \cdot B)}$ for $\prod_{k=1}^n e^{ix_k B_k}$. We will use the following well-known functional calculus(see [3] for the case $n = 1$) Define a bounded operator $f(B)$ by

$$f(B) \equiv \int_{\mathbf{R}^n} e^{i(x \cdot B)} \mathcal{F}f(x)\, dx,$$

where $\mathcal{F}$ is the Fourier transform, whenever $\mathcal{F}f(x) \in L^1(\mathbf{R}^n)$. The map $f \mapsto f(B)$ is an algebra homomorphism and clearly

$$\|f(B)\| \leq \|\mathcal{F}f\|_1,$$

for all such f.

Let $\mathcal{S}(\mathbf{R}^n)$ be the Schwartz space of rapidly decreasing smooth functions on $\mathbf{R}^n$.

We will write B for $(B_1, \cdots, B_n)$, B^α for $B_1^{\alpha_1} \cdots B_n^{\alpha_n}$. If p is a polynomial in n variables, $p(x) \equiv \sum_{|\alpha| \leq k} c_\alpha x^\alpha$ and $x \in \mathcal{D}_0 \equiv \bigcup_{f \in \mathcal{S}(\mathbf{R}^n)} Im(f(B))$, then

$$p(B)x \equiv \sum_{|\alpha| \leq k} c_\alpha B^\alpha x$$

and $p(B)$ is defined to be the closure of $p(B)|_{\mathcal{D}_0}$.

In particular, the operator D_k will be $i\frac{\partial}{\partial x_k}$, with D and the general constant coefficient differential operator $p(D)$ defined as above.

The operator $|B|^2 \equiv \sum_{k=1}^n B_k^2$, the generator of the strongly continuous semigroup

$$e^{-t|B|^2} \equiv f_t(B),\ f_t(x) \equiv e^{-t|x|^2}.$$

Matrix Operator Terminology 5.2. Throughout this section, $(p_{i,j})_{i,j=1}^m$ will be an $m \times m$ matrix of complex-valued polynomials in n variables. The number N will be $\max_{i,j}\{$ degree of $p_{i,j}\}$.

For a matrix $(p_{i,j})_{i,j=1}^m$, we define the operator $(p_{i,j}(B))$, on X^m, as follows. Let $\mathcal{P}$ be the restriction of $(p_{i,j}(B))$, such that $\mathcal{D}(\mathcal{P}) \equiv \mathcal{D}(|B|^N)^m$. Then, by [7, Theorem 5.1], an extension of $\mathcal{P}$ generates a regularized semigroup. This implies that $\mathcal{P}$ is closable; $(p_{i,j}(B))$ is defined to be the closure of $\mathcal{P}$.

If $\mathcal{F}f_{i,j} \in L^1(\mathbf{R}^n)$, for $1 \leq i,j \leq m$, and $\mathcal{G} = (f_{i,j})$, we define the bounded operator $\mathcal{G}(B)$ similarly.

When A is an operator on X, we will also write A for AI_m, the operator on X^m with domain $(\mathcal{D}(A))^m$.

When B is an $m \times m$ matrix of complex numbers, we will write $n.r.(B)$ for the *numerical range* of B,

$$n.r.(B) \equiv \{< Bx, x > \,|\, x \in \mathbf{C}^m\},$$

where $< \cdot >$ is the inner product in $\mathbf{C}^m$.

Definition 5.3. $\mathcal{F}(\omega, N)$ will be the set of all $m \times m$ matrices of polynomials in n variables $(p_{i,j})_{i,j=1}^m$ such that the degree of $p_{i,j}$ is less than or equal to N, for all i,j and

$$\sup\{Re(z) \,|\, z \in n.r.((p_{i,j}(x))_{i,j=1}^m\} \leq \omega,$$

for all $x \in \mathbf{R}^n$.

Note that, for $m = 1$, $\mathcal{F}(\omega, N)$ is the set of all operators $p(B)$ such that p is a polynomial in n variables of degree less than or equal to N, and $Re(p(x)) \leq \omega, \forall x \in \mathbf{R}^n$.

For $M > 0$, let $\mathcal{F}(\omega, N, M)$ be the set of all matrices $(p_{i,j}) \in \mathcal{F}(\omega, N)$ such that

$$|D^\alpha p_{i,j}(x)| \leq M(1 + \|x\|^2)^{\frac{1}{2}(N-|\alpha|)},$$

for all $x \in \mathbf{R}^n, 1 \leq i,j \leq m, |\alpha| \leq [\frac{n}{2}] + 1$.

Theorem 5.4. *Let* $\mathcal{A}_{\omega,N,M} \equiv \{(p_{i,j}(B)) \mid (p_{i,j}) \in \mathcal{F}(\omega, N, M)\}$. *Suppose* $r > \frac{k}{2}(N-1) + \frac{n}{4}$, *where* $k \equiv [\frac{n}{2}] + 1$.

Then $\forall \mu > \omega, \exists$ *a simultaneous* $(1 + |B|^2)^{-r}$*-existence family for* $\mathcal{A}_{\omega,N,M}$ *that is* $O(e^{\mu|\vec{t}|})$.

Remark 5.5. There exists a bounded, injective C, with dense range, commuting with B_j, for all j, such that the collection of *all* matrices of polynomials of B has a simultaneous C-existence family. This will appear in a future paper. This produces a Frechet space, in which the image of C is continuously embedded, on which all such matrices generate a strongly continuous, locally equicontinuous semigroup.

Remark 5.6. In [7], we show that, when $A \equiv (p_{i,j}(B))_{i,j=1}^m$ is a Petrovsky correct matrix of operators, then $\exists$ a $(1 + |B|^2)^{-r}$-existence family for A, when $2r > k(N-1) + \frac{n}{2} + N(m-1)$. If

$$Re(sp(p_{i,j}(x)) \leq \omega,$$

$\forall x \in \mathbf{R}^n$, then this existence family is $O(e^{\mu t}), \forall \mu > \omega$.

The proof of Theorem 5.4 will use the following lemmas, where $k \equiv [\frac{n}{2}] + 1$.

Lemma 5.7.

$$I(s) \equiv \int_{\mathbf{R}^n} (1 + \|x\|^2)^{-s}\, dx < \infty,$$

whenever $s > \frac{n}{2}$.

Lemma 5.8. *(see [17]) There exists* $M_1 < \infty$ *such that*

$$\int_{\mathbf{R}^n} \|(\mathcal{F}f)(x)\|\, dx \leq M_1 \sum_{|\alpha| \leq k} \|D^\alpha f\|_2,$$

for all $f \in H^k(\mathbf{R}^n)$.

The following lemma is a straightforward calculation, using the product rule and the fact that $\|x^\alpha\| \leq \|x\|^\alpha$.

Lemma 5.9. *There exists* K *such that*

$$\sum_{|\alpha| \leq k} \left\|D^\alpha \left((\prod_\ell e^{(p_{i,j,\ell}(x))_{i,j=1}^m}) g(x) \right)\right\| \leq K \sup_{1 \leq i,j \leq m} \sup_{1 \leq |\alpha| \leq k} \left(1 + \sum_\ell |D^\alpha p_{i,j,\ell}(x)|\right)^k \cdot$$

$$\left(\sup_{|\alpha| \leq k} |D^\alpha g(x)\| \right) \| \prod_\ell e^{(p_{i,j,\ell}(x))} |),$$

for any finite sequence $\{(p_{i,j,\ell})_{i,j=1}^m\}_\ell$, $x \in \mathbf{R}^n, g \in C^k(\mathbf{R}^n)$.

Proof of Theorem 6.9: Without loss of generality, we may assume that $\mu = 0, \omega < 0$.

Fix finite sequences $\{t_\ell\}_\ell \subseteq [0,\infty)$, $\{(p_{i,j,\ell})_{i,j=1}^m\}_\ell \subseteq \mathcal{A}_{\omega,N,M}$. Let $\mathcal{G}(x) \equiv \prod_\ell e^{t_\ell(p_{i,j,\ell}(x))}(1+\|x\|^2)^{-r}$, for $x \in \mathbf{R}^n$.

We will obtain a uniform bound for the norm of $\mathcal{G}(B)$, independently of $\{t_\ell\}_\ell, \{(p_{i,j,\ell})_{i,j=1}^m\}_\ell$.

Letting $g(x) \equiv (1+\|x\|^2)^{-\frac{k}{2}(N-1)}$, in Lemma 5.9, we have, for any $x \in \mathbf{R}^n$,

$$\sum_{|\alpha|\leq k} \|D^\alpha \left(\prod_\ell e^{(t_\ell p_{i,j,\ell}(x))_{i,j=1}^m} g(x)\right)\| \leq K\left(1+\sum_\ell (t_\ell + t_\ell^k)M(1+\|x\|^2)^{\frac{1}{2}(N-1)}\right)^k \cdot$$

$$(1+\|x\|^2)^{-\frac{k}{2}(N-1)} e^{\omega \sum_\ell t_\ell}$$

$$= \sum_{i=0}^k \binom{k}{i} (\sum_\ell (t_\ell + t_\ell^k)^i M^i (1+\|x\|^2)^{\frac{1}{2}(N-1)(i-k)} e^{\omega \sum_\ell t_\ell}$$

$$\leq (1+M^k)\sum_{i=0}^k \binom{k}{i}\left((\sum_\ell t_\ell) + (\sum_\ell t_\ell)^k\right)^i e^{\omega \sum_\ell t_\ell}$$

$$\leq K(\omega) \equiv (1+M^k)\sup_{s\geq 0} \sum_{i=0}^k \binom{k}{i}(s+s^k)^i e^{\omega s}.$$

By Lemmas 5.7 and 5.8, $\|\mathcal{G}(B)\|$ is uniformly bounded.

Straightforward calculations, using dominated convergence and the fact that $f \mapsto f(B)$ is an algebra homomorphism, now imply that

$$\mathcal{G}(B) = \prod_\ell e^{t_\ell(p_{i,j,\ell}(B))}(1+|B|^2)^{-r}.$$

Thus we have the desired existence family. ▮

Example 5.10. When $B_j \equiv i\frac{\partial}{\partial x_j}$, on $L^p(\mathbf{R}^n)(1 \leq p < \infty)$, then $|B|^2 = \Delta$, the Laplacian, and we have

$$[W^{2r,p}(\mathbf{R}^n)]^m \hookrightarrow Z_\mu(\mathcal{A}) \hookrightarrow [L^p(\mathbf{R}^n)]^m,$$

for r, μ as in Theorem 5.4, with $(p_{i,j}(B))|_{Z_\mu(\mathcal{A})}$ generating a strongly continuous semigroup, $\forall (p_{i,j}) \in \mathcal{F}(\omega, N, M)$.

One obtains similar results on any space where translation is strongly continuous and uniformly bounded.

VI. *C*-EVOLUTION SYSTEMS. It is clear how one may define, analogously to the definition of a C-existence family, the following generalization of an evolution system.

Definition 6.1. Suppose C is a bounded operator. We will say that $U(t,s)$ is a *C-evolution system* for $\{A(t)\}_{0\leq t\leq T}$ if $u(t) \equiv U(t,s)y$ is a solution of (1.2) with $x = Cy, \forall y \in X$.

When $CA(t) \subseteq A(t)C, \forall 0 \leq t \leq T$, we may define this algebraically and call it a *C-regularized evolution system* for $\{A(t)\}_{0\leq t\leq T}$(see [17]), as follows.

(1)

$$U(t,s)U(s,r) = U(t,r)C(t \geq s \geq r \geq 0),\ U(t,t) = C(0 \leq t \leq T).$$

(2) The map $t \mapsto U(t,s)x$, from $[s,T]$ into X, is continuous, $\forall x \in X, 0 \leq s \leq T$.

This definition is clear if you think of $U(t,s)$ as $e^{\int_s^t A(r)\,dr}C$.

In this section, we will give a simple illustration of how a C-evolution system may be thought of as a special case of a simultaneous C-existence family. In a future paper, we will give a noncommuting analogue of Theorem 6.3 and apply it to time dependent, space independent *systems* of partial differential initial value problems.

Definition 6.2. The complex number λ is in $\rho_C(A)$, the *C-resolvent set of A*, if $(\lambda - A)$ is injective and $Im(C) \subseteq Im(\lambda - A)$.

We will use Theorem 4.2 to create the following sufficient condition for having a C-regularized evolution system. Note that, when $\omega = 0$, we are producing bounded solutions of (1.2), for $x \in Im(C)$.

Theorem 6.3. *Suppose $\omega \in \mathbf{R}, \{A(t)\}_{0\leq t\leq T}$ is a family of closed operators on X, C is a bounded operator on X and $\mathcal{D} \subseteq \bigcap_{0\leq t\leq T} \mathcal{D}(A(t))$, such that*

(1) *The map $t \mapsto A(t)x$, from $[0,T]$ into X, is continuous, $\forall x \in \mathcal{D}$.*

(2) *$\mathcal{D}$ is dense in X.*

(3) *$(\omega,\infty) \subseteq \rho_C(A(t))$, for $0 \leq t \leq T$ and*

$$\sup\{\|\prod_k(\lambda_k - \omega)(\lambda_k - A(t_k))^{-1}C\| \mid \lambda_k > \omega, 0 \leq t_k \leq T\}$$

is finite.

(4) *All C-resolvents commute with C and each other, that is, $C(\lambda - A(t))^{-1}C = (\lambda - A(t))^{-1}C^2$ and $(\lambda - A(t))^{-1}C(\mu - A(s))^{-1}C = (\mu - A(s))^{-1}C(\lambda - A(t))^{-1}C, \forall 0 \leq s,t \leq T, \lambda,\mu > 0$.*

Then there exists a C-regularized evolution system $\{U(t,s)\}_{s,t\geq 0}$, for $\{A(t)\}_{0\leq t\leq T}$, and a constant M such that $\|e^{-\omega(t-s)}U(t,s)x\| \leq M\|x\|$, for $0 \leq s,t \leq T, x \in X$.

Example 6.4. As in Section V, let $D \equiv i(\frac{\partial}{\partial x_1}, \cdots, \frac{\partial}{\partial x_n})$ and for $0 \leq t \leq T$, let p_t be a polynomial in n variables.

We may apply Theorem 6.3 to time dependent, space independent partial differential initial value problems

$$\frac{d}{dt}u(t,x) = p_t(D)u(t,x),\ (x \in \mathbf{R}^n, 0 \leq t \leq T)$$

on $L^p(\mathbf{R}^n)(1 \leq p < \infty)$, or any space of functions on $\mathbf{R}^n$ where translation is strongly continuous and uniformly bounded, to obtain an $(i+\Delta)^{-r}$-regularized evolution system, for appropriate r, for $\{p_t(D)\}_{0\leq t\leq T}$, under the following hypotheses.

There exist N, ω such that

(1) The map $t \mapsto p_t(x)$, from $[0,T]$ into $\mathbf{C}$, is continuous, $\forall x \in \mathbf{R}^n$.

(2) $deg(p_t) \leq N, \forall 0 \leq t \leq T$.

(3) $Re(p_t(x)) \leq \omega, \forall 0 \leq t \leq T, x \in \mathbf{R}^n$.

See also [17, Theorem 3.4], for a similar result on $L^p(\mathbf{R}^n)(1 < p < \infty)$, where an inhomogeneous term is also included.

Remark 6.5. Note that hypothesis (3) of Theorem 6.3, with $C = I$, is saying that $\{A(t)\}_{0\leq t\leq T}$ is a *stable* family of operators. This is a common hypothesis for dealing with time-dependent evolution equations(see [15, Definition 2.1]).

Proof of Theorem 6.3: We will merely outline the proof. By translating, we may assume $\omega = 0$. Informally, we want to construct $\left(\int_s^t A(r)\,dr\right) C$ as the limit of Riemann sums, $\left(\sum_{k=1}^n \frac{1}{n} A(\frac{k}{n}(t-s)+s)\right) C$, so that $U(t,s)$ will be

$$e^{\int_s^t A(r)\,dr} C \equiv \lim_{n\to\infty} \left(\prod_{k=1}^n e^{\frac{1}{n}A(\frac{k}{n}(t-s)+s)} \right) C.$$

Let $Z \equiv Z_0(\{A(t)\}_{0\leq t\leq T})$.

By Theorem 4.2, $[Im(C)] \hookrightarrow Z$, thus, for any $n \in \mathbf{N}, 0 \leq s,t \leq T$, we may define a bounded operator on X by

$$U_n(t,s) \equiv \left(\prod_{k=1}^n e^{\frac{1}{n}A(\frac{k}{n}(t-s)+s)|Z} \right) C.$$

As in the case when $C = I$(see [9, 13.2]), we may show that, for any $x \in \mathcal{D}, \lim_{n\to\infty} U_n(t,s)x$ exists.

There exists M such that $\|y\|_Z \leq M\|y\|_{[Im(C)]}, \forall y \in Im(C)$. Thus, for any $x \in X$, $n \in \mathbf{N}, 0 \leq t,s \leq T$,

$$\|U_n(t,s)x\| \leq \|U_n(t,s)x\|_Z \leq \|Cx\|_Z \leq M\|Cx\|_{[Im(C)]} \leq M\|x\|.$$

Thus, since $\mathcal{D}$ is dense in X,

$$U(t,s)x \equiv \lim_{n\to\infty} U_n(t,s)x$$

exists, $\forall x \in X$. ∎

REFERENCES

[1] W. Arendt, F. Neubrander, and U. Schlotterbeck, *Interpolation of semigroups and integrated semigroups,* Semigroup Forum 45 (1992), 26-37.

[2] G. Da Prato, *Semigruppi regolarizzabili,* Ricerche Mat. 15 (1966), 223–248.

[3] E.B. Davies, "One-Parameter Semigroups," Academic Press, London, 1980.
[4] R. deLaubenfels, *Existence and uniqueness families for the abstract Cauchy problem,* J. London Math. Soc. 44 (1991), 310-338.
[5] R. deLaubenfels, *C-semigroups and strongly continuous semigroups,* Israel J. Math., to appear.
[6] R. deLaubenfels, *Automatic well-posedness,* J. London Math. Soc., to appear.
[7] R. deLaubenfels, *Matrices of operators and regularized semigroups,* Math. Z., to appear.
[8] R. deLaubenfels and S. Kantorovitz, *Laplace and Laplace-Stieltjes spaces,* J. Func. An., to appear.
[9] J.A. Goldstein, "Semigroups of Operators and Applications," Oxford, New York, 1985.
[10] M. Hieber, A. Holderrieth and F. Neubrander, *Regularized semigroups and systems of linear partial differential equations,* Ann. Scuola Norm. di Pisa, to appear.
[11] S. Kantorovitz "Spectral Theory of Banach Space Operators", Lecture Notes Math., Vol. 1012, Springer, Berlin Heidelberg New York(1983).
[12] S. Kantorovitz *The Hille-Yosida space of an arbitrary operator,* J. Math. An. and Appl. 136 (1988), 107-111.
[13] S. G. Krein, G. I. Laptev and G. A. Cretkova, *On Hadamard correctness of the Cauchy problem for the equation of evolution,* Soviet Math. Dokl. 11 (1970), 763-766.
[14] I. Miyadera and N. Tanaka, *A remark on exponentially bounded C-semigroups,* Proc. Japan Acad. 66 Ser. A (1990), 31-35.
[15] R. Nagel (ed.), "One-Parameter Semigroups of Positive Operators," Lect. Notes Math. 1184, 1986, Springer-Verlag.
[16] A. Pazy, "Semigroups of Linear Operators and Applications to Partial Differential Equations," Springer, New York, 1983.
[17] E. M. Stein, "Singular Integrals and Differentiability Properties of Functions", Princeton University Press, New Jersey, 1970.
[18] N. Tanaka, *Linear evolution equations in Banach spaces,* Proc. London Math. Soc. 63 (1991), 657-672.
[19] J.A. van Casteren, "Generators of strongly continuous semigroups", Research Notes in Math., 115, Pitman, 1985.

A Hilbert–Schmidt Property of Resolvent Differences of Singularly Perturbed Generalized Schrödinger Operators

M. DEMUTH University of Potsdam, Potsdam, Germany

J. A. van CASTEREN University of Antwerp, Antwerp, Belgium

Abstract

Let K_0 be the self-adjoint generator of a Feller semi-group in $L^2(E,m)$, let V be a Kato-Feller potential and let Σ be an appropriate open subset of the locally compact second countable Hausdorff space E. Conditions are given in order that differences of (powers) of resolvents of the form $(aI+K_0\dot{+}V)^{-q}-J^*(aI+(K_0\dot{+}V)_\Sigma)^{-q}J$ are Hilbert-Schmidt operators. Here Jf is the restriction of the function f to Σ and J^*g extends the function with 0 on the complement of Σ. The operator $(K_0\dot{+}V)_\Sigma$ is the generator of the Dirichlet semigroup in $L^2(E,m)$ generated by $K_0\dot{+}V$, but killed on the complement of Σ.

1. INTRODUCTION AND EXAMPLES

The main purpose of this paper is an exhibition of a number of conditions, guaranteeing that certain differences of powers of resolvents of singularly perturbed Schrödinger operators consist of Hilbert-Schmidt operators. We will consider so-called singular perturbations. This kind of properties has some spectral theoretical consequences like stability of the essential spectrum. In what follows we give some examples to which our results are applicable. For this reason we mention the kind of operators that generate (self-adjoint) Feller semi-groups in spaces of the form $C_\infty(E)$ or $L^2(E,m)$, where E is a locally compact second countable Hausdorff space. The formal definition of Feller semi-group will be given in BASSA, section 2.

EXAMPLE 1. Certain operators K_0 of the form

$$K_0 = -\frac{1}{2}\sum_{i,j=1}^{\nu} a_{ij}(x)\frac{\partial^2}{\partial x_i \partial x_j} + \sum_{j=1}^{\nu} b_j(x)\frac{\partial}{\partial x_j} + c(x) \tag{1.1}$$

generate self-adjoint semi-groups in $L^2(\mathsf{R}^\nu)$; for details and conditions to be imposed see Kochubeĭ [32, Theorem 2.].

Under some appropriate assumptions the operator K_0 generator generates a Feller semi-group and hence a Markov process

$$\{(\Omega, \mathcal{F}, \mathsf{P}_x), (X(t) : t \geq 0), (\vartheta_t : t \geq 0), (E, \mathcal{E})\}$$

with state space $E = \mathsf{R}^\nu$. Moreover the one-dimensional distribution $\mathsf{P}_x(X(t) \in A)$ are given by $\mathsf{P}_x(X(t) \in A) = \int_A p_0(t,x,y)dy$. If $c \equiv 0$ and the functions a_{ij} and b_j, $1 \leq i,j \leq \nu$ are uniformly bounded then the corresponding Markov process has infinite lifetime. If the coefficients (a_{ij}) are unbounded, then the corresponding heat kernels cannot be estimated in terms of the classical Gaussian kernel, see [13, Example 2.14]. On the other hand we do have $p_0(t,x,y) \leq Ct^{-\nu/2}$ for all $t > 0$ and for all x and y in R^ν. In [41] Taira considers operators of the form (1.1) on (open) subsets of R^ν, but now with boundary conditions (Neumann, Dirichlet, Wentzell).

In the following two examples we consider relativistic Hamiltonians, which were introduced by Ichinose (see e.g. [20, 21, 22, 23, 24]). For systems without electromagnetic fields we use the notation of Carmona, Masters and Simon [6].

EXAMPLE 2. The present example is taken from Carmona, Masters and Simon [6]. Let μ be non-negative measure on $\mathbf{R}^\nu$ with the property that $\int_{\mathbf{R}^\nu} \min\left(1, |x|^2\right) d\mu(x) < \infty$, let a and b be a vector in $\mathbf{R}^\nu$ and let C be a square $\nu \times \nu$ matrix. In addition let $h : \mathbf{R}^\nu \to \mathbf{R}^\nu$ be a function of compact support with the property that $h(x) = x$ for all x in a neighborhood of the origin. Define the negative-definite function $F : \mathbf{R}^\nu \to \mathbf{R}^\nu$ by

$$F(p) = a + ib.p + p.Cp - \int_{\mathbf{R}^\nu} \left[e^{ip.x} - 1 - ip.h(x)\right] d\mu(x).$$

Then there exists a generator K_0 of a semi-group $\{\exp(-tK_0) : t \geq 0\}$ with the property that $\exp(-tK_0)$ is given by

$$\left[\exp(-tK_0)f\right](x) = \int_{\mathbf{R}^\nu} f(x+y)dm_t(y),$$

where f belongs to $C_\infty(\mathbf{R}^\nu)$. Here the family $\{m_t : t \geq 0\}$ is a vaguely continuous convolution semi-group of probability measures on $\mathbf{R}^\nu$ with the property that the Fourier transform $\widehat{m_t}$ of m_t is given by $\widehat{m_t}(p) = \exp(-tF(p))$, $p \in \mathbf{R}^\nu$. If, in addition the integrals $\int_{\mathbf{R}^\nu} \exp(-tF(p))dp$ are finite, then the function $p_0(t,x,y)$ defined by

$$p_0(t,x,y) = \frac{1}{(2\pi)^\nu} \int_{\mathbf{R}^\nu} \exp\left(-tF(p) + ip.(x-y)\right) dp$$

defines the density of the corresponding Markov process. In case $F(p) \equiv |p|^\alpha$, $0 < \alpha \leq 2$ (stable case), the following inequalities can be proved:

$$\frac{c_1}{t^{\nu/\alpha}|x-y|^{\nu+\alpha}} \leq p_0(t,x,y) \leq \frac{c_2}{t^{\nu/\alpha}|x-y|^{\nu+\alpha}}, \qquad |x-y| \geq 1$$

if $\alpha \neq 2$. In particular, if $F(p) = \sqrt{p^2+m^2} - m$, m fixed, then $K_0 = \sqrt{-\Delta + m^2} - m$. In this case the density is given by

$$\begin{aligned} & p_0(t,x,y) \\ &= \frac{1}{(2\pi)^\nu} \frac{t}{\sqrt{|x-y|^2 + t^2}} \int_{\mathbf{R}^\nu} \exp\left(mt - \sqrt{\left(|x-y|^2 + t^2\right)(p^2+m^2)}\right) dp. \\ &= \int_0^\infty \frac{\exp(mt - m^2 u)}{(4\pi u)^{\nu/2}} \exp\left(-\frac{|x-y|^2}{4u}\right) \frac{t}{u\sqrt{u\pi}} \exp\left(-\frac{t^2}{4u}\right) du. \end{aligned}$$

The previous results can be generalized for negative definite functions F defined on a locally compact, second countable, abelian group G. In that case the variable p

varies over the dual group. It is also noticed that these results fit in the theory of Lévy processes.

EXAMPLE 3. The following example is related to the previous one and can be found in Ichinose [19]. Under suitable conditions on the vector field A, the following pseudodifferential operator:

$$\begin{aligned} K_0 &:= [H_A^m f](x) - mf(x) \\ &= \frac{1}{(2\pi)^\nu} \int\int_{\mathbf{R}^\nu \times \mathbf{R}^\nu} \exp(i(x-y).p)\, h_A^m\left(p, \frac{x+y}{2}\right) f(y)dydp, \end{aligned}$$

where h_A^m is the function

$$h_A^m(p,x) = \sqrt{(p - A(x))^2 + m^2}, \quad p \in \mathbf{R}^\nu, x \in \mathbf{R}^\nu,$$

is than the self-adjoint generator of a strongly continuous semi-group in $L^2(\mathbf{R}^\nu)$. Let the density function $p_0(t,x,y)$ be as Example 2. Then the integral kernel of the operator $\exp(-t(H_A^m - mI))$ is given by an imaginary path integral of the form

$$\exp(-t(H_A^m - mI))(x,y) = \lim_{s\uparrow t} \mathsf{E}_x \left(\exp(-iS(s))\, p_0(t-s, X(s), y)\right), \tag{1.2}$$

where $\{S(s) : s \geq 0\}$ is some additive process in terms of the Markov process generated by $H_0^m - mI$. If $A(x) \equiv 0$, then the density in (1.2) coincides with $p_0(t,x,y)$. For more details see Ichinose and Tamura [23].

EXAMPLE 4. This example is due to N. Jacob [26, 28]. A closely related example can be found in [18]. Let $p : \mathbf{R}^\nu \times \mathbf{R}^\nu \to \mathbf{R}$ be a continuous function such that for fixed $\xi \in \mathbf{R}^\nu$ the function $x \mapsto p(x,\xi)$ is a bounded C^∞-function with bounded derivatives of all orders. Suppose that for fixed $x \in \mathbf{R}^\nu$, the function $\xi \mapsto p(x,\xi)$ is negative definite. In addition we assume that there exists a continuous negative function $a : \mathbf{R}^\nu \mapsto \mathbf{R}$ such that, for some $0 < r \leq 2$,

$$c_0 c_1 \left(1 + |\xi|^2\right)^{r/2} \leq c_1 \left(1 + a(\xi)^2\right)^{1/2} \leq p(x,\xi) \leq c_2 \left(1 + a(\xi)^2\right)^{1/2}.$$

Define the Sobolev spaces $H^{q,a}(\mathbf{R}^\nu)$, $q \geq 0$, as follows:

$$H^{q,a}(\mathbf{R}^\nu) = \left\{ u \in L^2(\mathbf{R}^\nu) : \|u\|_{q,a} < \infty \right\},$$

where $\|\cdot\|_{q,a}$ is given by:

$$\|u\|_{q,a}^2 = \int_{\mathbf{R}^\nu} \left(1 + a(\xi)^2\right)^q |\widehat{u}|^2 \, d\xi.$$

If $a(\xi) \equiv |\xi|$, then we just write $H^q(\mathbf{R}^\nu)$ instead of $H^{q,a}(\mathbf{R}^\nu)$ and $\|\cdot\|_q$ replaces $\|\cdot\|_{q,a}$. Put $H^\infty(\mathbf{R}^\nu) := \bigcap_{q\geq 0} H^q(\mathbf{R}^\nu)$ and let the pseudodifferential operator $p(x,D)$ be defined by

$$[p(x,D)u](x) = \frac{1}{(2\pi)^{n/2}} \int_{\mathbf{R}^\nu} e^{ix.\xi} p(x,\xi)\widehat{u}(\xi)d\xi.$$

Again under some appropriate assumptions the operator $-p(x,D)$ generates a Feller semi-group in $C_\infty(\mathbf{R}^\nu)$.

EXAMPLE 5. It is perhaps interesting to recall Theorem 10.3. in Ikeda and Watanabe [25], stating that under appropriate conditions the P_x-distribution of the solution $(X(t) : t \geq 0)$ of the stochastic differential equation $dX(t) = \sigma(X(t))dB(t) + b(X(t))dt$, $X(0) = x$, defines a Markov process

$$\{(\Omega, \mathcal{F}, \mathsf{P}_x), (X(t) : t \geq 0), (\vartheta_t : t \geq 0), (E, \mathcal{E})\}$$

with the property that, for every compact subset K, there exists a constant C_K such that, for appropriate n, $\|1_K P(t)\|_{1,\infty} \leq C_K t^{-n/2}$, $t \geq 0$. Here $P(t)f(x) = \mathsf{E}_x(f(X(t)),\ f \in C_\infty(\mathbf{R}^\nu)$. Suppose that all the coefficients σ_{ij} and b_j are bounded and have bounded derivatives of all orders. Also suppose that Hörmander's condition is satisfied. Then the family $\{P(t) : t \geq 0\}$ is a Feller semi-group with C^∞-density $p_0(t,x,y)$. For more details we refer the reader to Ikeda and Watanabe [25, Chapter V].

EXAMPLE 6. The previous example has its counterpart for Riemannian manifolds. In fact instead of the Laplace operator on $\mathbf{R}^\nu$ we can also consider the Laplace-Beltrami operator on a Riemannian manifold. For details we refer the reader to Elworthy [15], [14], Azencott et al [2], Bismut [4] and to several others e.g. [40]. The authors also establish existence results for and bounds on the corresponding heat kernels. Recent and interesting paper [10] and [8] written by Davies. It provides the reader with much insight into the behavior of heat kernels. Of course his book [9] should be consulted also.

EXAMPLE 7. In this example we consider so-called hyper-singular integrals. Define the operators Δ_h^ℓ on $C(\mathbf{R}^\nu)$ as follows:

$$\left[\Delta_h^\ell f\right](x) = \sum_{k=0}^{\ell} (-1)^k \binom{\ell}{k} f(x - kh).$$

In [33] Kochubeĭ proves that operators K_0 of the form

$$[K_0 f](x) = -\beta \sum_{i,j=1}^{\nu} a_{ij}(x) \frac{\partial^2 f}{\partial x_i \partial x_j}(x) + \frac{1-\beta}{d_{n,\ell}(\gamma)} \int_{\mathbf{R}^\nu} \Omega\left(x, \frac{h}{|h|}\right) \frac{\left[\Delta_h^\ell f\right](x)}{|h|^{\nu+\gamma}} dh$$
$$+ \sum_{k=1}^{m} \frac{1}{d_{n,\ell}(\gamma_k)} \int_{\mathbf{R}^\nu} \Omega_k\left(x, \frac{h}{|h|}\right) \frac{\left[\Delta_h^\ell f\right](x)}{|h|^{\nu+\gamma_k}} dh + \sum_{j=1}^{\nu} b_j(x) \frac{\partial f}{\partial x_j}$$

generates a Feller semi-group in $C_\infty(\mathbf{R}^\nu)$ provided that the following conditions are satisfied:

(a) The functions Ω and Ω_k, $1 \le k \le m$, are non-negative and continuous on $\mathbf{R}^\nu \times S^{\nu-1}$. They are also even: $\Omega(x,\sigma) = \Omega(x,-\sigma)$, $x \in \mathbf{R}^\nu$, $\sigma \in S^{\nu-1}$.

(b) The orders of homogeneity γ, γ_k, $1 \le k \le m$, verify: $0 < \gamma_k < \gamma \le 2$. If $\gamma = 2$, then $\beta = 1$ and if $\gamma < 2$, then $\beta = 0$. If $\gamma = 1$, then $b_j \equiv 0$, $1 \le j \le \nu$.

(c) Some ellipticity conditions on (a_{ij}) are also required. In fact, the inequality $\operatorname{Re}\sum_{i,j=1}^{\nu} a_{ij}(x)\xi_i\overline{\xi}_j \ge a_0 |\xi|^2$, for all $\xi_j \in \mathbf{C}$, $1 \le j \le \nu$. Here a_0 is some strictly positive real number.

(d) The constants $d_{n,\ell}(\gamma)$ have to be chosen suitably. In fact they are chosen in such a way that the expression

$$\widetilde{\Omega}(x,\xi) = \frac{1}{d_{\nu,\ell}(\gamma)} \int_{\mathbf{R}^\nu} \frac{(1-\exp(-i\xi.h))^\ell}{|h|^{\nu+\gamma}} \Omega\left(x, \frac{h}{|h|}\right) dh,$$

called the *symbol* of the hyper-singular integral $D_\Omega^\gamma f$, does not depend on the particular choice of ℓ, where $\ell > \alpha$.

(e) The characteristics Ω and Ω_k, $1 \le k \le m$, are supposed to be non-negative and symmetric in the second variable, i.e. $\Omega(x,\sigma) = \Omega(x,-\sigma)$ for all $x \in \mathbf{R}^\nu$ and for all $\sigma \in S^{\nu-1}$.

Moreover the life time of the corresponding Markov process is ∞. In [33] Kochubeĭ proves that the corresponding Markov processes posses transition densities $p_0(t,x,y)$ verifying inequalities of the form

$$p_0(t,x,y) \le C\left\{ \frac{t}{\left[t^{1/\gamma} + |x-y|\right]^{\nu+\gamma}} + \sum_{k=1}^{m} \frac{t}{\left[t^{1/\gamma} + |x-y|\right]^{\nu+\gamma_k}} \right\}.$$

EXAMPLE 8. In this example we consider the generator $K_0 := -\frac{1}{2}\Delta + x.\nabla$ of the so-called Ornstein-Uhlenbeck process in $L^2\left(\mathbf{R}^\nu, \exp\left(-|y|^2\right)\pi^{-\nu/2}dy\right)$. Its integral kernel $p_0(t,x,y)$ is given by

$$p_0(t,x,y) = \frac{1}{(1-\exp(-2t))^{\nu/2}} \exp\left(-\frac{\exp(-2t)|x|^2 + \exp(-2t)|y|^2 - 2\exp(-t)\langle x,y\rangle}{1-\exp(-2t)} \right).$$

The semi-group in $L^2\left(\mathbf{R}^\nu, \exp\left(-|y|^2\right)\pi^{-\nu/2}dy\right)$ is given by

$$\begin{aligned}
\left[\exp(-tK_0)f\right](x) &= \int p_0(t,x,y) f(y) \exp\left(-|y|^2\right) \frac{dy}{\pi^{\nu/2}} \\
&= \int f\left(\exp(-t)x + \sqrt{1-\exp(-2t)}\,y\right) \exp\left(-|y|^2\right) \frac{dy}{\pi^{\nu/2}}.
\end{aligned}$$

For more details the reader is referred to e.g. Simon [39].

For more details the reader is referred to e.g. Simon [39].

EXAMPLE 9. In this example we consider the generator $K_0 := -\frac{1}{2}\Delta + \frac{1}{2}|x|^2$ of the oscillator process. The integral kernel of the corresponding semi-group $\exp(-tK_0)(x,y)$ may be written as (again see Simon [39])

$$\exp(-tK_0)(x,y) = \exp\left(-\frac{1}{2}|x|^2\right)\frac{1}{(2\pi\sinh t)^{\nu/2}}\exp\left(-\frac{|\exp(-t)x-y|^2}{1-\exp(-2t)}\right)\exp\left(\frac{1}{2}|y|^2\right)$$

$$= \exp\left(-\frac{1}{2}|x|^2\tanh t\right)\frac{1}{(2\pi\sinh t)^{\nu/2}}\exp\left(-\frac{1}{2\sinh t}\left|\frac{x}{(\cosh t)^{1/2}} - y(\cosh t)^{1/2}\right|^2\right).$$

It follows that the corresponding semi-group $\{\exp(-tK_0) : t \geq 0\}$ is given by

$$[\exp(-tK_0)f](x)$$
$$= \frac{\exp\left(-\frac{1}{2}|x|^2\tanh t\right)}{(2\pi\cosh t)^{\nu/2}}\int \exp\left(-\frac{1}{2}|y|^2\right) f\left(\frac{x}{\cosh t} + \sqrt{\tanh t}\,y\right)dy.$$

2. STOCHASTIC SPECTRAL ANALYSIS (BASSA)

There are different ways of introducing semi-groups with perturbed generators. The analytic way starts with the unperturbed semi-group and uses the Trotter-product formula to find a Feynman-Kac representation of the perturbed semi-group. The semi-analytic or semi-stochastic manner begins again with the unperturbed semi-group. Then the potentials are introduced stochastically by verifying the sensibility and the semi-group property of the Feynman-Kac formula.

In order to introduce semi-groups with perturbed generators we employ a purely stochastic approach in the sense that we begin with the process, or what is equivalent, with the transition density function. Our aim is to formulate all assumptions on the process or its generator in terms of assumptions on the density. An advantage is that we can consider a large class of generators, containing the examples in the introduction.

The objective of this paper is to present some Hilbert-Schmidt properties of differences of semi-groups generated by these operators. We start with the basic assumptions on the transition density function, which form the foundations of this theory. This theory will be called "Stochastic Spectral Analysis". The state space (or configuration space) will be a second countable locally compact Hausdorff space E with Borel field $\mathcal{E}$. A non-negative Radon measure m (reference measure) on $\mathcal{E}$ is given. Instead of $dm(x)$ or $m(dx)$ we usually write dx.

Basic Assumptions of Stochastic Spectral Analysis (BASSA).
In what follows the function $p_0(t,x,y)$ defined on $(0,\infty)\times E\times E$ will be a continuous density function with the following properties:

A1. It is non-negative and it verifies the Chapman-Kolmogorov identity, i.e.

$\int p_0(s,x,z)p_0(t,z,y)dz = p_0(s+t,x,y)$, s, $t>0$, x, $y\in E$, and its total mass is less than or equal to 1, i.e. $\int p_0(t,x,y)dy \le 1$, $t>0$, $x\in E$;

A2. (Feller property) For every $f\in C_\infty(E)$ the function $x\mapsto \int f(y)p_0(t,x,y)dm(y)$ belongs to $C_\infty(E)$;

A3. (continuity) For every $f\in C_\infty(E)$ and for every $x\in E$ the following identity is true: $\lim_{t\downarrow 0}\int f(y)p_0(t,x,y)dm(y)=f(x)$;

A4. The function $p_0(t,x,y)$ is symmetric: $p_0(t,x,y)=p_0(t,y,x)$ for all $t>0$ and for all x and y in E.

Sometimes we shall need a *boundedness assumption* of the following form:

B. There exists finite constants m, b and c such that $0\le p_0(t,x,y)\le ct^{-m}\exp(bt)$ for all $t>0$ and for all x, $y\in E$.

Remark 1. It is well-known that there exists a strong Markov process

$$\{(\Omega,\mathcal{F},\mathsf{P}_x),(X(t):t\ge 0),(\vartheta_t:t\ge 0),(E,\mathcal{E})\}$$

(see e.g. Blumenthal and Getoor [5]) with the following properties. The one-dimensional distributions are given by $\mathsf{P}_x(X(t)\in B)=\int_B p_0(t,x,y)dy$, $t>0$, B Borel subset of E. Its sample paths are P_x-almost surely right continuous and possess P_x-almost sure left limits in E on its life time. In other words the process $\{X(t),\mathsf{P}_x\}$ is cadlag on its life time. Moreover we may assume that the closure of the (random) set $\{X(s):0\le s<t\}$ is a compact subset of E, whenever $X(t-)$ belongs to E. In other the process does not re-enter E once it has hit δ, the point at infinity.

Remark 2. It is not necessarily true that densities are available. In principle one may formulate the basic assumptions (BASSA) in terms of the transition function $P(t,x,B):=\mathsf{P}_x(X(t)\in B)$, $t\ge 0$, $x\in E$, $B\in\mathcal{E}$, where $\mathcal{E}$ is the collection of Borel subsets of E.

This is perhaps the right place to fix some notation and insert an interesting inequality. Let K_0 be the L^2-generator of the Markov process above and let a be a strictly positive real number. For any Borel function g, defined on E, we write $[\exp(-sK_0)g](x)=\mathsf{E}_x(g(X(s)))=\int p_0(s,x,y)g(y)dy$ and $\left[(aI+K_0)^{-1}g\right](x)=\int_0^\infty e^{-as}[\exp(-sK_0)g](x)ds=\int_0^\infty e^{-as}p_0(s,x,y)g(y)dy$ whenever these expressions make sense.

Before in the next section we actually give some estimates on the norms of differences of semi-groups and resolvents, we insert a convenient inequality for the unperturbed resolvent. This inequality will among others be used in Theorem 2.4. Its proof will be omitted, but we refer to van Casteren [42, Theorem 6.4. p. 116-117] for a proof of a similar statement.

2.1. PROPOSITION. Let $g:E\to\mathsf{R}$ be a Borel measurable function and let a and η be strictly positive real numbers. The following inequalities are valid:

$$\left(1-e^{-a\eta}\right)\left\|(aI+K_0)^{-1}|g|\right\|_\infty \le \sup_{x\in E}\int_0^\eta \mathsf{E}_x(|g(X(s))|)ds \le e^{a\eta}\left\|(aI+K_0)^{-1}|g|\right\|_\infty.$$

For a concise formulation of our results we introduce the following definitions.

2.2. DEFINITION. Let $V : E \to [0, \infty]$ be a Borel measurable function on E.
(a) The function V is said to belong to $K(E)$ if

$$\limsup_{t\downarrow 0} \left\| \int_0^t P_0(s)V ds \right\|_{\infty,\infty} = \limsup_{t\downarrow 0} \sup_{x\in E} \int_0^t \left(\int p_0(s,x,y)V(y)dm(y) \right) ds = 0.$$

(b) The Borel measurable function $V : E \to [0, \infty]$ belongs to $K_{\text{loc}}(E) = K_{\text{loc}}(E, A_0)$ if $1_K V$ belongs to $K(E)$ for all compact subsets K of E.
(c) The Borel measurable function $V = V_+ - V_-$ is said to be a Kato-Feller potential if its positive part $V_+ = \max(V, 0)$ belongs to $K_{\text{loc}}(E)$ and if its negative part $V_- = \max(-V, 0)$ belongs to $K(E)$.

If a non-negative function W is a member of $K(E)$, then W is said to belong to *Kato's class* and if W is a member of $K_{\text{loc}}(E)$, then W is said to belong to *Kato's class locally*. The following general result can be proved. For details in the symmetric case see [42], [45], [44] and [43]. For the Gaussian semi-group the reader may consult Simon [38], [39] and Aizenman and Simon [1].

2.3. THEOREM. Suppose that $V = V_+ - V_-$ is a Borel measurable function defined on E such that V_- belongs to $K(E)$ and such that V_+ belongs to $K_{\text{loc}}(E)$.

(a) There exists a closed, densely defined linear operator $K_0 \dot{+} V$ in $C_\infty(E)$, extending $K_0 + V$, which generates a strongly continuous positivity preserving semi-group $\{\exp(t(K_0 \dot{+} V)) : t \geq 0\}$ in $C_\infty(E)$. Every operator $\exp(-t(K_0 \dot{+} V))$, $t > 0$, is of the form

$$\left[\exp(-t(K_0 \dot{+} V))f\right](x) = \int \exp(-t(K_0 + V))(x, y)f(y)dm(y), \quad f \in C_\infty(E),$$

where $\exp(-t(K_0 + V))(x, y)$ is a continuous function which verifies, for $t > 0$ and x, $y \in E$, the identity of Chapman-Kolmogorov:

$$\exp(-t(K_0 + V))(x, y) = \int \exp(-s(K_0 + V))(x, z) \exp(-t(K_0 + V))(z, y)dz.$$

(b) The semi-group $\{\exp(-t(K_0 \dot{+} V)) : t \geq 0\}$ also acts as a strongly continuous semi-group in $L^p(E, m)$, $1 \leq p < \infty$.

(c) If $\exp(-tK_0)$ maps $L^1(E, m)$ into $L^\infty(E, m)$ for all $t > 0$ (i.e. if $\sup\{p_0(t, x, y) : x, y \in E\} < \infty$ for all $t > 0$), then $\exp(-t(K_0 \dot{+} V))$, $t > 0$, maps $L^p(E, m)$ into $L^q(E, m)$, for $1 \leq p \leq q \leq \infty$. If $t > 0$ and if $1 \leq p \leq q < \infty$, then $\exp(-t(K_0 + V))$ maps $L^p(E, m)$ into $L^q(E, m) \cap C_\infty(E)$.

(d) In $L^2(E, m)$ the family $\{\exp(-t(K_0 \dot{+} V)) : t \geq 0\}$ is a self-adjoint positivity preserving strongly continuous semi-group with a self-adjoint generator.

(e) The Feynman-Kac semi-group in $L^2(E,m)$ coincides with the semi-group corresponding to the quadratic form Q with $D(Q) = D\left(K_0^{1/2}\right) \cap D\left(V_+^{1/2}\right)$ and defined by

$$Q(f,g) = \left\langle K_0^{1/2} f, K_0^{1/2} g\right\rangle - \left\langle V_-^{1/2} f, V_-^{1/2} g\right\rangle + \left\langle V_+^{1/2} f, V_+^{1/2} g\right\rangle,$$

where f and g belong to $D(Q)$.

Remark 1. From the general assumptions it follows that, for $t > 0$, the operator $\exp(-tK_0)$ maps $L^1(E,m)$ in $C_\infty(E)$. As indicated in (c), then we may prove that, always for $t > 0$, the operator $\exp(-t(K_0\dot{+}V))$ maps $L^p(E,m)$ in $L^q(E,m) \bigcap C_\infty(E)$, provided that $1 \leq p \leq q \leq \infty$, $p \neq \infty$. This is explained in [44], in [42] and in [45]. In fact the integral kernel $\exp\left(-t(K_0+V)\right)(x,y)$ is given by

$$\exp\left(-t(K_0+V)\right)(x,y) = \lim_{\tau\uparrow t} \mathsf{E}_x\left(\exp\left(-\int_0^\tau V(X(s))ds\right) p_0(t-\tau, X(\tau), y)\right).$$

Remark 2. Let K be a self-adjoint generator in a Hilbert space with a lower bound. Let ω_0 be the smallest number ω with the property that $\langle Kf, f\rangle \geq -\omega\langle f,f\rangle$ for all $f \in D(K)$. Then ω_0 is called the *type* of the semi-group $\{\exp(-tK) : t \geq 0\}$ generated by K. In fact it follows that $\|\exp(-tK)\| \leq \exp(\omega_0 t)$, $t \geq 0$.

Next we want to discuss the way in which the generator of the Feynman-Kac semi-group $\left\{\exp(-t(K_0\dot{+}V)) : t \geq 0\right\}$ is related to the Friedrichs' extension of K_0+V. We also are interested in "core"-type problems. Theorem 2.4. is closely related to the well-known KLMN-theorem: see Reed and Simon [35, Theorem X.17, p. 167]. The fact that $D\left(K_0^{1/2}\right) \cap D\left(V_+^{1/2}\right)$ is automatically dense implies that the Trotter-Lie product is available; see Kato [30, Theorem 1, p. 694].

2.4. THEOREM. Suppose that for every function $f \in D\left(K_0^{1/2}\right) \cap D\left(V_+^{1/2}\right)$ there exists a sequence $\{f_n : n \in \mathbb{N}\}$ in $D(K_0) \cap D(V)$ with the following properties:
(a) $\lim_{n\to\infty} \|f_n - f\|_2 = 0$;
(b) $\lim_{m,n\to\infty} \langle K_0(f_n - f_m), f_n - f_m\rangle = 0$;
(c) $\lim_{m,n\to\infty} \langle V_+(f_n - f_m), f_n - f_m\rangle = 0$.
Then the Feynman-Kac generator $K_0\dot{+}V$ is the Friedrichs' extension of $K_0 + V$.

It is noticed that the hypotheses in Theorem 2.4. can be rephrased as "the subspace $\mathrm{dom}(K_0) \cap \mathrm{dom}(V)$ is a core for the operator $K_0^{1/2} + V_+^{1/2}$" or, equivalently, "the subspace $\mathrm{dom}(K_0) \cap \mathrm{dom}(V)$ is a form core for $K_0 + V_+$".

PROOF. Fix a number a that is strictly larger then the type of the Feynman-Kac semi-group $\left\{\exp\left(-t(K_0\dot{+}V)\right) : t \geq 0\right\}$. The quadratic form $Q^{a,V}$ associated to the Feynman-Kac semi-group is given by

$$\begin{aligned} Q^{a,V}(f,g) &:= \left\langle \left(aI + K_0\dot{+}V\right)^{1/2} f, \left(aI + K_0\dot{+}V\right)^{1/2} g\right\rangle \\ &= \left\langle \left(aI + K_0\right)^{1/2} f, \left(aI + K_0\right)^{1/2} g\right\rangle + \left\langle V_+^{1/2} f, V_+^{1/2} g\right\rangle \\ &\quad - \left\langle V_-^{1/2}\left(aI + K_0\right)^{-1/2}\left(aI + K_0\right)^{1/2} f, V_-^{1/2}\left(aI + K_0\right)^{-1/2}\left(aI + K_0\right)^{1/2} g\right\rangle, \end{aligned}$$

where f and g belong to the domain $D(Q^{a,V}) = D\left(K_0^{1/2}\right) \cap D\left(V_+^{1/2}\right)$. Moreover, for a large enough, we have by Proposition 2.1. together with the definition of Kato-Feller potential,

$$\left\| V_-^{1/2}(aI+K_0)^{-1/2} h \right\|_2 \leq \left\| V_-^{1/2}(aI+K_0)^{-1/2} \right\|_{2,2} \|h\|_2$$

$$= \left\| V_-^{1/2}(aI+K_0)^{-1} V_-^{1/2} \right\|_{2,2}^{1/2} \|h\|_2 \leq \left\| (aI+K_0)^{-1} V_- \right\|_{\infty}^{1/2} \|h\|_2 \leq \sqrt{\varepsilon(a)}\, \|h\|_2,$$

where h belongs to $L^2(E,m)$. Since the negative part V_- of V is supposed to belong to Kato's class (because V is supposed to be a Kato-Feller potential), from Proposition 2.1. it follows that $\varepsilon(a) < 1$ for $a > 0$ large enough. From (a) it follows that the domain of $S := K_0 + V$ is dense in $L^2(E,m)$. This is so because the domain $D(Q^{a,V})$ is dense in $L^2(E,m)$. Since, in addition, the Feynman-Kac generator $K_0 \dot{+} V$ extends $K_0 + V$, we see that the operator $K_0 + V$ is closable. Let $\overline{S}$ denote this closure. We also write $\widetilde{S}$ for the Feynman-Kac generator $K_0 \dot{+} V$. Furthermore we define the operators T_1 and T_2 as follows:

$$T_1 := S^* \big|_{D\left(Q_{\overline{S}}^{a,V}\right) \cap D(S^*)}, \quad T_2 := S^* \big|_{D\left(Q_{\widetilde{S}}^{a,V}\right) \cap D(S^*)}.$$

Here $Q_{\overline{S}}^{a,V}$ is the quadratic form associated to $\overline{S}$ and $Q_{\widetilde{S}}^{a,V}$ is the quadratic form associated to $\widetilde{S}$, the so-called Feynman-Kac or Schrödinger form. Then, from Theorem 5.38 in Weidmann [46, p. 123], it follows that T_1 is the Friedrichs' extension of S. Since K_0 and V are both self-adjoint, the operator S is symmetric and so $S \subset S^*$. We also have $S \subset \widetilde{S}$ and hence $S^* \supset \widetilde{S}^* = \widetilde{S} \supset \overline{S} \supset S$. From the definition T_1 it is clear that $T_1 \subset S^*$ and hence $T_1 = T_1^* \supset \overline{S}$. We also readily see $T_2 \supset \widetilde{S}$ and thus $T_2^* \subset \widetilde{S}$. Since $D\left(Q_{\overline{S}}^{a,V}\right) \subseteq D\left(Q_{\widetilde{S}}^{a,V}\right)$, we also have $T_1 \subset T_2 \subset S^*$. A combination of these inclusions yields:

$$S \subseteq \overline{S} \subseteq T_2^* \subseteq \widetilde{S} \subseteq T_2 \subseteq S^* \quad \text{and} \quad S \subseteq \overline{S} \subseteq T_2^* \subseteq T_1 \subseteq T_2 \subseteq S^*.$$

From (a), (b) and (c) it follows that $D(S)$ forms a core for $Q_{\widetilde{S}}^{a,V}$. For let $f_0 \in D\left(Q_{\widetilde{S}}^{a,V}\right)$ be such that $\left\langle \left(aI+\widetilde{S}\right)^{1/2} f, \left(aI+\widetilde{S}\right)^{1/2} f_0 \right\rangle = 0$ for all $f \in D(S)$. Then $\langle (aI+S)f, f_0 \rangle = 0$ for all $f \in D(S)$. By properties (a), (b) and (c), there exists a sequence $(f_n)_{n\in\mathbb{N}}$ in $D(S)$ such that

$$\lim_{n\to\infty} \left\langle \left(aI+\widetilde{S}\right)^{1/2}(f_0 - f_n), \left(aI+\widetilde{S}\right)^{1/2}(f_0 - f_n) \right\rangle = 0.$$

Consequently

$$\left\| \left(aI+\widetilde{S}\right)^{1/2} f_0 \right\|_2^2 = \lim_{n\to\infty} \left\langle \left(aI+\widetilde{S}\right)^{1/2} f_n, \left(aI+\widetilde{S}\right)^{1/2} f_0 \right\rangle$$

$$= \lim_{n\to\infty} \left\langle \left(aI+\widetilde{S}\right) f_n, f_0 \right\rangle = \lim_{n\to\infty} \langle (aI+S) f_n, f_0 \rangle = 0.$$

Hence $f_0 = 0$. Next we are going to show that

$$D(S^*) \cap D\left(Q^{a,V}_{\widetilde{S}}\right) = D\left(\widetilde{S}\right) \cap D\left(Q^{a,V}_{\widetilde{S}}\right) = D\left(\widetilde{S}\right).$$

This is true, because let $f \in L^2(E,m)$ be such that the functional $g \mapsto \left\langle (aI+S)\,g, \left(aI+\widetilde{S}\right)^{-1/2} f\right\rangle$, $g \in D(S)$, is continuous. Since $D(S)$ is a core for $Q^{a,V}_{\widetilde{S}}$, it follows that the functional $g \mapsto \left\langle \left(aI+\widetilde{S}\right) g, \left(aI+\widetilde{S}\right)^{-1/2} f\right\rangle$, $g \in D\left(\widetilde{S}\right)$, is continuous as well. We infer that $R\left(\left(aI+\widetilde{S}\right)^{-1/2}\right) \cap D(S^*) = R\left(\left(aI+\widetilde{S}\right)^{-1/2}\right) \cap D\left(\widetilde{S}^*\right)$, or putting it differently

$$D\left(Q^{a,V}_{\widetilde{S}}\right) \cap D(S^*) = D\left(Q^{a,V}_{\widetilde{S}}\right) \cap D\left(\widetilde{S}\right) = D\left(\widetilde{S}\right)$$

and hence

$$T_2 = S^* \mid_{D\left(Q^{a,V}_{\widetilde{S}}\right) \cap D(S^*)} = S^* \mid_{D\left(\widetilde{S}\right)} = \widetilde{S}.$$

It follows that $T_2 = \widetilde{S}$ and hence $T_2 = T_2^* = T_1$. This shows that $\widetilde{S}$ is the Friedrichs' extension of S.

The result in Theorem 2.4. has its local counterpart. In fact, let Γ be a Borel subset of the second countable locally compact Hausdorff space E. In relation to the set Γ we shall be employing the following stopping times:

$$S = \inf\left\{s > 0 : \int_0^s 1_\Gamma(X(\sigma))d\sigma > 0\right\}, \quad T = \inf\{s > 0 : X(s) \in \Gamma\}.$$

It readily follows that $S \geq T$, P_x-almost surely, for all $x \in E$. The following proposition gives a sufficient condition on Γ, in order that, for all $x \in E$, $S = T$, P_x-almost surely. A point $x \in E$ belongs to Γ^r if $\mathsf{P}_x(T = 0) = 1$. Some authors call the time S the *penetration time*: see e.g. Herbst and Zhongxin Zhao [17]. Suppose $\Gamma^r = (\operatorname{int}(\Gamma))^r$. Then $S = T$ P_x-almost surely for all $x \in E$.

2.5. DEFINITION. Let S be the penetration time of Γ. The integral kernel $\exp(-\lambda K_\Sigma)(x,y)$ is given by

$$\exp(-tK_\Sigma)(x,y) = \lim_{t'\uparrow t} \mathsf{E}_x\left(\exp\left(-\int_0^{t'} V(X(\sigma))d\sigma\right) p_0(t-t', X(t'), y) : S > t'\right).$$

In the results below we let $\Sigma = E^\triangle \setminus \Gamma$ be an open subset of E and $\left(K_0 \dot{+} V\right)_\Sigma$ denotes the Feynman-Kac generator of the semi-group killed in the complement of

Σ, i.e. the semi-group $\{\exp(-t(K_0\dot{+}V)_\Sigma) : t \geq 0\}$, defined by

$$[\exp(-t(K_0\dot{+}V)_\Sigma)f](x) = \mathsf{E}_x\left(\exp\left(-\int_0^t V(X(s))ds\right) f(X(t)) : S > t\right)$$

$$= \int \exp(-t(K_0 + V)_\Sigma)(x, y) f(y)dy.$$

If $\Gamma = (\mathrm{int}(\Gamma))^r$, then the penetration time S may be replaced with the exit time T. The complement of Σ is supposed to be regular, which yields the fact that the Feynman-Kac semi-group, killed on the complement of Σ, leaves $C_\infty(\Sigma)$ invariant. Proofs are not given; they follow the same lines as the ones given above.

2.6. THEOREM. Suppose that for every function $f \in D\left((K_0)_\Sigma^{1/2}\right) \cap D\left(\left(V_+^{1/2}\right)_\Sigma\right)$ there exists a sequence $\{f_n : n \in \mathbb{N}\}$ in $D((K_0)_\Sigma) \cap D((V)_\Sigma)$ with the following properties:
(a) $\lim_{n\to\infty} \|f_n - f\|_2 = 0$;
(b) $\lim_{m,n\to\infty} \langle (K_0)_\Sigma(f_n - f_m), f_n - f_m\rangle = 0$;
(c) $\lim_{m,n\to\infty} \langle V_+(f_n - f_m), f_n - f_m\rangle = 0$.
Then the Feynman-Kac generator $(K_0\dot{+}V)_\Sigma$ is the Friedrichs' extension of $(K_0)_\Sigma + (V)_\Sigma$.

2.7. COROLLARY. If for all sufficiently large a, $a > 0$, the range of the operator $aI + (K_0)_\Sigma + (V)_\Sigma$ is dense in $L^2(\Sigma, m)$, then the operator $(K_0)_\Sigma + (V)_\Sigma$ is essentially self-adjoint and its closure generates the Feynman-Kac semigroup, killed on the complement of Σ.

Next we want to compare the operators $(K_0\dot{+}V)_\Sigma$ and $K_0\dot{+}V$. The operator H_Σ^{a+V} is defined as follows. Its domain is $D(K_0\dot{+}V)$ and its action is given by

$$\left[H_\Sigma^{a+V} f\right](x) = \mathsf{E}_x\left(\exp\left(-\int_0^S (a + V(X(s)))\, ds\right) f(X(S)) : S < \infty\right), \tag{2.1}$$

where f belongs to $D(K_0\dot{+}V)$. Intuitively, the function $H_\Sigma^{a+V} f$ is a function, that on $\Gamma = E \setminus \Sigma$ coincides with f and that on Σ is "$a + V$-harmonic". The operator J_Σ restricts functions defined on E to Σ and its dual J_Σ^* extends functions, defined on Σ, with 0 in Γ. The operator H_Σ^{a+V} as defined in (2.1) is a priori an operator defined on bounded continuous functions. It is not clear at all that it is defined on $L^2(E, m)$. In fact the latter does not seem to be true. However, in a natural way it is defined on the domain $D(K_0\dot{+}V)$ and a little more thought will show that it has a well-defined meaning on the domain of the corresponding quadratic form given by $D\left(aI + K_0\dot{+}V\right)^{1/2}$. For details see Proposition 2.8. and Corollary 2.9. below.

2.8. PROPOSITION. (a) Let $a > 0$ be large enough. The following identity holds:

$$\left(aI + (K_0\dot{+}V)_\Sigma\right) J_\Sigma \left(I - H_\Sigma^{a+V}\right) = J_\Sigma \left(aI + K_0\dot{+}V\right), \tag{2.2}$$

in the sense of domains and of equality of operators.
(b) The identity $(K_0 \dot{+} V)_\Sigma = J_\Sigma (K_0 \dot{+} V) J_\Sigma^*$ is valid in the sense of domains and of equality of operators.

PROOF. (a) First let f belong to $D\left((K_0 \dot{+} V)_\Sigma\right)$. Then define the function $g \in D(K_0 \dot{+} V)$ by the equality $J_\Sigma^* \left(aI + (K_0 \dot{+} V)_\Sigma\right) f = (aI + K_0 \dot{+} V)\, g$. For $x \in \Sigma$ we have

$$
\begin{aligned}
& g(x) - f(x) \\
&= \left[\left\{(aI + K_0 \dot{+} V)^{-1} - \left(aI + (K_0 \dot{+} V)_\Sigma\right)^{-1} J_\Sigma\right\} (aI + K_0 \dot{+} V)\, g\right](x) \\
&= \int_0^\infty ds \mathsf{E}_x \left(\exp\left(-\int_0^s (a + V(X(u)))\, du\right) (aI + K_0 \dot{+} V)\, g(X(s)) : S \le s\right) \\
&= \mathsf{E}_x \left(\int_S^\infty ds \exp\left(-\int_0^s (a + V(X(u)))\, du\right) (aI + K_0 \dot{+} V)\, g(X(s)) : S < \infty\right) \\
&= \mathsf{E}_x \left(\exp\left(-\int_0^S (a + V(X(u)))\, du\right)\right. \\
&\quad \times \left.\int_0^\infty ds \exp\left(-\int_0^s (a + V(X(u+S)))\, du\right) (aI + K_0 \dot{+} V)\, g(X(s+S)) : S < \infty\right)
\end{aligned}
$$

(Markov property)

$$
\begin{aligned}
&= \mathsf{E}_x \left(\exp\left(-\int_0^S (a + V(X(u)))\, du\right)\right. \\
&\quad \times \left.\int_0^\infty ds \mathsf{E}_{X(S)} \left(\exp\left(-\int_0^s (a + V(X(u)))\, du\right) (aI + K_0 \dot{+} V)\, g(X(s))\right) : S < \infty\right) \\
&= \mathsf{E}_x \left(\exp\left(-\int_0^S (a + V(X(u)))\, du\right) g(X(S)) : S < \infty\right).
\end{aligned}
$$

Consequently $g(x) - f(x) = \left[H_\Sigma^{a+V} g\right](x)$, $x \in \Sigma$. Conversely, let g belong to $D(K_0 \dot{+} V)$ and define $f \in D\left((K_0 \dot{+} V)_\Sigma\right)$ by the identity:

$$
\left(aI + (K_0 \dot{+} V)_\Sigma\right) f = J_\Sigma (aI + K_0 \dot{+} V)\, g.
$$

For $x \in \Sigma$ we have as above $g(x) - f(x) = \left[H_\Sigma^{a+V} g\right](x)$. This proves Proposition 2.8(a).

(b) From formula (2.2) in Proposition 2.8(a) we infer ($a > 0$ large enough)

$$
\begin{aligned}
(K_0 \dot{+} V)_\Sigma &= (K_0 \dot{+} V)_\Sigma J_\Sigma J_\Sigma^* = \left(aI + (K_0 \dot{+} V)_\Sigma\right) J_\Sigma J_\Sigma^* - a J_\Sigma J_\Sigma^* \\
&= \left(aI + (K_0 \dot{+} V)_\Sigma\right) J_\Sigma \left(I - H_\Sigma^{a+V}\right) J_\Sigma^* - a J_\Sigma J_\Sigma^* = J_\Sigma (K_0 \dot{+} V) J_\Sigma^*.
\end{aligned}
$$

As a corollary we have the following. The result should be compared to the fundamental identity for so-called λ-potentials in Port and Stone [34, p. 41].

2.9. COROLLARY. Let $a > 0$ be sufficiently large. The following identity holds in $L^2(E, m)$:

$$\left(aI + K_0 \dot{+} V\right)^{-1} - J^* \left(aI + \left(K_0 \dot{+} V\right)_\Sigma\right)^{-1} J = H_\Sigma^{a+V} \left(aI + K_0 \dot{+} V\right)^{-1}. \quad (2.3)$$

In addition the operator $H_\Sigma^{a+V} \left(aI + K_0 \dot{+} V\right)^{-1}$ is self-adjoint and form positive. In fact the following identities are true:

$$\begin{aligned} &H_\Sigma^{a+V} \left(aI + K_0 \dot{+} V\right)^{-1} \\ &= \left(\left(aI + K_0 \dot{+} V\right)^{1/2} H_\Sigma^{a+V} \left(aI + K_0 \dot{+} V\right)^{-1}\right)^* \left(aI + K_0 \dot{+} V\right)^{1/2} H_\Sigma^{a+V} \left(aI + K_0 \dot{+} V\right)^{-1} \\ &= H_\Sigma^{a+V} \left(aI + K_0 \dot{+} V\right)^{-1/2} \left(H_\Sigma^{a+V} \left(aI + K_0 \dot{+} V\right)^{-1/2}\right)^*. \end{aligned} \quad (2.4)$$

Hence the operators

$$\begin{gathered} H_\Sigma^{a+V} \left(aI + K_0 \dot{+} V\right)^{-1}, \quad H_\Sigma^{a+V} \left(aI + K_0 \dot{+} V\right)^{-1/2} \\ \text{and also} \quad \left(aI + K_0 \dot{+} V\right)^{1/2} H_\Sigma^{a+V} \left(aI + K_0 \dot{+} V\right)^{-1} \end{gathered}$$

are bounded operators in $L^2(E, m)$. Moreover the operator H_Σ^{a+V} is self-adjoint in the space $D(Q^{a+V})$ equipped with the inner-product

$$Q^{a+V}(f, g) = \left\langle \left(aI + K_0 \dot{+} V\right)^{1/2} f, \left(aI + K_0 \dot{+} V\right)^{1/2} g \right\rangle, \quad (2.5)$$

where f and g belong to $D(Q^{a+V}) = D\left(\left(aI + K_0 \dot{+} V\right)^{1/2}\right)$. In particular the operator $\left(aI + K_0 \dot{+} V\right)^{1/2} H_\Sigma^{a+V} \left(aI + K_0 \dot{+} V\right)^{-1/2}$ is a self-adjoint projection in $L^2(E, m)$.
PROOF. Notice the identity $\left[H_\Sigma^{a+V} f\right](x) = f(x)$, for $x \in \Gamma^r$ and for $f \in D\left(K_0 \dot{+} V\right)$. Also notice the fact that the operator H_Σ^{a+V} is a projection in the sense that its square $H_\Sigma^{a+V} \circ H_\Sigma^{a+V}$ equals H_Σ^{a+V}. In fact (2.3) is a reformulation of (2.2). The identities in (2.4) follow because the operator $H_\Sigma^{a+V} \left(aI + K_0 \dot{+} V\right)^{-1}$ is self-adjoint. The same argument applies for the proof of the the self-adjointness of the operator H_Σ^{a+V} with respect to the inner-product in (2.5). The latter also implies the final statement in Corollary 2.9.

3. SOME HILBERT-SCHMIDT PROPERTIES OF RESOLVENT DIFFERENCES

3.1. HYPOTHESES. We denote by $\mathcal{C}_1$, $\mathcal{C}_2$ and $\mathcal{C}_\infty$ the collection of Hilbert-Schmidt, the collection of trace class operators and the collection of compact operators respectively. As in section 2 we place ourselves in the surroundings of the basic assumptions on stochastic spectral analysis (BASSA). In fact, let K_0 be the generator of a self-adjoint semi-group $\{\exp(-tK_0) : t \geq 0\}$ in $L^2(E, m)$ of the form:

$\exp\left[\left(-tK_0\right)f\right](x) = \int p_0(t,x,y)f(y)dm(y)$, where $p_0(t,x,y)$ is symmetric and continuous on $(0,\infty)\times E\times E$ and where m is some non-negative Borel measure on E. Briefly, assumptions A1-A4 are verified. Usually we write dy instead of $dm(y)$. As in section 2 the generator K_0 will be perturbed in two ways. First there will be a "regular" perturbation, being a multiplication operator V and secondly, there will be a potential barrier on Γ. In principle Γ will be a closed subset of E. The singularity projection operator is defined by: $[Pf](x) = 1_\Gamma(x)f(x)$, $f \in L^2(E,m)$. Put $\Sigma := E\setminus\Gamma$ and introduce the restriction operator J as follows: $Jf = f\mid_\Sigma$. Hence $J^* = Id_{L^2(\Sigma)\to L^2(E)}$, $J^*J = I - P$ and $JJ^* = I_{L^2(\Sigma)}$. Let $K := K_0\dot{+}V$ be the Feynman-Kac generator, $K_M = K + MP$ with domain $\text{dom}(K) = \text{dom}(K_M)$ and denote with $K_\Sigma := \left(K_0\dot{+}V\right)_\Sigma$ the Feynman-Kac generator of the semi-group $\left\{\exp\left(-t(K_0\dot{+}V)_\Sigma\right) : t\geq 0\right\}$, killed on Γ. From formula (2.2) in Proposition 2.8(a) we infer $\left(K_0\dot{+}V\right)_\Sigma = J\left(K_0\dot{+}V\right)J^*$. Also notice that most of the time we write J instead of J_Σ. Also see the remarks preceding Proposition 2.8.

We wish to establish a number of Hilbert-Schmidt properties of resolvent differences. We begin with a proposition on differences of powers of resolvents. The main ingredient of the proof is the observation that the process $\{M_V^t(\tau) : 0\leq\tau<t\}$ is a P_z-martingale on the interval $(0,t)$ for all $z\in E$ and for all $t>0$. Here $M_V^t(\tau)$ is defined by

$$M_V^t(\tau) = \exp\left(-\int_0^\tau V(X(u))du\right)\exp\left(-(t-\tau)(K_0+V)\right)(X(\tau),y).$$

3.2. PROPOSITION. Suppose that $a>0$ and $q\geq 1$ are chosen in such a way that the integral $\int_\Sigma dx\left[H_\Sigma^{a+V}(aI+K_0+V)^{-2q}(\cdot,x)\right](x)$ is finite. Then the operator

$$J\left(aI+K_0\dot{+}V\right)^{-q} - \left(aI+(K_0\dot{+}V)_\Sigma\right)^{-q}J \tag{3.1}$$

is a Hilbert-Schmidt operator and

$$\begin{aligned}&\left\|J\left(aI+K_0\dot{+}V\right)^{-q} - \left(aI+(K_0\dot{+}V)_\Sigma\right)^{-q}J\right\|_{\mathrm{HS}}^2\\ &\leq \frac{\Gamma(2q-1)}{\Gamma(q)^2}\int_\Sigma dx\left[H_\Sigma^{a+V}(aI+K_0+V)^{-2q}(\cdot,x)\right](x).\end{aligned} \tag{3.2}$$

Here $S = \inf\left\{s>0 : \int_0^s 1_\Gamma(X(\sigma))d\sigma > 0\right\}$. The operator H_Σ^{a+V} is discussed in equality (2.3) of Corollary 2.9.

PROOF. Observe that the integral kernel of the operator in (3.1) is given by

$$\begin{aligned}&\frac{1}{\Gamma(q)}\int_0^\infty dt e^{-at}t^{q-1}\\ &\times\mathsf{E}_x\left(\exp\left(-\int_0^S V(X(u))du\right)\exp\left(-(t-S)\left(K_0\dot{+}V\right)\right)(X(S),y) : S<t\right),\end{aligned}$$

where x belongs to Σ and where y is in E. Hence from Chapman-Kolmogorov's identity it follows that

$$\int dy \left(J\left(aI + K_0 \dot{+} V\right)^{-q}(x,y) - \left(aI + \left(K_0 \dot{+} V\right)_\Sigma\right)^{-q} J(x,y)\right)^2$$

$$= \frac{1}{\Gamma(q)^2}\int_0^\infty dt_1 \int_0^\infty dt_2 e^{-a(t_1+t_2)}(t_1t_2)^{q-1}$$

$$\mathsf{E}_x \otimes \mathsf{E}_x\left((\omega,\omega') \mapsto \exp\left(-\int_0^{S(\omega)} V(X(u))(\omega)du\right) 1_{[0,t_1)}(S(\omega))\right.$$

$$\times \exp\left(-\int_0^{S(\omega')} V(X(u))(\omega')du\right) 1_{[0,t_2)}(S(\omega'))$$

$$\left.\times \exp\left(-\left(t_1+t_2-S(\omega)-S(\omega')\right)(K_0+V)\right)\left(X(S)(\omega), X(S)(\omega')\right)\right)$$

(apply Fubini, substitute $t_1 - S(\omega) = \tau_1$ and $t_2 - S(\omega') = \tau_2$ and apply Fubini again)

$$= \frac{1}{\Gamma(q)^2}\int_0^\infty d\tau_1 \int_0^\infty d\tau_2$$

$$\mathsf{E}_x \otimes \mathsf{E}_x\left((\omega,\omega') \mapsto \exp\left(-\int_0^{S(\omega)} \left(a + V(X(u))(\omega)\right) du\right) 1_{[0,\infty)}(S(\omega))\right.$$

$$\times \exp\left(-\int_0^{S(\omega')} \left(a + V(X(u))(\omega')\right) du\right) 1_{[0,\infty)}(S(\omega'))$$

$$\times \exp\left(-\left(\tau_1+\tau_2\right)(a+K_0+V)\right)\left(X(S)(\omega), X(S)(\omega')\right)$$

$$\left.\left(\tau_1 + S(\omega)\right)^{q-1}\left(\tau_2 + S(\omega')\right)^{q-1}\right)$$

(substitute $\tau_1 + \tau_2 = \tau$ and $\tau_1 = \sigma$)

$$= \frac{1}{\Gamma(q)^2}\int_0^\infty d\tau$$

$$\mathsf{E}_x \otimes \mathsf{E}_x\left((\omega,\omega') \mapsto \exp\left(-\int_0^{S(\omega)} \left(a + V(X(u))(\omega)\right) du\right) 1_{[0,\infty)}(S(\omega))\right.$$

$$\times \exp\left(-\int_0^{S(\omega')} \left(a + V(X(u))(\omega')\right) du\right) 1_{[0,\infty)}(S(\omega'))$$

$$\times \exp\left(-\tau(a+K_0+V)\right)\left(X(S)(\omega), X(S)(\omega')\right)$$

$$\left.\int_0^\tau d\sigma \left(\sigma + S(\omega)\right)^{q-1}\left(\tau - \sigma + S(\omega')\right)^{q-1}\right)$$

($2ab \le a^2 + b^2$ with $a = \left(\sigma + S(\omega)\right)^{q-1}$ and with $b = \left(\tau - \sigma + S(\omega')\right)^{q-1}$)

$$\leq \frac{1}{2(2q-1)}\frac{1}{\Gamma(q)^2}\int_0^\infty d\tau$$

$$\mathsf{E}_x \otimes \mathsf{E}_x\Bigg((\omega,\omega') \mapsto \exp\left(-\int_0^{S(\omega)} (a+V(X(u))(\omega))\,du\right) 1_{[0,\infty)}(S(\omega))$$

$$\times \exp\left(-\int_0^{S(\omega')} (a+V(X(u))(\omega'))\,du\right) 1_{[0,\infty)}(S(\omega'))$$

$$\times \exp\left(-\tau(a+K_0+V)\right)(X(S)(\omega), X(S)(\omega'))$$

$$\left((\tau+S(\omega))^{2q-1} + (\tau+S(\omega'))^{2q-1}\right)\Bigg)$$

(the roles of ω and ω' are interchangeble,
Fubini's theorem is applicable and $t = \tau + S(\omega')$ is substituted)

$$= \frac{1}{(2q-1)\Gamma(q)^2}\mathsf{E}_x\Bigg(\omega \mapsto \exp\left(-\int_0^{S(\omega)} (a+V(X(u))(\omega))\,du\right) 1_{[0,\infty)}(S(\omega))$$

$$\times \int_0^\infty dt t^{2q-1}\mathsf{E}_x\Bigg(\exp\left(-\int_0^{S(\omega')} (a+V(X(u))(\omega'))\,du\right) 1_{[0,t)}(S(\omega'))$$

$$\exp\left(-(t-S(\omega'))(a+K_0+V)\right)(X(S)(\omega), X(S)(\omega'))\Bigg)\Bigg)$$

(the process $\exp\left(-\int_0^\tau V(X(u))du\right)\exp\left(-(t-\tau)(K_0+V)\right)(y, X(\tau))$
is a martingale on the interval $(0,t)$)

$$= \frac{1}{(2q-1)\Gamma(q)^2}\int_0^\infty dt t^{2q-1}\mathsf{E}_x\Bigg(\exp\left(-\int_0^S (a+V(X(u)))\,du\right)$$

$$\times \exp\left(-t(a+K_0+V)\right)(X(S),x) : S<\infty\Bigg)$$

(definition of H_Σ^{a+V})

$$= \frac{\Gamma(2q)}{(2q-1)\Gamma(q)^2}\left[H_\Sigma^{a+V}(aI+K_0+V)^{-2q}(\cdot,x)\right](x). \tag{3.3}$$

Inequality (3.2) in the proposition follow upon integrating (3.3) with respect to x.

In what follows we write

$$C_\Sigma(t) = \sup_{z,w\in E} p_0(t/2,z,w)\int_\Sigma\left(\mathsf{E}_x\left(p_0(t,X(S),x) : S<\infty\right)\right)^{1/2} dx$$

and also $K_M := K_0\dot{+}V + M1_\Gamma$. We also recall that $K = K_0\dot{+}V$.

3.3. PROPOSITION. Suppose that there is a $q \in \mathbb{N}$, $q \geq 1$, such that for some constant $a_0 > 0$ and for some $q > 0$ the expression

$$\int_0^\infty t^{2q-1} e^{-a_0 t} C_\Sigma(t)dt \tag{3.4}$$

is finite. The following assertions hold true:

(i) $J(K_M - zI)^{-p} - (K_\Sigma - zI)^{-p} J \in \mathcal{C}_\infty\left(L^2(E,m), L^2(\Sigma,m)\right)$ for all $p \in \mathbb{N}$, for all $M \geq 0$ and for all $z \in \text{res}(K_M)\bigcap\text{res}(K_\Sigma)$.

(ii) Moreover $J(K_M - zI)^{-p} - (K_\Sigma - zI)^{-p} J$ belongs to $\mathcal{C}_2\left(L^2(E,m), L^2(\Sigma,m)\right)$, for all $p \geq q$, for all $M \geq 0$ and for all $z \in \text{res}(K_M)\bigcap\text{res}(K_\Sigma)$.

(iii) $\lim_{M\to\infty} \left\| J(K_M - zI)^{-p} - (K_\Sigma - zI)^{-p} J \right\|_r = 0$, for all $z \in \text{res}(K_\Sigma)$ for $r = 2$ if $p \geq q$ and for $r = \infty$ if $p = 1$.

(iv) The rate of convergence in (iii) is the same for all $z \in \text{res}(K_\Sigma)$.

(v) Suppose that the dimension of the semi-group $\{\exp(-tK_0) : t \geq 0\}$ is m, i.e. suppose $p_0(t,x,y) \leq c_1 t^{-m/2} e^{b_1 t}$. Also suppose that the inequality $\int dx \left(\mathsf{E}_x\left(p_0(t, X(S), x) : S < \infty\right)\right)^{1/2} \leq c_2 t^{-m/2} e^{b_2 t}$ is valid. Then, for $\text{Re}a > 0$ large,
$\left\| J\left(aI + K_0 \dot{+} V\right)^{-q} - \left(aI + (K_0 \dot{+} V)_\Sigma\right)^{-q} J \right\|_{\text{HS}} \leq \text{Constant} \times (\text{Re } a)^{-q+m/4}$.

Suppose $m < 4$. If $-\text{Re}z_0 = a$ is sufficiently large, then the following representation in the sense of Hilbert-Schmidt norm is valid:

$$\begin{aligned} &J(K_M - z_0 I)^{-1} - (K_\Sigma - z_0 I)^{-1} J \\ &= (q-1)\int_0^\infty t^{q-2}\left[J(K_M - (z_0 - t)I)^{-q} - (K_\Sigma - (z_0 - t)I)^{-q} J\right] dt. \end{aligned} \tag{3.5}$$

Remark 1. Representation (3.5) says that if the Hilbert-Schmidt property is true for some $q \in \mathbb{N}$, then it is true for all $q \in \mathbb{N}$.

Remark 2. Suppose that the penetration time S and the hitting T of Γ are equal P_x-almost surely. Since V is a Kato-Feller potential (i.e. $V_- \in K(E)$ and $V_+ \in K_{\text{loc}}(E)$) the operator K is a well-defined self-adjoint operator in $L^2(E,m)$. The spectrum of K is contained in $[-\gamma, \infty)$, $\gamma > 0$, or $\mathbb{C} \setminus [-\gamma,\infty) \subseteq \text{res}(K)$, the resolvent set of K. If the operator norm convergence u-$\lim_{M\to\infty} J(K_M - zI)^{-1} J^* = (K_\Sigma - zI)^{-1}$ can be established for some $z \in \text{res}(K_\Sigma)$, then this convergence is true for all $z \in \text{res}(K_\Sigma)$: see Kato [29, p. 211-212]. Because $\text{res}(K_\Sigma) = \bigcap_{M > M_0} \text{res}(K_M)$, for M_0 large enough, we also have $\mathbb{C} \setminus [-\gamma,\infty) \subseteq \text{res}(K_\Sigma)$.

PROOF of Proposition 3.3. Choose M_0 and C_0 in such a way that

$$\exp\left(-t(K_0 + V)\right)(x,y) \leq M_0 e^{C_0 t} p_0(t,x,y)^{1/2} \sup_{z,w\in E} p_0(t/2,z,w)^{1/2}.$$

Such constants M_0 and C_0 exist: see [44, p. 301]. For the proof we need the following property of functions V, that belong to the Kato-Feller class. For $a > 0$ sufficiently the following supremum $\sup_{x\in E} \mathsf{E}_x\left(\exp\left(-2aS + 2\int_0^S V_-(X(u))du\right) : S < \infty\right)$ is finite. A proof of this fact runs as follows. Fix $t_0 > 0$. From Khas'minskii's lemma it follows that $\sup_{y\in E} \mathsf{E}_y\left(\exp\left(2\int_0^{t_0} V_-(X(u))du\right)\right) < \infty$. Choose $a > 0$ so large that $e^{-at_0}\sup_{y\in E} \mathsf{E}_y\left(\exp\left(2\int_0^{t_0} V_-(X(u))du\right)\right) < 1$. From the Markov property it then follows that:

$$\begin{aligned}
&\mathsf{E}_x\left(\exp\left(-2aS + 2\int_0^S V_-(X(u))du\right) : S < \infty\right)\\
&\le \sum_{k=1}^{\infty} \mathsf{E}_x\left(\exp\left(-2a(k-1)t_0 + 2\int_0^{kt_0} V_-(X(u))du\right) : (k-1)t_0 < S \le kt_0\right)\\
&\le \sum_{k=1}^{\infty} \mathsf{E}_x\left(\exp\left(-2a(k-1)t_0 + 2\int_0^{(k-1)t_0} V_-(X(u))du\right)\right.\\
&\qquad \left.\mathsf{E}_{X((k-1)t_0)}\left(\exp\left(2\int_0^{t_0} V_-(X(u))du\right)\right) : (k-1)t_0 < S\right)\\
&\le \sum_{k=1}^{\infty} e^{-2a(k-1)t_0}\left(\sup_{y\in E}\mathsf{E}_y\left(\exp\left(2\int_0^{t_0} V_-(X(u))du\right)\right)\right)^k < \infty.
\end{aligned}$$

Henceforth we pick $a > C_0 + a_0$ so large that

$$\sup_{x\in E} \mathsf{E}_x\left(\exp\left(-2aS + 2\int_0^S V_-(X(u))du\right) : S < \infty\right) < \infty.$$

For such a it follows that

$$\begin{aligned}
&\mathsf{E}_x\left(\exp\left(-\int_0^S (a + V(X(u)))\,du\right)\exp\left(-t(K_0+V)\right)(X(S),x) : S < \infty\right)\\
&\le \left(\mathsf{E}_x\left(\exp\left(-2\int_0^S (a + V(X(u)))\,du\right) : S < \infty\right)\right)^{1/2}\\
&\quad \times \left(\mathsf{E}_x\left(\left(\exp\left(-t(K_0+V)\right)(X(S),x)\right)^2 : S < \infty\right)\right)^{1/2}\\
&\le M_0(a)\exp\left(C_0 t\right)\sup_{z,w\in E} p_0\left(t/2, z, w\right)^{1/2}\left(\mathsf{E}_x\left(p_0(t, X(S), x) : S < \infty\right)\right)^{1/2},
\end{aligned}$$

where

$$M_0(a) = M_0 \sup_{x\in E}\left(\mathsf{E}_x\left(\exp\left(-2\int_0^S (a + V(X(u)))\,du\right) : S < \infty\right)\right)^{1/2}.$$

Hence from this together with (3.4) it follows that

$$\begin{aligned}
&\left\| J\left(aI+K_0\dot{+}V\right)^{-q}-\left(aI+(K_0\dot{+}V)_\Sigma\right)^{-q}J\right\|_{\mathrm{HS}}^2\\
&\le \frac{1}{(2q-1)\Gamma(q)^2}\int_0^\infty dte^{-at}t^{2q-1}\int_\Sigma dx\mathrm{E}_x\Bigg(\exp\left(-\int_0^S (a+V(X(u)))\,du\right)\\
&\qquad \exp\left(-t(K_0+V)\right)(X(S),x):S<\infty\Bigg)\\
&\le \frac{M_0(a)}{(2q-1)\Gamma(q)^2}\int_0^\infty dtt^{2q-1}e^{-(a-C_0)t}C_\Sigma(t)<\infty. \qquad (3.6)
\end{aligned}$$

The proof of Proposition 3.3. begins with establishing the Hilbert-Schmidt property in (ii). Let z_0 be such that $\mathrm{Re}z_0=-a<-2A$. Suppose $p\ge q$. Then, as above, the Hilbert-Schmidt norm of the operator in (ii) can be estimated by

$$\begin{aligned}
&\left\|\left[J\left(K_M-z_0I\right)^{-p}-\left(K_\Sigma-z_0I\right)^{-p}J\right]\right\|_{\mathrm{HS}}^2\\
&\le \frac{M_0(a)}{(2p-1)\Gamma(p)^2}\int_0^\infty dtt^{2p-1}e^{-(a-C_0)t}C_\Sigma(t)\\
&\le \frac{M_0(a)}{(2p-1)\Gamma(p)^2}\int_0^\infty dtt^{2q-1}\max(1,t)^{2p-2q}e^{-(a-C_0)t}C_\Sigma(t)\\
&\le M_0'(a)\int_0^\infty dtt^{2q-1}e^{-a_0t}C_\Sigma(t),
\end{aligned}$$

where $M_0'(a)=\sup_{t>0}\left\{t^{2p-2q}e^{-(a-C_0-a_0)t}\right\}\dfrac{M_0(a)}{(2p-1)\Gamma(p)^2}$. So from (3.4) it follows that the operator in (ii) is a Hilbert-Schmidt operator for $-\mathrm{Re}z>C_0+a_0$ and $p\ge q$.

(i) Representation (3.5) always holds in the sense of the usual operator norm. Hence the compactness of $J\left(K_M-z_0I\right)^{-1}-\left(K_\Sigma-z_0I\right)^{-1}J$ follows for z_0 as in (i). But (i) and (ii) also hold for all other $z\in\mathrm{res}(K_M)\bigcap\mathrm{res}(K_\Sigma)$. Let d be the distance between z_0 and $\sigma(K)$. For $|z-z_0|<d$ one gets from the Neumann series:

$$\begin{aligned}
&\left\|J\left(K_M-zI\right)^{-1}-\left(K_\Sigma-zI\right)^{-1}J\right\|_r\\
&\le\sum_{k=0}^\infty(k+1)\frac{|z-z_0|^k}{d^k}\left\|J\left(K_M-z_0I\right)^{-1}-\left(K_\Sigma-z_0I\right)^{-1}J\right\|_r \qquad (3.7)
\end{aligned}$$

where $r=\infty$ or $r=2$. Consequently (i) and (ii) now follow.

From (3.7) together with the definition of the operator H_Σ^V it also follows that in the inequality (we always suppose $p\ge q$):

$$\begin{aligned}
&\left\|\left[J\left(K_M-z_0I\right)^{-p}-\left(K_\Sigma-z_0I\right)^{-p}J\right]\right\|_{\mathrm{HS}}^2\\
&\le\frac{\Gamma(2p-1)}{\Gamma(p)^2}\int_\Sigma dx\left[H_\Sigma^{a+V}\left(aI+K_0+V+M1_\Gamma\right)^{-2p}(\cdot,x)\right](x)
\end{aligned}$$

we may apply Lebesgue's theorem on dominated convergence. In fact fix x and $y \in E$ and let $\mathsf{E}_{y,0}^{x,t}$ be the conditional expectation that pins the process $\{X(s) : s \geq 0\}$ in y at time 0 and in x at time t. More precisely $\mathsf{E}_{y,0}^{x,t}$ is determined by the property that for all $0 < s < t$ and for all $A \in \mathcal{F}_s$ $\mathsf{E}_{y,0}^{x,t}(1_A)p_0(t,y,x) = \mu_{y,0}^{x,t}(A)$, where $\mu_{y,0}^{x,t}(A) = \mathsf{E}_y\left(p_0(t-s_1, X(s_1), x), A\right)$, with $s \leq s_1 < t$. Since the process $\{p_0(t-s, X(s), x) : 0 < s < t\}$ is a martingale on $(0,t)$, the measure $\mu_{y,0}^{x,t}$ is well-defined. It has the property that

$$\begin{aligned}
&\exp\left(-t(K_0 + V + M1_\Gamma)\right)(y,x) - \exp\left(-t(K_0+V)_\Sigma\right)(y,x) \\
&= \lim_{\tau \uparrow t} \mathsf{E}_y\left(\exp\left(-\int_0^\tau (V(X(u)) + M1_\Gamma(X(u)))\,du\right) p_0(t-\tau, X(\tau), x)\right) \\
&\quad - \lim_{\tau \uparrow t} \mathsf{E}_y\left(\exp\left(-\int_0^\tau (V(X(u)) + M1_\Gamma(X(u)))\,du\right) p_0(t-\tau, X(\tau), x), S > \tau\right) \\
&= \lim_{\tau \uparrow t} \mathsf{E}_y\left(\exp\left(-\int_0^\tau (V(X(u)) + M1_\Gamma(X(u)))\,du\right) p_0(t-\tau, X(\tau), x), S \leq \tau\right) \\
&= \int \exp\left(-\int_0^t (V(X(u)) + M1_\Gamma(X(u)))\,du\right) 1_{\{S<t\}}\,d\mu_{y,0}^{x,t}. \qquad (3.8)
\end{aligned}$$

Hence

$$\begin{aligned}
&\lim_{M\to\infty} \exp\left(-t(K_0 + V + M1_\Gamma)\right)(y,x) - \exp\left(-t(K_0+V)_\Sigma\right)(y,x) \\
&= \lim_{M\to\infty} \int \exp\left(-\int_0^t (V(X(u)) + M1_\Gamma(X(u)))\,du\right) 1_{\{S<t\}}\,d\mu_{y,0}^{x,t} \\
&= \int \exp\left(-\int_0^t V(X(u))du\right) 1_{\left\{\int_0^t 1_\Gamma(X(u))du = 0\right\}} 1_{\{S<t\}}\,d\mu_{y,0}^{x,t} \\
&= \int \exp\left(-\int_0^t V(X(u))du\right) 1_{\{S\geq t\}} 1_{\{S<t\}}\,d\mu_{y,0}^{x,t} = 0. \qquad (3.9)
\end{aligned}$$

Inserting the result of (3.9) in (3.8) yields

$$\begin{aligned}
&\lim_{M\to\infty} \left\| \left[J\,(K_M - z_0 I)^{-p} - (K_\Sigma - z_0 I)^{-p}\, J \right] \right\|_{\mathrm{HS}}^2 \\
&\leq \frac{1}{(2p-1)\Gamma(p)^2} \int_0^\infty dt e^{-at} t^{2p-1} \int_\Sigma dx \mathsf{E}_x\left(\exp\left(-\int_0^S (a + V(X(u)))\,du\right)\right. \\
&\qquad \left. \lim_{M\to\infty} \exp\left(-t(K_0 + V + M1_\Gamma)\right)(X(S), x) : S < \infty\right) \\
&= \frac{1}{(2p-1)\Gamma(p)^2} \int_0^\infty dt e^{-at} t^{2p-1} \int_\Sigma dx \mathsf{E}_x\left(\exp\left(-\int_0^S (a + V(X(u)))\,du\right)\right. \\
&\qquad \left. \exp\left(-t(K_0+V)_\Sigma\right)(X(S), x) : S < \infty\right) = 0.
\end{aligned}$$

because, on $\{S < \infty\}$, $X(S)$ is P_x-almost surely in Γ and for $y \in \Gamma$ the expression $\exp(-t(K_0 + V)_\Sigma)(y, x)$ vanishes. All this is true provided $\mathrm{Re}z_0 = -a < -2A$, i.e. for $-\mathrm{Re}z_0$ positive and large enough. An argument as in (3.7) yields the same result, not only for $-\mathrm{Re}z_0$ large, but for all $z \in \mathrm{res}(K_\Sigma)$. Instead of the Neumann series we write ($|z - z_0| < d$)

$$(K_M - zI)^{-p} = \sum_{k=0}^{\infty} \binom{k+p-1}{k} (z - z_0)^k (K_M - z_0 I)^{-k-1} .$$

This shows (iii) and also (iv), except for the convergence in operator norm for $p = 1$. For this we again take $-\mathrm{Re}z_0$ large enough and we use representation (3.5) for the operator norm. In fact we have

$$\begin{aligned}&\left\| J(K_M - z_0 I)^{-1} - (K_\Sigma - z_0 I)^{-1} J \right\| \\ &\leq (q-1) \int_0^\infty t^{q-2} \left\| J(K_M - (z_0 - t)I)^{-q} - (K_\Sigma - (z_0 - t)I)^{-q} J \right\| dt\end{aligned}$$

and, since the Hilbert-Schmidt norm dominates the operator norm, we know that $\lim_{M\to\infty} \left\| J(K_M - (z_0 - t)I)^{-q} - (K_\Sigma - (z_0 - t)I)^{-q} J \right\| = 0$. This proves (iii) for $p = 1$ and for the operator norm replacing the Hilbert-Schmidt norm.

Next we shall prove (v). It suffices to take for a a large real positive number. From (3.6) we infer

$$\begin{aligned}&\left\| J\left(aI + K_0 \dot{+} V\right)^{-p} - \left(aI + (K_0 \dot{+} V)_\Sigma\right)^{-p} J \right\|_{\mathrm{HS}} \\ &\leq \frac{M_0(a)}{(2p-1)\Gamma(p)^2} \int_0^\infty dt t^{2p-1} e^{-(a-C_0)t} C_\Sigma(t) \\ &\leq \frac{M_0(a)}{(2p-1)\Gamma(p)^2} \int_0^\infty dt t^{2p-1} e^{-(a-C_0)t} c_1^{1/2} 2^{m/4} t^{-m/4} \exp\left(\frac{1}{2} b_1 t\right) c_2^{1/2} t^{-m/4} \exp(b_2 t) \\ &\leq \frac{M_0(a) c_1^{1/2} c_2 2^{m/4}}{(2p-1)\Gamma(p)^2} \int_0^\infty dt t^{2p-1-m/2} \exp\left(-\left(a - C_0 - \frac{1}{2} b_1 - b_2\right) t\right) \\ &= \frac{M_0(a) c_1^{1/2} c_2 2^{m/4}}{(2p-1)\Gamma(p)^2} \frac{\Gamma(2p - m/2)}{\left(a - C_0 - \frac{1}{2} b_1 - b_2\right)^{2p-m/2}} .\end{aligned}$$

This proves the first part of (v). For $-\mathrm{Re}z_0 = a$ sufficiently large and for $t > 0$, the Hilbert-Schmidt norm of $J(K_M - (z_0 - t)I)^{-q} - (K_\Sigma - (z_0 - t)I)^{-q} J$ is dominated by a constant times $(a + t)^{-q+m/4}$. Since the mapping $t \mapsto \left\| J(K_M - (z_0 - t)I)^{-q} - (K_\Sigma - (z_0 - t)I)^{-q} J \right\|_{\mathrm{HS}}$ and since, upon employing the first claim of (v), the Hilbert-Schmidt norm of the right hand side of (3.5) can be estimated by ($q > 1 > m/4$)

$$\leq \text{Constant} \times (q-1) \int_0^\infty \frac{t^{q-2}}{(a+t)^{-q+m/4}} dt = \text{Constant} \times \frac{\Gamma(q)\Gamma(1 - m/4)}{a^{1-m/4}\Gamma(q - m/4)},$$

the assertion in (v) follows.

Next we turn our attention to resolvent differences on the whole space.

3.4. PROPOSITION. Suppose that the BASSA hypotheses A1-A4 are satisfied. Also suppose that $m(\text{bdr}(\Gamma)) = 0$ and assume that the boundedness condition B is verified. Then the following assertions are valid.

(i) For every $q \in \mathbb{N}$, for every $M \geq 0$ and for every $z \in \text{res}(K_M)$, the operator $P(K_M - zI)^{-q}$ is compact.

(ii) In fact, for every $q \in \mathbb{N}$, $q > m/2$, for every $M \geq 0$ and for every $z \in \text{res}(K_M)$ the operator $P(K_M - zI)^{-q}$ is a Hilbert-Schmidt operator.

(iii) $\lim_{M\to\infty} \left\| P(K_M - zI)^{-1} \right\| = 0$ for every $z \in \text{res}(K_\Sigma)$.

PROOF. It suffices to prove these assertions for $z_0 \in \text{res}(K)$, $\text{Re}z_0 = -a < -2A$. Assertion (ii) follows because

$$\begin{aligned}
&\left\| P(K_M - z_0 I)^{-q} \right\|_2^2 \\
&= \frac{1}{((q-1)!)^2} \int_\Gamma dx \int_E dy \left| \int_0^\infty d\lambda e^{z_0\lambda} \lambda^{q-1} \exp(-\lambda K_M)(x,y) \right|^2 \qquad (3.10)\\
&= \frac{1}{((q-1)!)^2} \int_\Gamma \int_0^\infty \int_0^\infty d\lambda d\mu e^{z_0(\lambda+\mu)} \lambda^{q-1} \mu^{q-1} \exp(-(\lambda+\mu)K_M)(x,x) \\
&\leq Cm(\Gamma) \int_0^\infty d\lambda \int_0^\infty d\mu e^{-(a-A)(\lambda+\mu)} \lambda^{q-1} \mu^{q-1} \sup_{x,y\in E} p_0\left(\frac{\lambda+\mu}{2}, x, y\right) C_1 m(\Gamma).
\end{aligned}$$

Here, the constant C_1 is finite, provided the integral kernel is of dimension $2m$, where $2q > m$, i.e. provided that $p_0(t,x,y) \leq Ct^{-m}$ for all $0 < t < 1$. Then (i) is a consequence of

$$P(K_M + aI)^{-1} = (q-1) \int_0^\infty dt t^{q-2} P(K_M + aI + tI)^{-q}. \qquad (3.11)$$

The next step shows that the equality in (3.10) yields:

$$\lim_{M\to\infty} \left\| P(K_M + aI)^{-q} \right\|_2^2 = 0, \quad \text{if} \quad q > \frac{m}{2}. \qquad (3.12)$$

Therefore we estimate:

$$\begin{aligned}
&((q-1)!)^2 \left\| P(K_M + a)^{-q} \right\|_2^2 \\
&= \int_\Gamma dx \int_E dy \left| \int_0^\infty d\lambda e^{-a\lambda} \lambda^{q-1} \exp(-\lambda K_M)(x,y) \right|^2 \\
&= \int_0^\infty d\lambda \int_0^\infty d\mu e^{-a\lambda - a\mu} \lambda^{q-1} \mu^{q-1} \int_\Gamma \exp(-(\lambda+\mu)K_M)(x,x) dx \\
&= \int_0^\infty d\lambda \int_0^\infty d\mu e^{-a\lambda - a\mu} (\lambda\mu)^{q-1} \int_\Gamma \Big(\exp(-(\lambda+\mu)K_M)(x,x) - \exp(-(\lambda+\mu)K_\Sigma)(x,x) \Big) dx \\
&\quad + \int_0^\infty d\lambda \int_0^\infty d\mu e^{-a\lambda - a\mu} (\lambda\mu)^{q-1} \int_\Gamma \exp(-(\lambda+\mu)K_\Sigma)(x,x) dx. \qquad (3.13)
\end{aligned}$$

Here

$$\exp(-\lambda K_\Sigma)(x,y) = \lim_{M\to\infty} \exp(-\lambda K_M)(x,y)$$
$$= \lim_{\lambda'\uparrow\lambda} \mathsf{E}_x\left(\exp\left(-\int_0^{\lambda'} V(X(\sigma))d\sigma\right) p_0(\lambda-\lambda', X(\lambda'), y) : S > \lambda'\right), \quad (3.14)$$

where $S = \inf\{s > 0 : \int_0^s 1_\Gamma(X(\sigma))d\sigma > 0\}$. The equality in (3.12) will follow from (3.13) together with (3.14), as soon as we have shown that the integral $\int_\Gamma \exp(-\lambda K_\Sigma)(x,x)dm(x)$ vanishes. Put $\widetilde{\Gamma}^r = \{x \in E : \mathsf{P}_x(S=0) = 1\}$. Then $\Gamma \setminus \widetilde{\Gamma}^r$ is contained in the boundary of Γ. Consequently $m(\Gamma \setminus \widetilde{\Gamma}^r) = 0$ and hence $\int_\Gamma \exp(-\lambda K_\Sigma)(x,x)dm(x) = 0$.

In the same way we have, for $t \geq 0$, $\lim_{M\to\infty} \|P(K_M + aI + tI)^{-q}\|_{\mathrm{HS}} = 0$. Hence (iii) follows by means of (3.11) and the dominated convergence theorem.

3.4. COROLLARY. Let the hypotheses be as in Proposition 3.3. The following assertions hold:

(i) For $M \geq 0$ and for $z \in \mathrm{res}(K_M) \cap \mathrm{res}(K_\Sigma)$ the resolvent difference $(K_M - zI)^{-1} - J^*(K_\Sigma - zI)^{-1} J$ is a compact operator.

(ii) The equality $\lim_{M\to\infty} \left\|(K_M - zI)^{-1} - J^*(K_\Sigma - zI)^{-1} J\right\| = 0$ is valid for $z \in \mathrm{res}(K_\Sigma)$.

PROOF. These assertions follow from the propositions 3.3 and 3.4 together with the identities:

$$J(K_M - zI)^{-1} - (K_\Sigma - zI)^{-1} J = J\left[(K_M - zI)^{-1} - J^*(K_\Sigma - zI)^{-1} J\right]$$

and

$$(K_M - zI)^{-1} - J^*(K_\Sigma - zI)^{-1} J = P(K_M - zI)^{-1}$$
$$- J^*\left[J(K_M - zI)^{-1} - (K_\Sigma - zI)^{-1} J\right].$$

Acknowledgement. The authors are grateful to R. Seiler, Technische Universität Berlin, for the support, which the second author received during his stay in Berlin (September 1990). The first author is grateful for the support given by the DFG for the project "Schrödinger operators" that he enjoyed together with Prof. W. Kirsch from Bochum. The second author is obliged to the University of Antwerp (UIA) and to the National Fund for Scientific Research (NFWO) for their material support. He is also indebted to the European Science Project: Evolutionary Systems, Deterministic and Stochastic Evolution Equations, Control Theory and Mathematical Biology.

REFERENCES

[1] M. Aizenman and B. Simon, Brownian motion and Harnack inequality for Schrödinger operators, Comm. Pure and Applied Math., Vol. XXXV, 1982, 209-273.

[2] R. Azencott, P. Baldi, A. Bellaiche, C. Bellaiche, P. Bougerol, M. Chaleyat-Maurel, L. Elie, J. Granara, Géodésiques et diffusions en temps petit, séminaire de probabilités, Université de Paris VII, Astérisque 84-85, Soc. Math. de France 1981.

[3] K. Baumgärtel and M. Wollenberg, Mathematical scattering theory, Birkhäuser, Basel 1983.

[4] J.-M. Bismut, Large deviations and the Malliavin Calculus, Birkhäuser 1984.

[5] R.M. Blumenthal and R.K. Getoor, Markov processes and potential theory, Academic Press 1986.

[7] H.L. Cycon, R.G. Froese, W. Kirsch and B. Simon, Schrödinger operators with applications to Quantum Mechanics and global geometry, Springer Verlag, Berlin 1987.

[8] E.B. Davies, Pointwise bounds on the space and time derivatives of heat kernels, *J. Operator Theory*, no. 21, 1989, 367-378.

[9] E.B. Davies, Heat kernels and spectral theory, Cambridge University Press, Cambridge 1989.

[10] E.B. Davies, Heat kernel bounds, conservation of probability and the Feller property, preprint London, 1991.

[11] E.B. Davies, Gaussian upper bounds for the heat kernels of some second-order operators on Riemannian manifolds, *J. of Funct. Analysis* **80**, 1988, 16-32.

[12] M. Demuth, On large coupling operator norm convergences of resolvent differences, SFB 237 - Preprint Nr. 83, Institut für Mathematik, Ruhr-Universität, Bochum, to be published in J. of Math. Physics.

[13] M. Demuth and J.A. van Casteren, On spectral theory for selfadjoint Feller generators, *Reviews in Mathematical Physics*, Vol 1, no. 4, 1989, 325-414.

[14] K.D. Elworthy, Geometric aspects of diffusions on manifolds, in École d'été de Probabilités de Saint-Flour XV-XVII, 1985-1987, editor P.L. Hennequin, Lecture Notes in Math. 1362, Springer-Verlag, Berlin 1988.

[15] K.D. Elworthy, Stochastic differential equations on manifolds, London Math. Soc. Lecture Notes in Mathematics **70**, Cambridge University Press 1982.

[16] K.D. Elworthy and A. Truman, Classical Mechanics, the diffusion (heat) equation, and the Schrödinger equation on a Riemannian manifold, *J. of Math. Physics* **22**, 1981, 2144-2166.

[17] I.W. Herbst and Zhongxin Zhao, Sobolov spaces, Kac regularity and the Feynman-Kac formula, in Seminar on Stochastic Processes 1987, edited by E. Cinlar, K.L. Chung, R.K. Getoor and J. Glover, Birkhäuser Basel 1988.

[18] W. Hoh and N. Jacob, Some Dirichlet forms generated by pseudo differential operators, preprint Universität Erlangen-Nürnberg 1991, *Bull. Sc. Math.*

[19] T. Ichinose, Essential selfadjointness of the Weyl quantized relativistic Hamiltonian, *Ann. Inst. Henri Poincaré*, Vol. 51, no. 3, 1889, 265-298.

[20] T. Ichinose, The nonrelativistic limit problem for a relativistic spinless particle in an electromagnetic field, *J. Functional Analysis* **73**, 233-257, 1987.

[21] T. Ichinose, Path integral for a Weyl quantized relativistic Hamiltonian and the nonrelativistic limit problem, in Differential Equations and Mathematical Physics, Lecture Notes in Mathematics **1285**, 205-210, 1988.

[22] T. Ichinose, Kato's inequality and essential self-adjointness for Weyl quantized relativistic Hamiltonian, *Proc. Japan Acad.* **64A**, 367-369, 1988.

[23] T. Ichinose and H. Tamura, Imaginary-time path integral for a relativistic spinless particle in a electromagnetic field, *Comm. in Math. Physics Vol.* **105**, 239-257, 1986.

[24] T. Ichinose and H. Tamura, Path integral for Weyl quantized relativistic Hamiltonian, *Proc. Japan Acad.* **62A**, 91-93, 1986.

[25] N. Ikeda and S. Watanabe, Stochastic differential equations and diffusion processes, second edition, North-Holland 1989.

[26] N. Jacob, Feller semigroups, Dirichlet forms, and pseudo differential operators, preprint Universität Erlangen, 1990, to appear in *Forum Math.*.

[27] N. Jacob, A class of elliptic differential operators generating symmetric Dirichlet forms, preprint Universität Erlangen, 1990.

[28] N. Jacob, Pseudo differential operators with negative definite functions as symbol: Applications in probability theory and Mathematical physics, Preprint University of Erlangen-Nürnberg.

[29] T. Kato, Perturbation theory for linear operators, Springer Verlag, Berlin 1976.

[30] T. Kato, On the Trotter-Lie product formula, *Proc. Japan Acad.*, **50**, 1974, 694-698.

[31] R.Z. Khas'minskii, On positive solutions of the equation $\mathcal{A}u + Vu = 0$, *Theory of Probability and its applications*, Vol. IV, no. 3, 1959, 309-318.

[32] A.N. Kochubeĭ, Singular parabolic equations and Markov processes, *Math. USSR Izvestiya*, Vol. **24**, no. 1, 1985, 73-97.

[33] A.N. Kochubeĭ, Parabolic pseudodifferential equations, hypersingular integrals and Markov processes, *Math. USSR Izvestiya*, Vol. **33**, no. 2, 1989, 233-259.

[34] S.C. Port and C.J. Stone, Brownian motion and classical potential theory, Academic Press, New York, 1978.

[35] M. Reed and B. Simon, Methods of Modern Mathematical Physics II: Fourier Analysis, Self-Adjointness, Academic Press, New York 1975.

[36] M. Reed and B. Simon, Methods of Modern Mathematical Physics III: Scattering Theory, Academic Press, New York 1979.

[37] M. Reed and B. Simon, Methods of Modern Mathematical Physics IV: Analysis of operators, Academic Press, New York 1978.

[38] B. Simon, Schrödinger semigroups, *Bulletin (New Series) of the Amer. Math. Soc.*, Vol. **7**, no. 3, 1982, p. 447-526.

[39] B. Simon, Functional integration and quantum physics, Academic Press, New York 1979.

[40] K.-Th. Sturm, Schrödinger semigroups on manifolds, Preprint, Univ. Erlangen-Nürnberg, 1991.

[41] K. Taira, On the existence of Feller semigroups with boundary conditions, Memoirs of the Am. Math. Soc. 267, Providence 1991.

[42] J.A. van Casteren, Generators of strongly continuous semigroups, Research Notes in Math. 115, Pitman, London 1985.

[43] J.A. van Casteren, Integral kernels and the Feynman-Kac formalism, in Aspects of Positivity in Functional Analysis, R. Nagel, U.Schlotterbeck and M.P.H. Wolff (editors), North-Holland, 1986, 179-185.

[44] J.A. van Casteren, A pointwise inequality for generalized Schrödinger semigroups, in the Proceedings of the Conference to be published by Teubner, Leipzig, in the Proceedings of the conference "Symposium Partial Differential Operators Holzhau 1988" held from April 24 until April 30, 1988, (B.W. Schulze and H. Triebel, editors), Teubner Texte zur Mathematik, Band 112, 298-312, Teubner Verlagungsgesellschaft, DDR-7010 Leipzig 1989.

[45] J.A. van Casteren, On generalized Schrödinger semigroups, in Markov Processes and Control Theory (H. Langer and V. Nollau, editors), Proceedings of the conference ISAM 88, "Markovsche Prozessen und Steuerungstheorie" held at Gaussig, DDR, 11-15 January 1988 under auspices of Technical University of Dresden, Mathematical Research Vol. 54, Akademie Verlag Berlin 1989, DDR, 16-39.

[46] J. Weidmann, Linear operators in Hilbert space, Graduate texts in Mathematics **68**, Springer-Verlag, Berlin 1980.

The "Cumulative" Formulation of (Physiologically) Structured Population Models

O. DIEKMANN CWI Amsterdam, The Netherlands, and Institute of Theoretical Biology, University of Leiden, Leiden, The Netherlands

MATS GYLLENBERG Luleå University of Technology, Luleå, Sweden

J. A. J. METZ Leiden University, Leiden, The Netherlands

HORST THIEME Arizona State University, Tempe, Arizona

1. STRUCTURED POPULATION MODELS: A CHALLENGE FOR SEMIGROUP THEORY?

The formulation of a physiologically structured population model starts at the individual level (*i*-level, for short). After the choice of finitely many *i*-state variables and the feasible *i*-state space $\Omega \subset \mathbb{R}^n$ one specifies how

- *i*-state change ('growth')
- probability per unit of time of dying
- probability per unit of time of giving birth
- distribution of *i*-state of offspring at birth
- influence on the environment ('consumption')

depend on the *i*-state of the individual concerned and on the state of the environment (see METZ & DIEKMANN, 1986, for a systematic exposition and a wealth of examples).

Lifting the *i*-model to the population level (*p*-level, for short) is, at least as long as one restricts attention to the deterministic (i.e. large numbers of individuals) approximation and to formal aspects, a matter of straightforward mathematical book-keeping. If one works with densities, the *p*-state space is $L_1(\Omega)$ and the Kolmogorov forward equation takes the form of a first order partial differential equation (pde) with non-local terms or boundary conditions to describe the reproduction. If one chooses the space of measures $M(\Omega)$ as p-state space it is easiest to write down the Kolmogorov backward equation, i.e. the pre-adjoint equation on $C_0(\Omega)$. Again one obtains a first order pde with non-local terms (see METZ & DIEKMANN, 1986, HEIJMANS 1986[a,b], DIEKMANN, to appear).

Thus the models are, completely in line with applied mathematical tradition, formulated in terms of rates and differential equations. The task for mathematicians is now to show that the pde's generate dynamical systems and to analyse how the qualitative and quantitative behaviour of the solution operators depends on the ingredients of the model (see WEBB, 1985, for the special case of age structure).

In order to keep models parameter-scarce, biologists will come up with 'idealizations' of reality which create discontinuities (e.g. individuals start reproducing exactly when their size passes a certain threshold value; size at birth has a fixed value). Such 'idealizations' severely complicate the task of describing the domain of definition of infinitesimal generators in the precise functional analytic sense.

Motivated to some, or even a large, extent by structured population models various linear and semi-linear methods have been developed in recent years to deal with refractory generators, in particular

- regularization via resolvent (DESCH & SCHAPPACHER 1984)
- integrated semigroups (ARENDT 1987, KELLERMAN & HIEBER 1989, NEUBRANDER 1988, THIEME 1990[a], ARENDT, NEUBRANDER & SCHLOTTERBECK to appear, LUMER 1990)
- restriction of maximal operator (GREINER 1987)
- $\odot$ * calculus (CLÉMENT et al. 1987, 1989, VAN NEERVEN 1992)
- non-densely defined operators (BÉNILAN, CRANDALL & PAZY 1988, DA PRATO & SINESTRARI 1987, THIEME 1990[b])

All of these methods work well when the i-state space is one dimensional and one restricts to autonomous linear or semilinear problems. For higher dimensional i-state space the best result in published form seems to be in TUCKER & ZIMMERMAN (1988), which is less functional analytic in spirit (and does not cover the case of a fixed birth-size).

It seems legitimate to ask why a general theory is still missing, despite the effort of several people (notably the authors) over a number of years. Are we asking for too much?

Let us concentrate for a moment on the case where the environmental variables are given functions of time (Mathematically speaking: the non-autonomous linear case). While an individual is 'growing', its i-state follows an orbit in Ω. In pde jargon these orbits are the (projected) characteristics. Due to the time-dependence in the 'growth' rates, these orbits change in time. At the p-level we have a density or measure translated along the orbits. The pde presupposes that we have differentiability, in some sense, along the changing orbits.

In the autonomous case the orbits are fixed. The Hille-Yosida theorem tells us that it is possible to unambiguously define a generator which completely characterizes the semigroup. In the non-autonomous case we have an evolutionary system of operators and there is no analogue of the Hille-Yosida theorem for those: it is questionable whether a generating family exists that completely characterizes the evolutionary system. In fact, the example of translating a function along a family of curves, which deform with time, nicely illustrates the difficulties involved.

The great advantage of differential equations as a modelling tool is that we can consider different mechanisms (like 'growth', death and reproduction) separately and then simply add their contributions since, in infinitesimal time intervals, they act independently. Yet there is a technical price (as another manifestation of technical difficulties, notice that birth rates may be undefined on time sets of measure zero, while the numbers, which one obtains by integration, are perfectly well defined). So do we insist on differential equations as the language to formulate the models?

In order to answer this question one has to take stock of alternatives. The aim of this paper is to demonstrate that a very natural and attractive alternative exists (and a very old and familiar one, in fact): integral equations.

2. MODEL INGREDIENTS AT THE i-LEVEL

In this and the next section we consider the situation in which the environmental variables are a given function of time, denoted by E. Even though all objects introduced below depend on E, we shall not express this dependence in our notation. In section 4 we shall describe how nonlinear problems are obtained by feedback through the environment and there we shall employ a more precise (but also more cumbersome) notation.

In this section we list one by one the modelling ingredients related to the various mechanisms of change in i-state or number of individuals. We do so in 'cumulative' terms, as opposed to rates. We first present the ingredients in the form they are needed at the p-level and then discuss how, in certain special situations, these ingredients themselves can be decomposed into more elementary building-blocks.

(i) *i-movement and survival*

$u(t, t_0, y_0)(\omega)$ = probability that an individual that is in state y_0 at time t_0 will still be alive at time t and then have i-state in the (measurable) subset ω of Ω

The interpretation requires that u satisfies the consistency relation

$$u(t+s, t_0, y_0)(\omega) = \int_\Omega u(t+s, t, y)(\omega)u(t, t_0, y_0)(dy) \tag{2.1}$$

which is often called the Chapman-Kolmogorov equation. The survival function $\mathcal{F}$ is defined by

$$\mathcal{F}(t, t_0, y_0) = u(t, t_0, y_0)(\Omega) = \int_\Omega u(t, t_0, y_0)(dy) \tag{2.2}$$

Taking $\omega = \Omega$ in (2.1) we find that the relation

$$\mathcal{F}(t+s, t_0, y_0) = \int_\Omega \mathcal{F}(t+s, t, y)u(t, t_0, y_0)(dy) \tag{2.3}$$

should hold.

In the special case of deterministic growth one postulates the existence of a function $Y(t, t_0, y_0)$, giving the i-state at time t, given that the i-state was y_0 at time t_0 and given survival. One then takes the survival function $\mathcal{F}$ as a second building-block and puts

$$u(t, t_0, y_0) = \delta_{Y(t,t_0,y_0)}\, \mathcal{F}(t, t_0, y_0) \tag{2.4}$$

where, as usual, δ_y denotes the unit measure concentrated in the point y. Substituting (2.4) into (2.1) we find that Y should have the semigroup property

$$Y(t + s, t_0, y_0) = Y(t + s, t, Y(t, t_0, y_0)) \tag{2.5}$$

while $\mathcal{F}$ should satisfy the consistency condition

$$\mathcal{F}(t + s, t_0, y_0) = \mathcal{F}(t + s, t, Y(t, t_0, y_0))\, \mathcal{F}(t, t_0, y_0). \tag{2.6}$$

Most often both Y and $\mathcal{F}$ will be derived from a differential equation, respectively

$$\frac{dY}{dt} = v(Y, E) \quad , \quad Y(t_0, t_0, y_0) = y_0, \tag{2.7}$$

and

$$\frac{d\mathcal{F}}{dt} = -\mu(Y,E)\mathcal{F} \quad , \quad \mathcal{F}(t_0, t_0, y_0) = 1, \tag{2.8}$$

and then (2.5) and (2.6) will automatically hold. Note that we can write

$$\mathcal{F}(t, t_0, y_0) = \exp(-\int_{t_0}^{t} \mu(Y(\tau, t_0, y_0), E(\tau))d\tau) \tag{2.9}$$

since (2.7) is decoupled from (2.8). When we start from data, rather than from model equations, $\mathcal{F}$ is the measured quantity and μ is introduced as minus its logarithmic derivative.

(ii) *reproduction*

$\Lambda(t, t_0, y_0)(\omega)$ = expected total number of direct offspring (i.e. children but not grand children, great grand children etc.), with state-at-birth in the (measurable) subset ω of Ω, of an individual having i-state y_0 at time t_0, in the time-interval $[t_0, t]$

Following the terminology of branching processes (JAGERS, 1989, 1991) we shall call Λ the *reproduction kernel*. Note that Λ is unconditional (i.e. we have not conditioned on survival of the individual till time t). Consequently the appropriate additive consistency relation is

$$\Lambda(t + s, t_0, y_0) = \Lambda(t, t_0, y_0) + \int_{\Omega} \Lambda(t + s, t, y)u(t, t_0, y_0)(dy). \tag{2.10}$$

In the special case of deterministic i-movement (2.10) reduces to

$$\Lambda(t + s, t_0, y_0) = \Lambda(t, t_0, y_0) + \Lambda(t + s, t, Y(t, t_0, y_0))\mathcal{F}(t, t_0, y_0). \tag{2.11}$$

Very often the components

$\lambda(y,e)$ = expected rate at which an individual in i-state y, currently living under environmental conditions e, gives birth

and

$p(y, e)(\omega)$ = probability that a neonate born from a mother with i-state y under environmental conditions e has itself i-state in the (measurable) subset ω of Ω

are used to define Λ by the formula

$$\Lambda(t, t_0, y_0) = \int_{t_0}^{t} \int_{\Omega} p(y, E(\tau))\lambda(y, E(\tau))u(\tau, t_0, y_0)(dy)d\tau \tag{2.12}$$

which in the special case of deterministic i-movement reduces to

$$\Lambda(t, t_0, y_0) = \int_{t_0}^{t} p(Y(\tau, t_0, y_0), E(\tau))\lambda(Y(\tau, t_0, y_0), E(\tau))\mathcal{F}(\tau, t_0, y_0)d\tau. \tag{2.13}$$

In certain pathological cases such a formula does not produce an unambiguous result. For instance, when λ has discontinuities as a function of λ, these have to be crossed transversally by Y, in the sense that the set $\{\tau{:}\lambda(\cdot, E(\tau))$ is discontinuous at $Y(\tau, t_0, y_0)\}$ has measure zero, in order that the right hand side of (2.13) yields a well-defined number. In a pde formulation, involving the ingredients p, λ, μ and v (see (2.7), (2.8) and (2.13)), one has to face this difficulty (which is rather hidden in the pde!) when analysing population behaviour. In the present approach, possible pathologies of specific models are dealt with in the phase of modelling i-behaviour, in particular when one concentrates on establishing the relationship between Λ and its constituents. On making, in the next section, the step from the i- to the p-level, we simply assume that Λ is well-defined and has suitable properties. We hope that this digression clarifies the advantage of the 'cumulative' (or renewal equation, see next section) approach as compared to the 'rates' (or pde) approach.

3. BOOK-KEEPING AT THE p-LEVEL

Let the population size and composition at some time t_0 be described by a (Borel) measure m. We shall call this group of individuals the zero'th generation. Let us, for a moment, disregard reproduction. At time $t>t_0$ both the size and composition are changed as a result of i-state change and death. Our description in terms of u at the i-level is immediately lifted to the p-level to yield the *generation development operators* U_0 defined by

$$(U_0(t, t_0)m)(\omega) = \int_{\Omega} u(t, t_0, y_0)(\omega)m(dy_0) \tag{3.1}$$

The Chapman-Kolmogorov equation (2.1) guarantees that U_0 is an evolutionary system:

$$U_0(t+s, t_0) = U_0(t+s, t)U_0(t, t_0). \tag{3.2}$$

Next, let's look at the *direct* offspring of the zero'th generation. The (cumulative) *direct offspring operators* are defined by

$$(K(t, t_0)m)(\omega) = \int_{\Omega} \Lambda(t, t_0, y_0)(\omega)m(dy_0) \tag{3.3}$$

and they yield the expected cumulative number of direct offspring in the time interval $[t_0, t]$, as distributed with respect to the i-state at birth (whence their name). The consistency condition (2.10) translates into the relation

$$K(t+s, t_0) = K(t, t_0) + K(t+s, t)U_0(t, t_0) \tag{3.4}$$

which is the non-autonomous counterpart of the defining relation for a *'cumulative output family'* as introduced in an autonomous setting by DIEKMANN, GYLLENBERG & THIEME (to appear).

So far everything is explicit, i.e. both (3.1) and (3.3) are explicit formulas in terms of the ingredients at the i-level. But now we have to pay attention to offspring of offspring, and so on indefinitely. In other words, we have to iterate the operator family $K(t, t_0)$ with due care for the time structure. As an equivalent alternative for such an infinite sequence expansion we can introduce an abstract Stieltjes renewal equation as follows.

Let the (cumulative) *total offspring operators* $R(t, t_0)$ be the analogues of $K(t, t_0)$ when considering the total clan, i.e. including offspring of offspring of ... . More precisely, $(R(t, t_0)m)(\omega)$ is the expected cumulative number of *all* births, with i-state at birth in the set $\omega \subset \Omega$, in the time interval $[t_0, t]$, given that the population at time t_0 was described by the measure m. Then consistency requires that

$$R(t, t_0) = K(t, t_0) + \int_{t_0}^{t} K(t, \tau)R(d\tau, t_0) \tag{3.5}$$

since any newborn is either the offspring of an individual already present at time t_0 or of an individual born after time t_0. Solving this equation by successive approximations, we obtain the generation expansion alluded to above.

We need a final step to convert i-state at birth to i-state at current time, given the time of birth, while accounting for the possibility of death. Fortunately the operators describing that transformation are already at our disposal, in the form of the U_0-family. We define the *population development operators* U by

$$U(t, t_0) = U_0(t, t_0) + \int_{t_0}^{t} U_0(t, \tau)R(d\tau, t_0) \tag{3.6}$$

which is an explicit expression, once we consider $R(t, t_0)$ as known. So $U(t, t_0)$ tells us how the population size and composition at time t derives from the same information at some earlier time t_0. Hence it should be an evolutionary system, i.e. the algebraic relation

$$U(t+s, t_0) = U(t+s, t)U(t, t_0) \tag{3.7}$$

should hold. The key to a verification of this property is the observation that R should be a cumulative output family for U, i.e.

$$R(t+s, t_0) = R(t, t_0) + R(t+s, t)U(t, t_0). \tag{3.8}$$

Note that (3.8) has exactly the same interpretation as (3.4), the only difference being that now we consider *all* offspring rather than direct offspring only. Relation (3.8) follows from (3.4), the definition (3.6) and the uniqueness of solutions of the renewal equation (3.5).

For completeness we give the expression for U_0 in the special case of deterministic i-state change (see (2.4)):

$$(U_0(t, t_0)m)(\omega) = \int_{\Omega \cap Y(t_0, t, \omega)} \mathcal{F}(t, t_0, y_0)m(dy_0) \tag{3.9}$$

Let us recapitulate the situation. The model ingredients at the i-level introduced in section 2 allow us to define operator families U_0 and K which satisfy the relations (3.2) and (3.4) or, in words, U_0 is an evolutionary system and K is a corresponding cumulative output family. Only one equation figures in our theory, the renewal equation (3.5), which can be solved by successive approximations. Once we have the solution of (3.5) we can write down an explicit expression, (3.6), for the object that we are after, the evolutionary system U that tells us how p-states at some time are mapped onto p-states at later times.

4. FEEDBACK THROUGH THE ENVIRONMENT

Let us begin by looking at an example. Suppose that substrate concentration S is one of the environmental variables and that substrate is consumed at a per capita rate $\gamma(y, S)$. If substrate dynamics follows the logistic differential equation in the absence of consumers, we add the equation

$$\frac{dS}{dt} = rS\left(1 - \frac{S}{K}\right) - \int_{\Omega} \gamma(y, S) n(dy) \tag{4.1}$$

to the description of the system, where n is the measure describing the consumer population and where the time argument is suppressed in the notation. Substituting for n in this equation $U^E(t, t_0)m$, where m is the initial condition at time t_0 and where we now have incorporated the dependence on E by writing a super-index, we are left with a *functional* differential equation for S (recall that E involves S as one of its components), which we have to solve. To verify that one can use a contraction mapping argument amounts to studying the dependence of $\langle \gamma(\cdot, S), U^E(t, t_0)m \rangle$ on the environmental component S, as a given (continuous, say) function on $[t_0, t]$. Here special properties of either $\gamma(\cdot, S)$ or m may contribute to the Lipshitz estimates. The technicalities are somewhat involved and will be dealt with in a future paper.

If all environmental variables are derived from a deterministic dynamical system we have, in general, the ode

$$\frac{dE}{dt} = F(E, L^E(n)) \tag{4.2}$$

where L^E is a continuous linear mapping from the space of measures on Ω into $\mathbf{R}^k$, for some k, with suitable smoothness conditions for both F and $E \mapsto L^E$. Quasi steady state assumptions produce an algebraic equation

$$E = G(E, L^E(n)) \tag{4.3}$$

as an alternative to (4.2).

It may be that only some of the environmental variables are described by an equation of the form (4.2) or (4.3), while others, such as temperature or light intensity, are considered as given (i.e. to be measured or experimentally controlled) functions of time (one does not want to make weather prediction a component of a population dynamics model!). The presence of such variables makes even the resulting non-linear dynamical system non-autonomous. But if such variables are absent, one wants to show that the resulting non-linear dynamical system is autonomous. The key to a proof of this property are the relations

$$\Lambda^{E_{-s}}(t + s, t_0 + s, y_0) = \Lambda^E(t, t_0, y_0) \tag{4.4}$$

and

$$u^{E_{-s}}(t + s, t_0 + s, y_0) = u^E(t, t_0, y_0) \tag{4.5}$$

where

$$E_{-s}(t) := E(t-s) \tag{4.6}$$

We intend to elaborate on this topic in the near future.

Finally we remark that in some models the environment is infinite dimensional, rather than finite dimensional as we have assumed in this paper for ease of formulation.

5. CONCLUSIONS

General physiologically structured population models can be mathematically described by a renewal equation and a feedback law. The main advantage of such a formulation is that it avoids the use of unbounded operators (the generating family) for which a precise description of the domain of definition is a technically hard and unpleasant task, if possible at all.

Our plan is to use this formulation as the starting point for a qualitative theory for nonlinear problems, dealing in particular with stability and bifurcation.

We think that even in the linear autonomous case the formulation has advantages. In that case (3.5) reduces to a convolution equation (see DIEKMANN, GYLLENBERG & THIEME, to appear) and Laplace transformation yields information about the asymptotic behaviour. This information is equivalent to a spectral analysis of the generator, but can be obtained without deriving a precise characterization of the domain of definition of the generator. Moreover, quite often a reduction in the 'size' of the problem is obtained, since the support of $\Lambda(t, t_0, y_0)$ is usually much smaller than $M(\Omega)$ (e.g., in the case of age, everybody is born with age zero!).

Acknowledgement. The research of H.R. Thieme was partially supported by NSF grant DMS-9101979. The research of M. Gyllenberg was partially supported by The Bank of Sweden Tercentenary Foundation, the Swedish Council for Forestry and Agricultural Research and the Swedish Cancer Foundation.

REFERENCES

ARENDT, W. (1987). Vector-valued Laplace transforms and Cauchy problems, Israel J. Math. **59**: pp. 327-352.

ARENDT, W., NEUBRANDER, F. and SCHLOTTERBECK, U. (to appear), Interpolation of semigroups and integrated semigroups, Semigroup Forum.

BÉNILAN, PH., CRANDALL, M.G. and PAZY, A. (1988). Bonnes solutions d'un problème d'évolution semi-linéaire. C.R. Acad. Sc. **306**: pp. 527-530.

CLÉMENT, PH., DIEKMANN, O., GYLLENBERG, M., HEIJMANS, H.J.A.M. and THIEME, H.R. (1987). Perturbation theory for dual semigroups. Part I. The sun-reflexive case, Math. Ann. **277**: pp. 709-725.

CLÉMENT, PH., DIEKMANN, O., GYLLENBERG, M., HEIJMANS, H.J.A.M. and THIEME, H.R. (1989). Perturbation theory for dual semigroups. Part IV. The intertwining formula and the canonical pairing, Semigroup Theory and Applications, (Ph. Clément, S. Invernizzi, E. Mitidieri, I.I. Vrabie, eds.) Marcel Dekker, New York, pp. 95-116.

Da Prato, G. and Sinestrari, E. (1987). Differential operators with non-dense domain, Ann. Sc. Norm. Pisa **14**: pp. 285-344.

Desch, W. and Schappacher, W. (1984). On relatively bounded perturbations of linear C_0-semigroups, Ann. Sc. Norm. Super. Pisa Cl. Sci., IV Ser. **11**: pp. 327-341.

Diekmann, O. (to appear). Dynamics of structured populations: biological modelling and mathematical analysis, Proceedings of Journées sur les Equations Différentielles, Marrakech, 1989.

Diekmann, O., Gyllenberg, M. and Thieme, H.R. (to appear). Perturbing semi-groups by solving Stieltjes renewal equations, Diff. Int. Equ.

Greiner, G. (1987). Perturbing the boundary conditions of a generator, Houston J. Math., **13**: pp. 213-229.

Heijmans, H.J.A.M. (1986)[a]. Structured populations, linear semigroups and positivity, Math. Z., **191**: pp. 599-617.

Heijmans, H.J.A.M. (1986)[b]. Markov semigroups and structured population dynamics, Aspects of Positivity in Functional Analysis, (R. Nagel, U. Schlotterbeck, M.P.H. Wolff, eds.), Elsevier, Amsterdam, pp. 199-208.

Jagers, P. (1989). The Markov structure of population growth, Evolution and Control in Biological Systems, (A.B. Kurzhanski & K. Sigmund, eds.) Kluwer, Dordrecht, pp. 103-114.

Jagers, P. (1991). The growth and stabilization of populations, Statistical Science, **6**: pp. 269-283.

Kellerman, H. and Hieber, M. (1989). Integrated semigroups, J. Funct. Anal., **84**: pp. 160-180.

Lumer, G., (1990). Solutions généralisées et semi-groupes intégrés, C.R. Acad. Sci. Paris, **310**: pp. 577-582.

Metz, J.A.J. and Diekmann, O. (eds) (1986). The Dynamics of Physiologically Structured Populations, Lect. Notes in Biomath. **68**, Springer, Berlin etc.,

van Neerven, J.M.A.M. (1992). The Adjoint of a Semigroup of Linear Operators, Ph. D. Thesis, Leiden University.

Neubrander, F. (1988). Integrated semigroups and their applications to the abstract Cauchy problem, Pacific J. Math. **135**: pp. 111-155.

THIEME, H.R. (1990)[a]. Integrated semigroups and integrated solutions to abstract Cauchy problems, J. Math. Anal. Appl. **152**: pp. 416-447.

THIEME, H.R. (1990)[b]. Semiflows generated by Lipschitz perturbations of non-densely defined operators. Diff. Int. Eq. **3**: pp. 1035-1066.

TUCKER, S.L. and ZIMMERMAN, S.O. (1988). A nonlinear model of population dynamics containing an arbitrary number of continuous variables, SIAM J. Appl. Math. **48**: pp. 549-591.

WEBB, G.F. (1985). Theory of Nonlinear Age-Dependent Population Dynamics, Marcel Dekker, New York.

Quasilinear Diffusions

J. R. DORROH and GISÈLE RUIZ GOLDSTEIN Louisiana State University, Baton Rouge, Louisiana

Nonlinear diffusion equations occur frequently in many applications, including engineering, physics, mathematical biology, and geometry. We shall focus on equations of the form

$$u_t = \varphi(t, x, u, \nabla u)\Delta u + \psi(t, x, u, \nabla u) \tag{1.1}$$

for $t \in [0, T]$, $x \in \Omega$, where Ω is a bounded domain in $\mathbb{R}^n$, $n \geq 1$. If $\varphi(t, x, p, q) > \delta > 0$ for all $x \in \Omega$, $p \in \mathbb{R}$, $q \in \mathbb{R}^n$, then the problem (1.1) is said to be *uniformly parabolic*. Uniformly parabolic problems have been studied extensively, and there is a vast amount of literature about them, see for example Friedman (1964), Ladyzhenskaya et al. (1968).

If φ vanishes for some $x \in \overline{\Omega}$, $p \in \mathbb{R}$, $q \in \mathbb{R}^n$, then the problem (1.1) is *degenerate parabolic*, or *singular*. Problems of this type arise naturally in applications as well, but less is known about the degenerate case. The equation

$$u_t = \Delta(u^m), \qquad m > 1 \tag{1.2}$$

describes the flow of fluids through a porous medium. Here the nonnegative function $u(t, x)$ represents a constant multiple of the density of a fluid at the point $x \in \Omega$. Formally, this equation can be written in the form (1.1), where the parabolicity degenerates when u vanishes. Aronson, with various coauthors, has studied existence, uniqueness, and regularity properties of solutions for the porous medium equation, see Aronson (1986), Aronson and Bénilan (1979), Aronson et al. (1982, 1985). Techniques from semigroups of nonlinear operators can also be used to study (1.2), see for example Bénilan et al. (1975).

Degenerate equations also arise naturally in problems from the calculus of variations. Steady-state solutions of the parabolic problem

$$u_t = \nabla \cdot (|\nabla u|^p \nabla u) \tag{1.3}$$

in a bounded domain Ω are solutions of the Euler-Lagrange equation for the problem

$$\min\{\Phi(u) : u \in W^{1,p+2}(\Omega),\ u - h \in W_0^{1,p+2}(\Omega)\}$$

where $h : \mathbb{R}^n \to \mathbb{R}$ and $\Phi(u) = \int_\Omega |\nabla u|^{p+2} dx$. Again, formally, (1.3) is a degenerate parabolic problem which can be written in the form (1.1). However, here φ degenerates when ∇u vanishes. These problems and properties of their solutions have been studied by many authors, see Evans (1982), Ladyzhenskaya and Ural'tseva (1968).

Singular parabolic problems also appear in mathematical biology. In the 1950's, Feller studied such equations as

$$u_t = x(1-x)u_{xx} \quad , \quad x \in [0, 1]. \tag{1.4}$$

The semigroup associated with (1.4) describes the diffusion limit of the Wright-Fisher model in genetics. This is a model involving an n-state Markov chain, described by a semigroup acting on the unit cube in $\mathbb{R}^n$. The limiting semigroup as $n \to \infty$ acts on

$C[0,1]$, and the corresponding operator is $x(1-x)(d^2/dx^2)$ with Wentzel boundary conditions, see for example Norman (1975).

The equation (1.4) is of the form (1.1); the diffusion coefficient vanishes at the spatial boundary of this one-dimensional problem. Goldstein and Lin (1987, 1989, 1991) began a program of studying the equation

$$u_t = \varphi(x,u,u_x)u_{xx} + \psi(x,u,ux) \tag{1.5}$$

where either $\varphi(0,p,q)$ or $\varphi(1,p,q)$ could vanish for some or all $p \in \mathbb{R}$, $q \in \mathbb{R}^n$. They assumed that the diffusion coefficient was independent of u, that is

$$\varphi(x,u,u_x) = \varphi(x,u_x) \tag{1.6}$$

and that ψ had a special form. If we let $Au = \varphi(\cdot,u')u'' + \psi(\cdot,u,u')$, then, with sufficient restrictions on φ and ψ and suitable boundary conditions, the operator A is the generator of a contraction semigroup in $C[0,1]$ equipped with the supremum norm. Specifically, they show that this operator A (with a suitable domain) is m-dissipative, that is $\|(I-\lambda A)f-(I-\lambda A)g\| \geq \|f-g\|$ for all $f,g \in \mathcal{D}(A)$, and $\mathcal{R}(I-\lambda A) = C[0,1]$ for some (and hence all) $\lambda > 0$. Thus, the problem can be solved by means of the Crandall-Liggett theorem. This technique allows a variety of boundary conditions to be used:

$$\begin{aligned} u(t,0) &= \alpha u_x(t,0) \\ u(t,1) &= -\beta u_x(t,1) \end{aligned} \tag{LBC}$$

for $\alpha, \beta \geq 0$, or

$$\begin{aligned} u(t,0) &= u(t,1) \\ u_x(t,0) &= u_x(t,1) \end{aligned} \tag{PBC}$$

or

$$u_x(t,0) \in \beta_0(u(t,0)) \tag{NBC}$$

$$u_x(t,1) \in -\beta_1(u(t,1))$$

for $\beta_0, \beta_1 : \mathbb{R} \to \mathbb{R}$ strictly increasing maximal monotone graphs with $\beta_i(0) \ni 0$ for $i = 0, 1$.

If the assumption that φ is independent of u is *not* made, the problem becomes far more difficult. At the level of the range condition, this u-dependence makes the corresponding elliptic problem more difficult to solve. However, the main (and technically very difficult) difference is the lack of global dissipativity, so that the Crandall-Liggett theorem no longer applies. This problem, as well as questions of regularity of solutions of (1.5), both with and without u-dependence in φ, has been studied in one dimension by Dorroh and Rieder ((1991) and (in preparation)) in the case of linear time-dependent boundary conditions

$$\begin{aligned} u(t,0) - \alpha u_x(t,0) &= g_1(t), \\ u(t,1) - \alpha u_x(t,1) &= g_2(t), \end{aligned} \tag{LBC$'$}$$

for $\alpha, \beta \geq 0$, and by Rieder ((1991) and (in preparation)) in the case of nonlinear boundary conditions (NBC) and periodic-like boundary conditions

$$\begin{aligned} u(t,0) - u(t,1) &= h_1(t), \\ u_x(t,0) - u_x(t,1) &= h_2(t). \end{aligned} \tag{PBC$'$}$$

In section 2 we state the main results, and in section 3 we briefly sketch the proof. Finally, we mention some open problems in the area.

2. Main Results

Consider the following initial value problem:

$$
\begin{aligned}
&u_t(t,x) = \varphi(t,x,u,u_x)u_{xx} + \psi(t,x,u,u_x),\\
&u(0,x) = u_0(x),
\end{aligned} \tag{2.1}
$$

together with either linear, nonlinear, or periodic boundary conditions. (We restrict ourselves to one space dimension.) We make the following assumptions on φ and ψ.

$$\varphi(t,x,p,q) \geq \varphi_0(x) \geq 0 \quad \textit{for } 0 \leq x \leq 1, \qquad \varphi(t,x,p,q) > 0 \quad \textit{for } 0 < x < 1, \tag{2.2}$$

where $\varphi_0 \in C[0,1]$ *and* $\varphi_0^{-1} \in L^1[0,1]$,

$$|\varphi(t,x,p,q) - \varphi(s,x,\tilde{p},q)| \leq \mathcal{K}_R(q)\varphi_0(x)\{|f(t)-f(s)| + |p-\tilde{p}|\}, \tag{2.3}$$

$$|\psi(t,x,p,q) - \psi(s,x,\tilde{p},q)| \leq \mathcal{L}_R(q)\{|f(t)-f(s)| + |p-\tilde{p}|\}, \tag{2.4}$$

and

$$|\psi(t,x,p,q)| \leq \mathcal{M}(|p|)(1+\varphi(t,x,p,q))\mathcal{N}(|q|) \tag{2.5}$$

for $|p|, |\tilde{p}| \leq R$, *where* $\mathcal{K}_R$, $\mathcal{L}_R$, $\mathcal{M}$, $\mathcal{N}$ *and* f *are nonnegative, nondecreasing functions, and* $\mathcal{N}$ *has at most linear growth.*

We define the operator $A(t)$ in $C[0,1]$ equipped with the supremum norm by

$$A(t)u = \varphi(t,\cdot,u,u_x)u_{xx} + \psi(t,\cdot,u,u_x)$$

with any *one* of the following domains.

$$\mathcal{D}_1(A(t)) = \{u \in C^1[0,1] \cap C^2(0,1) : A(t)u \in C[0,1] \text{ and } u \text{ satisfies } (LBC')\},$$

$$\mathcal{D}_2(A(t)) = \{u \in C^1[0,1] \cap C^2(0,1) : A(t)u \in C[0,1] \text{ and } u \text{ satisfies } (NBC)\},$$

$$\mathcal{D}_3(A(t)) = \{u \in C^1[0,1] \cap C^2(0,1) : A(t)u \in C[0,1] \text{ and } u \text{ satisfies } (PBC')\}.$$

We define the corresponding *generalized domains* by

$$\widehat{\mathcal{D}}_i(A(t)) = \{u \in C[0,1] : |||A(t)u||| < \infty\}$$

where

$$|||A(t)||| = \inf\{M : \exists\{u_n\} \subset \mathcal{D}_i(A(t)) \text{ such that } u_n \to u \text{ and } \|A(t)u_n\| \leq M\ \forall n\}.$$

THEOREM 1. *Assume (2.2)-(2.5), and let* $i \in \{1,2,3\}$.

(a)Then for each $u_0 \in \widehat{\mathcal{D}}_i(A(t))$, *the problem (2.1) has a unique limit solution on* $[0,T^*]$ *for some* $T^* \in (0,T]$ *which satisfies*

i) $u \in C([0,T^*]; C^1[0,1])$

ii) $u(t) \in \widehat{\mathcal{D}}_i(A(t))$,

iii) u satisfies the boundary conditions,

iv) $u_x \in C([0,T^*] \times [0,1])$.

v) u is Lipschitz continuous from $[0,T^*]$ *into* $C[0,1]$,

vi) If in addition, $\varphi_0^{-1} \in L^p[0,1]$ *for some p,* $1 < p < \infty$, *then* $u_x(t) \in C^{1/q}[0,1]$, *where* $p^{-1} + q^{-1} = 1$.

(b)If φ *satisfies (1.6), then the solution u of the problem (2.1) is a global solution defined on* $[0,T]$ *which satisfies i)-vi).*

The result as stated here is a combination of results found in Dorroh and Rieder ((1991) and (in preparation)) and Rieder ((1991) and (in preparation)). It should be noted that even under the hypothesis (1.6), part (b) extends Goldstein and Lin (1987) and Lin (to appear) by allowing more general functions ψ and by the addition of regularity results. In the case of nonlinear boundary conditions, Goldstein and Lin (1987) obtained results only for $\varphi_0^{-1} \in L^2[0,1]$. The notion of limit solution will be defined in section 3. Briefly, a limit solution is the limit of a sequence of convergent difference schemes, and it is known , see Bénilan (1972) and Evans (1975), that the notion of limit solution satisfies the basic requirements for an acceptable sense of weak solution.

We say that the operator $A(t)$ is in *divergence form* if, in addition to satisfying (2.2)-(2.5), we have

$$A(t)u = \rho(x)\frac{d}{dx}\Phi(t,x,u(x),u'(x)), \tag{2.6}$$

and

$$\Phi(t,1,\cdot,q) \textit{is nonincreasing, } \Phi(t,0,\cdot,q) \textit{nondecreasing for all } t,q. \tag{2.7}$$

where $\rho \in C[0,1]$, $\rho(x) > 0$ *for* $0 < x < 1$, *and* $\rho^{-1} \in L^1[0,1]$.

The assumption that $A(t)$ satisfies (2.2)-(2.6) is restrictive, but there are many nontrivial examples.

THEOREM 2. *Assume (2.2)-(2.7). Then for* $u_0 \in \widehat{\mathcal{D}}_i(A(0))$, *the problem*

$$u'(t) \in \overline{A(t)}u(t), \qquad (2.8)$$
$$u(0) = u_0$$

has a unique global $L^1([0,1], \rho^{-1}dx)$ *limit solution defined on* $[0,T]$. *In the case of Dirichlet or periodic boundary conditions, (2.7) is not needed.*

Here, the bar indicates closure in $L^1([0,1], \rho^{-1}dx)$. We note that the operator $A(t)$ is closed in $C[0,1]$, but not in $L^1([0,1], \rho^{-1}dx)$.

THEOREM 3. *Assume the hypothesis of Theorem 2. Then the solution u of (2.1) given by Theorem 1 is a strong* $L^1([0,1], \rho^{-1}dx)$ *solution. That is, the solution is Lipschitz continuous and differentiable a.e. as a function from* $[0,T^*]$ *into* $L^1([0,1], \rho^{-1}dx)$.

3. Sketch of Proofs

The main idea is to use the theory of semigroups of nonlinear operators. An operator A in a Banach space X is *dissipative* (respectively ω-*dissipative* or *quasidissipative*) if $I - \lambda A$ is invertible, and $(I - \lambda A)^{-1}$ is Lipschitz continuous with Lipschitz norm not exceeding 1 for $\lambda > 0$ (respectively, $\leq (1 - \lambda\omega)^{-1}$ for some $\omega \geq 0$ and $\lambda \in (0, \omega^{-1})$). A is *m-dissipative* (respectively *m-quasidissipative*) if A is dissipative (respectively quasidissipative) and $\mathcal{R}(I - \lambda A) = X$ for some $\lambda > 0$ (respectively, for some $\lambda \in (0, \omega^{-1})$). A is *essentially m-dissipative* if $\overline{\mathcal{R}(I - \lambda A)} = X$. The Crandall-Liggett-Bénilan theorem (Crandall and Liggett 1971 and Bénilan 1972) asserts that if A is m-dissipative (or m-quasidissipative), then for $u_0 \in \widehat{\mathcal{D}}(A(0))$, there is a unique limit solution to the problem

$$u' = Au, \qquad u(0) = u_0, \qquad (3.1)$$

If A is only essentially m-dissipative and even possibly multivalued, then there is still a unique limit solution to the differential inclusion $u' \in \bar{A}u$, $u(0) = u_0$. The analogue of this theorem for time dependent operators is due to Crandall and Pazy (1972) and Evans (1975, 1977).

We will now define the notion of limit solution. A *backward difference scheme* for

$$u'(t) = A(t)u(t), \qquad u(0) = u_0 \tag{3.2}$$

with $u_0 \in \widehat{\mathcal{D}}(A(0))$ is a sequence of triples $\{(t_k^n, x_k^n, f_k^n) : k = 0, 1, \ldots, N(n);\ n = 0, 1, \ldots\}$ such that: $\{t_n^k\}$ is a sequence of partitions of $[0, T]$ with mesh approaching zero as $n \to \infty$, $x_k^n, f_k^n \in X$ for each n, k, $\sup_{n,k} \|x_k^n\| < \infty$, $\sum_{k=0}^{N(n)} \|f_k^n\| \, |t_k^n - t_{k-1}^n| \to 0$ as $n \to \infty$, $x_0^n \to u_0$ as $n \to \infty$, and

$$\frac{x_k^n - x_{k-1}^n}{t_k^n - t_{k-1}^n} = A(t_k^n)x_k^n + f_k^n.$$

We define a sequence of step functions

$$u^n(t) = \begin{cases} x_k^n & \text{for} \quad t \in (t_{k-1}^n, t_k^n] \\ x_0^n & \text{for} \quad t = 0. \end{cases}$$

If the sequence $\{u^n\}$ converges uniformly to a continuous function u, then u is , by definition, a *limit solution*, or *mild solution*, of (3.2). It is known that limit solutions are unique, that a strong solution is a limit solution, that a limit solution which is Lipschitz continuous and differentiable a.e. is a strong solution, and that limit solutions are strong solutions if X is reflexive (under some restrictions on the t-dependence of the operator $A(t)$).

First, we note that a change of variables analogous to that for classical solutions reduces the problem with time dependent boundary conditions to an equivalent problem with time independent boundary conditions even in the case of limit solutions. Next, if φ is independent of u (the case considered in Goldstein and Lin (1987)), then an easy calculation shows that the corresponding operator $A(t)$ is dissipative. Demonstrating the range equation is equivalent to solving a degenerate elliptic boundary value problem in $C[0, 1]$. However, for general φ, there is no global dissipative estimate for $A(t)$. Our strategy is to first fix a $v \in C[0, 1]$ and solve

$$u - \lambda A^v(t)u = h \tag{3.3}$$

for $h \in C[0,1]$, where $A^v(t)u = \varphi(t,\cdot,v,u')u'' + \psi(t,\cdot,v,u')$. One can do this by approximating (3.3) by the uniformly elliptic problem

$$u - \lambda A_m^v(t)u = h, \tag{3.4}$$

where $A_m^v(t)$ is the operator $A^v(t)$ with φ replaced by $\varphi_m = \max\{\varphi, m^{-1}\}$. We then obtain a solution u_m of (3.4) by Green's function techniques for (LBC) or (PBC) and by a method due to Serizawa for (NBC). We obtain estimates on $\|u_m\|$, $\|u_m'\|$, and $\|u_m''\|_{L^1}$ which depend only on $\|h\|$ and various parameters that are independent of m. Passing to the limit as $m \to \infty$, we obtain a solution of (3.3). Next we define the operator $H_\lambda^h : C[0,1] \to C[0,1]$ for each $h \in C[0,1]$ by

$$H_\lambda^h v = J_\lambda^v(t)h,$$

where $J_\lambda^v(t) = (I - \lambda A^v(t))^{-1}$ is the resolvent of $A^v(t)$. Using the Schauder fixed point theorem, we prove H_λ^h has a fixed point; this is equivalent to solving

$$u - \lambda A(t)u = h.$$

The lack of a global dissipative estimate is a technically very difficult problem. The basic idea in this section of the proof is to restrict the domain of $A(t)$ to a smaller set on which $A(t)$ is dissipative and to obtain many estimates about the behavior of the resolvents $J_\lambda(t) = (I - \lambda A(t))^{-1}$ in combination with $A(s)$ and $J_\lambda(s)$ where s is not necessarily equal to t. Making careful use of these estimates, we choose sets $D_\lambda(t)$ in such a way that

$$J_\lambda(t) : D_\lambda(t) \to D_\lambda(s)$$

for $0 \le t \le s \le T^*$ for some $T^* \in (0,T]$. In this way we are able to construct a backward difference scheme and the corresponding limit solution. The details are quite complicated, see Dorroh and Rieder ((1991) and (in preparation)) and Rieder ((1991) and (in preparation)).

Theorem 2 is a consequence of the fact that an operator in divergence form is dissipative in $L^1([0,1], \rho^{-1}dx)$ and the fact that $\mathcal{R}(I - \lambda A) = C[0,1]$, which is dense in $L^1([0,1], \rho^{-1}dx)$.

To prove Theorem 3, we note that by Theorem 1, our solution is Lipschitz continuous into $C[0,1]$, and hence into $L^2([0,1],\rho^{-1}dx)$. Since L^2 has the Radon-Nikodym property, it follows that u is differentiable a.e. in L^2, and hence a.e. in L^1.

4. Open Problems

The main open problem is when can one assert that (2.1) has a global solution. It is trivial that blowup can occur for $\psi(t,x,p,q) = p^r$ for Dirichlet boundary conditions and $r > 1$. What restrictions on ψ and the boundary conditions will guarantee global solutions?

In the case of φ independent of u, Goldstein and Lin studied more rapid spatial degeneracy in A. This led naturally to the use of the Wentzel boundary condition. It is not known whether or not such techniques can also handle the dependence of the diffusion coefficient on u.

Finally, a problem of great interest is to find a method for unifying the study of degenerate equations in order to handle degeneracy in time, the spatial variables, the unknown function, and its gradient.

Acknowledgemet: The second named author gratefully acknowledges the partial support from a U. S. National Science Foundation grant.

REFERENCES

1. D. G. Aronson, The porous medium equation, *Nonlinear Diffusion Problems*, Lecture Notes in Math. 1229, Springer Verlag, Berlin, 1-46 (1986).
2. D. G. Aronson and Ph. Bénilan, Regularite des solutions de l'equations des milieux poreaux dans $\mathbf{R}^N$, *C. R. Acad Sc. Paris*, **288**: 103-105 (1979).
3. D. G. Aronson, L. A. Caffarelli, and J. L. Vasquez, Interface with a corner point in one-dimensional porous medium flow, *Comm. Pure and Appl. Math.*, **38**: 375-404 (1985).
4. D. G. Aronson, M. G. Crandall, and L. A. Peltier, Stabilizations of solutions of a degenerate nonlinear diffusion problem, *Nonlinear Anal. TMA*, **6**: 1001-1033 (1982).

5. Ph. Bénilan, *Equations d'Evolution dans un Espace de Banach Quelconque et Applications*, Ph. D. Thesis, Univ. of Paris (1972).
6. Ph. Bénilan, H. Brezis, and M. G. Crandall, A semilinear elliptic equation in $L^1(\mathbb{R}^N)$, *Ann. Scuola Norm. Sup. Pisa*, **2**: 523-555 (1975).
7. Ph. Bénilan and M. G. Crandall, The continuous dependence on φ of solutions of $u_t - \Delta\varphi(u) = 0$, *Indiana Univ. Math. J.*, **30**: 161-177 (1981).
8. M. G. Crandall and T. M. Liggett, Generation of semigroups of nonlinear transformations on general Banach spaces, *Amer. J. Math.*, **93**: 265-298 (1971).
9. M. G. Crandall and A. Pazy, Nonlinear evolution equations in Banach spaces, *Israel J. Math*, **11**: 67-94 (1972).
10. J. R. Dorroh and G. R. Rieder, A singular quasilinear parabolic problem in one space dimension, *J. Diff. Equations*, **91**: 1-23 (1991).
11. J. R. Dorroh and G. R. Goldstein, Existence and regularity of solutions of singular parabolic problems, in preparation.
12. J. R. Dorroh and G. R. Goldstein, A singular quasilinear parabolic problem in n dimensions, in preparation.
13. L. C. Evans, "Nonlinear Evolution Equations in an Arbitrary Banach Space," Math. Res. Center Tech Summary Report No. 1568, Madison (August 1975).
14. L. C. Evans, Nonlinear evolution equations in Banach spaces, *Israel J. Math.*, **26**: 1-42 (1977).
15. L. C. Evans, A new proof of local $C^{1,\alpha}$ regularity for solutions of certain degenerate elliptic P.D.E.'s, *J. Diff. Equations*, **45**: 356-373 (1982).
16. W. Feller, "Diffusion processes in genetics," Proceedings Second Berkeley Symposium Math. Stat. Prob.: 227-246 (1951).
17. A. Friedman, *Partial Differential Equations of Parabolic Type*, Prentice-Hall, Englewood Cliffs (1964).
18. J. A. Goldstein and C.-Y. Lin, Singular nonlinear parabolic boundary value problems in one space dimension, *J. Differential Equations*, **68**: 429-443 (1987).
19. J. A. Goldstein and C.-Y. Lin, An L^p-semigroup approach to degenerate parabolic boundary value problems, *Ann. Mat. Pura Appl.*, **159**: 211-227 (1991).

20. J. A. Goldstein and C.-Y. Lin, Highly degenerate parabolic boundary value problems, *Diff. & Int. Eqns.*, **2**: 216-227 (1989).

21. O. A. Ladyzhenskaya, V. Solonnikov, and N. N. Ural'tseva, *Linear and Quasilinear Equations of Parabolic Type*, Translations of Math. Monographs, American Math. Soc., Providence (1968).

22. O. A. Ladyzhenskaya and N. N. Ural'tseva, *Linear and Quasilinear Elliptic Equations*, Academic Press, New York (1968).

23. C.-Y. Lin, Degenerate nonlinear parabolic boundary value problems, *Nonlinear Anal. TMA*, to appear.

24. M. F. Norman, Approximation of stochastic processes by Gaussian diffusions and applications to Wright-Fisher genetic models, *SIAM J. Appl. Math.*, 29: 225-242 (1975).

25. G. R. Rieder, Spatially degenerate diffusion with periodic-like boundary conditions, *Differential Equations with Applications in Biology, Physics, and Engineering* (ed. by J. A. Goldstein, F. Kappel, and W. Schappacher): 301-312, Lecture Notes in Pure and Applied Mathematics, Marcel Dekker, New York (1991).

26. G. R. Rieder, Nonlinear singular diffusion with nonlinear boundary conditions, in preparation.

27. H. Serizawa, M-Bowder accretiveness of a quasilinear differential operator, *Houston J. Math.*, **10**: 147-152 (1984).

28. N. N. Ural'tseva, Degenerate quasilinear elliptic systems, *Zap. Naučn. Sem. Leningrad Otdel. Mat. Inst. Steklov*, **7**: 184-222 (1968) [in Russian].

29. A. D. Wentzel, On boundary conditions for multidemensional diffusion processes, *Theory Prob. Appl.*, **4**: 164-177 (1959).

The Coercive Solvability of Abstract Parabolic Equations in Banach Spaces

E. V. EMEL' ANOVA and V. VASILEV Pedagogical Institute, Belgorod, Russia

1. Let us consider the initial value problem

$$\frac{du}{dt} + Au = f, \quad u(0)=u_0, \quad (0\leqslant t\leqslant T), \tag{1}$$

in a Banach space E. Here A is a densly defined generator of an analytic semigroup with exponentialy decreasing norm; f(t) and $u_0 \in E$ are given.

We shall call the problem (1) coercively solvable (sometimes used for the maximal regularity property of the solution of the problem (1)) in the pair of the spaces $(E_1, \tilde{E})$ if there exists a constant M such that for any f(t), u(t) and u_0 satisfing (1) the following inequality is true

$$\|u'\|_{\tilde{E}} + \|Au\|_{\tilde{E}} + \|u(t)\|_{C([0,T],E_1)} \leqslant M(\|f\|_{\tilde{E}} + \|u_0\|_{E_1}), \tag{2}$$

This inequality is called the coercive inequality.

Here $E_1 \in E$ and $\tilde{E}$ - some space of the functions which are definded on [0,T].

The coercive solvability of the problem (1) means in fact that both terms of the left side of (1) have the same regularity as their sum and thus their regularity are "maximal". This property is very useful in the investigation of integro-differential equations, nonlinear problems et. (see [8]).

It is possible to investigate the coercive solvability of the abstract parabolic Cauchy problem in the case when the operator A depends of t:

$$\frac{du}{dt} + A(t)u = f, \quad u(0)=u_0, \quad (0\leqslant t\leqslant T), \tag{3}$$

The coercive inequality in this case takes the form

$$\|u'\|_{\tilde{E}} + \|A(t)u\|_{\tilde{E}} + \|u(t)\|_{C([0,T],E_1)} \leqslant M(\|f\|_{\tilde{E}} + \|u_0\|_{E_1}), \tag{4}$$

At first the coercive solvability of (1) in a Banach space was proved in [1] for the case $E_1 = D(A)$ with graph norm, $\tilde{E} = C_0^\alpha$ - the space of the functions with norm

$$\|f\|_{C_0^\alpha} = \sup_{t\in[0,T]} \|f(t)\| + \sup_{0\leqslant t,\tau,t+\tau\leqslant T} \frac{\|f(t+\tau)-f(t)\|}{\tau^\alpha} t^\alpha.$$

Then the similar results were obtained for the cases

a) (see [2])

$$E_1 := \{v\in D(A^{1-(\beta-\alpha)})\ ;\ \|v\|_{E_1} = \|v\| + \|A^{1-(\beta-\alpha)}v\|\ \},$$

$$\tilde{E} := \{f:\ \sup_{t\in[0,T]} \|f(t)\| t^{\beta-\alpha} + \sup_{0\leqslant t,\tau,t+\tau\leqslant T} \frac{\|f(t+\tau)-f(t)\|}{\tau^\alpha} t^\beta\ \}..$$

where $0\leqslant\alpha\leqslant\beta<1$.

b) (see [3])

$$E_1 := \{v\in D(A^{1-(\beta-\alpha)})\ :\ A^{1-(\beta-\alpha)}v \in E_\gamma\ ,$$

$$\|v\|_{E_1} := \|v\| + \|A^{1-(\beta-\alpha)}v\|_{E_\gamma}\ \},$$

$$\tilde{E} := \{f\ :\ \sup_{t\in[0,T]} \|f(t)\|_{E_\gamma} t^{\beta-\alpha} + \sup_{0\leqslant t,\tau,t+\tau\leqslant T} \frac{\|f(t+\tau)-f(t)\|}{\tau^\alpha} t^{\beta-\gamma}\},$$

where $0\leqslant\gamma\leqslant\alpha\leqslant\beta<1$, $\alpha>0$, $E_\gamma = (E,D(A))_{\gamma,\infty}$ - the interpolation space with the norm (see [4])

$$\|v\|_{E_\gamma} = \|v\|_\gamma := \sup_{z>0} z^{1-\gamma}\|A\ e^{-zA}v\|, \tag{5}$$

The similar results (for the case $\alpha=\beta=\gamma$) were proved in [8], where the following estimate

$$\|u'\|_{C^\alpha([0,T],E)} + \|Au\|_{C^\alpha([0,T],E)} \leqslant M\left[\|Au_0+f(0)\|_\alpha + \|f\|_{C^\alpha([0,T],E)}\right]$$

and inclusions $Au(t)+f(t)\in E_\alpha$ for all $t\geqslant 0$ are obtained. Here

$$\|f\|_{C^\alpha([0,T],E)} = \sup \frac{\|f(t)-f(\tau)\|}{(t-\tau)^\alpha}\ .$$

The same estimate is given in [8] with little Holder spaces $h^\alpha([0,T],E)$ instead $C^\alpha([0,T],E)$. The norm in $h^\alpha([0,T],E)$ is the same as in C^{α} and

$$h^{\alpha}([0,T],E) = \{f(t): \lim_{\tau\to t} \frac{\|f(t)-f(\tau)\|}{(t-\tau)^{\alpha}} = 0\}.$$

For the case when A - is an elliptic operator of second order in a bounded domain with Dirichlet boundary condition the space E_{γ} coincides with the Holder space $C^{2\gamma}$ with zero boundary condition (see [5]).

c) (see [6])

$$\tilde{E} := L_p([0,T],E_{\alpha,\beta}) \qquad (0<\alpha<1,\ 1\leqslant p\leqslant\infty\ ),$$

where $E_{\alpha,\beta}$ - is the interpolation space constructed by the real interpolation method - see [4].

E_1 - relative trace space.

d) (see [7])

$$\tilde{E} := \{f: \sup_{t\in[0,T]} \|f(t)\|_{\alpha-\gamma} + \sup_{0\leqslant t,\tau,t+\tau\leqslant T} \frac{\|f(t+\tau)-f(t)\|_{\alpha-\beta}}{\tau^{\beta}} (t+\tau)^{\gamma},$$

$E_1 := D(A_{\alpha-\gamma})$ with the norm $\|v\|_1 := \|Av\|_{\alpha-\gamma}$

for $0\leqslant\gamma\leqslant\beta\leqslant\alpha<1$, $\alpha>0$.

e) (see [11])

$\tilde{E}$ - the functional space with the norm:

$$\|f\|_{\tilde{E}} = \sup_{t\in[0,T]} \|f(t)\|_{\gamma}\, t^{\varkappa} + \sup_{t,\tau} \frac{\|f(t+\tau)-f(t)\|_{\beta-\alpha}}{\tau^{\alpha}} (t+\tau)^{\beta-\gamma} t^{\varkappa}$$

with $0\leqslant\alpha\leqslant\beta$, $0\leqslant\varkappa$, $0\leqslant \beta-\gamma+\varkappa \leqslant \alpha$, $\gamma\leqslant\beta$.

$E_1 = \{v\in E : A^{1-\varkappa}v\in E_{\gamma};\ \|v\|_{E_1} = \|A^{1-\varkappa}v\|_{\gamma}\ \}$,

This theorem remanes true for the space $\tilde{E}$ with the norm

$$\|f\|_{\tilde{E}} = \sup_{t\in[0,T]} \|f(t)\|_{\gamma}\, t^{\varkappa} + \sup_{t,\tau} \frac{\|f(t+\tau)-f(t)\|_{\beta-\alpha}}{\tau^{\alpha}}\, t^{\beta-\gamma+\varkappa}$$

with the same space E_1 (see also [11]).

f) (Anosov)

$\tilde{E}$ - the functional space with the norm:

$$\|f\|_{\tilde{E}} = \left\{ \int_0^1 \int_0^1 \frac{\|f(t)-f(\tau)\|^p}{|t-\tau|^{1+p\alpha}} \, dt d\tau + \int_0^1 \|f(t)\|^p dt \right\}^{1/p},$$

E_1 - is the trace space for $\tilde{E}$.

In all these cases the coercive solvability of (4) has place when the operators $A(t)$ satisfy the following conditions:
α) The operators $A(t)$ have the constant domain and

$$\|(A(t)-A(s))A^{-1}(r)\| \leqslant |t-s|^{\varepsilon},$$

β) There exist the constant C such that

$$\|(\lambda I+A(t))^{-1}\| \leqslant C\,(1+|\lambda|)^{-1} \text{ for all } \lambda \text{ with } \operatorname{Re}\lambda \geqslant 0$$

(the case c) were considered in [13]).

The interesting case when the domain of operator $A(t)$ depends of t were investigated in [12] where the condition α) were replaced by
α') For any $t,\tau \in [0,T]$, $u \in D(A(t))$, $v \in D(A^*(\tau))$ and some $\beta_i, \gamma_i \in [0,1]$, $\beta_i+\gamma_i \leqslant 1$, $\gamma_i < \varepsilon_i \leqslant 1$

$$|(A(t)u,v)-(u,A^*(\tau))| \leqslant \sum_{i=1}^{m} C_i |t-\tau|^{\varepsilon_i} \|A(t)u\|^{\beta_i} \|u\|^{1-\beta_i} \|A^*(\tau)v\|_*^{\gamma_i} \|v\|_*^{1-\gamma_i}.$$

The full solutions of the problem of maximal L^p-regularity for the problem (1) is unknown to the authors. The maximal L^p-regularity for problem (1) depends on the joint structure of space E and operator A. There is the following conditional result:

If E is reflexive and the problem (1) is coercively solvable for some $q \in [1,\infty)$ then the maximal L^p-regularity for (1) holds for every $p \in [1,\infty]$ with $E_1 = (E,D(A))_{1-1/p,p}$.
For most general exposition of this question see [14].

Some maximal regularity theorems were obtained without the assumption of density of $D(A)$ in E - see [9,10].

Literature.

1. Соболевский П.Е. Неравенства коэрцитивности для абстракт ных параболических уравнений. ДАН СССР.--1964.--157.-I.

2. Гудкин В.П., Дымент Д.А., Матвеев В.А. Коэрцитивная разрешимость абстрактных параболических уравнений в пространствах Гельдера с весом. Труды конференции ХАБИЖТ.-1973.

3. Васильев В.В. О коэрцитивной разрешимости абстрактных параболических уравнений с постоянным оператором. Дифф.уравнения. -Минск.-1978.-14.-р.1507-1510.

4. Трибель Х. Теория интерполяции.Функциональные пространства. Дифференциальные операторы. Москва.Мир.-1980.-664 р.

5. Lunardi A. Interpolation spaces between domains of elliptic operators and spaces of continuous functions with applications to nonlinear parabolic equations//Math.Narch.-1985. -121.-p.295-318.

6. Соболевкий П.Е. Некоторые свойства решений дифференциальных уравнений в дробных пространствах. Труды НИИМ ВГУ.-1974. -14.-р.68-74.

7. Ашыралиев А. О коэрцитивной разрешимости параболических уравнений в пространствах гладких функций. Изв. АН Турк.ССР. -сер.физ.-техн.,хим. и геолог. наук.-1989.-3.-р.3-13.

8. Da Prato G. Abstract differential equations, maximal regularity, and linearization//Proc.Symp.Pure Math.-1986.-45. -1.-p.359-370.

9. Da Prato G., Grisvard P. Equations d'evolution abstraites non lineaires de type parabolique//Ann.Mat.Pura Appl.-1979. -120.-4.-p.329-396.

10. Sinestrari E. On the abstract Cauchy problem of parabolic type in the spaces of continuous functions//J.Math.Anal. Appl.-1985.-107.-p.16-66.

11. Емельянова Е.В., Васильев В.В. Коэрцитивная разрешимость абстрактных параболических уравнений в весовых пространствах. Белгород.БГПИ.-Рукопись депонирована в ВИНИТИ, 15.04.91.-N 1590-В91.

12. Рудецкий В.А. Коэрцитивная разрешимость в интерполяционных пространствах параболических уравнений с переменным оператором, имеющим переменную область определения. Владивосток. -Дальневосточный политехнический институт.-1987.-Рукопись депонирована в ВИНИТИ N 9022-В87.-16 р.

13. Рудецкий В.А.Оценки решений параболических уравнений в нормах пространства следов. Владивосток.-Дальневосточный политехнический институт.-1988.-Рукопись депонирована в ВИНИТИ N-6621-В88.-34 р.

14. Cannarsa P., Vespri V. On maximal L^p regularity for the abstract Cauchy problem. Boll.U.M.I.-1986.-(6) 5-B.-p.165-175.

Smooth Solutions of Nonlinear Elliptic Systems with Dynamic Boundary Conditions

JOACHIM ESCHER University of Zürich, Zürich, Switzerland

1. Introduction and main results

In this note we consider the following problem

$$\begin{array}{llll}
(1) & \lambda u - \partial_j(a_{jk}\partial_k u) + a_j\partial_j u = f(\cdot, u) & \text{in} & \Omega \times (0,\infty)\,, \\
(2) & a_{jk}\nu^j\gamma_1\partial_k u = g(\cdot, \gamma_1 u) & \text{on} & \Gamma_1 \times (0,\infty)\,, \\
(3) & \gamma_2 u = 0 & \text{on} & \Gamma_2 \times (0,\infty)\,, \\
(4) & \partial_t(\gamma_3 u) + a_{jk}\nu^j\gamma_3\partial_k u = h(\cdot, \gamma_3 u) & \text{on} & \Gamma_3 \times (0,\infty)\,, \\
(5) & \gamma_3 u(\cdot, 0) = z_0 & \text{on} & \Gamma_3\,.
\end{array}$$

Let us briefly introduce some notation. Ω denotes a bounded domain in $\mathbb{R}^n$ with smooth boundary $\partial\Omega$. We assume that $\partial\Omega$ possesses a decomposition of the form $\partial\Omega = \Gamma_1 \cup \Gamma_2 \cup \Gamma_3$, where $\Gamma_i \cap \Gamma_j = \emptyset$ for $i \neq j$ and where each Γ_i is a finite union of connected subsets of $\partial\Omega$. The outer normal on $\partial\Omega$ is denoted by $\nu = (\nu^1, \ldots, \nu^n)$ and γ_i stands for the trace operator with respect to Γ_i, $i = 1, 2, 3$. λ is a real parameter.

In equations (1)-(5) we have used a vector notation. This means the unknown $u = (u^1, \ldots, u^N)$, $N \geq 1$, is a N-vector valued function, the coefficents $a_{jk}(x)$, $a_j(x) \in \mathcal{L}(\mathbb{R}^N)$ are real $N \times N$-matrices depending smoothly on the spatial variable $x \in \overline{\Omega}$ and satisfying appropriate ellipticity conditions. Also the right-hand sides $f = (f^1, \ldots, f^N)$, g, h and z_0 are smooth N-vector valued functions.

Thus (1) is a strongly coupled system of semilinear elliptic equations, which r-th component is given by

$$\lambda u^r - \partial_j(a_{jk}^{rs}\partial_k u^s) + a_j^{rs}\partial_j u^s = f^r(\cdot, u),$$

where we use the standard summation convention.

On each part of the decomposition $\partial\Omega = \Gamma_1 \cup \Gamma_2 \cup \Gamma_3$ we have different kinds of boundary conditions: Equations (2) and (3) are called nonlinear Neumann type and homogeneous Dirichlet type boundary condition, respectively.

On the portion Γ_3 of the boundary we consider a so called **nonlinear dynamic boundary condition**. Of course we assume that Γ_3 is not empty.

We are interested in unique classical solvability of problem (1)-(5), as well as in the dynamic qualities of the corresponding solutions like global existence, blow up phenomena or stability properties of stationary solutions.

We mention that the existence of a unique weak solution of problem (1)-(5) has already been proved in [5]. Thereby a function $u \in C(J, H_p^1(\Omega))$ is called a **weak solution** of (1)-(5) iff

$$\int_0^T \{-\int_{\Gamma_3}([\gamma_3\varphi]^{\cdot}|\gamma_3 u)d\sigma + \int_\Omega (\partial_j\varphi|a_{jk}\partial_k u) + (\varphi|a_j\partial_j u)dx\}dt =$$
$$\int_0^T \{\int_\Omega (\varphi|f(\cdot,u))dx + \int_{\Gamma_1}(\gamma_1\varphi|g(\cdot,\gamma_1 u))d\sigma + \int_{\Gamma_3}(\gamma_3\varphi|h(\cdot,\gamma_3 u))d\sigma\}dt + \int_{\Gamma_3}([\gamma_3\varphi](0)|z_0)d\sigma$$

for every $T \in \dot{J} := J \setminus \{0\}$ and every $\varphi \in C([0,T], H_{p'}^1(\Omega))$ satisfying $\gamma_2\varphi = 0$, $[\gamma_3\varphi](T) = 0$ and $\gamma_3\varphi \in C^1([0,T], B_{p'}^0(\Gamma_3))$.

In the following we denote for $s \in \mathbf{R}$ and $q \in (1,\infty)$ by $H_q^s(\Omega) := H_q^s(\Omega, \mathbf{R}^N)$ and by $B_q^s(\Gamma_i) := B_{qq}^{s-1/q}(\Gamma_i, \mathbf{R}^N)$ the standard Bessel potential spaces over Ω and the Besov spaces over Γ_i, respectively (see e.g. [4, 9]). Furthermore, $q' := 1 - 1/q$ stands for the dual exponent of q.

In this paper we are mainly interested in **classical solutions** of problem (1)-(5), i.e. to find $u \in C(J, H_p^1(\Omega)) \cap C(\dot{J}, C^\infty(\overline{\Omega}))$ with $\gamma_3 u \in C^1(\dot{J}, C^\infty(\Gamma_3))$ which satisfies (1)-(4) pointwise on $\dot{J} \times \overline{\Omega}$ and which fulfils the initial condition (5).

To give a precise formulation of our results we introduce the following regularity assumption for the coefficents and inhomogeneities.

$$\begin{aligned} &a_j,\ a_{jk} \in C^\infty(\overline{\Omega}, \mathcal{L}(\mathbf{R}^N)) \text{ for all } 1 \le j,\, k \le n, \\ &f \in C^\infty(\overline{\Omega} \times \mathbf{R}^N, \mathbf{R}^N),\ g \in C^\infty(\Gamma_1 \times \mathbf{R}^N, \mathbf{R}^N),\ h \in C^\infty(\Gamma_3 \times \mathbf{R}^N, \mathbf{R}^N). \end{aligned} \tag{6}$$

Furthermore, we assume that the so called Legendre-Hadamard condition is satisfied, i.e.

$$a_{jk}^{rs}(x)\xi^j\xi^k\eta_r\eta_s > 0 \quad \text{for all } \ x \in \overline{\Omega},\ \xi \in \mathbf{R}^n \setminus \{0\},\ \eta \in \mathbf{R}^N \setminus \{0\}. \tag{7}$$

Then we have the following *local* existence, uniqueness and regularity result for problem (1)-(5).

Theorem 1: *Suppose that $p > n$. Then there is a $\tilde{\lambda}_0 \geq 0$ such that for each $z_0 \in B_p^1(\Gamma_3)$ and each $\lambda \geq \tilde{\lambda}_0$ there exists a $\delta > 0$ and a unique classical solution*

$$u \in C([0,\delta], H_p^1(\Omega)) \cap C((0,\delta], C^\infty(\overline{\Omega})) \quad \textit{with} \quad \gamma_3 u \in C^1((0,\delta], C^\infty(\Gamma_3))$$

of problem (1)-(5).

The local solutions of Theorem 1 can be continued to a maximal smooth solution (in the sense that there is no proper extension), provided the following growth restriction for (f, g) is satisfied.

$$\begin{aligned}&\text{There exists a constant } M \geq 0 \text{ such that}\\ &|\partial_2 f(x,\xi)| + |\partial_2 g(y,\xi)| \leq M \text{ for } x \in \overline{\Omega},\ y \in \Gamma_1,\ \xi \in \mathbb{R}^N.\end{aligned} \tag{8}$$

Theorem 2: *Suppose that $p > n$ and that (8) holds.*

a) Then there is a $\lambda_0 \geq 0$ such that for each $z_0 \in B_p^1(\Gamma_3)$ and each $\lambda \geq \lambda_0$ problem (1)-(5) possesses a unique maximal classical solution

$$u \in C(J, H_p^1(\Omega)) \cap C(\dot{J}, C^\infty(\overline{\Omega})) \quad \textit{with} \quad \gamma_3 u \in C^1(\dot{J}, C^\infty(\Gamma_3)).$$

b) The maximal interval of existence $J = [0, t^+)$ is right open and

$$\limsup_{t \to t^+} |u(t)|_{H_p^1(\Omega)} = \infty\,, \quad \textit{provided } t^+ < \infty.$$

Conversely, u is a global solution, i.e. $t^+ = \infty$, if the trace $\gamma_3 u$ of u is bounded in $B_p^1(\Gamma_3)$.

Let us add some remarks and generalizations to Theorems 1 and 2.

Remarks:

a) Note that in Theorem 1 we don't require any growth condition with respect to f, g and h. Moreover, Theorem 1 remains true, even if we allow nonlinearities of the form $f(x, u, \nabla u)$, provided f is polynomially bounded with respect to the gradient ∇u.

b) In Theorem 2 only growth restriction with respect to f and g but not with respect to h are needed.

c) It is possible to assume one (or both) of the parts Γ_1 and Γ_2 of the boundary $\partial\Omega$ to be empty.

d) The proof of Theorem 2 shows that we obtain a *strict* classical solution of problem (1)-(5), i.e.

$$u \in C(J, C^\infty(\overline{\Omega})) \quad with \quad \gamma_3 u \in C^1(J, C^\infty(\Gamma_3))$$

for initial values $z_0 \in C^\infty(\Gamma_3)$.

e) Analogous results can be proved for time dependent problems (see [5]) and it is of course possible to consider less regular coefficents and right-hand sides like

$$a_{jk} \in C^{2+\alpha}(\overline{\Omega}, \mathcal{L}(\mathbf{R}^N)), \; a_j \in C^{1+\alpha}(\overline{\Omega}, \mathcal{L}(\mathbf{R}^N)) \quad \text{for all} \quad 1 \le j, k \le n,$$
$$f \in C^2(\overline{\Omega} \times \mathbf{R}^N, \mathbf{R}^N), \; g \in C^2(\Gamma_1 \times \mathbf{R}^N, \mathbf{R}^N), \; h \in C^2(\Gamma_3 \times \mathbf{R}^N, \mathbf{R}^N).$$

If $\alpha > 0$ the corresponding solution for such data possesses the following regularity

$$u \in C((0,t^+), C^2(\overline{\Omega})), \quad \gamma_3 u \in C^1((0,t^+), C^1(\Gamma_3)).$$

f) It should be observed that neither in the domain Ω nor in the part Γ_1 of the boundary a smoothing effect in time takes place, provided f or g depend on u.
If f and g are independent of u then one can prove that

$$u \in C^\infty((0,t^+), C^\infty(\overline{\Omega})).$$

2. Mild Solutions

In this section we introduce the concept of mild solutions of problem (1)-(5). For this we need some basic facts about the so called Lions-Magenes extension of elliptic boundary value problems. Secondly we introduce a family of pseudo-differential operators which generat strongly continuous analytic semigroups on the Besov spaces $B_p^s(\Gamma_3)$ for $s \in \mathbf{R}$.
Given $u \in H_p^2(\Omega)$, let us define

$$\mathcal{A}u := -\partial_j(a_{jk}\partial_k u) + a_j\partial_j u, \quad \mathcal{B}_i u := a_{jk}\nu^j\gamma_i\partial_k, \quad i = 1, 2, 3,$$

where $\gamma_i \in \mathcal{L}(H_p^s(\Omega), B_p^s(\Gamma_i))$ denotes for $s > 1/p$ the trace operator with respect to Γ_i.
Due to the regularity assumption (6) and the ellipticity condition (7) it is well know (see for instance [1, 2, 6]) that there exists a $\lambda_* \in \mathbf{R}$ such that

$$(\lambda + \mathcal{A}, \mathcal{B}_1, \gamma_2, \gamma_3) \in Isom(H_p^2(\Omega), L_p(\Omega) \times B_p^1(\Gamma_1) \times B_p^2(\Gamma_2) \times B_p^2(\Gamma_3)) \tag{9}$$

for all $\lambda \geq \lambda_*$.

Moreover, it is shown in [8] (cf. also [3]) that there is a continuously embedded subspace $\mathcal{D}(\Omega)$ of $L_p(\Omega)$ and extensions

$$(\overline{\mathcal{A}}, \overline{\mathcal{B}}_i, \overline{\gamma}_i) \in \mathcal{L}(\mathcal{D}(\Omega), L_p(\Omega) \times B_p^{-1}(\Gamma_i) \times B_p^0(\Gamma_i)), \quad i = 1, 2, 3 \tag{10}$$

of $(\mathcal{A}, \mathcal{B}_i, \gamma_i)$ such that

$$(\lambda + \overline{\mathcal{A}}, \overline{\mathcal{B}}_1, \overline{\gamma}_2, \overline{\gamma}_3) \in Isom(\mathcal{D}(\Omega), L_p(\Omega) \times B_p^{-1}(\Gamma_1) \times B_p^0(\Gamma_2) \times B_p^0(\Gamma_3))$$

for all $\lambda \geq \lambda_*$.

We now define

$$\begin{aligned} \mathcal{R} &:= (\lambda + \overline{\mathcal{A}}, \overline{\mathcal{B}}_1, \overline{\gamma}_2, \overline{\gamma}_3)^{-1} | L_p(\Omega) \times B_p^{-1}(\Gamma_1) \times \{0\} \times B_p^0(\Gamma_3), \\ \mathcal{S} &:= \mathcal{R} | L_p(\Omega) \times B_p^{-1}(\Gamma_1) \times \{0\}, \quad \mathcal{T} := \mathcal{R} | \{0\} \times B_p^0(\Gamma_3), \end{aligned} \tag{11}$$

where $\lambda \geq \lambda_*$ is fixed.

Denote by $[\cdot,\cdot]_\theta$, $\theta \in (0,1)$, the standard complex interpolation functor (cf. [4, 9]). Then it is known that (see e.g. [4, Theorem 6.4.5])

$$[L_p(\Omega), H_p^2(\Omega)]_\theta = H_p^{2\theta}(\Omega) \quad \text{and} \quad [B_p^s(\Gamma_i), B_p^t(\Gamma_i)]_\theta = B_p^{t\theta + s(1-\theta)}(\Gamma_i)$$

Thus it follows from $\mathcal{D}(\Omega) \hookrightarrow L_p(\Omega)$, (9) and (11) by interpolation that

$$\begin{aligned} &\mathcal{R} \in \mathcal{L}(L_p(\Omega) \times B_p^{2\theta-1}(\Gamma_1) \times B_p^{2\theta}(\Gamma_3), H_p^{2\theta}(\Omega)), \\ &\mathcal{S} \in \mathcal{L}(L_p(\Omega) \times B_p^{2\theta-1}(\Gamma_1), H_p^{2\theta}(\Omega)), \quad \mathcal{T} \in \mathcal{L}(B_p^{2\theta}(\Gamma_3), H_p^{2\theta}(\Omega)), \quad \theta \in [0,1]. \end{aligned} \tag{12}$$

With these notation we define a (unbounded) linear operator in $B_p^1(\Gamma_3)$ by

$$\mathbf{B}_{1/2} : B_p^2(\Gamma_3) \subset B_p^1(\Gamma_3) \to B_p^1(\Gamma_3), \quad \mathbf{B}_{1/2} z := \mathcal{B}_3 \mathcal{T} z, \quad z \in B_p^2(\Gamma_3).$$

It is shown in [7] that $\mathbf{B}_{1/2}$ generates a strongly continuous analytic semigroup in $B_p^1(\Gamma_3)$ and that

$$(\mu_* + \mathbf{B}_{1/2}) \in Isom(B_p^2(\Gamma_3), B_p^1(\Gamma_3)) \quad \text{for some} \quad \mu_* > 0.$$

To handle the nonlinear problem (1)-(5) we need an appropriate extension of $\mathbf{B}_{1/2}$. This can be done by considering the formal adjoint analogue of the construction above and by using some interpolation techniques. In fact, one can show the following result (see [5, Lemmas 1.3, 1.4 and Theorem 1.5]).

Theorem 3: *Let $-\infty < \alpha < \beta < \infty$ and $\sigma < 1 < \tau$ be given. Then the following assertions hold:*

a) The operator $\mathbb{B}_{1/2}$ is closable in $B_p^\sigma(\Gamma_3)$.

b) Denote by $\mathbb{B}_{\sigma-1/2}$ the closure of $\mathbb{B}_{1/2}$ in $B_p^\sigma(\Gamma_3)$ and by $\mathbb{B}_{\tau-1/2}$ the part of $\mathbb{B}_{1/2}$ in $B_p^\tau(\Gamma_3)$. Then

$$dom(\mathbb{B}_{\alpha-1/2}) = B_p^{\alpha+1}(\Gamma_3)$$

and $-\mathbb{B}_{\alpha-1/2}$ generates a strongly continuous analytic semigroup $\{e^{-t\mathbb{B}_{\alpha-1/2}}, t \geq 0\}$ in $\mathbb{B}_p^\alpha(\Gamma_3)$.

c) For all $t \geq 0$ we have

$$e^{-t\mathbb{B}_{\alpha-1/2}} = e^{-t\mathbb{B}_{\beta-1/2}} | B_p^\alpha(\Gamma_3).$$

In the following J denotes a nontrivial subinterval of $[0,\infty)$ with $0 \in J$. For $w \in C(J, H_p^1(\Omega))$ and $t \in J$ we then set

$$F(w)(t) := h(\gamma_3 w(t)) - \overline{\mathcal{B}}_3 \mathcal{S}(f(w(t)), g(\gamma_1 w(t))).$$

Observe that for $p > n$ we have the embeddings

$$H_p^1(\Omega) \hookrightarrow C(\overline{\Omega}) \hookrightarrow L_p(\Omega), \quad B_p^1(\Gamma_i) \hookrightarrow C(\Gamma_i) \hookrightarrow B_p^\eta(\Gamma_i) \quad \text{if} \quad \eta < 1/p.$$

Hence it follows from the regularity assumption (6), the mean value theorem and the trace theorem that

$$(f, g \circ \gamma_1, h \circ \gamma_3) \in C^{1-}(H_p^1(\Omega), L_p(\Omega) \times B_p^\eta(\Gamma_1) \times B_p^\eta(\Gamma_3)), \quad \eta < 1/p. \tag{13}$$

Consequently we obtain from (12) that

$$F(w) \in C(J, B_p^\eta(\Gamma_3)) \quad \text{for} \quad \eta \in [0, 1/p).$$

Due to this regularity for $F(w)$ it follows from Theorem 3 and [3, Theorem 9.3] that for each $\alpha < \eta$ and each $z_0 \in B_p^{\alpha+1}(\Gamma_3)$ the abstract linear Cauchy problem

$$\dot{z} + \mathbb{B}_{\alpha-1/2} z = F(w), \quad t \in \dot{J}, \quad z(0) = z_0$$

possesses a unique solution

$$z \in C(J, B_p^{\alpha+1}(\Gamma_3)) \cap C(\dot{J}, B_p^{\alpha'+1}(\Gamma_3)) \cap C^1(J, B_p^\alpha(\Gamma_3)) \cap C^1(\dot{J}, B_p^{\alpha'}(\Gamma)), \tag{14}$$

where $\alpha' \in (\alpha, \eta)$. This solution is given by

$$z(t) := z(t, w, z_0) = e^{-t\mathbb{B}_{\alpha-1/2}} z_0 + \int_0^t e^{-(t-s)\mathbb{B}_{\alpha-1/2}} F(w)(\tau)\, d\tau\,, \quad t \in J\,. \tag{15}$$

Finally, we define on $C(J, H_p^1(\Omega))$ the nonlinear operator

$$K(w)(t) := \mathcal{R}(f(w(t)), g(w(t)), z(t, w, z_0))\,, \quad w \in C(J, H_p^1(\Omega))\,, \quad t \in J\,.$$

Note that due to (12), (13) and (14) we have

$$K \in C(C(J, H_p^1(\Omega))\,,\, C(J, H_p^1(\Omega)))\,.$$

Using the above notation we are now able to introduce the notion of mild solutions. A function u is said to be a **mild solution on** J of problem (1)-(5) iff u is a fixed point of K in $C(J, H_p^1(\Omega))$. Suppose u is a mild solution on J satisfying

$$u \in C(\dot{J}, C^\infty(\overline{\Omega})) \quad \text{and} \quad \gamma_3 u \in C^1(\dot{J}, C^\infty(\Gamma_3)). \tag{16}$$

Then we have

$$u(t) = K(u)(t) = \mathcal{S}(f(u(t)), g(u(t))) + \mathcal{T} z(t, u, z_0)\,, \quad t \in J \tag{17}$$

where $z(t, u, z_0)$ is given by (15).
Observing the identities $(\lambda + \overline{\mathcal{A}}, \overline{\mathcal{B}}_1, \overline{\gamma}_2)\mathcal{S} = 1_{L_p(\Omega) \times B_p^{-1}(\Gamma_1) \times B_p^0(\Gamma_2)}$ and $(\lambda + \overline{\mathcal{A}}, \overline{\mathcal{B}}_1, \overline{\gamma}_2)\mathcal{T} = 0$, we conclude that

$$(\lambda + \mathcal{A}, \mathcal{B}_1, \gamma_2) u(t) = (\lambda + \overline{\mathcal{A}}, \overline{\mathcal{B}}_1, \overline{\gamma}_2) K(u(t)) = (f(u(t)), g(u(t)), 0) \quad \text{for} \quad t \in \dot{J}\,.$$

On the other side we have $\gamma_3 \mathcal{S} = 0$ and $\gamma_3 \mathcal{T} = 1_{B_p^1(\Gamma_3)}$. Therefore it follows from (17) and (15):

$$\begin{aligned} \partial_t(\gamma_3 u) + \mathcal{B}_3 u(t) &= \partial_t z(t, u, z_0) + \mathcal{B}_3 \mathcal{T} z(t, u, z_0) + \mathcal{B}_3 \mathcal{S}(f(u(t)), g(u(t))) \\ &= F(u)(t) + \mathcal{B}_3 \mathcal{S}(f(u(t)), g(u(t))) \\ &= h(\gamma_3 u(t))\,. \end{aligned}$$

Here we used the definition of $z(\cdot, u, z_0)$ and the fact that

$$\mathcal{B}_3 \mathcal{T} z(t, u, z_0) = \mathbb{B}_{-1/2} z(t, u, z_0).$$

This last identity is a consequence of Theorem 3 and our assumption (16).
Finally, due to (15) we note that $(\gamma_3 u)(0) = z(0, u, z_0) = z_0$.

Summing up, we find that every mild solution of problem (1)-(5) satisfying (16) is in fact a classical solution.

Conversely, suppose that $u \in C(J, H_p^1(\Omega)) \cap C(\dot{J}, C^\infty(\overline{\Omega}))$ is a classical solution of (1)-(5). Let $w(t) := \mathcal{S}(f(u(t)), g(u(t)))$ and $z(t) := \gamma_3 u(t)$ for $t \in J$. We conclude from (1)-(3) that $u = w + \mathcal{T}z$. Furthermore, equations (4), (5) and once again Theorem 3 show that z is a solution of the linear Cauchy problem

$$\dot{z} + \mathrm{B}_{-1/2} z = F(u), \quad z(0) = z_0 .$$

By uniqueness we find that $z = z(\cdot, u, z_0)$. Therefore

$$u = w + \mathcal{T}z(\cdot, u, z_0) = K(u),$$

which shows that u is a mild solution of problem (1)-(5). Collecting the above considerations we find

Proposition 1: *Every classical solution of problem (1)-(5) is a mild solution. Conversely, a mild solution of problem (1)-(5) satisfying (16) is already a classical solution of (1)-(5).*

3. Proof of Theorems 1 and 2

Suppose that $p > n$ and that the regularity and ellipticity assumption (6) and (7), respectively are satisfied.

a) Let us first mention that in [5, Theorem 1] we proved the existence and uniqueness of a local mild solution u of problem (1)-(5). Moreover, given growth restriction (8) u can be continued to a maximal mild solution of (1)-(5) (see Theorem 2 in [5]). Thus by Proposition 1 it suffices to show that the mild solution $u \in C(J, H_p^1(\Omega))$ satisfies

$$u \in C(\dot{J}, C^\infty(\overline{\Omega})) \quad \text{and} \quad \gamma_3 u \in C^1(\dot{J}, C^\infty(\Gamma_3)).$$

This follows from well known Sobolev embedding theorems, provided we can verify that for each $m \in \mathbb{N}$ we have

$$u \in C(\dot{J}, H_p^{m+1}(\Omega)) \quad \text{and} \quad \gamma_3 u \in C^1(\dot{J}, B_p^m(\Gamma_3)). \tag{18}$$

b) Notice that due to our regularity assumptions on the coefficents a_{jk}, a_j and on the domain Ω it follows that $(\mathcal{A}, \mathcal{B}_1, \gamma_2, \gamma_3, \Omega)$ is a regular boundary value problem of class C^∞ and order 2 in the sense of Agmon-Douglis-Nierenberg. Thus we obtain from [2, Theorem 13.1]:

$$\mathcal{R} \in \mathcal{L}(H_p^{2s}(\Omega) \times B_p^{1+2s}(\Gamma_1) \times B_p^{2+2s}(\Gamma_3)\,,\, H_p^{2+2s}(\Omega)) \quad \text{for} \quad s \geq 0. \tag{19}$$

Now interpolation between (11) and (19) yields

$$\mathcal{R} \in \mathcal{L}(H_p^{s}(\Omega) \times B_p^{s}(\Gamma_1) \times B_p^{1+s}(\Gamma_3)\,,\, H_p^{1+s}(\Omega)) \quad \text{for} \quad s \geq 0$$

and in particular

$$\mathcal{S} \in \mathcal{L}(H_p^{s}(\Omega) \times B_p^{s}(\Gamma_1)\,,\, H_p^{1+s}(\Omega))\,, \quad \mathcal{T} \in \mathcal{L}(B_p^{s}(\Gamma_3), H_p^{s}(\Omega))\,, \quad \text{for} \quad s \geq 0. \tag{20}$$

c) Next set $\delta := \frac{1}{2}(1 - \frac{n}{p}) > 0$ and observe that due to Sobolev's embedding theorem we have

$$H_p^{1+s}(\Omega) \hookrightarrow C^{2\delta+s}(\overline{\Omega}) \hookrightarrow H_p^{\delta+s}(\Omega) \quad \text{for} \quad s \geq 0. \tag{21}$$

On the other hand it follows from the mean value theorem that

$$(f, g \circ \gamma_1, h \circ \gamma_3) \in C^{1-}(C^\alpha(\overline{\Omega})\,,\, C^\alpha(\overline{\Omega}) \times C^\alpha(\Gamma_1) \times C^\alpha(\Gamma_3)) \quad \text{for} \quad \alpha \geq 0.$$

Thus the trace theorem and (21) imply

$$[u \mapsto (f(u), g(\gamma_1 u), h(\gamma_3 u))] \in C^{1-}(H_p^{1+s}(\Omega)\,,\, H_p^{\delta+s}(\Omega) \times B_p^{\delta+s}(\Gamma_1) \times B_p^{\delta+s}(\Gamma_3))\,,\ s \geq 0. \tag{22}$$

Furthermore note that, due to (19), (10) and once again the trace theorem, we have

$$\overline{\mathcal{B}}_3 \mathcal{S} \in \mathcal{L}(H_p^{s}(\Omega) \times B_p^{s}(\Gamma_1)\,,\, B_p^{s}(\Gamma_3)) \quad \text{for} \quad s \geq 0.$$

Consequently, we obtain from (22):

$$F = h(\gamma_3 \cdot) - \overline{\mathcal{B}}_3 \mathcal{S}(f(\cdot), g(\gamma_1 \cdot)) \in C^{1-}(H_p^{1+s}(\Omega), B_p^{\delta+s}(\Gamma_3)) \quad \text{for} \quad s \geq 0\,. \tag{23}$$

d) Now put $\delta' := \frac{\delta}{2} > 0$ and assume that for some $k \in \mathbb{N}$ we already proved that

$$u \in C(\dot{J}, H_p^{1+k\delta'}(\Omega)) \quad \text{and} \quad \gamma_3 u \in C^1(\dot{J}, B_p^{k\delta'}(\Gamma_3))\,. \tag{24}$$

Then it follows from (23) that

$$F(u) \in C(\dot{J}, B_p^{\delta+k\delta'}(\Gamma_3))\,.$$

Moreover, given $\tau \in \dot{J}$ we conclude from (24) and Theorem 3 that $z := \gamma_3 u = z(\cdot, u, z_0)$ is the solution of

$$\dot{z} + \mathbb{B}_{k\delta'-1/2} z = F(u), \quad \tau < t \leq T, \quad z(\tau) \in B_p^{1+k\delta'}(\Gamma_3).$$

Since $\tau \in \dot{J}$ was arbitrarily chosen it follows from (14) with $\alpha = k\delta'$, $\eta = k\delta' + \delta$ and $\alpha' = (k+1)\delta'$ that

$$z \in C(\dot{J}, B_p^{1+(k+1)\delta'}(\Gamma_3)) \cap C^1(\dot{J}, B_p^{(k+1)\delta'}(\Gamma_3)). \tag{25}$$

From (22) and (20) we see that $\mathcal{S}(f(u), g(u)) \in C(\dot{J}, H_p^{1+k\delta'+\delta}(\Omega))$. Since

$$u = K(u) = \mathcal{S}(f(u), g(u)) + \mathcal{T} z(\cdot, u, z_0),$$

the assertions (20) and (25) imply that

$$u \in C(\dot{J}, H_p^{1+(k+1)\delta'}(\Omega)) \quad \text{and} \quad \gamma_3 u \in C^1(\dot{J}, B_p^{(k+1)\delta'}(\Gamma_3)).$$

By induction we now obtain (18). □

4. References

[1] S. AGMON, A. DOUGLIS, L. NIERENBERG: Estimates near the boundary for solutions of elliptic partial differential equations satisfying general boundary conditions, II. Comm. Pure Appl. Math. **XVII** (1964), 35-92.

[2] H. AMANN: Existence and regularity for semilinear parabolic evolution equations. Annali Scuola Norm. Sup. Pisa Ser. IV XI (1984), 593-676.

[3] H. AMANN: Parabolic evolutions equations and nonlinear boundary conditions. J. Diff. Equs. **72** (1988), 201-269.

[4] J. BERGH & J. LÖFSTRÖM: Interpolation Spaces. An Introduction. Springer-Verlag, 1976.

[5] J. ESCHER: Nonlinear elliptic systems with dynamic boundary conditions. Math. Z., to appear.

[6] G. GEYMONAT & P. GRISVARD: Alcuni risultati di teoria spettrale per i problemi ai limiti lineari ellittici. Rend. Sem. Mat. Univ. Padova, **38** (1967), 121-173.

[7] T. HINTERMANN: Evolution equation with dynamic boundary conditions. Proc. Roy. Soc. Edinburgh Sect. A **105** (1989), 101-115.

[8] J.-L. LIONS & E. MAGENES: Problemi ai limiti non omogenei V. Annali Scuola Norm. Sup. Pisa **XVI** (1962), 1-44.

[9] H. TRIEBEL: Interpolation Theory, Function spaces, Differential Operators. North-Holland, Amsterdam, 1978.

Relaxed Controls, Differential Inclusions, Existence Theorems, and the Maximum Principle in Nonlinear Infinite Dimensional Control Theory

H. O. FATTORINI University of California, Los Angeles, California

§1. INTRODUCTION

Consider the semilinear control system

$$y'(t) = Ay(t) + f(t, y(t), u(t)), \qquad y(0) = \zeta. \tag{1.1}$$

A the infinitesimal generator of a strongly continuous semigroup in a Banach space E and $f(t, y, u)$ defined in $[0, T] \times E \times U$, where U is an arbitrary set called the *control set.* The system (1.1) is equipped with a space $U_{\text{ad}}(0, T; U)$ of *ordinary* controls taking values in U,

$$u(t) \in U \qquad (0 \le t \le T). \tag{1.2}$$

The space $U_{\text{ad}}(0, T; U)$ satisfies

(I) For every $y(\cdot) \in C(0, T; E)$ and every $u(\cdot) \in U_{\text{ad}}(0, T; U)$ the function $t \to f(t, y(t), u(t))$ belongs to $L^1(0, T; E)$. Solutions of (1.1) are, by definition, solutions of the (Lebesgue-Bochner) integral equation

$$y(t) = S(t)\zeta + \int_0^t S(t-\sigma)f(\sigma, y(\sigma), u(\sigma))d\sigma. \tag{1.3}$$

This work was supported in part by the NSF under grant QED-9902001.

We denote by $y(t, u)$ the solution (if any) of (1.1) corresponding to $u = u(\cdot) \in U_{\mathrm{ad}}(0, T; U)$.

Given a real valued *cost functional* $y_0(t, u)$, we consider the optimal control problem of minimizing $y_0(t, u)$ among all controls $u \in U_{\mathrm{ad}}(0, T; U)$ whose trajectories $y(t, u)$ satisfy the *target condition*

$$y(t, u) \in Y = \textit{target set} \subseteq E. \tag{1.4}$$

The *terminal time* t where (1.4) is required may or may not be fixed.

Existence theorems for optimal control problems (similarly to other problems of calculus of variations) are proved by taking limits of minimizing sequences. A *minimizing sequence* is a sequence $\{u^n(\cdot)\} \in U_{\mathrm{ad}}(0, t_n; U)$ such that the corresponding trajectory $y(t, u^n)$ is defined in $0 \le t \le t_n$ and satisfies

$$\limsup_{n\to\infty} y_0(t_n, u^n) \le m, \qquad \lim_{n\to\infty} \operatorname{dist}(y(t_n, u^n), Y) = 0$$

where m is the minimum of $y_0(t, u)$ subject to the target condition (1.4). To take limits we write (1.3) for each u^n,

$$y(t, u^n) = S(t)\zeta + \int_0^{t_n} S(t_n - \sigma) f(\sigma, y(\sigma, u^n), u^n(\sigma))d\sigma. \tag{1.5}$$

If $\{t_n\}$ is bounded, we may assume taking a subsequence that $t_n \to \bar{t}$. Typically, $\{u^n(\cdot)\}$ is not convergent, although if U is a closed, bounded convex subset of a dual Banach space, a subsequence will be convergent to an element $\overline{u}(\cdot) \in U_{\mathrm{ad}}(0, \bar{t}; U)$, say, in weak L^p topologies. It is sometimes possible to show that (a subsequence of) the sequence of trajectories $\{y(\cdot, u^n)\}$ converges; in finite dimensional systems, this is achieved using the Arzelá-Ascoli theorem [22] [24], while in infinite dimensional systems compactness of the semigroup $S(\cdot)$ or of the nonlinear term $f(t, y, u)$ must be used [10]. However, taking limits in (1.5) to obtain an optimal control $\overline{u}(\cdot)$ may fail in three ways:

(i) $\overline{u}(\cdot)$ may not satisfy the control constraint (1.2).

(ii) $y(t, \overline{u})$ may not satisfy the target condition (1.4).

(iii) the cost functional $y_0(t, u)$ may not satisfy

$$y(\bar{t}, \overline{u}) \le \limsup_{n\to\infty} y(t_n, u^n). \tag{1.6}$$

The three examples below illustrate these possibilities.

EXAMPLE 1.1. $E = \mathbf{R}^1$, $U = \{-1\} \cup \{1\}$, $\bar{t} = 1$, $Y = \{0\}$,

$$y'(t) = u(t), \qquad y(0) = 0$$

$$y_0(t, u) = \int_0^t y(\sigma)^2 d\sigma.$$

The sequence

$$u^n(t) = \begin{cases} 1 & (2k/2n \le t < (2k+1)/2n, \\ -1 & ((2k+1)/2n \le t < (2k+2)/2n, \end{cases} \qquad (k = 0,\ 1, \dots, n-1) \tag{1.7}$$

is a minimizing sequence for this problem with $y(\bar{t},\ u^n)$ uniformly convergent, but optimal controls do not exist; in fact, the "optimal control" $\overline{u}(t) \equiv 0$ (the weak limit of $\{u^n(\cdot)\}$) does not satisfy the control constraint (1.2).

EXAMPLE 1.2. $E = \mathbf{R}^2$, $U = [-1,\ 1]$, $\bar{t} = 1$, $Y = \{0,0\}$,

$$\begin{aligned} x'(t) &= u(t), & x(0) &= 0 \\ y'(t) &= (u(\sigma)^2 - 1)^2, & y(0) &= 0 \\ y_0(t, u) &= \int_0^t x(\sigma)^2\, d\sigma. \end{aligned}$$

The sequence (1.7) is a minimizing sequence with $\{x(t, u^n),\ y(t, u^n)\}$ uniformly convergent, but optimal controls do not exist; in fact, if $\overline{u}(t)$ is an optimal control, the first equation implies $x(t, \overline{u}) \equiv 0$, which in turn implies $\overline{u}(t) \equiv 0$; then $y(t, \overline{u})$ does not satisfy the target condition.

EXAMPLE 1.3. $E = \mathbf{R}^1$, $U = [-1,\ 1]$, $\bar{t} = 1$, $Y = \{0\}$,

$$\begin{aligned} y'(t) &= u(t), \qquad y(0) = 0 \\ y_0(t,\ u) &= \int_0^t \{y(\sigma)^2 + (u(t)^2 - 1)^2\} d\sigma. \end{aligned}$$

The sequence (1.7) is again a minimizing sequence with $\{y(\bar{t},\ u^n)\}$ uniformly convergent. However, there are no optimal controls; if $\overline{u}(\cdot)$ is optimal, then $y(t, \overline{u}) \equiv 0$, which means $\overline{u} = 0$. Condition (1.6) is violated for the minimizing sequence $\{u^n(\cdot)\}$; in fact,

$$y(\bar{t},\ \overline{u}) = 1, \ \limsup_{n \to \infty} y(\bar{t},\ u^n) = 0.$$

Nonexistence difficulties for finite dimensional systems described by a vector differential equation

$$y'(t) = f(t,\ y(t),\ u(t)), \qquad y(0) = \zeta \tag{1.8}$$

with a cost functional of the form

$$y_0(t,\ u) = \int_0^t f_0(\sigma,\ y(\sigma,\ u),\ u(\sigma)) d\sigma + g_0(t,\ y(t,\ u)) \tag{1.9}$$

were independently surmounted by Filippov [15], Warga [20], [21] and Gamkrelidze [17], who showed that in order to obtain existence of optimal controls the class of trajectories of (1.8) must be extended. Warga's extended class corresponds to controls whose values are probability measures $\mu(t,\ du)$ on the control set U (called *relaxed* controls); the original system (1.8) is replaced by the *relaxed* system

$$y'(t) = \int_U f(t,\ y(t),\ u)\mu(t,\ du) = \mathbf{f}(t,\ y(t))\mu(t) \tag{1.10}$$

and the cost functional by the *relaxed cost functional*

$$\mathbf{y}_0(t,\ \mu) = \int_0^t \int_U f_0(t,\ y(t,\ \mu),\ u)\mu(t,\ du) + g_0(t,\ y(t,\ \mu)). \tag{1.11}$$

Relaxed controls are a natural generalization of **Young measures** [23], [24] from calculus of variations to control theory. For expositions of finite dimensional relaxed control theory see [22], [24]; we limit ourselves to solve the problems in Examples 1.1, 1.2 and 1.3. In Example 1.1 each $\mu(t)$ is a probability measure based on $\{-1\} \cup \{1\}$, that is, a convex combination $\alpha(t)\delta(u-1)+(1-\alpha(t))\delta(u+1)$. The relaxed equation is

$$y'(t) = (-1)\mu(t,\ \{-1\}) + \mu(t,\ \{1\}).$$

A relaxed optimal control is given by

$$\mu(t) = \frac{1}{2}\ \delta(u-1) + \frac{1}{2}\ \delta(u+1). \tag{1.12}$$

In Example 1.2, each $\mu(t)$ is a probability measure in $[0,1]$. The relaxed system is

$$x'(t) = \int_0^1 u\mu(t, du),$$

$$y'(t) = \int_0^1 (u^2-1)^2\mu(t, du)$$

thus (1.12) is also an optimal control for this problem. In both examples it is unnecessary to relax the cost functional. In Example 1.3, each $\mu(t)$ is a probability measure in $[0,\ 1]$, the relaxed equation is

$$y'(t) = \int_0^t u\mu(t,\ du)$$

and the relaxed cost functional

$$y_0(t,\ u) = \int_0^t y(\sigma)^2 d\sigma + \int_0^t \int_{-1}^1 (u^2-1)^2\mu(t,\ du)dt$$

so that (1.12) is an optimal control.

Warga's approach to relaxation has been extended to infinite dimensional nonlinear systems in [1], [2] emphasizing countable additivity of the measures $\mu(t)$. A different class of measure-valued relaxed controls was introduced in [9], where compactness assumptions in the control set U and weak measurability assumptions on $f(t, y, u)$ are avoided using finitely additive measures. These relaxed controls (as Young's measures of variational calculus and Warga's relaxed controls for finite dimensional systems) are an attempt to fulfill Hilbert's dictum [24, p. 123] that "every problem of the calculus of variations has a solution, provided the word 'solution' is suitably understood". Existence theorems in this setting are considered in [10].

Generally speaking, any class of relaxed controls for (1.1) should pass four tests:

(a) Every ordinary trajectory is also a relaxed trajectory.

(b) Relaxed trajectories should be the same as the solutions of the differential inclusion

$$y'(t) \in Ay(t) + \overline{\text{conv}}\ f(t,\ y(t),\ U).$$

(c) For every relaxed trajectory $\mathbf{y}(t)$ and every $\varepsilon > 0$ there exists an ordinary trajectory such that

$$\|y(t) - \mathbf{y}(t)\| \leq \varepsilon \quad (0 \leq t \leq T).$$

(d) If $\{\mathbf{y}_n(t)\}$ is a convergent sequence of relaxed trajectories then $\mathbf{y}(t) = \lim \mathbf{y}_n(t)$ is a relaxed trajectory.

Test (a) is not decisive: if the original class $U_{\text{ad}}(0,\ T;\ U)$ does not guarantee existence, there is no reason why it should not be restricted to fulfill (a). Test (b) is just one of simplicity. On the other hand, failure of (c) would mean that driving the system with relaxed controls cannot even be approximately duplicated with ordinary controls, so that the relaxed control system is not closely related to the original system. Finally, (d) is the basis of all existence theorems.

§2. SPACES OF RELAXED CONTROLS AND THE RELAXED CONTROL SYSTEM IN REFLEXIVE SEPARABLE SPACES

Construction of relaxed controls is based on the dual of the space $L^1(0,\ T;\ E)$ of all (equivalence classes of) strongly measurable E-valued functions $f(\cdot)$ such that $\|f\|_1 = \int \|f(t)\|dt < \infty$ endowed with $\|\cdot\|_1$, where E is an arbitrary Banach space. If E^* is the dual of E, an E^*-valued function $g(\cdot)$ is *E-weakly measurable* if and only if $\langle g(\cdot),\ y\rangle$ is measurable for each $y \in E$. The space of all such $g(\cdot)$ with

$$|\langle g(t),\ y\rangle| \leq C\|y\| \qquad \text{a. e. in } 0 \leq t \leq T \tag{2.1}$$

("a. e." depending on y) is named $L^\infty_w(0,\ T;\ E^*)$ and equipped with the norm $\|g(\cdot)\|$ = infimum of all C such that (2.1) holds. A function $g(\cdot) \in L^\infty_w(0,\ T;\ E^*)$ produces an element in $L^1(0,\ T;\ E)^*$ through

$$\langle g(\cdot),\ f(\cdot)\rangle = \int_0^T \langle g(\sigma),\ f(\sigma)\rangle d\sigma \quad (f(\cdot) \in L^1(0,\ T;\ E)). \tag{2.2}$$

This functional only depends on the equivalence class of $g(\cdot)$ and its norm coincides with the $L_w^\infty(0,\ T;\ E^*)$-norm of $g(\cdot)$. Conversely, it has been shown in [4], [5] that every element of $L^1(0,\ T;\ E)^*$ admits the representation (2.2) with $g(\cdot) \in L_w^\infty(0,\ T;\ E^*)$, thus $L^1(0,\ T;\ E)^* = L_w^\infty(0,\ T;\ E^*)$ algebraically and metrically. The proof is a consequence of the Dunford-Pettis theorem [6], [7, Theorem 6, p. 503 and Lemma 8, p. 504], [19, Corollary 1, p. 89], which provides additional details: for instance, there exists a linear operator $S : L_w^\infty(0,\ T;\ E^*) \to L_w^\infty(0,\ T;\ E^*)$ such that Sg belongs to the equivalence class of g, $t \to \|(Sg)(t)\|$ is measurable in $0 \le t \le T$ and $\sup_{0\le t\le T} \|(Sg)(t)\| = \|g\|$.

We introduce three definitions of relaxed controls corresponding to increasingly stringent assumptions on the control set U. The first assumes nothing of U and is based on the Banach space $B(U)$ of all bounded functions in U equipped with the supremum norm, whose dual $B(U)^*$ is isometrically isomorphic to the space $\Sigma_{ba}(U,\ \Phi)$ of all bounded, finitely additive measures defined in the ring Φ of all subsets of U, $\Sigma_{ba}(U,\ \Phi)$ equipped with the total variation norm [7, pp. 258]. The duality pairing is

$$\langle \mu,\ f\rangle = \int_U f(u)\mu(du). \tag{2.3}$$

The space $V_b(0,\ T;\ U)$ of relaxed controls consists of all $\mu(\cdot) \in L^1(0,\ T;\ B(U))^* = L_w^\infty(0,\ T;\ \Sigma_{ba}(U,\ \Phi))$ that satisfy: (i)

$$\|\mu(\cdot)\| \le 1 \tag{2.4}$$

($\|\cdot\|$ the norm in $L_w^\infty(0,\ T;\ \Sigma_{ba}(U,\ \Phi))$ (ii) if $f(\cdot) \in L^1(0,\ T;\ B(U))$ is such that $f(t,\ u) \ge 0$ for $u \in U$ a. e. in $0 \le t \le T$ then

$$\int_0^T \int_U f(t,\ u)\mu(t,\ du)dt \ge 0, \tag{2.5}$$

(iii) if $\chi(\cdot)$ is the characteristic function of a measurable set $e \subseteq [0,\ T]$ and $\mathbf{1}$ is the function identically 1 in U, then

$$\int_0^T \int_U (\chi(t) \otimes \mathbf{1})\mu(t,\ du)dt = \ \text{meas}(e). \tag{2.6}$$

The second definition is for a normal topological space U and uses the Banach space $BC(U)$ of all bounded continuous functions in U equipped with the supremum norm, whose dual $BC(U)^*$ is isometrically isomorphic to the space $\Sigma_{rba}(U,\ \Phi_c)$ of all bounded, finitely additive regular measures defined in the ring Φ_c generated by the closed sets of U, $\Sigma_{rba}(U,\ \Phi_c)$ equipped with the total variation norm [7, p. 262]. The duality pairing is (2.4). The space $V_{bc}(0,\ T;\ U)$ of relaxed controls consists of all $\mu(\cdot) \in L^1(0,\ T;\ BC(U))^* = L_w^\infty(0,\ T;\ \Sigma_{rba}(U,\ \Phi_c))$ that satisfy (i), (ii), (iii).

The third space of relaxed controls is the classical one in [20], [21]: it assumes U to be a compact metric space and uses the Banach space $C(U)$ of all continuous functions in U equipped with the supremum norm, whose dual $C(U)^*$ is isometrically isometric to the space $\Sigma_{rca}(U, \Phi_b)$ of all bounded, countably additive regular measures defined in the σ-ring Φ_b of Borel sets of U (the σ-ring generated by the closed sets), $\Sigma_{rca}(U, \Phi_b)$ endowed with the total variation norm [7, p. 265]. The space $V_c(0, T; U)$ of relaxed controls consists of all $\mu(\cdot) \in L^1(0, T; C(U))^* = L^\infty_w(0, T; \Sigma_{rba}(U, \Phi_b))$ that satisfy (i), (ii), (iii).

In all three cases, (i) is a consequence of (ii) and (iii). In view of the comments after (2.2), each equivalence class in any of the spaces $V_b(0, T; U)$, $V_{bc}(0, T; U)$, $V_c(0, T; U)$ contains an element satisfying $\|\mu(t)\| \le 1$ in $0 \le t \le T$. This (and the conditions defining relaxed controls) implies that

$$\mu(t) \ge 0, \quad \mu(t, U) = \|\mu(t)\| = 1 \quad (0 \le t \le T) \tag{2.7}$$

thus relaxed controls (modulo equivalence) can be thought of as "taking probability measure values"; for the first two spaces, the measures are finitely additive. For a proof of (2.7) see [11, Lemma 2.2].

We define the relaxed version of (1.1) assuming E reflexive and separable (for nonreflexive spaces see §7). For $V_b(0, T; U)$ we assume

(II_b) (a) $f(t, y, u)$ is bounded in U for every t, y fixed; for every compact $K \subseteq E$ there exists $\alpha(\cdot) = \alpha(K, \cdot) \in L^1(0, T)$ such that

$$\|f(t, y, u)\| \le \alpha(t) \qquad (0 \le t \le T, \ y \in K, \ u \in U). \tag{2.8}$$

(b) If $y(\cdot) \in C(0, T; E)$ and $y^* \in E^*$ then $t \to \langle y^*, f(t, y(t), \cdot)\rangle$ is a strongly measurable $B(U)$-valued function.

We note that $t \to \langle y^*, f(t, y(t), \cdot)\rangle \in L^1(0, T; B(U))$, in fact, we have $\|\langle y^*, f(t, y(t), \cdot)\rangle\|_{B(U)} \le \alpha(K, t)\|y^*\|$ where $K = \{y(t);\ 0 \le t \le T\}$.

The *relaxed* system corresponding to $V_b(0, T; U)$ is

$$y'(t) = Ay(t) + \mathbf{f}(t, y(t))\mu(t), \ y(0) = \zeta \tag{2.9}$$

where $\mathbf{f} : [0, T] \times E \times \Sigma_{ba}(U, \Phi) \to E$ is defined thusly: $\mathbf{f}(t, y)\mu$ is the unique element of E satisfying

$$\langle y^*, \mathbf{f}(t, y)\mu\rangle = \int_U \langle y^*, f(t, y, u)\rangle \mu(du) \tag{2.10}$$

for all $y^* \in E^*$. We have

$$\|\mathbf{f}(t, y)\mu\| \le \alpha(K, t)\|\mu\|_{\Sigma_{ba}(U)} \quad (0 \le t \le T, \ y \in K). \tag{2.11}$$

Let $\mu(\cdot) \in L^\infty_w(0, T; \Sigma_{ba}(U, \Phi))$, $y(\cdot) \in C(0, T; E)$ and $y^* \in E^*$. By (b), the function

$$\langle y^*, \mathbf{f}(\cdot, y(\cdot))\mu(\cdot)\rangle = \int_U \langle y^*, f(\cdot, y(\cdot), u)\rangle \mu(\cdot, du) \tag{2.12}$$

belongs to $L^1(0, T)$, thus $\mathbf{f}(\cdot, y(\cdot))\mu(\cdot)$ is E^*-weakly measurable and, since E is separable, it is strongly measurable. Since $\mathbf{f}(t, y, u)$ satisfies (I) for $V_b(0, T; U)$, we interpret the differential equation (2.9) (the same as (1.1)) as the integral equation

$$y(t) = S(t)\zeta + \int_0^t S(t-\sigma)\mathbf{f}(\sigma, y(\sigma))\mu(\sigma)d\sigma. \tag{2.13}$$

The relaxed systems corresponding to the relaxed control space $V_{bc}(0, T; U)$ (resp. $V_c(0, T; U)$) are similarly defined; the assumptions on $f(t, y, u)$ are (II_{bc}) (resp. (II_c)) obtained interchanging the spaces $B(U)$, $\Sigma_{ba}(U, \Phi)$ by the spaces $BC(U)$, $\Sigma_{rba}(U, \Phi_c)$ (resp. $C(U)$, $\Sigma_{rca}(U, \Phi_b)$).

To check test (a) in §1 we must replicate the action of an ordinary control $u(\cdot)$ by a relaxed control $\mu(\cdot)$. For this, we define

$$\mu(t) = \delta(\cdot - u(t)) \tag{2.14}$$

where δ is the Dirac delta. If $\mu(\cdot)$ is to belong to $V_b(0, T; U)$, we need the function

$$t \to y(u(t)) \tag{2.15}$$

to be measurable for every $u(\cdot) \in U_{ad}(0, T; U)$. This essentially forces the elements of $U_{ad}(0, T; U)$ to be countably valued, measurable functions. On the other hand, if $V_{bc}(0, T; U)$ is used, we only need measurability of (2.15) for $y \in BC(U)$; this holds if U is a subset of a Banach space and the controls $u(\cdot)$ are merely strongly measurable. To include all spaces of relaxed controls in our results, we shall assume from now on that $U_{ad}(0, T; U)$ consists of all countably valued U-valued functions, that is, the set of all functions that take countably many values in measurable sets.

§3. APPROXIMATIONS OF MEASURES, RELAXED CONTROLS AND RELAXED TRAJECTORIES

We denote by $\Pi_b(U, \Phi)$ the subspace of $\Sigma_{ba}(U, \Phi)$ consisting of all probability measures, that is, of all μ such that $\mu \geq 0$, $\mu(U) = 1$. The sets $\Pi_{rba}(U, \Phi_c)$, $\Pi_{rca}(U, \Phi_b)$) are correspondingly defined. D is the set of all Dirac measures $\delta(\cdot - u)$, $u \in U$, which is a subset of $\Pi_b(U, \Phi)$, $\Pi_{bc}(U, \Phi_c)$, $\Pi_c(U, \Phi_b)$.

LEMMA 3.1. The convex hull conv(D) is dense in $\Pi_b(U, \Phi)$ in the weak $B(U)$-topology. Corresponding results hold for $\Pi_{bc}(U, \Phi_c)$, $\Pi_c(U, \Phi_b)$.

The proof is the same as that of [7, Lemma 6, p. 441]; for details, see [11]. Lemma 3.1 implies approximation properties for control spaces. Denote by $U_{ad}(0, T; U)$ the space of (countably valued) ordinary controls realized as relaxed controls as in (2.14). Then we have

APPROXIMATION THEOREM 3.2. The space $U_{ad}(0, T; U)$ is dense in $V_b(0, T; U)$ in the $L^1(0, T; B(U))$-weak topology of $L^\infty_w(0, T; \Sigma_{ba}(U, \Phi))$. Corresponding results hold for $V_{bc}(0, T; U)$, $V_c(0, T; U)$.

See [11, Theorem 3.2]. The approximation theorem can be used to obtain *relaxation theorems*, that is to verify test (b) in §1 under the following hypotheses.

(III_b) (a) $f(t, y, u)$ is bounded in U for every t, y fixed; moreover, for every $B > 0$ there exists $\alpha(\cdot) = \alpha(B, \cdot) \in L^1(0, T)$ such that

$$\|f(t, y, u)\| \le \alpha(t) \qquad (0 \le t \le T,\ \|y\| \le B,\ u \in U). \tag{3.1}$$

(b) If $y(\cdot) \in C(0, T; E)$ and $y^* \in E^*$ then $t \to \langle y^*, f(t, y(t), \cdot)\rangle$ is a strongly measurable $B(U)$-valued function.
(c) For every $B > 0$ there exists $\beta(\cdot) = \beta(B, \cdot) \in L^1(0, T)$ such that

$$\|f(t, y', u) - f(t, y, u)\| \le \beta(t)\|y' - y\| \tag{3.2}$$
$$(0 \le t \le T,\ \|y\| \le B,\ u \in U).$$

Part (a) (resp. part (c)) of Assumption (III_b) implies a local boundedness condition (resp. local Lipschitz condition) for $\mathbf{f}(t, y)\mu$. Under these, the integral equation (2.13) can be locally solved by successive approximations.

RELAXATION THEOREM 3.3. Let E be reflexive and separable. Assume that $S(t)$ is compact for $t > 0$ and that Assumption (III_b) holds. Let $\mu(\cdot) \in V_b(0, \bar{t}; U)$ be such that the solution $y(t, u)$ of (2.9) exists in $0 \le t \le \bar{t}$, $\varepsilon > 0$. Then there exists $u(\cdot) \in U_{ad}(0, T; U)$ such that $y(t, u)$ exists in $0 \le t \le \bar{t}$ and

$$\|y(t, u) - y(t, \mu)\| \le \varepsilon \qquad (0 \le t \le T). \tag{3.3}$$

For a proof, see [11, Theorem 5.2]. Theorem 3.3 holds without the assumption of compactness of $S(t)$ under a reinforced version of (b) of Assumption (III_b).

RELAXATION THEOREM 3.4. Let E be reflexive and separable. Let Assumption (III_b) hold with (b) replaced by:

(b') If $y(\cdot) \in C(0, T; E)$ then $t \to f(t, y(t), \cdot)$ is a strongly measurable $B(U; E)$-valued function.

Let $\mu(\cdot) \in V_b(0, \bar{t}; U)$ be such that the solution $y(t, \mu)$ of (2.9) exists in $0 \le t \le \bar{t}$, and let $\varepsilon > 0$. Then there exists $u(\cdot) \in U_{ad}(0, T; U)$ such that $y(t, u)$ exists in $0 \le t \le \bar{t}$ and (3.3) holds.
See [11, Theorem 5.2] and [14] for the Navier-Stokes equations. Relaxation theorems for differential inclusions are proved in [16]. For finite dimensional systems and the space $V_c(0, T; U)$ the results here are classical: ([22] and references therein). Finally, see [1], [2] for relaxation theory in a different setting.

§4. DIFFERENTIAL INCLUSIONS AND FILIPPOV'S THEOREM

If $F(t, y)$ is a set-valued function, an E-valued continuous function $y(t)$ is a *solution* of the differential inclusion (see [16])

$$y'(t) \in Ay(t) + F(t, y(t)), \qquad y(0) = \zeta \tag{4.1}$$

in $0 \leq t \leq T$ if and only if there exists a function $g(\cdot) \in L^1(0, T; E)$ such that $g(t) \in F(t, y(t))$ a. e. in $0 \leq t \leq T$ and

$$y(t) = S(t)\zeta + \int_0^t S(t-\sigma)g(\sigma)d\sigma \qquad (0 \leq t \leq T). \tag{4.2}$$

The pair $(y(\cdot), g(\cdot))$ is a *trajectory* of (4.1). The result below states that the control space $V_b(0, T; U)$ passes test (b) in §1 under Assumption (II_b). Similar results hold for $V_{bc}(0, T; U)$, $V_c(0, T; U)$.

THEOREM 4.1. Assume E is reflexive and separable and that Assumption (II_b) holds. Let $y(t, \mu)$ be a solution of (2.9) for $\mu(\cdot) \in V_b(0, T; U)$. Then $\mathbf{f}(t, y(t, \mu))\mu(t) \in \overline{\mathrm{conv}} f(t, y(t, \mu), U)$ a.e. in $0 \leq t \leq \bar{t}$ ($\overline{\mathrm{conv}}$ = closed convex hull) so that $(y(t, \mu), \mathbf{f}(t, y(t, \mu))\mu(t))$ is a trajectory of the differential inclusion

$$y'(t) \in Ay(t) + \overline{\mathrm{conv}}\, f(t, y(t), U) \tag{4.3}$$

in $0 \leq t \leq \bar{t}$. Conversely, let $(y(\cdot), g(\cdot))$ be a trajectory of (4.3). Then there exists $\mu(\cdot) \in V_b(0, T; U)$ such that

$$\mathbf{f}(t, y(t))\mu(t) = g(t) \text{ a. e. in } 0 \leq t \leq T \tag{4.4}$$

so that $y(t) = y(t, \mu)$ is a trajectory of (2.9).

For a proof, see [11, Theorem 4.1 and Theorem 4.3]. The second implication of Theorem 4.1 is a consequence of the following

LEMMA 4.2. Let E be a reflexive separable Banach space, U an arbitrary set, $\phi : [0, T] \times U \to E$ a function such that

(i) $\phi(t, u)$ is bounded in U for every t fixed.

(b) For each $y^* \in E^*$ the function $t \to \langle y^*, \phi(t, \cdot)\rangle$ is a strongly measurable $B(U)$-valued function.

Define $\boldsymbol{\phi}(t)\mu$ by

$$\langle y^*, \boldsymbol{\phi}(t)\mu\rangle = \int_U \langle y^*, \phi(t, u)\rangle \mu(du) \tag{4.5}$$

for $y^* \in E^*$, and let $g(\cdot)$ be a strongly measurable function such that

$$g(t) \in \overline{\mathrm{conv}}\phi(t, U) \text{ a. e. in } 0 \leq t \leq T. \tag{4.6}$$

Then there exists $\mu(\cdot) \in V_b(0,\ T;\ U)$ such that

$$\phi(t)\mu(t) = g(t) \quad \text{a. e. in } 0 \le t \le T. \tag{4.7}$$

Lemma 4.2 is an infinite dimensional generalization of the second half of Filippov's Theorem as formulated in [24, (34.7), p. 297], that is, of the statement dealing with relaxed controls, not ordinary controls. For the generalization corresponding to ordinary controls see [18].

The modifications necessary for the spaces $V_{bc}(0,\ T;\ U), V_c(0,\ T;\ U)$ are by now familiar: for $V_{bc}(0,\ T;\ U)$ we assume U a normal topological space and replace "bounded" by "continuous and bounded" in the hypotheses. Since these observations apply equally to all forthcoming results, we shall restrict ourselves to the space $V_b(0,\ T;\ U)$ in the rest of this paper and leave to the reader the modifications needed to deal with the other spaces. We shall use $\Sigma_{ba}(U)$ as shorthand for $\Sigma_{ba}(U,\ \Phi)$.

§5. EXISTENCE THEOREMS FOR OPTIMAL CONTROL PROBLEMS

We call the optimal control problem in §1 the *original* problem, with $m = \inf y_0(\bar{t},\ u)$, the infimum taken over all $u \in U_{ad}(0,\ \bar{t};\ U)$ whose trajectories $y(\cdot,\ u)$ satisfy (1.4). We assume that $-\infty < m < \infty$, which means that there exists some control $u \in U_{ad}(0,\ \bar{t};\ U)$ such that $y(\cdot,\ u)$ satisfies the target condition (1.4), and that we cannot reach the target set with arbitrarily low values of the cost functional; the latter is assumed of the form (1.9). The *relaxed control problem* is that described by (2.9), with controls $\mu(\cdot) \in V_b(0,\ T;\ U)$ and the *relaxed cost functional* is (1.11), or

$$y_0(t,\ \mu) = \int_0^t \mathbf{f}_0(\sigma,\ y(\sigma,\ \mu))\mu(\sigma)d\sigma + g_0(t,\ y(t,\ \mu)), \tag{5.1}$$

where $\mathbf{f}_0$ is defined by

$$\mathbf{f}_0(t,\ y)\mu = \int_U f_0(t,\ y,\ u)\mu(du). \tag{5.2}$$

The target condition is the same. The companion of assumption (II_b) giving sense to (5.1) is

(II_{b0}) (a) $f_0(t,\ y,\ u)$ is bounded in U for $t,\ y$ fixed.

(b) If $y(\cdot) \in C(0,\ T;\ E)$ then $t \to f_0(t,\ y(t),\ \cdot) \in L^1(0,\ T;\ B(U))$.

Corresponding to the relaxed control problem, $\mathbf{m} = \inf y_0(t,\ \mu)$, the infimum taken over all $\mu \in V_b(0, \bar{t};\ U)$ whose trajectories $y(\cdot,\ \mu)$ exist in $0 \le t \le \bar{t}$ and satisfy the target condition (5.2). The observations about m apply to $\mathbf{m}$ as well. Since there are more relaxed than ordinary controls, we will always have $\mathbf{m} \le m$. In principle, strict inequality is possible, including situations where $\mathbf{m} < \infty$, $m = \infty$ (the target may be hit by a trajectory $y(\cdot, \mu)$ of the relaxed system but not by a trajectory $y(\cdot, u)$ of the original system). It is desirable that $\mathbf{m} = m$, for if $\mathbf{m} < m$ it could be maintained that the relaxed problem is not really related to the original problem.

The basis of all our existence theorems is the *saturation theorem* below.

THEOREM 5.1. Let $\{y(t, \mu_n)\}$ be a sequence of trajectories of the relaxed system (2.9) in $0 \le t \le \bar{t}$. Assume that (i) there exists $y(\cdot) \in C(0, T; E)$ such that $y(t, \mu_n) \to y(t)$ in E for each t, $0 \le t \le \bar{t}$, (ii)

$$\langle y^*, f(\cdot, y(\cdot, \mu_n), \cdot)\rangle \to \langle y^*, f(\cdot, y(\cdot), \cdot)\rangle,$$
$$f_0(\cdot, y(\cdot, \mu_n), \cdot) \to f_0(\cdot, y(\cdot), \cdot),$$

in $L^1(0, T; B(U))$ for every $y^* \in E^*$. Then $y(t)$ is a trajectory of (2.9), that is, there exists a relaxed control $\mu(\cdot) \in V_b(0, T; U)$ such that $y(t) = y(t, \mu)$; moreover, $y_0(\bar{t}, \mu) = \lim_{n\to\infty} y_0(\bar{t}, \mu_n)$.

Theorem 5.1 is simply a scheme or "template" for the construction of existence theorems. For the finite dimensional system (1.8), under y-continuity of $f(t, y, u)$, $f_0(t, y, u)$ assumption (ii) follows from convergence of $\{y(t, u^n)\}$, which is guaranteed by a priori bounds and the Arzelà-Ascoli theorem. For the infinite dimensional system (1.1), one must rely on compactness of the semigroup $S(t)$ or of the nonlinear term $f(t, y, u)$. Compactness of $S(t)$ is used via Lemma 5.2 below, where E may be an arbitrary Banach space.

LEMMA 5.2. The operator

$$(\Lambda g)(t) = \int_0^t S(t-\sigma)g(\sigma)d\sigma$$

is bounded from $L^1(0, T; E)$ into $C(0, T; E)$. If $S(t)$ is compact for $t > 0$ and $\{g_n(\cdot)\}$ is a sequence in $L^1(0, T; E)$ such that the integrals of $\|g_n(\cdot)\|$ are equicontinuous in $0 \le t \le T$, then $\{\Lambda g_n(\cdot)\}$ has a subsequence convergent in $C(0, T; U)$.

For a proof, see [10], [12]. Lemma 4.2 is put to work under the following assumptions on f and f_0, which add a condition to (II_b), (II_{b0}).

(IV_b) (a) and (b) as in (II_b). (c) If $\{y_n(\cdot)\} \subset C(0, T; E)$ is such that $y_n(\cdot) \to y(\cdot)$ in $C(0, T; E)$ then

$$\langle y^*, f(\cdot, y_n(\cdot), \cdot)\rangle \to \langle y^*, f(\cdot, y(\cdot), \cdot)\rangle$$

in $L^1(0, T; B(U))$.

(IV_{b0}) (a) and (b) as in (II_{b0}). (c) If $\{y_n(\cdot)\} \subset C(0, T; E)$ is such that $y_n(\cdot) \to y(\cdot)$ in $C(0, T; E)$ then

$$f_0(\cdot, y_n(\cdot), \cdot) \to f_0(\cdot, y(\cdot), \cdot)$$

in $L^1(0, T)$.

THEOREM 5.3. Let E be reflexive and separable, $S(t)$ a compact semigroup, and let the target set Y be closed. Assume that (IV_b) and (IV_{b0}) hold and that there exists a minimizing sequence $\{\mu_n(\cdot)\}$, $\mu_n(\cdot) \in V_b(0, t_n; U)$ of relaxed controls with $\{t_n\}$ bounded, $\{y(t, \mu_n)\}$ uniformly bounded in $0 \le t \le t_n$ and $\|f(t, y(t, \mu_n), \cdot)\| \le \gamma(t)$, $(u \in U)$, where $\gamma(\cdot) \in L^1(0, T)$. Then there exists a solution $\overline{\mu}(\cdot)$ of the relaxed optimal control problem.

For the proof, see [10, Theorem 5.2].

Many compact semigroups (such as those generated by uniformly elliptic partial differential operators) are also **holomorphic**. Modulo a translation we may assume that the origin belongs to the resolvent set of the infinitesimal generator A, so that fractional powers $(-A)^\alpha$ can be defined for α real: $(-A)^\alpha S(t)$ is bounded in $t > 0$ for all $\alpha > 0$ and

$$\|(-A)^\alpha S(t)\| \le C_\alpha t^{-\alpha} e^{\omega t} \quad (t \ge 0).$$

Writing the integral equation (2.13) in the form

$$y(t) = S(t)\zeta + \int_0^t (-A)^\alpha S(t-\sigma)(-A)^{-\alpha} \mathbf{f}(t, y(\sigma))\mu(\sigma) d\sigma$$

we may prove an analogue of Theorem 5.3 where hypothesis (IV_b) is required of $(-A)^{-\alpha} f(t, y, u)$ rather than of $f(t, y, u)$. See also [13], [14], where related tricks are applied to the Navier-Stokes equations.

In cases where $S(t)$ is not compact, we rely on the nonlinear term $f(t, y, u)$. This works for instance for the semilinear wave equation

$$\begin{aligned} y_{tt}(t, x) = \sum_{j=1}^m \sum_{k=1}^m \partial^j (a_{jk}(x) \partial^k y(t, x)) - \\ - \phi(y(t, x), u(t)) \ (x \in \Omega), \ y(t, x) = 0 \ (x \in \Gamma) \end{aligned} \tag{5.3}$$

in an arbitrary domain Ω with boundary Γ in m-dimensional Euclidean space $\mathbf{R}^m$; here $x = (x_1, x_2, \ldots, x_m)$, $\partial^j = \partial/\partial x_j$. We assume that $a_{jk} = a_{kj}$ and that the operator A is uniformly elliptic: $\Sigma\Sigma a_{jk}(x)\xi_j\xi_k > \kappa|\xi|^2 (\xi \in \mathbf{R}^m, x \in \overline{\Omega}, \kappa > 0)$. The nonlinear term $\phi(y, u)$ is defined in $\mathbf{R} \times U$. To be able to solve (5.3), we assume that for each $y \in \mathbf{R}$, $\phi(y, \cdot) \in B(U)$ and that the conditions below hold in function of the dimension m.

Dimension $m > 2$. Let $\alpha = m/(m-2)$. Then

$$|\phi(y, u)| \le C(1 + |y|^\alpha) \quad (y \in \mathbf{R}^m, u \in U) \tag{5.4}$$

$$|\phi(y', u) - \phi(y, u)| \le K(1 + |y|^{\alpha-1} + |y'|^{\alpha-1})|y' - y| \tag{5.5}$$

$$(y, y' \in \mathbf{R}^m, u \in U).$$

Dimension $m = 2$. There exists some $\alpha > 0$ such that (5.4) and (5.5) hold.

Dimension $m = 1$. For each $B > 0$ there exist $C = C(B),\ K = K(B)$ such that

$$|\phi(y,\ u)| \le C \qquad\qquad (|y| \le B,\ u \in U)$$
$$|\phi(y',\ u) - \phi(y,\ u)| \le K|y' - y| \qquad (|y|,\ |y'| \le B,\ u \in U).$$

Under these conditions, the equation (5.3) has a unique global solution and existence of solutions to optimal control problems can be shown for $m \le 2$; for $m > 2$ we need a slightly stronger assumption.

THEOREM 5.4. Assume ϕ satisfies (5.4), (5.5) with $\alpha < m/(m-2)$ in case $m > 2$, and let the cost functional satisfy (II_{b0}). Let $\{\mu_n(\cdot)\}$ be a minimizing sequence with $\{t_n\}$ bounded. Then there exists a relaxed solution to the optimal control problem.

For a proof, see Theorem 6.1 in [10], where (5.3) is reduced to a first order equation in the energy space in the customary way. Note that the condition on α in the case $m > 3$ is in excess of what is needed for global existence, where nonstrict inequality would be sufficient. This is due to the need for the Rellich-Kondrachev theorem to show compactness of the nonlinearity.

§6. THE MAXIMUM PRINCIPLE

Control systems of the forms (1.1) or (2.9) fit the model in [10] with minor modifications. To apply the nonlinear programming theory in [8] we equip $V_b(0,\ T;\ U)$ with the metric

$$d(\mu(\cdot),\ \nu(\cdot)) = \lambda\{t \in [0,\ T];\ \mu(t) \neq \nu(t)\},$$

where λ is the outer measure generated by the Lebesgue measure on the real line, and define as equivalent elements of $V_b(0,\ T;\ U)$ that lie at distance zero; this equivalence relation is more restrictive than that of $L^\infty_w(0,\ T;\ \Sigma_{ba}(U))$, which we discard. The space of (relaxed) controls is now called $V_b(0,\ T;\ U)_d$, and is complete under d.

Pontryagin's maximum principle for the control system (2.9) is obtained by application of the results on the problem

$$\text{minimize } f_0(u) \text{ subject to } f(u) \in Y$$

in [8], where $f : V \to E$ (V a complete metric space, E a Banach space) and $f_0 : V \to \mathbf{R}$. For fixed terminal time $\bar{t}$, we take $V = V_b(0,\ \bar{t};\ U)_d$ and

$$f(\mu) = y(\bar{t},\ \mu), \qquad f_0(\mu) = y_0(\bar{t},\ \mu).$$

The optimal control $\bar{\mu}$ is assumed to exist or is provided by one of the existence theorems in §5. Without aiming for maximum generality, we place conditions on

E, U, f and f_0 that legitimize all statements that follow. We assume that E is reflexive and separable and that $f(t, y, u)$ satisfies

(V_b) (a) $f(t, y, u)$ is continuous with respect to t, y in $[0, T] \times E \times U$ uniformly with respect to u, and for every $B > 0$ there exists $\alpha(\cdot) = \alpha(B, \cdot) \in L^1(0, T)$ such that

$$\|f(t, y, u)\|_E \le \alpha(t) \qquad (0 \le t \le T,\ \|y\| \le B,\ u \in U).$$

(b) $f(t, y, u)$ has a Fréchet derivative $\partial_y f(t, y, u)$ with respect to y in $[0, T]\times E\times U$ uniformly with respect to u, i.e.

$$f(t, y + h, u) = f(t, y, u) + \partial_y f(t, y, u)h + \rho(t, y, h, u),$$

where for each t, y we have $\|\rho(t, y, h, u)\|/\|h\| \to 0$ as $h \to 0$ uniformly with respect to $u \in U$. (c) $\partial_y f(t, y, u)$ is strongly continuous with respect to t, y in $[0, T] \times E \times U$ uniformly with respect to u and for every $B > 0$ there exists $\beta(\cdot) = \beta(B, \cdot)$ such that

$$\|\partial_y f(t, y, u)\|_{L(E,E)} \le \beta(t) \quad (0 \le t \le T,\ \|y\| \le B,\ u \in U).$$

These properties of $f(t, y, u)$ imply that $\mathbf{f}(t, y)\mu$ is continuous in $[0, T]\times E\times\Sigma_{ba}(U)$ and

$$\|\mathbf{f}(t, y)\mu\| \le \alpha(t)\|\mu\|_{\Sigma_{ba}(U)} \quad (0 \le t \le T,\ \|y\| \le B,\ \mu \in \Sigma_{ba}(U));$$

moreover, the function $t \to \mathbf{f}(t, y(t))\mu(t)$ is strongly measurable for every $y(\cdot) \in C(0, T; E)$ and $\mu(\cdot) \in L^\infty_w(0, T; \Sigma_{ba}(U))$ and $\mathbf{f}(t, y)\mu$ has a Fréchet derivative $\partial_y \mathbf{f}(t, y)\mu$, given by

$$\langle y^*, (\partial_y \mathbf{f}(t, y)\mu)h\rangle = \int_U \langle y^*, \partial_y f(t, y, u)h\rangle \mu(du) \quad (y^* \in E^*).$$

Finally, $\partial_y \mathbf{f}(t, y)\mu$ is strongly continuous in $[0, T] \times E \times \Sigma_{ba}(U)$ and

$$\|\partial_y \mathbf{f}(t, y)\mu\|_{L(E,E)} \le \beta(t)\|\mu\|_{\Sigma_{ba}(U)} \quad (0 \le t \le T,\ \|y\| \le B,\ \mu \in \Sigma_{ba}(U)).$$

Existence of the Fréchet derivative and the mean value theorem imply

LEMMA 6.1. Let $\bar\mu(\cdot) \in L^\infty_w(0, \bar t; \Sigma_{ba}(U))$ be such that the trajectory $y(t, \bar\mu)$ exists in $0 \le t \le \bar t$. Then there exists $\rho > 0$ such that if $d(\mu, \bar\mu) \le \rho$ then the trajectory $y(t, \mu)$ exists in the same interval and

$$\|y(t, \mu) - y(t, \bar\mu)\| \le C \int_{[\{t;\mu(t)\ne\bar\mu(t)\}]} \alpha(\sigma)d\sigma.$$

The notation is: given a set e, $[e]$ is a **measurable envelope** of e (that is, a measurable set with $e \subseteq [e]$, $\lambda(e) =$ meas $[e]$).

The companion of assumption (V) for f_0 is:

(V_0) (a) $f_0(t, y, u)$ is continuous with respect to t, y in $[0, T] \times E \times U$ uniformly with respect to u, and for every $B > 0$ there exists $\alpha_0(\cdot) = \alpha_0(B, \cdot) \in L^1(0, T)$ such that

$$\|f(t, y, u)\|_E \leq \alpha_0(t) \qquad (0 \leq t \leq T, \ \|y\| \leq B, \ u \in U).$$

(b) $f(t, y, u)$ has a Fréchet derivative $\partial_y f_0(t, y, u)$ with respect to y in $[0, T] \times E \times U$ uniformly with respect to u. (c) $\partial_y f_0(t, y, u)$ is continuous with respect to t, y in $[0, T] \times E \times U$ uniformly with respect to u and for every $B > 0$ there exists $\beta_0(\cdot) = \beta_0(B, \cdot) \in L^1(0, T)$ such that

$$\|\partial_y f_0(t, y, u)\|_{E^*} \leq \beta_0(t) \qquad (0 \leq t \leq T, \ \|y\| \leq B, \ u \in U).$$

The assumption implies that $\mathbf{f}_0(t, y)\mu$ is continuous in $[0, T] \times E \times \Sigma_{ba}(U)$ and

$$|\mathbf{f}_0(t, y)\mu| \leq \alpha_0(t)\|\mu\|_{\Sigma_{ba}(U)} \qquad (0 \leq t \leq T, \ \|y\| \leq B, \ \mu \in \Sigma_{ba}(U));$$

moreover, the function $t \to \mathbf{f}_0(t, y(t))\mu(t)$ is measurable for every $y(\cdot) \in C(0, T; E)$ and $\mu(\cdot) \in L^\infty_w(0, T; \Sigma_{ba}(U))$ and $\mathbf{f}_0(t, y)\mu$ has a Fréchet derivative $\partial_y \mathbf{f}_0(t, y)\mu$ given by

$$(\partial_y \mathbf{f}_0(t, y)\mu)h = \int_U \partial_y f_0(t, y, u)h\mu(du).$$

Finally, $\partial_y \mathbf{f}_0(t, y)\mu$ is continuous in $[0, T] \times E \times \Sigma_{ba}(U)$ and

$$\|\partial_y \mathbf{f}_0(t, y)\mu\|_{E^*} \leq \beta_0(t)\|\mu\|_{\Sigma_{ba}(U)} \quad (0 \leq t \leq T, \ \|y\| \leq C, \ \mu \in \Sigma_{ba}(U)).$$

As a very particular consequence of the properties of $\mathbf{f}_0$, we obtain using Lemma 6.1 that the function $\mu \to y_0(\bar{t}, \mu)$ is continuous in $B(\bar{\mu}, \rho)$, the ball of center $\bar{\mu}$ and radius ρ in $V_b(0, \bar{t}; U)_d$.

The proof of all results in this section can be found in [10].

THEOREM 6.2 (Pontryagin's maximum principle). Let $\bar{\mu}(\cdot)$ a solution of the relaxed optimal control problem in $0 \leq t \leq \bar{t}$. Then there exists $(z_0, z) \in \mathbf{R} \times E^*$ such that $z_0 \leq 0$,

$$\begin{aligned} z_0\mathbf{f}_0(t, y(t, \bar{\mu}))\bar{\mu}(t) + \langle z(t), \mathbf{f}(t, y(t, \bar{\mu}))\bar{\mu}(t)\rangle = \\ = \max_{\nu \in \Sigma_{ba}(U,\Phi)} \{z_0\mathbf{f}_0(t, y(t, \bar{\mu}))\nu + \langle z(t), \mathbf{f}(t, y(t, \bar{\mu}))\nu\rangle\} \end{aligned}$$

a. e. in $0 \leq t \leq \bar{t}$, where $z(t)$ is the solution of the final value problem

$$\begin{aligned} z'(t) = -\{A^* + \partial_y \mathbf{f}(t, y(t, \bar{\mu}))\bar{\mu}(t))^*\}z(t) - \\ - z_0\partial_y \mathbf{f}_0(t, y(t, \bar{\mu}))\bar{\mu}(t), \ z(\bar{t}) = z. \end{aligned}$$

Under additional assumptions, the multiplier $(z_0,\ z)$ is nontrivial. We use the condition in [8, Corollary 2.14]:

LEMMA 6.3. Assume that, for every sequence $\{y^n\} \subseteq Y$ and every sequence $\{\mu_n\} \in V_r(0,\ \bar{t};\ U)_d$ such that $y^n \to \bar{y} = y(\bar{t},\ \bar{\mu}),\ \mu_n \to \bar{\mu}$ fast enough there exists a compact set Q such that

$$\bigcap_{n=1}^{\infty} \{R_n(\bar{t}) - T_Y(y^n) + Q\}$$

contains an interior point, where $T_Y(y^n)$ is the tangent cone to Y at y^n and $R_n(\bar{t})$ is the reachable space of the system

$$\begin{aligned} &z'(t) = \{A + \partial_y \mathbf{f}(t,\ y(t,\ \mu_n))\mu_n(t)\}z(t) + \\ &\mathbf{f}_0(t,\ y(t,\ \mu_n))\nu(t) - \mathbf{f}_0(t,\ y(t,\ \mu_n))\mu_n(t),\ z(0) = 0, \end{aligned}$$

$\nu(\cdot) \in V_b(0,\ \bar{t};\ U)_d$. Then $(z_0,\ z) \neq 0$.

The time optimal problem needs a special treatment but the final result may be included in Theorem 6.2; the result guarantees a multiplier $(z_0,\ z) = (0,\ z)$ with $z \neq 0$.

The conditions of Lemma 6.3 are always satisfied if the target set Y is "large" (for instance, a ball). For small target sets (say, $Y = \{y\}$) they are satisfied by some hyperbolic systems. They are also satisfied automatically when E is finite dimensional; we may take Q = unit ball of E. For comments on this condition for abstract parabolic equations see [8, §6].

§7. ABSTRACT PARABOLIC SYSTEMS IN NONREFLEXIVE SPACES

There are systems (for instance, reaction-diffusion equations in spaces of continuous functions or in L^1 spaces, see [8]) where the reflexivity assumption on E is unnatural. We indicate how the results in §3, §4 and §5 can be extended to nonreflexive spaces; existence theorems in this setting have been proved in [10, §7].

Let E be a Banach space, $S(t)$ a strongly continuous semigroup in E. Denote by $E^{\odot} \subseteq E^*$ the closure of the domain of $D(A^*)$ in E^* or, equivalently, the maximal subspace where the semigroup $S(t)^*$ is strongly continuous. The restriction of the semigroup $S(\cdot)^*$ to $E^{\odot}$ is a strongly continuous semigroup $S^{\odot}(t)$ called the *Phillips adjoint* of $S(t)$. The infinitesimal generator $A^{\odot}$ of $S^{\odot}(t)$ is the restriction of A^* with domain $D(A^{\odot}) = \{y \in D(A^*);\ A^*y \in E^{\odot}\}$. We shall assume that

(i) $S(t)E \subseteq D(A)$ and $AS(t)$ is continuous in the operator norm in $t > 0$.

(ii) E and $E^{\odot}$ are separable. E is ${}^{\odot}$*-reflexive* with respect to $S(\cdot)$, that is, $E^{\odot\odot} = E$.

The second condition obviously implies that $S^{\odot\odot}(t) = S(t)$, $A^{\odot\odot} = A$. Also, we have $S(t)^*E^* \subseteq E^{\odot}$, $S^{\odot}(t)^*(E^{\odot})^* \subseteq E$, $S(t)^*|_{E^{\odot}} = S^{\odot}(t)$, $S^{\odot}(t)^*|_E = S(t)$.

This abstract setup is justified by one key example, elliptic differential operators in the space $E = C(\overline{\Omega})$ of continuous functions in the closure $\overline{\Omega}$ of a domain Ω in Euclidean space. In this case, E^* is the space $\Sigma(\overline{\Omega})$ of regular Borel measures in $\overline{\Omega}$ with the total variation norm and $E^{\odot} = L^1(\Omega)$, $(E^{\odot})^* = L^{\infty}(\Omega)$. The same scheme applies to operators in $L^1(\Omega)$; here $L^1(\Omega)^* = L^{\infty}(\Omega)$, $L^1(\Omega)^{\odot} = C(\overline{\Omega})$. For further details on this spaces and operators see [8, §3].

The spaces $V_b(0,\ T;\ E)$, $V_{bc}(0,\ T;\ U)$, $V_c(0,\ T;\ U)$ are defined as in §2. To define the relaxed system corresponding to (1.1) we require the following less stringent version of Assumption (II_b):
($\mathrm{II}_b^{\odot}$) (a) Same as (a) of (II_b). (b) Same as (b) of (II_b) but with $y^* \in E^{\odot}$.
The relaxed system corresponding to (1.1) is

$$y'(t) = Ay(t) + \mathbf{f}(t,\ y(t))\mu(t), \qquad y(0) = \zeta \tag{7.1}$$

where $\mathbf{f}(t,\ y)\mu$ is the unique element of $(E^{\odot})^* \supseteq E$ that satisfies

$$\langle y^*,\ \mathbf{f}(t,\ y)\ \mu\rangle = \int_U \langle y^*,\ f(t,\ y,\ u)\rangle\mu(du) \tag{7.2}$$

for every $y^* \in E^{\odot}$. The corresponding integral equation is

$$y(t) = S(t)\zeta + \int_0^t S^{\odot}(t-\sigma)^*\mathbf{f}(t,\ y(\sigma))\mu(\sigma)d\sigma \tag{7.3}$$

if $\zeta \in E$; the integrand takes values in E. We may also take $\zeta \in (E^{\odot})^*$, in which case $S(t)\zeta$ is replaced by $S^{\odot}(T)^*\zeta$. The integral equation (7.3) is interpreted using

LEMMA 7.1. Let $g(\cdot)$ be a $E^{\odot}$-weakly measurable $(E^{\odot})^*$-valued bounded function defined in $0 \le t \le T$. Then the E-valued function

$$\sigma \to S^{\odot}(t-\sigma)^*g(\sigma)$$

is strongly measurable in $0 \le \sigma \le t$. If $\|g(\cdot)\|_{(E^{\odot})^*} \in L^1(0,\ T)$ the integral

$$y(t) = \int_0^t S^{\odot}(t-\sigma)^*g(\sigma)d\sigma$$

exists, is E-valued and continuous in $0 \le t \le T$.

We review the results in Sections 3 and 4 under the present hypotheses. Theorem 3.3 remains true under Assumption ($\mathrm{III}_b^{\odot}$) below:

($III_b^{\odot}$) (a) Same as (a) of (III_b). (b) Same as (b) of (III_b) but with $y^* \in E^{\odot}$. (c) Same as (c) of (III_b).

Solutions of the differential inclusion (4.1) are defined as in §4, but we only assume that $g(\cdot)$ is a $(E^{\odot})^*$-valued $E^{\odot}$-weakly measurable function with $\|g(\cdot)\| \in L^1(0, T)$. The integral (4.2) is interpreted through Lemma 7.1. Theorem 4.1 is valid under Assumption ($II_b^{\odot}$), but in $\overline{\text{conv}}$ the closure is taken *in the weak $E^{\odot}$-topology of* $(E^{\odot})^*$. Lemma 4.2 remains true; in Assumption (b), $y^* \in E^{\odot}$; in (4.6) $\overline{\text{conv}}$ is interpreted as above.

§8. GENERALIZATIONS AND CONCLUSIONS

In certain control systems modeling fluid flow [13], [14] one is forced to consider systems of the type of (1.1) but where $f(t, y, u)$ and $f_0(t, y, u)$ are unbounded in U, a situation where the use of the spaces $B(U)$ or $BC(U)$ to generate relaxed controls would render the definition (2.10) of $\mathbf{f}(t, y)\mu$ and that of $\mathbf{f}_0(t, y)\mu$ senseless. We need to impose "growth at infinity" conditions on the relaxed control spaces to be sure of convergent integrals. For instance, in [13], [14] relaxed controls satisfy

$$\int_U \kappa(u)^2 \mu(t, du) \in L^1(0, T).$$

where $\kappa(u)$ is positive, unbounded and bounded below.

A general facetious comment on all spaces of relaxed controls is: although their construction requires some esoteric, nonconstructive mathematics (such as the Hahn-Banach theorem) once they have been "installed" in the equation and cost functional, we can forget about them; the relaxed system works in the same way as the original system. In a sense, running a system with relaxed controls is the same as to upgrade software; if the upgrade is well designed, the software will function as before, but it will crash less often ("crashing" means here to deduce theorems such as the maximum principle for nonexistent controls). We don't even have to be interested in the "source code" (the esoteric mathematics).

REFERENCES

1. Ahmed, N. U. (1983). Properties of relaxed trajectories for a class of nonlinear evolution equations in a Banach space, *SIAM J. Control and Optimization*, **21**: 953-967.
2. Ahmed, N. U. (1986). Existence of optimal controls for a class of systems governed by differential inclusions in Banach spaces, *Jour. Optimization Theory Appl.*, **50**: 213-237.
3. Diestel, J., and Uhl, J. J. (1977). **Vector Measures**, Mathematical Surveys **15**, Amer. Math. Soc., Providence.
4. Dieudonné, J. (1947-48). Sur le théorème de Lebesgue - Nikodym (III), *Ann. Université Grenoble*, **23**: 25-53.
5. Dieudonné, J. (1951). Sur le théorème de Lebesgue - Nikodym (IV), *J. Indian Math. Soc.*, **22**: 77-86.

6. Dunford, N., and Pettis, B. J. (1940). Linear operators on summable functions, *Trans. Amer. Math. Soc.*, **47**: 323-392.
7. Dunford, N., and Schwartz, J. T. (1958). **Linear Operators**, part 1, Interscience, New York.
8. Fattorini, H. O. Optimal control problems for distributed parameter systems in Banach spaces, to appear in Applied Math. Optimization.
9. Fattorini, H. O. (1991). Relaxed controls in semilinear infinite dimensional systems, *Int. Series Num. Math.*, **100**: Birkhäuser, Basel (1991) 115-128.
10. Fattorini, H. O. Existence theory and the maximum principle for relaxed infinite dimensional optimal control problems, to appear in SIAM Jour. Control & Optimization.
11. Fattorini, H. O. Relaxation theorems, differential inclusions and Filippov's theorem for relaxed controls in infinite dimensional systems, to appear in Jour. Differential Equations.
12. Fattorini, H. O. Relaxation in semilinear infinite dimensional control systems, to appear in L. Markus Festschrift, Lecture Notes in Pure and Applied Mathematics, Marcel Dekker.
13. Fattorini, H. O., and Sritharan, S. S. Optimal chattering controls for viscous flow, to appear.
14. Fattorini, H. O., and Sritharan, S. S. Relaxation in viscous flow control problems, to appear.
15. Filippov, A. F. (1959). On certain questions in the theory of optimal control, *Vestnik Moskov. Univ. Ser. Mat. Mech. Astronom.*, **2**: 25-32. English translation: *SIAM J. Control*, **1** (1962) 76-84.
16. Frankowska, H. (1990). A priori estimates for operational differential inclusions, *Jour. Diff. Eq.*, **84**: 100-128.
17. Gamkrelidze, R. V. (1962). On sliding optimal states, *Dokl. Acad. Nauk SSSR*, **143**: 1243-1245.
18. Himmelberg, C. J., Jacobs, M. Q., and Van Vleck, F. S. (1969). Measurable multifunctions, selectors and Filippov's Implicit Functions Lemma, *Jour. Math. Anal. Appl.*, **25**: 276-284.
19. Ionescu Tulcea, A., and Ionescu Tulcea, C. (1969). **Topics in the Theory of Lifting**, Springer, Berlin.
20. Warga, J. (1962). Relaxed variational problems, *J. Math. Anal. Appl.*, **4**: 111-128.
21. Warga, J. (1962). Necessary conditions for minimum in relaxed variational problems, *J. Math. Anal. Appl.*, **4**: 129-145.
22. Warga, J. (1971). **Optimal Control of Differential and Functional Equations**, Academic Press, New York.
23. Young, L. C. (1937). Generalized curves and the existence of an attained absolute minimum in the calculus of variations, *Comptes Rendus Soc. des Sciences et des Lettres de Varsovie*, classe III, **30**: 212-234.
24. Young, L. C. (1969). **Lectures on the Calculus of Variations and Optimal Control Theory**, W. B. Saunders, Philadelphia.

On Second Order Implicit Differential Equations in Banach Spaces

A. FAVINI University of Bologna, Bologna, Italy

A. YAGI Himeji Institute of Technology, Himeji, Hyogo, Japan

1. Introduction and Preliminaries

It is the purpose of this paper to establish some existence results on second order differential equations in Banach space, of the type

$$\frac{d}{dt}(Cu') + Bu' + Au = f, \tag{1}$$

where $u' = du / dt$ and nothing is assumed on the invertibility of the closed linear operator C.

There is a very extensive literature on this subject and we refer to [1] for large bibliography and methods.

Since the results that we shall obtain heavily depend upon the ones in [3,4] on first order differential equations, we think it opportune to recall the main definitions and theorems of those papers.

Let A be a multivalued linear operator in the complex Banach space X. We are given a continuous function $f: [0,T]\to X$ and $v_0 \in \mathcal{D}(A)$, the domain of A.

We seek a function v from $[0,T]$ into X such that

$$\text{(P)} \qquad \begin{cases} v'(t)+Av(t) \ni f(t), & 0<t\leq T, \\ v(0) = v_0. \end{cases}$$

under the assumptions

(a) $\rho(A) = \{z\in\mathbb{C}: \exists\, (z+A)^{-1}\in\mathcal{L}(X)\}\supseteq$

$$\supseteq \Sigma = \{z : \text{Re } z \geq -c(1+|\text{Im } z|)^{\eta}\},\ c > 0,$$

and

(b) $$\|(z + A)^{-1}; \mathcal{L}(X)\| \leq M(1+|z|)^{-\beta},\ z \in \Sigma,$$

where

$$0 < \beta \leq \eta \leq 1.$$

We precise what is intended by a solution to (P).

DEFINITION 1. A function $v:[0,T] \to X$ is a CLASSICAL solution of (P) if $v \in C^1((0,T];X)$, $v(t) \in \mathcal{D}(A)$, $0 < t \leq T$, and (P) holds.

We have [3]:

PROPOSITION 1. Assume (a) - (b), and $2\eta + \beta > 2$.

Let $(2 - \eta - \beta)/\eta < \sigma \leq 1$, $v_0 \in \mathcal{D}(A)$.

Then for any $f \in C^{\sigma}([0,T];X)$, problem (P) has a unique classical solution.

Proposition 1 has an immediate application to degenerate Cauchy problems, as we see in a moment.

Let L, M be two closed linear operators from Y into X, where X and Y are complex Banach spaces, $\mathcal{D}(L) \subseteq \mathcal{D}(M)$, $L^{-1} \in \mathcal{L}(X)$, (so that $ML^{-1} \in \mathcal{L}(X)$), $f: [0,T] \to X$ strongly continuous, $u_0 \in \mathcal{D}(L)$.

DEFINITION 2. The function $u:(0,T] \to Y$ is a CLASSICAL solution of (E):

(E) $$\begin{cases} \frac{d}{dt}(Mu(t)) + Lu(t) = f(t), & 0<t\leq T, \\ Mu(0) = Mu_0, \end{cases}$$

if $u(t) \in \mathcal{D}(L)$, $t \in (0,T]$, $Mu \in C^1((0,T];X)$, $Lu \in C((0,T];X)$ and (E) holds, where the initial condition reads

$$\lim_{t \downarrow 0} \|M[u(t) - u_0]; X\| = 0.$$

In view of Proposition 1, we easily have

PROPOSITION 2. Assume

(a)' $$\{z \in \mathbb{C};\ zM + L \text{ has a bounded inverse}\} \supseteq \Sigma_{\eta}$$

(b)' $$\|M(zM + L)^{-1}; \mathcal{L}(X)\| \leq C(1 + |z|)^{-\beta},\ z \in \Sigma_{\eta},\ 0<\beta\leq\eta\leq 1.$$

If $2\eta + \beta > 2$ and $(2 - \eta - \beta)/\eta < \sigma \leq 1$, then for any $f \in C^{\sigma}([0,T];X)$ and all $u_0 \in \mathcal{D}(L)$, problem (E) has a unique classical solution.

More information on the regularity of the solution u to (E) can be achieved if $\eta = 1$, according.

DEFINITION 3. The function $u:[0,T]\to Y$ is a STRICT solution to (E) if $u(t)\in \mathcal{D}(L)$ for $0\le t\le T$, $Lu\in C([0,T];X)$, $Mu\in C^1([0,T];X)$ and

$$\begin{cases}\frac{d}{dt}(Mu(t)) + Lu(t) = f(t), & 0\le t\le T,\\ Mu(0) = Mu_0.\end{cases}$$

In [4] we proved

PROPOSITION 3. Assume (a)', (b)', $\eta = 1$.

If $\sigma\in(1-\beta, 1)$ and $f\in C^{\sigma}([0,T];X)$, $u_0\in\mathcal{D}(L)$ satisfy the compatibility condition

$$f(0) - Lu_0 \in \text{imm}\,(ML^{-1}),$$

then there exists a unique strict solution u to (E) such that

$$Lu(\cdot),\ \frac{d}{dt}Mu(\cdot)\in C^{\sigma+\beta-1}([0,T];X).$$

The remainder of the paper consists of two sections. In section 2 we shall consider equation (1) in the simpler case $\mathcal{D}(B)\subseteq\mathcal{D}(A)$.

Our result shall allow us to handle various initial-boundary-value problems and to obtain solutions to them more regular than those known in literature.

Developping and extending a device by S. Yu. Yakubov [8], (see also [5,pp.277-278]), in section 3 we shall treat the more complicated case $\mathcal{D}(A)\subset\mathcal{D}(B)$.

Furthermore, we shall give a condition under which our main assumptions are guaranteed and show that it is verified in a concrete example of partial differential equations.

2. Equation (1) in the case $\mathcal{D}(B)\subseteq\mathcal{D}(A)$

Let us consider the problem

$$(2)\qquad \begin{cases}\frac{d}{dt}(Cu') + Bu' + Au = f = f(t), & 0<t\le T,\\ u(0) = u_0,\\ Cu'(0) = Cu_1,\end{cases}$$

under the following hypotheses on the operators A,B,C :

$$(3)\qquad \begin{cases}\text{A,B,C are closed linear operators from the Banach space X into itself},\\ \mathcal{D}(B)\subseteq\mathcal{D}(C),\ \mathcal{D}(B)\subseteq\mathcal{D}(A).\end{cases}$$

If one puts $u' = v$ and

$$M = \begin{bmatrix} I & 0\\ 0 & C\end{bmatrix},\qquad L = \begin{bmatrix} 0 & -I\\ A & B\end{bmatrix},$$

the operators M and L are closed linear operators from the Banach space $\mathcal{D}(B)\times X$ into itself, where $\mathcal{D}(B)$ is endowed with the graph norm.

In view of our further assumptions, it is not restrictive to suppose that B has a bounded inverse.

Problem (2) is trasformed into

$$(4)\qquad \begin{cases} \frac{d}{dt}(Mz(t)) + Lz(t) = F(t), & 0<t\leq T, \\ Mz(0) = Mz_0, \end{cases}$$

where $z(t) = (u(t),v(t))$, $z_0 = (u_0,u_1)$, $F(t) = (0,f(t))$, $0 < t \leq T$.

Our aim is to apply Proposition 2 and Proposition 3 and to this end we introduce the operator pencil

$$P(z) = z^2C + zB + A, \quad z \text{ a complex number.}$$

We first observe that, at least formally,

$$P(z) = (z + AB^{-1})(zC + B) - z\,AB^{-1}C = (z + AB^{-1})\,Q(z)(zC + B),$$

where

$$Q(z) = I - z\,(z + AB^{-1})^{-1}AB^{-1}C(zC + B)^{-1}.$$

This remark accounts for the type of assumptions that we shall make on the operators B and C, that is,

$$(5)\qquad \begin{cases} zC + B \text{ has a bounded inverse for all } z\in\Sigma_\eta, \\ \|C(zC + B)^{-1}; \mathscr{L}(X)\| \leq k(1 + |z|)^{-\beta},\ z\in\Sigma_\eta, \\ \text{with } 0<\beta\leq\eta\leq 1. \end{cases}$$

In virtue of (3) and (5), we deduce that then for all $z \in \Sigma_\eta$, $zM + L$ has a bounded inverse and $M(zM + L)^{-1} = [A_{ij}(z)]$, $i,j = 1,2$, where

$$A_{11}(z) = z^{-1}(I - P(z)^{-1}A),\ A_{12}(z) = P(z)^{-1},\ A_{21}(z) = CP(z)^{-1}A,\ A_{22}(z) = zCP(z)^{-1}.$$

Furthermore, there is $k' > 0$ such that

$$\|M(zM + L)^{-1}; \mathscr{L}(D(B)\times X)\| \leq k'(1 + |z|)^{-\beta},\ z\in\Sigma_\eta.$$

In a way analogous to the one for first order equations, we introduce the definitions of CLASSICAL or STRICT solutions to problem (2), that we omit for brevity.

Then it is not too hard to prove

THEOREM 1. Assume (3) and (5).

If $u_0, u_1 \in \mathcal{D}(B)$ and $f \in C^\sigma([0,T];X)$, where

$$2\eta + \beta > 2,\ (2 - \eta - \beta)/\beta < \sigma \leq 1,$$

then problem (2) has a unique classical solution u with

$$u' \in C((0,T];D(B)),\ Cu' \in C^1((0,T];X),\ Au \in C((0,T];X).$$

THEOREM 2. Suppose that (3) and (5) hold with $\eta = 1$, and let $1 - \beta < \sigma < 1$.

If $u_0, u_1 \in \mathcal{D}(B)$, $f \in C^\sigma([0,T];X)$ and $f(0) - Au_0 - Bu_1 \in C(D(B))$, then problem (2) has one and only one strict solution u such that

$$Bu' , \frac{d}{dt}(Cu') \in C^{\sigma+\beta-1}([0,T];X).$$

We quickly give two examples of application of the preceding Theorems.

EXAMPLE 1.
Let us consider the problem

$$(6)\quad \begin{cases} \dfrac{\partial}{\partial t}(m(x)\dfrac{\partial u}{\partial t}) - \Delta\dfrac{\partial u}{\partial t} + \alpha(x)\Delta u = f, & \text{in } (0,T]\times\Omega, \\ u = \dfrac{\partial u}{\partial t} = 0, \ \text{in } (0,T]\times\partial\Omega, \\ u(0,x) = u_0(x), \ x\in\Omega, \\ m(x)\dfrac{\partial u}{\partial t}(t,x) \to m(x)u_1(x) \ \text{per } t\to 0+, \ x\in\Omega, \end{cases}$$

where Ω is a bounded domain in R^n with a smooth boundary $\partial\Omega$, $n\in\mathbb{N}$, α is a scalar-valued continuous function on $\bar{\Omega}$ and $m\in L^\infty(\Omega)$ is non negative on Ω.

Problems of the type (6) have been studied very much and we refer to [1] for the literature.

Usually, one works in the space $H^{-1}(\Omega)$, since in this space one has the best estimates for the modified resolvent, that is, $\eta = \beta = 1$.

But if we want to treat problem (6) in the space $X = L^2(\Omega)$, and hence

$$\mathcal{D}(B)=H^2(\Omega)\cap H^1_0(\Omega), \quad B=-\Delta,$$

$$C = \text{multiplication by } m(x)$$

$$\mathcal{D}(A) = \mathcal{D}(B), \quad Au = \alpha\cdot u, \quad u\in\mathcal{D}(A),$$

then only the worse estimate

$$\|C(zC+B)^{-1}; \mathcal{L}(X)\| \le k\,|z|^{-1/2}, \ z\ne 0, \ z\in\Sigma_1.$$

is available ([3]).

One can verify that if $m,\alpha\in C^1(\bar{\Omega})$ and $|\nabla m(x)|\le Cm(x)^\rho$, with $0<\rho\le 1$, then the preceding estimate can be improved to

$$\|C(zC+B)^{-1};\mathcal{L}(X)\| \le k|z|^{-\gamma}, \quad z\ne 0, \quad z\in\Sigma_1,$$

where $\gamma = (2-\rho)^{-1}$.

EXAMPLE 2. For sake of simplicity, we consider the interval $\Omega = (0,1)\subset\mathbb{R}$. Let $C_0>0$ and consider

$$
(7)\begin{cases}
\dfrac{\partial}{\partial t}(m(x)\dfrac{\partial u}{\partial t}) - \dfrac{\partial^3 u}{\partial x^2 \partial t} + C_0 \dfrac{\partial u}{\partial t} + \dfrac{\partial^2 u}{\partial x^2} = f = f(t,x), \quad 0 < t \le T,\ 0 < x < 1, \\
u(t,0) = u(t,1) = \dfrac{\partial u}{\partial t}(t,0) = \dfrac{\partial u}{\partial t}(t,1) = 0, \qquad 0 < t \le T, \\
\lim_{t \to 0} u(t,x) = u_0(x), \quad 0<x<1, \\
\lim_{t \to 0} m(x)\dfrac{\partial u}{\partial t}(t,x) = m(x)\, u_1(x), \quad 0<x<1.
\end{cases}
$$

This time we take $X = L^p(0,1)$, $1 < p < \infty$,

$\mathcal{D}(B) = W^{2,p}(0,1) \cap W_0^{1,p}(0,1)$, $(Bu)(x) = -u''(x) + C_0 u(x)$, $u \in \mathcal{D}(B)$, $m \in L^\infty(0,1)$, $m(x) \ge 0$.

In virtue of some results in [4], one has

$$
\|C(zC + B)^{-1}; \mathcal{L}(X)\| \le k|z|^{-1/p}, \quad z \neq 0, \quad z \in \Sigma_1.
$$

Hence, also problem (7) can be considered with the aid of Theorems 1 and 2.

3. The case $\mathcal{D}(A) \subseteq \mathcal{D}(B)$

In order to treat problem (2) under the assumption $\mathcal{D}(A) \subseteq \mathcal{D}(B)$ we extend the method of [8], (see [5,pp.276-279]), where (2) is studied with C equal to the identity operator.

In short, we multiply the equation in (4) by an operator matrix to reduce the term in front of z(t) to a triangular form, more suitable to be handled.

As it is well-known, such an approach has been already used with success in various fields of operator theory and to this purpose we refer to the recent monograph [6].

To begin with, we observe that problem (2) can be equivalently formulated in the space $X \times X$ by means of

$$
(8)\begin{cases}
\dfrac{d}{dt}\left(\begin{bmatrix} B & C \\ 0 & C \end{bmatrix}\begin{bmatrix} u(t) \\ v(t) \end{bmatrix}\right) + \begin{bmatrix} A & 0 \\ A & B \end{bmatrix}\begin{bmatrix} u(t) \\ v(t) \end{bmatrix} = \begin{bmatrix} f(t) \\ f(t) \end{bmatrix}, \quad 0 < t \le T, \\
\begin{bmatrix} B & C \\ 0 & C \end{bmatrix}\begin{bmatrix} u(0) \\ v(0) \end{bmatrix} = \begin{bmatrix} B & C \\ 0 & C \end{bmatrix}\begin{bmatrix} u_0 \\ u_1 \end{bmatrix}.
\end{cases}
$$

As our aim is to apply Proposition 1 to (8), we need consider the operator C^{-1}, that is multivalued in general.

Let

$$
AB^{-1} = U, \quad BC^{-1} = V,
$$

so that

$$
\begin{bmatrix} A & 0 \\ A & B \end{bmatrix}\begin{bmatrix} B^{-1} & -B^{-1} \\ 0 & C^{-1} \end{bmatrix} = \begin{bmatrix} U & -U \\ U & -U+V \end{bmatrix} = \mathcal{Q},
$$

To establish the invertibility of $\mathcal{Q} + z$, where z is a complex number, first of all we introduce the notation Σ'_{η} for the set

$$
\Sigma'_{\eta'} = \{z\ ;\ \mathrm{Re}\, z \ge -c'(1 + |\mathrm{Im}\, z|)^{\eta'}\}, \quad c' > 0, \quad 0 < \eta' \le 1.
$$

and then we formulate our hypotheses as follows.

(a)" $\|C(zC+B)^{-1}; \mathcal{L}(X)\| \le k(1+|z|)^{-\beta}$, $z\in\Sigma_\eta$, $0<\beta\le\eta\le 1$,

(b)" $\|B(zB+A)^{-1}; \mathcal{L}(X)\| \le k(1+|z|)^{-\alpha}$, $z\in\Sigma'_{\eta'}$, $0<\alpha\le\eta'\le 1$,

(c)" $\mathcal{D}(V)\ (=\mathcal{D}(BC^{-1}))\subseteq\mathcal{D}(U) = \mathcal{D}(AB^{-1})$ and there exists a positive constant δ such that

$$\|U(z+V)^{-1}; \mathcal{L}(X)\| \le k\,|z|^{-\delta}, \text{ for all } z\in\Sigma_\eta, |z| \text{ large.}$$

For our aim it is not restrictive to suppose $\eta\le\eta'$ and $c = c'$.
One then obtains

THEOREM 3. Assume (a)", (b)", (c)", $\eta\le\eta'$ and $\alpha+\delta$, $\alpha+\beta>1$, $2\eta+\alpha+\beta>3$.
If

$$(3-\eta-\alpha-\beta)/\eta < \sigma \le 1,$$

then for any $u_0\in\mathcal{D}(A)$, $u_1\in\mathcal{D}(B)$ and $f\in C^\sigma([0,T];X)$, problem (2) has a unique classical solution.

THEOREM 4. Suppose (a)", (b)", (c)", $\eta=\eta'=1$, and $0<\alpha,\beta\le 1$, $\alpha+\beta,\alpha+\delta>1$, $2-\alpha-\beta<\sigma<1$.

Then for any $u_0, u_1\in\mathcal{D}(A)$ and $f\in C^\sigma([0,T];X)$ satisfying

$$f(0) - Au_0 - Bu_1\in C(\mathcal{D}(B)),$$

there is a unique strict solution u to (2) such that

$$Au,\ Bu',\ (Cu')'\in C^{\sigma+\alpha+\beta-2}([0,T];X).$$

Clearly, the most restrictive condition consists in hypothesis (c)", that involves a strong connection between the operators.
Next result provides a criterion (rather simple and useful for PDEs problems) to have (c)".

THEOREM 5. Assume (a)", (b)". Furthermore,

i) $\mathcal{D}(B)$ is invariant under the operator C,

ii) $\|BCu;X\| \le k\,\|Bu;X\|$ for all $u\in\mathcal{D}(B)$, where $k>0$,

iii) B is a positive operator with bounded imaginary powers B^{is}, $s\in\mathbb{R}$, $\|B^{is};\mathcal{L}(X)\| \le$ Constant, for all real s, $|s|\le\rho$, some positive ρ,

iv) there exists $\tau\in(0,1)$ such that

$$\mathcal{D}(B^{1+\tau})\subseteq\mathcal{D}(A)$$

and

$$\|Au;X\| \le k'\,\|B^{1+\tau}u;X\|,\quad u\in\mathcal{D}(B^{1+\tau}),\ k'>0.$$

Then, if $0<\tau<\beta$, condition (c)" holds.

PROOF. Condition ii) assures that

$$\|B(z + BC^{-1})^{-1}f;X\| = \|BC(zC + B)^{-1}f;X\| \le k\,\|B(zC + B)^{-1}f;X\| \le$$

$$\le k_1(1 + |z|^{1-\beta})\|f;X\|, \quad z\in\Sigma_{\eta'}, \quad f\in X.$$

On the other hand, ([7],p.103]), in virtue of iii), for any $\theta\in(0,1)$, $\mathcal{D}(B^{\theta})$ coincides with the complex interpolation space $[X,\mathcal{D}(B)]_{\theta}$, and thus

$$\|C(zC + B)^{-1};\mathcal{L}(X;\mathcal{D}(B^{\tau}))\| = \|(z + BC^{-1})^{-1};\mathcal{L}(X;\mathcal{D}(B^{\tau}))\| \le$$

$$\le k''|z|^{-(\beta-\tau)}, \quad z\in\Sigma, \quad |z| \text{ large. \#}$$

We finish this work by giving an application to a partial differential equation.

EXAMPLE 3. Here we only consider the open set $\Omega = (a,b) = I\subset\mathbb{R}$, even if some extensions to $\mathbb{R}^n$, with $n > 1$, are allowed and in L^p spaces, $1 < p < \infty$ too, in view of de Laubenfels' results [2].

Let $X = L^2(I)$ and denote by C the multiplication operator by $m(x)$ in the space X, where $m(\cdot)$ is a non negative function on $\bar{I}$ which is also sufficiently regular.

Let $K{:}X\to X$ be the operator defined by

$$\mathcal{D}(K) = H^2(I)\cap H^1_0(I), \quad Ku = -u'', \quad u\in D(K).$$

Then

$$A = K^{m+q}, \quad B = K^m,$$

where m, q are two positive integers and $q < m$.

It is not too hard to recognize that, in view of the boundary values assigned to $u\in\mathcal{D}(B)$, condition (a)" holds with $\beta = 1/2$ and $\eta = 1$. Moreover, (b)" is verified with $\alpha = \eta' = 1$.

To apply Theorem 5 we need further regularity to the function $m(\cdot)$; precisely,

$$m(\cdot)\in C^{(2m)}(\bar{I}) \text{ e } m^{(2j+1)}(a) = m^{(2j+1)}(b) = 0, \; j = 0,1,\dots, m-1.$$

Then the assumptions i), ii) in Theorem 5 are satisfied.

Since iii) holds and iv) is true with $\tau = q/m$, we conclude that if $2q < m$, our last result applies.

Hence we are allowed to treat the initial-boundary-value problem connected with the partial differential equation

$$\frac{\partial}{\partial t}\left(m(x)\frac{\partial u}{\partial t}\right) + (-1)^m\frac{\partial^{2m+1}u}{\partial x^{2m}\partial t} + (-1)^{m+q}\frac{\partial^{2(m+q)}u}{\partial x^{2(m+q)}} = f(t,x), \quad 0<t\le T, \quad a<x<b.$$

REFERENCES

[1] R.W.Carroll-R.E.Showalter, "Singular and Degenerate Cauchy Problems", *Academic Press*, 1976.

[2] R.de Laubenfels, Powers of generators of holomorphic semigroups, *Proceed. AMS* 99(1987), 105-108.

[3] A.Favini-A.Yagi, Multivalued linear operators and degenerate evolution problems, to appear on *Annali Mat.Pura Appl.*

[4] A.Favini-A.Yagi, Space and time regularity for degenerate evolution equations, to appear on *J.Math.Soc.Japan* .

[5] S.G.Krein, "Linear Differential Equations in Banach Space", *AMS*, 1971.

[6] L.Rodman, "An Introduction to Operator Polynomials", *Birkhäuser,* 1989.

[7] H.Triebel, "Interpolation Theory, Function Spaces, Differential Operators", *North-Holland*, 1978.

[8] S.Ya.Yakubov, A nonlocal boundary value problem for a class of Petrovskii well posed equations, *Math.Sb.(N.S.)*, 118(60),(1982), 252-261; *Math.USSR-SB*, 46(1983), 255-265.

Bounded Imaginary Powers of Abstract Differential Operators

MARCO FUHRMAN Polytechnico di Milano, Milan, Italy

Abstract. We prove that differential operators of the form

$$(Pu)(t) := a_0 u^{(n)}(t) + a_1 u^{(n-1)}(t) +\ldots+ a_n u(t) + \mu\, u(t)$$

with $a_0 = \mp 1$ properly chosen and $a_i \in \mathbb{C}$, have bounded imaginary powers for large $\mu > 0$ in the space $L^p(\mathbb{R},E)$, where $p \in (1,\infty)$ and E is a UMD Banach space.

1 INTRODUCTION

The aim of this paper is to prove that some differential operators have bounded imaginary powers in the space $L^p(\mathbb{R},E)$, where E is a UMD Banach space and $p \in (1,\infty)$ (precise definitions are given below). The importance of the operators with bounded imaginary powers has increased after (Dore-Venni, 1987). Since then, Dore and Venni's results have found several applications (see e.g. (Clément-Prüss, 1990), (Prüss-Sohr, 1990) and the examples and references given there). We only recall that the boundedness of imaginary powers, in L^p spaces of complex functions of several real variables, has been proved for elliptic differential operators with smooth coefficients (Seeley, 1971), for negative generators of special semigroups (Prüss-Clément, 1990), and for second order elliptic operators with Hölder-continuous coefficients (Prüss-Sohr, 1991). We deal with differential operators in L^p spaces of vector-valued functions of a real variable. Some cases have already been considered in (Dore-Venni, 1987) (the operator d/dt), and (Prüss-Clément, 1990) (with applications to integro-differential equations). Here we consider higher order differential operators. Our techniques are fairly simple: we use an extension of Michlin's multiplier theorem, and a perturbation result (Dore-Venni, 1990) in order to deal with lower order terms. Deeper perturbation theorems would allow to extend our results to a more general class of operators: the reader may consult (Prüss-Sohr, 1991). I wish to thank Professor Jan Prüss for pointing out to me the paper (Prüss-Sohr, 1991).

2 NOTATIONS AND TERMINOLOGY

Throughout this paper X (or E) is a complex Banach space with norm $\|\cdot\|$. We denote by $L^p(\mathbb{R},X)$, $p \in [1,\infty)$, the usual space of strongly measurable X-valued functions $f(t)$ such that $t \mapsto \|f(t)\|^p$ is integrable over $\mathbb{R}$. $\mathcal{D}(\mathbb{R},X)$ is the space of test functions. The vector-valued Sobolev spaces $W^{m,p}(\mathbb{R},E)$ are defined in analogy with the scalar case.

X is said to have the UMD property if the Hilbert transform

$$(Hf)(t) := \int_{\mathbb{R}} s^{-1}\, f(t-s)\, ds$$

defined for $f \in \mathcal{D}(\mathbb{R},X)$, has an extension to a bounded operator in $L^2(\mathbb{R},X)$. We refer to (Zimmermann, 1989) and the references given there for equivalent characterizations and further properties of UMD Banach spaces. We only recall that if E is a UMD Banach space, so is $L^p(\mathbb{R},E)$, $\forall p \in (1,\infty)$.

For a linear operator A in X let D(A), R(A), N(A), $\rho(A)$ denote its domain, range, kernel, resolvent set respectively. $\mathcal{L}(X)$ denotes the set of bounded linear operators in X.

Let $\vartheta \in (0,\pi)$. We say that a linear operator A in X belongs to the class BIP (ϑ) (Bounded Imaginary Powers) if there exist $M > 0$, $K > 0$ such that

1) $N(A) = 0$, $D(A)$ and $R(A)$ are dense in X, $(-\infty,0) \subset \rho(A)$,

$$\|(A+\lambda)^{-1}\|_{\mathcal{L}(X)} \le M\,\lambda^{-1}, \quad \forall\lambda > 0. \tag{1}$$

2)
$$\|A^{i\tau}\|_{\mathcal{L}(X)} \le K\, e^{|\tau|\vartheta}, \quad \forall\tau \in \mathbb{R}\setminus\{0\}. \tag{2}$$

For an operator satisfying (1) complex powers (in general unbounded) are defined for any $z \in \mathbb{C}$ (see (Komatsu, 1966) for details). Here we only recall the definition of (2) and some useful formulas. For $x \in D(A) \cap R(A)$ and $\mathrm{Re}(z) \in (-1,1)$, $z \ne 0$, we define

$$A^z x := \pi^{-1}\sin(\pi z)\left\{\frac{x}{z} - \frac{A^{-1}x}{1+z} + \int_0^1 \lambda^{z+1}(A+\lambda)^{-1}A^{-1}x\,d\lambda + \int_1^\infty \lambda^{z-1}(A+\lambda)^{-1}Ax\,d\lambda\right\} \tag{3}$$

$A^z x$ is an X-valued analytic function of z in $\mathrm{Re}(z) \in (-1,1)$. The operator $x \mapsto A^z x$ is closable, and its closure is denoted by A^z. For $x \in D(A)$ and $\mathrm{Re}(z) \in (0,1)$ we also have

$$A^z x := \pi^{-1}\sin(\pi z)\int_0^\infty \lambda^{z-1}(A+\lambda)^{-1}Ax\,d\lambda. \tag{4}$$

In the sequel we will use Fourier transform defined for $u(t) \in L^1(\mathbb{R},E)$ by $\hat{u}(\xi) := \int_{\mathbb{R}} e^{-i\xi t}\,u(t)\,dt$. The inverse Fourier transform is denoted by $\check{u}(t) = (1/(2\pi))\int_{\mathbb{R}} e^{i\xi t}\,u(\xi)\,d\xi$. We call $T : \mathcal{D}(\mathbb{R},E) \to L^p(\mathbb{R},E)$ a multiplier transformation if there exists a complex function $m(\xi)$ such that

$$(Tu)(t) = (m(\xi)\hat{u}(\xi))^{\vee}(t), \quad \forall u \in \mathcal{D}(\mathbb{R},E).$$

We now state two results that we will use later. Their proofs can be found in (Zimmermann, 1989) and (Dore-Venni, 1990) respectively.

THEOREM 1. *Let* E *be a* UMD *Banach space,* $p \in (1,\infty)$, $m(\xi) \in C^1(\mathbb{R}\setminus\{0\})$ *and*

$$\gamma_1(m) := \sup_{\xi\in\mathbb{R}} \max\left\{|m(\xi)|,\ |\xi\, m'(\xi)|\right\} < \infty. \tag{5}$$

Then the multiplier transformation $(Tu)(t) = (m(\xi)\hat{u}(\xi))^{\vee}(t)$ *satisfies*

$$\|Tu\|_{L^p(\mathbb{R},E)} \le c\,\gamma_1(m)\,\|u\|_{L^p(\mathbb{R},E)}, \quad \forall u \in \mathcal{D}(\mathbb{R},E), \tag{6}$$

where c *depends only on* p *and* E.

THEOREM 2. *Let* X *be a Banach space,* A *and* B *closed linear operators in* X, *with* $0 \in \rho(B)$. *Assume*

1) there exists $\vartheta \in (0,\pi)$ *such that* $B \in BIP(\vartheta)$; (7)

2) there exists $\eta \in (0,1)$ *such that* $D(B^{\eta}) \subset D(A)$. (8)

Then for any $\varepsilon \in (0,\pi-\vartheta)$ *there exists* $\mu_0 > 0$ *such that*

$$A + B + \mu \in BIP(\vartheta+\varepsilon), \quad \forall\mu > \mu_0. \tag{9}$$

3 MAIN RESULTS

LEMMA 3. *Let* E *be a* UMD *Banach space,* $p \in (1,\infty)$.

In the space $X = L^p(\mathbb{R},E)$ *consider the differential operators*

$(B_1u)(t) := (-1)^m u^{(2m)}(t)$, $D(B_1) = W^{2m,p}(\mathbb{R},E)$, $m = 1,2,3,\ldots$

$(B_2u)(t) := \sigma\, u^{(2m+1)}(t)$, $D(B_2) = W^{2m+1,p}(\mathbb{R},E)$, $m = 0,1,2,\ldots$, $\sigma = +1$ *or* -1.

Then

$$B_1 \in BIP(\varepsilon), \qquad \forall\varepsilon \in (0,\pi); \tag{10}$$

$$B_2 \in BIP(\pi/2 + \varepsilon), \qquad \forall\varepsilon \in (0,\pi/2). \tag{11}$$

More precisely, there exists $c = c(p,m,E) > 0$ *such that*

$$\|B_1^{i\tau}\|_{\mathcal{L}(X)} \le c\,(1+|\tau|)\,, \qquad \forall\tau \in \mathbb{R}\setminus\{0\}, \tag{12}$$

$$\|B_2^{i\tau}\|_{\mathcal{L}(X)} \le c\,(1+|\tau|)\,e^{|\tau|\pi/2}, \qquad \forall\tau \in \mathbb{R}\setminus\{0\}. \tag{13}$$

Proof. We only consider B_2 with $\sigma = (-1)^m$. All the other cases are similar. First remark that $N(B_2) = 0$, $D(B_2)$ and $R(B_2)$ are dense in $L^p(\mathbb{R},E)$, for $p \in [1,\infty)$. Fix $f \in \mathcal{D}(\mathbb{R},E)$, $\lambda > 0$, $p \in [1,\infty)$. We want to solve the equation $(B_2+\lambda)u = f$, i.e.

$$(-1)^m\, u^{(2m+1)}(t) + \lambda\, u(t) = f(t), \qquad u \in W^{2m+1,p}(\mathbb{R},E). \tag{14}$$

Applying Fourier transformation we obtain formally

$$\hat{u}(\xi) = (i\xi^{2m+1} + \lambda)^{-1}\,\hat{f}(\xi). \tag{15}$$

Call $K_\lambda(t)$ the functions whose Fourier transform is $(i\xi^{2m+1} + \lambda)^{-1}$. $K_\lambda(t)$ can be explicitly computed as follows. Let

$$\alpha_k = \lambda^{\frac{1}{2m+1}} \exp\left(i\,\frac{\pi}{2}\,\frac{1}{2m+1} + \frac{2k\pi i}{2m+1}\right), \qquad k = 0,\ldots,2m$$

be the roots of order 2m+1 of $i\lambda$. Then

$$(i\xi^{2m+1} + \lambda)^{-1} = -i \prod_{k=0}^{2m} (\xi - \alpha_k)^{-1}$$

Clearly there exist positive constants c(k,m) such that

$$|\mathrm{Im}(\alpha_k)| = c(k,m)\, \lambda^{\frac{1}{2m+1}} \neq 0$$

for every k. Moreover

$$((\xi - \alpha_k)^{-1})^{\vee}(t) = i \exp(i\alpha_k t)\chi_{(0,\infty)}(t), \qquad \text{if } \mathrm{Re}(i\alpha_k) = -\mathrm{Im}(\alpha_k) < 0,$$

$$((\xi - \alpha_k)^{-1})^{\vee}(t) = -i \exp(i\alpha_k t)\chi_{(-\infty,0)}(t), \qquad \text{if } \mathrm{Re}(i\alpha_k) = -\mathrm{Im}(\alpha_k) > 0,$$

(χ denotes the characteristic function) so that

$$\|((\xi - \alpha_k)^{-1})^{\vee}\|_{L^1(\mathbb{R})} = |\mathrm{Im}(\alpha_k)|^{-1},$$

for k = 0,...2m. It follows that

$$K_\lambda = -i\, ((\xi - \alpha_0)^{-1})^{\vee} * ((\xi - \alpha_1)^{-1})^{\vee} * \ldots * ((\xi - \alpha_{2m})^{-1})^{\vee} \in L^1(\mathbb{R})$$

$$\|K_\lambda\|_{L^1(\mathbb{R})} \le \prod_{k=0}^{2m} |\mathrm{Im}(\alpha_k)|^{-1} = \prod_{k=0}^{2m} c(k,m)^{-1}\, \lambda^{-\frac{1}{2m+1}} = c(m)\, \lambda^{-1}.$$

Then

$$u(t) := (K_\lambda * f)(t) \tag{16}$$

is easily verified to be the unique solution of (14). By Young's inequality for convolutions

$$\|u\|_{L^p(\mathbb{R},E)} \le \|K_\lambda\|_{L^1(\mathbb{R})} \|f\|_{L^p(\mathbb{R},E)} \le c(m)\, \lambda^{-1} \|f\|_{L^p(\mathbb{R},E)} \tag{17}$$

as can be checked by explicit calculation. Since $\mathcal{D}(\mathbb{R},E)$ is dense in $L^p(\mathbb{R},E)$, an approximation argument shows that (14) has a unique solution u for any $f \in L^p(\mathbb{R},E)$ and (17) gives

$$\|(B_2 + \lambda)^{-1} f\|_{L^p(\mathbb{R},E)} \le c(m)\, \lambda^{-1} \|f\|_{L^p(\mathbb{R},E)}, \qquad \forall f \in L^p(\mathbb{R},E),\ \forall \lambda > 0. \tag{18}$$

Therefore (1) holds for B_2. Remark that (18) also holds for p = 1.

Now choose $f \in \mathcal{D}(\mathbb{R},E)$, $\sigma \in (0,1)$, $\tau \in \mathbb{R}\setminus\{0\}$, $p \in (1,\infty)$. By (4)

$$B^{\sigma+i\tau} f = \pi^{-1}\sin(\pi(\sigma+i\tau)) \int_0^\infty \lambda^{\sigma+i\tau-1}\, (B_2(B_2+\lambda)^{-1} f)\, d\lambda \tag{19}$$

Remark that $h(\lambda,t) := \lambda^{\sigma+i\tau-1}\, (B_2(B_2+\lambda)^{-1} f)\,(t)$ belongs to $L^1((0,\infty) \times \mathbb{R},E)$, as can be easily seen using (18). So taking Fourier transforms in (19), in view of (15)

$$(B_2^{\sigma+i\tau} f)^{\wedge}(\xi) = \pi^{-1}\sin(\pi(\sigma+i\tau)) \int_0^\infty \lambda^{\sigma+i\tau-1} \frac{i\xi^{2m+1}}{i\xi^{2m+1} + \lambda}\, d\lambda\, \hat{f}(\xi) \tag{20}$$

By the change of variable $\lambda = |\xi|^{2m+1} v^{1/2}$ and recalling that

$\int_0^\infty \frac{v^{z-1}}{v+1}\,dv = \frac{\pi}{\sin(\pi z)}$ for Re(z) $\in$ (0,1), one easily obtains

$$\int_0^\infty \lambda^{\sigma+i\tau-1} \frac{1}{i\xi^{2m+1}+\lambda}\,d\lambda = |\xi|^{(2m+1)(\sigma+i\tau-1)}(\pi/2)\cdot \tag{21}$$

$$\cdot\left[\frac{1}{\cos((\sigma+i\tau)\pi/2)} - i\,\frac{\operatorname{sgn}(\xi)}{\sin((\sigma+i\tau)\pi/2)}\right]$$

So (20) becomes

$$(B_2^{\sigma+i\tau}f)^{\wedge}(\xi)=|\xi|^{(2m+1)(\sigma+i\tau)}\left[\cos((\sigma+i\tau)\pi/2) + i\operatorname{sgn}(\xi)\sin((\sigma+i\tau)\pi/2)\right]\hat{f}(\xi)$$

so that $(B_2^{\sigma+i\tau}f)^{\wedge} \in L^1(\mathbb{R},E)$. It follows that

$$(B_2^{\sigma+i\tau}f)(t) = \left\{|\xi|^{(2m+1)(\sigma+i\tau)}\left[\cos((\sigma+i\tau)\pi/2) + \right.\right. \tag{22}$$

$$\left.\left. + i\operatorname{sgn}(\xi)\sin((\sigma+i\tau)\pi/2)\right]\hat{f}(\xi)\right\}^{\vee}(t),\quad \text{a.e. } t\in\mathbb{R}$$

Now we let $\sigma \to 0$. Then for any $\xi \in \mathbb{R}$

$$|\xi|^{(2m+1)(\sigma+i\tau)}\left[\cos((\sigma+i\tau)\pi/2) + i\operatorname{sgn}(\xi)\sin((\sigma+i\tau)\pi/2)\right]\hat{f}(\xi) \to$$

$$\to |\xi|^{(2m+1)i\tau}\left[\cos(i\tau\pi/2) + i\operatorname{sgn}(\xi)\sin(i\tau\pi/2)\right]\hat{f}(\xi) := m(\xi)\,\hat{f}(\xi) \text{ and}$$

$$\left\|\, |\xi|^{(2m+1)(\sigma+i\tau)}\left[\cos((\sigma+i\tau)\pi/2) + i\operatorname{sgn}(\xi)\sin((\sigma+i\tau)\pi/2)\right]\hat{f}(\xi)\,\right\| \le$$

$$\le (1+|\xi|)^{(2m+1)}\left[\,|\cos(i\tau\pi/2)| + |\sin(i\tau\pi/2)|\,\right]\|\hat{f}(\xi)\| \in L^1(\mathbb{R},E).$$

By the dominated convergence theorem $(B_2^{\sigma+i\tau}f)^{\wedge}(\xi) \to m(\xi)\hat{f}(\xi)$ in $L^1(\mathbb{R},E)$, so that by (22)

$$\lim_{\sigma\to 0}(B_2^{\sigma+i\tau}f)(t) = (m(\xi)\hat{f}(\xi))^{\vee}(t),\qquad \text{a.e. } t\in\mathbb{R}.$$

If, in addition, $f \in R(B_2)$, we have $B_2^{\sigma+i\tau}f \to B^{i\tau}f$ in $L^p(\mathbb{R},E)$, so finally

$$(B_2^{i\tau}f)(t) = (m(\xi)\hat{f}(\xi))^{\vee}(t),\qquad \text{a.e. } t\in\mathbb{R},\ \forall f\in\mathcal{D}(\mathbb{R},E)\cap R(B_2).$$

Therefore $B_2^{i\tau}$ is a multiplier transformation corresponding to the multiplier

$$m(\xi) = |\xi|^{(2m+1)i\tau}\left[\cos(i\tau\pi/2) + i\operatorname{sgn}(\xi)\sin(i\tau\pi/2)\right].$$

It is easy to verify that $\gamma_1(m) \le c(m)\,(1+|\tau|)e^{|\tau|\pi/2}$ (see (5)), so by (6)

$$\|B_2^{i\tau}f\|_{L^p(\mathbb{R},E)} \le c(m,p,E)\,(1+|\tau|)e^{|\tau|\pi/2}\,\|f\|_{L^p(\mathbb{R},E)},\quad \forall f\in\mathcal{D}(\mathbb{R},E)\cap R(B_2),\ \forall\tau\in\mathbb{R}\setminus\{0\}$$

which yields

$$\|B_2^{i\tau}\|_{\mathcal{L}(L^p(\mathbb{R},E))} \le c(m,p,E)\,(1+|\tau|)e^{|\tau|\pi/2},\qquad \forall\tau\in\mathbb{R}\setminus\{0\}.$$

THEOREM 4. *Let* E *be a* UMD *Banach space,* n *a positive integer,* $a_0 = +1$ *or* -1, $a_1, \dots, a_n \in \mathbb{C}$, $p \in (1,\infty)$. *In* $L^p(\mathbb{R},E)$ *consider the differential operator*

$$(Pu)(t) := a_0 u^{(n)}(t) + a_1 u^{(n-1)}(t) + \dots + a_n u(t) + \mu\, u(t)$$

with domain $W^{n,p}(\mathbb{R},E)$. *Then for any* ε *satisfying respectively*

$\varepsilon \in (0,\pi/2)$, *if* n *is odd;*

$\varepsilon \in (0,\pi)$, *if* n *is even and* $a_0 = (-1)^{n/2}$;

there exists $\mu_0 \in \mathbb{R}$ *such that respectively*

$P \in BIP(\pi/2 + \varepsilon)$, $\quad \forall \mu > \mu_0$, *if* n *is odd;*

$P \in BIP(\varepsilon)$, $\quad \forall \mu > \mu_0$, *if* n *is even and* $a_0 = (-1)^{n/2}$.

Proof. In the space $X = L^p(\mathbb{R},E)$ define

$(P_n u)(t) := a_0 u^{(n)}(t)$, $\quad$ with domain $D(P_n) := W^{n,p}(\mathbb{R},E)$;

$(Au)(t) := a_1 u^{(n-1)}(t) + \dots + a_n u(t)$ $\quad$ with domain $D(A) := W^{n-k,p}(\mathbb{R},E)$,

where $k \in \{1,\dots,n\}$ is the least integer such that $a_k \neq 0$. We now apply theorem 2. We define in X

$$B := P_n + I, \qquad \text{with domain } D(B) := D(P_n),$$

so that $0 \in \rho(B)$. By lemma 3, P_n has bounded imaginary powers. Since E is a UMD Banach space and $p \in (1,\infty)$, X is also a UMD Banach space. Together with theorem 3 of (Prüss-Sohr, 1990) this gives

$B \in BIP(\pi/2 + \varepsilon)$, $\quad \forall \varepsilon \in (0,\pi/2)$, if n is odd;

$B \in BIP(\varepsilon)$, $\quad \forall \varepsilon \in (0,\pi)$, if n is even and $a_0 = (-1)^{n/2}$.

Finally we verify (8). By theorem 1.15.3 of (Triebel, 1978) we have $D(B^\eta) = [X, D(B)]_\eta$ (complex interpolation space), so that

$$D(B^\eta) = [L^p(\mathbb{R},E),\ W^{n,p}(\mathbb{R},E)]_\eta \subset W^{n-k,p}(\mathbb{R},E) = D(A)$$

if $\eta \in (0,1)$ is sufficiently close to 1. So (8) holds and the theorem follows by (9) from $A + B = P + I$ and the arbitrariness of ε.

REMARK. The same proof can be adapted to yield results in case the coefficients $a_1, \dots, a_n$ depend on t and belong to various function spaces, and a_0 is a complex number. In order to consider a variable top order coefficient $a_0 = a_0(t)$ one needs more powerful perturbation results for the class BIP. The reader may consult (Prüss-Sohr, 1991).

4 AN EXAMPLE

THEOREM 5. *Let* E *be a* UMD *Banach space,* n *an odd positive integer,* $a_1,\ldots,a_n \in \mathbb{C}$, $\mu > 0$. *Consider the differential equation, with unknown* u,

$$u^{(n)}(t) + a_1 u^{(n-1)}(t) + \ldots + a_n u(t) + \mu\, u(t) + \Lambda u(t) = f(t), \qquad t \in \mathbb{R}, \quad (23)$$

where $f \in L^p(\mathbb{R},E)$, $p \in (1,\infty)$. *Suppose there exists* $\vartheta_\Lambda \in (0,\pi/2)$ *such that* $\Lambda \in BIP(\vartheta_\Lambda)$ *in the space* E. *Then* (23) *has a unique solution in* $W^{n,p}(\mathbb{R},E)$, *provided* μ *is sufficiently large.*

Proof. Set $X = L^p(\mathbb{R},E)$ and define

$$(Bu)(t) := u^{(n)}(t) + a_1 u^{(n-1)}(t) + \ldots + a_n u(t) + \mu\, u(t), \qquad D(B) = W^{n,p}(\mathbb{R},E)$$

$$(Au)(t) := \Lambda u(t),\ D(A) = \{u \in X : u(t) \in D(\Lambda) \text{ a.e., } t \mapsto \Lambda u(t) \text{ belongs to } X\}$$

Then $A \in BIP(\vartheta_\Lambda)$ in X and $B \in BIP(\pi/2 + \varepsilon)$, by theorem 4, provided μ is sufficiently large. So theorem 5 follows from the results in (Dore-Venni, 1987).

REFERENCES

Clément, Ph., Prüss, J. (1990). Completely positive measures and Feller semigroups, *Math. Ann.*, **287**: 73-105.

Dore, G., Venni, A. (1987). On the closedness of the sum of two closed operators, *Math. Z.*, **196**: 189-201.

Dore, G., Venni, A. (1990). Some results about complex powers of closed operators, *J. Math. Anal. Appl.*, **149**: 124-136.

Komatsu, H. (1966). Fractional Powers of Operators, *Pacific Journal of Mathematics*, **19**, no. 2: 285-346.

Prüss, J., Sohr, H. (1990). On operators with bounded imaginary powers in

Banach spaces, *Math. Z.*, **203**: 429-452.

Prüss, J., Sohr, H. (1991). Imaginary powers of elliptic second order differential operators in L^p spaces, to appear.

Seeley, R. (1971). Norm and domains of the complex powers A_B^z, *Amer. J. Math.*, **93**: 299-309.

Triebel, H. (1978). *Interpolation theory, function spaces, differential operators*, North Holland, Amsterdam.

Zimmermann, F. (1989). On vector-valued Fourier multiplier theorems, *Studia Mathematica*, **93**: 201-222.

Regularity and Decay for Nonlinear Parabolic Boundary Problems

GISÈLE RUIZ GOLDSTEIN and JEROME A. GOLDSTEIN Louisiana State University, Baton Rouge, Louisiana

§1. Introduction

A key feature of parabolic partial differential equations is that they are generally regularity improving. In the linear (autonomous) theory, this is expressed by noting that the governing semigroup is an analytic one and so maps the whole space into the domain of each power of the generator. Also, if the underlying spatial domain is bounded the semigroup maps L^1 initial data to an L^∞ (in space) solution for positive time. In the nonlinear case these considerations are more subtle. Here we are concerned with *nonlinear* parabolic equations (in divergence form) with *nonlinear* boundary conditions, and we wish to allow the ellipticity to degenerate at the spatial boundary.

The purpose of this note is to extend some results of L.C. Evans [6] along these lines. The first of our two theorems establishes the L^2 norm decay to zero of the time derivative of the solution (as $t \to \infty$). In this result a mild degeneracy is allowed at the boundary. Next we assume uniform ellipticity and show that the solution at any positive time is bounded if the initial data is merely in L^1. Evans [6] deduced these results for the usual heat equation $u_t = \Delta u$ with nonlinear boundary conditions. He also obtained supremum norm estimates of $\partial u/\partial t$, but we have been unable to obtain the corresponding result in our context.

§2 Background and Main Results

Of concern is the n dimensional parabolic partial differential equation

$$\frac{\partial u}{\partial t} = \sum_{i=1}^{n} \frac{\partial}{\partial x_i}(\psi_i(x, \nabla u)) \tag{2.1}$$

for $t \geq 0$ and $x \in \Omega$, a smooth bounded domain in $\mathbb{R}^n$. The boundary condition is

$$-\psi(x, \nabla u) \cdot \nu = \beta(x, u) \qquad \text{for } (x,t) \in \partial\Omega \times [0, \infty]. \tag{2.2}$$

Here ν is the outer unit normal to $\partial\Omega$ at x, and β is a monotone increasing function of u; (2.2) may be expressed as

$$-\sum_{i=1}^{n} \psi_i(x, \nabla u)\nu_i = \beta(x, u) \tag{2.3}$$

for $(x,t) \in \partial\Omega \times \mathbb{R}^+$. The initial condtion is

$$u(x, 0) = f(x) \tag{2.4}$$

for $x \in \Omega$. In one dimension, by parabolicity, the equation and boundary conditions reduce to

$$\begin{aligned} \frac{\partial u}{\partial t} &= \frac{\partial}{\partial x}(\psi(x, \frac{\partial u}{\partial x})) && \text{for } (x,t) \in \Omega \times \mathbb{R}^+ \\ (-1)^j \frac{\partial u}{\partial x} &\in \beta_j(u) && \text{for } x = j \in \{0, 1\} \text{ and } t \in \mathbb{R}^+. \end{aligned} \tag{2.5}$$

(In one dimension we allow for multivalued functions β_j.)

Under suitable hypotheses, these equations are governed by a strongly continuous contraction semigroup on all of the spaces X_p for $1 \leq p \leq \infty$ where $X_p = L^p(\Omega)$ for $1 \leq p < \infty$ and $X_\infty = C[0,1]$.

The hypotheses are as follows. First, we consider $n = 1$.

$$\psi \in C([0,1] \times \mathbb{R}) \cap C^1((0,1) \times \mathbb{R}),\ \psi(x, 0) \equiv 0; \tag{2.6}$$

$$\frac{\partial}{\partial \xi}\psi(x, \xi) > 0 \textit{ for } x \in (0,1) \textit{ and } \frac{\partial}{\partial \xi}\psi(x, \xi) \geq \varphi_0(x) \textit{ for all } (x, \xi) \in [0,1] \times \mathbb{R}$$

where

$$0 \leq \varphi_0(x) \in C[0,1] \textit{ with } \varphi_0^{-1} \in L^1[0,1].$$

(2.7) *β_j is a maximal monotone graph in* $\mathbb{R} \times \mathbb{R}$ *with* $\beta_j(0) \ni 0$ *and β_j strictly increasing.*

(Strict monotonicity means that if $x_1 < x_2$ and $\beta_i(x_i) \ni y_i$, then $y_1 < y_2$.)

The hypotheses for the case $n > 1$ are as follows.

(2.8) $$\psi_i \in C^{2+\epsilon}(\Omega \times \mathbb{R}^n) \cap C^{\epsilon}(\bar{\Omega} \times \mathbb{R}^n) \text{ for some } \epsilon > 0 \text{ and } \psi_i(x,0) \equiv 0.$$

(2.9) *There are numbers $\tau > -1$ and $q > n$ and a nonnegative function $\varphi_0 \in C(\bar{\Omega})$ such that for all $(x, \xi, \eta) \in \bar{\Omega} \times \mathbb{R}^n \times \mathbb{R}^n$,*

$$\sum_{i_j j=1}^{n} \frac{\partial \psi_i}{\partial \xi_i}(x,\xi)\eta_i\eta_j \geq \varphi_0(x)|\eta|^2(1 + |\xi|^2)^{\tau}$$

where $\frac{\partial \psi_i}{\partial \xi_i}(x, \xi) > 0$ *for* $x \epsilon (0,1)$, *and* $\varphi_0^{-1} \epsilon L^q(\Omega)$.

(2.10) *For some positive constant C, all $(x, \xi) \in \Omega \times \mathbb{R}^n$ and all i, j,*

$$|\frac{\partial \psi_i}{\partial \xi_j}|(1 + |\xi|^2) + (|\psi_i| + |\frac{\partial \psi_i}{\partial x_j}|)(1 + |\xi|) \leq C(| + |\xi|)^{\tau+2}$$

and

$$|\frac{\partial \psi_i}{\partial x_j}| \leq C(1 + |\xi|)^2.$$

(Usually we can take $\tau = 0$.)

Define the operator A by

$$Au = \sum_{i=1}^{n} \frac{\partial}{\partial x_i}(\psi_i(x, \nabla u))$$

for $u \epsilon D$ where

$$D = \{u \epsilon C^2(\Omega) \cap C^1(\bar{\Omega}) : Au \in C(\bar{\Omega}) \text{ and } -\sum_{i=1}^{n} \psi_i(x, \nabla u)\nu_i = \beta(x,u) \text{ on } \partial\Omega\}.$$

If $n = 1$, we define the domain by

$$D = \{u \in C^2(0,1) \cap C^1[0,1] : Au \in C[0,1] \text{ and } (-1)^j u'(j) \epsilon \beta_j(u(j)) \text{ for } j = 0,1\}.$$

Our theorems also hold if the nonlinear boundary conditions are replaced by Dirichlet conditions . In fact, if we let the boundary be the union of two disjoint parts, Γ_1 and Γ_2, and we let $D = \{u \in C^2(0,1) \cap C^1[0,1] : Au \in C[0,1], u = 0 \text{ on } \Gamma_1 \text{ and } -\psi(x, \nabla u) \cdot \nu = \beta(x,u) \text{ on } \Gamma_2\}$, then our theorems are still valid.

Let A_p be the closure of A on X_p; note that A_∞ is itself closed on $C(\bar{\Omega})$.

THEOREM 1. *Assume (2.6) and (2.7) if $n = 1$, or assume (2.8)-(2.10) if $n > 1$. Then the operator A_p on X_p is m-dissipative, that is A_p uniquely determines a semigroup $\{T(t) : t \geq 0\}$ on X_p, and there exists a unique mild solution $u \in C([0,\infty); X_p)$ of*

$$u'(t) = Au(t)$$
$$U(0) = f$$

for $f \epsilon X_p$ if $1 \leq p < \infty$ or for $f \in \bar{D}$ if $p = \infty$. Furthermore, $u(t) = T(t)$ where $T(t)$ is a strongly continuous contraction semigroup on X_p for $1 \leq p < \infty$ or on $\bar{D}$ for $p = \infty$. If $1 < p < \infty, u$ is a strong solution on $(0, \infty)$; if in addition $f \epsilon D$, then u is a strong solution on $[0, \infty]$.

Theorem is proved under more severe restrictions by J. Goldstein and Lin [9]. The above version can be proved by noting that the proof in [9] works in a more general context and by using techniques which are developed in G. Goldstein [9]. We omit the details of the existence proof. For background on dissipative operators and nonlinear semigroups see for example [1], [2], [4], [5], [8].

We also note that $-A_2$ is the subdifferential of a lower semicontinuous convex functional J on X_2. We explain this formally. Let

$$J(u) = \int_\Omega \Psi(x, \nabla u(x))dx + \int_{\partial\Omega} j(x, u(x))dx$$

where $j(x,\xi)$ is continuous for $(x,\xi) \in \partial\Omega \times \mathbb{R}$ and convex in ξ for each $x \in \partial\Omega$, and $\Psi(x,\xi)$ is continuous on $\bar{\Omega} \times \mathbb{R}^n$ and convex in ξ for each $x \in \bar{\Omega}$. Then the derivative $J'(u)$ is defined by

$$J(u+h) = J(u) + < J'(u), h > + o(h).$$

But

$$\begin{aligned} J(u+h) &= \int_\Omega \Psi(x, \nabla u + \nabla h)dx + \int_{\partial\Omega} j(x, u+h)dS \\ &= J(u) + \int_\Omega \nabla_2\Psi(x, \nabla u) \cdot \nabla h dx + \int_{\partial\Omega} \partial_2 j(x,u)hdS + o(h) \\ &= J(u) + \int_\Omega \nabla \cdot (\nabla_2 \Psi(x, \nabla u)h)dx - \int_\Omega \sum_{i=1}^n \frac{\partial}{\partial x_i}\psi_i(x, \nabla u)hdx \\ &+ \int_{\partial\Omega} \partial_2 j(x,u)hdx + o(h) \end{aligned}$$

if

$$\nabla_2 \Psi = (\psi_1, \ldots, \psi_n). \tag{2.11}$$

Then by the divergence theorem,

$$J(u+h) = J(u) - < \sum_{i=1}^{n} \frac{\partial}{\partial x_i} \psi_i(x, \nabla u), h > + o(h)$$

provided

$$\nabla_2 \Psi(x, \nabla u) \cdot \nu = -\partial_2 j(x, u) \qquad \text{on } \partial\Omega$$

which is the boundary condition (2.2) when $\partial_2 j(x, \cdot) = \beta(x, \cdot)$. We *assume* (2.11). This always holds in one dimension and is a restriction in higher dimensions.

We note that in the special case

$$\Psi(x, \nabla u) = \frac{1}{2} \Phi(x) |\nabla u|^2$$

the operator A defined by $Au = -J'(u)$ becomes

$$Au(x) = \Phi(x) \Delta(x) + \nabla \Phi(x) \cdot \nabla u(x).$$

The boundary condition becomes

$$-\Phi(x) \nabla u \cdot \nu = \beta(x, u) \qquad \text{on } \partial\Omega.$$

This reduces to $-\frac{\partial u}{\partial \nu} = \beta(u)$ when $\Phi \equiv 1, Au = \Delta u$, and β is independent of x. This is the case studied by Evans [6].

We now state the two main theorems.

Theorem 2: *Let $n > 1$, and assume (2.8) - (2.11). Let $u \in C([0, \infty); L^2(\Omega))$ be the unique mild solution of $u'(t) = Au(t), u(0) = f \in L^2(\Omega)$ given by Theorem 1. Then, in a certain sense, u satisfies the problem (2.1), (2.2), (2.4), $u(t)$ is differentiable a.e. in $L^2(\Omega)$, and*

$$\| \frac{du}{dt} \|_2 \leq \frac{C}{t} \|f\|_2 \tag{2.12}$$

for all $t > 0$ where

$$C = \sqrt{\frac{11 + 5\sqrt{5}}{2}} \approx 3.33.$$

If $n = 1$ and we assume (2.6), (2.7), then the same conclusions hold for the appropriately modified versions of (2.1) and (2.2).

Theorem 3: *Assume the hypotheses of Theorem 2, except only assume $f \in L^1(\Omega)$ Suppose $\psi_i(x, \nabla u)$ depends only on $\frac{\partial u}{\partial x_i}$ and that A is uniformly elliptic, that is*

$$\psi_i = \psi_i(x, \frac{\partial u}{\partial x_i}), \text{ and}$$

(2.13)

$$\sum_{i=1}^{n} \psi_i(x, \xi_i)\xi_i \geq xi_o|\xi|^2$$

holds for some $c_0 > 0$ and for all $\xi \in \mathbb{R}^n$, and $x \in \bar{\Omega}$. Then, the mild solution u of (2.1), (2.2), (2.4) exists and is in $L^\infty(\Omega)$ for each $t > 0$, and there exist constants C_1 and C_2 (independent of t and f)

$$(2.14) \qquad ||u(t)||_\infty \leq (C_1 + \frac{C_2}{t})||f||_1 \qquad for\ all\ T > 0.$$

We shall prove Theorem 2 in Section 3. The proof of Theorem 3 is considerably more complicated and will be given elsewhere. It is based on Evans' proof [6] and requires estimates similar to those to be obtained in the following proof as well as the Nash-Moser iteration technique.

§3. Proof of Theorem 2.

Let B_λ be the Yosida approximation of A_2, that is, $A_2 = -\partial J$ and $B_\lambda = -\partial J_\lambda$ where

$$(3.1) \qquad J(u) = \int_\Omega \Psi(x, \nabla u(x))dx + \int_{\partial\Omega} j(x, u(x)dS(x)$$

(or $+\infty$ if u is not in the domain of J) and

$$J_\lambda(u) = \inf_{v \in X_2} \left\{ \frac{||u - v||_2^2}{2\lambda} + J(v) \right\}.$$

Then J and J_λ are both lower semicontinuous and convex on X_2. Also, J_λ is defined on all of X_2, is Lipschitz continuous, and

$$||J_\lambda(u)||_2^2 = \frac{||u - (I - \lambda A)^{-1}u||_2^2}{2\lambda} + J((I - \lambda A)^{-1}u)$$

for all $u \in X_2$ and $\lambda > 0$. Moreover, for the solution $u_\lambda(t)$ of

(3.2) $$\frac{du_\lambda}{dt} = -J_\lambda(u_\lambda), \quad u_\lambda(0) = f,$$

we have that $u_\lambda(t)$ converges as $\lambda \to 0^+$ uniformly for t in compacta to the solution $u(t)$ of $\frac{du}{dt} = A_2 u, u(0) = f$.

Let $T(t)$ (or $T_p(t)$) be the contraction semigroup on X_p generated by A_p. Then $||u(t)||_p$ is a nonincreasing function of t for $1 \le p \le \infty$ since $T(t)0 = 0$ and $T(t)$ is a contraction. If $1 < p < \infty$ and $u(0) \epsilon D(A_p)$, then u is a strong solution, and for $0 \le r < s$ and $t > 0$,

$$||u(s+t) - u(r+t)||_p \le ||u(s) - u(r)||_p.$$

Dividing by $s - r$ and letting $s - r \to 0$, we see that $||u'(t)||_p$ is nonincreasing.

For $f \,\epsilon\, L^2(\Omega)$ let u_λ be the solution of (3.2), and let $k > -1$. We shall suppress the x dependence to simplify the notation. Define

(3.3) $$I_1 = \int_0^t \int_\Omega \tau^{k+2} \big(\frac{d}{d\tau} u_\lambda(\tau)\big)^2 dx d\tau.$$

Then

(3.4) $$I_1 = \int_0^t \int_\Omega \tau^{k+2} u'_\lambda(\tau)(-J'_\lambda(u_\lambda(\tau)) dx d\tau.$$

Note that $\int_\Omega u'_\lambda(J'_\lambda(u_\lambda)) dx = \frac{d}{d\tau} J_\lambda(u_\lambda)$, so that we can integrate by parts (in τ) in (3.4) to conclude

$$\begin{aligned} I_1 &= (k+2) \int_0^t \tau^{k+1} J_\lambda(u_\lambda) d\tau - [\tau^{k+2} J_\lambda(u_\lambda)]|_0^t \\ &\le (k+2) \int_0^t \int_\Omega \tau^{k+1} J'_\lambda(u_\lambda) u_\lambda d\tau \end{aligned}$$

since $J_\lambda \ge 0$ and $J_\lambda(u_\lambda) \le \langle J'_\lambda(u_\lambda), u_\lambda \rangle = \int_\Omega J'_\lambda(u_\lambda) u_\lambda dx$. Then

$$\begin{aligned} I_1 &\le (k+2) \int_0^t \int_\Omega \left[\tau^{\frac{k+2}{2}} (-u'_\lambda(\tau)) \frac{1}{\sqrt{k+2}} \right] \left[\tau^{\frac{k}{2}} u'_\lambda(\tau) \sqrt{k+2} \right] dx d\tau \\ &\le \frac{1}{2} \int_0^t \int_\Omega \tau^{k+2} u'_\lambda(\tau)^2 dx d\tau + \frac{(k+2)^2}{2} \int_0^t \int_\Omega \tau^k u'_\lambda(\tau)^2 dx d\tau \end{aligned}$$

by the Schwarz inequality. Consequently

(3.5) $$I_1 \le (k+2)^2 \int_0^t \int_\Omega \tau^k u'_\lambda(\tau)^2 dx d\tau.$$

By a theorem of Brezis [3], $u(t) \in D(A_2)$ even if $f \in X_2$. Using the facts that u is a strong solution of $u' = Au$ and that $||u(t)||_2$ is nonincreasing and letting $\lambda \downarrow 0$ in (3.5) we see that

$$||u'(t)||_2^2 t^{k+3} \leq (k+3)I_1 \leq (k+3)(k+2)^2 \int_0^t \int_\Omega \tau^k u'(\tau)^2 \, dx d\tau$$
$$\leq (k+3)(k+2)^2(k+1)^{-1} t^{k+1} ||f||_2^2.$$

Thus,

$$||u(t)||_2 \leq \frac{C_k}{t} ||f||_2$$

where

$$C_k^2 = (k+3)(k+2)^2(k+1)^{-1}$$

and C_k is independent of t. An easy calculation shows that the minimum of C_k is

$$C = \frac{11 + 5\sqrt{5}}{2}.$$

Acknowledgements

This work was begun in the spring of 1990, when we enjoyed the hospitality of MSRI in Berkeley, and when we benefitted from the helpful presence of Craig Evans. Both authors gratefully acknowledge the partial support of NSF grants. Finally we thank Gunter Lumer, Philippe Clement, and the organizers for arranging the splendid conference at Han-sur-Lesse at which this work was presented.

References

1. V. Barbu, *Nonlinear Semigroups and Differential Equations in Banach Space,* Noordhoff, Leiden (1976).
2. Ph. Benilan, M.G. Crandall, and A. Pazy, *Nonlinear Evolution Governed by Accretive Operators,* in preparation.
3. H. Brezis, Propriétes regularisantes de certains semi-groupes non lineáires, *Israel J. Math* 9: 513-534 (1971).
4. H. Brezis, *Operateurs Maximaux Monotones et Semi-Groupes de Contractions dans Espaces de Hilbert,* North Holland, Amsterdam (1973).
5. M. G. Crandall, Nonlinear semigroups and evolution governed by accretive operators, *Proceedings of Symposia in Pure Mathematics*, Vol 45, Part 1, Amer. Math Society, Providence, 305-337 (1986).
6. L.C. Evans, Regularity properties for the heat operation subject to nonlinear boundary constraints, *Nonlinear Anal TMA* 1: 593-602(1977).
7. G. R. Goldstein, Nonlinear singular diffusion with nonlinear boundary conditions, in preparation.
8. J. A. Goldstein, *Semigroups of Nonlinear Operators and Applications,* in preparation.
9. J. A. Goldstein and C.-Y. Lin, An L^p-semigroup approach to degenerate parabolic boundary value problems, *Ann. Mat. Pura Appl.* 159: 211-227 (1991).

Boundary Control of Cracked Domains

P. GRISVARD University of Nice, Nice, France

1. Introducing the problem

Let Ω be a bounded open subset of $\mathbb{R}^n$ with boundary $\Gamma=\partial\Omega$. We denote by Q the cylinder $\Omega\times]0,T[$ and by $\Sigma=\Gamma\times]0,T[$ its boundary. The problem of exact boundary controllability for the wave equation is as follows.

Given $(u_0,u_1)\in L_2(\Omega)\times H^{-1}(\Omega)$ we look for $T>0$ and $v\in L^2(\Sigma)$, such that u the (weak) solution of

$$u''=\Delta u \text{ in } Q,$$
$$u(0)=u_0,\ u'(0)=u_1,$$
$$u|_\Sigma=v,$$

vanishes at time T and subsequent times. The unknown v is the Dirichlet data and is called the "control". The problem is more realistic when the control is the Neumann data but exposition is easier on Dirichlet control.

One also endeavors to minimize Σ_0 the support of v.

Known results, when Ω is smooth, have been proven by Triggiani (1988) and Lions (1988). Optimal, i.e. minimal T, has been obtained by Bardos-Lebeau-Rauch (1988). Here I want to consider non smooth domains Ω and in particular 2d domains with cracks.

2. Hilbert Uniqueness Method

A systematic approach, called HUM (i.e. Hilbert Uniqueness Method) has been put forward by Lions (1988). Exact controllability is achieved as a consequence of the existence of a constant C such that the "dual" inequality

$$(1)\quad \|\varphi_0\|^2_{H^1_0(\Omega)}+\|\varphi_1\|^2_{L^2(\Omega)}\leqslant C\int_{\Sigma_0}\left|\frac{\partial\varphi}{\partial\nu}\right|^2 dt\,d\sigma$$

holds for every $(\varphi_0,\varphi_1)\in H^1_0(\Omega)\times L^2(\Omega)$. Here φ is the (finite energy) solution of

$$\varphi'' = \Delta\varphi \text{ in } Q,$$
$$\varphi(0) = \varphi_0, \varphi'(0) = \varphi_1,$$
$$\varphi|_\Sigma = 0,$$

and thus φ fulfils a homogeneous Dirichlet boundary condition.

When (1) holds the control is supported by Σ_0.

It has been proven for smooth or polygonal or polyhedral Ω that

$$\frac{\partial\varphi}{\partial\nu} \in L^2(\Sigma).$$

This is no longer true on domains with cracks.

The basic inequality (1) is obtained by calculating by parts the following integral

$$(2) \quad \int_Q (\varphi'' - \Delta\varphi)\mathbf{m}.\nabla\varphi \, dtdx$$

where $\mathbf{m}$ is a suitable multiplier. Formally (i.e. disregarding the convergence of the involved integrals) one gets

$$\int_\Omega \Delta\varphi \mathbf{m}.\nabla\varphi \, dx =$$

$$- \int_\Omega \nabla\varphi.\nabla(\mathbf{m}.\nabla\varphi) \, dx + \int_\Gamma \mathbf{m}.\nu \left|\frac{\partial\varphi}{\partial\nu}\right|^2 ds =$$

$$- \sum_{k,l} \int_\Omega D_k m_l D_k\varphi D_l\varphi \, dx - \frac{1}{2} \int_\Omega \mathbf{m}.\nabla|\nabla\varphi|^2 \, dx + \int_\Gamma \mathbf{m}.\nu \left|\frac{\partial\varphi}{\partial\nu}\right|^2 ds =$$

$$- \sum_{k,l} \int_\Omega D_k m_l D_k\varphi D_l\varphi \, dx + \frac{1}{2} \int_\Omega \operatorname{div}\mathbf{m} \, |\nabla\varphi|^2 \, ds + \frac{1}{2} \int_\Gamma \mathbf{m}.\nu \left|\frac{\partial\varphi}{\partial\nu}\right|^2 ds.$$

On the other hand one has

$$\int_Q \varphi'' \mathbf{m}.\nabla\varphi \, dtdx = \int_\Omega \varphi' \mathbf{m}.\nabla\varphi \, dx \Big|_0^T - \int_Q \varphi' \mathbf{m}.\nabla\varphi' \, dtdx =$$

$$\int_\Omega \varphi' \mathbf{m}.\nabla\varphi \, dx \Big|_0^T + \frac{1}{2} \int_Q \operatorname{div}\mathbf{m} \, \varphi'^2 \, dtdx$$

It follows that

$$(3) \quad 0 = \int_Q (\varphi'' - \Delta\varphi)\mathbf{m}.\nabla\varphi \, dtdx =$$

$$\int_\Omega \varphi' \mathbf{m}.\nabla\varphi\, dx \Big|_0^T + \frac{1}{2}\int_Q \mathrm{div}\,\mathbf{m}\{\varphi'^2 - |\nabla\varphi|^2\}\, dtdx + \sum_{k,l}\int_Q D_k m_l D_k\varphi D_l\varphi\, dxdt -$$

$$\frac{1}{2}\int_\Sigma \mathbf{m}.\nu\left|\frac{\partial\varphi}{\partial\nu}\right|^2 dsdt$$

since $\varphi''=\Delta\varphi$.

Now let us pay attention to the convergence of the involved integrals. The boundary integral that comes out is proportional to

$$(4) \qquad \int_\Sigma \mathbf{m}.\nu\left|\frac{\partial\varphi}{\partial\nu}\right|^2 dsdt$$

We recall that a stationary solution of the wave equation may behave as $\sqrt{r}$ where r is the distance to any crack tip. This is actually true of all solutions and $\frac{\partial\varphi}{\partial\nu}$ behaves as $\frac{1}{\sqrt{r}}$. See Grisvard (1985). The boundary term (4) will remain meaningful only if we assume that $\mathbf{m}.\nu$ vanishes at all crack tips, a first geometric assumption.

3. Elementary multiplier technique

Lions results are obtained by assuming

$$\mathbf{m} = x - x_0$$

with a suitable x_0. In 2d the above identity (3) simplifies into

$$(5) \qquad \int_\Omega \varphi' \mathbf{m}.\nabla\varphi\, dx \Big|_0^T + \int_Q |\varphi'|^2\, dxdt = \frac{1}{2}\int_\Sigma \mathbf{m}.\nu\left|\frac{\partial\varphi}{\partial\nu}\right|^2 dsdt$$

since $\mathrm{div}\,\mathbf{m} = 2$ and $D_k m_l = \delta_{kl}$. The first integral is easily bounded by E the energy of φ. Also the second integral is seen to dominate as much energy as we wish by chosing large enough T and writing

$$\int_Q |\varphi'|^2\, dxdt = \frac{1}{2}\int_Q \{\varphi'^2 + |\nabla\varphi|^2\}\, dtdx + \frac{1}{2}\int_Q \{\varphi'^2 - |\nabla\varphi|^2\}\, dtdx =$$

$$TE - \frac{1}{2}\int_Q (\varphi'' - \Delta\varphi)\varphi\, dxdt + \frac{1}{2}\int_\Omega \varphi'\varphi\, dx \Big|_0^T =$$

$$TE + \frac{1}{2}\int_\Omega \varphi'\varphi\, dx \Big|_0^T .$$

Therefore (1) holds for large enough T with Σ_0 defined by $\mathbf{m}.\nu > 0$, on a smooth Ω at least.

In a cracked domain the above geometric assumption means that all cracks point to, or from, a same point x_0. This is a drastic restriction.

Under this restriction integration by parts has actually to be performed in a truncated Ω_ε obtained from Ω by cutting away a disc of radius ε centered at every crack tip. If we denote by γ_ε the "new" boundary, i.e. $\Gamma_\varepsilon\backslash\Gamma$, and calculate (2) on $Q_\varepsilon=\Omega_\varepsilon\times]0,T[$ we obtain the additional boundary terms

$$(6)\quad \frac{1}{2}\int_{\gamma_\varepsilon\times]0,T[}\{m.\nu[-(\varphi')^2+(\frac{\partial\varphi}{\partial\tau})^2-(\frac{\partial\varphi}{\partial\nu})^2]-m.\tau\frac{\partial\varphi}{\partial\tau}\frac{\partial\varphi}{\partial\nu}\}\,dtd\sigma$$

The worst possible behavior of the integrand is like $\frac{1}{\varepsilon}$ and since the length of γ_ε is proportional to ε this boundary integral obviously remains bounded as $\varepsilon\to 0$. It actually has a limit whose sign depends on the sign of $\mathbf{m}.\tau$. As a consequence inequality (1) holds provided $\mathbf{m}.\tau\geqslant 0$ at all crack tips, another geometric assumption. See Grisvard (1989).

4 More general multipliers

More general multipliers have been introduced by Triggiani (1988) who only assumes that $m\in C^2(\overline{\Omega})^2$ and that

$$Dm+Dm^*>0$$

everywhere on $\overline{\Omega}$. In performing the above integrations by parts we obtain the additional integral

$$\int_Q (\nabla\mathrm{div}(m).\nabla\varphi)\varphi\, dtdx$$

which no longer vanishes since div$\mathbf{m}$ is no longer constant. However it is easily bounded by

$$C\int_Q |\varphi|^2\, dtdx$$

for some constant C. Instead of inequality (1) one obtains the weaker inequality

$$(1')\quad \|\varphi_0\|^2_{H^1_0(\Omega)}+\|\varphi_1\|^2_{L^2(\Omega)}\leqslant C\{\int_{\Sigma_0}|\frac{\partial\varphi}{\partial\nu}|^2\, dtd\sigma+\int_Q|\varphi|^2\, dtdx\}$$

with possibly another constant C for large enough T.

5. Applying compactness

In order to conclude one usually proves the existence of yet another constant C such that the inequality

$$(7) \quad \int_Q |\varphi|^2 \, dtdx \leqslant C \int_{\Sigma_0} |\frac{\partial\varphi}{\partial\nu}|^2 \, dtd\sigma$$

holds. By contradiction, if (7) were not true, there would exist a sequence of functions φ_m solutions with finite energy of

$$\varphi_m'' = \Delta\varphi_m \text{ in } Q,$$
$$\varphi_m|_\Sigma = 0,$$

such that

$$\int_Q |\varphi_m|^2 \, dtdx = 1$$

and such that

$$\int_{\Sigma_0} |\frac{\partial\varphi_m}{\partial\nu}|^2 \, dtd\sigma \longrightarrow 0$$

Taking weak limits, and making use of (1'), this implies the existence of a solution with finite energy φ of

$$\varphi'' = \Delta\varphi \text{ in } Q,$$
$$\varphi|_\Sigma = 0,$$
$$\frac{\partial\varphi}{\partial\nu}|_{\Sigma_0} = 0$$

with, in addition,

$$\int_Q |\varphi|^2 \, dtdx = 1.$$

This is contradictory by Holmgren's uniqueness theorem when Ω is smooth.

Following Bardos-Rauch (1991) we shall dodge this uniqueness theorem. Let us denote by N the space of the solutions with finite energy of

$$\varphi'' = \Delta\varphi \text{ in } Q,$$
$$\varphi|_\Sigma = 0,$$
$$\frac{\partial\varphi}{\partial\nu}|_{\Sigma_0} = 0.$$

We shall show that $N = \{0\}$.

Inequality (1') implies, by conservation of energy, that the $H^1(Q)$ and $L^2(Q)$ norms are equivalent on N. Then the compactness of the injection from $H^1(Q)$ into $L^2(Q)$ implies that N is finite dimensional.

We observe that N is invariant by $\frac{\partial}{\partial t}$.

If we assume, by contradiction, that dim N $\geqslant$ 1, the operator $\frac{\partial}{\partial t}$ has at least one non zero eigenvector : $u \in N$, $u \neq 0$ and $u' = \lambda u$. Consequently $u = e^{\lambda t}\psi$ where $\Delta\psi = \lambda\psi$. Since $\psi = \frac{\partial\psi}{\partial\nu} = 0$ on Γ_0, an open subset of Γ, one has $\psi \equiv 0$ on an open subset ω of Ω provided Γ is piecewise analytic. If Ω is connected we conclude that $u \equiv 0$, a contradiction.

Then if (7) were not true this would imply dimN $\geqslant$ 1. From (1') and (7) we deduce (1) and exact controllability.

6. Conclusion

The conclusion is as follows. The four assumptions

(i) $D\mathbf{m} + D\mathbf{m}^* > 0$ on $\overline{\Omega}$

(ii) Ω is connected and Γ is piecewise analytic (or even C^2)

(iii) $\mathbf{m}.\boldsymbol{\nu} = 0$ at crack tips,

(iv) $\mathbf{m}.\boldsymbol{\tau} > 0$ at crack tips

imply exact controllability. If in addition $\mathbf{m}.\boldsymbol{\nu} \leqslant 0$ along the cracks, the control v vanishes on the cracks as one would reasonably expect.

The extra flexibility allowed by these possible choices of **m** yields more general distribution of the cracks than the very particular choice of **m** considered in the above §3. Various examples of such multipliers **m** are given in Triggiani (1988).

References

Bardos-Lebeau-Rauch (1988) Appendice 2 in Lions (1988) refered below.

Bardos-Rauch (1991) Observation and control of low frequency waves (Preprint).

Grisvard (1985) Elliptic problems in non smooth domains, Monographs and studies in Mathematics, 24, Pitman, London.

Grisvard (1989) Contrôlabilité exacte des solutions de l'équation des ondes en présence de singularités, J. de Mathématiques Pures et Appliquées, 68, p.215/259.

Lions (1988) Contrôlabilité exacte, perturbations et stabilisation de systèmes distribués, tome 1, RMA n°8, Masson, Paris.

Triggiani (1988) Exact boundary controllability on $L_2(\Omega) \times H^{-1}(\Omega)$ of the wave equation with Dirichlet boundary control acting on a portion of the boundary $\partial\Omega$ and related problems, Applied Mathematics and Optimisation, 18, p.241/277.

Differential Equations on Branched Manifolds

K. P. HADELER and THOMAS HILLEN Tübingen University, Tübingen, Germany

Summary: Vector fields on two-dimensional branched manifolds can be seen as caricatures of three-dimensional problems. The corresponding semiflows are easy to visualize locally because of dimension two, on the other hand they are not continuous. With suitable transversality conditions one can obtain information on qualitative behavior and limit sets of trajectories. Particular attention is given to the fact that limit sets, in general, need not be invariant. Typical bifurcations are studied. The phenomena and applications are illustrated by graphical and numerical examples.

Introduction

The qualitative behavior of differential equations on two-dimensional compact manifolds is relatively well understood (see, e.g., Palis and di Melo [11]) whereas the behavior of differential equations in three dimensions can be very complicated and difficult to visualize. Therefore branched manifolds have been used as two-dimensional caricatures of three-dimensional systems. Williams [15] has developed a general concept of branched manifolds and he has studied caricatures of the Lorenz attractor ([16], [17], see also Guckenheimer and Holmes [5] and Sparrow [13]). One can make a similar construction for the Rössler attractor ([12], see also Jetschke [7]). Hadeler and Shonkwiler [6] have used branched manifolds in an epidemiological model. There is a rather close connection between differential equations on branched manifolds and differential equations with reset conditions or, in other terminology, differential equations with discontinuous right hand sides. Filippov [4] has developed a theory of such differential equations and has collected a vast bibliography. His view is essentially restricted to local qualitative analysis whereas we shall try to take a global view. In an abstract setting differential equations with discontinuous right hand sides can also be seen as differential inclusions (see, e.g., Aubin and Cellina [2]) or systems for set-valued functions.

Mostly a branched manifold has been seen as a set of planes (with boundaries) joined together along certain edges to form a geometric object on which vector fields are studied. At the edges transition conditions have been defined ad hoc ("if the trajectory ... "). In this view the connectedness and, in particular, the embedding in three dimensional space play a major role. Here we follow a more abstract approach which provides a rigorous definition of transition conditions. The appropriate construction requires some effort. Later we shall return to the familiar object by way of identifications.

Definition of a branched manifold

The purpose of branched manifolds is, of course, to study vector fields and their trajectories. In our approach (as in Williams [15]) we define a branched manifold as a geometric object independent of any vector fields or flows. Roughly speaking a branched manifold is a collection $\tilde{\mathcal{M}}$ of smooth manifolds M_i with transition conditions. Trajectories "run" on one manifold, "arrive" at some submanifold, and "continue" on another manifold.

If the transition conditions are used to identify points on the manifolds M_i one obtains an object $\mathcal{M}$ which heuristically can be seen as a set of surfaces glued together along lines. Then the transition conditions ensure that at each point of $\mathcal{M}$ there is a well-defined tangent space.

In this approach one can, on a fixed branched manifold, study vector fields depending on parameters and related bifurcations. In another view, used in [6], one can start from given vector fields on the M_i and study changes in the transition conditions.

In applications the constituting M_i will be mostly spheres S^2, and the N_{ij} will be spheres S^1.

In realistic applications the manifolds will be simply connected planar domains, and the submanifolds will be line segments. However, these manifolds are topologically equivalent to spheres. As usual in differential equations the assumption of compact manifolds without boundary merely leads to some simplification of the representation and some unification, it makes a coherent theory possible.

Let M_i, $i = 1,\ldots,r$, be two-dimensional compact C^1 manifolds without boundary.

Let N_{ij}, $j = 1,\ldots,s_i$, be one-dimensional (compact) C^1 submanifolds of M_i. Assume

$$N_{ij} \cap N_{ik} = \emptyset \quad \text{for} \quad j \neq k. \tag{1}$$

We define

$$M_i' = M_i \setminus \bigcup_{j=1}^{s_i} N_{ij}. \tag{2}$$

Now we define transition conditions. It is useful to introduce an index set $J = \{(i,j) : i = 1,\ldots,r;\ j = 1,\ldots,s_i\}$. One can visualize J as a matrix with rows of different lengths. Furthermore denote by L the set $L = \{1,\ldots,r\}$. Let $V : J \to L$ be a function.

For $(i,j) \in J$ let g_{ij} be a C^1 mapping $g_{ij} : N_{ij} \to M_l$ where $l = V(i,j)$. By construction the image $g_{ij}(N_{ij}) \subset M_l$ is compact.

The main hypothesis on the various manifolds is the following nonintersection property. For $l \in L$ we require

$$\left(\bigcup_{V(i,j)=l} g_{ij}(N_{ij}) \right) \bigcap \left(\bigcup_{j=1}^{s_l} N_{lj} \right) = \emptyset. \tag{3}$$

The object $\tilde{\mathcal{M}} = \{M_i, N_{ij}, g_{ij} : (i,j) \in J\}$ is called a branched manifold. In addition to $\tilde{\mathcal{M}}$ we shall need the underlying point set $M = \cup_{i=1}^{r} M_i$.

Example 1: $r = 1$, $s = 1$. Thus V maps $(i,j) = (1,1)$ into $l = 1$. There is only one manifold $M = M_1 = S^2$, and only one submanifold $N = N_{11}$, and $g : N \to M$. Condition (3) reduces to $g(N) \cap N = \emptyset$. This example corresponds to the classical reset problem. The trajectory runs on M until it meets N. Then it is reset to $g(N)$ and starts again (see Fig.1a). The caricature of the dynamics of the Rössler attractor as given by Jetschke [7] fits into this scheme (Fig.1b).

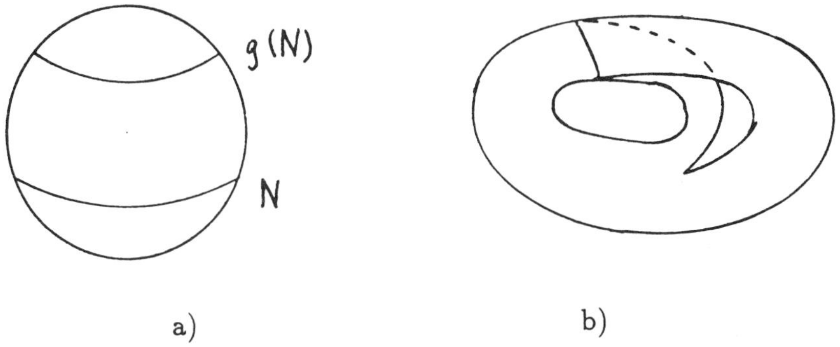

Fig.1: The reset problem as a branched manifold

Example 2: $r = 1$, $s = 2$, $M_1 = S^2$, $V(1,1) = 1$, $V(1,2) = 1$. N_1, N_2 are disjoint sets S^1, and the images $g_1(N_1)$, $g_2(N_2)$ coincide (as sets). $g_1(N_1) = g_2(N_2)$ is a sphere S_1 disjoint from $N_1 \cup N_2$ (Fig.2a). The Fig.2b shows a flow on M. The shaded area in Fig.2b is equivalent to William's caricature of the Lorenz attractor (Fig.2c).

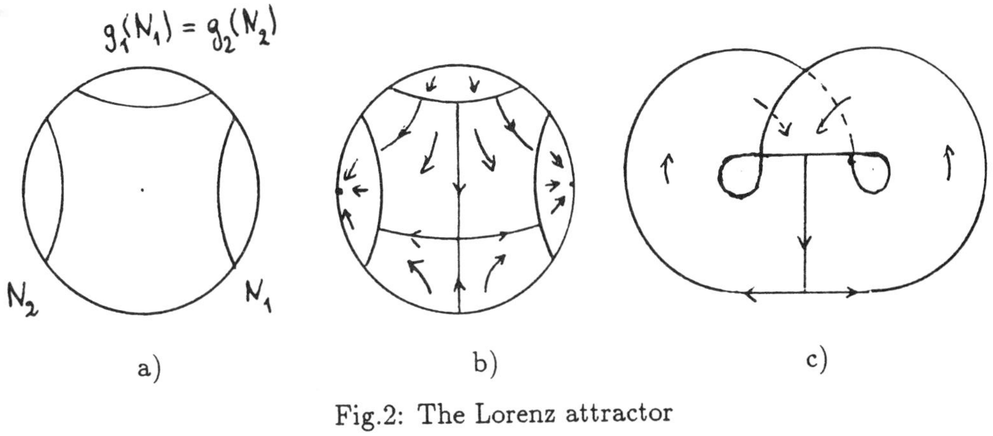

Fig.2: The Lorenz attractor

Example 3: $r = 2$, $s_1 = s_2 = 1$, $M_1 = S^2$, $M_2 = S^2$, $V(1,1) = 2$, $V(2,1) = 1$. This setting is the common situation in problems of pest control or epidemic control ([6]) where there is a switch between strategies. Fig.2a presents the general situation for spheres. In the case studied in [6] one cap of each sphere can be discarded since no trajectory ever returns to these caps. Then the part of the branched manifold which contains the essential dynamics can be represented as in Fig.2b.

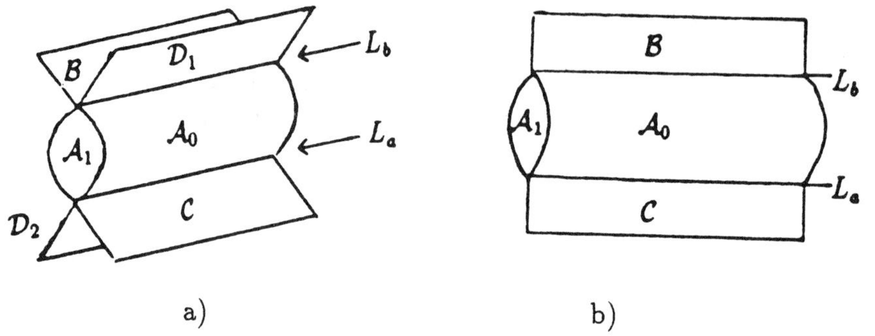

Fig.3: Epidemic control problem as a branched manifold

On $\tilde{\mathcal{M}}$ (i.e., on M endowed with the structure of $\tilde{\mathcal{M}}$) we introduce an equivalence relation $\sim$ in the following way. Let $\mathcal{G}$ be the set of functions $\mathcal{G} = \{g_{ij} : (i,j) \in J\} \cup \{id\}$ where id is the identity on M. For points $\tilde{x}, \tilde{y} \in \tilde{\mathcal{M}}$ we define $\tilde{x} \sim \tilde{y}$ if there are functions $f, g \in \mathcal{G}$ such that $f(\tilde{x}) = g(\tilde{y})$. One can visualize the identification by $\sim$ as glueing together the M_i along N_{ij} and $g(N_{ij})$, respectively.

Proposition 1: *The relation $\sim$ is an equivalence relation.*

Proof: $\tilde{x} \sim \tilde{x}$ since $id \in \mathcal{G}$. $\tilde{x} \sim \tilde{y}$ implies $\tilde{y} \sim \tilde{x}$ by symmetry of definition. Assume $\tilde{x} \sim \tilde{y}$ and $\tilde{y} \sim \tilde{z}$. Then there are functions $f, g, h, k \in \mathcal{G}$ such that $f(\tilde{x}) = g(\tilde{y})$ and $h(\tilde{y}) = k(\tilde{z})$. In view of (1) there are only two cases.
Case 1: $g = h$. Then $f(\tilde{x}) = h(\tilde{z})$, $\tilde{x} \sim \tilde{z}$.
Case 2: $g = id$ (or $h = id$). Then $f(\tilde{x}) = \tilde{y}$ and $h(\tilde{y}) = k(\tilde{z})$. $h = id$ would lead to Case 1. Thus assume $h \neq id$. From (3) it follows that $f = id$, $\tilde{x} = \tilde{y}$, $h(\tilde{x}) = k(\tilde{z})$, $\tilde{x} \sim \tilde{z}$.

Now we define the object $\mathcal{M} = \tilde{\mathcal{M}}/\sim$. The set $\mathcal{M}$ will be endowed with the quotient topology and thus it becomes a compact Hausdorff space. We shall call both objects $\mathcal{M}$ and $\tilde{\mathcal{M}}$ the branched manifold. We shall denote by π the natural projection $\pi : \tilde{\mathcal{M}} \to \mathcal{M}$.

The object $\mathcal{M}$ looks rather similar to the ramified spaces which have been studied by Lumer [8], v.Below [3], Nicaise [9], Ali-Mehmeti [1] in connection with diffusion equations. We underline that they are indeed different. Ramified spaces are also constructed from smooth manifolds by way of identifications, but in their construction there is no directional information. Of course the present notion of branched manifold is closely related to the notion of Williams [15]. In Williams' definition of a branched manifold there is, by definition, a well-defined tangent space at every point. The natural visualization of Williams' branched manifold is a set of two-dimensional surfaces in $\mathbb{R}^3$ which pass through a curve such that at each point of this curve all surfaces have the same tangent plane. In the present context the appropriate visualization is given by Figs.1 and 2. Here the uniqueness of the tangent space is a consequence of the construction as it will be shown in the next proposition.

Proposition 2: *In each equivalence class $x \in \mathcal{M}$ there is one and only one point $\tilde{x}$ with the property:*

$$\left\{\tilde{y} \in \tilde{\mathcal{M}},\ f \in \mathcal{G},\ f(\tilde{x}) = \tilde{y}\right\} \Rightarrow f = id. \tag{4}$$

Proof: If x contains only one point $\tilde{x}$ then $\tilde{x}$ has property (4). Suppose x contains two points $\tilde{y}, \tilde{z} \in M$ with $\tilde{y} \neq \tilde{z}$. Then there are $f, g \in \mathcal{G}$ such that $g(\tilde{y}) = h(\tilde{z})$. Put $\tilde{x} = g(\tilde{y})$. Then $\tilde{x} \in x$ because $id \in \mathcal{G}$. The point $\tilde{x}$ has property (4) in view of condition (3). Now assume there are $\tilde{x}_1, \tilde{x}_2 \in x$ with property (4). Then there are functions $h_1, h_2 \in \mathcal{G}$ such that $h_1(\tilde{x}_1) = h_2(\tilde{x}_2)$. Then, again by (4), $h_1 = h_2 = id$ and $\tilde{x}_1 = \tilde{x}_2$.

Definition: For each equivalence class $x \in \mathcal{M}$ the point $\tilde{x} \in \mathcal{M}$ with property (4) is called the actual point $\alpha(x)$. The tangent space at $x \in \mathcal{M}$ is given by $T_x\mathcal{M} = T_{\alpha(x)}\tilde{\mathcal{M}}$.

Thus $\alpha : \mathcal{M} \to \tilde{\mathcal{M}}$ is the map which attributes to each point on $\mathcal{M}$ the actual point in $\tilde{\mathcal{M}}$. An equivalence class or point $x \in \mathcal{M}$ is called trivial if it contains only one point of $\tilde{\mathcal{M}}$. Otherwise the point is called a branch point. The set of trivial equivalence classes is open in $\mathcal{M}$. The function α is continuous on this open set. It should be underlined that π does not induce a natural projection of $T\tilde{\mathcal{M}}$ to $T\mathcal{M}$.

Flows

Suppose that on each manifold M_i, $i = 1, \ldots, r$, there is a C^1 vector field f_i. This collection of vector fields defines a C^1 vector field $\tilde{f}$ on M and $\tilde{\mathcal{M}}$. By integrating the vector fields f_i we obtain flows $\Phi_i(t, x)$, $i = 1, \ldots, r$ which exist for all $t \in \mathbb{R}$. The function $\Phi_i(t, x_0)$ is the solution of

$$\dot{x}(t) = f_i(x(t)), \quad x(0) = x_0, \tag{5}$$

i.e.,

$$\begin{aligned} \frac{\partial \Phi_i(t, x)}{\partial t} &= f_i(\Phi_i(t, x)), \\ \Phi_i(0, x) &= x. \end{aligned} \tag{6}$$

We want to construct a semiflow $\tilde{\Phi}$ on $\tilde{\mathcal{M}}$ which for small $t > 0$ has the property

$$\tilde{\Phi}(t, \tilde{x}) = \begin{cases} \Phi_i(t, \tilde{x}) & \text{if } \tilde{x} \in M_i', \\ \Phi_l(t, g_{ij}(\tilde{x})) & \text{if } \tilde{x} \in N_{ij},\ V(i, j) = l. \end{cases} \tag{7}$$

Theorem 3:
i) There is a unique semiflow $\tilde{\Phi}$ on $\tilde{\mathcal{M}}$ which has the property (7).
ii) The function

$$\Phi(t,x) = \pi\tilde{\Phi}(t,\alpha(x)) \tag{8}$$

defines a semiflow Φ on $\mathcal{M}$ for which $\tilde{\Phi}(t,\alpha(x)) = \alpha(\Phi(t,x))$ has the property (7).

Proof: Define, for $\tilde{x} \in M_i$

$$\tau_i(\tilde{x}) = \inf\left\{t \geq 0 : \Phi_i(t,\tilde{x}) \in \cup_{j=1}^{s_i} N_{ij}\right\}. \tag{9}$$

Furthermore define

$$\delta = \inf_{l\in L}\left(\inf\left\{\tau_l(\tilde{x}) : \tilde{x} \in \cup_{V(i,j)=l}\, g_{ij}(N_{ij})\right\}\right). \tag{10}$$

The number δ is positive in view of (3) and compactness. For $0 \leq t < \delta$ define a local semiflow by

$$\tilde{\Phi}(t,\tilde{x}) = \begin{cases} \Phi_i(t,\tilde{x}) & \text{if } \tilde{x} \in M_i \text{ and } 0 \leq t \leq \tau_i(\tilde{x}), \\ \Phi_l(t-\tau_i(\tilde{x}), g_{ij}(\Phi_i(\tau_i(\tilde{x}),\tilde{x}))) & \text{if } \tilde{x} \in M_i,\ \Phi_i(\tau_i(\tilde{x}),\tilde{x}) \in N_{ij}, \\ & V(i,j) = l, \text{ and } \tau_i(\tilde{x}) < t < \delta. \end{cases} \tag{11}$$

The function $\tilde{\Phi}$ is well-defined and satisfies

$$\tilde{\Phi}(t+s,\tilde{x}) = \tilde{\Phi}(s,\tilde{\Phi}(t,\tilde{x})), \quad \tilde{\Phi}(0,\tilde{x}) = \tilde{x} \tag{12}$$

for $t, s \geq 0$ with $t+s < \delta$. Since δ is uniform and $\tilde{\mathcal{M}}$ is compact, this local semiflow can be continued to a semiflow.

These properties carry over to Φ as defined by (8).

Corollary 4: *The trajectory $t \mapsto \Phi(t,x)$ is continuous (in the topology of $\mathcal{M}$). The trajectory $t \mapsto \tilde{\Phi}(t,\tilde{x})$ has at most one discontinuity (in the topology of $\tilde{\mathcal{M}}$) in any given time interval of length δ.*

The flow $\tilde{\Phi}$ can be recovered from Φ as

$$\tilde{\Phi}(t,\tilde{x}) = \lim_{\substack{s\to t\\ s<t}} \alpha(\Phi(s,x)), \quad \tilde{x} \in x, \quad \text{for} \quad t > 0,$$

$$\tilde{\Phi}(0,\tilde{x}) = \tilde{x}.$$

It is evident that both $\tilde{\Phi}(t,\tilde{x})$ and $\Phi(t,x)$ are not continuous in $\tilde{x}$ or x, respectively. There is no sensible way to make these functions continuous. In some sense this is the sacrifice one has to make in replacing smooth three-dimensional vector fields by vector fields on branched manifolds. Some continuity can be recovered under transversality assumptions.

The idea of the construction can be explained as follows. For a given nontrivial point $x \in \mathcal{M}$ there are several points $\tilde{x} \in M_i$ for appropriate $i \in L$. Thus there are several candidates $f_i(\tilde{x})$ for a tangent vector. By the principle of the actual point one of these tangent vectors is selected. Thus at every point of $\mathcal{M}$ there is a unique tangent vector. This vector field is piecewise smooth though it may not be smooth if $\mathcal{M}$ is embedded into some space of higher dimension.

The construction of Williams is somewhat different. Geometrically speaking, his construction assumes that the different manifolds glued together, embedded into some space of higher dimension, are tangent to each other.

Transversality

Suppose N is a closed C^1 curve in M_i. The vector field f_i is called transversal to N at $\tilde{x} \in N$ if $\tilde{x}$ is a simple point of N and $f_i(\tilde{x})$ is not tangent to N at $\tilde{x}$. The vector field f_i is called transversal to N if it is transversal to N at every point of N.

A point $\tilde{x} \in \tilde{\mathcal{M}}$ is called transversal if either $\tilde{x} \in M_i^!$ or $\tilde{x} \in N_{ij}$ and f_i is transversal to N_{ij} at $\tilde{x}$. The vector field $\tilde{f}$ is called transversal if all points of $\tilde{\mathcal{M}}$ are transversal.

A point $x \in \mathcal{M}$ is called transversal if all $\tilde{x} \in x$ are transversal.

Suppose $\Phi(t,x)$, $t \geq 0$, is a trajectory in $\mathcal{M}$. Suppose $\Phi(t_0,x)$ is transversal for some $t_0 > 0$. Then $\Phi(t,x)$ is transversal for $t_0 \leq t < t_0 + \delta$ where δ is defined by (9).

Theorem 5: *Suppose $x \in \mathcal{M}$ is trivial and $\Phi(t,x)$ is transversal for all $t \geq 0$. Then for every $T > 0$ there is a neighborhood $U_T \subset \mathcal{M}$ of x such that Φ is continuous in $[0,T] \times U_T$.*

Proof: The orbit $\Phi(t,x)$ through x has only countably many transitions $\tau_1 < \tau_2 < \cdots$. By assumption $\alpha(x) \in M_i^!$. Hence there is a neighborhood U_0 of x and a $\delta_0 > 0$, $\delta_0 < \delta$, such that $\Phi(t,y)$ contains only one point for $(t,y) \in [0,\delta_0) \times U_0$, and consequently Φ is continuous in this set.

Now suppose it has been shown that there is a δ_k, $0 < \delta_k < \delta$, and a neighborhood U_k of x such that $\Phi(t,y)$ is continuous in $[0,\tau_k + \delta_k) \times U_k$. Since $\Phi(\tau_k - \varepsilon, x)$ is transversal for small $\varepsilon > 0$ by assumption, by Ważewski's theorem [14] and assumption (3) there is a δ_{k+1}, $0 < \delta_{k+1} < \delta$, and a neighborhood $U_{k+1} \subset U_k$ such that $\Phi(t,y)$ is continuous in $[0,\tau_{k+1} + \delta_{k+1}) \times U_{k+1}$. This argument can be repeated.

We shall need a similar assertion for the flow $\tilde{\Phi}$.

Corollary 6: *Let $\tilde{x} \in \tilde{\mathcal{M}}$, $\tilde{x} \in M_i^!$, and suppose that $\tilde{\Phi}(t,\tilde{x})$ is transversal for all $t \geq 0$. Then for every $T > 0$ there is a neighborhood $\tilde{U}_T \subset M_i$, $\tilde{U}_T \ni \tilde{x}$, such that $\tilde{\Phi}$ is continuous in $[0,t] \times \tilde{U}_T$.*

The proof is essentially the same as that of Theorem 5.

The assumption of transversality everywhere is much too restrictive. In order to have something concrete at hand we define a class of vector fields which have some generic properties.

Suppose $x \in \mathcal{M}$ is a branch point which is not transversal. Then there is $\tilde{x} \in x$ such that $\tilde{x} \in N_{ij} \subset M_i$, and f_i is tangent to N_{ij}. We call a branch point x a generic contact point if for all $\tilde{x} \in x$ the following is true: If $\tilde{x} \in N_{ij}$ then the contact of $\Phi_i(t, x)$ and N_{ij} is only of first order. Then locally $\Phi_i(t, \tilde{x})$ stays on one side of N_{ij} (see Fig.4).

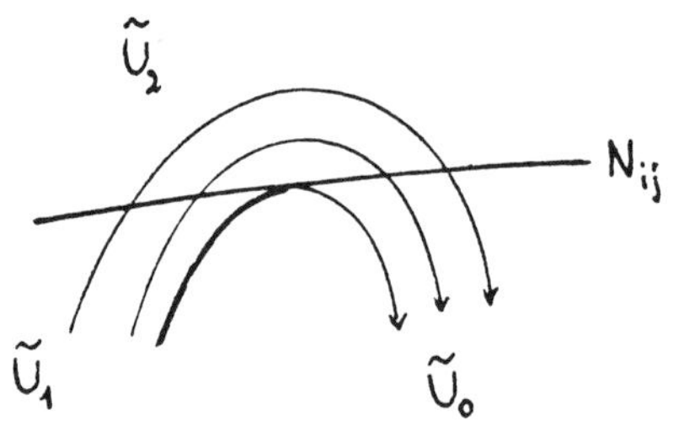

Fig.4: Generic contact point

If x is a generic contact point then at each $\tilde{x} \in x$, $\tilde{x} \in N_{ij}$, there is a neighborhood $\tilde{U} \subset M_i$ of $\tilde{x}$ such that the trajectory $\Phi_i(t, \tilde{x})$ and N_{ij} define three domains $\tilde{U}_0$, $\tilde{U}_1$, $\tilde{U}_2$ (see Fig.4). Trajectories of $\tilde{\Phi}$ starting in $\tilde{U}_0$ stay in M_i as long as they stay in $\tilde{U}$, trajectories in $\tilde{U}_1$ and $\tilde{U}_2$ leave M_i. That is why $\tilde{\Phi}$ is not continuous at these points.

We call a vector field f on $\tilde{\mathcal{M}}$ a field of generic contact if there are only finitely many points which are not transversal and if all these points are generic contact points. The vector fields in [6] have this property.

At least in topologically simple cases strong transversality properties provide equivalences between branched manifolds and unbranched manifolds of known topological structure. In the following examples 4a,b,c we assume that the vector field f_i is transversal to all curves N_{ij} and also to all curves $g_{jl}(N_{jl})$ with $V(j, l) = i$. We say that the Poincaré-Bendixson property holds for a trajectory when the trajectory eventually remains in some S^2 (or disc).

Example 4a: M is a 2-sphere, N and $g(N)$ are disjoint circles. These curves define an annulus A and two discs D_0 (bounded by N) and D_1 (bounded by $g(N)$). Since the vector field is transversal on N and $g(N)$ there are just four qualitatively different situations which can be presented as follows.
a) D_0 and D_1 are positively invariant. Then a trajectory either stays in D_0 or D_1, or it stays in A, or it leaves A to stay in D_1. Hence the Poincaré-Bendixson property holds.

b) Trajectories from the annulus never arrive at N. There is no reset. The Poincaré-Bendixson property holds.
c) D_0 is positively invariant, but D_1 is negatively invariant. Then the two discs can be discarded. The two curves can be identified, a 2-torus remains.

Example 4b: M_1, M_2 are 2-spheres, N_{11}, N_{21} and their images are disjoint circles. On M_i these curves define an annulus and two discs. There are 16 qualitatively different cases. If we exchange M_1 and M_2 then ten cases remain. Among these all cases are trivial where either N_{11} or N_{21} cannot be reached from the interior of the annulus. In these cases every trajectory has at most one transition from one sphere to the other. Then four cases remain. In three of these cases one sees easily, as in Example 1, that the Poincaré-Bendixson property holds. In the remaining case one can disregard the four discs, and connect the two annuli to a 2-torus. Thus we conclude that the dynamics of a transversal vector field on this branched manifold can be essentially represented on a 2-sphere or on a 2-torus.

Example 4c: M_1, M_2 are 2-spheres with $s_1 = 2$ and $s_2 = 1$. The six resulting curves and the direction of the transversal vector field are shown in Fig.5. This structure cannot be reduced to a smooth (classical) manifold. A similar observation holds for the Lorenz manifold of Williams [16], [17].

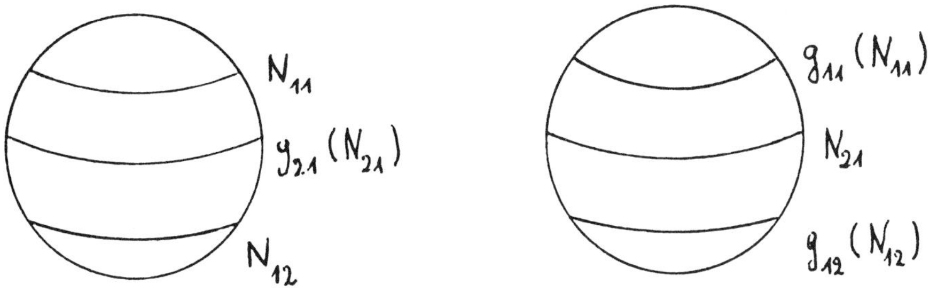

Fig.5: Illustration of Example 4c

Limit sets

As usual the limit set (ω limit set) of a trajectory $\Phi(t, x)$ is defined as

$$\omega(x) = \{z : \exists t_1 < t_2 < \cdots,\ t_k \to \infty,\ \Phi(t_k, x) \to z\}.$$

As in the classical case one proves the following proposition.

Proposition 7: *The set $\omega(x)$ is*
i) nonempty,
ii) compact in the topology of $\mathcal{M}$,
iii) connected in the topology of $\mathcal{M}$.

Proof:
i) Choose any sequence t_k. $\mathcal{M}$ is compact. Hence there is an accumulation point.

ii) Let $y \in \mathcal{M}$, $y \notin \omega(x)$. Then there is a neighborhood U of y and a $T > 0$ such that $U \cap \{\Phi(t,x) : t \geq T\} = \emptyset$. Hence $\omega(x)$ is closed and thus compact.
iii) Suppose $\omega(x)$ is not connected in the topology of $\mathcal{M}$. Then there are open sets U and V such that $U \cap V = \emptyset$, $\omega(x) \subset U \cup V$, and $\omega(x) \cap U \neq \emptyset$, $\omega(x) \cap V \neq \emptyset$. Define $K = \mathcal{M} \setminus (U \cup V)$. K is compact. For every $T > 0$ there are $t_1 > T$, $t_2 > T$ such that $\Phi(t_1, x) \in U$, $\Phi(t_2, x) \in V$, hence also $t > T$ such that $\Phi(t,x) \in K$. Hence there is a sequence $t_k \to \infty$ such that $\Phi(t_k, x)$ converges to some point $y \in K$. Hence $y \in \omega(x)$. This gives a contradiction.

Limit sets on branched manifolds need not be positively invariant. A simple counterexample: In Example 1 above assume that a trajectory in $M \setminus N$ approaches a stationary point on N. Then the stationary point is the limit set, but if the stationary point is chosen as an initial condition then the trajectory continues on $g(N)$.

Even if the vector fields f_i are structurally stable (see, e.g., [10]) and all critical elements are transversal to the N_{ij} there may be limit sets which are not positively invariant. This fact is shown by Example 5b. However, one can prove the following result.

Theorem 8: *Assume that all points of $\omega(x)$ are transversal. Then $\omega(x)$ is positively invariant.*

Proof: Let $y \in \omega(x)$. By assumption there is a sequence $t_k \to \infty$ such that $\Phi(t_k, x) \to y$. We have to show $\Phi(s, y) \in \omega(x)$ for $s \geq 0$. We can use the flow property

$$\Phi(t_k + s, x) = \Phi(s, \Phi(t_k, x)).$$

If Φ were continuous in x then $\Phi(t_k + s, x) \to \Phi(s, y)$ would follow immediately. Since Φ is not continuous in general, we have to consider the problem in more detail.

Case 1: y is trivial. In view of Theorem 5, for every $T > 0$, there is a neighborhood $U_T \subset \mathcal{M}$, $U_T \ni y$, such that Φ is continuous in $[0,T] \times U_T$. Choose $T > s$. For large k we have $\Phi(t_k, x) \in U_T$. We use

$$\Phi(s + t_k, x) = \Phi(s, \Phi(t_k, x)).$$

For $k \to \infty$ the right hand side has the limit $\Phi(s, y)$ whereas the left hand side, by definition of the limit set, coverges to some point in $\omega(x)$.

Case 2: y is not trivial. Let $\alpha(y) = \tilde{y} \in M_l$ be the actual point. The sequence $\Phi(t_k, x)$ may contain nontrivial points.

Case 2a: There is an infinite subsequence such that the actual points are in M_l. We can assume that the given sequence has already this property, i.e., that $\alpha(\Phi(t_k, x)) = \tilde{y}_k \in M_l$. By Corollary 6, for every $T > 0$ there is a neighborhood

$\tilde{U}_T \subset M_l$, $\tilde{U}_T \ni \tilde{y}$ such that $\tilde{\Phi}$ is continuous in $[0,T] \times \tilde{U}_T$. Choose $T > s$. For large k we have $\tilde{\Phi}(t_k, \alpha(x)) \in \tilde{U}_T$. Then

$$\tilde{\Phi}(s + t_k, \alpha(x)) = \tilde{\Phi}(s, \tilde{\Phi}(t_k, \alpha(x)),$$

and thus

$$\Phi(s + t_k, x) = \Phi(s, \Phi(t_k, x)).$$

For $k \to \infty$ we have $\tilde{\Phi}(t_k, \alpha(x)) \to \tilde{y}$, thus $\Phi(t_k, x) \to y$. Now we can continue as in Case 1.

Case 2b: There is no such sequence as in Case 2a. We choose, if necessary, a subsequence, and have the following situation. There is $\hat{y} \in y$, $\hat{y} \neq \alpha(y)$, $\hat{y} \in N_{ij} \subset M_i$. There is an open neighborhood $U \subset M_i$ of $\hat{y}$ such that $N_{ij} \cap U$ is homeomorphic to an interval, and $N_{ij} \cap U$ separates U into two open sets U_- (where trajectories of $\tilde{\Phi}$ leave M_i) and U_+ (where trajectories of $\tilde{\Phi}$ stay in M_i) and $U = U_+ \cup U_- \cup (N_{ij} \cap U)$. We can assume that Φ_i is transversal to N_{ij} along $N_{ij} \cap U$.

Case 2bα: There is an infinite subsequence such that the actual points are in U_+. We can assume that the original sequence has this property, i.e., $\alpha(\Phi(t_k, x)) \in U_+$. For large k the trajectory of Φ_i (or $\tilde{\Phi}$ through $\alpha(\Phi(t_k, x))$ crosses N_{ij} close to $\hat{y}$. In U_+ this trajectory is a trajectory of Φ. Hence for each k there is an s_k such that $\alpha(\Phi(t_k - t, x)) \in U_+$ for $0 \leq t < s_k$, and $\Phi_i(-s_k, \alpha(\Phi(t_k, x)) \in N_{ij}$. This is a contradiction to the fact that there is a transition to M_l at N_{ij}. Hence Case 2bα is impossible.

Case 2bβ: There is an infinite subsequence such that $\alpha(\Phi(t_k, x)) \in U_-$. We can assume that the given sequence has this property. Choose $\varepsilon > 0$ such that $\Phi_i(t, \hat{y}) \in U_-$ for $-\varepsilon \leq t < 0$. Then for k sufficiently large there is s_k such that $\Phi(t_k + s_k + t, x) \in U_-$ for $-\varepsilon \leq t < 0$. Then $\Phi(t_k + s_k + t, x) \to \Phi_i(t, \tilde{y})$ uniformly in $-\varepsilon \leq t < 0$ for $k \to \infty$. Hence $\Phi_i(t, \tilde{y}) \in \omega(x)$ for $-\varepsilon \leq t < 0$. Now choose any of these points and proceed as in Case 1.

We add some comments on attractors and basins. Consider a vector field on S^2 with finitely many critical elements. Then the basins of the attractors form finitely many open domains and their boundaries are formed by trajectories, hence are piecewise differentiable curves. These boundaries may contain repellers and saddle points. The basin boundaries are themselves invariant sets. On branched manifolds we have a different situation. In Example 5b there are finitely many stationary points and periodic orbits, every trajectory converges to one of these. The basin of each attractor contains an open set such that the closures of these sets cover the whole manifold. But the basins need not be open.

Return mappings and global behavior

The global behavior of the dynamical system defined by $(\mathcal{M}, f)$ can be studied by return mappings (Poincaré mappings). The idea is that trajectories which have only finitely many transitions stay eventually in one of the M_i and hence are "trivial". Thus one chooses a curve L (preferably one of the curves N_{ij} or $g_{ij}(N_{ij})$) and one follows all trajectories starting from L. For $x \in L$ (in the following we shall omit all tildes) define

$$\tau(x) = \inf\{t > 0 : \Phi(t, x) \in L\}.$$

Then define

$$L_\infty = \{x \in L : \tau(x) = \infty\}.$$

Introduce the symbol $\emptyset$ for "empty". Define a mapping φ on $L \cup \{\emptyset\}$ by

$$\varphi(x) = \begin{cases} \Phi(\tau(x), x) & \text{if } x \in L \setminus L_\infty, \\ \emptyset & \text{if } x \in L_\infty, \\ \emptyset & \text{if } x = \emptyset. \end{cases}$$

The mapping φ is called the return mapping. At least in cases where the number of the N_{ij} is small one can get a global view of the asymptotic behavior by studying φ and its iterates.

If x is such that $\varphi^k(x) = \emptyset$ for some k then the trajectory $\Phi(t, x)$ meets L only finitely often. Fixed points of φ other than $\emptyset$ correspond to periodic orbits.

Here we study a type of branched manifold closely related to Example 3. Let $M_1 = \mathbb{R}^2$, $M_2 = \mathbb{R}^2$, each endowed with the same cartesian coordinate system (x, y). Let $N_{11} = \{x = b\}$ and $N_{21} = \{x = a\}$, and let $g_{11}(x) = x$, $g_{21}(x) = x$ (with respect to the identical coordinate system). For convenience we assume $a < b$. If we discard the sets $\{x \in M_1 : x > b\}$ and $\{x \in M_2 : x < a\}$ then we arrive at the manifold studied in [6].

As in [6] we introduce the sets $\mathcal{A}_1 = \{(x, y) \in M_1 : a < y < b\}$, $\mathcal{A}_2 = \{(x, y) \in M_2 : a < y < b\}$, $\mathcal{B} = \{(x, y) \in M_2 : y > b\}$, $\mathcal{C} = \{(x, y) \in M_1 : y < a\}$. Furthermore we introduce the lines $L_a = \{y = a\}$ and $L_b = \{y = b\}$ as subsets of M_1 and M_2. It is sufficient to consider trajectories which meet N_{11} and N_{21} infinitely often. Choose $L = L_a$. Define a mapping φ_1 as follows. For any point $x \in L_a \subset M_1$ define $\tau_1(x) = \inf\{t > 0 : \Phi(t, x) \in L_b \subset M_1\}$. Define $\varphi_1(x) = \Phi(\tau_1(x), x)$ if $\tau_1(x)$ is finite and $\varphi_1(x) = \emptyset$ otherwise. Extend the definition of φ_1 by putting $\varphi_1(\emptyset) = \emptyset$. Define τ_2 and φ_2 using M_2 instead of M_1. Then $\varphi = \varphi_2 \circ \varphi_1$ is the return mapping.

To have something concrete at hand we assume that the trajectories in $\mathcal{A}_1 \cup \mathcal{C}$ look as in Fig.6a, whereas the trajectories in $\mathcal{A}_2 \cup \mathcal{B}$ look as in Fig 6b.

Apparently there are four interesting points called P, Q, R, S as indicated. We assume that P and S do not coincide on $\mathcal{M}$, neither do Q and R. The points P and Q are generic contact points. The points R and S are just points where f_l is not transversal to $g_{ij}(N_{ij})$.

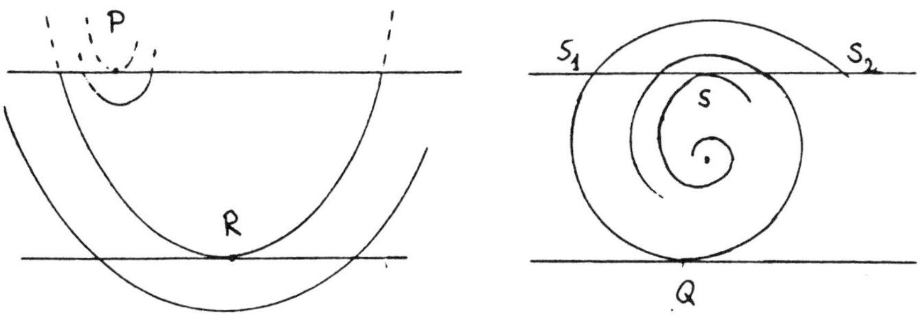

a) b)

Fig.6: The flows on M_1 and M_2

Then the function φ_1 is continuous everywhere, it is decreasing for $x < R$ and increasing for $x > R$. Hence it attains its minimum at R (see Fig 7a). Let S_1, S_2 be the two preimages of Q with respect to the flow in $\mathcal{A}_2$. Then φ_2 increases for $x < S_1$, decreases for $x > S_2$, and φ_2 takes the interval (S_1, S_2) to $\emptyset$ (see Fig.7b).

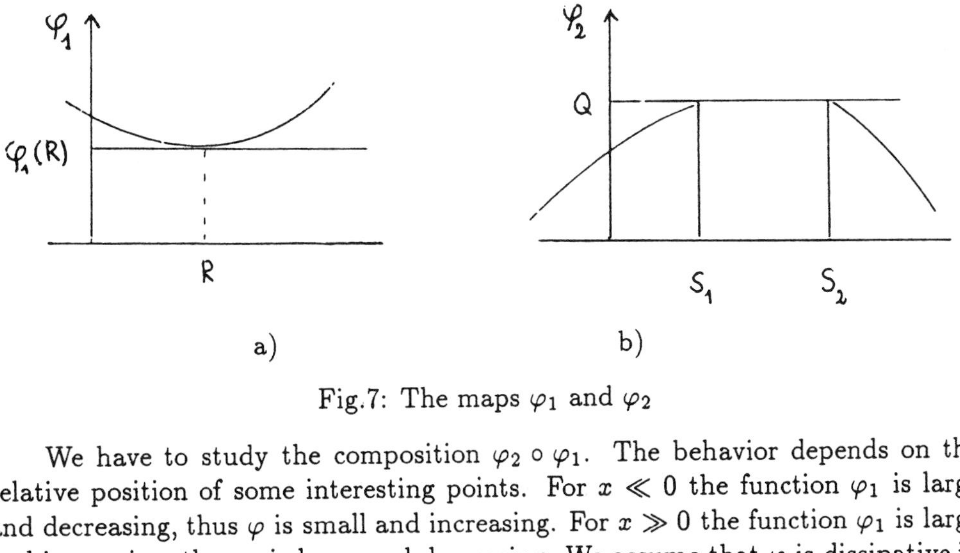

a) b)

Fig.7: The maps φ_1 and φ_2

We have to study the composition $\varphi_2 \circ \varphi_1$. The behavior depends on the relative position of some interesting points. For $x \ll 0$ the function φ_1 is large and decreasing, thus φ is small and increasing. For $x \gg 0$ the function φ_1 is large and increasing, thus φ is large and decreasing. We assume that φ is dissipative in the sense that $\varphi(x) < x$ for $x \gg 0$ and $\varphi(x) > x$ for $x \ll 0$.

The minimum $\varphi_1(R)$ of φ may be located below S_1, between S_1 and S_2, or above S_2 (we do not discuss limit cases).

Case 1: $\varphi_1(R) > S_2$. Then the function φ does not assume the value $\emptyset$, it is first increasing, then decreasing. The maximum is assumed at R. If $\varphi(R) < R$ then there is an odd number of fixed points, all located below R, these are alternatingly stable and unstable, every trajectory (of φ) converges to one of these fixed points. If $\varphi(R) > R$ then there is an even number of fixed points in $(-\infty, R]$, and a

single fixed point in $[R, +\infty)$. The latter may be unstable and give rise to period doubling or other complex behavior.
Case 2: $\varphi_1(R) \in (S_1, S_2)$. There are two values R_2 and R_2' with $R_2 < R < R_2'$ and $\varphi_1(R_2) = \varphi_1(R_2') = S_2$. The function φ is increasing in $(-\infty, R_2]$, decreasing in $[R_2', +\infty)$, and it carries (R_2, R_2') into $\emptyset$. Furthermore, $\varphi(R_2) = \varphi(R_2') = Q$. Now there are again three cases.
If $Q < R_2$ then there is an odd number of fixed points in $(-\infty, R_2]$, and no fixed point in $[R_2', +\infty)$. Trajectories cannot enter the interval (R_2, R_2'), all trajectories end up in $(-\infty, R_2]$, and approach one of the fixed points.
There is a largest fixed point in $(-\infty, R_2]$, and this fixed point is stable (in a generic situation, otherwise it is stable from above). This fixed point corresponds to a stable periodic orbit of Φ. The interval (R_2, R_2') corresponds to trajectories of Φ which approach the attractor in $\mathcal{A}_2 \cup \mathcal{B}$. Trajectories of φ cannot enter (R_2, R_2'). The trajectory of Φ starting from R_2 is the boundary between the basins of the stable periodic orbit and the attractor in $\mathcal{A}_2 \cup \mathcal{B}$. The boundary itself approaches the attractor.
If $Q \in (R_2, R_2')$ then there is an even number of fixed points in $(-\infty, R_2]$, and no fixed point in $[R_2', +\infty)$. Some trajectories (of φ) can end up in (R_2, R_2') (all trajectories will end up in this interval, if φ has no fixed points.
If $Q > R_2'$ then there is an even number of fixed points in $(-\infty, R_2]$, and exactly one fixed point in $[R_2', +\infty)$. The interval (R_2, R_2') will attract some trajectories.

Case 3: $\varphi_1(R) < S_1$. There are four values $R_2 < R_1 < R < R_1' < R_2'$ such that $\varphi_1(R_2) = \varphi_1(R_2') = S_2$, $\varphi(R_1) = \varphi(R_1') = S_1$. There are five cases depending on where the point Q is located in relation to these numbers. We shall list some essential features.
$Q < R_2$. In $(-\infty, R_2]$ an odd number of fixed points, no fixed points otherwise.
$R_2 < Q < R_1$. In $(-\infty, R_2]$ there is an even number of fixed points, no fixed points otherwise.
$R_1 < Q < R_1'$. An even number of fixed points in $(-\infty, R_2]$, an odd number of fixed points in (R_1, R_1'), no fixed point in $[R_2', +\infty)$.
$R_1' < Q < R_2'$. An even number of fixed points in $(-\infty, R_2]$ and in (R_1, R_1'), no other fixed points.
$R_2' < Q$. An even numer of fixed points in $(-\infty, R_2]$ and in (R_1, R_1'), exactly one fixed point in $[R_2', +\infty)$.

It should be underlined that the return map φ descr ibes all trajectories $\Phi(t, x)$ of the original system which meet the curve L_a. All other trajectories have trivial behavior insofar as they stay eventually either in M_1 or in M_2.

Theorem 9: *Under the assumptions stated above there are two types of limit sets: limit sets of the Poincaré-Bendixson type in $\mathcal{A}_0 \cup \mathcal{C}$ and $\mathcal{A}_1 \cup \mathcal{B}$, and periodic orbits which meet L_a and L_a.*

Numerical examples

On the manifolds M_i of the preceding section consider two vector fields f_i, in polar coordinates,

$$f_i: \quad \begin{aligned} \dot{r} &= r(1 - r/R_i), \\ \dot{\varphi} &= 1. \end{aligned}$$

where $i = 1, 2$. Hence the problem depends on four parameters a, b, R_0, R_1. In the numerical study we keep R_1, R_2 fixed and vary a, b. The numerical calculations indicate that all non-constant solutions converge to periodic solutions. The number of periodic solutions varies between one and two (and not between one and three, as one might think). The transition between the different cases shows the saddle-node bifurcation of periodic orbits which has been found in [6].

Example 5: Let $r = 2$, $s_1 = 1$, $s_2 = 1$, $V(1,1) = 2$, $V(2,1) = 1$, $M_1 = S^2 = \mathbb{R}^2 \cup \{\infty\}$, $M_2 = S^2 = \mathbb{R}^2 \cup \{\infty\}$. Define the g_{ij} with the obvious identifications in cartesian coordinates
$N_{11} = \{(x,y) : y = b\}$, $g_{11} : N_{11} \to M_2$, $x \mapsto x$,
$N_{21} = \{(x,y) : y = a\}$, $g_{21} : N_{21} \to M_1$, $x \mapsto x$.
a) $R_1 = 1$, $R_2 = 0.5$. $a = -0.4$, and b is ranging from 0.1 to 0.9. Then there are two unstable stationary points on $\mathcal{M}$, and the two circles are not periodic orbits on $\mathcal{M}$. Numerical evidence shows that there is a unique "large" periodic orbit, which is globally stable (with the exception of the stationary points). Fig.8 shows the case $b = 0.4$.

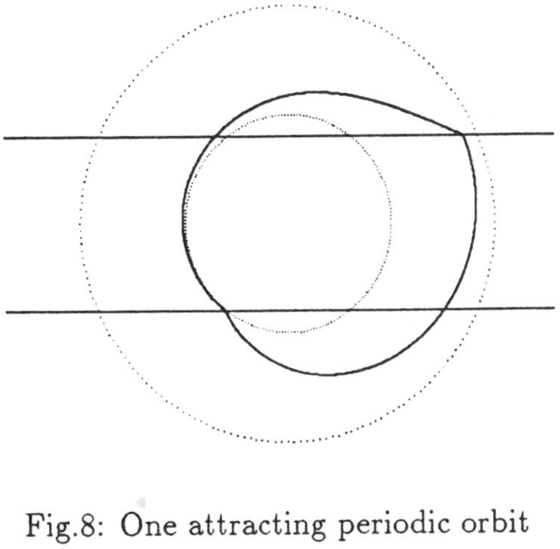

Fig.8: One attracting periodic orbit

b) $R_1 = 1$, $R_2 = 0.3$, $b = 0.2$, and a ranges from -0.4 to -0.1. There are two unstable stationary points on $\mathcal{M}$, and the smaller circle is a locally stable periodic orbit on $\mathcal{M}$ for $-0.4 \leq a < 0.3$.

For $-0.4 \leq a \leq -0.3$ the smaller circle is a limit set. But for $a = -0.3$ the smaller circle is not a periodic orbit. The trajectory starting at $x = 0$, $y = -0.3$ leaves the limit set. Hence the limit set is not positively invariant.

Numerical evidence shows that the smaller circle is globally stable (with the exception of the stationary points) for a close to -0.4. On the other hand, for $a \in [-0.3, -0.1)$ we are in the situation of Example 5a, there is a single "large" periodic orbit which changes between M_1 and M_2.

However, for a in between something interesting happens. At $a = \hat{a} \approx -0.302$ there is a saddle-node bifurcation of periodic orbits which results in a "large" stable periodic orbit and a basin boundary. For $a \in (\hat{a}, -0.3)$ the two stable orbits coexist. The boundary of the basins is defined by the trajectory of f_2 which is tangent to the line L_a. The boundary trajectory itself approaches the large periodic orbit. Hence the basin is not open. Fig.9 shows the situation for $a = -0.301$.

c) $R_1 = 1$, $R_2 = 0.2$, $a = 0.5$, $b = 0.7$. There are two unstable stationary points and one stable periodic orbit which shows a complicated behavior (Fig.10).

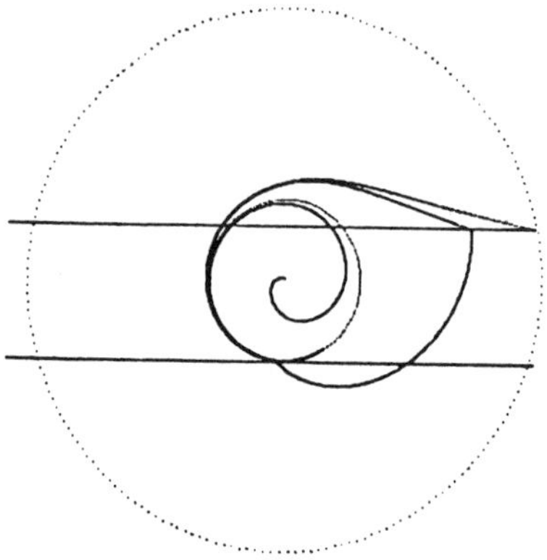

Fig.9: Coexistence of two stable periodic orbits

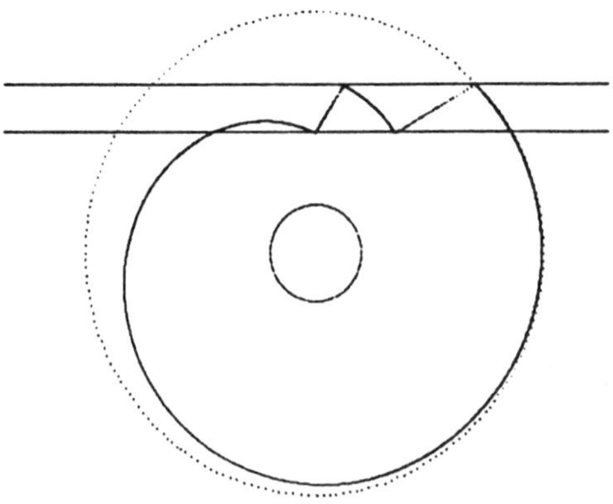

Fig.10: Complicated orbit

References

[1] Ali-Mehmeti, F., Linear and nonlinear transmission and interaction problems. Semesterbericht Funktionalanalysis Tübingen 13 (1987/88)

[2] Aubin, J.P., Cellina, A., Differential Inclusions. Springer Verlag 1984

[3] Below, J.v., Classical solvability of linear parabolic equations on networks. J. Diff. Equ. 72, 316-337 (1988)

[4] Filippov, A.F., Differential equations with discontinuous right hand sides. Kluwer Acad. Publ. 1988

[5] Guckenheimer, J., Holmes, Ph., Nonlinear Oscillations, Dynamical Systems, and Bifurcation of Vector Fields. Springer Verlag 1983

[6] Hadeler, K.P., Shonkwiler, R., An implicit differential equation related to epidemic models. Proc. Internat. Conference on Differ. Equ., Edinburg, Texas, Pitman, to appear.

[7] Jetschke, G., Mathematik der Selbstorganisation. Vieweg Verlag, Braunschweig 1989

[8] Lorenz, E.N., Deterministic nonperiodic flow. J. Atmospheric Sciences 20, 130-141 (1963)

[9] Lumer, G., Connecting local operators and evolution equations on networks. Potential Theory Copenhagen 1979. Lect. Notes Math. 787 , Springer 1980

[10] Nicaise, S., Diffusion sur les espaces ramifiés. Thèse du Doctorat, Mons 1986

[11] Palis, J., di Melo, W., Geometric Theory of Dynamical Systems. Springer 1982

[12] Rössler, O.E., Different types of chaos in two simple differential equations. Zeitschr. f. Naturforschung 31a, 1664-1670 (1978)

[13] Sparrow, C., The Lorenz equations: Bifurcations, Chaos, and Strange Attractors. Springer Verlag 1982

[14] Ważewski, T., Sur une méthode topologique de l'examen de l'allure asymptotiquoe des intégrales des équations différentielles. Proc. Int. Congr. Math., Vol III, 132-139, Amsterdam 1954

[15] Williams, R.F., Expanding attractors. Publ. Math. I.H.E.S. 43, 169-203 (1974)

[16] Williams, R.F., The structure of Lorenz attractors. In: Turbulence Seminar, p. 93-113, P.Bernard (ed.), Lect. Notes in Math. 615, Springer 1977

[17] Williams, R.F., Structure of Lorenz attractors. Publ. Math. I.H.E.S. 50, 59-72 (1980)

Regulator Problem for Linear Distributed Control Systems with Delay in Outputs

HERNÁN R. HENRÍQUEZ University of Santiago, Santiago, Chile

Abstract

In this paper we are concerned with distributed linear control systems with delays in the observed outputs and constants disturbances. We introduce a finite dimensional controller which regulates the system. That is, the controller stabilizes the system and the controlled output $z(t) \to z_r, \quad t \to \infty$, where z_r is a reference signal.

1 Introduction.

In the last years much effort has been made to generalize the control theory for finite dimensional system to distributed control systems.

In particular several papers have been devoted to the problem of track an input signal in spite of disturbances inputs. We refer to the works of Curtain (1985), Kobayashi (1983), Pohjolainen (1982) and Schumacher (1983) and also to their references. In all these works the observed output in the time t is a function of the state in the same time t.

The aim of this paper is to study the regulation problem for distributed linear control systems which presents delays in the observed output. Certainly this model represents a more realistic situation.

We will extend the ideas of Kobayashi (1983) to define a control law which stabilizes and regulates the system. Moreover in our approximation the stabilization of the system is obtained by using a finite dimensional compensator, which allows us to avoid some smoothness conditions on the semigroup associated to the system, used in Kobayashi (1983) and Pohjolainen (1982). Our generalization is a consequence of some stability properties of abstract functional differential equations.

This work was supported in part by FONDECYT, Project 89-0744 and by DICYT, Project 8933 HM.

This paper is organized as follows. In Section 2 we introduce the class of systems which will be considered. In Section 3 we will prove the existence of a regulator and in the Appendix we collected some results on functional differential equations.

Our terminology is the generally used in functional analysis. Throughout this paper we use the same symbol $\|\cdot\|$ to indicate the norm in all of the Banach spaces considered. Given two Banach spaces X and Y, we denote by $\mathcal{B}(X;Y)$ the space of all bounded linear operators from X into Y endowed with the norm of operators and we will abbreviate this notation to $\mathcal{B}(X)$ when $X = Y$.

In this paper we will consider retarded functional differential equations with values in a Banach space X and whose initial conditions are continuous functions. For a fixed $h > 0$, we will denote by $C([-h,0];X)$ for the space of continuous functions $\varphi : [-h,0] \to X$ endowed with the norm of uniform convergence. Moreover, as usual in the theory of functional equations, if $x : [-h,\infty) \to X$ is a continuous function then, for each $t \geq 0$, $x_t \in C([-h,0];X)$ denotes the function defined by $x_t(\Theta) = x(t+\Theta)$, for $-h \leq \Theta \leq 0$.

Finally, given Banach spaces $X_1, \cdots, X_n$ we will consider their product $X_1 \times \cdots \times X_n$ endowed with the norm

$$\|(x_1, x_2, \cdots, x_n)^T\| = \sum_{i=1}^{n} \|x_i\| \tag{1.1}$$

2 System Description

We consider a first order linear control system described by the following equations:

$$\dot{x}\,(t) = Ax(t) + Bu(t) + w, \qquad t \geq 0, \tag{2.1}$$

$$y(t) = \Lambda(x_t) \tag{2.2}$$

$$z(t) = Dx(t). \tag{2.3}$$

The variable $x(t) \in X$ denotes the state, and the state space X is a complex Banach space. We designate by $w \in X$ an unknown constant which represents an external disturbance. The variables $u(t) \in U$, $y(t) \in Y$ and $z(t) \in Z$ denote the control, the observed output, and the controlled output, respectively. We will restrict us to consider only systems where U, Y and Z are finite-dimensional spaces. This situation includes most of the control systems of practical interest.

In the abstract equation (2.1) we assume that $A : D(A) \subseteq X \to X$ is the infinitesimal generator of a strongly continuous semigroup $T(t)$ on X and that $B : U \to X$ is a bounded linear operator. The control function $u(\cdot)$ will be at least locally integrable and we consider the solutions of (2.1) in the mild sense (Curtain and Pritchard, 19788). Thus, the solution of (2.1) is given by

$$x(t) = T(t)x(0) + \int_0^t T(t-s)Bu(s)ds, \qquad t \geq 0. \tag{2.4}$$

It is well known that $x(\cdot)$ is a continuous function. By this reason in our model we may consider $\Lambda : C([-h,0];X) \to Y$ as a linear bounded operator, where $h > 0$ represent the delay of the observation. In the sequel we will denote by I the interval $[-h, 0]$.

In (2.3) we will assume that $D : X \to Z$ is a bounded linear operator.

Our objective in this work is to design a finite order compensator such that the resulting closed - loop system with the perturbation $w = 0$ will be uniformly stable and the controlled output $z(t)$ will be regulated to an arbitrary reference signal $z_r \in Z$. This means that $\lim_{t\to\infty} z(t) = z_r$, and this regulation property should occurs for all initial function $x_0 \in C(I;X)$ and for every constant perturbation $w \in X$. This compensator will be called a regulator of the system (2.1) - (2.3).

In order to obtain a finite order regulator, we need some additional assumptions on the system. We essentially need that there exists an appropiate decomposition of the space X using the modes of the operator A.

Specifically we consider control systems (2.1) - (2.3) which satisfy the following conditions:

Assumption I

a) There exists a constant $\alpha > 0$ such that the set $\sigma_1 = \{\lambda \in \sigma(A) : Re(\lambda) > -\alpha\}$ is finite.

b) For every $\varepsilon > 0$, there exists $\beta > \alpha$ such that the set $\sigma_2 = \{\lambda \in \sigma(A) : -\beta < Re(\lambda) \leq -\alpha\}$ is finite.

If we set $\sigma_3 = \{\lambda \in \sigma(A) : Re(\lambda) \leq -\beta\}$ then the space X can be decomposed in the form $X = X_1 \oplus X_2 \oplus X_3$ corresponding to the decomposition of $\sigma(A)$ into the spectral sets σ_1, σ_2 and σ_3 (Nagel 1986, theorem A - III, 3.3). In this decomposition each X_i is an invariant space under A. Let A_i be the restriction of A to X_i, $i = 1,2,3$. Then each A_i is the infinitesimal generator of a strongly continuous semigroup T_i on X_i and the operators A_1 and A_2 are bounded.

We will assume that the spaces X_1 and X_2 have finite dimension and that the operator A_3 satisfies the Spectrum Determined Growth Assumption (see Curtain and Pritchard (1978) and Triggiani (1975)). This means that the growth bound $\omega(T_3)$ and the spectral bound $s(A_3)$ coincide.

Clearly we may consider the space $C(I;X_i)$ as a closed subspace of $C(I:X)$, $i = 1,2,3$, and from the decomposition of X it follows that

$$C(I;X) = C(I;X_1) \oplus C(I;X_2) \oplus C(I;X_3).$$

c) If D_i denotes the restriction of D to X_i and Λ_i represents the restriction of Λ to $C(I;X_i)$, $i = 1,2,3$, we will assume that, for each $\varepsilon > 0$, it is possible to choose β so that $\|D_3\| \leq \varepsilon$ and $\|\Lambda_3\| \leq \varepsilon$.

There exists a large class of systems for which these conditions hold. This class includes the systems modelled by parabolic partial differential equation on bounded domains and the systems described by retarded functional differential equations.

The condition c) allows us to reduce a problem of infinite-dimensional nature into a finite-dimensional one, disregarding the effect of the component of state in the space X_3. This is the spillover phenomenon of Balas (1983).

We will denote by P_i, $i = 1, 2, 3$, the projections determined by the direct sum decomposition of X and we will introduce the notations $x_i(t) = P_i x(t)$, $B_i = P_i B$ and $w_i = P_i w$, for $i = 1, 2,$. Moreover, we will represent by $x_{i,t}$ the functions defined by $x_{i,t}(\Theta) = x_i(t + \Theta)$, for $-h \leq \Theta \leq 0$ and $i = 1, 2, 3.$,

With these notations we can write equations (2.1) - (2.3) under the form

$$\dot{x}_i\,(t) = A_i x_i(t) + B_i u(t) + w_i, \quad t \geq 0, \quad i = 1, 2, 3, \tag{2.5}$$

$$y(t) = \Lambda_1(x_{1,t}) + \Lambda_2(x_{2,t}) + \Lambda_3(x_{3,t}), \tag{2.6}$$

$$z(t) = D_1 x_1(t) + D_2 x_2(t) + D_3 x_3(t) \tag{2.7}$$

with initial conditions $x_{i,0} \in C(I; X_i)$.

On the other hand, as X_1 is a finite-dimensional space we can apply the Riesz Representation Theorem to conclude that there exists a normalized bounded variation operator function $\eta_1 : I \to \mathcal{B}(X_1, Y)$ such that

$$\Lambda_1(\varphi) = \int_{-h}^{0} [d\eta_1(\Theta)]\varphi(\Theta), \tag{2.8}$$

for every $\varphi \in C(I; X_1)$. Let us introduce the operator C_1 defined by

$$C_1 = \int_{-h}^{0} [d\eta_1(\Theta)]e^{A_1\Theta}, \tag{2.9}$$

and the operator $K_1 : L^1(I; X_1) \to Y$ defined by the formula

$$K_1(\psi) = \int_{-h}^{0} [d\eta_1(\Theta)]e^{A_1(\Theta)} \int_{\Theta}^{0} e^{-\tau A_1}\psi(\tau)d\tau. \tag{2.10}$$

From this expression it is clear that K_1 is a bounded linear operator. Furthermore, using the variation of constant formula to obtain the solution $x_1(t)$ of equation (2.5) for $i = 1$, and substituing C_1 and K_1 we may write

$$\Lambda_1(x_{1,t}) = C_1 x_1(t) - K_1(B_1 \otimes u_t) + q_1(t), \quad t \geq 0, \tag{2.11}$$

where $q_1(t)$ is a continuous function which depends on the initial condition $x_{1,0}$ and that is vanish for $t > h$ and $B \otimes u_t$ denotes the function defined by $(B \otimes u_t)(\Theta) = Bu_t(\Theta)$, for $\Theta \in I$, with the convention that $u(s) = 0$, for $s < 0$.

3 Construction of the Regulator

In this section we will show that there exists a regulator of finite order for the system considered in Section 2.

First we introduce a finite order observer. The idea of this type of observer goes to Watanabe and Ito (1981) and Klamka (1982), who studied it for finite dimensional systems. The equations of this system are:

$$\begin{aligned} \dot{v}_1(t) &= A_1 v_1(t) + GC_1 v_1(t) - Gy(t) - GK_1(B_1 \otimes u_t) \\ &+ G\mathcal{A}_2(v_{2,t}) + B_1 u(t) \end{aligned} \tag{3.1}$$

$$\dot{v}_2(t) = A_2 v_2(t) + B_2 u(t) \tag{3.2}$$

with variables $v_i(t) \in X_i$ and initial function $v_{2,0} \in C(I; X_2)$.

Since A_2 is a bounded operator, the solution of equation (3.2) corresponding to a continuous control function $u(\cdot)$ is a continuously differentiable function. Moreover, since $\sigma(A_2) = \sigma_2$ the operator A_2 is invertible. Thus we may define the variable $\xi(t) \in Z$ by the equation

$$\dot{\xi}(t) = Dx(t) - D_2 A_2^{-1} \dot{v}_2(t) - z_r. \tag{3.3}$$

We will consider the following feedback control law,

$$u(t) = F_0 \xi(t) + F_1 v_1(t) \tag{3.4}$$

In these formulas, $F_0 \in \mathcal{B}(Z; U)$, $F_1 \in \mathcal{B}(X_1, U)$ and $G \in \mathcal{B}(Y, X_1)$ are unknown operators that we will must determine in order to obtain the regulation property.

We observe that such as in the case studied in Kobayashi (1983) and Pohjolainen (1982), a PI control law is proposed to regulate the system. The proportional part is used to stabilize the X_1–component of the system and the purpose of the integral or servo-compensator part is to modify the system steady state in order to obtain the output regulation.

If we write $e_i(t)$ for the error signals

$$e_i(t) = v_i(t) - x_i(t) \tag{3.5}$$

for $i = 1, 2,$ then from (2.5) - (2.7) and (3.11) - (3.5) we obtain the following closed loop system

$$\dot{\xi}(t) = -R_2 F_0 \xi(t) + (D_1 - R_2 F_1) x_1(t) - R_2 F_1 e_1(t) - D_2 e_2(t) + D_3 x_3(t) - z_r, \tag{3.6}$$

$$\dot{x}_1(t) = B_1 F_0 \xi(t) + (A_1 + B_1 F_1) x_1(t) + B_1 F_1 e_1(t) + w_1, \tag{3.7}$$

$$\dot{e}_1(t) = (A_1 + GC_1) e_1(t) + G\mathcal{A}_2(e_{2,t}) - G\mathcal{A}_3(x_{3,t}) - Gq_1(t) - w_1 \tag{3.8}$$

$$\dot{x}_2(t) = B_2 F_0 \xi(t) + B_2 F_1 x_1(t) + B_2 F_1 e_1(t) + A_2 x_2(t) + w_2, \tag{3.9}$$

$$\dot{x}_3(t) = B_3F_0\xi(t) + B_3F_1x_1(t) + B_3F_1e_1(t) + A_3x_3(t) + w_3, \tag{3.10}$$

$$\dot{e}_2(t) = A_2e_2(t) - w_2, \tag{3.11}$$

where we have abbreviated $R_2 = D_2A_2^{-1}B_2$. Solving equation (3.11) by means of the variation of constants formula we obtain

$$e_2(t) = e^{A_2t}e_2(0) - A_2^{-1}(e^{A_2t} - I)w_2$$

and since A_2 is a bounded operator then the operator e^{A_2t} satisfies the estimate

$$\|e^{A_2t}\| \le M_2e^{-\alpha t}, \qquad t \ge 0, \tag{3.12}$$

for some $M_2 \ge 1$. It is follows that

$$\lim_{t\to+\infty} e_2(t) = A_2^{-1}w_2 \tag{3.13}$$

and from this last relation it is easy to see that

$$\lim_{t\to+\infty} e_{2,t} = A_2^{-1}w_2 \tag{3.14}$$

where, as it is usual, we have identified the vector $A_2^{-1}w_2$ with the corresponding constant function.

We now consider the augmented system (3.6) - (3.10) in the product space

$$X_e = Z \times X_1 \times X_1 \times X_2 \times X_3.$$

Introducing the vector

$$x_e(t) = (\xi(t), x_1(t), e_1(t), x_2(t), x_3(t))^T,$$

we can write the equations (3.6) - (3.10) in the abbreviated form

$$\dot{x}_e(t) = A_ex_e(t) + \Lambda_e(x_{e,t}) + f(t), \tag{3.15}$$

where the operators A_e and Λ_e have the following block form

$$A_e = \begin{bmatrix} -R_2F_0 & D_1 - R_2F_1 & -R_2F_1 & 0 & D_3 \\ B_1F_0 & A_1 + B_1F_1 & B_1F_1 & 0 & 0 \\ 0 & 0 & A_1 + GC_1 & 0 & 0 \\ B_2F_0 & B_2F_1 & B_2F_1 & A_2 & 0 \\ B_3F_0 & B_3F_1 & B_3F_1 & 0 & A_3 \end{bmatrix} \tag{3.16}$$

$$\Lambda_e = \begin{bmatrix} 0 & 0 & 0 & 0 & 0 \\ 0 & 0 & 0 & 0 & 0 \\ 0 & 0 & 0 & 0 & -G\Lambda_3 \\ 0 & 0 & 0 & 0 & 0 \\ 0 & 0 & 0 & 0 & 0 \end{bmatrix} \tag{3.17}$$

and the function f is defined by

$$f(t) = (-D_2 e_2(t) - z_r,\ w_1,\ G\Lambda_2(e_{2,t}) - Gq_1(t) - w_1,\ w_2,\ w_3)^T \tag{3.18}$$

It is clear from these expressions that the operator A_e is the infinitesimal generator of a strongly continuous semmigroup on X_e and $\Lambda_e : C(I; X_e) \to X_e$ is a bounded linear operator. The abstract equations of type (3.15) have been studied by several authors. We refer to the works of Travis and Webb (1974) and Nagel (1986), part B -IV, 3. Furthermore in the appendix we state some stability properties which we shall need in the sequel. Our first result shows that if the homogeneous Abstract Cauchy Problem

$$\dot{x}_e\,(t) = A_e x_e(t) + \Lambda_e(x_{e,t}), \quad t \geq 0, \tag{3.19}$$

is uniformly stable then the original control system has been regulated. Specifically we shall now prove.

Theorem 1 *Suppose that the semigroup* T_e *generated by* A_e *satisfies the estimate*

$$\|T_e(t)\| \leq M_e e^{\omega t}, \qquad t \geq 0 \tag{3.20}$$

for some $\omega < 0$ *and that*

$$\frac{M_e \|\Lambda_e\|}{-\omega} e^{-\omega h} < 1. \tag{3.21}$$

Then the system (3.1) - (3.2) with the control law defined by (3.4) is a regulator of the system (2.1) - (2.3).

Proof: Let us define the bounded linear operator L on X_e by

$$L(x_e) = \Lambda_e(\bar{x}_e)$$

where $\bar{x}_e$ denotes the constant function $\bar{x}_e(\theta) = x_e$, for every $\theta \in I$.

We may apply lemma 2 in the apprendix to conclude that for every initial condition $x_{e,o} \in C(I; X_e)$ the solution of equation (3.15) is convergent as $t \to \infty$ and

$$\lim_{t\to\infty} x_e(t) = -(A_e + L)^{-1} \lim_{t\to\infty} f(t).$$

Now, using (3.13) and (3.14) we obtain that

$$f(t) \to (-D_2 A_2^{-1} w_2 - z_r, w_1 - GA_2(A_2^{-1} w_2) - w_1, w_2, w_3)^T,$$

as $t \to \infty$. Introducing tha notations

$$\psi = \lim_{t \to \infty} x_e(t) = (\psi_0, \psi_1, \tilde{\psi}_1, \psi_2, \psi_3)^T$$

and $f_\infty = \lim_{t\to\infty} f(t)$, we may write

$$\psi = -(A_e + L)^{-1} f_\infty$$

which implies that $\psi \in D(A_e)$ and

$$(A_e + L)\psi = -f_\infty. \tag{3.22}$$

A simple computation applying the operator $[I, D_2 A_2^{-1}, 0, 0, 0,]$ to relation (3.22) yields

$$D_1\psi_1 + D_2\psi_2 + D_3\psi_3 = z_r.$$

Hence we infer that

$$\lim_{t\to\infty} z(t) = D \lim_{t\to\infty} x(t) = z_r.$$

□

Therefore, such as in the case treated by Pohjolainen (1982) and Kobayashi (1983) if the augmented system can be stabilized, then the regulation occurs, for every constant perturbation.

Now we will prove that assuming some additional condition, there exist operators F_0, F_1 and G that verify the conditions of theorem 1.

Let $X = X_1 \oplus X_0$ be the spectral decomposition of X corresponding to the spectral sets

$$\sigma_1 = \{\lambda \in \sigma(A) : Re(\lambda) > -\alpha\}$$

and

$$\sigma_0 = \{\lambda \in \sigma(A) : Re(\lambda) \le -\alpha\}.$$

Furthermore we denote by P_1 the projection with range space X_1 and Kernel X_0. It is well known that this proyection is given by

$$P_1 = \frac{1}{2\Pi i}\int_\Gamma R(\lambda, A)d\lambda \tag{3.23}$$

where Γ denotes a positively oriented closed curve containing σ_1 in its interior and σ_0 in its exterior. Let $P_0 = I - P_1$ be the projection on X_0. We will denote by A_0 and D_0 the restrictions of A and D, respectively, on X_0. Then

A_0 generates a semigroup T_0 which coincides with the restriction of the semigroup T on X_0. Consequently, T_0 satisfies the Spectrum Determined Growth Assumption, which implies that there exists a constant $M_0 \geq 1$ such that the following estimate holds

$$\|T_0(t)\| \leq M_0 \bar{e}^{\alpha t}, \qquad t \geq 0. \tag{3.24}$$

Let us introduce the notation $B_0 = P_0$ and $R_0 = D_0 A_0^{-1} B_0$. Let $S \in \mathcal{B}\,(U \times X_1,\ Z \times X_1)$ be the operator defined by

$$S = \begin{bmatrix} -R_0 & D_1 \\ B_1 & A_1 \end{bmatrix}.$$

Our next result establish sufficient conditions to regulate the system.

Theorem 2 *Suppose that the control system (2.1) - (2.3) satisfies the assumption I. If the pair (A_1, B_1) is controllable, the pair (C_1, A_1) is observable and the operator S is surjective then there exist operators F_0, F_1 and G so that the system (3.1) - (3.2) with the control law (3.4) regulates the control system (2.1) - (2.3).*

Proof: We have only to show that it is possible to choose F_0, F_1 and G so that the conditions of theorem 1 hold. Initially we observe that the controllability of (A_1, B_1) and the surjectivity of S combined with the Hautus test of controllability (see Kailath (1980)) allow us to conclude that the system

$$\left(\begin{bmatrix} 0 & D_1 \\ 0 & A_1 \end{bmatrix}, \begin{bmatrix} -R_0 \\ B_1 \end{bmatrix} \right)$$

with state space $Z \times X_1$ and control space U is controllable.

On the other hand, since the pair (C_1, A_1) is observable we can assert that for every $\mu > 0$ there exist matrices F_0, F_1 and G such that the spectrum of matrix

$$\tilde{A}_0 = \begin{bmatrix} -R_0F_0 & D_1 - R_0F_1 & -R_0F_1 \\ -B_1F_0 & A_1 - B_1F_1 & -B_1F_1 \\ 0 & 0 & A_1 + GC_1 \end{bmatrix} \tag{3.25}$$

is included in the set $\{\lambda \in \mathbf{C} : Re(\lambda) \leq -\mu\}$. Consequently, there exists a constant $N \geq 1$ such that

$$\|e^{\tilde{A}_0 t}\| \leq N e^{-\mu t}, \qquad t \geq 0, \tag{3.26}$$

Let us fix a constant $0 < \delta < \alpha$ in such a way that $-\alpha + \delta$ represents the degree of stability of the closed-loop system (3.19). And, let us take the number μ such that $\mu > \alpha$. Let ε be a constant such that

$$0 < \varepsilon < min\left\{\frac{\mu-\alpha}{4\cdot N\|A_0^{-1}\|\ \|B\|\cdot k},\ \frac{\delta}{M_e(1+\|G\|e^{\alpha h})}\right\} \tag{3.27}$$

where k and M_e are constants given by

$$k = max\{\|F_0\|,\ \|F_1\|\}, \tag{3.28}$$

and

$$M_e = max\left\{M_0, N\left(1+\frac{4\|B_0\|M_0 k}{\mu-\alpha}\right)\right\}. \tag{3.29}$$

By the Assumption I we may choose $\beta > \alpha$ so that the spectral decomposition $X_0 = X_2 \oplus X_3$ corresponding to the spectral sets

$$\sigma_2 = \{\lambda \in \sigma_0 : -\beta < Re(\lambda) \leq -\alpha\}$$

and

$$\sigma_3 = \{\lambda \in \sigma_0 : Re(\lambda) \leq -\beta\}$$

satisfies the conditions $\|D_3\| \leq \varepsilon$ and $\|A_3\| \leq \varepsilon$.

Since $P_0 = P_2 + P_3$ then

$$\begin{aligned} R_0 u &= D_0 A_0^{-1} B_0 u \\ &= D_0 A_0^{-1}(P_2+P_3)Bu \\ &= D_2 A_2^{-1} B_2 u + D_3 A_3^{-1} B_3 u, \end{aligned}$$

for every $u \in U$. Thus $R_0 = R_2 + \Delta R$ with

$$\|\Delta R\| \leq \varepsilon \cdot \|A_0^{-1}\| \cdot \|B\|.$$

If we introduce matrix

$$A_{e,11} = \begin{bmatrix} -R_2F_0 & D_1 - R_2F_1 & -R_2F_1 \\ -B_1F_0 & A_1 + B_1F_1 & -B_1F_1 \\ 0 & 0 & A_1 + GC_1 \end{bmatrix}$$

then from (3.25) we obtain that $A_{e,11} = \tilde{A}_0 + A_r$ where the perturbation A_r is given by

$$A_r = \begin{bmatrix} \Delta R\ F_0 & \Delta R\ F_1 & \Delta R\ F_1 \\ 0 & 0 & 0 \\ 0 & 0 & 0 \end{bmatrix}$$

Evaluating the norm of operators defined on product spaces using the norm defined by (1.1) we obtain that

$$\|A_r\| \leq \|\Delta R\| max\{\|F_0\|,\ \|F_1\|\}.$$

Then, applying the perturbation theory of semigroups (Pazy (1983), theorem 3.1.1) combined with the choice of ε we obtain that the semigroup generated by $A_{e,11}$ satisfies the estimate

$$\begin{aligned} \|e^{A_{e,11}t}\| &\leq Ne^{(-\mu+N\|A_r\|)t} \\ &\leq Ne^{-\nu t}, \end{aligned} \tag{3.30}$$

for $t \geq 0$, where $\nu = \mu - \varepsilon N \|A_0^{-1}\|\ \|B\| \cdot k > \alpha$.

On the other hand, if we introduce the operator

$$\tilde{A}_e = \begin{bmatrix} & A_{e,11} & & & 0 \\ \\ B_2F_0 & B_2F_1 & B_2F_1 & A_2 & 0 \\ \\ B_3F_0 & B_3F_1 & B_3F_1 & A_3 & 0 \end{bmatrix}$$

then from Schumacher (1981), proposition 4.7, it follows that the semigroup $\tilde{T}_e(t)$ generated by the operator $\tilde{A}_e$ has the following block triangular from

$$\tilde{T}_e(t) = \begin{bmatrix} e^{A_{e,11}t} & & 0 & \\ \\ \tilde{T}_{e,21}(t) & T_2(t) & & 0 \\ \\ & & 0 & T_3(t) \end{bmatrix} \tag{3.31}$$

where the operator $\tilde{T}_{e,21}(t) : Z \times X_1 \times X_1 \to X_2 \times X_3$ is defined pointwise by the formula

$$\tilde{T}_{e,21}(t) = \int_0^t \begin{bmatrix} T_2(t-s) & 0 \\ \\ 0 & T_3(t-s) \end{bmatrix} \begin{bmatrix} B_2F_0 & B_2F_1 & B_2F_1 \\ \\ B_3F_0 & B_3F_1 & B_3F_1 \end{bmatrix} e^{A_{e,11}s} ds \tag{3.32}$$

Observing that $T_i(t)$ is the restriction of $T_0(t)$ on X_i, $i = 2, 3$, we may assert that

$$\|T_i(t)\| \leq M_0 e^{-\alpha t},\ \ t \geq 0, \tag{3.33}$$

which combined with (3.30) and (3.32) allows us to obtain the estimate

$$\begin{aligned} \|\tilde{T}_{e,21}(t)\| &\leq \textstyle\int_0^t M_0 e^{-\alpha(t-s)} \left\| \begin{bmatrix} B_2F_0 & B_2F_1 & B_2F_1 \\ \\ B_3F_0 & B_3F_1 & B_3F_1 \end{bmatrix} \right\| Ne^{-\nu s} ds \\ &\leq \tfrac{2 \cdot M_0 \cdot N \|B_0\| \cdot k}{\nu - \alpha} e^{-\alpha t}. \end{aligned}$$

Now, from (3.31) a simple calculation yields

$$\|\tilde{T}_e(t)\| \leq M_e e^{-\alpha t}, \qquad t \geq 0 \tag{3.34}$$

with the constant M_e defined in (3.29).

We observe now that

$$A_e = \tilde{A}_e + \begin{bmatrix} 0 & \cdot & D_3 \\ & & \\ 0 & \cdot & 0 \end{bmatrix}$$

so that the operator A_e can be considered as a perturbation of $\tilde{A}_e$. Therefore, the same argument used previously shows that the semigroup $T_e(t)$ generated by A_e satisfies the following estimate

$$\|T_e(t)\| \leq M_e e^{(-\alpha+\varepsilon M_e)t} \tag{3.35}$$

for every $t \geq 0$, which implies that the semigroup $T_e(t)$ is uniformly stable and that the condition (3.20) holds with $\omega = -\alpha + \varepsilon M_e$.

Finally, it follows from (3.17) that $\|\Lambda_e\| \leq \|G\| \cdot \varepsilon$ and accordingly to the choice of ε we obtain

$$\frac{\varepsilon M_e \|G\| e^{(\alpha-\varepsilon M_e)h}}{\alpha - \varepsilon M_e} \leq \frac{\varepsilon M_e \|G\| e^{\alpha h}}{\alpha - \varepsilon M_e} < 1,$$

which shows that condition (3.21) holds. This completes the proof.

□

We remark that the lemma 1 in the appendix and the above argument show that the semigroup associated to equation (3.19) has growth bound $-\alpha + \varepsilon M_e + \varepsilon M_e \|G\| e^{\alpha h} < -\alpha + \delta$.

4 Appendix

In this appendix we collect some stability properties of abstract retarded diferential equations.

Let T be a strongly continuous semigroup defined on a Banach space X and let A be its infinitesimal generator. Let Λ be a bounded linear operator from $Y = C([-h,0];X)$ into X, for some $h > 0$, and let f be a continuous function with values in X. The initial value problem

$$\dot{x}(t) = Ax(t) + \Lambda(x_t) + f(t), \quad t \geq 0, \tag{4.1}$$

$$x_0 = \varphi \tag{4.2}$$

has been studied by several authors (see Nagel (1986) and Travis and Webb (1974)). It is known that there exists a unique continuous solution $x(t)$ of (4.1) - (4.2) on $[0, \infty)$ in the sense that x solves

$$x(t) = T(t)\varphi(0) + \int_0^t T(t-s)A(x_s)ds + \int_0^t T(t-s)f(s)ds \tag{4.3}$$

for $t \geq 0$ and $x_0 = \varphi$.

If $x(t, \varphi)$ denotes the solution to this equation with $f = 0$ (homogeneous equation), then the operator $U(t) : Y \to Y$, $U(t)\varphi = x_t(\cdot, \varphi)$ defines a strongly continuous semigroup on Y. We will denote by A_U the infinitesimal generator of this semigroup.

We say that the Abstract Cauchy Problem (4.1) - (4.2) is uniformly stable if the semigroup $U(t)$ is uniformly stable.

Lemma 1 *Suppose that*

$$\|T(t)\| \leq Me^{\omega t}, \qquad t \geq 0, \tag{4.4}$$

for $t \geq 0$ *and some constant* $\omega < 0$. *Then*

$$\|U(t)\| \leq Me^{\omega h}e^{\mu t},$$

where

$$\mu = \omega + M\|A\|e^{-\omega h}.$$

This lemma follows from Travis and webb (1974), proposition 3.1. Consequently, if

$$\frac{M\|A\|}{-\omega}e^{-\omega h} < 1 \tag{4.5}$$

then the system (4.1) - (4.2) is uniformly stable.

□

Now, we are interested in the solutions of non-homogeneous equation (4.3) which have a limit in infinity. It is a well known fact that if a semigroup T satisfies a stability condition (4.4) then the spectrum set $\sigma(A)$ is included in $\{\lambda \in \mathbf{C} : Re(\lambda) \leq \omega\}$. Thus, we see that the inverse operator A^{-1} exists in $\mathcal{B}(X)$. By the same reason, if (4.5) holds then the operator A_U^{-1} exists in $\mathcal{B}(Y)$. Let us introduce the operator $L : X \to X$ defined by

$$L(x) = A(\bar{x}),$$

where $\bar{x}$ denotes the constant function $\bar{x}(\Theta) = x$, for $-h \leq \Theta \leq 0$. Using the characterization of $\sigma(A_U)$ established in Nagel (1986), proposition B-IV, 3.4 we obtain that $(A + L)^{-1}$ exists in $\mathcal{B}(X)$. In these conditions we may state and prove the following result.

Lemma 2 *Suppose that (4.4) and (4.5) hold. If* $\lim_{t\to\infty} f(t) = f_0$, *then for every* $\varphi \in C([-h,0];X)$ *the solution* $x(\cdot)$ *of (4.3) satisfies*

$$\lim_{t\to\infty} x(t) = -(A+L)^{-1} f_0 \tag{4.6}$$

Proof: Let φ be an arbitrary initial function. Let us consider the space C_L of continuous functions $x : [0,\infty) \to X$ such that $x(0) = \varphi(0)$ and $x(t)$ converges to some value x^o as $t \to \infty$. It is easy to see that C_L is a closed subspace of the space $C_b([0,\infty);X)$ of bounded continuous functions endowed with the norm of uniform convergence.

Consider now the map τ defined on C_L by

$$(\tau x)(t) = T(t)\varphi(0) + \int_0^t T(t-s)\Lambda(x_s)ds + \int_0^t T(t-s)f(s)ds, \tag{4.7}$$

for $t \geq 0$.

Since $x(t) \to x^o$, as $t \to \infty$, then $x_t \to \bar{x}^o$, as $t \to \infty$, in the space of continuous functions. In view of Λ is a continuous operator, it is clear that $\Lambda(x_t) \to \Lambda(\bar{x}^o) = L(x^o)$, as $t \to \infty$. Furthermore, since $T(t)$ is a uniformly stable semigroup, from the semigroup theory (c.f. Pazy (1983), theorem 4.4.4) we obtain that

$$\lim_{t\to\infty} (\tau x)(t) = -A^{-1}L(x^o) - A^{-1}f_0. \tag{4.8}$$

Therefore, $\tau : C_L \to C_L$ is a well-defined map. Moreover, from (4.7) it is easy to obtain the estimate

$$\|\tau x - \tau y\|_\infty \leq \frac{M\|\Lambda\|}{-\omega}\|x - y\|_\infty$$

for every $x, y \in C_L$.

In view of (4.5) we may conclude that τ is a contractive map. Thus, the fixed point of τ coincides with the unique solution $x(t) = x(t,\varphi)$ of equation (4.3). Hence we infer that

$$x^o = \lim_{t\to\infty} x(t) = -A^{-1}L(x^o) - A^{-1}f_0$$

from which we obtain that $x^o \in D(A)$ and

$$(A+L)x^o = -f_0$$

and since the operator $A+L$ has inverse, this completes the proof of the assertion.

□

References

[1] Balas, M. J. (1983). *The mathematical structure of the Feedback control problem for linear distributed parameter systems with finite-dimensional controllers.* **Lect. Notes in Control and Inf. Sci., 54:** 1-34.

[2] Curtain, R. (1985). *Pole assignment for distributed systems by finite-dimensional control*, **Automatica, Vol. 21** N^o **1:** 57-67.

[3] Curtain, R. and Pritchard, A. J. (1978). *Infinite Dimensional Linear Systems Theory.* **Lect. Notes in Control and inf. Sciences, Vol. 8**, Springer-Verlag. New York.

[4] Kailath,T. (1980). *Linear Systems.* Prentice-Hall.

[5] Klamka, J. (1982). *Observer for linear feedback control of systems with distributed delays in controls and outputs*, **Systems and Control letter 1,** N^o **5:** 326-331.

[6] Kobayshi, T. (1983). *Regulator problem for infinite-dimensional systems*, **Systems and Control letter 3,** N^o **1:** 31-39.

[7] Nagel, R. (editor), (1986). *One-parameter Semigroups of Positive Operator*, **Lect. Notes in Mathematics 1184**, Springer-Verlag.

[8] Pazy, A. (1983). *Semigroup of linear Operators and Applications to Partial Differential Equations*, Springer-Verlag, New York.

[9] Pohjolainen, S. A. (1982). *Robust Multivariable PI-Controller for Infinite Dimensional Systems*, **IEEE Trans. Aut. Contr., Vol. AC-27** N^o **1:** 17-30.

[10] Schumacher, J. M. (1981). *Dynamic Feedback in Finite and Infinite-Dimensional linear Systems*, Mathematical Centre Tracts N° 143, Amsterdam.

[11] Schumacher, J. M. (1983). *Finite-dimensional regulators for a class of infinite-dimensional systems*, **Systems and Control Letters 3,** N^o **1:** 7-12.

[12] Travis, C. C. and Webb, G. F. (1974). *Existence and Stability for Partial Functional Differential Equations*, **Trans. Am.Math.Soc. Vol. 200**: 395-418.

[13] Triggiani, R. (1975). *On the stabilization problem in Banach spaces*, **J. Math. Anal. Appl. Vol. 52,** N^o **3:** 383-403.

[14] Watanabe, K. and Ito, M. (1981). *An observer for linear feedback control laws of multivariable systems with multiple delays in controls and outputs*, **Systems and Control letter, 1,** N^o **1:** 54-59.

On the L^p Theory of Systems of Linear Partial Differential Equations

MATTHIAS HIEBER University of Zürich, Zürich, Switzerland

1. Introduction

The aim of this paper is to present a brief survey of a certain operator-theoretical approach to systems of linear partial differential equations of the form

$$\frac{\partial u}{\partial t} = \sum_{|\alpha|\leq m} A_\alpha D^\alpha u, \qquad u(0) = u_0. \tag{1.1}$$

Here $u = u(t,\cdot)$ is a function defined on $\mathbb{R}^n$ which takes values in $\mathbb{C}^N$ and A_α are constant $N \times N$ matrices. In particular, we shall discuss various concepts of wellposedness and examine the question whether the differential operator, associated with the right hind side of (1.1) in the natural way, generates a C_0-semigroup, an integrated semigroup or a regularized semigroup on an L^p space.

We emphasize that the approach we shall discuss is strongly related to the theory of Fourier multipliers. We therefore restrict ourselves to constant coefficient operators and choose as underlying Banach space a function space such as $L^p(\mathbb{R}^n)^N$.

The first major contribution to the subject was perhaps the outstanding 1937-paper by Petrovskii [P]. Among many other publications we mention the fundamental results by Friedrichs [Fr], Kreiss [K] and Brenner [Br].

For the basic properties of integrated and regularized semigroups, respectively, we refer to [A],[HHN],[Lu],[N], and [deL].

2. Preliminaries

We denote by $\mathcal{S}^N$ the space of all functions from $\mathbb{R}^n$ to $\mathbb{C}^N$ having each component in $\mathcal{S}$, the space of all rapidly decreasing functions. The dual space $(\mathcal{S}^N)'$ of $\mathcal{S}^N$ is called the space of tempered distributions. We denote by $M_N(\mathbb{C})$ the ring of all $N \times N$-matrices. The Fourier transform $\mathcal{F}$ of a matrix-valued tempered distribution is defined by applying the transform elementwise. An L^∞ function $a : \mathbb{R}^n \to M_N(\mathbb{C})$ is called a Fourier-multiplier for $L^p(\mathbb{R}^n)^N (1 \leq p \leq \infty)$ if $\mathcal{F}^{-1}(a\mathcal{F}\phi) \in L^p(\mathbb{R}^n)^N$ for all $\phi \in \mathcal{S}^N$ and if

$$\|a\|_{\mathcal{M}_p^N} := \sup\{\|\mathcal{F}^{-1}(a\mathcal{F}\phi)\|_{L^p(\mathbb{R}^n)^N}; \phi \in \mathcal{S}^N, \|\phi\|_{L^p(\mathbb{R}^n)^N} \leq 1\} < \infty.$$

It is well known that the space of all L^∞ functions satisfying the above condition is a (non-commutative) Banach algebra, which we denote by $\mathcal{M}_p^N$. The norm of $\mathcal{M}_p^N$ is the above supremum, written $\|\cdot\|_{\mathcal{M}_p^N}$.

Let us also note that $\mathcal{M}_p^N = \mathcal{M}_q^N$, $(\frac{1}{p}+\frac{1}{q}=1, 1\le p\le\infty)$ and that for $1\le p\le\infty$ we have

$$\sup_\xi \|a(\xi)\| = \|a\|_{\mathcal{M}_2^N} \le \|a\|_{\mathcal{M}_p^N} \le \|a\|_{\mathcal{M}_1^N}.$$

For details and proofs, see [Hö] and [S].

With a given differential operator $\sum_{|\alpha|\le m} A_\alpha D^\alpha$ and its symbol a defined by $a(\xi) := \sum_{|\alpha|\le m} A_\alpha (i\xi)^\alpha$ we associate a linear operator $\mathcal{A}_p$ on $L^p(\mathbb{R}^n)^N$ as follows. Set

$$\begin{aligned} &D(\mathcal{A}_p) := \{f\in L^p(\mathbb{R}^n)^N; \mathcal{F}^{-1}(a\mathcal{F}f)\in L^p(\mathbb{R}^n)^N\} \qquad \text{and define} \\ &\mathcal{A}_p f := \mathcal{F}^{-1}(a\mathcal{F}f) \quad \text{for all} \quad f\in D(\mathcal{A}_p). \end{aligned} \tag{2.1}$$

Then $\mathcal{A}_p$ is closed. Here α is a multiindex and D^α is defined by $D^\alpha = (\frac{\partial}{\partial x_1})^{\alpha_1}\dots(\frac{\partial}{\partial x_n})^{\alpha_n}$ $(|\alpha| = \sum_{j=1}^n)$.

Consider now the initial value problem for a system with constant coefficients

$$\frac{\partial}{\partial t}u(x,t) = \sum_{|\alpha|\le m} A_\alpha D^\alpha u(x,t), \qquad x\in\mathbb{R}^n,\ t\ge 0, \qquad u(0,x)=u_0(x). \tag{2.2}$$

After taking Fourier transforms with respect to x we obtain a system of ordinary differential equations with the solution

$$\hat{u}(t,\xi) = e^{ta(\xi)}\widehat{u_0}(\xi),$$

at least for $u_0\in\mathcal{S}^N$. The wellposedness of (2.2) will therefore depend on multiplier properties of e^{ta}. More precisely, the following basic result is well known.

Proposition 2.1. *The operator $\mathcal{A}_p$ generates a C_0-semigroup on $L^p(\mathbb{R}^n)^N$ if and only if $e^{ta}\in\mathcal{M}_p^N$ and $\|e^{ta}\|_{\mathcal{M}_p^N}\le Me^{\omega t}$ for all $t\ge 0$ and suitable constants $M,\omega\in\mathbb{R}$. In particular, $\mathcal{A}_2$ generates a C_0-semigroup on $L^2(\mathbb{R}^n)^N$ if and only if there exist constants $M,\omega>0$ such that*

$$\sup_{\xi\in\mathbb{R}^n} \|e^{ta(\xi)}\| \le Me^{\omega t} \qquad \textit{for all} \quad t\ge 0. \tag{2.3}$$

Symbols satisfying (2.3) were completely characterized by Kreiss [K]. Indeed, he proved the following fundamental result.

Theorem 2.2. *Let $a : \mathbb{R}^n \to M_N(\mathbb{C})$ be given by $a(\xi) := \sum_{|\alpha|\leq m} A_\alpha(i\xi)^\alpha$ and let $\mathcal{A}_2$ be defined as in (2.1). Then the following assertions are equivalent.*

a) The operator $\mathcal{A}_2$ generates a C_0-semigroup on $L^2(\mathbb{R}^n)^N$.

b) There exists a constant $C_1 > 0$ such that the resolvent $R(\lambda, \mathcal{A}_2)$ of $\mathcal{A}_2$ satisfies

$$\|R(\lambda, \mathcal{A}_2)\| \leq \frac{C_1}{Re\lambda} \qquad (\lambda \in \mathbb{C}, Re\lambda > 0).$$

c) There exist constants $C_2, C_3 > 0$ and for every $\xi \in \mathbb{R}^n$ a nonsingular matrix $S(\xi)$ such that $\sup_\xi\{\|S(\xi)\|, \|S^{-1}(\xi)\|\} < C_2$ and such that $S(\xi)a(\xi)S^{-1}(\xi) = B(\xi) = (b_{ij}(\xi))$ is an upper triangular matrix with elements satisfying

$$Reb_{NN}(\xi) \leq \ldots \leq Reb_{11}(\xi) \leq C_3 \quad \text{and}$$
$$|b_{ij}(\xi)| \leq C_3(1 + |Reb_{ii}(\xi)|) \quad (1 \leq i < j \leq N).$$

In general, it is quite difficult to decide whether or not a given matrix-valued function belongs to $\mathcal{M}_p^N$. Mere existence and uniqueness results for solutions of (2.2) can be obtained more directly from the knowledge of the behavior of the real parts of the eigenvalues of $a(\xi)$. For an $N \times N$-matrix A with eigenvalues $\lambda_1 \ldots \lambda_N$, we define the spectral bound $\Lambda(A)$ of A by $\Lambda(A) := \max_j Re\lambda_j$.

Petrovskii [P] proved that e^{ta} satisfies an estimate of the form $\|e^{ta(\xi)}\| \leq Me^{\omega t}(1+|\xi|^q)$ for all $\xi \in \mathbb{R}^n$ and suitable M, ω, q if and only if

$$\Lambda(a(\xi)) \leq C_1 \log(1 + |\xi|) + C_2 \qquad (\xi \in \mathbb{R}^n)$$

for suitable constants C_1 and C_2. Later Gårding proved that one can choose $C_1 = 0$. Hence, Parseval's relation yields that the solution of (2.2) exists for $u_0 \in H^q$ and satisfies $\|u(t,\cdot)\|_{L^2} \leq Me^{\omega' t}\|u_0\|_{H^q}$ (q, ω' suitable) if and only if there exists a constant ω such that

$$\Lambda(a(\xi)) \leq w \qquad \text{for all} \quad \xi \in \mathbb{R}^n. \tag{2.4}$$

Systems satisfying condition (2.4) are called Petrovskii-correct. We shall see in section 4 that in an L^p space, Petrovskii-correct systems are exactly those systems which solutions are governed by C-regularized semigroups where C is a certain power of $R(1, \Delta)$ (Δ denotes the Laplace operator).

As a third concept, "lying in between" the two preceding ones, we introduce the theory of integrated semigroups. For the abstract theory concerning integrated semigroups we refer to [A] and [N].

In our particular situation of constant coefficient differential operators, generators of integrated semigroups can be characterized in terms of Fourier multipliers.

Proposition 2.3. *Let $1 \leq p < \infty$ and $k \in \mathbb{N}$. Then the operator $\mathcal{A}_p$ defined as in (2.1) generates a k-times integrated semigroup on $L^p(\mathbb{R}^n)^N$ if and only if $\int_0^t \frac{(t-s)^{k-1}}{(k-1)!} e^{sa} ds \in \mathcal{M}_p^N$ for $t \geq 0$ and there exist constants $M, \omega > 0$ such that $\| \int_0^t \frac{(t-s)^{k-1}}{(k-1)!} e^{sa} ds \|_{\mathcal{M}_p^N} \leq M e^{\omega t}$ for all $t \geq 0$.*

3. First order systems

Consider a first order system

$$\frac{\partial u}{\partial t} = \sum_{j=1}^{n} A_j \frac{\partial u}{\partial x_j}, \qquad u(0) = u_0, \tag{3.1}$$

where A_j are constant $N \times N$-matrices over $\mathbb{C}$. We define the operator $\mathcal{A}_p$ on $L^p(\mathbb{R}^n)^N$ with symbol $a(\xi) := i \sum_{j=1}^n A_j \xi_j$ as in (2.1). Then, as a consequence of the Kreiss matrix theorem 2.2 we obtain the following result (see [K,Thm.5].

Proposition 3.1. *The operator $\mathcal{A}_2$ generates a C_0-semigroup on $L^2(\mathbb{R}^n)^N$ if and only if for any real ξ there exists a nonsingular matrix $S(\xi)$ such that $\sup_\xi \{\max(\|S(\xi)\|, \|S^{-1}(\xi)\|)\} < \infty$ and $S(\xi)a(\xi)S^{-1}(\xi)$ is a diagonal matrix with purely imaginary eigenvalues.*

These conditions are evidently satisfied whenever the A_j are symmetric; $S(\xi)$ can be chosen to be orthogonal. For arbitrary matrices A_j we make the following hypothesis (H). For each $\xi \in \mathbb{R}^n \backslash \{0\}$, all eigenvalues of $a(\xi)$ are non-zero.
Then, the resolvent condition b) of Theorem 2.2 together with an homogeneity argument implies the following result (see [Be] and [Hi2]).

Proposition 3.2. *Suppose (H) is satisfied. Then the operator $\mathcal{A}_2$ generates a C_0-semigroup on $L^2(\mathbb{R}^n)^N$ if and only if $a(\xi)$ is diagonalizable for all $\xi \in \mathbb{R}^n$ and $\sigma(a(\xi)) \subset i\mathbb{R}$ for all $\xi \in \mathbb{R}^n$.*

In case $a(\xi)$ is not diagonalizable for some $\xi \in \mathbb{R}^n$, the solution of (3.1) can be described in terms of integrated semigroups. Indeed, denote by $mult(a)$ the maximum multiplicity of the roots of the minimum polynomial of $a(\xi)$ for $\xi \in \mathbb{R}^n \backslash \{0\}$. Observe that in general the multiplicity of the minimum polynomial of $a(\xi)$ may vary with ξ. Then, by [Hi2;Prop.3.3], the following result holds true.

Proposition 3.3. *Suppose (H) is satisfied. Let $l, k \in \mathbb{N}$ and assume that $\sigma(a(\xi)) \subset i\mathbb{R}$ for all $\xi \in \mathbb{R}^n \backslash \{0\}$. If $mult(a) = k$, then $\mathcal{A}_2$ generates an l-times integrated semigroup on $L^2(\mathbb{R}^n)^N$ whenever $l \geq \min\{k+1, N-1\}$.*

In the special case of one space variable, i.e. $a(\xi) = i\xi A$, $A \in M_N(\mathbb{C})$, the order of integration can be characterized in terms of $mult(a)$. In fact, assume that $0 \notin \sigma(A)$. Then $\mathcal{A}_2$ generates a $(k-1)$-times integrated semigroup on $L^2(\mathbb{R})^N$ $(k \geq 1)$ if and only if $\sigma(a(\xi)) \subset i\mathbb{R}$ for all $\xi \in \mathbb{R}\backslash\{0\}$ and $mult(a)$ is at most k.

The above condition (H) guarantees the existence of the resolvent $R(\lambda, \mathcal{A}_2)$ of $\mathcal{A}_2$ for λ in a right halfplane. More generally, assuming only the existence of the resolvent of $\mathcal{A}_2$ in a right halfplane, Holderrieth [Ho,Thm.3] proved the following fact.

Proposition 3.4. *Assume that the resolvent set of $\mathcal{A}_2$ contains a right halfplane. Then the operator $\mathcal{A}_2$ generates a k-times integrated semigroup on $L^2(\mathbb{R}^n)^N$ whenever $k \geq 2N$.*

Without assuming the existence of the resolvent $R(\lambda, \mathcal{A}_2)$ of $\mathcal{A}_2$ for λ in a right halfplane we obtain the following result, see [HHN,Thm.4.1].

Proposition 3.5. *The operator $\mathcal{A}_2$ generates a C-regularized semigroup on $L^2(\mathbb{R}^n)^N$ with $C = R(1, \Delta)^{N-1}$ if and only if there exists a constant ω such that $\Lambda(a(\xi)) \leq \omega$.*

An example of an operator satisfying the above condition and having empty resolvent set is given by $\mathcal{A}_2 = \begin{pmatrix} D^{(1,0)} & D^{(0,1)} \\ 0 & D^{(1,0)} \end{pmatrix}$ defined on $L^2(\mathbb{R}^2)^2$.

The corresponding situation in L^p-spaces for $p \neq 2$ is completely different. Indeed, consider the symmetric hyperbolic system

$$\frac{\partial u}{\partial t} = \sum_{j=1}^{n} A_j \frac{\partial u}{\partial x_j}, \qquad A_j^* = A_j, \tag{3.2}$$

and recall from Proposition 3.1 that the corresponding initial value problem is well posed in L^2. For $p \neq 2$, Brenner [Br] proved the following theorem.

Theorem 3.6. *The initial value problem for (3.2) is well posed in $L^p(\mathbb{R}^n)^N, (1 \leq p < \infty), p \neq 2$, if and only if*

$$A_j A_k = A_k A_j, \qquad j, k = 1 \ldots n.$$

This result implies in particular that the Cauchy problem for the wave equation, Maxwell's equations or the Dirac equation is not well posed in L^p for $p \neq 2$. A similar result for the wave equation was proved previously by Littman[Li].

In case the matrices A_j don't commute, the solution of (3.2) can be described in terms of regularized semigroups, see [Hi2,Thm.5.1].

Proposition 3.7. *Let $1 \leq p < \infty$ and let $a : \mathbb{R}^n \to M_N(\mathbb{C})$ be given by $a(\xi) := i\sum_{j=1}^n A_j\xi_j$, where A_j are constant, symmetric $N \times N$-matrices. Let $q_p \in \mathbb{R}$ such that $q_p > n|\frac{1}{2} - \frac{1}{p}|$. Then the operator $\mathcal{A}_p$ generates a C-regularized semigroup on $L^p(\mathbb{R}^n)^N$ with $C = R(1,\Delta)^{\frac{q_p}{2}}$.*

For the wave equation the above constant q_p can be improved. In fact, the operator corresponding to the wave equation generates a k-times integrated semigroup on $L^p(\mathbb{R}^n)^{n+1}$ whenever $k \geq (n-1)|\frac{1}{2} - \frac{1}{p}|$ (cf.[Hi1]).

4. Systems of arbitrary order.

Let $\mathcal{A}_p$ be the operator of order m defined as in (2.1). Then necessary and sufficient conditions for $\mathcal{A}_p$ to be the generator of a C_0-semigroup on $L^p(\mathbb{R}^n)^N$ were in principal given in Proposition 2.1. In many cases, however, it is quite difficult to verify that e^{ta} belongs to $\mathcal{M}_p^N$.

A helpful criterion for estimating the exponential of an arbitrary $N \times N$-matrix A is given in the following lemma (cf.[Fd,p.169]).

Lemma 4.1. *If A is an $N \times N$-matrix , then $\|e^A\| \leq e^{\Lambda(A)} \sum_{j=0}^{N-1}(2\|A\|)^j$.*

For parabolic systems, the situation is fairly easy to describe. In fact, suppose that there exist constants δ and C such that

$$\Lambda(a(\xi)) \leq -\delta|\xi|^m + C, \qquad \xi \in \mathbb{R}^n.$$

Then, the above lemma implies that $\xi \mapsto e^{ta(\xi)} \in \mathcal{S}^N$ and that $\mathcal{F}^{-1}(e^{ta}) \in L^1$. Therefore, the corresponding operator $\mathcal{A}_p$ generates a C_0-semigroup on $L^p(\mathbb{R}^n)^N$ for all p with $1 \leq p < \infty$.

For general systems the situation is more complicated. A typical "negative" result for semigroup generators on L^p spaces ($p \neq 2$) goes back to Hörmander [Hö].

Theorem 4.2. *Let a be a not identically vanishing real homogeneous polynomial on $\mathbb{R}^n$ of degree $m > 1$. Then e^{ia} does not belong to $\mathcal{M}_p$ for $p \neq 2$.*

Starting from this situation, we analyze now in further detail the L^p-behavior of operators which generate C_0-semigroups only on L^2. To this end, define a real number q_p by

$$q_p := mn\left|\frac{1}{2} - \frac{1}{p}\right| \quad (1 < p < \infty) \quad \text{and let} \quad q_1 > \frac{mn}{2}.$$

A proof of the following result can be found in [Hi2,Thm.5.1].

Theorem 4.3. *Let $a : \mathbb{R}^n \to M_N(\mathbb{C})$ be given by $a(\xi) := \sum_{|\alpha|\leq m} A_\alpha(i\xi)^\alpha$ and let $\mathcal{A}_p$ be defined as in (2.1). If $\mathcal{A}_2$ generates a C_0-semigroup on $L^2(\mathbb{R}^n)^N$, then $\mathcal{A}_p$ generates a C-regularized semigroup on $L^p(\mathbb{R}^n)^N$ $(1 \leq p < \infty)$ with $C = R(1,\Delta)^{\frac{q_p}{2}}$.*

If in addition the symbol of the resolvent of $\mathcal{A}_2$ satisfies certain growth conditions, then the solution of (1.1) is governed by an integrated semigroup. More precisely, assume that there exist constants $M, L, \omega > 0$ such that the inverse of $(\lambda - a(\xi))$ satisfies an inequality of the form

$$\|(\lambda - a(\xi))^{-1}\| \leq M|\xi|^{-m}$$

for $\lambda > \omega$ and for $\xi \in \mathbb{R}^n$ with $|\xi| \geq L$. Then the operator $\mathcal{A}_p$ generates a k-times integrated semigroup on $L^p(\mathbb{R}^n)^N$ $(1 < p < \infty)$ whenever $k \geq n|\frac{1}{2} - \frac{1}{p}|$ and an l-times integrated semigroup on $L^1(\mathbb{R}^n)^N$ whenever $l > \frac{n}{2}$.

Moreover, the example of the operator $\mathcal{A}_p = i\Delta$ on $L^p(\mathbb{R}^n)$ shows that the order of integration in Theorem 4.3 is optimal, see [Hi1;Thm.4.3].

We give two typical examples of classes of operators which fulfill the above assumptions.

a) Elliptic operators on $L^p(\mathbb{R}^n)$.

Let $a : \mathbb{R}^n \to \mathbb{C}$ be an elliptic polynomial satisfying $Re a(\xi) \leq C$ for all $\xi \in \mathbb{R}^n$ and suitable $C > 0$. Then $\mathcal{A}_p$ generates a k-times integrated semigroup on $L^p(\mathbb{R}^n)$ whenever $k \geq n|\frac{1}{2} - \frac{1}{p}|$ $(1 < p < \infty)$ and an l-times integrated semigroup on $L^1(\mathbb{R}^n)$ whenever $l > \frac{n}{2}$. This holds in particular for the operator $i\Delta$. We remark that recently Boyadzhiev and deLaubenfels [BdeL] obtained similar results by a different approach.

b) First order systems on $L^p(\mathbb{R}^n)^N$.

Assume that condition (H) of Section 3 is satisfied. Suppose moreover that $a(\xi)$ is diagonalizable for $\xi \in \mathbb{R}^n\backslash\{0\}$ and has only purely imaginary eigenvalues. Then $\mathcal{A}_p$ generates a k-times integrated semigroup on $L^p(\mathbb{R}^n)^N$ $(1 < p < \infty)$ for all $k \geq n|\frac{1}{2} - \frac{1}{p}|$ and an l-times integrated semigroup on $L^1(\mathbb{R}^n)^N$ for all $l > \frac{n}{2}$.

Coming back to the concept of Petrovskii-correct systems it turns out that these systems can be characterized in terms of regularized semigroups. In fact, let q_p be a real number such that

$$q_p > (N-1)m + mn|\frac{1}{2} - \frac{1}{p}|.$$

Then, according to [HHN,Thm.4.1] or [deL] the following result holds true.

Theorem 4.4. *Let* $a : \mathbb{R}^n \to M_N(\mathbb{C})$ *be given by* $a(\xi) := \sum_{|\alpha|\le m} A_\alpha(i\xi)^\alpha$ *and let* $\mathcal{A}_p$ *be defined as in (2.1). Then* $\mathcal{A}_p$ *generates a* C*-regularized semigroup on* $L^p(\mathbb{R}^n)^N$, $(1 \le p < \infty)$ *with* $C = R(1,\Delta)^{\frac{q_p}{2}}$ *if and only if there exists a constant* ω *such that* $\Lambda(a(\xi)) \le \omega$.

It follows easily in the proof of Theorem 4.4 that the estimate

$$\Lambda(a(\xi)) \le C_1 + C_2 \log(1+|\xi|)$$

is a necessary condition for $\mathcal{A}_p$ to be the generator of such a regularized semigroup. Using the Seidenberg-Tarski theorem, one can show that this estimate implies Petrovskii's condition. For details, we refer to [Fd;Sect.14,15].

References

[A] W. Arendt, *Vector valued Laplace transforms and Cauchy problems.* Israel J. Math. 59 (1987), 327-352.

[Be] R. Beals, *Hyperbolic equations and systems with multiple characteristics.* Arch. Rat. Math. Anal. 48 (1972), 123-152.

[BdeL] K. Boyadzhiev, R. deLaubenfels, *Boundary values of holomorphic semigroups.* preprint 1991.

[Br] P. Brenner, *The Cauchy problem for symmetric hyperbolic systems in* L^p. Math. Scand. 19 (1966), 27-37.

[deL] R. deLaubenfels, *C-semigroups and the Cauchy problem.* J. Func. Anal. 111 (1993), 44-61.

[Fd] A. Friedman, *Generalized functions and partial differential equations.* Prentice-Hall, New Jersey, 1963.

[Fr] K.O. Friedrichs, *Symmetric hyperbolic linear differential equations.* Comm. Pure Appl. Math. 7 (1954), 345-392.

[Hi1] M. Hieber, *Integrated semigroups and differential operators on* L^p *spaces.* Math. Ann. 291 (1991), 1-16.

[Hi2] M. Hieber, *On linear hyperbolic systems with multiple characteristics.* Differential Integral Equations. (to appear).

[HHN] M. Hieber, A. Holderrieth, F. Neubrander, *Regularized semigroups and systems of linear partial differential equations.* Ann. Scoula Norm. Pisa, Vol.XIX, (1992), 363-379 .

[Ho] A. Holderrieth, *Matrix multiplication operators generating one parameter semigroups.* Semigroup Forum 42 (1991), 155-166.

[Hö] L. Hörmander, *Estimates for translation invariant operators in L^p spaces.* Acta Math. 104 (1960), 93-140.

[K] H. O. Kreiss, *Über sachgemäße Cauchyprobleme.* Math. Scand. 13 (1963), 109-128.

[Li] W. Littman, *The wave operator and L^p-norms.* J. Math. Mech. 12 (1963),55-68.

[Lu] G. Lumer, *Solutions généralisees et semi-groupes intégrés.* C. R. Acad. Sci. Paris, 310, sér.I, (1990), 577-582.

[N] F. Neubrander, *Integrated semigroups and their applications to the abstract Cauchy problem.* Pacific J. Math. 135 (1988), 111-155.

[P] I.G. Petrovskii, *Über das Cauchysche Problem für ein System linearer partieller Differentialgleichungen.* Rec. Math. (Mat. Sb.) 2 (1937) , 814-868.

[S] E. M. Stein, *Singular Integrals and Differentiability Properties of Functions.* Princeton University Press, New Jersey, 1970.

Commuting Multiplicative Perturbations

ALBRECHT HOLDERRIETH Tübingen University, Tübingen, Germany

In [deL2] R. deLaubenfels considered commuting multiplicative perturbations of generators of bounded strongly continuous groups. He showed in examples that even if the perturbing operator is "nice", i.e., the numerical range is a compact subset of the positive real axis, the perturbed operator is in general no longer the generator of a strongly continuous semigroup. On the other hand he proved that the perturbed operator still generates an integrated semigroup. Here we use different methods to show results on the multiplicative perturbation of generators of strongly continuous semigroups, integrated semigroups and regularized semigroups. For the theory of integrated semigroups see [Ar], [Ar-K] or [Ne] and for the theory of regularized or C-semigroups we refer to [DaP], [D-P], [deL1] or [Hi-Ho-Ne].

Before stating the results we introduce some notation. If A is a linear operator on a Banach space X we denote its domain by $D(A)$, the resolvent set by $\rho(A)$, the spectrum by $\sigma(A)$ and the resolvent for some $\lambda \in \rho(A)$ by $R(\lambda, A)$. Furthermore we write "$f(t) = O(t^n)$" if the positive real valued function f "grows like t^n", i.e., if there exists $M > 0$ such that $f(t) \leq M(1 + |t|)^n$ for all $t \in \mathbb{R}$.

Note that 0-times integrated semigroups are exactly strongly continuous semigroups. Hence the following result includes the case of generators of a strongly continuous semigroups.

Theorem 1. Let $(A, D(A))$ be the generator of an n-times integrated semigroup $(S(t))_{t\geq 0}$, for some $n \in \mathbb{N} \cup \{0\}$ on a Banach space X satisfying

$$\|S(t)\| \leq Me^{\omega t}$$

for all $t \geq 0$ and suitable constants $M > 0, \omega \in \mathbb{R}$. Assume for the bounded operator $B \in \mathcal{L}(X)$ that $\sigma(B) \subset [\alpha, \beta]$ where $0 < \alpha < \beta$ and that B commutes with $(S(t))_{t\geq 0}$. If there exists an $m \in \mathbb{N} \cup \{0\}$ such that

$$\|e^{itB^{-1}}\| = O(t^m)$$

then the operator BA with domain $D(BA) = D(A)$ is the generator of an n+m+4-times integrated semigroup. If in addition $D(A)$ is dense, then BA generates even an n+m+2-times integrated semigroup.

Proof. First we assume $\beta < 1$.

Since $\sigma(B^{-1}) \subset [\beta^{-1}, \alpha^{-1}]$ there exists $M_1 > 0$ such that $\|e^{-tB^{-1}}\| \leq M_1 e^{-t}$ for all $t \geq 0$. Therefore

$$R(\lambda)x := (\lambda B^{-1})^n \int_0^\infty e^{-\lambda B^{-1} t} S(t)x dt$$

exists for all $x \in X$ and $\operatorname{Re}\lambda > \omega$ and defines a bounded operator. Since A is a closed operator commuting with B, we obtain for $x \in D(A)$
if $n = 0$:

$$\begin{aligned}
AR(\lambda) &= \int_0^\infty e^{-\lambda B^{-1}t} AS(t)x\,dt = \int_0^\infty e^{-\lambda B^{-1}t} S'(t)x\,dt \\
&= \left[e^{-\lambda B^{-1}t} S(t)x\right]_0^\infty + \int_0^\infty \lambda B^{-1} e^{-\lambda B^{-1}t} S(t)x\,dt \\
&= -x + \lambda B^{-1} \int_0^\infty e^{-\lambda B^{-1}t} S(t)x\,dt = \left(\lambda B^{-1} R(\lambda) - I\right)x,
\end{aligned}$$

if $n \geq 1$:

$$\begin{aligned}
AR(\lambda)x &= (\lambda B^{-1})^n \int_0^\infty e^{-\lambda B^{-1}t} AS(t)x\,dt \\
&= (\lambda B^{-1})^n \int_0^\infty e^{-\lambda B^{-1}t} \left(S'(t)x - \frac{1}{(n-1)!} t^{n-1} x\right) dt \\
&= (\lambda B^{-1})^n \left[e^{-\lambda B^{-1}t} S(t)x\right]_0^\infty + (\lambda B^{-1})^n \int_0^\infty \lambda B^{-1} e^{-\lambda B^{-1}t} S(t)x\,dt \\
&\quad - (\lambda B^{-1})^n \int_0^\infty \frac{1}{(n-1)!} t^{n-1} e^{-\lambda B^{-1}t} x\,dt.
\end{aligned}$$

By repeated integration by parts one obtains

$$\int_0^\infty \frac{1}{(n-1)!} t^{n-1} (\lambda B^{-1})^n e^{-\lambda B^{-1}t} x\,dt = x.$$

Therefore

$$\begin{aligned}
AR(\lambda) &= 0 + \lambda B^{-1} (\lambda B^{-1})^n \int_0^\infty e^{-\lambda B^{-1}t} S(t)x\,dt - x \\
&= \left(\lambda B^{-1} R(\lambda) - I\right)x.
\end{aligned}$$

Since B and A commute we have for all $x \in D(A)$

$$(\lambda B^{-1} - A)R(\lambda)x = R(\lambda)(\lambda B^{-1} - A)x = x.$$

If $D(A)$ is dense the closedness of $\lambda B^{-1} - A$ implies that $(\lambda B^{-1} - A)R(\lambda)x = x$ for all $x \in X$, i.e. $(\lambda B^{-1} - A)^{-1}$ exists and equals $R(\lambda)$ for all $\operatorname{Re}\lambda > \omega$.
If $D(A)$ is not dense we define $S_1(t) := \int_0^t S(s)ds$. Then A is the generator of the n+1-times integrated semigroup $(S_1(t))_{t \geq 0}$. Furthermore $S_1(t)x \in D(A)$ for $t \geq 0$ and the mapping $t \to S_1(t)x$ is continuous into $D(A)$ equipped with the graph norm for all $x \in X$ (see [Ar2]). Now define for $x \in X$ and $\operatorname{Re}\lambda > \omega$

$$R_1(\lambda)x := (\lambda B^{-1})^{n+1} \int_0^\infty e^{-\lambda B^{-1}t} S_1(t)x\,dt.$$

Since B leaves $D(A)$ invariant we conclude that $R_1(\lambda)x \in D(A)$ for all $x \in X$. By the same arguments as above we obtain that $R_1(\lambda)(\lambda B^{-1} - A)x = x$ for $x \in D(A)$ and $(\lambda B^{-1} - A)R_1(\lambda)x = x$ for all $x \in X$. Hence $(\lambda B^{-1} - A)^{-1}$ exists and equals $R_1(\lambda)$.

Since

$$(\lambda - BA) = B(\lambda B^{-1} - A)$$

we have that $\rho(BA) \supset \{z \in \mathbb{C} : \operatorname{Re} z > \omega\}$. Moreover the resolvent of BA is given by

$$R(\lambda, BA)x = B^{-1}R(\lambda)x = \lambda^n B^{-(n+1)} \int_0^\infty e^{-\lambda B^{-1}t} S(t)x dt$$

if $D(A)$ is dense. Otherwise

$$R(\lambda, BA)x = B^{-1}R_1(\lambda)x = \lambda^{n+1} B^{-(n+2)} \int_0^\infty e^{-\lambda B^{-1}t} S_1(t)x dt$$

for all $\operatorname{Re}\lambda > \omega$. If $D(A)$ is dense there exist constants $C_1, C_2 > 0$ such that

$$\begin{aligned}
\|R(\lambda, BA)\| &\leq \|B^{-(n+1)}\||\lambda^n| \int_0^\infty \|e^{-\lambda B^{-1}t}\|\|S(t)\|dt \\
&\leq \|B^{-(n+1)}\||\lambda^n| \int_0^\infty \|e^{-i\operatorname{Im}\lambda B^{-1}t}\|\|e^{-\operatorname{Re}\lambda B^{-1}t}\|\|S(t)\|dt \\
&\leq C_1|\lambda^n| \int_0^\infty (1 + |\operatorname{Im}\lambda t|)^m e^{(\omega - \operatorname{Re}\lambda)t} dt \\
&\leq C_2(1 + |\lambda|)^{n+m}
\end{aligned}$$

for $\operatorname{Re}\lambda > \omega + 1$. By a well-known theorem on integrated semigroups (see e.g. [A-K], Section 3) this implies that BA generates a n+m+2-times integrated semigroup.
If $D(A)$ is not dense we obtain similarly for a suitable $C > 0$

$$\|R(\lambda, BA)\| \leq C(1 + |\lambda|)^{n+m+1}.$$

This shows that BA generates a n+m+4-times integrated semigroup.
If $\beta \geq 1$ there exists a $\delta > 0$ such that $\delta\beta < 1$ and the statements of the theorem hold for the operator δBA, hence for BA. □

Remarks. 1. If one applies the theory of α-times integrated semigroups (see [Hi1]), the statement of Theorem 1 reads as follows: If $D(A)$ is dense then the operator BA generates an n+m+1 + ϵ-times integrated semigroup, otherwise a n+m+3 + ϵ-times integrated semigroup for every $\epsilon > 0$.
2. The assumption that $\|e^{itB^{-1}}\| = O(t^m)$ implies that B is a "generalized spectral operator" (see e.g. [C-F]). Furthermore by [C-F], Lemma 4.3.2 and Theorem 5.4.5 it follows that B is a so called "C^{m+2}-self adjoint operator", i.e., it posses a C^{m+2}-functional calculus.

Examples. 1. Let $C_b^k(J)$ be the space of all k-times continuously differentiable functions on the intervall J such that all derivatives up to the order k are bounded equipped with norm $\|g\|_k := \sup_{0 \leq i \leq k} \|g^{(i)}\|_\infty$. Furthermore let $X := C_0\left([0,\infty), C_b^k(J)\right)$ be the Banach space of all continuous functions from $[0,\infty)$ into $C_b^k(J)$ which vanish at infinity equipped with the supremum norm. Define the strongly continuous semigroup $(T(t))_{t\geq 0}$ by $T(t)f(x,y) := f(x+t,y)$ for $x \in [0,\infty)$, $y \in J$ and let A its generator. Assume that the bounded function $b : J \to \mathbb{R}_+$ is k-times continuously differentiable and bounded away from 0. If the k-th derivative is not vanishing then the operator bA defined as $bAf(x,y) := b(y)Af(x,y)$ is not the generator of a strongly continuous semigroup. Since $\|e^{itb^{-1}}\| = O(t^k)$ Theorem 1 shows that bA generates an at least $2+k$-times integrated semigroup .

2. Let $i\Delta_x$ be the Schrödinger operator on $L^p(\mathbb{R}^n)$, $1 \leq p < \infty$, with natural domain. Define the operator A on $L^p(\mathbb{R}^n \times \mathbb{R}^m)$ by $Af(x,y) := i\Delta_x f(x,y)$. Then it follows by [Hi2] that A generates a k-times integrated semigroup for every $k > n\left|\frac{1}{2} - \frac{1}{p}\right|$. Assume that the bounded function $b : \mathbb{R}^m \to \mathbb{R}_+$ is measurable and bounded away from 0. Then the operator bA defined as $bAf(x,y) := b(y)i\Delta_x f(x,y)$ generates a $k+2$-times integrated semigroup on $L^p(\mathbb{R}^{n+m})$.

Next we present the corresponding result for regularized semigroups. We will follow the notation and definitions in [Hi-Ho-Ne].

Theorem 2. Let the closed operator $(A, D(A))$ be the generator of a regularized semigroup $(S(t))_{t\geq 0}$ with regularizing operator C. Assume that $\|S(t)\| \leq Me^{\omega t}$ for all $t \geq 0$ and some $M > 0$, $\omega \in \mathbb{R}$. Assume further that the bounded operator B commutes with $(S(t))_{t\geq 0}$ and C and that $\sigma(B) \subset [\alpha, \beta]$ for $0 < \alpha < \beta$. If there exists an $m \in \mathbb{N} \cup \{0\}$ such that

$$\|e^{itB^{-1}}\| = O(t^m)$$

then the operator BA is the generator of an exponentially bounded regularized semigroup.

Proof. As in the proof of Theorem 1 we assume without loss of generality that $\beta < 1$. Then we define for $x \in X$ and $\operatorname{Re}\lambda > \omega$ the operator

$$R(\lambda)x := \int_0^\infty e^{-\lambda B^{-1}t} S(t)x dt.$$

Now one can show that $R(\lambda)$ is the C-regularized inverse of $(\lambda B^{-1} - A)C$. Again by similar considerations as above one concludes that BA generates an m+2-times integrated regularized semigroup with regularizing operator C if A is the densely defined. Otherwise one has to integrate $S(t)$ once and with similar arguments using the integrated regularized semigroup generated by A one obtains that BA generates an m+4-times integrated regularized semigroup. The assertion then follows from [Hi-Ho-Ne], Theorem 3.2. □

In case $(A, D(A))$ generates an exponentially stable strongly continuous semigroup our method also applies to unbounded perturbations.

Theorem 3. Let $(A, D(A))$ be the generator of a strongly continuous semigroup $(T(t))_{t\geq 0}$ with negative growth bound, i.e. there exist $M_1 > 0$, $\omega < 0$ such that $\|T(t)\| \leq M_1 e^{\omega t}$ for all $t \geq 0$. Assume that B is a densely defined closed operator satisfying $\sigma(B) \subset [\alpha, \infty)$ for some $0 < \alpha$ such that B^{-1} commutes with $(T(t))_{t\geq 0}$. If

$$\|e^{itB^{-1}}\| = O(t^m)$$

then the operator BA with domain $D(BA) := \{x \in D(A) : Ax \in D(B)\}$ is the generator of an m+2-times integrated semigroup.

Proof. Since B^{-1} is bounded the operator BA with domain $D(BA) := \{x \in D(A) : Ax \in D(B)\}$ is closed. From $D(BA) = A^{-1}B^{-1}X = B^{-1}A^{-1}X = B^{-1}D(A)$ it follows that $D(BA)$ is dense in X.
The assumptions on A and B imply that the integral

$$R(\lambda)x := \int_0^\infty e^{-\lambda B^{-1}t}T(t)x\,dt$$

exists for all $\mathrm{Re}\,\lambda > 0$ and all $x \in X$. As in the proof of Theorem 2 one has for $x \in D(A)$

$$\begin{aligned} AR(\lambda)x &= \int_0^\infty e^{-\lambda B^{-1}t}AT(t)x\,dt \\ &= \left[e^{-\lambda B^{-1}t}T(t)x\right]_0^\infty + \lambda B^{-1}\int_0^\infty e^{-\lambda B^{-1}t}T(t)x\,dt \\ &= \left(-I + \lambda B^{-1}R(\lambda)\right)x. \end{aligned}$$

Since $\lambda B^{-1} - A$ with domain $D(A)$ is closed we have that $\left(\lambda B^{-1} - A\right)^{-1}$ exists and coincides with $R(\lambda)$ for all $\mathrm{Re}\,\lambda > 0$. Now $\lambda - BA = B(\lambda B^{-1} - A) = BR^{-1}(\lambda)$ and we obtain that $\rho(BA) \supset \{z \in \mathbb{C} : \mathrm{Re}\,z > 0\}$. Moreover

$$R(\lambda, BA)x := B^{-1}\int_0^\infty e^{-\lambda B^{-1}t}T(t)x\,dt$$

for all $x \in X$ and $\mathrm{Re}\,\lambda > 0$. As in the proof of Theorem 1 we have for suitable constants $C_1, C_2 > 0$

$$\begin{aligned} \|R(\lambda, BA)\| &\leq \|B^{-1}\| \int_0^\infty \|e^{-i\mathrm{Im}\,\lambda B^{-1}t}\|\|e^{-\mathrm{Re}\,\lambda B^{-1}t}\|\|T(t)\|dt \\ &\leq C_1 \int_0^\infty (1 + |t\mathrm{Im}\,\lambda|)^m e^{\omega t}dt \\ &\leq C_2 (1 + |\lambda|)^m \end{aligned}$$

for all $\mathrm{Re}\,\lambda > 1$. Again by [A-K] we conclude that BA generates an m+2-times integrated semigroup. □

Examples. 1. Let $X := C_0([0,1), C_b^k((0,1])$ defined analogously as Example 1 above and A be the generator of the nilpotent semigroup $(T(t))_{t\geq 0}$ defined by

$$T(t)f(x,y) := \begin{cases} f(x+t,y) & \text{if } x+t<1 \\ 0 & \text{if } x+t\geq 1 \end{cases}$$

for $x \in [0,1)$, $y \in (0,1]$ and $t \geq 0$. If we define $b(y) = \frac{1}{y}$ for $y \in (0,1]$ then the operator A and the multiplication operator by b fulfill the assumptions of Theorem 3 with $m = k$. Therefore bA generates an $k+2$-times integrated semigroup on X.

2. Let X be the Sobolev space $W^{k,p}(\mathbb{R} \times \mathbb{R})$, $1 \leq p < \infty$, $k > 0$. For $\omega > 0$ define the strongly continuous semigroup $T(t)f(x,y) := e^{-\omega t} f(x+t,y)$ for $t \geq 0$ and $x, y \in \mathbb{R}$. Then $(T(t))_{t\geq 0}$ has negative growth bound $-\omega$. Denote the generator of $(T(t))_{t\geq 0}$ by A. Assume that the function $b : \mathbb{R} \to \mathbb{R}_+$ is k-times continuously differentiable and bounded away from 0 and take the operator $bAf(x,y) := b(y)Af(x,y)$. Then Theorem 3 shows that bA generates an k+2-times integrated semigroup.

References

[Ar1] W. Arendt, *Vector valued Laplace transforms and Cauchy problems*, Israel J. Math. **59** (1987), 327-352.

[A-K] W. Arendt, H. Kellermann, *Integrated solutions of Volterra integro-differential equations and applications*, in: Volterra Integrodifferential Equations in Banach Spaces and Applications. G. Da Prato, M. Iannelli (eds.), Pitman 1989.

[C-F] I. Colojoară and C. Foiaş, Theory of generalized spectral operators, Gordon and Breach, New York, 1968.

[DaP] G. Da Prato, *Semigruppi regolarizzabili*, Ricerche Mat. **15** (1966), 223-246.

[D-P] E.B. Davies and M.M. Pang, *The Cauchy problem and a generalization of the Hille-Yosida Theorem*, Proc. London Math. Soc. **55** (1987), 181-208.

[deL1] R. deLaubenfels, *C-semigroups and the Cauchy problem*, J. Func. Anal., to appear.

[deL2] R. deLaubenfels, *Bounded, commuting multiplicative perturbations of strongly continuous group generators*, Houston J. Math. **17**, (1991), 299–310.

[Hi1] M. Hieber, *Laplace transforms and α-times integrated semigroups*, Forum Math. **3** (1991), 595–612.

[Hi2] M. Hieber, *Integrated semigroups and differential operators on L^p spaces*, Math. Ann. **291** (1991), 1–16.

[Hi-Ho-Ne] M. Hieber, A. Holderrieth and F. Neubrander, *Regularized semigroups and systems of partial differential equations*, Lecture Notes in Functional Analysis and Partial Differential Equations, Lousiana State University 1991, 108–128.

[Ne] F. Neubrander, *Integrated semigroups and their applications to the abstract Cauchy problem*, Pacific J. Math. **135** (1988), 111-155.

Superstable C_0-Semigroups on Banach Spaces

SENZHONG HUANG* Tübingen University, Tübingen, Germany, and Nankai University, Tianjin, China

F. RÄBIGER Tübingen University, Tübingen, Germany

Abstract. From a result of Arendt and Batty (1988) and Lyubich and Phong (1988) it follows that a bounded C_0-semigroup $\mathcal{T}$ with generator A on a reflexive Banach space E is stable if the spectrum $\sigma(A)$ of A intersected with the imaginary axis is countable. In this paper we will characterize the countability of $\sigma(A) \cap i\mathbb{R}$ by means of a compactness property of the ultrapowers of $\mathcal{T}$. The analogous result for discrete semigroups has recently been obtained by Nagel and Räbiger (1992).

1. Stable C_0-Semigroups

Let $\mathcal{T} = (T(t))_{t\geq 0}$ be a strongly continuous semigroup (shortly C_0-semigroup) of bounded linear operators on a Banach space E and denote with A its generator and with $D(A)$ the domain of A (cf. Nagel (1986), A-I, for these notions).

This paper is part of a research project supported by the DFG.

* Financially supported by DAAD.

DEFINITION 1.1 The semigroup $\mathcal{T} = (T(t))_{t\geq 0}$ is called *stable* if $\{T(t)x : t \geq 0\}$ is relatively compact in E for all $x \in E$.

Instead of 'stable' some authors use the notion 'almost periodic' (see e.g. Glicksberg and deLeeuw (1961), Lyubich (1988), Lyubich and Phong (1990)).

Independently, Arendt and Batty (1988) and Lyubich and Phong (1988) obtained the following sufficient condition for stability which we only state for reflexive Banach spaces.

THEOREM 1.2 Let $\mathcal{T} = (T(t))_{t\geq 0}$ be a bounded C_0-semigroup with generator A on a reflexive Banach space E. If $\sigma(A) \cap i\mathbb{R}$ is countable and contains no eigenvalues, then $\lim_{t\to\infty} \|T(t)x\| = 0$ for all $x \in E$.

If $\mathcal{T} = (T(t))_{t\geq 0}$ is a bounded C_0-semigroup on a reflexive Banach space E one can apply the Jacobs-Glicksberg-deLeeuw decomposion theorem to $\mathcal{T}$ and obtains the following decomposition of E (cf. Krengel (1985), 2.4.4, 2.4.5, Glicksberg and deLeeuw (1961), Thm.4.11):

$$E = E_0 \oplus E_r, \quad \text{where}$$
$$E_0 = \{x \in E : 0 \text{ is in the weak closure of } \{T(t)x : t \geq 0\}\} \quad \text{and}$$
$$E_r = \overline{\text{lin}}\,\{x \in E : \text{there is } \gamma : \mathbb{R}_+ \to \mathbb{C} \text{ such that } |\gamma(t)| = 1 \text{ and } T(t)x = \gamma(t)x \text{ for all } t \geq 0\}.$$

Moreover the restriction of the weak operator closure of $\mathcal{T}$ to E_r is a group.

Actually one can show that

$$\begin{aligned} E_r &= \overline{\text{lin}}\,\{x \in E : T(t)x = e^{\lambda t}x \text{ for } \lambda \in i\mathbb{R} \text{ and for all } t \geq 0\} \\ &= \overline{\text{lin}}\,\{x \in E : Ax = \lambda x \text{ for } \lambda \in i\mathbb{R}\} \end{aligned}$$

(cf. Krengel (1985), p.107, Remark, Nagel (1986), A-III, 6.4). Obviously, E_0 and E_r are $T(t)$-invariant for all $t \geq 0$ and the restriction of $\mathcal{T}$ to E_r has relatively compact orbits. If A_0 denotes the generator of $\mathcal{T}|_{E_0}$, then $\sigma(A_0) \cap i\mathbb{R} \subseteq \sigma(A) \cap i\mathbb{R}$ and A_0 has no eigenvalues on the imaginary axis (cf. Nagel (1986), A-III, 4.3, 6.4). Hence, from Theorem 1.2 we obtain the following result (cf. Lyubich and Phong (1990), Thm.3).

THEOREM 1.3 Let $\mathcal{T} = (T(t))_{t\geq 0}$ be a bounded C_0-semigroup with generator A on a reflexive Banach space E. If

$$(ABLP) \qquad \sigma(A) \cap i\mathbb{R} \quad \text{is countable,}$$

then $\mathcal{T}$ is stable.

The translation semigroup $T(t)f(x) := f(x+t)$, $t \geq 0$, on $L^2(\mathbb{R}_+)$ shows that condition $(ABLP)$ is not necessary for stability. In fact this semigroup has the generator $A = \frac{d}{dx}$

and the spectrum of A is the left half plane $\{\lambda \in \mathbb{C}: \operatorname{Re}\lambda \le 0\}$ (cf. Nagel (1986), A-III, 2.4). Hence, $\sigma(A) \cap i\mathbb{R} = i\mathbb{R}$ is uncountable but $\lim_{t\to\infty} \|T(t)f\| = 0$ for every $f \in L^2(\mathbb{R}_+)$.

The aim of this paper is to characterize condition $(ABLP)$ (for bounded C_0-semigroups on superreflexive Banach spaces) through a compactness property of the ultrapower extensions of the given semigroup. An analogous result for discrete semigroups $\{T^n : n \in \mathbb{N}\}$ generated by a bounded linear operator T has recently been obtained by Nagel and Räbiger (1992).

2. Ultrapowers and Superstable C_0-Semigroups

We briefly recall some facts on ultrapowers of Banach spaces and operators. For details and a variety of applications we refer to Heinrich (1980), Schaefer (1974), Sims (1982) and Stern (1978).

Let E be a Banach space and $\mathcal{U}$ a (free) ultrafilter on $\mathbb{N}$. Consider, $l^\infty(E)$, the space of all bounded E-valued sequences equipped with the norm $\|(x_n)\| := \sup_n \|x_n\|$ and let $c_\mathcal{U}(E)$ be the closed subspace of all bounded sequences converging to zero along $\mathcal{U}$. The quotient space $E_\mathcal{U} := l^\infty(E)/c_\mathcal{U}(E)$ is the *$\mathcal{U}$-power* of E and the norm on $E_\mathcal{U}$ is given by the formula

$$\|(x_n) + c_\mathcal{U}(E)\| = \lim_{\mathcal{U}} \|x_n\|. \tag{1}$$

For elements $(x_n)+c_\mathcal{U}(E) \in E_\mathcal{U}$ we also use the notation $[x_n]_\mathcal{U}$. The space E is isometrically embedded into $E_\mathcal{U}$ via $x \mapsto [x]_\mathcal{U}$. Moreover, if F is a closed subspace of E then $F_\mathcal{U}$ can be considered as a closed subspace of $E_\mathcal{U}$ by virtue of the isometric embedding $(x_n)+c_\mathcal{U}(F) \mapsto (x_n) + c_\mathcal{U}(E)$.

Every bounded linear operator $T \in \mathcal{L}(E)$ has a norm preserving extension $T_\mathcal{U} \in \mathcal{L}(E_\mathcal{U})$ given by $T_\mathcal{U}[x_n]_\mathcal{U} := [Tx_n]_\mathcal{U}$. The mapping $T \mapsto T_\mathcal{U}$ from $\mathcal{L}(E)$ into $\mathcal{L}(E_\mathcal{U})$ is an algebra homomorphism. In particular this implies equality of the spectra

$$\sigma(T_\mathcal{U}) = \sigma(T)$$

and yields the formula

$$R(\lambda, T_\mathcal{U}) = R(\lambda, T)_\mathcal{U} \quad \text{for} \quad \lambda \in \rho(T) := \mathbb{C} \setminus \sigma(T),$$

where $R(\lambda, T) := (\lambda - T)^{-1}$ denotes the resolvent of T at λ.

Let now $\mathcal{T} = (T(t))_{t\ge 0}$ be a C_0-semigroup of operators in $\mathcal{L}(E)$. Then $(T(t)_\mathcal{U})_{t\ge0}$ is a semigroup in $\mathcal{L}(E_\mathcal{U})$ which, in general, is not strongly continuous. Actually one can show that $(T(t)_\mathcal{U})_{t\ge0}$ is a C_0-semigroup if and only if $\mathcal{T}$ is uniformly continuous (cf. Huang (1992)). In order to obtain a C_0-semigroup one has to restrict $(T(t)_\mathcal{U})_{t\ge0}$ to an appropriate (and for our purpose sufficiently large) subspace of $E_{\mathcal{U}}$. We briefly recall this construction which has its origin in a paper of Derndinger (1980) (cf. also Nagel (1986), A-I, 3.6, A-III, 4.5).

Consider the closed subspace

$$l_{\mathcal{T}}^{\infty}(E) := \{(x_n) \in l^{\infty}(E) : \lim_{t\to 0} \sup_n \|T(t)x_n - x_n\| = 0\}$$

of $l^{\infty}(E)$. The quotient space

$$E_{\mathcal{U}}^{\mathcal{T}} := l_{\mathcal{T}}^{\infty}/c_{\mathcal{U}}(E) \cap l_{\mathcal{T}}^{\infty}(E)$$

is called the *$\mathcal{U}$-power of E with respect to $\mathcal{T}$* and the norm on $E_{\mathcal{U}}^{\mathcal{T}}$ is given by

$$\|(x_n) + c_{\mathcal{U}}(E) \cap l_{\mathcal{T}}^{\infty}(E)\| = \lim_{\mathcal{U}} \|x_n\|.$$

By (1), $E_{\mathcal{U}}^{\mathcal{T}}$ can be considered as a closed subspace of $E_{\mathcal{U}}$, namely

$$E_{\mathcal{U}}^{\mathcal{T}} = \{[x_n]_{\mathcal{U}} : (x_n) \in l_{\mathcal{T}}^{\infty}(E)\}.$$

Obviously, this subspace is $T(t)_{\mathcal{U}}$-invariant for all $t \geq 0$ and $(T(t)_{\mathcal{U}}|_{E_{\mathcal{U}}^{\mathcal{T}}})_{t\geq 0}$ is strongly continuous. The C_0-semigroup

$$\mathcal{T}_{\mathcal{U}} = (T_{\mathcal{U}}(t))_{t\geq 0} := (T(t)_{\mathcal{U}}|_{E_{\mathcal{U}}^{\mathcal{T}}})_{t\geq 0}$$

in $\mathcal{L}(E_{\mathcal{U}}^{\mathcal{T}})$ is called the *ultrapower of $\mathcal{T}$*.

If $(A, D(A))$ and $(A_{\mathcal{U}}, D(A_{\mathcal{U}}))$ denote the generator of $\mathcal{T}$ and $\mathcal{T}_{\mathcal{U}}$, respectively, then the following holds (cf. Nagel (1986), A-I, 3.6):

$$(2) \qquad D(A_{\mathcal{U}}) = \{[x_n]_{\mathcal{U}} : (x_n), (Ax_n) \in l_{\mathcal{T}}^{\infty}(E)\} \quad \text{and} \quad A_{\mathcal{U}}[x_n]_{\mathcal{U}} = [Ax_n]_{\mathcal{U}}.$$

Moreover, the spectra of A and $A_{\mathcal{U}}$ are related as follows (cf. Nagel (1986), A-III, 4.5):

$$(3) \qquad \sigma(A_{\mathcal{U}}) = \sigma(A) \quad \text{and} \quad P\sigma(A_{\mathcal{U}}) = A\sigma(A_{\mathcal{U}}) = A\sigma(A),$$

where $P\sigma(.)$ and $A\sigma(.)$ denote the point spectrum and the approximate point spectrum, respectively.

In Nagel and Räbiger (1992) an operator $T \in \mathcal{L}(E)$ is called *superstable* if, for every ultrafilter $\mathcal{U}$ on $\mathbb{N}$ and every $x \in E_{\mathcal{U}}$, the orbit $\{T_{\mathcal{U}}^n x : n \in \mathbb{N}\}$ is relatively compact in $E_{\mathcal{U}}$. The following definition is the analogue for C_0-semigroups.

DEFINITION 2.1 A C_0-semigroup $\mathcal{T} = (T(t))_{t\geq 0}$ on a Banach space E is called *superstable* if $\mathcal{T}_{\mathcal{U}} = (T_{\mathcal{U}}(t))_{t\geq 0}$ is stable for every ultrafilter $\mathcal{U}$ on $\mathbb{N}$.

Recall that a Banach space E is called *superreflexive* if one (and hence all) ultrapower(s) of E are reflexive (cf. Heinrich (1980)). Among such spaces are the L^p-spaces for $p \in (1, \infty)$. The coincidence of the spectra $\sigma(A)$ and $\sigma(A_{\mathcal{U}})$ implies that A satisfies $(ABLP)$ if and only if $A_{\mathcal{U}}$ satisfies $(ABLP)$ for any ultrafilter $\mathcal{U}$. Thus our next result is an immediate consequence of Theorem 1.3.

THEOREM 2.2 Let $\mathcal{T} = (T(t))_{t \geq 0}$ be a bounded C_0-semigroup with generator A on a superreflexive Banach space E. If

$$(ABLP) \qquad \sigma(A) \cap i\mathbb{R} \quad \text{is countable,}$$

then $\mathcal{T}$ is superstable.

In our main result we will show (see Section 4) that superstability already characterizes $(ABLP)$. For this we need an iteration property of semigroup-ultrapowers which is the subject of the next section.

3. Iteration of Semigroup-Ultrapowers

It is shown by Stern (1978) that the formation of ultrapowers is closed with respect to finitely many iterations. Precisely, the following holds (cf. Nagel and Räbiger (1992), Lemma 3.1, Stern (1978), Prop.2.1).

PROPOSITION 3.1 Let E be a Banach space and let $\mathcal{U}$, $\mathcal{V}$ be ultrafilters on $\mathbb{N}$. Then there exists an ultrafilter $\mathcal{W}$ on $\mathbb{N}$ and an isometric bijection $\Phi : (E_{\mathcal{U}})_{\mathcal{V}} \mapsto E_{\mathcal{W}}$ such that for every $T \in \mathcal{L}(E)$ the following diagram commutes:

$$\begin{array}{ccc} E_{\mathcal{W}} & \xrightarrow{\ T_{\mathcal{W}}\ } & E_{\mathcal{W}} \\ \Phi \uparrow & & \uparrow \Phi \\ (E_{\mathcal{U}})_{\mathcal{V}} & \xrightarrow[(T_{\mathcal{U}})_{\mathcal{V}}]{} & (E_{\mathcal{U}})_{\mathcal{V}} \end{array}$$

The corresponding result for semigroup-ultrapowers needs some preparation. Let A be the generator of a C_0-semigroup $\mathcal{T} = (T(t))_{t \geq 0}$ on a Banach space E and let $\mathcal{U}$ be an ultrafilter on $\mathbb{N}$. If $\lambda \in \rho(A)$, then $R(\lambda, A)_{\mathcal{U}} E_{\mathcal{U}}^{\mathcal{T}} \subseteq E_{\mathcal{U}}^{\mathcal{T}}$ since $R(\lambda, A)$ commutes with every $T(t), t \geq 0$. Then from (2) it follows that

$$(4) \qquad R(\lambda, A_{\mathcal{U}}) = R(\lambda, A)_{\mathcal{U}}|_{E_{\mathcal{U}}^{\mathcal{T}}} \quad \text{for} \quad \lambda \in \rho(A).$$

Hence,

$$R(\lambda, A)_{\mathcal{U}} E_{\mathcal{U}}^{\mathcal{T}} = D(A_{\mathcal{U}}) \subseteq E_{\mathcal{U}}^{\mathcal{T}}.$$

The next lemma shows that we have even more.

LEMMA 3.2 Let $\mathcal{T} = (T(t))_{t\geq 0}$ be a C_0-semigroup with generator A on a Banach space E and let $\mathcal{U}$ be an ultrafilter on $\mathbb{N}$. If $\lambda \in \rho(A)$, then $R(\lambda, A)_{\mathcal{U}} E_{\mathcal{U}}$ is a dense subspace of $E_{\mathcal{U}}^{\mathcal{T}}$.

Proof. Let $\lambda \in \rho(A)$ and $(x_n) \in l^\infty(E)$. Set $y_n := R(\lambda, A)x_n, n \in \mathbb{N}$. From

$$\begin{aligned}\|T(t)y_n - y_n\| &\leq \|T(t)y_n - e^{\lambda t}y_n\| + |e^{\lambda t} - 1|\,\|y_n\| \\ &= \|\int_0^t e^{\lambda(t-s)}T(s)(\lambda - A)y_n\, ds\| + |e^{\lambda t} - 1|\,\|y_n\|\end{aligned}$$

(cf. Nagel (1986), A-I, 3.1) one obtains that

$$\lim_{t\to 0} \sup_n \|T(t)y_n - y_n\| = 0.$$

Hence, $[y_n]_{\mathcal{U}} = R(\lambda, A)_{\mathcal{U}}[x_n]_{\mathcal{U}} \in E_{\mathcal{U}}^{\mathcal{T}}$, *i.e.*, $R(\lambda, A)_{\mathcal{U}} E_{\mathcal{U}} \subseteq E_{\mathcal{U}}^{\mathcal{T}}$. Since $D(A_{\mathcal{U}}) \subseteq R(\lambda, A)_{\mathcal{U}} E_{\mathcal{U}}$ the space $R(\lambda, A)_{\mathcal{U}} E_{\mathcal{U}}$ is dense in $E_{\mathcal{U}}^{\mathcal{T}}$.

We are now in the position to establish the iteration property of semigroup-ultrapowers.

PROPOSITION 3.3 Let $\mathcal{T} = (T(t))_{t\geq 0}$ be a C_0-semigroup with generator A on a Banach space E and let $\mathcal{U}, \mathcal{V}$ be ultrafilters on $\mathbb{N}$. Then there exists an ultrafilter $\mathcal{W}$ on $\mathbb{N}$ and an isometric bijection $\Psi : (E_{\mathcal{U}}^{\mathcal{T}})_{\mathcal{V}}^{\mathcal{T}_{\mathcal{U}}} \to E_{\mathcal{W}}^{\mathcal{T}}$ such that the following diagram commutes:

$$\begin{array}{ccc} E_{\mathcal{W}}^{\mathcal{T}} & \xrightarrow{\ \mathcal{T}_{\mathcal{W}}\ } & E_{\mathcal{W}}^{\mathcal{T}} \\ \Psi \uparrow & & \uparrow \Psi \\ (E_{\mathcal{U}}^{\mathcal{T}})_{\mathcal{V}}^{\mathcal{T}_{\mathcal{U}}} & \xrightarrow[(\mathcal{T}_{\mathcal{U}})_{\mathcal{V}}]{} & (E_{\mathcal{U}}^{\mathcal{T}})_{\mathcal{V}}^{\mathcal{T}_{\mathcal{U}}} \end{array}$$

Proof. Let $\lambda \in \rho(A)$ and let $\mathcal{W}$ and $\Phi : (E_{\mathcal{U}})_{\mathcal{V}} \to E_{\mathcal{W}}$ be as in Propositon 3.1. Then

$$\begin{aligned} D((A_{\mathcal{U}})_{\mathcal{V}}) &= R(\lambda, A_{\mathcal{U}})_{\mathcal{V}}(E_{\mathcal{U}}^{\mathcal{T}})_{\mathcal{V}}^{\mathcal{T}_{\mathcal{U}}} \quad \text{(by (4))} \\ &\subseteq R(\lambda, A_{\mathcal{U}})_{\mathcal{V}}(E_{\mathcal{U}}^{\mathcal{T}})_{\mathcal{V}} \\ &= (R(\lambda, A)_{\mathcal{U}})_{\mathcal{V}}(E_{\mathcal{U}}^{\mathcal{T}})_{\mathcal{V}} \quad \text{(by (4))} \\ &\subseteq (R(\lambda, A)_{\mathcal{U}})_{\mathcal{V}}(E_{\mathcal{U}})_{\mathcal{V}} \\ &= (R(\lambda, A)_{\mathcal{U}})_{\mathcal{V}}\Phi^{-1}E_{\mathcal{W}} \\ &= \Phi^{-1}R(\lambda, A)_{\mathcal{W}}E_{\mathcal{W}} \quad \text{(by Prop.3.3)}. \end{aligned}$$

Since $(E_{\mathcal{U}}^{\mathcal{T}})_{\mathcal{V}}^{\mathcal{T}_{\mathcal{U}}}$ and $E_{\mathcal{W}}^{\mathcal{T}}$ are closed in $(E_{\mathcal{U}})_{\mathcal{V}}$ and $E_{\mathcal{W}}$, respectively, it follows from Lemma 3.2 that $\Phi((E_{\mathcal{U}}^{\mathcal{T}})_{\mathcal{V}}^{\mathcal{T}_{\mathcal{U}}}) = E_{\mathcal{W}}^{\mathcal{T}}$. If Ψ is the restriction of Φ to $(E_{\mathcal{U}}^{\mathcal{T}})_{\mathcal{V}}^{\mathcal{T}_{\mathcal{U}}}$, then the assertion follows from Proposition 3.1.

Since closed subspaces and ultrapowers of superreflexive Banach spaces are superreflexive again (cf. Heinrich (1980)) we have the following consequence.

COROLLARY 3.4 Let $\mathcal{T} = (T(t))_{t\geq 0}$ be a C_0-semigroup on a superreflexive Banach space E. Then $\mathcal{T}$ is superstable if and only if $\mathcal{T}_{\mathcal{U}}$ is superstable for every ultrafilter $\mathcal{U}$ on $\mathbb{N}$.

4. Spectral Characterization of Superstable C_0-Semigroups

We now present the main result of this paper. The discrete analogue has been shown by Nagel and Räbiger (1992).

THEOREM 4.1 Let $\mathcal{T} = (T(t))_{t\geq 0}$ be a bounded C_0-semigroup with generator A on a superreflexive Banach space E. Then

$$(ABLP) \qquad\qquad \Lambda := \sigma(A) \cap i\mathbb{R} \quad \text{is countable}$$

if and only if $\mathcal{T}$ is superstable.

Proof. The 'only if' part is shown in Theorem 2.2. Conversely, let $\mathcal{T}$ be superstable and suppose that Λ is uncountable. We will show that this is impossible. By considering an equivalent norm and passing to an ultrapower of $\mathcal{T}$ we may assume that $\mathcal{T}$ is a semigroup of contractions and $\Lambda \subseteq P\sigma(A)$ (see (3) and Corollary 3.4). We now apply the Jacobs-Glicksberg-deLeeuw decomposition theorem (see Section 1) and note that the restriction of the weak operator closure of $\mathcal{T}$ to $E_r = \overline{\text{lin}}\{x \in E : Ax = \lambda x \text{ for some } \lambda \in \Lambda\}$ is a group of contractions. Hence, $T(t)|_{E_r}$ is an isometry for every $t \geq 0$. Choose now $n_0 \in \mathbb{N}$ such that $\Omega := \Lambda \cap i[-n_0, n_0]$ is uncountable. Then Ω contains an uncountable $\mathbb{Q}$-linearly independent subset G. Let $(\lambda_n)_{n\in\mathbb{N}}$ be dense in G. The $\mathbb{Q}$-linear independence of $(\lambda_n)_{n\in\mathbb{N}}$ implies that

$$\overline{\{(e^{2\pi m\lambda_1}, \ldots, e^{2\pi m\lambda_n}) : m \in \mathbb{N}\}} = \Gamma^n \quad \text{for each} \quad n \in \mathbb{N},$$

where $\Gamma := \{z \in \mathbb{C} : |z| = 1\}$ denotes the unit circle (cf. Hewitt and Ross (1979), p.408). Choose now $x_n \in D(A)$, $\|x_n\| = 1$, such that $Ax_n = \lambda_n x_n$. Then $T(2\pi)x_n = e^{2\pi\lambda_n}x_n$ (cf. Nagel (1986), A-III, 6.4). Hence, $F := \overline{\text{lin}}\{x_n : n \in \mathbb{N}\} \subseteq E_r$ is $T(2\pi)$-invariant. If $S := T(2\pi)|_F$, then S is an isometry and $\sigma(S) \cap \Gamma \supseteq \overline{\{e^{2\pi\lambda_n} : n \in \mathbb{N}\}}$ is uncountable. By Nagel and Räbiger (1992), Thm.3.7, the operator S is not superstable, *i.e.*, there exists an ultrafilter $\mathcal{U}$ on $\mathbb{N}$ and $x \in F_{\mathcal{U}} \subseteq E_{\mathcal{U}}$ such that $\{T(2\pi n)_{\mathcal{U}}x : n \in \mathbb{N}\}$ is not relatively compact in $E_{\mathcal{U}}$ (observe that $F_{\mathcal{U}}$ is a closed subspace of $E_{\mathcal{U}}$). To derive the desired contradiction it suffices to show that $F_{\mathcal{U}} \subseteq E^{\mathcal{T}}_{\mathcal{U}}$. Let $(y_n) \in l^\infty(F)$. Then there exist $z_n = \sum_{k=1}^{m_n} \alpha_{kn} x_k \in D(A)$, $n \in \mathbb{N}$, such that $\lim_n \|y_n - z_n\| = 0$. As in the proof of Proposition 3.3 in Nagel and Räbiger (1992) one can show that, for every $n \in \mathbb{N}$ and every choice of scalars (α_k) and (β_k) with $|\alpha_k| \leq |\beta_k|$, $1 \leq k \leq n$, one has

$$\|\sum_{k=1}^{n} \alpha_k x_k\| \leq \|\sum_{k=1}^{n} \beta_k x_k\|.$$

Thus

$$\|Az_n\| = \|\sum_{k=1}^{m_n} \alpha_{kn} \lambda_k x_k\| \leq \|\sum_{k=1}^{m_n} n_0 \alpha_{kn} x_k\| = n_0 \|z_n\|$$

for every $n \in \mathbb{N}$, *i.e.*, (Az_n) is bounded. From the identities

$$T(t)z_n - z_n = \int_0^t T(s)Az_n \, ds$$

(cf. Nagel (1986), A-I, 1.6) it follows that

$$\lim_{t \to 0} \sup_n \|T(t)z_n - z_n\| = 0.$$

Hence $[y_n]_{\mathcal{U}} = [z_n]_{\mathcal{U}} \in E^{\mathcal{T}}_{\mathcal{U}}$ and thus the theorem is proved.

We conclude with an example which shows that in Theorem 4.1 superreflexivity cannot be omitted. We quote the example in Nagel and Räbiger (1992), Example 2.6 b).

EXAMPLE. Let $E := l^p(l^\infty_n)$, $p \in (1, \infty)$, be the l^p-sum of the spaces l^∞_n. The space E is reflexive, but not superreflexive. Choose an increasing real sequence $0 < \alpha_n \uparrow 1$. The multiplication operators $A_n : (\xi_1, ..., \xi_n) \mapsto (\xi_1 \ln \alpha_1, ..., \xi_n \ln \alpha_n)$ on l^∞_n induce a bounded operator $A \in \mathcal{L}(E)$ given by $A(x_n) := (A_n x_n)$. Then $\sigma(A) = \{\ln \alpha_n : n \in \mathbb{N}\} \cup \{0\}$. Consider $T(t) := \exp(tA)$ for $t \geq 0$. Then $\mathcal{T} = (T(t))_{t \geq 0}$ is a stable C_0-semigroup on E with bounded generator A. Hence, for any ultrafilter $\mathcal{U}$ on $\mathbb{N}$ one has $E^{\mathcal{T}}_{\mathcal{U}} = E_{\mathcal{U}}$. But, the $\mathcal{U}$-power $\mathcal{T}_{\mathcal{U}}$ of $\mathcal{T}$ is not stable since there exists $z \in E_{\mathcal{U}}$ such that $\{T_{\mathcal{U}}(1)^n z : n \in \mathbb{N}\} = \{T(1)^n_{\mathcal{U}} z : n \in \mathbb{N}\}$ is not relatively compact (cf. Nagel and Räbiger (1992), Example 2.6 b)).

ACKNOWLEDGEMENT. The authors wish to thank Rainer Nagel for many helpful discussions.

REFERENCES

1. Arendt, W. and Batty, C.J.K. (1988). Tauberian theorems and stability of one parameter semigroups, *Trans. Amer. Math. Soc.* **306**, 837–852.
2. Derndinger, R. (1980). Über das Spetrum positiver Generatoren, *Math. Z.* **172**, 281–293.
3. Glicksberg, I. and deLeeuw, K. (1961). Applications of almost periodic compactifications, *Acta Math.* **105**, 63–97.
4. Heinrich, S. (1980). Ultraproducts in Banach space theory, *J. Reine Angew. Math.* **313**, 72–104.
5. Hewitt, E. and Ross, K.A. (1979). Abstract Harmonic Analysis I. 2nd Edition, Springer-Verlag, Berlin-Heidelberg-New York.
6. Huang, S. (1992). In preparation.
7. Krengel, U. (1985). Ergodic Theorems, De Gruyter, Berlin-New York.
8. Lyubich, Yu.I. (1988). Introduction to the Theory of Banach Representations of Groups, Birkhäuser-Verlag, Basel-Boston-Berlin.
9. Lyubich, Yu.I. and Phong, V.Q. (1988). Asymptotic stability of linear differential equations in Banach spaces, *Studia Math.* **88**, 37–42.
10. Lyubich, Yu.I. and Phong, V.Q. (1990). A spectral criterion for the almost periodicity of one parameter semigroups, *J. Soviet Math.* **48**, 644–647.
11. Nagel, R. (ed.) (1986). One-parameter Semigroups of Positive Operators, Lecture Notes in Mathematics **1184**, Springer-Verlag, Berlin-Heidelberg-New York-Tokyo.
12. Nagel, R. and Räbiger, F. (1992). Superstable operators on Banach spaces, submitted.
13. Schaefer, H.H. (1974). Banach Lattices and Positive Operators, Grundl. Math. Wiss. **215**, Springer-Verlag, Berlin-Heidelberg-New York.
14. Sims, B. (1982). Ultra-Techniques in Banach Spaces Theory, Queen's papers in pure and applied mathematics **60**, Queen's University, Kingston, Ontario.
15. Stern, J. (1978). Ultrapowers and local properties of Banach spaces, *Trans. Amer. Math. Soc.* **240**, 231–252.

Gradient Blow-Ups and Global Solvability After the Blow-Up Time for Nonlinear Parabolic Equations

N. KUTEV Bulgarian Academy of Sciences, Sofia, Bulgaria, and University of Heidelberg, Heidelberg, Germany

The aim of this paper is to investigate the global solvability of the Dirichlet and Neumann problem for one-dimensional nonlinear parabolic equations

$$u_t - u_{xx} = f(t, x, u, u_x) \quad \text{in } Q = (a, b) \times (0, \infty) \tag{1}$$

$$u(a,t) = A(t)\ , \quad u(b,t) = B(t)\ , \quad u(x,0) = \psi(x) \tag{2}$$

$$u_x(a,t) - \sigma u(a,t) = A_1(t)\ , \quad u_x(b,t) + \tau u(b,t) = B_1(t)\ , \quad u(x,0) = \psi(x) \tag{3}$$

where $\sigma, \tau = \text{const} > 0$, $\psi \in C^{2,\alpha}[a, b]$, $A_1, B_1 \in C^{1,\alpha}[0, \infty)$, $A, B \in C^{2,\alpha}[0, \infty)$, $0 < \alpha < 1$. We propose precise conditions which guarantee global solvability for $t \geq 0$ of the Dirichlet and Neumann problem (1)–(3). However, without some additional assumptions as quadratic growth of the right-hand side $f(t, x, u, p)$ with respect to p the solutions of the Dirichlet problem do not satisfy the classical boundary value problem. Under super quadratic growth of f in a neighbourhood of a or b the problem (1), (2) has a unique, global viscosity solution (in the sense of Crandall and P.-L. Lions [4]). In fact the solution is a classical one, i.e. $C^2(Q) \cap C(\bar{Q})$ smooth, and up to the first boundary gradient blow up time satisfies the classical Dirichlet problem i.e. it attains continuously its boundary data. After the blow up time the solution detaches from the data at one of the points a or b, when the difference of the data $|B(t) - A(t)|$ is sufficiently large and satisfies the Dirichlet problem in viscosity sense.

In the multidimensional case the quadratic growth of the lower order term has to be replaced with a more general condition depending on the geometry of the domain and the growth of the coefficients with respect to the gradient at infinity. This precise condition defined by means of the so-called Bernstein-Serrin function (see [16]) guarantees classical solvability of the Dirichlet problem or gradient blow up for a finite time and global solvability but only in viscosity sense of the boundary value problem after the blow up time.

This new phenomenon i.e. detachment of the solution from the data sheds light on the well-known nonexistence results for the Dirichlet problem for nonlinear elliptic and parabolic equations. In fact in the beginning of this century, Bernstein [1] first pointed out that the Dirichlet problem for uniformly elliptic equations in the plane has no classical solution when the lower order term has more than quadratic growth with respect to the gradient. For the minimal surface equation, which is a nonuniformly elliptic one and was considered also by Bernstein, the nonexistence result is even simpler. If the domain Ω is not convex in the two-dimensional case, or is not mean convex in the multidimensional case, then there exist infinitely smooth data with arbitrarily small oscillation for which the Dirichlet problem has no $C^2(\Omega) \cap C(\bar{\Omega})$ solution. The results of Bernstein were extended in the multidimensional case by Serrin [22] for elliptic equations and by Lieberman [17] for parabolic equations. As we will show this nonexistence phenomena which lead to detachment of the solution from the boundary data do not appear for the Neumann problem (1), (3). However, in the multidimensional case, except the Monge-Ampère type equations for which the same result was proved in [19] the problem is still open.

In order to formulate the main results we need the following conditions for the right-hand side of f. Suppose that

$$\begin{gathered} f(t,x,u,p)\,\mathrm{sign}(u) \le g(x)/h(p) \\ \int_a^b g(x)\,dx < \int_{-\infty}^{+\infty} h(p)\,dp \end{gathered} \tag{4}$$

for $(t,x) \in \bar{Q}$, $|u| \ge m$, $p \in \mathbb{R}$, for some positive constant m and for some positive functions $g \in L^1[a,b]$, $h \in L^1_{\mathrm{loc}}(\mathbb{R})$.

This condition guarantees a priori estimates for the amplitude of the solutions of the Dirichlet and Neumann problem when the source term in the right-hand side can be eliminated and is the best possible one. Indeed, as Theorem 3 shows, if equalities hold in (4) then the amplitude of the solutions of (1)-(3) blows up at infinity. The case of strictly opposite inequalities in (4) is quite complicated since some new effects appear and is considered for mean curvature equations in [14].

Let us recall that the amplitude blow ups of the solutions of weakly nonlinear parabolic equations depending only on the unknown function were investigated by many mathematicians in the last twenty years (see for example the references in [3]). However, only in the papers of Chipot and Weissler [3], Kawohl and Peletier [15], Galaktionov and Posashkov [7] (see also [6] and [21]) were considered equations with gradient nonlinearities (with subquadratic growth or with gradient diffusion) and was found out the precise dependence between the source and dumping term in order to have global solvability or amplitude blow ups of the solutions. Note that in all of the above papers the amplitude of the solution blows up for a finite time because of the source term, while in the simplest case (4) the blow up is at infinity. In spite of that, condition (4) is necessary and sufficient for amplitude blow ups in the class of nonlinear equations for which the dependence of u can be eliminated.

As for the gradient estimates of the solutions, only sufficient conditions are well known and it is not clear whether they are necessary too. In fact, the behaviour of some complexes of derivatives with respect to the gradient at infinity is important for the validity of the gradient a priori estimates. More precisely, when the Bernstein function or the trace of the matrix of the coefficients before the second order derivatives (which for equation (1) are p^2 and p, resp.) dominate over the derivatives $f_u + f_x/p$, $pf_p - f$, or some of the above complexes have a constant sign and dominate over the other invariants then the gradient estimates for the solutions of (1)-(3) hold. That is why we suppose that at least one of the following conditions is fulfilled:

$$|f(t,x,u,p)| \le Cp^2 \; ; \tag{5)$_{\rm i}$ }$$

$$f_u + f_x/p \le 0 \; ; \qquad (5)_{\rm ii}$$

$$f_u + f_x/p \le \frac{|pf_p - f|}{2(C+1)\operatorname{osc}(u)} \quad \text{and } pf_p - f \text{ has a constant sign } ; \qquad (5)_{\rm iii}$$

$$pf_p - f \text{ or } f - pf_p \le Cp^2 \,, \quad f_u + f_x/p \le p^2/2[C+1-C\exp\left(-(C+1)\operatorname{osc}(u)\right)] \quad (5)_{\rm iv}$$

for $(t,x) \in \bar{Q}$, $\inf u \le u \le \sup u$, $|p| \ge M$, for some positive constants M, C.

As for the Dirichlet problem (1), (2) we will need some additional assumptions in order to clarify when the solution satisfies the boundary value problem in classical or in viscosity sense (see definition 7.4 in [5]). Let $\varphi, \Phi \in C[0,\infty)$ be positive nondecreasing functions which satisfy the growth conditions

$$\int_1^\infty \frac{dt}{t\varphi(t)} = \infty \,, \quad \int_1^\infty \frac{dt}{t\Phi(t)} < \infty \qquad (6)$$

(for example $\varphi(t) = \ln(2+t)$, $\Phi(t) = \ln^{1+\varepsilon}(2+t)$ for some $\varepsilon > 0$). Moreover, suppose that in sufficiently small neighbourhoods N_a of a or N_b of b the function f satisfies either the conditions

$$|f(t,x,u,p)| \le \varphi^{1+\beta}(p^2)\,(x-a)^\beta (1+|p|)^{2+\beta} \quad \text{for } x \in N_a \qquad (7)_{\rm a}$$

$$|f(t,x,u,p)| \le \varphi^{1+\beta}(p^2)\,(b-x)^\beta (1+|p|)^{2+\beta} \quad \text{for } x \in N_b \qquad (7)_{\rm b}$$

for $t \ge 0$, $\inf u \le u \le \sup u$, $p \in \mathbb{R}$, for some nonnegative constant β and for some function φ satisfying (6), or

$$|f(t,x,u,p)| \ge \Phi^{1+\beta}(p^2)\,(x-a)^\beta (1+|p|)^{2+\beta} \quad \text{for } x \in N_a \qquad (8)_{\rm a}$$

$$|f(t,x,u,p)| \ge \Phi^{1+\beta}(p^2)\,(b-x)^\beta (1+|p|)^{2+\beta} \quad \text{for } x \in N_b \qquad (8)_{\rm b}$$

for $t \ge 0$, $u \in \mathbb{R}$, $p \in \mathbb{R}$, for some nonnegative constant β and for some function Φ satisfying (6).

THEOREM 1: Let $f \in C^1$ satisfy (4) and at least one of the conditions (5). Suppose $\psi \in C^{2,\alpha}[a,b]$, $A_1, B_1 \in C^{1,\alpha}[0,\infty)$, $0 < \alpha < 1$, and the compatibility conditions $\psi'(a) - \sigma\psi(a) = A_1(0)$, $\psi'(b) + \tau\psi(b) = B_1(0)$ hold.

Then the Neumann problem (1), (3) has a unique bounded global $C^{2+\alpha,1+\alpha/2}([a,b] \times [0,\infty))$ solution.

THEOREM 2: Let $f \in C^1$, $f_u \le 0$ satisfy (4) and at least one of the conditions (5) hold. Suppose $\psi \in C^{2,\alpha}[a,b]$, $A, B \in C^{2,\alpha}[0,\infty)$, $0 < \alpha < 1$, and the compatibility conditions $\psi(a) = A(0)$, $\psi(b) = B(0)$, $A'(0) - \psi''(a) - f(0,a,\psi(a),\psi'(a)) = 0$, $B'(0) - \psi''(b) - f(0,b,\psi(b),\psi'(b)) = 0$ hold.

Then: (i) the Dirichlet problem (1), (2) has a unique $C^{2+\alpha,1+\alpha/2}([a,b] \times [0,\infty))$ solution which satisfies the boundary value problem in viscosity sense (see definition 7.4 in [5]);

(ii) if additionally $(7)_{\rm a}$ and $(7)_{\rm b}$ hold then the solution of (1), (2) satisfies the classical Dirichlet problem;

(iii) if either $(8)_a$ or $(8)_b$ holds then for all constant data A, B whose difference is less than some critical difference $\Delta^*(t)$, i.e. $|B-A| \leq \Delta^*(t)$, and for all initial data $\psi(x)$ with sufficiently small C^1 norm the solution $u(x,t)$ satisfies the classical Dirichlet problem;

(iv) if $(8)_a$ or $(8)_b$ holds, but $|B(t)-A(t)| > \Delta^*(t)$, then the gradient of the solution $u(x,t)$ blows up on the boundary for a finite time T_*. Up to the blow up time T_* the solution of (1), (2) satisfies the boundary value problem in classical sense and after the blow up time in viscosity sense.

The idea of the proof of Theorem 1 and conclusions (ii), (iii) in Theorem 2 is to find out a priori estimates for the amplitude as well as for the gradient of the solutions in $\bar{Q}$ and then to apply the method of continuity. Using (4), the amplitude a priori estimates for the solutions of both the Dirichlet and Neumann problem follow by means of sub and supersolutions v which are linear functions i.e. $v_\pm = \pm Rx + D$. The constants R, D are chosen suitably, under control, in order to have an equality in the first condition of (4) and $v_\pm$ to be supporting lines of the solution.

As for the global gradient estimates $|u_x|$, the proof is also one and the same for both of the boundary value problems and is based on the original idea of Bernstein to differentiate equation (1) with respect to x and multiplying with u_x to obtain a new equation for the new variable $v = u_x^2$. The rest of the proof follows from (5) by means of the weak maximum principle or by means of the generalized maximum principle (see [20]), i.e. applying the weak maximum principle for the function $u_x^2/w(u)$, with suitable choice of the positive auxiliary function $w(u)$.

In case of the classical Dirichlet problem i.e. conclusions (ii), (iii) in Theorem 2, we need additionally boundary gradient estimates at the end points a and b. For this purpose, from $(7)_a$ and $(7)_b$ local barriers can be found out of the form $v(x-a) + A(t) + (B(t)-A(t))(x-a)/(b-a)$ in a neighbourhood of a, and analogously in a neighbourhood of b, where $v \in C^2[0,\delta]$, $\delta > 0$, has the following properties: $v(0) = 0$, $v' \geq M$, $v(\delta) \geq K$, $v''(s) = -\varphi^{1+\beta}(v'^2)\, s^\beta (1+|v'|)^{2+\beta}$, for some sufficiently large positive constants M, K and φ satisfying (6).

The proof of the existence part in conclusion (i) in Theorem 2 is based on Perron's method for viscosity solutions introduced by Ishii [8]. By means of (4), global viscosity sub and supersolutions are found out and the rest of the proof follows from the procedure in [5] (see also [2], [8]–[10]). The higher regularity is a consequence of the gradient estimates obtained from conditions (5).

The uniqueness result is derived from the comparison principle for viscosity solutions as in [2], [5], [11]–[13].

As for the blow up result and the detachment of the solutions in (iv), it is obtained by means of local viscosity sub or supersolutions near the end points a or b of the form $v(x-a) + \psi(x) + C(t-T)$, where C is a suitably chosen positive constant, $v(0) = 0$, $v'(0) = \infty$ and $v''(s) = -\Phi^{1+\beta}(v'^2)\, s^\beta (1 + |v'|)^{2+\beta}$, for some constant $\beta \geq 0$ and Φ satisfying (6).

The following theorem shows that condition (4) is necessary for amplitude estimates for the solutions of the Dirichlet and Neumann problem.

THEOREM 3: Suppose $f \in C^1$ satisfies at least one of the conditions (5), the following conditions

$$f(t,x,u,p) = g(x)/h(p) \ , \quad \int_a^b g(x)\,dx = \int_{-\infty}^{+\infty} h(p)\,dp \tag{9}$$

as well as the compatibility conditions in Theorem 1 or 2.

Then: (i) the Neumann problem (1), (3) has a unique global $C^{2+\alpha,1+\alpha/2}([a,b]\times[0,\infty))$ solution whose amplitude blows up at infinity;

(ii) the Dirichlet problem (1), (2) has a unique global $C^{2+\alpha,1+\alpha/2}([a,b]\times[0,\infty))$ solution which satisfies the boundary value problem in viscosity sense, the amplitude of which blows up at infinity;

(iii) if additionally $(7)_a$ and $(7)_b$ or $(8)_a$ and $(8)_b$ hold then the conclusions (i)–(iii) in Theorem 2 are true and the amplitude of the solution of the Dirichlet problem (in classical or viscosity sense) (1), (2) blows up at infinity.

Sketch of the proof: Let $\varepsilon > 0$ be an arbitrary positive constant. Then the auxiliary boundary value problem

$$w_t - w_{xx} = (g(x) - \varepsilon)/h(w_x) \quad \text{in } Q \tag{10}$$

with Dirichlet or Neumann conditions (2), (3) fulfils all assumptions of Theorem 1 and 2 and consequently (10) has a unique global viscosity solution $w_\varepsilon(t,x)$ which satisfies (2) in classical sense and (3) in classical or viscosity sense. Now the proof of Theorem 3 follows from the fact that $w_\varepsilon(t,x) \to \infty$ when $t \to \infty, \ \varepsilon \to 0$.

ACKNOWLEDGEMENT: This work was accomplished during the author's research period as A. v. Humboldt fellow at the University of Heidelberg.

REFERENCES

1. S. Bernstein, Conditions nécessaires et suffisantes pour la possibilité du problème de Dirichlet, *C. R. Acad. Sci. Paris, 150:* 514–515 (1910).
2. Yun-Gang Chen, Yoshikazu Giga and Shun'ichi Goto, Uniqueness and existence of viscosity solutions of generalized mean curvature flow equations, *J. Diff. Geometry, 33:* 749–786 (1991).
3. M. Chipot and F. B. Weissler, Some blow up results for a nonlinear parabolic equation with a gradient term, *SIAM J. Math. Anal., 20:* 886–907 (1989).
4. M. Crandall and P.-L. Lions, Viscosity solutions of Hamilton-Jacobi equations, *Trans. Amer. Math. Soc., 277:* 1–42 (1983).
5. M. Crandall, H. Ishii and P.-L. Lions, User's guide to viscosity solutions of second-order partial differential equations, *Bull. A. M. S., 27:* 1–67 (1992).
6. M. Fila, Remarks on blow up for a nonlinear parabolic equation with a gradient term, *Proc. Amer. Math. Soc., 111:* 795–801 (1991).
7. V. Galaktionov and S. Posashkov, Single point blow-up for n-dimensional quasilinear equations with gradient diffusion and source, *Indiana Univ. Math. J., 40:* 1041–1060 (1991).

8. H. Ishii, Perron's method for Hamilton-Jacobi equations, *Duke Math. J., 55:* 369–384 (1987).
9. H. Ishii, On uniqueness and existence of viscosity solutions of fully nonlinear second-order elliptic p.d.e's, *Comm. Pure Appl. Math., 42:* 15–45 (1989).
10. H. Ishii and P.-L. Lions, Viscosity solutions of fully nonlinear second-order elliptic partial differential equations, *J. Diff. Eq., 83:* 26–78 (1990).
11. R. Jensen, The maximum principle for viscosity solutions of fully nonlinear second-order partial differential equations, *Arch. Rat. Mech. Anal., 101:* 1–27 (1988).
12. R. Jensen, Uniqueness criteria for viscosity solutions of fully nonlinear elliptic partial differential equations, *Indiana Univ. Math. J., 38:* 629–667 (1989).
13. R. Jensen, P.-L. Lions and P. E. Souganidis, A uniqueness result for viscosity solutions of second-order fully nonlinear partial differential equations, *Proc. Amer. Math. Soc., 102:* 975–978 (1988).
14. B. Kawohl and N. Kutev, Global behaviour of solutions to a parabolic mean curvature equation, preprint.
15. B. Kawohl and L. A. Peletier, Observations on blow up and dead cores for nonlinear parabolic equations, *Math. Z., 202:* 207–217 (1989).
16. N. Kutev, On the solvability of Dirichlet's problem for a class of nonlinear elliptic and parabolic equations, *Proc. Conf. Equadiff*, 1991, Barcelona, to appear.
17. G. Lieberman, The first initial-boundary value problem for quasilinear second order parabolic equations, *Ann. Scuola Norm. Sup. Pisa, 13:* 347–387 (1986).
18. P.-L. Lions, Generalized solutions of Hamilton-Jacobi equations, *Pitman*, London, 1982.
19. P.-L. Lions, N. Trudinger and J. Urbas, The Neumann problem for equations of Monge-Ampère type, *Comm. Pure Appl. Math., 39:* 539–563 (1986).
20. M. Protter and H. Weinberger, Maximum principles in differential equations, *Prentice-Hall*, New Jersey.
21. P. Quittner, Blow-up for semilinear parabolic equations with a gradient term, *Math. Methods in the Appl. Sci., 14* (1991).
22. J. Serrin, The problem of Dirichlet for quasilinear elliptic differential equations with many independent variables, *Philos. Trans. Roy. Soc. London, Ser. A 264:* 413–496 (1969).

Derivative Nonlinear Schrödinger Equations

HORST LANGE University of Cologne, Cologne, Germany

1 INTRODUCTION

In this note we consider some mathematical problems in connection with nonlinear Schrödinger equations which contain derivatives in the nonlinear term. The equations we look for are (in the simplest 1-1-dimensional case) of type

$$iu_t = -u_{xx} + f(\gamma + i\partial_x,\ |u|^2) \cdot u. \tag{1.1}$$

Nonlinear Schrödinger equations of this form can be found in many physical situations (see e.g. Bass and Nasonov (1990), Kosevich, Ivanov and Kovalev (1990), Makhankov and Fedyanin (1984)). The aim of this paper is to treat some specific cases where mathematical problems arise in the presence of derivatives in the nonlinearity.

In Section 2 we treat the case when we have

$$f = i\alpha(\partial_x|u|^2) \cdot u + i\beta|u|^2\partial_x u \tag{1.2}$$

This case is usually called derivative nonlinear Schrödinger equation; it is important in plasma physics, fluid dynamics, theory of magnetic crystals, electromagnetic spin waves (see e.g. Pathria and Morris (1990), Kaup and Newell (1978), Mio, Ogino, Minami and Takeda (1976), Mjølhus (1976)).

In Section 3 we state some results on the case where f involves a second order derivative term in the nonlinearity namely

$$f = \beta(\partial_{xx}|u|^2) \cdot u\ ; \tag{1.3}$$

this nonlinearity has applications in plasma physics, superfluidity, Heisenberg ferromagnets, fluid mechanics (see e.g. Kosevich, Ivanov and Kovalev (1990), Makhankov, and Fedyanin (1984), Litvak and Sergeev (1978), Nakamura (1977), Kurihara (1981), Quispel and Capel (1982), Takeno and Homma (1981)).

Finally in Section 4 we treat an intermediate case where

$$f = \beta\phi(u; x, t) \cdot u\ ; \tag{1.4}$$

here ϕ is a functional coming from a potential equation with nonlinear boundary conditions which are coupled to (1.1). In some sense the nonlinearity corresponds to the case (1.2) with a real constant (instead of the pure imaginary in (1.2)); this is the reason for some difficulties with uniqueness of solutions to (1.1) with nonlinearity (1.4);

(1.4) arises in fluid mechanics (theory of deep water waves; see e.g. Lo and Mei (1985), Dysthe (1979)).

We use the following notation:
$H = L^2(0,1)$, $S = [0,T]$, $Q = [0,1] \times S$, $(u,v) = \int_0^1 u(x)\overline{v(x)}dx$ (or different domain of integration); furthermore $V^m = \{u \in H^m(0,1) \mid u^{(j)}(0) = u^{(j)}(1), \quad j = 0,\ldots,m-1\}$ and $V := V^1$; V^{-m} should be the antidual space of V^m; we use the continuous and dense embeddings

$$V \subset H \subset V^{-1}$$

which are obtained by identifying H with its dual. Also $L : V \to V^{-1}$ denotes the continous continuation of $-\partial_x^2 : V^2 \to V^{-1}$ defined by

$$(Lu, v) = \int_0^1 u_x \overline{v}_x dx \quad (u, v \in V);$$

$C_w(S, V^m)$ is the space of weakly-continuous functions from S to V^m $(m \in \mathbb{Z})$; by $h_k(x) = \exp(2\pi ikx)$ $(k \in \mathbb{Z})$ we denote the elements of the complete orthogonal system $\{h_k \mid k \in \mathbb{Z}\}$ in H . If $u_0 \in H$ $u_{0,n}$ is the n-th Fourier-sum of u_0, i.e. $u_{0,n} = \sum_{k=-n}^{n}(u_0, h_k)h_k$. $\|u\|$ always means the L^2-norm.

2 FIRST ORDER DERIVATIVE NONLINEAR SCHRÖDINGER EQUATION

In this section we consider a nonlinear Schrödinger equation of type

$$iu_t = -u_{xx} + i\alpha(\partial_x|u|^2) \cdot u \; ; \tag{2.1}$$

for simplicity we deleted the term $i\beta|u|_x^2 u$ in (1.2) which is simpler to treat. We investigate the space periodical initial-boundary value problem

$$u(x,0) = u_0(x) \tag{2.2}$$
$$u(x+1,t) = u(x,t) \tag{2.3}$$

for (2.1) where $u_0 \in H$.

The Cauchy problem for (2.1) has been treated by Tsutsumi and Fukuda (1979), (1980), (1981), Hayashi (1991), Lee (1989), Hayashi and Ozawa (1992); usually one has small initial data.

By an H^1-solution of problem (2.1) - (2.3) on $S = [0,T]$ we mean a function $u \in C(S,H) \cap C_w(S,V)$ such that $u_t \in C_w(S,V^{-1})$, (2.2) is satisfied, and (2.1) in the sense

$$(iu_t - Lu - i\alpha|u|_x^2 u, v) = 0$$

for all $t \in S, v \in V$.

THEOREM 2.1 *Let* $u_0 \in V^1$. *Then there is an* H^1*-solution of* (2.1) - (2.3) *with* $u \in C(Q)$. *If* $u_0 \in V^2$ *there is a unique* H^1*-solution of* (2.1) - (2.3) *such that*

$$u, u_x \in C(Q), u \in L^\infty(S, V^2), u_t \in L^\infty(S, H),$$

and equation (2.1) *is satisfied* a.e. *on* Q .

We present here a condensed version of the proof of *Theorem 1* which uses the standard method of *Galerkin sequences* approximating the solution; also we treat only the case $u_0 \in V^2$ to involve the uniqueness result. The n-th Galerkin approximation of (2.1) - (2.3) is a function

$$u_n(t) = \sum_{k=-n}^{n} b_k^n(t) h_k \in C^1(S, V^m)$$

(any $m \in \mathbb{Z}$) which is the unique solution of the problem

$$(iu_n' + u_{n,xx} - i\alpha |u_n|_x^2 u_n, h_\ell) = 0 \tag{2.4}$$

$$u_n(0) = u_{0,n} \tag{2.5}$$

($-n \le \ell \le n$). One sees that (2.4), (2.5) is equivalent to the first order ordinary differential system for the coefficients $b_k^n(t)$ given by

$$\frac{d}{dt} b_\ell^n(t) = g_\ell(b_{-n}^n(t), \ldots, b_n^n(t), \overline{b_{-n}^n(t)}, \ldots, \overline{b_n^n(t)}) \tag{2.6}$$

$$b_\ell^n(0) = (u_0, h_\ell) \qquad (-n \le \ell \le n)$$

(together with the complex conjugate equations $\frac{d}{dt}\overline{b_\ell^n}(t) = \overline{g_\ell}(\ldots), \overline{b_\ell^n(0)} = \overline{(u_0, h_\ell)}$ for $\overline{b_\ell^n}$)) where

$$g_\ell(b_{-n}^n, \ldots, \overline{b_n^n}) = i(2\pi\ell)^2 b_\ell^n(t) + 2\pi i\alpha \sum_{k,r,s=-n}^{n} (\ell - k)\alpha_{krs}^\ell b_k^n(t) b_r^n(t) \overline{b_s^n(t)}$$

with $\alpha_{krs}^\ell = 1$ for $k + r = s + \ell, \alpha_{krs}^\ell = 0$ otherwise. One often needs the following lemmata:

LEMMA 2.2 *For all* $u \in H^1(0,1)$ *one has* $\|u\|_{C[0,1]} = \|u\|_\infty \le \|u\|(\|u\| + 2\|u_x\|)$. (This can be found e.g. in Kastenholz, Lange und Nieder (1989)).

LEMMA 2.3 *Let* X *be a Banach space,* X_0, X_1 *reflexive Banach spaces such that* $X_0 \subset X \subset X_1$ *with compact embedding* of X_0 *in* X *and continuous and dense embedding of* X *in* X_1. *Then for* $1 < p, q < \infty$ *the space*

$$W = \{u \in L^p(S, X_0) \; |u' \in L^q(S, X_1)\}$$

with norm $\|u\|_{L^p(S,X_0)} + \|u'\|_{L^q(S,X_1)}$ *is a Banach space which is compactly embedded in* $L^p(S, X)$. (This is Aubin's embedding theorem and can be found in Lions (1969)).

The main lemma for the existence proof of *Theorem 2.1* is

LEMMA 2.4 *For $u_0 \in V^2$ problem (2.1) - (2.3) has a unique Galerkin sequence $u_n(t)$, and there is a constant $C > 0$ such that*

$$\|u_n\|_{L^\infty(Q)} \le C\,, \qquad \|u_n\|_{L^\infty(S,V^2)} \le C$$
$$\|u_{n,x}\|_{L^\infty(Q)} \le C\,, \qquad \|u_{n,t}\|_{C(S,V^{-1})} \le C$$

for all $n \in \mathbb{N}$, $t \in S$ and furthermore (u_n) is compact in $L^2(Q)$.

Instead of proving *Lemma 2.4* in full (and much technical) detail we give a proof of the following result which contains in analogous form the main step in the conclusions of *Lemma 2.4* namely to get an a priori bound on the Galerkin sequence (or for the solution itself!) by using conservation laws for equation (2.1):

LEMMA 2.5 *Let u be an H^1-solution of (2.1) - (2.3), $u_0 \in V$ such that equation (2.1) is satisfied a.e. on Q, and $u \in L^\infty(S, V^2)$, $u_t \in L^2(S, V^{-1})$. Then one has*

$$\int_0^1 |u(x,t)|^2 dx = \int_0^1 |u_0(x)|^2 dx, \tag{2.6}$$

$$E(u;t) := \int_0^1 \{|u_x|^2 + \alpha|u|^2 Im(u\overline{u}_x) + \frac{\alpha^2}{3}|u|^6\}dx = E(u_0) \tag{2.7}$$

for $t \in S$.

Proof (of *Lemma 2.5*) By using (2.1) we have

$$\begin{aligned}\partial_t|u|^2 &= 2\,\mathrm{Re}\,(u_t\overline{u}) = 2\,\mathrm{Im}\,(iu_t\cdot\overline{u}) = \\ &= -2\,\mathrm{Im}\,(u_{xx}\cdot\overline{u}) + 2\alpha\,\mathrm{Re}\,(|u|_x^2|u|^2) \\ &= \partial_x[2\,\mathrm{Im}\,(u\overline{u}_x) + \alpha|u|^4].\end{aligned} \tag{2.8}$$

By integration we get (2.6). Analogously, on could use the relation

$$\partial_t\left\{|u_x|^2 + \alpha|u|^2 Im(u\overline{u}_x) + \frac{\alpha^2}{3}|u|^6\right\} = \partial_x A \tag{2.9}$$

where

$$\begin{aligned}A = {}& 2\,\mathrm{Im}\,(u\overline{u}_{xx} + \frac{3}{4}\alpha(|u|_x^2)^2 + \alpha\,[\mathrm{Im}\,(u\overline{u}_x)]^2 + \\ &+ \frac{\alpha^3}{2}|u|^8 + 2\alpha^2\,\mathrm{Im}\,(u\overline{u}_x) + \alpha|u|^2|u_x|^2 \\ &- \alpha|u|^2\,\mathrm{Re}\,(u\overline{u}_{xx})\end{aligned}$$

which can be proved by direct calculation using equation (2.1), (2.8) and

$$\begin{aligned}\partial_t|u_x|^2 &= 2\,\mathrm{Im}\,(u_x\overline{u}_{xx})_x + \frac{\alpha}{2}(|u|_x^2)_x^2 + 2\alpha|u|_x^2|u_x|^2\,, \\ \partial_t\,\mathrm{Im}\,(u\overline{u}_x) &= \frac{1}{2}|u_x|_x^2 + 2\alpha|u|_x^2\,\mathrm{Im}\,(u\overline{u}_x) - \mathrm{Re}\,(u\overline{u}_{xxx})\,, \\ \partial_t|u|^6 &= 6|u|^4\,\mathrm{Im}\,(u\overline{u}_x)_x + 3\alpha|u|^4|u|_x^2\,.\end{aligned}$$

For this one needs some further regularity, but by directly looking at $\partial_t E$ one can use the same calculation and integration by parts to avoid the appearance of third order derivatives.

q.e.d.

Now, applying the analogue of *Lemma 2.5* (and *Lemma 2.2*) to the Galerkin sequence one gets *Lemma 2.4*. Using *Lemma 2.3* by extracting some subsequences one finally gets an appropriate subsequence of the Galerkin sequence which converges in various topologies (including the strong topology of $L^2(S,H)$ and the weak topology of $L^2(S,V)$) to an H^1-solution of (2.1) - (2.3).

For the uniqueness proof one needs some further regularity, namely that $u \in L^\infty_{\text{loc}}(S,H^2)$ (which then gives $u \in L^\infty(S,H^2)$, also). This is proved by considering a local a priori bound for $u_{n,xx}$ in the L^2-norm. Again, to avoid technical overloading in the proof of this property we proceed by just looking at $\partial_t \|u_{xx}\|^2$ (for the H^1-solution u) assuming already enough regularity for u_{xx} (which $u_{n,xx}$ has !). This gives

$$\begin{aligned}
\partial_t \int |u_{xx}|^2\,dx &= 2\,\text{Re}\int u_{xxt}\overline{u}_{xx}\,dx = \\
&= -2\alpha\,\text{Re}\int |u|^2_{xx} u\overline{u}_{xxx}\,dx - 2\alpha\,\text{Re}\int |u|^2_x u_x \overline{u}_{xxx}\,dx \\
&=: I_1 + I_2\ ,
\end{aligned}$$

$$\begin{aligned}
\int |u|^2_{xx}\,\text{Re}\,(u\overline{u}_{xxx})\,dx &= 2\int |u_x|^2\,\text{Re}\,(u\overline{u}_{xxx})\,dx + \\
&+ 2\int \text{Re}\,(u\overline{u}_{xx})\,\text{Re}\,(u\overline{u}_{xxx})\,dx = -4\int |u_x|^2_x\,\text{Re}\,(u\overline{u}_{xx})\,dx \\
&- \int |u_x|^2 |u_x|^2_x\,dx - 2\int \text{Re}\,(u\overline{u}_{xxx})\,\text{Re}\,(u\overline{u}_{xx})\,dx\ .
\end{aligned}$$

This leads to

$$I_1 = 12\alpha \int |u_x|^2_x\,\text{Re}\,(u\overline{u}_{xx})\,dx$$

by partial integration. Analogously

$$\begin{aligned}
I_2 &= 2\alpha\int |u|^2_{xx}\,\text{Re}\,(u\overline{u}_{xx})\,dx + \alpha\int |u|^2_{xx}|u|^2_x\,dx \\
&= 4\alpha\int |u_x|^2\,\text{Re}\,(u\overline{u}_{xx})\,dx + 4\alpha\int [\text{Re}\,(u\overline{u}_{xx})]^2\,dx\ .
\end{aligned}$$

With *Lemma 2.2* and the boundedness of $\|u\|_\infty$, $\|u_x\|$ (by *Lemma 2.5*) one gets by using $\|u_x\|^2_\infty \le C\|u_{xx}\|$

$$|I_1| \le C\|u_{xx}\|^{\frac{5}{2}} + C_1\ ,\ |I_2| \le C\|u_{xx}\|^2 + C_1\ . \tag{2.10}$$

By using a local Gronwall lemma of type

$$\varphi(t) \le A + B\varphi(t)^\beta \qquad (t \in [0,T))$$

(for a continuous function $\varphi : [0,T) \to \mathbb{R}^+$, $A, B > 0$, $\beta > 1$) for the function

$$\varphi(t) = \sup_{s \le t} \|u_{xx}(s)\| \qquad (t \in [t_0 - \delta,\ t_0 + \delta])$$

one gets $u \in L^\infty_{\text{loc}}(S, H^2)$ by choosing δ small enough, since by (2.10) one arrives at

$$\begin{aligned}\varphi(t) &\le (\varphi(t_0) + C_1) + C \int_{t_0}^{t} \varphi(s)^{\frac{5}{2}}\, ds \\ &\le C_2 + C\delta\varphi(t)^{\frac{5}{2}}\end{aligned}$$

on $[t_0 - \delta, t_0 + \delta]$.

To prove uniqueness of the H^1-solutions (where $u_0 \in H^2$) one uses the local H^2-bound for the solutions, namely if $u(t)$, $v(t)$ are two H^1-solutions of (2.1) - (2.3), and if we set $w(t) = u(t) - v(t)$ we have

$$\begin{aligned} iw_t &= -w_{xx} + i\alpha(|u|_x^2 u - |v|_x^2 v) , \\ w_t &= iw_{xx} + \alpha|v|_x^2 w + \alpha(|u|^2 - |v|^2)_x u . \end{aligned} \tag{2.11}$$

This implies:

$$\begin{aligned}\partial_t \int |w|^2\, dx &= -2\alpha \int (|u|^2 - |v|^2) \operatorname{Re}\, (u\overline{w})_x\, dx \\ &\quad + 2\alpha \int |v|_x^2 |w|^2\, dx = R_1 + R_2 ,\end{aligned}$$

$$\begin{aligned}|R_1| &= 2|\alpha| \cdot \left| \int (u\overline{w} + \overline{w}v) \operatorname{Re}\, (u_x\overline{w} + u\overline{w}_x)\, dx \right| \\ &\le C[\|w\|^2 + \|w_x\|^2] , \\ |R_2| &\le C\|w\|^2 .\end{aligned}$$

This means that $\partial_t \|w(t)\|^2 \le C[\|w(t)\|^2 + \|w_x(t)\|^2]$. Furthermore we have by using (2.11)

$$\begin{aligned}\partial_t \int |w_x|^2\, dx &= -2 \operatorname{Re} \int \overline{w}_{xx} w_t\, dx = \\ &= -2\alpha \operatorname{Re} \int (|u|^2 - |v|^2)_x \cdot u\overline{w}_{xx}\, dx - 2\alpha \int |v|_x^2 \operatorname{Re}\, (w\overline{w}_{xx})\, dx \\ &=: S_1 + S_2 ,\end{aligned}$$

$$\begin{aligned}|S_2| &= 2|\alpha| \cdot \left| \int \{|v|_{xx}^2 \operatorname{Re}\, (w\overline{w}_x) + |v|_x^2\, |w_x|^2\}\, dx \right| \\ &\le C \int \{|w|\, |w_x|(|v_x|^2 + |v||v_{xx}|) + |v|_x^2 |w_x|^2\}\, dx \\ &\le C[\|w\|^2 + \|w_x\|^2] ,\end{aligned}$$

since $\int |v|\ |v_{xx}|\ |w|\ |w_x|\ dx \le C\|w\|_\infty\ \|v_{xx}\|\ \|w_x\| \le C_1[\|w\|_\infty^2 + \|w_x\|^2] \le C_2[\|w\|^2 + \|w_x\|^2]$ by *Lemma 2.2.* Furthermore by writing $|u|^2 - |v|^2 = \mathrm{Re}\ (w\overline{u} + v\overline{w})$ we get

$$
\begin{aligned}
S_1 &= 2\alpha \int (|u|^2 - |v|^2)_{xx}\, \mathrm{Re}\ (u\overline{w}_x)\ dx + 2\alpha \int (|u|^2 - |v|^2)_x\, \mathrm{Re}\ (u_x\overline{w}_x) \\
&= 4\alpha \int \mathrm{Re}\ (u\overline{w}_x)\ (u_x\overline{w}_x + \overline{v}_x w_x + \overline{v}_{xx} w)\ dx \\
&\quad + 4\alpha \int \mathrm{Re}\ (u\overline{w}_x)\, \mathrm{Re}\ (u\overline{w}_{xx})\ dx \\
&\quad + 4\alpha \int \mathrm{Re}\ (u_x\overline{w}_x)\ (u\overline{w}_x + \overline{v}_x w)\ dx =: T_1 + T_2 + T_3\ .
\end{aligned}
$$

One readily gets $|T_1|,\ |T_3| \le C[\|w\|^2 + \|w_x\|^2]$. For T_2 one observes that

$$
T_2 = -4\alpha \int \mathrm{Re}\ (u_x\overline{w}_{xx})\, \mathrm{Re}\ (u\overline{w}_x)\ dx - 8\alpha \int \mathrm{Re}\ (u\overline{w}_x)\, \mathrm{Re}\ (u_x\overline{w}_x)\ dx\ ,
$$

which gives $T_2 = 4\alpha \int \mathrm{Re}\ (u_x\overline{w}_x)\, \mathrm{Re}\ (u\overline{w}_x)\ dx$, $|T_2| \le C[\|w\|^2 + \|w_x\|^2]$, also. Since $w(0) = 0$ Gronwall's lemma implies $w(t) \equiv 0$ on $[0, T]$.

q.e.d.

REMARK By the uniqueness of H^1-solutions if follows that the Galerkin sequence $(u_n(t))$ of problem (2.1) - (2.3) converges to the unique solution $u(t)$ in various norms (and not only a subsequence!). With further regularity assumptions one can also prove error estimates of type

$$
\|u - u_n\| = 0(n^{-\gamma}) \qquad (\gamma > 0)
$$

in various norms.

3 SECOND ORDER DERIVATIVE NONLINEAR SCHRÖDINGER EQUATION

The simplest nonlinear Schrödinger equation of physical relevance which contains second order derivative nonlinear terms is the equation

$$
iu_t = -u_{xx} + \beta |u|^2_{xx} \cdot u \tag{3.1}
$$

(β real). Here we consider the whole space initial-boundary value problem

$$
u(x, 0) = u_0(x)\ . \tag{3.2}
$$

It turns out that this problem (even local existence of solutions!) is much harder than problem (2.1) - (2.3). The reason for that is the lack of an appropriate local a priori H_2-bound on the local solution or any approximative sequence. One is able to prove a priori bounds on $\|u\|$ and $\|u_x\|$ (and then also for $\|u\|_\infty$) by using the conservation laws

$$
\int_{\mathbb{R}} |u(x,t)|^2\ dx = const. \tag{3.3}
$$

$$
\int_{\mathbb{R}} \left\{ |u_x|^2 - \frac{\beta}{2}(|u|^2_x)^2 \right\}\ dx = const. \tag{3.4}
$$

for sufficient regular solutions of (3.1); this follows from the local conservation law

$$\partial_t |u|^2 = \partial_x [2\,\mathrm{Im}\,(u\overline{u}_x)] \tag{3.5}$$

and by computation from

$$\begin{aligned}\partial_t \int |u_x|^2\,dx &= -2\,\mathrm{Re}\int \overline{u}_{xx} u_t\,dx \\ &= 2\beta\,\mathrm{Im}\int u\overline{u}_{xx}\cdot |u|^2_{xx}\,dx = -\beta\int \partial_t |u|^2\,|u|^2_{xx}\,dx \\ &= \frac{\beta}{2}\partial_t \int (|u|^2_x)^2\,dx\,.\end{aligned}$$

For any sequence (u_n) of approximating solutions which converge strongly to a function u in $L^2(S, L^2(\mathbb{R}))$ and weakly in $L^2(S, H^1(\mathbb{R}))$ (which can be provided by (3.3) and (3.4) for e.g. negative β) by compactness arguments similar to those in *Section 2* one cannot identify the limit u as a solution since $\int_0^T \int_{\mathbb{R}} |u_n|^2_{xx} u_n \varphi\,dxds$ (for an arbitrary test function φ) may not converge to $\int_0^T \int_{\mathbb{R}} |u|^2_{xx} u\varphi\,dxds$ (unless $u_{n,x}$ converges strongly to u_x in $L^2(S, L^2(\mathbb{R}))$). But using the theory of infinite order Sobolev spaces

$$W = W^\infty\{a_n, p_n, r_n\} = \left\{ u \in C^\infty(\mathbb{R}) \mid \sum_{n=0}^{\infty} a_n \|\partial_x^n u\|^{p_n}_{L^{r_n}} < \infty \right\}$$

for some parameter sequences $a_n \geq 0,\ a_0 > 0,\ p_n \geq 1,\ r_n \geq 1$ one can show that there exists an appropriate constant $C (C > 1)$ such that the following theorem is valid:

THEOREM 3.1 *Let* $u_0 \in W^\infty\{a_n,\ p_n,\ r_n\}$ *with* $a_n = C^{n2^n},\ p_n = 2^{n+1},\ r_n = 2$. *Then* (3.1) - (3.2) *has a unique classical solution* $u \in C^\infty(\mathbb{R} \times S)$ *such that* $u(\cdot, t) \in W^\infty\{a_n,\ p_n,\ r_n\}$ *for every* $t \in S$. *The solutions of* (3.1) - (3.2) *are unique within the class* $C^1(S, H^2(\mathbb{R}))$ *for* $\beta \leq 0$ *or small* $L^\infty(\mathbb{R} \times S)$*-solutions.*

The proof of *Theorem 3.1* will appear elsewhere (see Audonnet, Fleckinger, Lange (1992)), but we will give a brief sketch of it. When looking for a higher order conservation law (which we did not find!) one comes naturally to an expression of type

$$I_n(t) = \int_{\mathbb{R}} \left\{ |\partial_x^n u|^2 - \frac{\beta}{n+1} (\partial_x^n |u|^2)^2 \right\} dx$$

for which one can show (when assuming enough regularity for a local solution or an approximating sequence) that

$$I_n(t) \leq C_n \left(\int_{\mathbb{R}} |\partial_x^{n+1} u|^2\,dx \right)^2 \tag{3.6}$$

for any $n \geq 1$; here one has $C_n \sim C \cdot n \cdot 2^n$.

By setting

$$\varphi(t) = \sum_{n=0}^{\infty} a_n \|\partial_x^n u(t)\|^{2\beta_n}$$

one may show with (3.6) that

$$\dot{\varphi}(t) \leq K_1\varphi(t) + K_2 \qquad (t \in S)$$

which implies the existence of an a priori bound for $\varphi(t)$ on S. This is enough to construct a global solution within the class mentioned in *Theorem 3.1.*

The uniqueness result is proved by writing equation (3.1) as a system for real and imaginary part of $w = u - v$, where u, v are two solutions of (3.1) - (3.2) within the class of *Theorem 3.1* and with $u(0) = u_0$, $v(0) = u_0$. If $U = \begin{pmatrix} \mathrm{Re}\ u \\ \mathrm{Im}\ u \end{pmatrix} = \begin{pmatrix} u_1 \\ u_2 \end{pmatrix}$, $V = \begin{pmatrix} \mathrm{Re}\ v \\ \mathrm{Im}\ v \end{pmatrix} = \begin{pmatrix} v_1 \\ v_2 \end{pmatrix}$ and $W = U - V$ one get a system of type

$$W_t = MW_{xx} + NW_x + SW + R(W) \tag{3.7}$$

where

$$M = \begin{pmatrix} 2\beta u_1 u_2 & -1 + 2\beta u_2^2 \\ 1 - 2\beta u_1^2 & -2\beta u_1 u_2 \end{pmatrix}, \quad N = 4\beta \begin{pmatrix} u_2 u_{1,x} & u_2 u_{2,x} \\ -u_1 u_{1,x} & -u_1 u_{2,x} \end{pmatrix}$$

and S, R are some 2×2-matrices depending on U, V, W such that

$$|S| \leq C\{|U| + |U_x| + |U_{xx}| + |V| + |V_x| + |V_{xx}|\}\ ,$$
$$|R(W)| \leq K|W|^2.$$

Instead of looking for an a priori bound of $\|W\|^2 + \|W_x\|^2$ by computing $\partial_t \left[\|W\|^2 + \|W_x\|^2\right]$ one derives an inequality of type

$$\partial_t \tilde{E}(t) \leq C_1 \tilde{E}(t) \tag{3.8}$$

for

$$\tilde{E}(t) = \int_{\mathbb{R}} \{(W_x, CMW_x) + (W_x, C(N - \partial_x M)W) + a|W|^2\}\ dx$$

where $C = \begin{pmatrix} 0 & 1 \\ -1 & 0 \end{pmatrix}$ and a is a suitable positive constant; $\tilde{E}(t)$ is equivalent to $\int_{\mathbb{R}}(|W|^2 + |W_x|^2)\ dx$ since CM is positive definite and symmetric (e.g. for $\beta \leq 0$); the proof of (3.8) uses heavily that the matrix $C(N - \partial_x M)$ is antisymmetric!

REMARKS. 1.) A serious problem in connection with *Theorem 3.1* is the *nontriviality* of the space W, i.e. $W \neq \{0\}$. In case of the Cauchy problem the necessary and sufficient condition for the nontriviality of the space W used in *Theorem 3.1* is the existence of a number $q > 0$ such that

$$\sum_{n=0}^{\infty} a_n q^{2n\beta_n} < \infty\ ; \tag{3.9}$$

with our choice of $(a_n), (\beta_n)$ this is satisfied for s small enough q (see Dubinskij (1986)); The number q is an upper bound of the length of the support of the Fourier transform of the initial function u_0 (which must lie in $(-q, q)$). Thus (3.9) inherits a hidden smallness assumption on the initial data. Furthermore the function u_0 has to be an entire holomorphic function (see Dubinskij (1986)).

2.) The necessary and sufficient conditions which have to be satisfied such that the analogously defined Sobolev spaces for the space periodical initial boundary value or for the Dirichlet problem with homogenous boundary data are *not satisfied* for our choice of parameter sequences $(a_n), (p_n), (r_n)$ as in *Theorem 3.1*; thus in the cases of periodical or Dirichlet boundary conditions the Sobolev spaces of infinite order belonging to those problems are trivial.

4 AN INTERMEDIATE DERIVATIVE NONLINEAR SCHRÖDINGER PROBLEM

In this section we consider a nonlinear Schrödinger equation arising in the theory of deep water waves (see Section 1):

$$iu_t = -u_{xx} + \gamma \cdot \phi_u \cdot u \tag{4.1}$$

(γ real) where $\phi_u(x,t)$ is $\frac{\partial \phi}{\partial x}|_{y=0}$, and $\phi(x,y,t)$ satisfies the potential equation

$$\begin{cases} \phi_{xx} + \phi_{yy} &= 0 \qquad (-1 < y < 0) \\ \frac{\partial \phi}{\partial y}|_{y=-1} &= 0 \quad , \quad \frac{\partial \phi}{\partial y}|_{y=0} = \delta |u|_x^2 \; ; \end{cases} \tag{4.2}$$

$u = u(x,t)$ described the complex amplitude of the first harmonic of the Stokes wave in the evolution of a weakly deep water wave traveling predominantly in one direction, whereas $\phi(x,y,t)$ is the potential induced by the mean current (see Lo and Mei (1985)). If one looks for the space periodical initial boundary value problem

$$u(x,0) = u_0(x) \tag{4.2}$$

$$u(x+1,t) = u(x,t) \tag{4.3}$$

one can prove

THEOREM 4.1 *Let $u_0 \in V$ and $\gamma\delta < 0$. Then* (4.1) - (4.3) *has an H^1-solution u satisfying the conservation laws*

$$\int_0^1 |u(x,t)|^2 \, dx = \int_0^1 |u_0(x)|^2 \, dx \tag{4.4}$$

$$E(u,t) = \int_0^1 \{|u_x|^2 + \lambda |\psi_u|^2\} \, dx = E(u_0) \tag{4.5}$$

where $\psi_u(x,t) = \sum_{k \neq 0} \alpha_k(|u|^2, \ h_k) \, h_k, \alpha_k^2 = 2\pi k \cdot \subset th(2\pi k)$ $(k \in \mathbb{Z})$ *and* $\lambda = -\frac{\gamma\delta}{2}$.

The proof of *Theorem 4.1* (which can be found in Lange and Kühnen (1992), Kühnen (1992)) uses the same method as in Section 2, namely a Galerkin approximation procedure; one solves the potential equation (4.2) for $\phi(x,y,t)$ with the nonlinear Neumann conditions coupled to the wave function u explictly and inserts this into equation (4.1). To get a priori bounds as those in *Lemma 2.4* for the Galerkin sequence one uses the analogues of the conversation laws (4.4) and (4.5). With this method it seems not to be possible to prove uniqueness of H^1-solution even with higher order derivative boundedness results; the reason for this seems to be (as mentioned in the introduction) that the nonlinearity in (4.1) corresponds to a term of type $\gamma|u|_x^2 u$ with a real coefficient instead of a pure imaginary one as in Section 2.

REFERENCES

1. Audonnet, J., Fleckinger,J., and Lange, H. (1992). A Schrödinger equation with a second order derivate in the nonlinearity, *Preprint*, Mathematisches Institut, Univ. Köln.
2. Bass, F.G., and Nasonov, N.N. (1992). Nonlinear electromagnetic-spin waves, *Physics Reports, 189:* 162-223 .
3. Dubinskij, J.A. (1986). Sobolev spaces of infinite order and differential equations, D. Reidel Publ. Co., Dordrecht.
4. Dysthe, K.B. (1979). Note on a modification to the nonlinear Schrödinger equation for application to deep water waves, *Proc. Roy. Soc. London, A 369:* 105-114.
5. Hayashi, N. (1991). The initial value problem for the derivative nonlinear Schrödinger equation in the energy space, *Preprint.*
6. Hayashi, N., and Ozawa, T. (1992). On the derivative nonlinear Schrödinger equation, *Physica D 55:* 14-36.
7. Kastenholz, M., Lange H., and Nieder, D. (1989). Periodic solutions of nonlinear Schrödinger equations, *Portugaliae Mathematica 46:* 517-537 .
8. Kaup D.J., and Newell, A.C. (1978). An exact solution for a derivative nonlinear Schrödinger equation, *J. Math. Phys. 19:* 789-801.
9. Kosevich, A.M., Ivanov B.A., and Kovalev, A.S. (1990). Magnetic solitons. *Physics Reports 194:* 117-238.
10. Kühnen, C. (1992). Existenz von Lösungen eines nichtlinearen Schrödinger-Systems aus der Strömungsmechanik, *Diplom-Arbeit*, Mathematisches Institut, Univ. Köln.
11. Kurihara, S. (1982). Large-amplitude quasi-solitons in superfluid films, *J. Phys. Soc. Japan 50:* 3262-3267 .
12. Lange, H., and Kühnen C. (1992). A nonlinear Schrödinger system from the theory of deep water waves. *Preprint*, Mathematisches Institut, Univ. Köln.
13. Lee, J.H. (1989). Global solvability of the derivative nonlinear Schrödinger equation, *Trans. Amer. Math. Soc. 314:* 107-118.
14. Lions, J.L. (1969). Quelques méthodes de résolution des problèmes aux limites nonlinéaires, Dunod, Paris.
15. Litvak A.G., and Sergeev, A.M. (1978). One dimensional collapse of plasma waves, *JETP Letters 27:* 517-520 .
16. Lo E., and Mei, Ch.C. (1985). A numerical study of water-wave modulation based on a higher-order nonlinear Schrödinger equation, *J. Fluid Mech. 150:* 395-416.
17. Makhankov V.G., and Fedyanin, V.K. (1984). Non-linear effects in quasi-onedimensional models of condensed matter theory, *Physics Reports 104:* 1-86.

18. Mio, W., Ogino, T., Minami, K., and Takeda, S. (1976). Modified nonlinear Schrödinger equations for Alfven waves propagating along the magnetic field in cold plasmas, *J. Phys. Soc. Japan 41:* 265-271.
19. Mjølhus, E. (1976). On the modulational instability of hydromagnetic waves parallel to the magnetic field, *J. Plasma Phys. 16:* 321-334.
20. Nakamura, A. (1977). Damping and modification of exciton solitary waves, *J. Phys. Soc. Japan 42:* 1824-1835.
21. Pathria D., and Morris, J. LL. (1990). Pseudo-spectral solution of nonlinear Schrödinger equations, *J. Comp. Phys. 87:* 108-125.
22. Quispel G.R.W., and Capel, H.W. (1982). Equation of motion for the Heisenberg spin chain, *Physica 110 A:* 41-80.
23. Takeno S., and Homma, S. (1981). Classical planar Heisenberg ferromagnet, complex scalar field and nonlinear excitations, *Progr. Theoret. Phys. 65:* 172-189.
24. Tsutsumi M., and Fukuda, I. (1979). On solutions of some nonlinear dispersive wave equations, *Memoirs School of Sci. Engineering Waseda Univ. 43:* 109-146 .
25. Tsutsumi M., and Fukuda, I. (1980). On solutions of the derivative nonlinear Schrödinger equation. Existence and uniqueness theorem, *Funkcialaj Ekvacioj 23.* 259-277.
26. Tsutsumi M., and Fukuda, I. (1981). On solutions of the derivative nonlinear Schrödinger equation. II. *Funkcialaj Ekvacioj 24:* 85-94.

The Wave Equation Method in the Spectral Theory of Self-Adjoint Differential Operators

B. M. LEVITAN University of Minnesota, Minneapolis, Minnesota

1 INTRODUCTION

As it will be later shown in this paper, the knowledge of the structure of fundamental solutions of hyperbolic equation permit us to obtain important information about the asymptotic distribution of eigenfunctions and eigenvalues of self-adjoint differential operators (ordinary and with partial derivatives).

Besides the knowledge of the structure of the fundamental solution an important part in this investigation concerns some special Tauberian theorems for Fourier-type integrals.

One of the simplest Tauberian theorems of this kind is:

THEOREM (LEVITAN [1]) Let the function $\sigma(\lambda), -\infty < \lambda < \infty$, satisfy the following two conditions:

$$1)\ \mathrm{Var}_a^{a+1}\{\sigma(\lambda)\} = O(|a|^r), \quad r \geq 0, \quad a \to \infty. \tag{1}$$

2) For every smooth function $g(t)$ with finite support, $\operatorname{supp} g(t) \subset (-\delta, \delta)$, $\delta > 0$

$$\int_{-\infty}^{\infty} \hat{g}(\lambda)\, d\sigma(\lambda) = 0, \tag{2}$$

where

$$\hat{g}(\lambda) = \frac{1}{2\pi} \int_{-\infty}^{\infty} g(t) e^{i\lambda t}\, dt.$$

Then the estimate

$$\int_{-\nu}^{\nu}\left(1-\frac{\lambda^2}{\nu^2}\right)^s d\sigma(\lambda)=O(\nu^{r-s} \tag{3}$$

$s \geq 0, \nu \to \infty$ holds.

In particular, for $s > r$

$$\int_{-\nu}^{\nu}\left(1-\frac{\lambda^2}{\nu^2}\right)^s d\sigma(\lambda)=o(1), \quad \nu \to \infty.$$

REMARK 1 If the function $\sigma(\lambda)$ is odd, then we can choose the function $g(t)$ as even. The estimate (3) can be written as

$$\int_{o}^{\nu}\left(1-\frac{\lambda^2}{\nu^2}\right)^s d\sigma(\lambda)=O(\nu^{r-s}). \tag{3'}$$

REMARK 2 If in the estimate (1) O can be replaced by o then the same is true in the estimates (3) and (3').

REMARK 3 In many cases instead of the condition (2) it is more convenient to use the condition

$$\int_{-\infty}^{\infty} \hat{g}(\lambda)\, d\sigma(\lambda) = \int_{-\delta}^{\delta} g^{(n)}(t) f(t)\, dt \tag{2'}$$

where $f(t) \in \mathcal{L}(-\delta, \delta)$ and $n \leq [r]$.

PROOF OF THE ASSERTION OF REMARK 3 By means of the Parseval equation (usually for Fourier integrals) the condition (2') can be reduced to condition (2).

2 APPLICATION OF THE METHOD OF WAVE EQUATION TO THE SPECTRAL THEORY OF LAPLACE OPERATORS IN EUCLIDIAN SPACES

Consider the eigenvalues problem: $\mathcal{D}$-compact domain

$$\Delta u = -\lambda u, \quad x \in \mathcal{D}, \quad \mathcal{D} \in R^3; \qquad u/\Gamma = 0, \quad \Gamma = \partial\mathcal{D}. \tag{4}$$

Denote by $0 < \lambda_1 \leq \lambda_2 \cdots \leq \lambda_n \leq \cdots$ the eigenvalues and by $\varphi_o(x) = \text{const}$, $\varphi_1(x)$, $\varphi_2(x), \ldots, \varphi_n(x), \ldots, x \in \mathcal{D}$ the corresponding eigenfunctions of problem (4), normed in Hilbert space $\mathcal{L}_2^2(\mathcal{D})$ to one. Consider together with the spectral problem (4) the initial-boundary problem:

$$\frac{\partial^2 u}{\partial t^2} = \Delta u, u|_{t=0} = f(x), \quad \frac{\partial u}{\partial t}|_{t=0} = 0, \quad u|_r = 0, \tag{5}$$

where $f(x), x \in \mathcal{D} \subset \mathcal{R}^3$ is an arbitrary smooth function with finite support in $\mathcal{D}$.

By means of the Fourier method the solution of problem (5) can be expressed as

$$u(x,t) = \sum_{n=0}^{\infty} \cos \mu_n t \varphi_n(x) \int_{(\mathcal{D})} f(y)\varphi_n(y)\, dy \tag{6}$$

where $\mu_n = \sqrt{\lambda_n}$.

Here, it is very convenient to do a very important remark, which is crucial in all cases when using the wave equation method; the differential equation (5) is of hyperbolic type. But for the solutions of partial differential equations of hyperbolic type the finite domain defense of the initial data is valid.

Therefore, if x is a fixed inner point of the domain $\mathcal{D}$ and $\operatorname{dist}(x, \Gamma) = \epsilon > 0, \Gamma = \partial\mathcal{D}$ then for $0 < t \leq \epsilon$ the boundary condition in problem (5) can be neglected. Now, the solution of the pure Cauchy problem (5) is:

$$u(x,t) = \frac{1}{4\pi}\frac{\partial^2}{\partial t^2}\int_{r\leq\epsilon}\frac{f(y)}{r}\,dy, \quad r = |x-y|. \tag{7}$$

As the solution of the pure Cauchy problem is unique from (6) and (7) we have

$$\sum_{n=0}^{\infty}\cos\mu_n t\varphi_n(x)\int_{(\mathcal{D})}f(y)\varphi_n(y)\,dy = \frac{1}{4\pi}\int_{r\leq t}\frac{f(y)}{r}\,dy. \tag{8}$$

Now, let $g(t)$ be a smooth even function with finite support: $\operatorname{supp} g(t) \subset (-\epsilon, \epsilon)$, ϵ is the same as earlier. Let

$$\hat{g}(\mu) = \frac{2}{\pi}\int_0^{\epsilon} g(t)\cos\mu t\,dt$$

be the cos–Fourier transform of $g(t)$. Multiply both sides of equation (8) by $\frac{2}{\pi}g(t)$ and integrate. We obtain:

$$\sum_{n=0}^{\infty}\hat{g}(\mu_n)\varphi_n(x)\int_{(\mathcal{D})}f(y)\varphi_n(y)\,dy = \frac{1}{2\pi^2}\int_0^{\infty}g(t)\left[\frac{\partial^2}{\partial t^2}\int_{r\leq t}\frac{f(y)}{r}\,dy\right]dt. \tag{9}$$

As $\hat{g}(\mu)$ rapidly decreases we can invert the order of summation and integration in the left side of equation (9). The right side of the identity (9) we integrate once by part and then reverse the order of integration. The result will be:

$$\int_{(\mathcal{D})}f(y)\left[\sum_{n=0}^{\infty}\hat{g}(\mu_n)\varphi_n(x)\varphi_n(y)\right]dy$$
$$= \frac{1}{2\pi^2}\int_{r\leq\epsilon}f(y)\frac{g'(r)}{r}\,dy, \quad r = |x-y|.$$

As $f(y)$ is arbitrary from the last equation we obtain the identity:

$$\int_0^{\infty}\hat{g}(\mu)\,d_\mu\theta(x,y,\mu) = \begin{cases}\dfrac{1}{2\pi^2}\dfrac{g'(r)}{r}, & \text{if } |x-y|\leq\epsilon\\ 0, & \text{if } |x-y|\geq\epsilon\end{cases} \tag{10}$$

where

$$\theta(x,y;\mu) = \sum_{\mu_n<\mu}\varphi_n(x)\varphi_n(y).$$

Using the inversion formula for Fourier integrals we have:

$$\begin{aligned} g(r) &= \int_0^{\infty}\hat{g}(\mu)\cos\mu r\,d\mu, \quad g'(r) = -\int_0^{\infty}\mu\hat{g}(\mu)\sin\mu r\,d\mu,\\ \frac{g'(r)}{r} &= -\frac{1}{r}\int_0^{\infty}\mu\hat{g}(\mu)\sin\mu r\,d\mu = \int_0^{\infty}\hat{g}(\mu)d_\mu\left[\frac{\sin\mu r - \mu r\cos\mu r}{r^3}\right].\end{aligned} \tag{11}$$

From (10) and (11) follows:

$$\int_0^\infty \hat{g}(\mu)\, d_\mu \left[\theta(x,y;\mu) - \frac{1}{2\pi^2}\,\frac{\sin\mu r - \mu r\cos\mu r}{r^3}\right] = 0,$$

$$\text{if } 0 \le r \le \epsilon, \quad r = |x-y|. \tag{12}$$

Equation (12) means that the Fourier transform on variable t (in the sense of distribution) of the function, which stands under the square bracket in equation (12), is zero for $0 \le t \le \epsilon$. Using the Tauberian theorem of the introduction we obtain the estimate:

$$\left|\theta(x,y;\nu) - \frac{1}{2\pi^2}\,\frac{\sin\nu r - \nu r\cos\nu r}{r^3}\right| \le c\nu^2. \tag{13}$$

The most interesting case is $r = 0$, that is, $x = y$. In this case, we have from (13)

$$\sum_{\mu_n \le \nu} \varphi_n^2(x) = \frac{1}{6\pi^2}\,\nu^3 + O(\nu^2). \tag{14}$$

REMARK 1 The condition (1) in the Tauberian theorem of the introduction, that is, the preliminary estimate

$$\sum_{a<\mu_n\le a+1} \varphi_n^2(x) = O(a^2), \quad a \to \infty$$

can be obtained from the identity (10) by choosing a special $g(t)$ (see Levitan [2]).

REMARK 2 By the same method the asymptote of the spectral function of the Laplace operator in Euclidian space of arbitrary dimension can be studied (Levitan [2]).

3 THE ASYMPTOTIC BEHAVIOR OF THE SPECTRAL FUNCTION AND DISCRETE SPECTRUM OF LAPLACE–BELTRAMI OPERATOR ON LOBACHEVSKY'S SPACE

We restrict ourselves in this report to the case of Lobachevsky's space of two dimensions, but some of our results can be directly generalized to the case of arbitrary dimensions.

Let S be the Lobachevsky plane, realized as the upper half-plane $z = x + iy, y > 0$. S is a homogeneous space for the group $S\mathcal{L}(2,R)$ which acts in S by the transformation

$$z \to gz = \frac{az+b}{cz+d}, \quad g = \begin{pmatrix} a, & b \\ c, & d \end{pmatrix}, \quad ad - bc = 1, \quad a,b,c,d \in R.$$

The differential operator $A = y^2\left(\frac{\partial^2}{\partial x^2} + \frac{\partial^2}{\partial y^2}\right)$ and the measure $d\mu(z) = \frac{dx\,dy}{y^2}$ are invariant under the transformation g. Let Γ be a discrete subgroup of $S\mathcal{L}(2,R)$ satisfying the following conditions:

1) S/Γ is noncompact, but has a finite measure.

2) Γ contains only one parabolic subgroup. In this case, one can choose the fundamental domain $F = F_\Gamma$ for Γ so that:

a) F lies in the strip $-X < x < X, y > Y > 0$.

b) for some $d > 1$ the intersection $F \cap (y > d)$ coincides with the strip $-X_1 < x < X_1$. In the following we assume that the distance is chosen so that $X_1 = 1/2$.

c) The boundary of F is piecewise smooth and consist of geodesic segments with a finite number of angle points.

In the space $H = \mathcal{L}^2(F, d\mu)$, consider a symmetric operator $\tilde{\mathcal{L}}$ defined by the differential expression

$$-A - \frac{1}{4} = -y^2\left(\frac{\partial^2}{\partial x^2} + \frac{\partial^2}{\partial y^2}\right) - \frac{1}{4}$$

on the space of all sufficiently smooth and uniformly bounded on S functions which satisfy the authomorphic condition, that is, which are restrictions on F of smooth "periodic" functions $u : u(z) = u(\gamma z), z \in F, \gamma \in \Gamma$. The operator $\tilde{\mathcal{L}}$ has in H only one self-adjoint extension $\mathcal{L}$. The spectrum of $\mathcal{L}$ consists of many negative eigenvalues of finite multiplicity:

$$\lambda_{-n} \leq \lambda_{-n+1} \leq \cdots \leq \lambda_1 \leq 0,$$

also, maybe, of infinitely many positive eigenvalues $\lambda_l, l = 1, 2, \ldots$ each of finite multiplicity and of a simple branch of absolutely continuous spectrum $\lambda = \kappa^2, -\infty < \kappa < +\infty$.

We denote the eigenfunctions of the discrete spectrum by $\psi_l(z), l = -n, -n+1, \ldots, -1, 1, 2, \ldots$ and will suppose that they have norms equal to one.

The "eigenfunctions" $\varphi(z, \kappa)$ of the continuous spectrum are normalized by the asymptote: for $y \to +\infty$

$$\varphi(z, \kappa) = \varphi_o(z, \kappa) + \varphi_1(z, \kappa), \tag{15}$$

where

$$\begin{aligned} &\varphi_o(z, \kappa) = y^{1/2+i\kappa} + S(\kappa)y^{1/2-i\kappa} \\ &\varphi_1(z, \kappa) = O(1), \quad y \to +\infty. \end{aligned} \tag{15'}$$

By such norming the continuous spectrum identifies with the whole real κ-line and the spectral density by Lebesgue measure is $\frac{1}{4\pi}$. The function $S(\kappa)$ in the asymptote (15′) is called the coefficient of reflection and is defined uniquely by the condition (15).

By a spectral function of the operator $\mathcal{L}$ we call the function:

$$\theta(z, z') = \begin{cases} \sum\limits_{\lambda_l < \lambda} \psi_l(z)\overline{\psi_l(z')}, & \lambda < 0 \\ \sum\limits_{\lambda_l < \lambda} \psi_l(z)\overline{\psi_l(z')} + \dfrac{1}{4\pi}\displaystyle\int_0^\kappa [\varphi(z, \kappa)\overline{\varphi(z', \kappa)} + \overline{\varphi(z, \kappa)}\varphi(z', \kappa)]\, d\kappa & \\ \qquad \lambda > 0, \kappa = \sqrt{\lambda}. & \end{cases}$$

The Asymptotic Behavior of the Spectral Function

Consider the Cauchy problem:

$$\begin{aligned} &y^2 \Delta u + \frac{1}{4}u = \frac{\partial^2 u}{\partial t^2}, \quad u|_{t=0} = f(z) = f(x, y) \\ &\qquad \frac{\partial u}{\partial t}|_{t=0} = 0. \end{aligned} \tag{16}$$

Here $f(x, y)$ is an arbitrary smooth function with finite support in the domain F, extended as a "periodic" function on the whose upper plane. The problem (16) can be solved by two methods: by means of Kirchhoff-type formula and by means of the spectral expansion of the operator $\mathcal{L}$. Equating these two expressions we obtain the identity:

$$\sum_{l=-n}^{\infty} \cos\sqrt{\lambda_l}\, t\psi_l(x,y) \int_F (\xi,\eta)\overline{\psi(\xi,\eta)}\, d\mu(\xi,\eta)$$

$$+ \frac{1}{4\pi}\int_{-\infty}^{\infty} \cos\kappa t\varphi(x,y;\kappa) \int_F f(\xi,\eta)\overline{\varphi(\xi,\eta;\kappa)} d\mu(\xi,\eta)$$

$$= \frac{1}{4\pi}\sum_{\gamma\in\Gamma} \frac{\partial}{\partial t}\int_{r_\gamma\leq t} \chi_F(\xi,\eta)\frac{f(\xi,\eta)}{(sh^2\frac{t}{2} - sh^2\frac{r_\gamma}{2})^{1/2}}\, d\mu(\xi,\eta).$$

Here, $\chi_F(\xi,\eta)$ is the characteristic function of the domain F, r_γ is the geodesic distance between the points $\gamma(\xi,\eta), (\xi,\eta) \in F, \gamma \in \Gamma$ and the point $(x,y) \in F$.

Let $g(t)$ be an even, smooth function with finite support, $\operatorname{supp} g(t) \subset (-b,b), b > 0, b < \infty$ and let

$$\hat{g}(\mu) = \int_o^{\infty} g(t)\cos\mu t\, dt.$$

Multiplying both sides of the identity (17) by $g(t)$ and using the arbitrariness of the function we get $f(\xi,\eta)$:

$$\sum_{l=-n}^{\infty} \hat{g}(\sqrt{\lambda_l})\psi_l(x,y)\overline{\psi_l(\xi,\eta)} + \frac{1}{4\pi}\int_{-\infty}^{\infty} \hat{g}(\kappa)\varphi(x,y;\kappa)\overline{\varphi(\xi,\eta;\kappa)}\, d\kappa$$

$$= -\frac{1}{4\pi}\sum_{\gamma\in\Gamma}\int_{r_\gamma}^{\epsilon} \frac{g'(t)\, dt}{\left(sh^2\frac{t}{2} - sh^2\frac{r_\gamma}{2}\right)^{1/2}}. \tag{18}$$

For $(\xi,\eta) = (x,y)$ from (18) follows:

$$\sum_{l=-n}^{\infty} \hat{g}(\sqrt{\lambda_l})|\psi(x,y)|^2 + \frac{1}{4\pi}\int_{-\infty}^{\infty} \hat{g}(\kappa)|\varphi(x,y;\kappa)|^2\, d\kappa$$

$$= -\frac{1}{4\pi}\sum_{\gamma\in\Gamma}\int_{r_\gamma}^{\epsilon} \frac{g'(t)\, dt}{\left(sh^2\frac{t}{2} - sh^2\frac{r_\gamma}{2}\right)^{1/2}} \tag{19}$$

where $r_\gamma = \operatorname{dist}[(x,y),\gamma(x,y)], \gamma \in \Gamma$.

As $g(t)$ has a compact support, the right-hand side sum of (19) is finite, but the number of its terms depends on y. Choose a fixed $Y > 0$ and denote by F_Y the set of points of F, for which $y \leq Y$. Then, for $(x,y) \in F_Y$, the number of terms in the right-hand sum of (19) is bounded. (See Fig.1 on which the fundamental domain of the modular group is presented). Suppose that the domain F does not contain fixed points. Then, for $(x,y) \in F_Y$,

$$\min_{\gamma\in\Gamma/I} r_\gamma = \bar{r} > 0, \quad r_\gamma = d[(x,y),\gamma(x,y)].$$

If Y is sufficiently large then this minimum is equal to the geodesic distance between (x,Y) and $(x+1,Y)$. It can be shown that

$$\operatorname{dist}[(x,y),(x+1,y)] = \bar{r} = \frac{1}{y} + O\left(\frac{1}{y^2}\right), \quad y \to +\infty.$$

Thus, if $\operatorname{supp} g(t) \subset (-\frac{1}{Y}, \frac{1}{Y})$, then the right-hand side of the identity (19) contains only one term, which corresponds to the identity I of the discrete subgroup Γ. Using the Tauberian

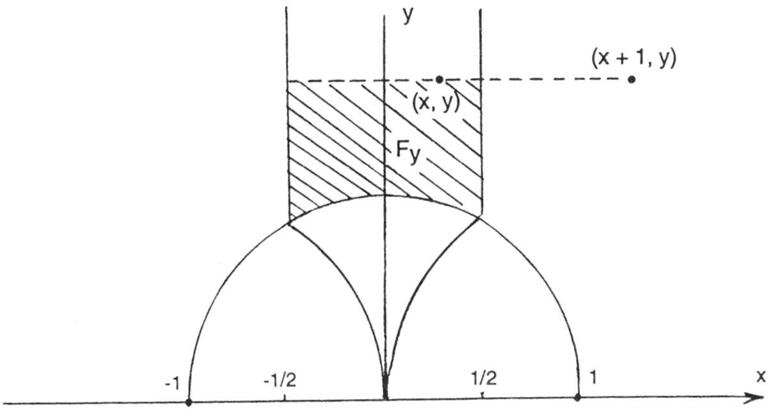

Figure 1.

theorem of the introduction one can obtain the following Theorem. There exists a constant C, independent of V and Y such that for $(x, y) \in F_Y$ and $V \to +\infty$

$$\left| \sum_{0 \le \sqrt{\lambda_l} \le V} |\psi_l(x,y)|^2 + \frac{1}{2\pi} \int_0^V |\varphi(x,y;\kappa)|^2 \, d\kappa - \frac{1}{4\pi} V^2 \right| \le C(VY + Y^2). \tag{20}$$

REMARK 1 In the estimate (20) V is relatively exact, but with respect to Y can be improved.

REMARK 2 If the fixed points exist then the asymptotic estimate (20) is valid for $(x, y) \in F$ outside some neighborhoods of fixed points (which are finite in number). Examination of the asymptotic behavior of the spectral function near the fixed points requires a more detailed analysis.

REMARK 3 Immediately in the fixed points the contribution of the elements of Γ which leave the fixed points immovable is the same as from the identity I of Γ. Therefore, in formula (20), the main term $\frac{1}{4\pi} V^2$ must be multiplied by some entire positive number.

3 THE ASYMPTOTIC BEHAVIOR OF THE DISCRETE SPECTRUM

In order to obtain from identity (19) the asymptotic behavior of the discrete spectrum we must integrate identity (19) over the domain F. It is natural at first to integrate over the domain F_Y and then tend Y to infinity. Here we must take into account that the geodesic distance between the points (x, Y) and $(x + 1, Y)$ tend to zero when Y tends to infinity. But in the Tauberian theorem we must have information about the Fourier transform in some fixed interval near zero. Therefore, as Y increases, the number of summations that one has to consider on the right-hand side of the formula (19) also increases infinitely with Y. Thus we have to calculate the contribution from some infinite sum. This calculation is analogous to the support in the Selberg's trace formula from the parabolic elements of Γ.

The final result is:

$$V(\lambda) = \sum_{\lambda_l < \lambda} 1 = \frac{1}{4\pi}|F|\lambda - \frac{1}{2\pi}[\sqrt{\lambda}\ln\lambda - \operatorname{Im} S(\lambda)] + O(\sqrt{\lambda}). \tag{21}$$

In the case of a modular group

$$S(\lambda) = \sqrt{\pi}\,\frac{\Gamma(i\lambda)}{\Gamma(\frac{1}{2} + i\lambda)} \cdot \frac{\zeta(2i\lambda)}{\zeta(2i\lambda + 1)}, \quad |F| = \frac{\pi}{3}.$$

Using the known asymptotic behavior of Γ and ζ functions and some improvement of the Tauberian theorem of the introduction (Marchenko [3]) we can obtain in the case of modular group the asymptote:

$$V(\lambda) = \frac{1}{12}\lambda - \frac{1}{\pi}\sqrt{\lambda}\ln\lambda + \frac{\sqrt{\lambda}}{\pi}(2 + \ln\pi - \ln 2) + \overline{o}(\sqrt{\lambda}).$$

REMARK The same method can be applied in the case of Dirichlet and Neumann boundary conditions on a fundamental domain [6].

REFERENCES

1. Levitan, B.M. On a special Tauberian theorem. Izv. Acad. Sci. USSR, Bd 17 (1953), 269–284 (in Russian).
2. Levitan, B.M. On eigenfunction expansion of Laplace operator. Math. Sbornic, Bd.35(77), #2 (1954), 267–316 (in Russian).
3. Marchenko, V.A. Theorems of Tauberian type in the spectral theory of differential operators. Izv. Acad. Sci. USSR, Bd. 19 (1955), 381–422 (in Russian).
4. Lax, S.D., and Phillips, R.S. Scattering Theory for Automorphic Functions. Princeton, NJ: Princeton University Press, 1976.
5. Levitan, B.M. On the asymptotic behavior of the spectral function and discrete spectrum of Laplace operator on a fundamental domain of Lobachevsky's plane. Rendiconty Math. Fis. Milano, vol. LII (1982), 195–219.
6. Levitan, B.M., and Parnovsky, L.B. On the asymptotic of discrete spectra for Dirichlet's and Neumann's problems on a regular polygon of Lobachevsky's space. Funct. Anal. Appl., Bd. 24:1 (1990), 21–28 (in Russian).

Nonlinear Boundary Value Problems for a Class of Hyperbolic Systems

RODICA LUCA Polytechnic Institute of Iaşi, Iaşi, Romania

GHEORGE MOROŞANU University of Iaşi, Iaşi, Romania

In this paper we shall investigate the following hyperbolic partial differential system

$$
(S) \qquad \begin{cases} \dfrac{\partial i_k}{\partial t} + \dfrac{\partial v_k}{\partial x} + \alpha_k(x, i_k) = 0 \\ \dfrac{\partial v_k}{\partial t} + \dfrac{\partial i_k}{\partial x} + \beta_k(x, v_k) = 0, \end{cases}
$$

$$
0 < x < 1, \ t > 0, \ k = \overline{1, n}
$$

with the boundary condition

$$
\begin{pmatrix} col\ \big(i_1(t,0), -i_1(t,1), \ldots, i_n(t,0), -i_n(t,1)\big) \\ S\Big(col\big(\frac{dw_1}{dt}(t), \ldots, \frac{dw_m}{dt}(t)\big)\Big) \end{pmatrix} \in
$$

$$
(BC) \qquad \in -G\begin{pmatrix} col\big(v_1(t,0), v_1(t,1), \ldots, v_n(t,0), v_n(t,1)\big) \\ col\big(w_1(t), \ldots, w_m(t)\big) \end{pmatrix} +
$$

$$
+\begin{pmatrix} col\big(b_1(t), b_2(t), \ldots, b_{2n-1}(t), b_{2n}(t)\big) \\ col\big(b_{2n+1}(t), \ldots, b_{2n+m}(t)\big) \end{pmatrix} \ , t > 0
$$

and the initial data:

$$(IC) \qquad \begin{cases} i_k(0,x) = i_{k0}(x), \quad v_k(0,x) = v_{k0}(x), \quad 0 < x < 1, \ k = \overline{1,n}, \\ w_j(0) = w_{j0}, \quad j = \overline{1,m} \end{cases}$$

Here $i_k(t,x)$, $v_k(t,x)$, $w_j(t)$ are the unknown functions, S is a positive diagonal matrix and α_k, β_k, G satisfy some appropriate assumptions (see $(A1)$, $(A2)$, $(A3)$ below).

This problem represents a model for a network of integrated circuits, consisting of a multiport with $2n+m$ ports at wich n distributed elements and m linear capacitors are connected. The case of time dependent sources (i.e., the functions b_j in (BC) depend on t) is investigated. For more details concerning this problem we refer the reader to Moroşanu and Luca (1992). We mention also that parabolic problems associated to integrated circuits were studied in recent years by Marinov and Lehtonen (1989), Marinov and Neittaanmäki (1988), Moroşanu (1990), Moroşanu (1991) and Moroşanu et al. (1991) (see also the references therein).

In all which follows we denote by (A1), (A2) and (A3) the following assumptions:

(A1) The functions $x \mapsto \alpha_k(x,p)$ and $x \mapsto \beta_k(x,p)$ are in $L^2(0,1)$ for any fixed $p \in \mathbf{R}$. Besides, the functions $p \mapsto \alpha_k(x,p)$ and $p \mapsto \beta_k(x,p)$ are continuous and nondecreasing from $\mathbf{R}$ into $\mathbf{R}$, for $a.e. x \in (0,1), (k = \overline{1,n})$.

(A2)

(a) $G : D(G) \subset \mathbf{R}^{2n+m} \to \mathbf{R}^{2n+m}$ is a maximal monotone (possibly multivalued) mapping. Moreover, G can be split in

$$G = \begin{pmatrix} G_{11} & G_{12} \\ G_{21} & G_{22} \end{pmatrix}$$

where $G_{11} : D(G_{11}) \subset \mathbf{R}^{2n} \to \mathbf{R}^{2n}$, $G_{12} : D(G_{12}) \subset \mathbf{R}^m \to \mathbf{R}^{2n}$, $G_{21} : D(G_{21}) \subset \mathbf{R}^{2n} \to \mathbf{R}^m$, $G_{22} : D(G_{22}) \subset \mathbf{R}^m \to \mathbf{R}^m$.

(b) There exists $k_0 > 0$ such that for all $x, y \in D(G)$ and all $w_1 \in G(x), w_2 \in G(y)$

$$\langle w_1 - w_2, x - y \rangle_{\mathbf{R}^{2n+m}} \geq k_0 \|x - y\|^2_{\mathbf{R}^{2n+m}}.$$

(A3) $\quad S = diag(s_1, s_2, ..., s_m), \ s_j > 0, \ j = \overline{1,m}.$

Remark Our assumption $(A2)(a)$ is a technical one. Obviously, it is automatically satisfied if G is a matrix. It would be interesting to consider the general case in which G is not necessarily an operator matrix as in assumption $(A2)(a)$.

We consider the following *spaces*: $X = \left(L^2(0,1)\right)^{2n}$, $\mathbf{R}^m$ and $Y = X \times \mathbf{R}^m$ with the following scalar products:

$$\langle f,g\rangle_X = \sum_{k=1}^{2n} \int_0^1 f_k(x)g_k(x)dx, \; f = col(f_1,...,f_{2n}), \; g = col(g_1,...,g_{2n}) \in X,$$

$$\langle x,y\rangle_s = \sum_{i=1}^{m} s_i x_i y_i, \; x,y \in \mathbf{R}^m,$$

$$\left\langle \begin{pmatrix} f \\ x \end{pmatrix}, \begin{pmatrix} g \\ y \end{pmatrix} \right\rangle_Y = \langle f,g\rangle_X + \langle x,y\rangle_s, \begin{pmatrix} f \\ x \end{pmatrix}, \begin{pmatrix} g \\ y \end{pmatrix} \in Y.$$

Now, let us define the *operators* A and B as folows: $A : D(A) \subset Y \to Y$ with $D(A) = \{col(i,v,w) \in Y;\; i_k,\; v_k \in H^1(0,1),\; k = \overline{1,n},\; \binom{\gamma_0 v}{w} \in D(G)$ *and* $\gamma_1 i \in -G_{11}(\gamma_0 v) - G_{12}(w)\}$,

where

$$\gamma_1 i = col(i_1(0), -i_1(1), ..., i_n(0), -i_n(1)), \; \gamma_0 v = col(v_1(0), v_1(1), ..., v_n(0), v_n(1)),$$

$$A\begin{pmatrix} i \\ v \\ w \end{pmatrix} = \begin{pmatrix} col(v_1', v_2', \ldots, v_n') \\ col(i_1', i_2', \ldots, i_n') \\ S^{-1}G_{21}(\gamma_0 v) + S^{-1}G_{22}(w) \end{pmatrix}$$

and $B : D(B) \subset Y \to Y$ with $D(B) = \{col(i,v,w) \in Y,\; B(col(i,v,w)) \in Y\}$,

$$B\begin{pmatrix} i \\ v \\ w \end{pmatrix} = \begin{pmatrix} col(\alpha_1(\cdot,i_1), \ldots, \alpha_n(\cdot,i_n)) \\ col(\beta_1(\cdot,v_1), \ldots, \beta_n(\cdot,v_n)) \\ 0 \end{pmatrix}$$

Remark Under the assumptions $(A1)$ and $(A2)a$, we can easily show that $D(A) \neq \emptyset$ and $D(A) \subset D(B)$.

LEMMA 1 If $(A2)$ and $(A3)$ hold, then the operator A is maximal monotone.

For the proof of Lemma 1, see Luca and Moroşanu [3].

LEMMA 2 If $(A1)$, $(A2)a$ and $(A3)$ hold, then the operator $A+B$ is maximal monotone.

Sketch of proof We shall consider without loss of generality that A and G are single-valued. It is easy to check that B is maximal monotone. To show that $A+B$ is maximal monotone it is sufficient to prove that for any fixed $col(p,q,r) \in Y$ equation

$$\begin{pmatrix} i \\ v \\ w \end{pmatrix} + A\begin{pmatrix} i \\ v \\ w \end{pmatrix} + B\begin{pmatrix} i \\ v \\ w \end{pmatrix} = \begin{pmatrix} p \\ q \\ r \end{pmatrix} \tag{1}$$

has (at least) a solution. To this end consider the approximate problem

$$(2)\qquad \begin{cases} \begin{pmatrix} i^\lambda \\ v^\lambda \\ w^\lambda \end{pmatrix} + A \begin{pmatrix} i^\lambda \\ v^\lambda \\ w^\lambda \end{pmatrix} + B_\lambda \begin{pmatrix} i^\lambda \\ v^\lambda \\ w^\lambda \end{pmatrix} = \begin{pmatrix} p \\ q \\ r \end{pmatrix} \\ \begin{pmatrix} i^\lambda \\ v^\lambda \\ w^\lambda \end{pmatrix} \in D(A), \qquad (\lambda > 0) \end{cases}$$

where B_λ is the Yosida approximate of B, that is

$$B_\lambda \begin{pmatrix} i \\ v \\ w \end{pmatrix} = \begin{pmatrix} col(\alpha_{1\lambda}(\cdot, i_1), \ldots, \alpha_{n\lambda}(\cdot, i_n)) \\ col(\beta_{1\lambda}(\cdot, v_1), \ldots, \beta_{n\lambda}(\cdot, v_n)) \\ 0 \end{pmatrix}$$

Here $\alpha_{k\lambda}(x, \cdot)$ and $\beta_{k\lambda}(x, \cdot)$ are the Yosida approximates of $\alpha_k(x, \cdot)$ and $\beta_k(x, \cdot)$. For any $\lambda > 0$ the problem (2) has a solution $col(i^\lambda, v^\lambda, w^\lambda) \in D(A)$. We intend to prove that $\{B_\lambda(col(i^\lambda, v^\lambda, w^\lambda));\ \lambda > 0\}$ is bounded in Y. This implies that eq. (1) is solvable (see Brezis et al., 1970). The problem (2) is equivalent with

$$(3)\qquad \begin{cases} i_k^\lambda + (v_k^\lambda)' + \alpha_{k\lambda}(x, i_k^\lambda) = p_k \\ v_k^\lambda + (i_k^\lambda)' + \beta_{k\lambda}(x, v_k^\lambda) = q_k, \ 0 < x < 1, \ k = \overline{1, n} \\ w_j^\lambda + [S^{-1}G_{21}(\gamma_0 v^\lambda) + S^{-1}G_{22}(w^\lambda)]_j = r_j, \ j = \overline{1, m} \\ \gamma_1 i^\lambda = -G_{11}(\gamma_0 v^\lambda) - G_{12}(w^\lambda). \end{cases}$$

For a fixed $col(i^0, v^0, w^0) \in D(A)$, denote

$$(4)\qquad \begin{pmatrix} i^0 \\ v^0 \\ w^0 \end{pmatrix} + A \begin{pmatrix} i^0 \\ v^0 \\ w^0 \end{pmatrix} + B_\lambda \begin{pmatrix} i^0 \\ v^0 \\ w^0 \end{pmatrix} =: \begin{pmatrix} \alpha^\lambda \\ \beta^\lambda \\ \delta^\lambda \end{pmatrix}$$

An easy computation shows that $\{B_\lambda(col(i^0, v^0, w^0))\}_{\lambda>0}$ is bounded in Y; hence, $\{\alpha_k^\lambda\}_{\lambda>0}$, $\{\beta_k^\lambda\}_{\lambda>0}$ are bounded in $L^2(0,1)$, $(k = \overline{1,n})$ and $\{\delta_j^\lambda\}_{\lambda>0}$ is bounded in $\mathbf{R}$, $(j = \overline{1,m})$. From (2) and (4) we obtain

$$\left\| \begin{pmatrix} i^\lambda - i^0 \\ v^\lambda - v^0 \\ w^\lambda - w^0 \end{pmatrix} \right\|_Y \le const., \ \forall \lambda > 0.$$

Therefore $\{i_k^\lambda\}_{\lambda>0}$ and $\{v_k^\lambda\}_{\lambda>0}$ are bounded in $L^2(0,1)$, $(k = \overline{1,n})$ and $\{w_j^\lambda\}_{\lambda>0}$ is bounded in $\mathbf{R}$, $(j = \overline{1,m})$. Now, using (3) and (4), we can show (by a computation similar to that used in Moroşanu and Petrovanu, 1985-86) that

$$(5)\quad \sum_{k=1}^n [\int_0^1 |(v_k^\lambda)'| dx + \int_0^1 |(i_k^\lambda)'| dx] \le -\sum_{k=1}^n [v_k^\lambda(1) - v_k^0(1)] \cdot [i_k^\lambda(1) - i_k^0(1)] +$$

$$+\sum_{k=1}^{n}[v_k^\lambda(0)-v_k^0(0)]\cdot[i_k^\lambda(0)-i_k^0(0)]+const.,\ \forall\lambda>0$$

and

$$(6)\qquad \|w^\lambda-w^0\|_s^2+\langle G_{21}(\gamma_0 v^\lambda)+G_{22}(w^\lambda)-G_{21}(\gamma_0 v^0)-G_{22}(w^0),w^\lambda-w^0\rangle_{\mathbf{R}^m}\le$$
$$\le const.,\ \forall\lambda>0.$$

By adding (5) and (6), we have:

$$\sum_{k=1}^{n}\Big[\int_0^1|(v_k^\lambda)'|dx+\int_0^1|(i_k^\lambda)'|dx\Big]+\|w^\lambda-w^0\|_s^2\le$$
$$\le-\Big\langle G\begin{pmatrix}\gamma_0 v^\lambda\\ w^\lambda\end{pmatrix}-G\begin{pmatrix}\gamma_0 v^0\\ w^0\end{pmatrix},\begin{pmatrix}\gamma_0(v^\lambda-v^0)\\ w^\lambda-w^0\end{pmatrix}\Big\rangle_{\mathbf{R}^{2n+m}}+const.\le const.$$

Because, besides, $\{i_k^\lambda\}_{\lambda>0}$ and $\{v_k^\lambda\}_{\lambda>0}$ are bounded in $L^2(0,1)$, it follows that $\{i_k^\lambda\}_{\lambda>0}$ and $\{v_k^\lambda\}_{\lambda>0}$ are bounded in $C([0,1])$. Now, on account of $(A1)$, we conclude that $\{\alpha_{k\lambda}(\cdot,i_k^\lambda)\}_{\lambda>0}$ and $\{\beta_{k\lambda}(\cdot,v_k^\lambda)\}_{\lambda>0}$ are bounded in $L^2(0,1)$. Q.E.D.

Now, we make the following change of functions:

$$i_k=\tilde{i}_k+\hat{i}_k,$$

where

$$\hat{i}_k(t,x)=(1-x)b_{2k-1}(t)-xb_{2k}(t),\ k=\overline{1,n}.$$

Then, our problem $(S),(BC),(IC)$ can be written as:

$$(\tilde{S})\qquad \begin{cases}\dfrac{\partial\tilde{i}_k}{\partial t}+\dfrac{\partial v_k}{\partial x}+\alpha_k(x,\tilde{i}_k+\hat{i}_k(t,x))=f_k(t,x)\\[2mm] \dfrac{\partial v_k}{\partial t}+\dfrac{\partial\tilde{i}_k}{\partial x}+\beta_k(x,v_k)=g_k(t,x),\end{cases}$$

with the boundary value condition

$$\begin{pmatrix}col\ (\tilde{i}_1(t,0),-\tilde{i}_1(t,1),\dots,\tilde{i}_n(t,0),-\tilde{i}_n(t,1))\\ S\Big(col(\frac{dw_1}{dt}(t),\dots,\frac{dw_m}{dt}(t))\Big)\end{pmatrix}\in$$

$$(\widetilde{BC})\qquad \in-G\begin{pmatrix}col(v_1(t,0),v_1(t,1),\dots,v_n(t,0),v_n(t,1))\\ col(w_1(t),\dots,w_m(t))\end{pmatrix}+$$

$$+\begin{pmatrix}col(0,\dots,0)\\ col(b_{2n+1}(t),\dots,b_{2n+m}(t))\end{pmatrix}$$

and the initial data:

$$(\widetilde{IC}) \qquad \begin{cases} \tilde{i}_k(0,x) = \tilde{i}_{k0}(x), \quad v_k(0,x) = v_{k0}(x), \quad k = \overline{1,n}, \\ w_j(0) = w_{j0}, \quad j = \overline{1,m} \end{cases}$$

where

$$\begin{cases} f_k(t,x) = -(1-x)b'_{2k-1}(t) + xb'_{2k}(t), \\ g_k(t,x) = b_{2k-1}(t) + b_{2k}(t) \\ \tilde{i}_{k0}(x) = i_{k0}(x) - (1-x)b_{2k-1}(0) + xb_{2k}(0) \end{cases}$$

Using the operators A and B, our problem $(\tilde{S}), (\widetilde{BC}), (\widetilde{IC})$ lead us to consider in the space Y the following time dependent Cauchy problem

$$(P) \qquad \begin{cases} (d/dt)\begin{pmatrix} \tilde{i} \\ v \\ w \end{pmatrix} + A\begin{pmatrix} \tilde{i} \\ v \\ w \end{pmatrix} + B\begin{pmatrix} \tilde{i} + \hat{i}(t) \\ v \\ w \end{pmatrix} = \begin{pmatrix} f(t,\cdot) \\ g(t,\cdot) \\ S^{-1}B_2(t) \end{pmatrix}, \quad t > 0 \\ \begin{pmatrix} \tilde{i}(0) \\ v(0) \\ w(0) \end{pmatrix} = \begin{pmatrix} \tilde{i}^0 \\ v^0 \\ w^0 \end{pmatrix} \end{cases}$$

where $f = col(f_1, \cdots, f_n)$, $g = col(g_1, \cdots, g_n)$, $B_2(t) = col(b_{2n+1}(t), \cdots, b_{2n+m}(t))$.

THEOREM 1 Assume that $(A1)$, $(A2)$ and $(A3)$ hold. If $b_k \in W^{1,2}(0,T)$, $k = \overline{1, 2n+m}$ (with $T > 0$ fixed), $i_{k0}, v_{k0} \in H^1(0,1)$ $(k = \overline{1,n})$, $w^0 \in \mathbf{R}^m$, $G\binom{\gamma_0 v^0}{w^0} \in D(G)$ and $B_1(0) \in \gamma_1 i^0 + G_{11}(\gamma_0 v^0) + G_{12}(w^0)$. Then, problem $(S), (BC), (IC)$ has a unique strong solution $col(i,v,w) \in W^{1,\infty}(0,T;Y)$. Moreover, $i_k, v_k \in L^\infty(0,T;H^1(0,1))$, hence $i_k, v_k \in L^\infty((0,T)\times(0,1))$, $k = \overline{1,n}$.

(We have denoted $B_1(t) = col(b_1(t), \cdots, b_{2n}(t))$).

Sketch of proof In a first stage, we assume that $b_k \in W^{2,\infty}(0,T)$, $k = \overline{1,2n}$, $b_k \in W^{1,\infty}(0,T)$, $k = \overline{2n+1, 2n+m}$ and $\alpha_k(x,\cdot)$ are Lipschits continuous with Lipschits constants independent on x. Consider the operators $K(t)$, $t \geq 0$ defined by: $D(K(t)) = D(A)$ and

$$K(t)\begin{pmatrix} \tilde{i} \\ v \\ w \end{pmatrix} = A\begin{pmatrix} \tilde{i} \\ v \\ w \end{pmatrix} + B\begin{pmatrix} \tilde{i} + \hat{i}(t) \\ v \\ w \end{pmatrix} - \begin{pmatrix} f(t,\cdot) \\ g(t,\cdot) \\ S^{-1}B_2(t) \end{pmatrix}.$$

According to Lemma 2, the operators $K(t)$ are maximal monotone in Y, $t \geq 0$. Besides, there exists $L > 0$ such that

$$\left\| K(t)\begin{pmatrix}\tilde{i}\\ v\\ w\end{pmatrix} - K(s)\begin{pmatrix}\tilde{i}\\ v\\ w\end{pmatrix}\right\|_Y \leq L|t-s|,\ \forall \begin{pmatrix}\tilde{i}\\ v\\ w\end{pmatrix} \in D(A),\ t,s\in[0,T].$$

Therefore, the family $\{K(t),\ t\geq 0\}$ satisfies Kato's conditions (see Kato, 1967). By our assumptions it follows that $col(\tilde{i}^0, v^0, w^0) \in D(A)$. Hence, problem (P) has a unique strong solution $col(\tilde{i}, v, w) \in W^{1,\infty}(0,T;Y)$, $col(\tilde{i}(t), v(t), w(t)) \in D(A)$, for all $t \in [0,T]$; Moreover, $col(\tilde{i}, v, w)$ is differentiable from the right on $[0,T]$. Therefore, $col(i,v,w)$, with $i = \tilde{i} + \hat{i}$, is solution of problem $(S),(BC),(IC)$. Now, consider α_k without Lipschits condition and replace $\alpha_k(x,\cdot)$ by their Yosida approximates $\alpha_{k\lambda}(x,\cdot)$, $\lambda > 0$, $k = \overline{1,n}$. From the reasoning above, it follows that problem (P), with $\alpha_{k\lambda}$ instead of α_k, has a unique strong solution $col(\tilde{i}^\lambda, v^\lambda, w^\lambda) \in W^{1,\infty}(0,T;Y)$, that is, $col(i^\lambda, v^\lambda, w^\lambda)$ verifies the following problem

$$(7)\quad \begin{cases} (d^+/dt)\begin{pmatrix}\tilde{i}^\lambda(t)\\ v^\lambda(t)\\ w^\lambda(t)\end{pmatrix} + A\begin{pmatrix}\tilde{i}^\lambda(t)\\ v^\lambda(t)\\ w^\lambda(t)\end{pmatrix} + B_\lambda\begin{pmatrix}\tilde{i}^\lambda(t)\\ v^\lambda(t)\\ w^\lambda(t)\end{pmatrix} = \begin{pmatrix}0\\ 0\\ S^{-1}B_2(t)\end{pmatrix}, \\ (\gamma_1 i^\lambda)(t) = -G_{11}(\gamma_0 v^\lambda)(t) - G_{12}(w^\lambda)(t) + B_1(t),\ 0\leq t < T \\ \begin{pmatrix}\tilde{i}^\lambda(0)\\ v^\lambda(0)\\ w^\lambda(0)\end{pmatrix} = \begin{pmatrix}i^0\\ v^0\\ w^0\end{pmatrix} \end{cases}$$

with

$$B_\lambda\begin{pmatrix}i\\ v\\ w\end{pmatrix} = \begin{pmatrix}col(\alpha_{1\lambda}(\cdot,i_1),\cdots,\alpha_{n\lambda}(\cdot,i_n))\\ col(\beta_1(\cdot,v_1),\cdots,\beta_n(\cdot,v_n))\\ 0\end{pmatrix}.$$

After some computations, we obtain from (7) and $(A2)b$

$$(d^+/dt)\left\|\begin{pmatrix}i^\lambda(t+h)-i^\lambda(t)\\ v^\lambda(t+h)-v^\lambda(t)\\ w^\lambda(t+h)-w^\lambda(t)\end{pmatrix}\right\|_Y^2 \leq const.\|B(t+h) - B(t)\|^2_{\mathbf{R}^{2n+m}},$$

for $0 \leq t < t+h < T$, $\lambda > 0$.

(We have denoted $B(t) = col(B_1(t), B_2(t))$).

This implies (see again (7)) that

$$\sup_{\lambda>0}\left\|(d/dt)\begin{pmatrix}i^\lambda(t)\\ v^\lambda(t)\\ w^\lambda(t)\end{pmatrix}\right\|_Y \leq const.,\quad 0<t<T.$$

Therefore

$$(8)\qquad \begin{cases} \{i_k^\lambda\}_{\lambda>0},\ \{v_k^\lambda\}_{\lambda>0} \quad \text{are bounded in} \quad L^\infty(0,T;L^2(0,1)), \\ \{w_j^\lambda\}_{\lambda>0} \quad \text{are bounded in} \quad L^\infty(0,T). \end{cases}$$

Using an argument similar to that we used in Lemma 2, we deduce that

$$(9)\qquad \left\{\frac{\partial i_k^\lambda}{\partial x}\right\}_{\lambda>0},\ \left\{\frac{\partial v_k^\lambda}{\partial x}\right\}_{\lambda>0} \quad \textit{are bounded in} \quad L^\infty(0,T;L^1(0,1)).$$

From (8) and (9) we obtain

$$\{i_k^\lambda\}_{\lambda>0} \ \textit{and}\ \{v_k^\lambda\}_{\lambda>0} \ \textit{are bounded in} \quad L^\infty((0,T)\times(0,1)).$$

This implies

$$\left\{B_\lambda\begin{pmatrix} i^\lambda \\ v^\lambda \\ w^\lambda\end{pmatrix}\right\}_{\lambda>0} \quad \textit{is bounded} \quad \textit{in} L^2(0,T;Y).$$

From the inequality

$$(d/dt)\left\|\begin{pmatrix} i^\lambda(t)-i^\mu(t) \\ v^\lambda(t)-v^\mu(t) \\ w^\lambda(t)-w^\mu(t)\end{pmatrix}\right\|_Y^2 \le -2\left\langle B_\lambda\begin{pmatrix} i^\lambda(t) \\ v^\lambda(t) \\ w^\lambda(t)\end{pmatrix} - B_\mu\begin{pmatrix} i^\mu(t) \\ v^\mu(t) \\ w^\mu(t)\end{pmatrix},\right.$$
$$\left.,\begin{pmatrix} i^\lambda(t)-i^\mu(t) \\ v^\lambda(t)-v^\mu(t) \\ w^\lambda(t)-w^\mu(t)\end{pmatrix}\right\rangle_Y, 0<t<T,\quad \lambda,\mu>0.$$

we deduce that

$$\left\|\begin{pmatrix} i^\lambda(t)-i^\mu(t) \\ v^\lambda(t)-v^\mu(t) \\ w^\lambda(t)-w^\mu(t)\end{pmatrix}\right\|_Y \le \ \textit{const.}(\lambda+\mu)^{1/2},\ 0\le t\le T,\quad ,\lambda,\mu>0.$$

This last inequality shows that $col(i^\lambda, v^\lambda, w^\lambda)$ converges to some $col(i,v,w)$ in $C([0,T];Y)$, as λ tends to 0. Next, by Lebesgue's Dominated Convergence Theorem we can prove that

$$B_\lambda\begin{pmatrix} i^\lambda \\ v^\lambda \\ w^\lambda\end{pmatrix} \longrightarrow B\begin{pmatrix} i \\ v \\ w\end{pmatrix},\ \textit{as}\lambda\to 0, \textit{strongly in}\ L^2(0,T;Y).$$

Since A and G are closed, by letting $\lambda \longrightarrow 0$ in (7), we conclude that $col(i,v,w)$ is a strong solution of problem $(S),(BC),(IC)$. In the next stage, we shall consider $b_k \in W^{1,2}(0,T)$, $k = \overline{1,2n+m}$. Let $col(i^0,v^0,w^0)$, $col(i_1^0,v_1^0,w_1^0) \in Y$ be such that $col(\tilde{i}^0,v^0,w^0), col(\tilde{i}_1^0,v_1^0,w_1^0) \in D(A)$ and $b_k, \tilde{b}_k \in W^{2,\infty}(0,T)$ $(k = \overline{1,2n})$, $b_j, \tilde{b}_j \in W^{1,\infty}(0,T)$ $(j = \overline{2n+1,2n+m})$. Then, the corresponding solutions satisfy the following inequality

(10) $$\left\| \begin{pmatrix} i(t) \\ v(t) \\ w(t) \end{pmatrix} - \begin{pmatrix} i_1(t) \\ v_1(t) \\ w_1(t) \end{pmatrix} \right\|_Y^2 \le \left\| \begin{pmatrix} i^0 \\ v^0 \\ w^0 \end{pmatrix} - \begin{pmatrix} i_1^0 \\ v_1^0 \\ w_1^0 \end{pmatrix} \right\|_Y^2 +$$

$$+const. \int_0^t \|B(s) - \tilde{B}(s)\|^2_{\mathbf{R}^{2n+m}} ds, \ 0 \le t \le T.$$

Now, we consider the sequences $\{b_k(j)\}_{j\ge1} \in W^{2,\infty}(0,T)$ such that $b_k(j) \longrightarrow b_k$, as $j \to \infty$, in $W^{1,2}(0,T)(k = \overline{1,2n})$ and $\{b_k(j)\}_{j\ge1} \in W^{1,\infty}(0,T)$ such that $b_k(j) \longrightarrow b_k$, as $j \to \infty$, in $W^{1,2}(0,T)(k = \overline{2n+1,2n+m})$. Then, fixing $col(i^0,v^0,w^0) \in Y$ such that $col(\tilde{i}^0,v^0,w^0) \in D(A)$, we can conclude that the sequence of the corresponding strong solutions $col(i(j),v(j),w(j))$ converges uniformly to some $col(i,v,w)$, which is a strong solution. The regularity properties $i_k, v_k \in L^\infty(0,T;H^1(0,1))$ $(k = \overline{1,2n})$ follow by a standard argument. Q.E.D.

THEOREM 2 Assume that $(A1)$, $(A2)$ and $(A3)$ hold. If $b_k \in L^2(0,T)$ $(k = \overline{1,2n+m})$, $i_{k0}, v_{k0} \in L^2(0,1)$ $(k = \overline{1,n})$, $w^0 \in \overline{D(G_{12} \cap D(G_{22})}$, then problem (S), $(BC),(IC)$ has a unique weak solution $col(i,v,w) \in C([0,T];Y)$.

The proof can be done by using a density argument involving estimate (10) (see Luca and Moroşanu, 1992).

Aknowledgement The authors are indebted to the referee for his valuable remarks and for suggesting some improvements.

REFERENCES

1. Brezis, H., Crandall, M. G. and Pazy, A. (1970). Perturbations of nonlinear maximal monotone sets in Banach spaces, *Comm. Pure Appl. Math.*, 23, 123-144.

2. Kato, T. (1967). Nonlinear semigroups and evolution equations, *J. Math. Soc. Japan*, 19(4), 508-520.

3. Luca, R. and Moroşanu, G. (1992). Hyperbolic problems in integrated circuits modelling, *Stud. Cerc. Mat. (Mathematical Reports)*, 44(5), 355-373.

4. Luca, R. and Moroşanu, G. On a class of nonlinear hyperbolic systems, *Memoriile Secţiilor Ştiinţifice ale Academiei Române* (in press).

5. Marinov, C. and Lehtonen, A. (1989). Mixed - type circuits with distributed and lumped parameters, *IEEE Transactions on Circuits and Systems, CAS*, 36(8), 1080-1086.

6. Marinov, C. and Neittaanmäki, P. (1988). A theory of electrical circuits with resistively coupled distributed structures: Delay time predicting, *IEEE Transactions on Circuits and Systems, CAS*, 35(2), 173-183.

7. Moroşanu, G. (1988). *Nonlinear Evolution Equations and Applications*, D. Reidel, Dordrecht.

8. Moroşanu, G. (1990). *Nonlinear Evolutions. Stability Theory and Applications*, Reports on Appl. Math. and Computing, No. 5, University of Jyväskylä, Finland.

9. Moroşanu, G. (1991). Parabolic problems associated to integrated circuits with time dependent sources, in vol. *Differential Equations and Control Theory* (V. Barbu, ed.), Pitman Research Notes in Math. Series 250, Longman Scientific and Technical, 208-216.

10. Moroşanu, G., Marinov, C. and Neittaanmäki, P. (1991). Well - posed nonlinear problems in integrated circuit modelling, *Circ. Syst. Sign. Proc.*, 10, 53-69.

11. Moroşanu, G. and Petrovanu, D. (1985-86). Nonlinear 7boundary value problems for a class of hyperbolic partial differential systems, *Atti Sem. Mat. Fis. Univ. Modena*, 34, 295-316.

Models for Diffusion-Type Phenomena with Abrupt Changes in Boundary Conditions in Banach Space and Classical Context. Asymptotics Under Periodic Shocks

GÜNTER LUMER University of Mons, Mons, Belgium

0. INTRODUCTION

The purpose of this paper is to treat, first in a quite general Banach space context, then in a classical PDE context, models for diffusion-type phenomena for which abrupt changes occur in the boundary conditions (such as a sudden drop or increase of heat on the boundary, or change in the boundary concentration of a solute, etc.) and where we have a "close to reality" description of the phenomenum (say via sup-norm, hence in particular avoiding to "hide" the abrupt change under a weakened norm).

First we shall show how such models give rise to evolution problems that can be correctly formulated, can be well posed as an "initial value-boundary value problem" in what must be a generalized sense (generalized sense indeed since the abrupt changes make it impossible for such a problem to be well posed in the usual sense in a "strict norm" as mentioned above). Then we show how such problems are solved, explicit formulas obtained for the solutions, some qualitative properties (useful in computer-graphics studies of these models) established, and asymptotics studied for periodic shocks in quite some detail. This is done first in a Banach space context, and then in a concrete PDE setting (periodic heat shocks).

To do what we have just mentioned, we first recall in section 1 the necessary basic items concerning generalized solutions (Lumer (1990)) and initial values in generalized sense (Lumer (1991a)), and we recall items concerning irregular analytic semigroups in section 2. In section 3 we describe how the earlier developed notions recalled in section 1 can now be somewhat extended and adapted, and then used to formulate properly the problems (models) in which abrupt changes on the boundary occur, and we establish the general existence and uniqueness result 3.1 for "optimal regular" solutions in that context. In section 4 we show how explicit formulas for the asymptotic values for solutions in "periodic shocks models" can be given at the Banach space level; we take up the same matters in the classical context in section 5. We discuss the inhomogeneous case (existence of a forcing term — distributed sources, etc.) in section 6. In section 7 we derive some general qualitative facts concerning periodic shocks (in the Banach space generality), which can be directly applied, as we shall do in a following paper, to computer analysis of such models in the classical context.

Since we must anyway refer to, and make use of, results of Lumer (1990), Lumer (1991a), we shall use freely in this paper the notations and notions of these articles, some of which we recall in the next section. We shall often use the abbreviations "s.g.", "i.s.g.", "loc.lip.", "lip.", for "semigroup(s)", "integrated semigroup(s)", "locally lipschitz", "lipschitz".

1. GENERALIZED SOLUTIONS AND INITIAL VALUES IN THE GENERALIZED SENSE

We refer to Lumer (1990) for the earlier introduced notions and notations of strong and mild generalized solution of order n (n-s.g.s. and n-m.g.s.) of

$$u' = Au + F(t) \quad , \quad u(0) = f \quad , \tag{1}$$

where $A \in \mathcal{L}(X)$, A is closed, X is a Banach space, $f \in X$, $F \in L^1_{loc}([0, +\infty[, X)$. A very essential role is played by the notions and notations concerning "$u(0) = f$ (g.s.)" and here more particularly "$u(0) = f$ (mild g.s.)" where "g.s." stands for "generalized sense". See Lumer (1991a). To say that u is a mild solution of $u' = Au + F(t)$ on $]0, +\infty[$ satisfying $u(0) = f$ (mild g.s.) means that, for some $n \geqslant 1$, $\exists$ an n-s.g.s. v_n of $u' = Au + F(t)$, $u(0) = f$, such that $v_n \in C^n(]0, +\infty[, X)$ and $v_n^{(n)} = u$ for $t > 0$, $w = v_n^{(n-1)}$ satisfying $w' = Aw + f + F_1(t)$ classically for $t > 0$ i.e. being a 1-s.g.s. on $]0, +\infty[$. $u = w'$ is then a mild generalized solution of order 0, 0-m.g.s., (from here on we say simply mild solution), on $]0, +\infty[$. For such a u we shall write

$$\begin{aligned} u' &= Au + F(t) \quad , \\ u(0) &= f \quad \text{(mild g.s.)} \quad , \end{aligned} \tag{2}$$

and shall often simply say (as an equivalent statement) u is a solution of (2), or u satisfies (2).

If A has the "uniqueness property" (i.e. "$u' = Au$, $u(0) = 0$" implies "$u = 0$") and is closed, as assumed in Lumer (1990), then if (2) is satisfied with two different values of n in v_n, say $n_1 < n_2 = n_1 + k$, we have necessarily $v_{n_2} = (v_{n_1})^{[k]}$ where $^{[k]}$ denotes "iterated integration from 0 to t, k times", and hence $u = v_{n_1}^{(n_1)} = v_{n_2}^{(n_2)}$ is uniquely determined by (2) without having to specify indeed which n is used.

We shall discuss later what happens without the mentioned assumption of uniqueness property, as will indeed be the case in the general situation below.

2. IRREGULAR ANALYTIC SEMIGROUPS

We refer to definitions and results in Sinestrari (1985), Da Prato and Sinestrari (1987), Lumer (1991a). Irregular bounded analytic semigroups can be defined by simply deleting in the usual definition of bounded analytic s.g. (say as given in Reed and Simon (1975)) the requirement about strong continuity at $z = 0$. It can be shown that these irregular semigroups (when considering only the nondegenerate ones) coincide essentially with the sometimes called "generalized analytic semigroups" introduced by E. Sinestrari, Sinestrari (1985), (see Lumer (1992)). These s.g. provide an essential ingredient for the construction of — actually the explicit expressions for — the solutions in the problems with shocks treated below. Next we recall some facts (Lumer (1992), Lumer (1991a), Sinestrari (1985), Da Prato and Sinestrari (1987)) and prove some properties of the mentioned s.g., properties directly needed below. We assume the reader has some knowledge of integrated semigroups.

PROPOSITION 2.1 (Lumer (1992)). If $Q(z)$ is an irregular bounded analytic s.g. of angle α ($Q(t)$ the restricion of $Q(z)$ to the real $t > 0$) then $Q(z) = S'(z)$ for $z \in \Gamma_\alpha = \{0 \neq z \in \mathbb{C} : |\arg z| < \alpha\}$, where $S(z)$ is a lip. analytic i.s.g. of angle α. ($Q(z)$ is nondegenerate iff $S(z)$ is nondegenerate.) $S(z)$ has the same generator (see Lumer (1992)) as $Q(z)$ if the latter is nondegenerate. $Q(z)$ (extended by 1 at $z = 0$) is strongly continuous on $\overline{D(A)}$ (converges strongly to 1 on $\overline{D(A)}$) as $z \to 0$ in any $\Gamma_{\alpha'}$, $0 < \alpha' < \alpha$.

Next we prove facts stated in part or in weaker form in Lumer (1992), Lumer (1991a), using results from Lumer (1992).

PROPOSITION 2.2 Let $Q(z)$ be an irregular bounded analytic nondegenerate s.g. of angle α and generator A.
Then $Q(z) : X \to D(A) \quad \forall z \in \Gamma_\alpha$. If $\exists\ A^{-1} \in \mathcal{B}(X)$ then $\exists\ \alpha > 0$, $M > 0$, such that

$$\|Q(t)\| \leqslant Me^{-\alpha t} \quad \forall\, t > 0 \quad . \tag{3}$$

(Remark. Using the identification proved in Lumer (1992) between the $Q(z)$ in 2.2 and essentially the generalized analytic s.g. introduced in Sinestrari (1985), 2.2 can also be derived from facts in Da Prato and Sinestrari (1987), Sinestrari (1985).)

Proof of 2.2. Let $S(z)$ be as in 2.1 above. Assume first $\exists\ A^{-1} \in \mathcal{B}(X)$. Then (Lumer (1992) and Lumer (1991a)) $S(z) = (e^{zA_0} - 1)A^{-1}$ for $z \in \Gamma_\alpha$, where A_0 is the part of A in $\overline{D(A)}$, $Q(z) = A_0 e^{zA_0} A^{-1}$ where e^{zA_0} is a (usual) analytic s.g. on $X_0 = \overline{D(A)}$. $\forall\ t > 0$ $\ e^{zA_0} : X_0 \to D(A_0)$; and since $\exists\ A_0^{-1} \in \mathcal{B}(X_0)$, $\|e^{tA_0}\| \leqslant M_0 e^{-\alpha t}$ for some M_0, $\alpha > 0$, all $t \geqslant 0$; hence choosing some $\epsilon > 0$ we have : for $t > \epsilon$, $\|Q(t)\| = \|A_0 e^{(t-\epsilon)A_0} e^{\epsilon A_0} A^{-1}\| = \|e^{(t-\epsilon)A_0} A_0 e^{\epsilon A_0} A^{-1}\| = \|e^{(t-\epsilon)A_0} Q(\epsilon)\| \leqslant M_0 e^{-\alpha(t-\epsilon)} C$ (since $\|Q(t)\| \leqslant$ some $C > 0$ for all $t > 0$). Hence $\|Q(t)\| \leqslant (Ce^{\alpha\epsilon} + CM_0 e^{\alpha\epsilon})e^{-\alpha t}$, so $\|Q(t)\| \leqslant Me^{-\alpha t} \quad \forall t > 0$ where $M = Ce^{\alpha\epsilon}(1 + M_0)$. This proves the part of our statement concerning (3).

Also it is clear from $Q(t) = e^{(t-\epsilon)A_0} Q(\epsilon)$, $t > \epsilon$, $\epsilon > 0$ otherwise arbitrary, that $Q(t)$ maps into $D(A_0) \subset D(A)$; and since in the general case we can reduce ourselves to the previous assumption about A^{-1} replacing A by some $A - \lambda_0$ (because $\exists\ R(\lambda, A)$ for $\lambda > 0$ $Q(t)$ being bounded), with the original $Q(t)$ then replaced by $e^{-\lambda_0 t} Q(t)$, it is clear that the other part of the statement also holds as claimed.

∎

The following result is quite important for what we do later on in dealing with periodic shocks.

LEMMA 2.3 Let $Q(z)$ be an irregular bounded analytic s.g. nondegenerate. Then if $\|Q(t)\| \leqslant C$ for a constant C, $\exists$ an equivalent norm $|\ \ |$ on X such that $Q(t)$, $t > 0$, becomes a s.g. of contractions, i.e. $|Q(t)| \leqslant 1 \quad \forall\, t > 0$. If $\exists\, A^{-1} \in \mathcal{B}(X)$ then we can find an equivalent norm on X, $|\ \ |$, and $\alpha > 0$, such that

$$|Q(t)| \leqslant e^{-\alpha t} \qquad \forall\, t > 0 \quad . \tag{4}$$

Proof. Let $S(z)$, $S'(z) = Q(z)$, be as in 2.1, 2.2, $S(\ )$ and $Q(\ )$ having generator A. We know from Kellemann-Hieber (1989), Arendt (1987), that a closed operator $G \in \mathcal{L}(X)$ is the generator of a loc.lip. integrated (nondegenerate) s.g. $S(t)$ satisfying $\|S(t) - S(s)\| \leqslant Me^{a\omega}|t - s|$ if $0 \leqslant t,\ s \leqslant a$ for any $a \geqslant 0$, with constants $M, \omega \geqslant 0$ independent of a, iff $\exists\ R(\lambda, G)$ for $\lambda > \omega$ with $\|R(\lambda, G)^n\| \leqslant M/(\lambda - \omega)^n$ for $n = 1, 2, \ldots$. Now if $\exists\ A^{-1} \in \mathcal{B}(X)$, then (3) of 2.2 holds and $Q^\#(t) = e^{\alpha t}Q(t)$ satifies $\|Q^\#(t)\| \leqslant M$ and has generator $A^\# = A + \alpha$, $S^\#(t) = \int_0^t Q^\#(s)ds$, $\|S^\#(t) - S^\#(s)\| \leqslant M|t - s|$. So for $\lambda > 0$, $\|\lambda^n R(\lambda, A^\#)^n\| \leqslant M$ for $n = 1, 2, \ldots$, and by a renorming lemma of Pazy (1983) (lemma 1.5.1), not requiring dense domain, $\exists\ |\ \ |$ equivalent such that $|\lambda R(\lambda, A^\#)| \leqslant 1$ for $\lambda > 0$. Using again the above Kellerman-Hieber, Arendt, results, $|S^\#(t) - S^\#(s)| \leqslant |t - s|$ in the new norm, and since $Q^\#(t) = S^{\#\prime}(t)$, $|e^{\alpha t}Q(t)| = |Q^\#(t)| \leqslant 1\ \forall\, t > 0$, hence $|Q(t)| \leqslant e^{-\alpha t}$. (The rest of the statement is also established in the proof above.)

■

One has at once from the equivalent renorming result of 2.3 the following

COROLLARY 2.4 Let $a_0 + a_1 z + a_2 z^2 + \cdots + a_n z^n + \cdots$ be a power series in the complex variable z with complex coefficients, and with radius of convergence $\geqslant 1$. Then for $Q(t)$ an irregular bounded analytic s.g. (nondegenerate) of generator A such that $\exists\ A^{-1} \in \mathcal{B}(X)$, and any $r > 0$,

$$a_0 + a_1 Q(r) + a_2 Q(r)^2 + \cdots + a_n Q(r)^n + \cdots \quad \text{converges in } \mathcal{B}(X)\text{-norm.} \tag{5}$$

3. THE BANACH SPACE SETUP FOR (MODELLING OF) DIFFUSION-TYPE PROBLEMS (PHENOMENA) WITH ABRUPT CHANGES IN BOUNDARY CONDITIONS

Roughly speaking, to set up an appropriate model, in a strict norm, accounting for sudden and essentially unrestricted changes in boundary values, one can no longer adequately include the description of such boundary behavior in the definition of the domain $D(\cdot)$ of an operator coming basically from the expression of the law of evolution of the phenomenum (system) in the interior. Instead we must use both an operator $\hat{A}$ with an appropriate sufficiently large domain to allow for the essentially arbitrary shocks, values, on the boundary (while having the mentioned action in the interior) and a boundary operator B by means of which we can specify exactly the abrupt change of boundary conditions occuring in a given shock. (So we shall be dealing with initial value-boundary value problems rather than simply initial value problems.)

We describe next the precise setup in the Banach space context of the model we treat here.

We make use of the notions recalled in section 1, but further extended to meet the present more general needs : (i) in defining generalized solutions, $u(0) = f$ in generalized sense, we no longer assume from here on, unless otherwise specified, that A has the uniqueness property (which is definitely not satisfied in general and in the most interesting cases for the operator $\hat{A}$ we consider below); because of the lack of uniqueness property for A, we no longer know the fact mentioned after (2) in section 1, i.e. that u is uniquely determined by (2); (ii) from now on the notions of generalized solutions, and initial values in generalized sense (in particular in this paper "$u(0) = f$ (mild g.s.)"), are also defined when using finite intervals ($[0, a[$, $]0, a[$, $[0, a]$, or $[t_0, t_1[$, etc.) in the same way than for $[0, +\infty[$, $]0, +\infty[$, so we can write "$u' = Au + F(t)$, $u(t_0) = f$ (g.s.) (or $u(t_0) = f$ (mild g.s.)), $t_0 < t \leqslant t_1$" i.e. holding for the interval $]t_0, t_1]$ ("for $t_0 < t \leqslant t_1$").

We set up now our model in X (the corresponding model in classical context is treated in section 5). We suppose : (I) we are given $A \in \mathcal{L}(X)$ generator of an irregular bounded analytic s.g. (nondegenerate) $Q(t)$, $\overline{D(A)} = X_0 \neq X$, such that $\exists\ A^{-1} \in \mathcal{B}(X)$. (II) We are given a linear subspace $H \subset X$ (one may intuitively think of H as a space of harmonic functions), and $B \in \mathcal{B}(X)$ called the boundary operator, such that $B : X \to H$, $B : H \to H$ identically, $B : X_0 \to 0$ (0 stands for $\{0\}$). (Notice that this implies $H \cap \overline{D(A)} = 0$, $\overline{D(A)} \oplus H$ is well defined in X and every $\hat{f} \in \overline{D(A)} \oplus H$ has a unique representation of the form $f + h$ with $f \in \overline{D(A)}, h \in H$. [1])

We then define $\hat{A} \in \mathcal{L}(X)$ by $D(\hat{A}) = D(A) \oplus H$, and for $\hat{f} = f + h \in D(\hat{A})$, $f \in D(A)$ and $h \in H$, $\hat{A}\hat{f} = Af$. (One sees immediately that H is a closed subspace, $\hat{A}$ a closed operator.)

Finally, under these circumstances, we shall say that in our model the "study of the evolution of our system (phenomenum) for $0 < t \leqslant t_1$ after the system reaches continuously (from the left) the state f at $t = 0$ (with some boundary value $\varphi_0 \in H$), under the shock (abrupt change) due to the boundary value going from φ_0 at $t = 0$ to $\varphi_1 = \varphi \in H$ for $0 < t \leqslant t_1$, and also under the action of a forcing term F" corresponds to the study of the following initial value-boundary value problem :

$$\begin{aligned} &u' = \hat{A}u + F(t) \quad , \\ &u(0) = f \quad \text{(mild g.s.)}, \\ &Bu(t) = \varphi \quad , \quad 0 < t \leqslant t_1 \quad . \end{aligned} \tag{6}$$

Also 0 can be replaced by any t_0 using simply "translation" to define things analoguously for a shock occuring at t_0, (6) then becomes

$$\begin{aligned} &u' = \hat{A}u + F(t) \quad , \\ &u(t_0) = f \quad \text{(mild g.s.)}, \\ &Bu(t) = \varphi_1 \quad , \quad t_0 < t \leqslant t_1 \,. \end{aligned} \tag{6'}$$

[1] Here the symbol $\oplus$ is used of course in the sense of "algebraic direct sum" as defined for instance in Goldberg (1966).

We must now clarify certain aspects concerning (6), ((6')), and establish basic properties of solutions. First notice that in the definition of "u satisfies (6) (or (6'))" there is involved a regularization via a $\hat{v}_n$ which is not unique (see section 1 and (i), (ii) of the third paragraph of section 3, $\hat{v}_n$ taking here, in (6) or (6'), the place of v_n just as $\hat{A}$ takes the place of the A of section 1 and of (i), (ii) of section 3) and we may consider for a given u satisfying (6) (or (6')) the smallest n in the $\hat{v}_n$ corresponding to that u, say n_0; we then say that this u has order of regularization n_0. A solution of (6') is called optimal regular iff there does not exist another solution of the same problem having a lower order of regularization. We shall show that the problem (6) (or (6')) is uniquely determined and in an appropriate sense well-posed (when considering optimal regular solutions), and show that its solution (mild solution) can be expressed with the help of the irregular s.g. $Q(t)$ generated by A.

Since A generates a loc.lip. i.s.g. (Kellerman and Hieber (1989), Lumer (1991b), Da Prato-Sinestrari (1987)), it has the uniqueness property, Arendt (1987), while this certainly does not hold in general for $\hat{A}$ the boundary condition $Bu(t) = \varphi$ being necessary to determine the problem. Notice however that, as we show later, the shocks problem (in $[0, t_1]$) for which $\varphi = 0$, i.e. (6) with $Bu(t) = 0$, is (in terms of optimal regular solutions) equivalent to the simple initial value problem $u' = Au + F(t)$, $u(0) = f$ (mild s.g.) (which models merely a shock of the type "a drop (or rise) of temperature on the boundary from φ_0 to 0 as we pass from $t = 0$ to $0 < t \leqslant t_1$"); in that situation no boundary operator B is needed, and A can be used instead of the extended $\hat{A} \supset A$.
In the general situation (6), one has

THEOREM 3.1 For any $f \in X, F \in L^1_{loc}([0, t_1], X)$, $\varphi \in H$, (6) has a unique optimal regular solution u, which is a mild solution of $u' = \hat{A}u + F(t)$ on $]0, t_1]$ and is given for $0 < t \leqslant t_1$ by

$$u(t) = \varphi + Q(t)(f - \varphi) + \int_0^t Q(t-s)F(s)ds \quad . \tag{7}$$

If $\varphi = 0$ (6) is (in terms of optimal regular solutions) equivalent to the mere initial value problem $u' = Au + F(t)$, $u(0) = f$ (mild g.s.), $0 < t \leqslant t_1$.

Proof. Consider any given f, F, as in the above statement. Let $S(t)$, $Q(t) = S'(t)$ be the i.s.g. and irregular s.g., respectively, generated by A (see 2.1 above). Then from the usual results on loc. lip. i.s.g., Kellerman and Hieber (1989), Lumer (1990), Lumer (1991b), one knows that $\exists$ in $C^1([0, t_1], X)$, $t \mapsto v_2(t, f) \in D(A)$, satisfying in $[0, t_1]$

$$v_2' = Av_2 + t(f - \varphi) + F_2(t), v_2(0) = 0 \quad , \tag{8}$$

where for $t \in [0, t_1]$

$$v_2'(t) = S(t)(f - \varphi) + \int_0^t S(t-s)F(s)ds \quad . \tag{9}$$

According to Da Prato and Sinestrari (1987) the integral term in the right hand side of (9) is C^1 for $0 \leqslant t \leqslant t_1$, and by 2.1 the other term is C^1 for $0 < t \leqslant t_1$, and (see Lumer (1991a), Lumer (1991b)) we have moreover

$$\text{for } 0 < t \leqslant t_1 \quad \exists v_2'' \text{ and}$$

$$v_2''(t) = Q(t)(t - \varphi) + \int_0^t Q(t-s)F(s)ds \quad . \tag{10}$$

Now set :

$$\hat{v}_2(t) = \hat{v}_2(t, f, F) = \frac{1}{2}t^2\varphi + v_2(t) \quad \text{for } 0 < t \leqslant t_1 \quad . \tag{11}$$

Then $\exists\ \hat{v}_2''$ for $0 < t \leqslant t_1$, and setting $\hat{v}_2'' = u$,

$$\begin{aligned} u(t) &= \varphi + v_2''(t) = \\ &\quad \varphi + Q(t)(f - \varphi) + \int_0^t Q(t-s)F(s)ds \quad , \\ &\quad \text{for } 0 < t \leqslant t_1. \end{aligned} \tag{12}$$

We have since $\varphi \in H \subset D(\hat{A})$, $\hat{A}\varphi = 0$, $\hat{A}\hat{v}_2 = Av_2$, from (11), (8), (12).

$$\begin{aligned} \hat{v}_2' &= t\varphi + v_2' = t\varphi + Av_2 + t(f - \varphi) + F_2(t) \\ &= \hat{A}\hat{v}_2 + tf + F_2(t), \hat{v}_2(0) = v_2(0) = 0 \quad , \\ &\quad \text{for } 0 \leqslant t \leqslant t_1 \quad ; \text{ and} \\ Bu(t) &= B\varphi + Bv_2''(t) = \varphi \quad , \quad \text{for } 0 < t \leqslant t_1. \end{aligned} \tag{13}$$

Also, since $v_2 \in C^2(]0, t_1], X)$, from (8) A being closed, $v_2'' = Av_2' + (f - \varphi) + F_1(t)$, $\hat{v}_2'' = \varphi + \hat{A}\hat{v}_2' + (f - \varphi) + F_1(t) = \hat{A}\hat{v}_2' + f + F_1(t)$, so setting for $0 < t \leqslant t_1$ $w(t) = \hat{v}_2'(t), w \in C^1(]0, t_1], X)$,

$$w' = \hat{A}w + f + F_1(t) \quad \text{for } 0 < t \leqslant t_1. \tag{14}$$

This says that w is a 1-s.g.s. of $u' = \hat{A}u + F(t)$ on $]0, t_1]$, and $u = w'$ a 0-m.g.s., i.e. a mild solution, on $]0, t_1]$, of the latter problem; also from (13) and (12), (11), we have that u is a mild solution of $u' = \hat{A}u + F(t)$ on $]0, t_1]$ satisfying $u(0) = f$ (mild g.s.). So finally the latter, (13), and (12), show that (6) is indeed satisfied with a u given by formula (7), which concludes the "existence part" of the proof.

In the opposite direction, let now $u^{\#}$ be any optimal regular solution of (6). Since we have just shown that there always exists a solution $u = \hat{v}_2''$ given by (7) our purpose will be to show that $u^{\#} = u$. Since $u^{\#}$ is optimal regular we must have $1 \leqslant n \leqslant 2$, and we may assume $n = 2$ since we can otherwise work with $\hat{v}_1^{\#[1]}$ (where $^{[k]}$ is defined in the second paragraph of section 1) because $(\hat{v}_1^{\#[1]})^{(2)} = \hat{v}_1^{\#(1)} = u^{\#}$ in that case. So we have $\hat{v}_2^{\#}$ in $C^1([0, t_1], X)$ satisfying

$$\begin{aligned} &\hat{v}_2^{\#\prime} = \hat{A}\hat{v}_2^{\#} + tf + F_2(t), \hat{v}_2^{\#}(0) = 0, 0 \leqslant t \leqslant t_1, \\ &\exists \text{ for } 0 < t \leqslant t_1 \quad \hat{v}_2^{\#\prime\prime} = u^{\#} \text{ continuous}, \\ &Bu^{\#}(t) = \varphi \ , \ 0 < t \leqslant t_1 \quad . \end{aligned} \tag{15}$$

Let us write $\hat{v}_2^{\#}(t) = v_2^{\#}(t) + h^{\#}(t)$ in the direct sum decomposition $D(\hat{A}) = D(A) \oplus H$, $0 \leqslant t \leqslant t_1$.

$$B\hat{v}_2^{\#}(t) = h^{\#}(t) \quad , \quad 0 \leqslant t \leqslant t_1 \quad , \tag{16}$$

so in particular $h^{\#}$, hence $v_2^{\#}$, are in $C^1([0,t_1],X)$. Now for $0 < t \leqslant t_1$,

$$B\hat{v}_2^{\#''} = h^{\#''} = Bu^{\#} = \varphi \quad , \tag{17}$$

and since $\hat{v}_2^{\#}(0) = \hat{v}_2^{\#'}(0) = 0$ by (15), from (16)

$$h^{\#}(0) = h^{\#'}(0) = 0 \quad . \tag{18}$$

Hence from (17) and (18),

$$h^{\#}(t) = \frac{1}{2}t^2\varphi \quad , \quad 0 \leqslant t \leqslant t_1 \quad . \tag{19}$$

Then going back with (19) to (15), we have

$$\hat{v}_2^{\#'} = v_2^{\#'} + t\varphi = Av_2^{\#} + tf + F_2(t) \quad , \quad \hat{v}_2^{\#}(0) = v_2^{\#}(0) = 0 \quad . \tag{20}$$

Hence :

$$v_2^{\#'} = Av_2^{\#} + t(f-\varphi) + F_2(t) \quad , \quad v_2^{\#}(0) = 0 \quad ,$$

which by the uniqueness property for the generator A, means that $v_2^{\#}$ coincides with the uniquely determined v_2 given by (9) above. So $\hat{v}_2^{\#} = v_2 + \frac{1}{2}t^2\varphi$, $u^{\#}(t) = \hat{v}_2^{\#''}(t) = \varphi + Q(t)(f-\varphi) + \int_0^t Q(t-s)F(s)ds$, $0 < t \leqslant t_1$, which proves the uniqueness of the optimal regular solution of problem (6).

Finally let us consider the case in which $\varphi = 0$, $Bu(t) = 0$, $0 < t \leqslant t_1$. Then the above facts show that $u(t) = v_2''(t)$, v_2 satisfying $v_2' = Av_2 + tf + F_2(t)$ with $v_2(0) = 0$, so in fact in that case problem (6) is (in terms of optimal regular solutions) equivalent to the initial value problem indicated in our statement.

■

Of course things are entirely similar for the problem (6') instead of (6).

From here on in this paper we shall only consider optimal regular solutions and shall thus omit the expression "optimal regular".

4. PERIODIC SHOCKS. ASYMPTOTICS

Consider the solution (i.e. mild solution) $u(t)$ describing the behavior of the system (model) studied here (section 3, 3.1) with a shock at $t = 0$. For $t > 0$, as long as no new abrupt change occurs, say in the context of 3.1 up to $\tilde{t}$ where $0 < \tilde{t} < t_1$, solutions of (6') can be glued together in the natural way, i.e. if we take the solution $u(t)$ of (6) up to $\tilde{t} < t_1$ and then the solution $\tilde{u}$ of (6') with initial value $\tilde{f} = u(\tilde{t})$ and unchanged boundary value $B\tilde{u}(t) = \varphi \quad \tilde{t} < t \leqslant t_1$, and the same F, then

$$\tilde{\tilde{u}}(t) = \begin{cases} u(t) & \text{for } 0 < t \leqslant \tilde{t} \\ \tilde{u}(t) & \text{for } \tilde{t} < t \leqslant t_1 \end{cases} \tag{21}$$

is the solution (as considered in (6), (6')) with initial value f, boundary value φ, forcing term $F(t)$, between 0 and t_1 i.e. for $0 < t \leqslant t_1$. We leave it to the reader to check this fact. (Of course in this paragraph initial values are meant (mild s.g.), but we omit the expression "(mild g.s.)" here, and also often below when it is not necessary given the context.)

If we reach a new abrupt change at $t = t_1$ we pass to the boundary value φ_2 for $t_1 < t \leqslant t_2$; then the uniquely determined solution of

$$\begin{aligned} &u' = \hat{A}u + F(t), \\ &u(0) = f \quad \text{(mild g.s.)}, \\ &Bu(t) = \varphi_{i+1} \text{ for } t \in]t_i, t_{i+1}], i = 0, 1, \varphi_{i+1} \in H, \end{aligned} \tag{22}$$

(where we write $t_0 = 0$, and the φ of the preceding paragraph is now written φ_1), is defined by gluing together as in the preceding paragraph containing (21) the corresponding solutions u, $\tilde{u}$, with initial value (mild g.s.) f at $t_0 = 0$, $\tilde{f} = u(t_1)$ at t_1, and respective boundary values φ_1, φ_2 (and same F; but since F is only L^1_{loc} this simply means we are giving F_{i+1} integrable on each $]t_i, t_{i+1}]$). This definition is consistent by what was said above, and also extends immediately to the case of any number of abrupt changes, at $t_0 < t_1 < t_2 < \cdots < t_n < \cdots$.[2]

$$\begin{aligned} &u' = \hat{A}u + F(t) \quad , \\ &u(t_0) = f \quad \text{(mild g.s.)}, \\ &Bu(t) = \varphi_{i+1} \quad \text{for } t \in]t_i, t_{i+1}], i = 0, 1, 2, \ldots, \\ &\varphi_{i+1} \in H \quad . \end{aligned} \tag{22'}$$

We shall now treat the following situation of periodic shocks of period $2r > 0$, where the abrupt changes are located at $t = 0, r, 2r, 3r, 4r, \ldots$, with $\varphi_0 = \varphi$, $\varphi_1 = 0$, $\varphi_2 = \varphi$, $\varphi_3 = 0$, $\varphi_4 = \varphi, \ldots, 0 \neq \varphi \in H$, and for the moment $F = 0$. We examine the asymptotic behavior starting from any initial value $f \in X$ at 0 (with $Bf = \varphi_0 = \varphi$).

By what we have seen, starting with f at 0 we have a solution u reaching $t_1 = r$ with the value $u(t_1) = u(r) = Q(r)f$, next by 3.1 (7), and what is said in the paragraph containing (22) and (22'), $u(2r) = \varphi + Q(r)(Q(r)f - \varphi) = \varphi - Q(r)\varphi + Q(r)^2 f$, $u(3r) = Q(r)u(2r) = Q(r)\varphi - Q(r)^2\varphi + Q(r)^3 f$, $u(4r) = \varphi + Q(r)(u(3r) - \varphi) = \varphi - Q(r)\varphi + Q(r)^2\varphi - Q(r)^3\varphi + Q(r)^4 f, \ldots$. By induction we see easily that at the end of the n-th odd numbered interval (i.e. $t = (2n-1)r$) we reach the value $(n > 1)$:

$$(Q(r) - Q(r)^2 + Q(r)^3 - \cdots (-1)^{2n-3} Q(r)^{2n-2})\varphi + Q(r)^{2n-1} f \quad , \tag{23}$$

and at the end of the n-th even numbered interval $(n > 0)$ we reach the value

$$(1 - Q(r) + Q(r)^2 - \cdots (-1)^{2n-1} Q(r)^{2n-1})\varphi + Q(r)^{2n} f \quad . \tag{24}$$

---

[2]It is important to notice that while we take $\hat{A}$ constant in (22') through the shocks, the general theory developed in the previous section and this section up to (22') (and also thereafter but for the latter part we should say "to a certain extent, with appropriate assumptions") can be applied (or adapted) also to situations where not only the values of φ vary (φ_{i+1} becomes $\varphi_{i+2}, \varphi_{i+3}, \ldots$, as we go through the shock points $t_{i+1}; t_{i+2}, \ldots$) but furthermore the $\hat{A}$ can change as we go through these shock points $t_{i+1}, t_{i+2}, \ldots$ (satisfying of course at least always the same basic assumptions made for $\hat{A}$ in section 3).

Now $Q(t)$ has exponential decay (2.2 (3)) so $Q(r)^k = Q(kr) \to 0$ as $k \to \infty$, hence $Q(r)^{2n}f$, $Q(r)^{2n-1}f \to 0$ as $n \to \infty$; but what about the terms $(Q(r) - Q(r)^2 + \cdots)\varphi$, $(1-Q(r)+Q(r)^2-\cdots)\varphi$ in (23), (24)? The latter converge indeed (uniformly in $\|\varphi\| \leqslant$ constant) by 2.4 since $1 - Q(r) + Q(r)^2 - \cdots$ converges in $\mathcal{B}(X)$-norm to $(1+Q(r))^{-1}$. Hence the expressions in (23), (24) converge to the values w_i, w_s, respectively (intuitively "lower value" and "upper value" thinking of φ as "> 0")

$$\begin{aligned} w_s &= (1+Q(r))^{-1}\varphi \\ w_i &= Q(r)(1+Q(r))^{-1}\varphi \quad . \end{aligned} \tag{25}$$

From (25) we have $w_s + w_i = (1+Q(r))^{-1}\varphi + Q(r)(1+Q(r))^{-1}\varphi = (1+Q(r))(1+Q(r))^{-1}\varphi = \varphi$, i.e. the useful relation

$$w_s + w_i = \varphi \quad . \tag{26}$$

Consider now our given periodic shocks situation (problem) when starting at 0 with $f = w_s$, i.e. with $u(0) = w_s$ (mild g.s.). Then in $]0, 2r]$ the solution u is of the form (since $Q(r)w_s = w_i$)$u^\star(t)$,

$$u^\star(t) = \begin{cases} Q(t)w_s & 0 < t \leqslant r \\ \varphi + Q(t-r)(w_i - \varphi) & r < t \leqslant 2r \quad . \end{cases} \tag{27}$$

But $u^\star(2r) = \varphi + Q(r)(w_i - \varphi) =$ (using (26)) $\varphi - Q(r)w_s = \varphi - w_i = w_s$; hence in $]2r, 4r]u(t) = u^\star(t-2r)$, $u(4r) = u^\star(2r) = w_s$, etc. Hence u is a periodic solution of the periodic shocks problem considered, obtained by extending $2r$-periodically $u^\star$ in (27) above to all the positive reals.

LEMMA 4.1 There can exist only one periodic solution (of period $2r$) for the periodic shocks problem considered above.

Proof. A periodic solution as mentioned in the statement must come from an initial value $f \in X$ satisfying $\varphi + Q(r)(Q(r)f - \varphi) = f$. Hence

$$(1 - Q(r))\varphi = (1 - Q(r)^2)f \quad . \tag{28}$$

As we have already seen, by 2.3 $|Q(r)| < 1$, $\exists\ (1 - Q(r))^{-1} \in \mathcal{B}(X)$, so from (28) multiplying by $(1 - Q(r))^{-1}$

$$\varphi = (1 + Q(r))f \quad . \tag{29}$$

Since also $\exists\ (1 + Q(r))^{-1} \in \mathcal{B}(X)$,

$$f = (1 + Q(r))^{-1}\varphi = w_s \quad , \tag{30}$$

which proves the stated uniqueness.

∎

As we have seen, for any $f \in X$ there is a unique global solution (mild solution) with initial value f (mild g.s.) in our given periodic shocks context (model). We call that solution "the orbit corresponding to f (in the given periodic shocks model)"[3].

From (24) and what follows thereafter it is clear that for any orbit $u(t) = u(t, f)$, $u(2nr) \to w_s$ as $n \to \infty$; using this, the existence of the periodic orbit u_p defined by $2r$-periodic extension from (27) and 4.1, it is seen at once that $u(t, f)$ tends asymptotically to u_p in the following sense : $\forall\, \epsilon > 0 \;\; \exists\, t_\epsilon > 0$ such that $\|u(t, f) - u_p(t)\| \leqslant \epsilon$ as soon as $t > t_\epsilon$. Summarizing we have :

THEOREM 4.2 For the periodic shocks model (with period $2r > 0$) described above in the paragraph following (22'), there exists a unique periodic orbit u_p, corresponding to the initial value $w_s = (1 + Q(r))^{-1}\varphi$ (and given in $]0, 2r]$ by (27) above). If $u(t, f)$ is any orbit, then $u(t, f)$ tends to $u_p(t)$ asymptotically; in this sense one can say that u_p is an asymptotically attracting (periodic) orbit.

5. MODEL IN THE CLASSICAL CONTEXT. PERIODIC HEAT SHOCKS

Let Ω be a bounded domain in R^N with C^∞ boundary (a very regular domain in the sense of Lions (1965)). We are given a uniformly elliptic (as defined in Friedman (1969)) operator $-A(x, D)$,

$$A(x, D) = \sum_{i,j=1}^{N} \frac{\partial}{\partial x_i} a_{ij}(x) \frac{\partial}{\partial x_j} + c_0(x) \quad , \tag{31}$$

with C^∞ coefficients, real, $a_{ij} = a_{ji}$, $c_0 \leqslant 0$, all defined on $\overline{\Omega}$. To $A(x, D)$ we associate two operators : A in $X = C(\overline{\Omega})$, and $\mathcal{A}$ in $\mathcal{X} = L^2(\Omega)$, as follows

$$\begin{aligned} D(A) &= \{f \in C_0(\Omega) \cap W^{2,p}(\Omega) : A(x, D)f \in C(\overline{\Omega})\} \quad , \\ Af &= A(x, D)f \quad \text{for } f \in D(A) \quad , \end{aligned} \tag{32}$$

where $p > N$ (in which case it can be shown that A does not depend on p). (See Lumer (1977), Paquet (1978), Pazy (1983), [4].)

$$\begin{aligned} D(\mathcal{A}) &= H^2(\Omega) \cap H_0^1(\Omega) \quad , \\ \mathcal{A}f &= A(x, D)f \in L^2(\Omega) \quad \text{for } f \in D(\mathcal{A}). \end{aligned}$$

It is easily shown that $A \subset \mathcal{A}$ in $L^2(\Omega) = \mathcal{X}$, and it follows from Stewart (1974), and arguments as used in Lumer and Paquet (1977, 1979), see 5.1 below, that A generates an

[3] In the terminology used here "the orbit associated to f" is not a set but the map $t \mapsto u(t, f)$, for $t > 0$; we also call sometimes the map $t \mapsto u(t, f)$ "the orbit corresponding to the initial value f" or "the orbit with initial value f".

[4] In these references one uses in turn results or techniques from Agmon, Douglis and Nirenberg (1959), Bers, John and Schechter (1974), Bony (1967).

irregular bounded analytic s.g. (nondegenerate). Also one has : $\exists\, A^{-1} \in \mathcal{B}(X)$, $\exists\, \mathcal{A}^{-1} \in \mathcal{B}(\mathcal{X})$. Moreover $\mathcal{A}^{-1}$ is self-adjoint, compact, in $\mathcal{X}$, $\mathcal{A} \leqslant 0$, and

$$e^{t\mathcal{A}} = \sum_n e^{-\lambda_n t}(\cdot, \varphi_n)\varphi_n \tag{33}$$

where $-\lambda_n, 0 < \lambda_1 \leqslant \lambda_2 \leqslant \lambda_3 \leqslant \cdots$, are the eigenvalues of $\mathcal{A}$, φ_n corresponding (normalized) eigenvectors. (See Friedman (1969), Showalter (1977).)

REMARK 5.1 The fact that A generates an irregular bounded analytic s.g. (nondegenerate) actually can be proved for much less regular bounded domains Ω than mentioned here or in the references, by using results of Stewart (1974) applied to $C_0(G)$, G very regular with $G \supset \overline{\Omega}$, and the comparison of resolvent techniques of Lumer and Paquet (1977, 1979).

THE CONCRETE HEAT SHOCKS MODEL 5.2 Our general Banach space model, and hence the results of sections 3, 4, apply to the present situation where for the concrete heat shocks model, and models of diffusion phenomena with shocks, (in particular the concrete version of the $2r$-periodic shocks model of section 4), $X = C(\overline{\Omega})$, A is as in (32) $\hat{A}$ being then defined for that A as in section 3, $H = \{f \in C(\overline{\Omega}) \cap C^\infty(\Omega) : A(x, D)f = 0 \text{ in } \Omega\}$, B being defined on $X = C(\overline{\Omega})$ by $Bf = H^\Omega(f|\partial\Omega)$ where "|" means "restriction to", $H^\Omega(\psi)$ is the solution in $C(\overline{\Omega})$ of the Dirichlet problem relative to $A(x, D)$ for $\psi \in C(\partial\Omega)$ (which exists under our assumptions, see Bony (1967)).

Concerning the comparison of the irregular s.g. $Q(t)$ generated by A on X, and the (usual) s.g. $\mathcal{Q}(t) = e^{t\mathcal{A}}$ generated by $\mathcal{A}$ on $\mathcal{X}$, we can compare for both cases the 2-s.g.s. $v_2 = v_2(t, f)$ in X, and $v_2^\# = v_2^\#(t, f)$ in $\mathcal{X}$, which satisfy respectively

$$\begin{aligned} v_2' &= Av_2 + tf \quad , \quad v_2(0) = 0 \quad , \\ v_2^{\#\prime} &= \mathcal{A}v_2^\# + tf \quad , \quad v_2^\#(0) = 0 \quad . \end{aligned} \tag{34}$$

Since $A \subset \mathcal{A}$ in $\mathcal{X}$, $X \hookrightarrow \mathcal{X}$, and we have uniqueness of solution for equations of the type of the second one in (34), we conclude that in $\mathcal{X}$

$$\begin{aligned} &v_2 = v_2^\# \text{ in } \mathcal{X} \quad \text{for } \ f \in X, \\ &v_2'(t, f) = Q(t)f = v_2^{\#\prime}(t, f) = \mathcal{Q}(t)f \quad \text{in } \mathcal{X} \ \text{ for } \ f \in X. \end{aligned} \tag{35}$$

$$Q(t) = e^{t\mathcal{A}} \ \text{ in } \mathcal{X}, \ \text{ on } X, \tag{35'}$$

and $e^{t\mathcal{A}}$ can be computed via (33).

6. SHOCKS WITH SOURCES (FORCING TERM $F(t)$)

We return to the context of section 4 and the periodic shocks problem treated there, but now with an arbitrary forcing term $F(\,) \in L^\infty([0, +\infty[, X)$ added, for which we assume $\exists$ a.e. $\lim_{t\to\infty} F(t) = c_0 = F(\infty)$.

Then taking up again, with the mentioned modification, the computations leading to (23), (24), we have for the first intervals, 0 to $t_1 = r$, and $t_1 = r$ to $t_2 = 2r$: $u(r) = Q(r)f + \int_0^r Q(r-s)F(s)ds$,

$$\begin{aligned} u(2r) &= \varphi + Q(r)(Q(r)f + \int_0^r Q(r-s)F(s)ds - \varphi) + \int_r^{2r} Q(2r-s)F(s)ds \\ &= \varphi - Q(r)\varphi + \int_0^{2r} Q(2r-s)F(s)ds + Q(r)^2 f \quad . \end{aligned}$$

By induction we find that simply the terms

$$\int_0^{(2n-1)r} Q((2n-1)r - s)F(s)ds \quad ,$$

$$\int_0^{2nr} Q(2nr - s)F(s)ds \quad ,$$

are added respectively to the expressions in (23), (24), to obtain the values at the end of the nth odd, even, numbered intervals. But according to Lumer (1991a)

$$\exists \lim_{t\to\infty} \int_0^t Q(t-s)F(s)ds = -A^{-1}c_0 = -A^{-1}F(\infty) \quad .$$

From what precedes and section 4, we immediately derive the following

THEOREM 6.1 In the context of 4.2 but adding an L^∞ forcing term $F(t)$ as specified above in this section, u_p being as in 4.2, $\exists$ a unique periodic orbit in $X : t \mapsto u(t) = u_p(t) - A^{-1}F(\infty)$ asymptotically attracting all orbits.

7. SOME QUALITATIVE PROPERTIES CONCERNING PERIODIC SHOCKS

We now examine the behavior of solutions in their asymptotic (steady-state) form for the periodic heat shocks model of section 5, in particular shortly after a shock, i.e. we shall study the qualitative behavior of the periodic asymptotically attracting orbits $u_{p,r}$ (depending on the period $2r$) in particular as $t \to 0_+$. In fact we shall derive some qualitative properties analytically in the Banach space generality of section 4, in close connection with the applications to, and interaction with, computer analysis of such problems which we treat in a following paper.

We use the notations introduced in section 4. For $r > 0$, we write $w_s = w_{s,r}$, $u_p = u_{p,r}$; for $0 < t \leqslant r$

$$\delta_r^s(t) = Q(t)w_{s,r} - w_{s,r} = u_{p,r}(t) - w_{s,r} \tag{36}$$

and

$$\delta_r^i(t) = u_{p,r}(r+t) - w_i \quad . \tag{37}$$

We first have the following simple fact :

LEMMA 7.1

$$\delta_r^s(t) = -\delta_r^i(t) \quad . \tag{38}$$

Proof. $\delta_r^i(t) = \varphi + Q(t)(Q(r)w_s - \varphi) - w_i$. Now $Q(r)w_s = w_i$ and ((26) section 4) $w_i + w_s = \varphi$. So $\delta_r^i(t) = \varphi - w_i - Q(t)w_s = -(Q(t)w_s - w_s) = -\delta_r^s(t)$.

■

The next result shows that the behavior "just after a shock" i.e. $\delta_r^s(t)$ for t small > 0 is essentially independent of r ("insensitive to the parameter r") while of course globally the shape of the periodic asymptotic orbit varies greatly. We have :

THEOREM 7.2 $\delta_r^s(t) = Q(t)\varphi - \varphi + \epsilon_r(t)$ where $\epsilon_r(t) \to 0$ as $t \to 0_+$, for any $r > 0$. Hence for $r_1, r_2 > 0$ $\|\delta_{r_1}(t) - \delta_{r_2}(t)\| \to 0$ as $t \to 0_+$.

Proof. $w_s = w_{s,r} = (1 + Q(r))^{-1}\varphi = \varphi - Q(r)\varphi + Q(r^2)\varphi - \cdots$ where the series converges in X, and all $Q(r) : X \to D(A)$ (by 2.4 and 2.2), so

$$w_s = \varphi + w_0 \quad , \quad w_0 \in \overline{D(A)} \quad . \tag{39}$$

By (39) $\delta_r^s(t) = Q(t)\varphi - \varphi + Q(t)w_0 - w_0 = Q(t)\varphi - \varphi + \epsilon_r(t)$, where $\epsilon_r(t) \to 0$ as $t \to 0$ since $Q(t)$ is strongly continuous at $t = 0$ on $\overline{D(A)}$ (see 2.1). Hence for any given $r_1, r_2 > 0$, $\delta_{r_1}^s(t) - \delta_{r_2}^s(t) = \epsilon_{r_1}(t) - \epsilon_{r_2}(t) \to 0$ as $t \to 0$.

■

This research was supported in part by the EC, through the Science Plan, project "Evolutionary systems", contract SC1*-CT90-0464.

REFERENCES

Agmon, S., Douglis, A. and Nirenberg, L. (1959). Estimates near the boundary for elliptic partial differential equations satisfying general boundary conditions, I, Comm. Pure Appl. Math., 12, 623-727.

Arendt, W. (1987). Vector-valued Laplace transforms and Cauchy problems, Israël J. Math., 59, 327-352.

Bers, L., John, F. and Schechter, M. (1974). Partial differential equations, A.M.S., Providence RI.

Bony, J.M. (1967). Principe du maximum dans les espaces de Sobolev, C.R. Acad. Sci. Paris, 265, série A, 333-336.

Da Prato, G. and Sinestrari, E. (1987). Differential operators with non dense domain, Ann. Scuola Normale Pisa, 14, 285-344.

Friedman, A. (1969). Partial differential equations, Holt, Rinehart and Winston, New York.

Goldberg, S. (1966). Unbounded linear operators, Mc Graw-Hill, New York.

Kellerman, H. and Hieber, M. (1989). Integrated semigroups, J. Funct. Analysis, 84, 160-180.

Lions, J.L. (1965), Problèmes aux limites dans les équations aux dérivées partielles, 2è ed., Presses de l'Université de Montréal.
Lumer, G. (1977). Equations d'évolution en norme uniforme pour opérateurs elliptiques. Régularité des solutions, C.R. Acad. Sci. Paris, 284, série A, 1435-1437.
Lumer, G. (1990). Solutions généralisées et semi-groupes intégrés, C.R. Acad. Sci. Paris, 310, série I, 557-582.
Lumer, G. (1991a) Problèmes dissipatifs et "analytiques" mal posés : solutions et théorie asymptotique, C.R. Acad. Sci. Paris, 312, série I, 831-836.
Lumer, G. (1991b). A (very) direct approach to locally lipschitz integrated semigroups and some new related results oriented towards applications, via generalized solutions, LSU Seminar Notes in Functional Analysis and PDES 1990-1991, Louisiana State University, Baton Rouge, 88-107.
Lumer, G. (1992). Semi-groupes irréguliers et semi-groupes intégrés : application à l'identification de semi-groupes irréguliers analytiques et résultats de génération, preprint, to appear in C.R. Acad. Sci. Paris.
Lumer, G. and Paquet, L. (1977). Semi-groupes holomorphes et équations d'évolution, C.R. Acad. Sci. Paris, 284, série A, 237-240.
Lumer, G. and Paquet, L. (1979). Semi-groupes holomorphes, produit tensoriel de semi-groupes et équations d'évolution, Séminaire de Théorie du Potentiel, Paris, n° 4, Lect. Notes in Math., Vol. 713, Springer-Verlag, 156-177.
Paquet, L. (1978). Equations d'évolution pour opérateurs locaux et équations aux dérivées partielles, C.R. Acad. Sci. Paris, 286, série A, 215-218.
Pazy, A. (1983). Semigroups of linear operators and applications to partial differential equations, Springer-Verlag.
Reed, M. and Simon, B. (1975). Methods of Modern Mathematical Physics : II Fourier analysis, Self-adjointness, Academic Press, New York.
Sinestrari, E. (1985). On the abstract Cauchy problem of parabolic type in spaces of continuous functions, J. of Math. Analysis and Appl., 107, 16-66.
Showalter, R.E. (1977). Hilbert space methods for partial differential equations, Pitman, London.
Stewart, B. (1974). Generation of analytic semigroups by strongly elliptic operators, Trans. Amer. Math. Soc., 199, 141-162.

On Uniqueness and Regularity in Models for Diffusion-Type Phenomena with Shocks

GÜNTER LUMER University of Mons, Mons, Belgium

0. INTRODUCTION

In a recent paper [4] we treat models for diffusion-type phenomena in which abrupt changes in the boundary conditions occur; we establish the governing equations, find solutions and develop applications to periodic shocks (in particular periodic heat shocks). The basic equations governing the evolution between two shocks occuring say at times t_0 and $t_1 > t_0$, when set up in Banach space context, are written (see [4])

$$\begin{aligned} &\frac{du}{dt} = u' = \hat{A}u + F(t) \quad , \\ &u(t_0) = f \quad \text{(mild g.s.)}, \\ &Bu(t) = \varphi, t_0 < t \leqslant t_1 \quad , \end{aligned} \tag{1}$$

$f \in X$, $\varphi \in H \subset X$, (see [4] section 3 for notations and all details)[1].

Such problems, models, as just mentioned above, bring up — make use of — mathematical objects behaving in several respects in a way quite different from what one encounters in usual diffusion-type models and more generally in the usual pure initial value problems (Cauchy problems). Concerning these aspects, the present brief article deals with : (i) nonuniqueness of non optimal regular solutions ([4] section 3) for the problem (1), (and lack of "uniqueness property" for $\hat{A}$); (ii) continuity/discontinuity behavior of solutions as one moves through shock points $t_0, t_1, \ldots$, in a sup-norm $C(\overline{\Omega})$ context. (i) is treated in sections 1 and 2 below, (ii) in section 3.

[1]See also footnote [2] in section 4 of [4]. The results in the present paper can also be used (adapted) for the more general situations mentioned there.

To help in keeping this article brief we use freely below the notions, notations and terminology of [4], and refer the reader to [4] for all necessary background material.

1. SOME REMARKS CONCERNING THE LACK OF UNIQUENESS PROPERTY FOR THE OPERATOR $\hat{A}$ IN (1)

Already physical intuition indicates, looking at the phenomena we are modeling governed by (1), ((6), (6') of [4]), that $\hat{A}$ (as defined precisely in section 3 of [4]) will not have in general the uniqueness property since, corresponding to the mentioned phenomena, the solutions of (1) should depend not only on the initial condition f but also on the imposed boundary value φ. Indeed mathematically we are dealing with initial value-boundary value problems, not initial value problems (Cauchy problems). In a rigorous way, the lack of uniqueness property for $\hat{A}$ can be seen from 3.1, (7), of [4]. More precisely and explicitely (in the context and with the notations of [4] in particular section 3) take any $\varphi \in H$, $Q(t)$ the irregular s.g. generated by A (see section 3 of [4]), and $S(t)$ the i.s.g. such that $S'(t) = Q(t)$ for $t > 0$ (2.1 of [4]); define moreover $S_2(t) = \int_0^t S(\sigma)d\sigma$ (strong integral) for $t \geqslant 0$.[2]. Set

$$v(t) = \frac{1}{2}t^2\varphi - S_2(t)\varphi \quad \text{for } t \geqslant 0. \tag{2}$$

Then $v \in C^1([0, +\infty[, X)$, and since $S_2(t)\varphi \in D(A)$, $v(t) \in D(\hat{A})$,

$$\hat{A}v(t) = -AS_2(t)\varphi = t\varphi - S(t)\varphi = v'(t); v(0) = 0 \quad . \tag{3}$$

(We have used of course in (3) the fact that $S(t)$ is an i.s.g. [1]).

Now $v(t) \equiv 0$ would imply, differentiating three times for $t > 0$, that $Q'(t)\varphi = AQ(t)\varphi = 0 \quad \forall t > 0$, [3], [6], so multiplying on the left by A^{-1} we would find $Q(t)\varphi = 0$ hence $\varphi = 0$ (because $Q(t)$ is nondegenerate). The latter and (3) tell us that the uniqueness property does not hold for $\hat{A}$ if $\exists 0 \neq \varphi \in H$ i.e. whenever $H \neq 0$. (On the other hand if $H = 0$ then $\hat{A} = A$, and since A generates the loc.lip. i.s.g. $S(t)$ the uniqueness property holds in that case for $\hat{A} = A$, as is well known [1].) Summarizing we have

1.1. PROPOSITION. The operator $\hat{A}$ in (1) (i.e. the operator $\hat{A}$ defined in section 3 of [4]) fails to have the uniqueness property iff $H \neq 0$ (or equivalently iff $B \neq 0$ where B is the boundary operator also defined in section 3 of [4]).

2. ON THE NEED OF POSING DIFFUSION-TYPE SHOCKS PROBLEMS IN TERMS OF OPTIMAL REGULAR SOLUTIONS.

From what we have seen in section 1 above, an evolution problem in terms of $\hat{A}$ will certainly not be well posed in general as a pure initial value problem; but will it be well

[2]It suffices to consider the case $t_0 = 0$ which is what we do below in the computations/proofs in this section and the next; also in the same occasions we do not mention t_1 (i.e. we write, say, $\forall t > 0$ instead of $\forall 0 < t \leqslant t_1$, etc.).

posed in an appropriate sense[3] as an initial value-boundary value problem formulated in the way written in (1) ((6), (6') of [4])? As shown in 3.1 of [4] the answer to the preceding question is yes indeed if we consider only optimal regular solutions. Below we show that if we do not restrict ourselves to optimal regular solutions the answer would be no in general (we establish below a somewhat more complete and precise result, 2.1).

We thus consider the general context of section 3 of [4] and problem (6) there (problem (1) above). As shown in 3.1 of [4] there is an optimal regular solution u for the mentioned problem, of the form $u(t) = \hat{v}''(t)$ for $t > 0$, where $\hat{v}(t) = v_2(t) + h(t)$, $t \geqslant 0$, with $v_2(t) \in D(A)$, $h(t) \in H$, and

$$\begin{aligned} \hat{v}' &= \hat{A}\hat{v} + tf + F_2(t), \hat{v}(0) = 0, \\ v_2' &= Av_2 + t(f-\varphi) + F_2(t), v_2(0) = 0. \end{aligned} \tag{4}$$

Let us now investigate if there can be another, non optimal regular, solution $\overline{u} \neq u$ for (6) of [4].

From (4) by integration $v_3 = v_2^{[1]}$ satisfies

$$v_3' = Av_3 + \frac{t^2}{2}(f-\varphi) + F_3(t), v_3(0) = 0. \tag{5}$$

We shall look for another solution $\overline{u}$ of the form $\hat{\overline{v}}_3^{(3)}$ where $\hat{\overline{v}}_3 = \overline{v}_3 + \overline{h}$ in the $D(A) \oplus H$ decomposition, with $\overline{h}$ of the form

$$\overline{h}(t) = \frac{1}{6}t^3\varphi + \frac{1}{2}t^2 h_2, \text{ some } h_2 \in H. \tag{6}$$

Set $w = v_3 - \overline{v}_3$; then we must have :

$$\begin{aligned} \hat{\overline{v}}_3' &= v_3' - w' + \frac{t^2}{2}\varphi + th_2 = Av_3 + \frac{t^2}{2}(f-\varphi) + F_3(t) - w' + \frac{t^2}{2}\varphi + th_2 \\ &= \hat{A}\hat{\overline{v}}_3 + \frac{t^2}{2}f + F_3(t) + Aw + th_2 - w' \\ &= \hat{A}\hat{\overline{v}}_3 + \frac{t^2}{2}f + F_3(t), \text{ and } \hat{\overline{v}}_3(0) = 0. \end{aligned} \tag{7}$$

The latter relations hold iff

$$w' = Aw + th_2\,, \; w(0) = 0 \quad . \tag{8}$$

Since A generates the i.s.g. $S(t)$, (8) has exactly the unique solution (using the notation $S_2(t)$ introduced in section 1 above) :

$$w(t) = S_2(t)h_2 \quad , \quad t \geqslant 0 \quad . \tag{9}$$

Now with w as in (9), we have for $t > 0$ using (6) and (12) of [4]

$$\begin{aligned} &\hat{\overline{v}}_3''' = v_3''' - w''' + \varphi = \\ &v_2'' + \varphi - w''' = \\ &u - w''' = u - Q'(t)h_2 = \overline{u} \quad . \end{aligned} \tag{10}$$

[3]Existence and uniqueness of solution for given data f, φ.

From (10) we have $B\overline{u}(t) = \varphi$ for $t > 0$, and since, with w given by (9), (7) holds $\overline{u}(t) = \hat{\overline{v}}_3'''(t)$ is indeed a solution of (6) of [4] ((1) above), and we shall show next that it is different from u iff $h_2 \neq 0$. Indeed, by (10) $\overline{u} = u$ iff, [6], [3], (analyticity),

$$Q'(t)h_2 = AQ(t)h_2 = 0 \qquad \forall t > 0 \quad . \tag{11}$$

Multiplying by A^{-1} on the left in (11) and using the fact that $Q(\cdot)$ is nondegenerate [3], [6], (11) is seen to be equivalent to $h_2 = 0$.

Considering the above facts and the fact that for $H = 0$ $\hat{A} = A$, $B = 0$, we see that we have proved the following

2.1. THEOREM. For any problem (1), ((6) of [4]) with arbitrary chosen data f, φ, there always exists besides the unique optimal regular solution (given by (7) of 3.1 of [4]) infinitely many other non optimal regular solutions, provided $H \neq 0$ (and only if $H \neq 0$).

It is thus essential indeed to remain within the class of optimal regular solutions, which also permits to continue these well determined, uniquely determined optimal regular solutions, across any number of shocks (see [4]) obtaining a uniquely determined global solution, (so we have a globally well posed initial value-boundary value problem, in an appropriate sense).

3. CONTINUITY EXCEPT AT THE BOUNDARY OF THE FLOW OF SOLUTIONS THROUGH THE SHOCK POINTS IN DIFFUSION-TYPE MODELS IN THE CLASSICAL SUP-NORM $C(\overline{\Omega})$ CONTEXT.

The purpose of this section is to prove in great generality the "nearly sup-norm continuous" behavior at shock points of solutions in $C(\overline{\Omega})$ context, suggested in the title of the section, behavior which one would intuitively expect for the phenomena in question at least in the more regular cases; (this behavior is also clearly "visible" in computer graphics studies of sufficiently simple cases). This behavior is of course in line with the various reasons why it is realistic to use the strict norm models (sup-norm models here) developed in [4] for the kind of phenomena we are considering (for instance an L^p-norm does not give a very faithful modeling of the sup-norm discontinuous yet somehow "nearly sup-norm continuous" behavior we establish below).

Now, again to keep this article short as already said, we shall do two things :
(i) we collect in the next lemma, 3.1 below, some facts needed in the proof of the main result of this section, 3.2, facts whose detailed proof will appear elsewhere [5] (the proof of the mentioned facts — which have indeed many uses besides their use in 3.2 — is based fundamentally on the maximum principles holding for elliptic operators like $-A(x, D)$ of section 5 of [4], or more generally for locally dissipative local operators under not very restrictive conditions (see 3.3 below)).
(ii) for simplicity we state 3.1 and prove 3.2 under the assumptions of section 5 of [4], while they both hold assuming much less regularity (in the sense of differentiability of the coefficients of the differential operator and of the boundary $\partial\Omega$); indications concerning the extensions of 3.1, 3.2, to the more general versions just mentioned are given in 3.3.

Anyhow the proof of the main result of this section, 3.2, given below is essentially the general one (i.e. essentially the same which can be used for the more general version) as explained with more precision in 3.3 below.

3.1. LEMMA. In the classical $C(\overline{\Omega})$ context of section 5 of [4], with $A(x, D)$, A (and the irregular s.g. $Q(t)$ generated by the latter), as considered there, one has (writing for $f \in C(\overline{\Omega})$ real "$f > 0$" iff "$f(x) \geqslant 0 \quad \forall x \in \Omega$ ") :

$$\begin{aligned} &Q(t) \geqslant 0 \quad (\text{i.e. } Q(t)f \geqslant 0 \text{ if } f \geqslant 0) \quad \forall t > 0 \quad , \\ &\lambda R(\lambda, A)h \leqslant h \,, \forall 0 \leqslant h \in H \,, \lambda > 0 \quad , \\ &Q(t)h \leqslant h \,, \forall 0 \leqslant h \in H \,, t > 0 \quad . \end{aligned} \tag{12}$$

We now establish the following

3.2. THEOREM. Consider the general $2r$-periodic shocks model in the classical $C(\overline{\Omega})$ context, as defined and under the assumptions of section 5 of [4]. Then for any orbit (solution) $u(t, f)$ starting at an arbitrary $f \in C(\overline{\Omega})$ at $t = 0$, $u(t, f)(x) \to f(x)$ as $t \to 0_+$, $x \in \Omega$, uniformly on compacta of Ω, and moreover $u(t, f)(x) \to u(kr, f)(x)$ as $t \to kr + 0$ $(k = 1, 2, \ldots)$, $x \in \Omega$, uniformly on compacta of Ω; $(t, x) \mapsto u(t, f)(x)$ is continuous at all points of $[0, +\infty[\times\Omega$ (i.e. across the shock points), while of course $u(t, f)(x)$ is in general not continuous on $[0, +\infty[\times\overline{\Omega}$, (not continuous on $[0, +\infty[\times\partial\Omega$ if $\varphi \neq 0$).

Proof. We consider $u(t, f)$ near $t = 0$. There for $t > 0 \quad u(t, f) = Q(t)f$. We proceed in several steps.

(a) We first take $0 \leqslant f = h \in H$. We shall then prove that $Q(t)h \uparrow h$ (monotone convergence) as $t \downarrow 0$, on Ω. Indeed, write $u(t, h)(x) = u(t, x, h) = u(t, x)$; then for $t > 0$, $x \in \overline{\Omega}$,

$$\exists \; \frac{\partial u}{\partial t}(t, x) = (Q'(t)h)(x) = (AQ(t)h)(x) \quad . \tag{13}$$

Now $Q(t) : X = C(\overline{\Omega}) \to D(A_0)$ see [4] section 2, and on $D(A_0) \quad A_\lambda = \lambda^2 R(\lambda, A) - \lambda$ for $\lambda > 0$ (the Yosida approximation) converges strongly (as $\lambda \to \infty$) to A_0 by the usual arguments. Hence

$$\begin{aligned} AQ(t)h = \lim_{\lambda \to \infty} A_\lambda Q(t)h = \\ Q(t)A_\lambda h \leqslant 0 \quad , \end{aligned} \tag{14}$$

since $A_\lambda h = \lambda(\lambda R(\lambda, A)h - h) \leqslant 0$ by 3.1 (12).
From (13) and (14)

$$\frac{\partial u}{\partial t}(t, x) \leqslant 0 \qquad \forall x \in \overline{\Omega} \,, 0 < t \leqslant r \quad . \tag{15}$$

Hence $\forall x \in \overline{\Omega}\quad \exists \lim_{t\downarrow 0_+} u(t,x) = \psi(x)$, and using again 3.1,

$$0 \leqslant \psi \leqslant h \quad . \tag{16}$$

Now take K compact $\subset \Omega$ and g real, $0 \leqslant g \in C_0(\Omega)$ such that $g \leqslant h$, $g = h$ on K. By (12) $0 \leqslant Q(t)g \leqslant Q(t)h \leqslant \psi$ for $t > 0$ small; but since $g \in \overline{D(A)} = C_0(\Omega)$, $Q(t)g \to g$ in $C(\overline{\Omega})$ as $t \downarrow 0_+$ (see [3]) and it then follows from $g = h$ on K, $Q(t)g \leqslant Q(t)h$ and (16), that $\psi = h$ on K. Hence

$$\lim_{t\downarrow 0_+} u(t,x) = \lim_{t\downarrow 0_+} u(t,h)(x) = h(x) \quad \text{for } x \in \Omega. \tag{17}$$

Since h is continuous and the convergence in (17) monotone, we have by Dini's theorem uniform convergence on compacta of Ω.

(b) Next, for any real $0 \leqslant f \in C(\overline{\Omega})$, we can write $f = h + f - h$ with $h \in H$, $f - h \in \overline{D(A)}$ (simply $h = Bf$), and we have $h \geqslant 0$. So from (a) applied to h, and also from the fact that since $f - h \in \overline{D(A)}\quad Q(t)(f-h) \to f - h$ in $C(\overline{\Omega})$,

$$\begin{aligned} &u(t,x,f) = u(t,x,h) + u(t,x,f-h) \\ &\to h(x) + (f-h)(x) = f(x) \,\text{on } \Omega, \\ &\text{as } t \to 0_+, \text{ uniformly on compacta of } \Omega. \end{aligned} \tag{18}$$

(c) For any $f \in C(\overline{\Omega})$ real, $f = f_+ - f_-$ with f_+ and $f_- \geqslant 0$, $f_+, f_- \in C(\overline{\Omega})$; $u(t,x,f) = u(t,x,f_+) - u(t,x,f_-)$, and so using (b) we have also $u(t,x,f) \to f_+(x) - f_-(x) = f(x)$ uniformly on compacta of Ω. Similarly the conclusion extends now to any $f \in C(\overline{\Omega})$ by decomposing f into real and imaginary parts.

(d) The rest of our statement follows now at once.

■

3.3. ON THE EXTENSION OF 3.2 TO SITUATIONS WITH LESS REGULARITY

As alrealy mentioned much less regularity is actually needed for the conclusion of 3.2 than is assumed above (i.e. in section 5 of [4]). In fact, once the conclusions of 3.1 are established under much less restrictive conditions (as we do in [5]), and these conditions furthermore modified slightly so as to also imply that (I) and (II) in (the fourth paragraph of) section 3 of [4] hold as well, then the proof of 3.2 given above will go through essentially without change. The mentioned less restrictive conditions, used in [5] to state and prove a correspondingly more general version of 3.1 above (and to prove the implications (I) and (II) mentioned above in this paragraph), are stated in terms of a locally dissipative local operator L (see [2] where L is denoted A) on some locally compact Hausdorff space $\tilde{\Omega}$ (with $\overline{\Omega}$ compact $\subset \tilde{\Omega}$) replacing the elliptic operator $-A(x,D)$ defined near $\overline{\Omega}$; the exact assumptions in that setup will be the ones introduced in the earlier paper [2] (as needed for the maximum principle 2.5 of that paper [2]) plus an additional assumption in the same context, again not very restrictive, garanteeing the validity of (I) and (II) mentioned above[4]. Details are in [5]. Extended in this way the more general version of 3.2 is then applicable with second order elliptic operators having merely continuous real coefficients (rather than C^∞ coefficients as above)[5], and domains Ω merely regular in the sense of the L-Dirichlet problem.

References

[1] Arendt, W.; Vector-valued Laplace transforms and Cauchy problems, Israël J. Math., 59 (1987), 327-352.

[2] Lumer, G.; Problème de Cauchy et fonctions surharmoniques, Séminaire de Théorie du Potentiel, Paris N° 2, Lect. Notes in Math. 563 (1976), Springer-Verlag, 202-218.

[3] Lumer, G.; Semi-groupes irréguliers et semi-groupes intégrés : application à l'identification de semi-groupes analytiques et résultats de génération, C.R. Acad. Sci. Paris, 314 série I (1992), 1033-1038.

[4] Lumer, G.; Models for difusion-type phenomena with abrupt changes in boundary conditions, in Banach space and classical context. Asymptotics under periodic shocks, preprint (1992).

[5] Lumer, G.; Aspects of qualitative and computer analysis for periodic heat/diffusion shocks, forthcoming.

[6] Sinestrari, E.; On the abstract Cauchy problem of parabolic type in spaces of continuous functions, J. of Math. Analysis and Appl., 107 (1985), 16-66.

---

[4]Specifically, in the terminology of [2] (with L instead of the A used there), one assumes that $\exists G \in \mathcal{R}$ containing $\overline{\Omega}$ such that the s.g. e^{tL_G} (which exists necessarily) in analytic.

[5]Instead of the $A(x,D)$ of [4], one will then be able to apply the results with an $A(x,D) = \sum_{|\alpha|\leqslant 2} c_\alpha(x)D^\alpha$, c_α real and merely continuous near $\overline{\Omega}$, $c_0 \leqslant 0$, and of course $-A(x,D)$ uniformly elliptic (with $c_\alpha = a_{ij} = a_{ji}$ for $|\alpha| = 2$).

Idempotent Semimoduli in Optimization Problems

V. P. MASLOV Russian Academy of Sciences, Moscow, Russia

This paper deals with new idempotent algebraic structures which can be used for solving certain problems of mathematical physics and control theory.

First we show how these structures appear in the process of constructing generalized solutions of the Cauchy problem for the simplest Hamilton-Jacobi equation [1, 2].

Consider the Bürgers equation

$$\frac{\partial w}{\partial t}+\frac{1}{2}\left(\frac{\partial w}{\partial x}\right)^2-\frac{h}{2}\frac{\partial^2 w}{\partial x^2}=0, \qquad x\in\mathbf{R},\ t\geqslant 0. \tag{1}$$

The change $u=\exp\{-w(x,t)/t\}$ transforms it into the following linear heat equation

$$\frac{\partial u}{\partial t}=\frac{h}{2}\frac{\partial^2 u}{\partial x^2}. \tag{2}$$

Therefore we have that equation (1) is also linear if in the space of functions the following semigroup operations of "addition"

$$a\oplus_h b=-h\ln\left(\exp\left\{-\frac{a}{h}\right\}+\exp\left\{-\frac{b}{h}\right\}\right) \tag{3}$$

and "multiplication"

$$a\odot b=a+b \tag{4}$$

are introduced.

Obviously, the ring obtained is isomorphic to the usual one. Consider the limit of this structure as $h\to 0$. Since the Hamilton-Jacobi equation

$$\frac{\partial w}{\partial t}+\frac{1}{2}\left(\frac{\partial w}{\partial x}\right)^2=0 \tag{5}$$

is the limit of equation (1) as $h\to 0$, equation (5) must be linear in this new structure. The operation (4) is independent of h, and it is easy to see that

$$\lim_{h\to 0} a\oplus_h b=\min(a,b),$$

namely, the limit of operation (3) is a semigroup operation

$$a \oplus b = \min(a, b).$$

It is well-known that for a smooth convex initial function $w_0(x)$ the Cauchy problem for equation (5) has the unique classical (i.e., everywhere smooth) solution which has the form

$$R_t w_0(x) = w(t, x) = \inf_{\xi} \left(w_0(\xi) + \frac{(x-\xi)^2}{2t} \right). \tag{6}$$

Now we use the linearity of the structure introduced in order to get this formula for unsmooth and nonconvex initial data. Namely, we note that the operator R_t is self-adjoint if we define the corresponding bilinear scalar product

$$\langle \varphi, \psi \rangle = \inf_x (\varphi(x) + \psi(x)) \tag{7}$$

on the space of functions with operations $\oplus = \min$, $\odot = +$. Actually,

$$\langle R_t \psi, \varphi \rangle = \inf_x \left\{ \inf_\xi \left(\psi(\xi) + \frac{(x-\xi)^2}{2t} \right) + \varphi(x) \right\} =$$
$$= \inf_{x,\xi} \left(\varphi + \psi(\xi) + \frac{(x-\xi)^2}{2t} \right) = \langle \psi, R_t \varphi \rangle.$$

Since the set of smooth convex functions is complete in the space of continuous functions (obviously, any function on $\mathbf{R}^n$ can be approximated by linear combinations in the sense of operations $\oplus = \min$, $\odot = +$ applied to convex smooth functions), the $R_t \varphi$ can be considered, precisely as in the linear theory, as a weak solution of the Hamilton-Jacobi equation (5). Similar constructions can be carried out for a more general equation

$$\frac{\partial S}{\partial t} + H\left(\frac{\partial S}{\partial x}, x\right) = 0, \tag{8}$$

in particular, for the Bellman equation

$$\frac{\partial S}{\partial t} + \max_u \left(f(x, u) \cdot \frac{\partial S}{\partial x} + g(x, u) \right) = 0.$$

These constructions are given in [3, 4].

Consider more precisely the method of constructing general solutions of the stationar Hamilton-Jacobi equation

$$H\left(x, \frac{\partial S}{\partial x}\right) = 0. \tag{9}$$

Suppose x belongs to a certain bounded domain $\Omega \subset \mathbf{R}^n$ and the first boundary-value problem must be solved for equation (9), i.e., a generalized solution coinciding with a given function $S|_{\partial\Omega}$ on the boundary must be constructed.

Let $B(\Omega, A)$ denote the space (semimodulus) of functions bounded from below on Ω with values in the semiring $A = \mathbf{R} \cup \{+\infty\}$ with operations $a \oplus b = \min(a, b)$, $a \odot b = a + b$ and neutral elements $\mathbf{0} = +\infty$ and $\mathbf{1} = 0$ assigned to the operations of addition $\oplus$ and multiplication $\odot$ respectively.

In $B(\Omega, A)$ the bilinear scalar product (7) is defined. We define the set Φ of basic functions as the set of real functions continuously differentiable in Ω and countinuous in $\overline{\Omega}$. Two functions f and g in $B(\Omega, A)$ will be called equivalent if $\langle f, \varphi\rangle = \langle g, \varphi\rangle$ for any $\varphi \in \Phi$. We shall say that the sequence of functions (or classes of equivalence) $\{f_i\}_{i=1}^{\infty}$ converges weakly to $f \in B(\Omega, A)$ (and write $f = w_0 - \lim f_i$) if for any $\varphi \in \Phi$ we have

$$\lim_{i\to\infty} \langle f_i, \varphi\rangle = \langle f, \varphi\rangle$$

By $\mathcal{P}^0$ we denote the set of classes of equivalence in $B(\Omega, A)$ defined by weak limits of sequences of functions from Φ. The convergence in $\mathcal{P}^0$ is weak by definition. Obviously, a continuous embedding i_0 of the space of continuous functions $C(\overline{\Omega}, \mathbf{R})$ into the semimodulus $\mathcal{P}^0$ is defined.

Let ξ be a vector field over $\overline{\Omega}$ defined in any chart $U \subset \Omega$ by the set $(\xi_1, \dots, \xi_k)$ of scalar functions (components). We shall say that two vector fields ξ and η over Ω are equivalent if for any local coordinate neighbourhood U their components ξ_i and η_i are equivalent. We denote by $\mathcal{P}^0(T\Omega)$ the space of classes of equivalence of vector fields with components from $\mathcal{P}^0$.

DEFINITION 1. The sequence of differentiable in X functions f_i is w_1-convergent to $f \in B(\overline{\Omega}, A)$ (we write $f = w_1 - \lim f_i$) if $f = w_0 - \lim_{i\to\infty} f_i$ and the vector field $w_0 - \lim Df_i$, where Df is the gradient of the function f, exists and belongs to $\mathcal{P}^0(T\Omega)$.

DEFINITION 2. A function (class of equivalence) f belongs to the space $\mathcal{P}^1(\overline{\Omega})$ if it is a w_1-limit of the sequence of differentiable functions. The convergence $\mathcal{P}^1(\overline{\Omega})$ is defined by the w_1-convergence.

Obviously, the elements from $\mathcal{P}^1(\overline{\Omega})$ are continuous in the sence that there is a unique continuous function in each class of equivalence. It is also clear that a natural continuous embedding is defined

$$i_1 : C^1(\overline{\Omega}, \mathbf{R}) \to \mathcal{P}^1(\overline{\Omega}),$$

where $C^1(\overline{\Omega}, \mathbf{R})$ are smooth functions, and we have the following diagram

$$\begin{array}{ccc} C^1(\overline{\Omega}, \mathbf{R}) & \xrightarrow{\ i_1\ } & \mathcal{P}^1(\overline{\Omega}) \\ \downarrow{\scriptstyle D} & & \downarrow{\scriptstyle D} \\ C(TX) & \xrightarrow{\ i_0\ } & \mathcal{P}^0(TX). \end{array}$$

It is easy to see that in $\mathcal{P}^0(\overline{\Omega})$ the operations $\oplus\colon \mathcal{P}^0 \times \mathcal{P}^0 \to \mathcal{P}^0$ and $\odot\colon A \times \mathcal{P}^0 \to \mathcal{P}^0$ are continuous, but this does not hold already in the semimodulus $\mathcal{P}^1(\overline{\Omega})$, as is shown in the following simple example. Let $f \equiv 0$ and $g_\varepsilon(x) = |x| - \varepsilon$. Then $D(f \oplus g_0) = D(f) \equiv 0$, but $w_0 - \lim_{\varepsilon \to 0}(D(f \oplus g_\varepsilon)) \not\equiv 0$.

THEOREM 1. *(Maslov, Samborskiĭ) In $\mathcal{P}^1(\overline{\Omega})$ there exists a semimodulus $\mathcal{A}(\Omega)$ maximal with respect to inclusion, containing all the differentiable functions, in which the operation $\oplus$ and the multiplication by numbers $\odot$ are continuous in the topology induced from $\mathcal{P}^1(\overline{\Omega})$. In this case $f \in \mathcal{A}(\Omega)$ if and only if for any point $x_0 \in \Omega$ and any nonnegative differentiable function φ the equality $(f, \varphi) = f(x_0)$ implies the differentiability of f in x_0.*

EXAMPLE. The function $|x|$ does not belong to $\mathcal{A}([-1, 1])$, and the function $-|x|$ belongs.

THEOREM 2. *(Maslov, Samborskiĭ) Suppose $H\colon \Omega \times \mathbf{R}^n \to \mathbf{R}$ is a continuous function such that*

1) the mapping $\mathcal{H}\colon \mathcal{P}^1(\overline{\Omega}) \to \mathcal{P}^0(\overline{\Omega})$ of the form $(\mathcal{H}S)(x) = H(x, \frac{\partial S}{\partial x})$ is continuous;

2) the set $\Lambda_x = \{p \in \mathbf{R}^n\colon H(x, p) \geqslant 0\}$ is not empty, compact and strictly convex for all $x \in X$;

3) there exists a function $\overline{S} \in \mathcal{P}^1(\overline{\Omega})$ such that

$$H\left(x, \frac{\partial \overline{S}}{\partial x}\right) \geqslant 0.$$

Then equation (9) has the solution $S \in \mathcal{A}(\Omega)$, such that the restrictions of S and $\overline{S}$ coincide on $\partial\Omega$.

The solution of equation (9) with given boundary conditions is, generally speaking, not unique. The set of these solutions forms a convex (in the sense of operations $\oplus$, $\odot$) subset in $\mathcal{A}(\Omega)$, i.e. if $_1$, S_2 are solutions, then $a_1 \odot S_1 \oplus a_2 \odot S_2$ is also a solution for any a_1, a_2, such that $a_1 \oplus a_2 = \mathbf{1} = 0$.

For example, on the interval $[0, 1]$, for the equation

$$((S')^2 - 1)((S')^2 - 4) = 0, \tag{10}$$

with boundary conditions $S(0) = S(1) = 0$, there are two "extreme" solutions $S_1 = \min(x, -x)$, $S_2 = \min(2x, 2 - 2x)$, and any linear combination $a \odot S_1 \oplus S_2 = \min(a + S_1, S_2)$ is also a solution for $a > 0$.

We note that the Hamilton-Jacobi equation describes, in particular, the logarithmic limits $\lim_{h \to 0} h \ln y$ of the corresponding pseudodifferential equations. In particular, the equation

$$(\Delta - 1)(\Delta - 4)y = 0, \tag{11}$$

where Δ is the Laplace operator, corresponds to equation (10). The logarithmic limit for the asymptotic solutions of (11) of the form $C\exp\{-\frac{S(x)}{h}\}$ with boundary conditions $y(0) = 1 + \exp\{-\frac{a}{h}\}$, $y'(0) = 2 + \exp\{-\frac{a}{h}\}$ (and symmetric conditions at the right end) will be obtained in the form of a linear combination $a \odot S_1 \oplus S_2$ mentioned above. A similar example can be presented in the multidimensional situation.

In conclusion we present a theorem on finite dimension for the set of solutions of a stationary equation in the whole space.

We define the solution of a stationary equation

$$H\left(x, \frac{\partial S}{\partial x}\right) = \lambda \tag{12}$$

as the eigenfunction of the resolving operator R_t of the Cauchy problem for the equation (8), i.e. as a function such that $S(x) + t\lambda$ is the generalized solution of the Cauchy problem for (8).

THEOREM 3. *(Kolokoltzov) Suppose the Lagrangian $L(x, v)$ of the problem (12), i.e., the Legendre transform of the function H with respect to the second argument, has a finite number of minima $(\xi_1, 0), \ldots (\xi_k, 0)$, and $L = 0$ at them. $L(x, v) \to +\infty$ as $\|x\|, \|v\| \to +\infty$. Then there exist k functions $S_j(x)$ tending to $+\infty$ as $\|x\| \to +\infty$, such that for $\lambda = 0$ any bounded from below solution of equation (12) has the form $\bigoplus_{j=1}^{k} a_j \odot S_j$ with certain constants a_j, and for $\lambda \neq 0$, there are no solutions.*

These stationary solutions describe the limits as $t \to \infty$ of stationary problems

$$S_0(\xi) + \int_0^t L(x, \dot{x})\, d\tau \quad \to \quad \inf,$$

where inf is taken over all ξ and over all the curves joining ξ to x during the time t. Namely, let $S(t, x)$ be a solution of such a problem. Then

$$\lim S(t, x) = \bigoplus_{j,k} \langle S_0, \tilde{S}_j \rangle \odot \langle \tilde{S}_j, S_k \rangle \odot S_k(x),$$

where $\tilde{S}_j$ are eigenfunctions of the adjoint equation

$$\frac{\partial S}{\partial t} + H\left(x, -\frac{\partial S}{\partial x}\right) = 0,$$

and the scalar product was defined in (7).

The methods of idempotent analysis also allow to consider differential equations in optimization problems with vector criterion of merit [5] and with stochastic and game control [6].

References

1. V. P. Maslov, *On a new superposition principle for optimazation problems*, Uspekhi Mat. Nauk **42:3** (1987), 39-48 (Russian); English transl. in Russian Math. Surveys.
2. V. P. Maslov, *Asymptotic methods for solving pseudodifferential equations*, Nauka, Moscow, 1988. (Russian)
3. V. N. Kolokoltsov, and V. P. Maslov, *Cauchy problem for homogeneous Bellman equation*, Dokl. Acad. Nauk SSSR **296:4** (1987), 796–800 (Russian); English transl. in Soviet Math. Dokl.
4. V. N. Kolokoltsov, and V. P. Maslov, *Idempotent analysis as an apparatus of control theory*, Funktsional. Anal. i Prilozhen. (1989); I **23:1**, 1–14; *II* **23:4**, 53–62 (Russian); English transl. in Functional Anal. Appl.
5. V. N. Kolokoltsov, and V. P. Maslov, *Differential Bellman equations and Pontryagin maximum principle for multicriterial optimization problems*, Dokl. Akad. Nauk USSR **324:1** (1992) (Russian); English transl. in Soviet Math. Dokl.
6. V. N. Kolokoltsov, *Stochastic Bellman equation as a nonlinear equation in Maslov spaces. Perturbation theory*, Dokl. Akad. Nauk SSSR **323:2** (1992) (Russian); English transl. in Soviet Math. Dokl.
7. V. P. Maslov, and S. N. Samborskiĭ, *Stationary Hamilton-Jacobi equations*, Dokl. Akad. Nauk SSSR (to apper).

Dual Semigroups and Functional Differential Equations

MIKLAVŽ MASTINŠEK University of Maribor, Maribor, Slovenia

1 INTRODUCTION

The purpose of this paper is to obtain a characterization of the dual solution semigroup of the delay differential equation (DDE):

$$\dot{x}(t) = Ax(t) + Bx(t) + Ax(t-h) + \int_{-h}^{0} a(s)Ax(t+s)ds, t \geq 0$$
$$x(0) = \phi^0, x(s) = \phi^1(s) \text{ a.e. on } [-h, 0), \tag{1.1}$$

where A is the infinitesimal generator of an analytic semigroup on a Hilbert space X, B is a densely defined closed operator in X, and $a(\cdot)$ is an element of $L^2(-h, 0; R)$.

Equations of this type were considered, for instance, by Di Blasio, et al. (5,6). They have developed a state space theory with initial values $\phi = (\phi^0, \phi^1)$ from the product space $Z = V \times L^2(-h, 0; D(A))$, where V is a real interpolation space between $D(A)$ and X, and where B belongs to $\mathcal{L}(D(A), V)$. Integrodifferential equations related to (1.1) were also considered by Desch, et al. (4,7).

Bernier, et al. in their study of DDE in R^n (see Refs. 2 and 3) have shown that the so-called structural operators can be employed to describe the influence of the delay part in (1.1) on the evolution of the trajectories as well as to characterize the adjoint semigroup of the solution semigroup of (1.1). Their results have recently been proved for the case of infinite-dimensional spaces, where bounded and unbounded operators act in the delayed part of the equation (see Refs. 11–13, 15), for example. Nakagiri (13) considered structural properties of functional differential equations with bounded operators acting on the delays.

Tanabe (15) considered structural operators for equation (1.1) where A is defined by a sesquilinear form, $B = 0$ and $a(\cdot)$ is Hölder continuous.

The paper is organized as follows. In Section 2, the solution semigroup $T(t)$ of the DDE (1.1) is given. By replacing the operator A acting in the delay part of (1.1) with its Yosida-type approximation A_λ and B with a bounded operator $B_\lambda = \lambda\, \mathrm{BR}(\lambda, A)$, the approximative delay differential equation (ADDE) is obtained. The solution semigroup $T_\lambda(t)$ of ADDE is defined for every $\lambda > 0$ and strong convergence $T_\lambda(t)\phi \to T(t)\phi$ is shown for every $\phi \in Z$. In Section 3, the transposed semigroups are introduced. These semigroups are defined as solution semigroups of dual equations, which are obtained by replacing all of the operators in DDE and ADDE with their respective adjoints. In Section 4, structural operators associated with DDE and ADDE are introduced. It is shown that they provide the same relationship between adjoint semigroups $T_\lambda^*(t)$ and $T^*(t)$ and their respective transposed semigroups $T_\lambda^T(t)$ and $T^T(t)$, as that given in Refs. 2 and 3.

The method given above can also be used for more general forms of DDEs.

The notation in this paper is rather standard. Let X denote a complex Banach space with norm $\|\cdot\|_X$. For real numbers $a < b, L^2(a,b;X)$ denotes the vector space (of equivalence classes) of strongly measurable functions x from $[a,b]$ to X such that $t \longrightarrow \|x(t)\|_X^2$ is Lebesgue integrable on $[a,b]$. If X is a Hilbert space with inner product $\langle .,. \rangle$ then $L^2(a,b;X)$ is a Hilbert space with inner product

$$\langle x,y\rangle_{L^2(a,b;X)} = \int_a^b \langle x(t), y(t)\rangle\, dt.$$

The space of continuous functions on $[a,b]$ with values in X is denoted by $C(a,b;X)$ and $W^{1,2}(a,b;X)$ is the space of absolutely continuous functions f from $[a,b]$ to X with $\dot{f} \in L^2(a,b;X)$. Given a function x from $[-h,\tau]$ to X and $t \in [0,\tau]$ the segment of a trajectory x is defined by $x_t(s) := x(t+s)$ for $s \in [-h,0]$.

If Y is a Banach space, then $\mathcal{L}(X,Y)$ is the space of all linear bounded operators from X to Y and the inclusion $X \circlearrowleft Y$ means that X is continuously and densely embedded in Y. If $A : D(A) \subset X \to X$ is an injective linear operator, $D(A)$ will be regarded as a normed space equipped with the graph norm $\|x\|_{D(A)} := \|Ax\|_X$. This norm is equivalent to the norm $\|x\|_{D_A} := \|x\|_X + \|Ax\|_X$ used in (5), when $0 \in \rho(A)$ – the resolvent set of A. As usual, $R(\lambda,A) = (\lambda I - A)^{-1}$, for every $\lambda \in \rho(A)$.

If A is the infinitesimal generator of an analytic semigroup $S(t)$, then the intermediate vector space V between $D(A)$ and X is defined as follows:

$$V := \left\{ v \in X \middle| \int_0^\infty \|AS(t)v\|^2\, dt < \infty \right\} \tag{1.2}$$

with norm

$$\|v\|_V := \|v\| + \left(\int_0^\infty \|AS(t)v\|^2\, dt \right)^{1/2}. \tag{1.3}$$

We have the following relations: $D(A) \circlearrowleft V \circlearrowleft X$. It is known that the space $L^2(0,\tau;D(A)) \cap W^{1,2}(0,\tau;X)$ is continuously embedded in $C(0,\tau;V)$ so that there exists c_0 such that

$$\|x\|C(0,\tau;V) \le c_0 \|x\|_{L^2(0,\tau;D(A)) \cap W^{1,2}(0,\tau;X)} \tag{1.4}$$

for each $x \in L^2(0,\tau;D(A)) \cap W^{1,2}(0,\tau;X)$. For details, see, e.g., Ref. 10, p. 23.

2 SOLUTION SEMIGROUPS

Let X be a Hilbert space with norm $\|\cdot\|$ and let $A : D(A) \subset X \to X$ be the infinitesimal generator of a bounded analytic semigroup $\{S(t); t \geq 0\}$ on X with $0 \in \rho(A)$. Let $B \in \mathcal{L}(D(A), V)$, where V is the intermediate space between $D(A)$ and X defined by (1.2) and (1.3).

We consider the following delay differential equation:

$$\begin{aligned} &\dot{x}(t) = Ax(t) + Bx(t) + L_1 x(t-h) + L_2 x_t, \\ &x(0) = \phi^0, x(s) = \phi^1(s) \text{ a.e. on } [-h, 0), \end{aligned} \tag{2.1}$$

where

$$\begin{aligned} &L_1 = A \\ &L_2 x_t = \int_{-h}^{0} a(s) A x(t+s) ds \end{aligned} \tag{2.2}$$

for almost every $t \in (0, \tau)$ and $\phi = (\phi^0, \phi^1)$ an element from the product space $M^2 := X \times L^2(-h, 0; X)$. We assume $h > 0, \tau > 0$ and $a \in L^2(-h, 0; R)$.

As mentioned before, this form of the DDE (2.1) was studied by Di Blasio, et al. in (5,6). By Theorems 3.3 and 4.1 in (5) we have the following result:

For every $\phi \in Z := V \times \mathcal{H} := V \times L^2(-h, 0; D(A))$ the solution x of (2.1) exists uniquely and the following estimate holds

$$\|x\|_{L^2(0,\tau;D(A)) \cap W^{1,2}(0,\tau;X)} \leq c_1(\|\phi^0\|_V + \|\phi^1\|_{\mathcal{H}}), \tag{2.3}$$

for some constant c_1 dependent on τ. Moreover, the family of operators $\{T(t); t \geq 0\}$ defined by

$$T(t)\phi := (x(t), x_t) \tag{2.4}$$

is a strongly continuous semigroup on Z.

By assumption A generates a bounded semigroup $S(t)$ on X, so there is a constant $M > 0$ such that

$$\|S(t)\|_{\mathcal{L}(X)} \leq M \text{ for } t \geq 0, \text{ and} \tag{2.5}$$

$$\|\lambda R(\lambda, A)\|_{\mathcal{L}(X)} \leq M \text{ for } \lambda > 0, \tag{2.6}$$

[see, e.g., (1, 14)]. For the proof of the following lemma, see, e.g., (5).

LEMMA 2.1 If $f \in L^2(0, \tau; X)$, then setting

$$(S * f)(t) = \int_0^t S(t-s) f(s) ds, \quad 0 \leq t \leq \tau \tag{2.7}$$

we have

$$(S * f) \in L^2(0, \tau; D(A)) \cap W^{1,2}(0, \tau; X), \tag{2.8}$$

$$\|S * f\|_{L^2(0,\tau;D(A)) \cap W^{1,2}(0,\tau;X)} \leq c_2 \|f\|_{L^2(0,\tau;X)} \tag{2.9}$$

with c_2 depending on M and τ. If $v \in V$, then the function $t \to S(t)v$ belongs to $L^2(0, \tau; D(A)) \cap W^{1,2}(0, \tau; X)$ and

$$\|S(.)v\|_{L^2(0,\tau;D(A))} \leq \|S(.)v\|_{W^{1,2}(0,\tau;X)} \leq c_3 \|v\|_V \tag{2.10}$$

with $c_3 = \max(M\sqrt{\tau}, 1)$. Moreover, there is a function $\gamma(\tau)$ such that $\lim_{\tau \to 0} \gamma(\tau) = 0$ and such that

$$\|S * f\|_{L^2(0,\tau;D(A))} \leq \gamma(\tau)\|f\|_{L^2(0,\tau;V)} \tag{2.11}$$

for each $f \in L^2(0, \tau; V)$.

The first objective of this paper is to approximate the unbounded operators in (2.1) with bounded operators so that the existing results on structural operators can be used for the characterization of the approximate dual solution semigroup. The second objective, then, is to show convergence of approximate dual solution semigroups to the dual solution semigroup of equation (2.1) and thus to obtain its characterization.

We let $A_\lambda := \lambda\, \mathrm{AR}(\lambda, A)$ be the Yoshida-type approximation of the operator A for $\lambda > 0$. For details, see, e.g., Refs. 1, 14. Moreover, we define bounded operators R_λ and B_λ in X by

$$R_\lambda = \lambda R(\lambda, A) \text{ and } B_\lambda = \mathrm{BR}_\lambda \tag{2.12}$$

for $\lambda > 0$. By (2.6) we have the estimate $\|R_\lambda\|_{(X)} \leq M$. We also have the equation

$$\lim_{\lambda \to \infty} \|R_\lambda x - x\| = 0 \quad \forall x \in X. \tag{2.13}$$

By using the fact that $D(A)\circlearrowleft V\circlearrowleft X$ the following estimates can be readily obtained

$$\|B_\lambda x\|_V \leq M \cdot \|B\|_{\mathcal{L}(D(A),V)} \cdot \|x\|_{D(A)}, \quad \forall x \in D(A) \tag{2.14}$$

$$\lim_{\lambda \to \infty} \|B_\lambda x - Bx\| = 0, \quad \forall x \in D(A). \tag{2.15}$$

We consider the following approximate delay differential equation (ADDE)

$$\begin{aligned} \dot{x}_\lambda(t) &= Ax_\lambda(t) + B_\lambda x_\lambda(t) + L_{1\lambda} x_\lambda(t-h) + L_{2\lambda} x_{\lambda t} \\ x_\lambda(0) &= \phi^0, \quad x_\lambda(s) = \phi^1(s) \text{ a.e. on } [-h, 0), \end{aligned} \tag{2.16}$$

where

$$L_{1\lambda} = A_\lambda$$

$$L_{2\lambda} x_t = \int_{-h}^{0} a(s) A_\lambda x(t+s) ds.$$

Since the operators in (2.16), excluding A, are bounded in X we can use Proposition 2.1 in (9) and get the following result:

For every $\phi \in M^2$ the mild solution x_λ of (2.16) exists uniquely; i.e., x_λ solves the integral equation

$$x_\lambda(t) = S(t)\phi^0 + \int_0^t S(t-s)[B_\lambda x_\lambda(s) + L_{1\lambda} x_\lambda(s-h) + L_{2\lambda} x_{\lambda s}] ds. \tag{2.17}$$

Moreover, the family of operators $\{T_\lambda(t); t \geq 0\}$ defined by

$$T_\lambda(t)\phi = (x_\lambda(t), x_{\lambda t}) \tag{2.18}$$

is a strongly continuous semigroup on M^2 for every $\lambda > 0$.

Next we will show that the restriction of $T_\lambda(t)$ to the subspace $Z \subset M^2$ is a strongly continuous semigroup in Z.

LEMMA 2.2 Let $x \in L^2(-h, \tau; D(A))$ and let us define functions $\hat{L}_{1\lambda} x$ and $\hat{L}_{2\lambda} x$ by: $(\hat{L}_{1\lambda} x)(t) = L_{1\lambda} x(t-h)$ and $(\hat{L}_{2\lambda} x)(t) = L_{2\lambda} x_t$ for a.e. $t \in (0, \tau)$. Then we have

$$\|B_\lambda x\|_{L^2(0,\tau;V)} \leq M \|B\|_{\mathcal{L}(D(A);V)} \|x\|_{L^2(0,\tau;D(A))} \tag{2.19}$$

$$\|\hat{L}_{1\lambda}x\|_{L^2(0,\tau;X)} \leq M\|x\|_{\mathcal{H}} \tag{2.20}$$

$$\|\hat{L}_{2\lambda}x\|_{L^2(0,\tau;X)} \leq M\|L_2\|_{\mathcal{L}(\mathcal{H};X)}\sqrt{\tau}\|x\|_{L^2(-h,\tau;D(A))}. \tag{2.21}$$

Proof. The results readily follow from (2.6) and (2.14).

PROPOSITION 2.3 For every $\phi \in Z$ the mild solution x_λ of ADDE (2.16) satisfies the relation

$$\|x_\lambda\|_{L^2(0,\tau;D(A))\cap W^{1,2}(0,\tau;X)} \leq c_4(\|\phi^0\|_V + \|\phi^1\|_{\mathcal{H}}), \tag{2.22}$$

where c_4 depends on τ and is independent of λ.

Proof. The idea for the proof was inspired by the method given in the proof of Theorem 3.3 in (5). Let x_λ be the solution of (2.17) and let $\tau \in (0, h)$ be fixed so that

$$\beta := \gamma(\tau) \cdot M\|B\| + c_2\sqrt{\tau}M\|L_2\| < 1. \tag{2.23}$$

The norms of the operators are always equal to those in Lemma 2.2, so we omit the subscripts. Note that from (2.6) it follows

$$\|A_\lambda v\| \leq M\|Av\|, \tag{2.24}$$

for every vector $v \in D(A)$. By using (2.10), (2.9), and (2.11) in equation (2.17) we get

$$\begin{aligned}\|x_\lambda\|_{L^2(0,\tau;D(A))\cap W^{1,2}(0,\tau;X)} &\leq c_3\|\phi^0\|_V + c_2\|\hat{L}_{1\lambda}x_\lambda + \hat{L}_{2\lambda}x_\lambda\|_{L^2(0,\tau;X)}\\ &\quad + \gamma(\tau)\|B_\lambda x_\lambda\|_{L^2(0,\tau;D(A))}.\end{aligned}$$

We note that $x_\lambda = \phi^1$ on $[-h, 0)$. Hence, by (2.19)–(2.21) it follows

$$\|x_\lambda\|_{L^2(0,\tau;D(A))\cap W^{1,2}(0,\tau;X)} \leq \beta_1(\|\phi^0\|_V + \|\phi^1\|_{\mathcal{H}}, \tag{2.25}$$

with

$$\beta_1 = (1-\beta)^{-1}[c_3 + c_2M(1 + \|L_2\|\sqrt{\tau}].$$

Thus (2.22) is verified on $(0, \tau)$ with $\tau < h$ satisfying (2.23). By uniqueness of the mild solution of (2.16) the relation (2.22) can be readily proved for an arbitrary $\tau > 0$. For details, see, e.g., Ref. 5.

COROLLARY 2.4 The restriction of $T_\lambda(t)$ to Z is a strongly continuous semigroup on Z and we have

$$\|T_\lambda(t)\phi\|_Z \leq c_5\|\phi\|_Z, \quad 0 \leq t \leq \tau, \tag{2.26}$$

for every $\phi \in Z$. The constant c_5 is independent of λ.

Proof. The statement is a consequence of the continuity of the shift operator in $L^2(-h, \tau; D(A))$.

PROPOSITION 2.5 For every $\phi \in Z$ the following equation holds

$$\lim_{\lambda\to\infty} \|T)_\lambda(t)\phi - T(t)\phi\|_Z = 0, \tag{2.27}$$

uniformly over bounded time intervals.

Proof. Let x_λ and x be mild solutions of (2.16) and (2.1) with the same initial value $\phi \in Z$. Again, let $\tau \in (0,h)$ be fixed, so that (2.2) holds. Let us denote $\mathcal{H}^0 = L^2(0,\tau;X)$. By (1.5), (2.14) and Lemma 2.2 we get the estimate

$$(1-\beta)\|x_\lambda - x\|_{L^2(0,\tau;D(A))} \tag{2.28}$$
$$\leq c_2[\|B_\lambda x - Bx\|_{\mathcal{H}^0} + \|\hat{L}_{1\lambda}\phi^1 - \hat{L}_1\phi^1\|_{\mathcal{H}^0} + \|\hat{L}_{2\lambda}x - \hat{L}_2 x\|_{\mathcal{H}^0},$$

where $(\hat{L}_1\phi^1)(t) = L_1\phi^1(t-h)$ and $(\hat{L}_2 x)(t) = L_2 x_t$. By (2.13), (2.15), and Lebesgue's dominated convergence theorem from (2.28) it follows thus

$$\lim_{\lambda\to\infty} \|x_\lambda - x\|_{L^2(0,\tau;D(A))} = 0.$$

Since by (1.4) we have

$$\|x_\lambda - x\|_{C(0,\tau;V)} \leq c_0\|x_\lambda - x\|_{L^2(0,\tau;D(A))},$$

it follows that

$$\lim_{\lambda\to\infty} \|T_\lambda(t)\phi - T(t)\phi\|_Z = 0, \text{ for } \phi \in Z,$$

uniformly in $t \in [0,\tau]$ and τ satisfying (2.23). By induction, (2.27) can be proved for an arbitrary time interval.

3 TRANSPOSED SEMIGROUPS

The objective of this section is to introduce transposed semigroups that allow characterization of the adjoint semigroup $\{T^*(t); t \geq 0\}$ of the solution semigroup $\{T(t); t \geq 0\}$ of DDE (2.1). It is well known that the adjoint operator A^* of A is the infinitesimal generator of the semigroup of adjoint operators $\{S^*(t); t \geq 0\}$ of $S(t)$. The spectrum of the adjoint A^* is just the conjugate of the spectrum of A. Therefore, the operator A^* is itself the generator of a bounded analytic semigroup and the same estimates as (2.5) and (2.6) hold (see, e.g., Ref. 14):

$$\|S^*(t)\|_{\mathcal{L}(X)} \leq M, \text{ for } t \geq 0 \text{ and}$$
$$\|\lambda R(\lambda, A^*)\|_{\mathcal{L}(X)} \leq M, \text{ for } \lambda > 0.$$

We consider the dual or transposed DDE associated with equation (2.1)

$$\begin{aligned} \dot{y}(t) &= A*y(t) + B*y(t) + A*y(t-h) + \int_{-h}^{0} a(s)A*y(t+s)ds \\ y(0) &= \psi^0, y(s) = \psi^1(s) \text{ a.e. on } [-h,0), \end{aligned} \tag{3.1}$$

for a.e. $t \in (0,\tau)$. The initial value $\psi = (\psi^0, \psi^1)$ is an element of the product space $Z_* := V_* \times L^2(-h,0;D(A^*))$, where

$$V_* := \left\{ w \in X \,\middle|\, \int_0^\infty \|A*S*(t)w\|^2\,dt < \infty \right\} \text{ with norm}$$

$$\|w\|_{V_*} = \|w\| + \left(\int_0^\infty \|A*S*(t)w\|^2\,dt\right)^{1/2}$$

is the intermediate space between $D(A*)$ and X and $D(A*)$ is, as usual, equipped with the graph norm. We assume that the operator $B*$ belongs to $\mathcal{L}(D(A*), V_*)$. Thus the operators appearing in (3.1) are of the same type as those given in the original equation (2.1). Therefore, by Theorems 3.3 and 4.1 in Ref. 5 we have the result:

For every $\psi \in Z_*$ there is a unique solution of (3.1). Moreover, the family of operators $\{T^T(t); t \geq 0\}$ defined by

$$T^T(t)\psi := (y(t), y_t) \tag{3.2}$$

is a strongly continuous semigroup on Z_*.

Let $A^*_\lambda = \lambda A * R(\lambda, A^*)$ be the Yoshida-type approximation of the operator A^* for $\lambda > 0$. We note that $R(\lambda, A^*) = (R(\lambda, A))^*$, so we can write $(A^*)_\lambda = (A_\lambda)^* = A^*_\lambda$. Moreover, $B^*_\lambda y = (\mathrm{BR}_\lambda)^* y = R^*_\lambda B^* y$ for every $y \in D(A^*)$. For details, see, e.g., Ref. 16, p. 195.

We consider the approximate dual DDE of (1.1):

$$\dot{y}_\lambda(t) = A^* y_\lambda(t) + B^*_\lambda y(t) + A^*_\lambda y_\lambda(t-h) + \int_{-h}^{0} a(s) A^*_\lambda y_\lambda(t+s)ds \tag{3.3}$$

$$y_\lambda(0) = \psi^0, \qquad y_\lambda(s) = \psi^1(s) \text{ a.e. on } (-h, 0),$$

for a.e. $t \in (0, \tau)$ and $\psi \in M^2$.

The approximate dual equation (3.3) is of the same type as ADDE (2.16). Therefore, the unique mild solution y_λ of (3.3) exists for every $\psi \in M^2$ and $\lambda > 0$. Also the family of approximate dual semigroups $\{T^T_\lambda(t); t \geq 0\}$ is defined by

$$T^T_\lambda(t)\psi := (y_\lambda(t), y)_{\lambda t}), \quad \forall \psi \in M^2. \tag{3.4}$$

PROPOSITION 3.1 For every $\psi \in Z_*$ the following equation holds

$$\lim_{\lambda \to \infty} \|T^T_\lambda(t)\psi - T^T(t)\psi\|_{Z_*} = 0, \tag{3.5}$$

uniformly over bounded time intervals.

Proof. The relation can be proved exactly in the same way as Proposition 2.5; hence, the proof is omitted.

4 STRUCTURAL OPERATORS AND ADJOINT SEMIGROUPS

In this section we introduce structural operators that provide the essential connection between the adjoint and the transposed semigroup associated with equations (2.1) and (2.16). These operators describe the structure of the delayed part of the delay differential equation. The properties of structural operators associated with DDEs (2.1) and (2.16) were previously studied (11–13). Hence we give here only definitions and refer to cited papers for additional information. First we define the structural operators $F_\lambda : M^2 \to M^2$ associated with the DDE (2.16):

$$F_\lambda \phi := (\phi^0, H_\lambda \phi^1), \qquad \text{for } \phi \in M^2, \tag{4.1}$$

where

$$(H_\lambda \phi^1)(s) := A_\lambda \phi^1(-h-s) + \int_{-h}^{s} a(r) A_\lambda \phi^1(r-s)dr,$$

for a.e. $s \in [-h, 0)$. We note that $F_\lambda \in \mathcal{L}(M^2)$ for $\lambda > 0$. By a change of variables it can be shown that the adjoint operator F^*_λ of F_λ is given by

$$F^*_\lambda \psi = (\psi^0, H^*_\lambda \psi^1), \text{ for } \psi \in M^2, \tag{4.2}$$

where

$$(H_\lambda^*\psi^1)(s) = A_\lambda^*\psi^1(-h-s) + \int_{-h}^{2} a(r)A_\lambda^*\psi^1(r-s)dr,$$

for a.e. $s \in [-h,0)$.

The product space M^2 is a Hilbert space, so that the elements of the topological dual $(M^2)'$ can be identified with the elements of M^2 itself. Therefore, the adjoint semigroup $\{T_\lambda^*(t); t \geq 0\}$ is a strongly continuous semigroup on M^2 for every $\lambda > 0$.

We note that the operator B_λ in the DDE (2.16) is bounded. By perturbation theory the operators' sum $A_0 = A + B_\lambda$ is itself the infinitesimal generator of a strongly continuous semigroup on X (see, e.g., Ref. 14). This means that the DDE (2.16) with bounded operators acting in the delays has the same form as the DDE studied in (11), Section 2. Therefore, the following characterization of the adjoint semigroup $T_\lambda^*(t)$ is a direct consequence of Theorem 2.1 in (11) (see also Theorem 4.2 in Ref. 13):

THEOREM 4.1 For $\lambda > 0$ let $\{T_\lambda^*(t); t \geq 0\}$ be the adjoint semigroup associated with DDE (2.16) and $\{T_\lambda^T(t); t \geq 0\}$ be the transposed semigroup associated with (3.3). Then we have

$$T_\lambda^*(t)F_\lambda^*\psi = F_\lambda^*T_\lambda^T(t)\psi, \qquad t \geq 0, \tag{4.3}$$

for every $\psi \in M^2$.

Next, let us define the structural operator $F : Z \to M^2$ associated with DDE (2.1) by

$$F\phi := (\phi^0, H\phi^1), \text{ for } \phi \in Z, \tag{4.4}$$

where

$$(H\phi^1)(s) := A\phi^1(-h-s) + \int_{-h}^{s} a(r)A\phi^1(r-s)dr.$$

We note that $F \in \mathcal{L}(Z, M^2)$. Moreover, let Z' and Z'_* denote dual spaces of Z and Z_*, respectively. Then we have the following relations:

$$Z \subset M^2 = (M^2)' \subset Z' \text{ and } Z_* \subset M^2 = (M^2)' \subset Z'_*.$$

Therefore, it follows that $F \in \mathcal{L}(Z, Z'_*)$ and $F^* \in \mathcal{L}(Z_*, Z')$.

By a change of variables the following characterization of the adjoint F^* can be obtained (for details, see Ref. 12):

LEMMA 4.3 For $\psi \in Z_*$ the adjoint F^* of the operator F is given by

$$F^*\psi = (\psi^0, H^*\psi^1) \text{ and}$$

$$(H^*\psi^1)(s) = A^*\psi^1(-h-s) + \int_{-h}^{s} a(r)A^*\psi^1(r-s)dr. \tag{4.5}$$

The following theorem shows that the structural operator F^* associated with the DDEs (2.1) and (3.1) provides the same relationship between the adjoint semigroup $T^*(t)$ and its transposed semigroup $T^T(t)$, as that given in Refs. 2 and 3.

THEOREM 4.4 Let $\{T^*(t); t \geq 0\}$ be the adjoint semigroup of DDE (2.1) and let $\{T^T(t); t \geq 0\}$ be the transposed semigroup associated with dual equation (3.1). Then the following equation holds for every $\psi \in Z_*$:

$$T^*(t)F^*\psi = F^*T^T(t)\psi, \quad \text{for } t \geq 0. \tag{4.6}$$

Proof. By using Theorem 4.1, Proposition 2.5, and Proposition 3.1 the relation (4.6) can be proved exactly in the same way as relation (4.12) in Ref. 12. We omit the details.

REFERENCES

1. A. Bensoussan, G. Da Prato, M. Delfour, S. Mitter, Representation and Control of Infinite-dimensional Systems, Vol. 1. Boston: Birkhauser, 1992.
2. C. Bernier and A. Manitius, On semigroups in $R^n \times L^p$ corresponding to differential equations with delays. Can. J. Math. 30 (1978), 897–914.
3. M.C. Delfour and A. Manitius, The structural operator F and its role in the theory of retarded systems. Part I: J. Math. Anal. Appl. 73 (1980), 466–490; Part II: J. Math. Anal. Appl. 74 (1980), 359–381.
4. W. Desch, R. Grimmer, and W. Schappacher, Some considerations for linear integrodifferential equations. J. Math. Anal. Appl. 104 (1984), 219–234.
5. G. DiBlasio, K. Kunisch, and E. Sinestrari, L^2-regularity for parabolic integrodifferential equations with delay in the highest order derivatives. J. Math. Anal. Appl. 102 (1984), 38–57.
6. G. DiBlasio, K. Kunisch, and E. Sinestrari, Stability for abstract linear functional differential equations. Isr. J. Math. 50 (1985), 231–263.
7. R.C. Grimmer and J. Prüss, On linear Volterra equations in Banach spaces. Comput. Math. Appl. 11 (1985), 189–205.
8. K. Kunisch and M. Mastinšek, Dual semigroups and structural operators for partial functional differential equations with unbounded operators acting on the delays. Diff. Int. Equations 3 (1990), 733–756.
9. K. Kunisch and W. Schappacher, Mild and strong solutions for partial differential equations with delay. Ann. Mat. Pur. Appl. 125 (1980), 193–219.
10. J.L. Lions and E. Magenes, Problèmes aux Limites non Homogènes et Applications, Vol. I. Paris: Dunod, 1968.
11. M. Mastinšek, Structural operators for abstract functional differential equations. In Semigroup Theory and Applications, edited by P. Clément, S. Invernizzi, E. Mitidieri, I.I. Vrabie, Lecture Notes in Pure and Applied Mathematics 116. New York: Marcel Dekker, Inc., 1989.
12. M. Mastinšek, Dual semigroups for delay differential equations with unbounded operators acting on the delays. Diff. Int. Equations (1993) (to appear).
13. S. Nakagiri, Structural properties of functional differential equations in Banach spaces. Osaka J. Math. 25 (1988), 353–398.
14. A. Pazy, Semigroups of Linear Operators and Applications to Partial Differential Equations. New York: Springer, 1983.
15. H. Tanabe, Structural operators for linear delay-differential equations in Hilbert space. Proc. Jpn Acad. 64, Ser. A (1988).
16. K. Yoshida, Functional Analysis. Berlin: Springer-Verlag, 1980.

Min–Max Game Theory and Algebraic Riccati Equations for Boundary Control Problems with Continuous Input-Solution Map, Part I: The Stable Case

C. M. McMILLAN[1] and R. TRIGGIANI[2] University of Virginia, Charlottesville, Virginia

Abstract

We consider the abstract dynamical framework [L-T.3, class (H.2)] which models a variety of mixed partial differential equation problems in a smooth bounded domain $\Omega \in \mathbb{R}^n$, with L_2-boundary control and with L_2-boundary disturbance. We then set and solve a min-max game theory problem with quadratic indefinite cost. The present Part I considers the stable case of the free dynamics where all relevant quantities of the min-max problem can be expressed directly and explicitly in terms of the problem data. A companion paper, Part II, considers the general case which requires a more complicated approach which, in particular, and unlike Part I, involves the Riccati operator $P_{0,\infty}$ which arises when there is no disturbance $w \equiv 0$.

[1]Research partially supported by an IBM Graduate Student Fellowship
[2]Research partially supported by the National Science Foundation under Grant NSF-DMS-8902811-01

1 Introduction

1.1 Problem setting

Let U (control) and Y (state) be separable Hilbert spaces. We introduce the following abstract state and output equations

$$\dot{y}(t) = Ay(t) + Bu(t) + Gw(t) \quad \text{in } [D(A^*)]'; \;\; y(0) = y_0 \in Y \tag{1.1.1}$$

$$z(t) = Ry(t) + Du(t). \tag{1.1.2}$$

Here, the function $u \in L_2(0, \infty; U)$ is the control and $w \in L_2(0, \infty; Y)$ is a deterministic disturbance. The dynamics (1.1.1), (1.1.2) is subject to the following assumptions, which will be maintained throughout the paper:

(H.1) $A : Y \subset D(A) \longrightarrow Y$ is the infinitesimal generator of a strongly continuous (s.c.) semigroup e^{At} on the Hilbert space Y, and A^{-1} is boundedly invertible: $A^{-1} \in \mathcal{L}(Y)$

(H.2) B: continuous $U \longrightarrow [D(A^*)]'$; or, equivalently, $A^{-1}B \in \mathcal{L}(U; Y)$, where $[D(A^*)]'$ denotes the dual of $D(A^*)$ with respect to the Y-topology, and A^* is the Y-adjoint of A;

(H.3) the following abstract trace regularity holds (see Remark 1.2.2): the (closable) operator $B^*e^{A^*t}$ admits a continuous extension, denoted by the same symbol, from $Y \longrightarrow L_2(0, T; U)$:

$$\int_0^T \|B^*e^{A^*t}x\|_Y^2 dt \le c_T \|x\|_Y^2 \quad \forall\, T < \infty; \tag{1.1.3}$$

where B^* is the dual of B, satisfies $B^* \in \mathcal{L}(D(A^*), U)$ after identifying $[D(A^*)]''$ with $D(A^*)$.

(H.4) G is an operator of the same class as B; i.e., G is continuous $U \longrightarrow [D(A^*)]'$ and

$$\int_0^T \|G^*e^{A^*t}x\|_Y^2 \le c_T \|x\|_Y^2 \quad \forall\, T < \infty \tag{1.1.4}$$

(H.5) the following simplifying assumptions hold:

$$R \in \mathcal{L}(Y), \quad D \in \mathcal{L}(U;Y) \text{ with } D^*D = I \text{ and } D^*R = 0 \tag{1.1.5a}$$

(see Remark 1.2.3), so that recalling (1.1.2)

$$\|z(t)\|_Y^2 = \|Ry(t)\|_Y^2 + \|u(t)\|_U^2 \tag{1.1.5b}$$

The min-max game theoretic problem with indefinite cost considered below in section 1.2, and leading to a characterization such as the one of the main Theorem 1.3.1 below in terms of an algebraic Riccati equation, may be solved for a general s.c. semigroup e^{At}: this requires the additional assumptions of "Finite Cost Condition" or "stabilizability" (for existence) and "detectability" (for uniqueness) of the Riccati operator. The case where e^{At} is actually a unitary group is the one which arises in the boundary control problems (mixed problems) for the canonical conservative partial differential equations such as those listed in Remark 1.2.2, which satisfy the trace regularity property (1.1.3). However, the general analysis of this min-max problem and the corresponding Riccati theory – which extends the quadratic control problem $w \equiv 0$ of [L-T.2, section 5], [F-L-T.1] – is complicated. Our treatment of this problem is contained in the companion paper, Part II, of the present article [M-T.1]. Another treatment, which likewise falls into the analysis of [F-L-T.1], is found in [B.1], where further references to the related H^∞-control problem are given in the case where A, B, and G are matrices. As it turns out, a much more simplified, short-cut treatment –which is also more informative, cleaner, and fully explicit –may be given in the case where the original dynamics, e^{At}, is assumed to be (exponentially) uniformly stable (in which case the assumptions of stabilizability and detectability are automatically satisfied). This is done in the present Part I, which provides the explicit characterizations of the relevant optimal quantities which are not possible in the general case. Applications include all the aforementioned conservative equations with an additional damping, see Remark 1.2.2, which constitute a large and physically significant class. All this justifies a separate Part I. Accordingly, in the present article we shall make, in addition to the hypotheses above, the further assumption that

(H.6) there exist constants $M \geq 1$ and $\delta > 0$, such that

$$\|e^{At}\|_{\mathcal{L}(Y)} \leq Me^{-\delta t}, \quad t \geq 0 \tag{1.1.6}$$

The solution to the state equation (1.1.1) is given explicitly by

$$y(t) = y(t; y_0) = e^{At}y_0 + (Lu)(t) + (Ww)(t) \tag{1.1.7}$$

where the operators L and W are defined in (2.1.2) and (2.1.4) below.

1.2 Game Theory Problem

For a fixed $\gamma > 0$, we associate with (1.1.1) and (1.1.2) the cost functional

$$J(u,w) = J(u,w,y(u,w)) = \int_0^\infty [\|Ry(t)\|_Y^2 + \|u(t)\|_U^2 - \gamma^2\|w(t)\|_Y^2]dt \tag{1.2.1a}$$

$$= \int_0^\infty [\|z(t)\|_Y^2 - \gamma^2\|w(t)\|_Y^2]dt \tag{1.2.1b}$$

where $y(t) = y(t; y_0)$ is given by (1.1.7), and where in going from (1.2.1a) to (1.2.1b) we have used (1.1.5b). The aim of this paper, is to study the following game-theory problem:

$$\sup_w \inf_u J(u,w) \tag{1.2.2}$$

where the infimum is taken over all $u \in L_2(0,\infty;U)$, for w fixed, and the supremum is taken over all $w \in L_2(0,\infty;Y)$. In P.D.E. mixed problems (see Remark 1.2.2), both control u ("good" player) and disturbance w ("bad" player) act on the boundary $\partial\Omega$ (or part thereof) of the spatial domain $\Omega \in \mathbb{R}^n$.

Remark 1.2.1: The above game-theory problem in a finite-dimensional setting has been shown by various authors (see, e.g. [B-B.1]) to be related to the following H^∞-problem with state feedback:

Find necessary and sufficient conditions for the existence of a stabilizing feedback $u = Fy$ such that the transfer function from $w(t)$ to $z(t)$:

$$T(s;F) = (R + DF)(sI - A - BF)^{-1}G \tag{1.2.3}$$

has H^∞-norm, $||T(s;F)||_\infty < \gamma, \quad \gamma > 0$. (Here, "$H^\infty$" stands for the Hardy space of all complex-valued functions which are analytic and bounded in the open half-plane, $\mathbb{C}^+$.)

Thus, this paper extends the H^∞-problem to the infinite-dimensional case which includes the various partial differential equation models described in Remark 1.2.2 below.

Remark 1.2.2: Assumption (H.3) = (1.1.3) is an abstract trace theory property. Over the past ten years, this property has been proved to hold true for many classes of partial differential equations by purely P.D.E.'s methods (energy methods either in differential or in pseudo-differential form), including: second order hyperbolic equations; Euler-Bernoulli, Kirchhoff, and Schroedinger equations; first order hyperbolic systems, etc., all in arbitrary space dimensions and on explicitly identified spaces; see e.g. [L-T.3, class (H.2)]. In order to satisfy the stability assumption (1.1.6), one may add damping to the aforementioned dynamics.

Remark 1.2.3: Assumption (H.5) is for simplicity only. It can be relaxed [B.1].

1.3 Statement of main results

Main Theorem 1.3.1 *Assume (H.1) - (H.6). Then there exists a (critical) value $\gamma_c > 0$ defined explicitly in terms of the problem data by Eq. (2.2.1) below such that:*

(i) if $0 < \gamma < \gamma_c$, then taking the supremum in w as in (1.2.2) leads to $+\infty$; i.e. there is no finite solution of the game theory problem (1.2.2);

(ii) if $\gamma > \gamma_c$, then

(ii_1) there exists a unique solution $\{u^*(\cdot\,;y_0); w^*(\cdot\,;y_0); y^*(\cdot\,;y_0)\}$ *of the game theory problem (1.2.2);*

(ii_2) the following pointwise feedback relation holds

$$u^*(t;y_0) = -B^*Py^*(t;y_0) \in L_2(0,\infty;U) \tag{1.3.1}$$

(see (5.1.2)), where P *is the unique bounded, nonnegative self-adjoint operator which satisfies the following Algebraic Riccati Equation for all* $x, y \in D(A)$*:*

$$(PAx,z)_Y + (Px,Az)_Y + (Rx,Rz)_Y = (B^*Px, B^*Pz)_U - \gamma^{-2}(G^*Px, G^*Pz)_Y \tag{1.3.2}$$

(see Theorem 5.2.1 below), with the property (see (5.2.10) below)

$$B^*P \in \mathcal{L}(D(A);U); \qquad G^*P \in \mathcal{L}(D(A);U) \tag{1.3.3}$$

(ii_3) the operator (F stands for "feedback")

$$A_F = A - BB^*P + \gamma^{-2}GG^*P \tag{1.3.4}$$

is the generator of a s.c. semigroup on Y *(see (5.2.4) below) and, in fact, for* $y_0 \in Y$ *(see (5.2.1) below):*

$$y^*(t;y_0) = e^{(A-BB^*P+\gamma^{-2}GG^*P)t}y_0 \in C_b([0,\infty];Y) \cap L_2(0,\infty;Y) \tag{1.3.5}$$

where, moreover, the semigroup is uniformly stable in Y*;*

(ii_4)
$$\gamma^2 w^*(t;y_0) = G^*Py^*(t;y_0) \in C_b([0,\infty];Y) \cap L_2(0,\infty;Y) \tag{1.3.6}$$

(see (5.1.3) below)

(ii_5) for any $y_0 \in Y$ *(see (5.1.5) below)*

$$(Py_0,y_0) = J^*(y_0) \equiv J(u^*(\cdot\,;y_0), w^*(\cdot\,;y_0), y^*(\cdot\,;y_0)) = \sup_w \inf_u J(u(\cdot\,;y_0), w(\cdot\,;y_0), y(\cdot\,;y_0)) \tag{1.3.7}$$

Additional results are given in the treatment below.

2 Minimization of J over u for w fixed

2.1 Existence of a unique optimal pair and optimality conditions

We return to the functional cost J in (1.2.1). In this section we consider the following problem: given a fixed but arbitrary $w \in L_2(0, \infty; Y)$: minimize

$$J(u, w, y_0) = \int_0^\infty [\|Ry(t)\|_Y^2 + \|u(t)\|_U^2 - \gamma^2 \|w(t)\|_Y^2] dt \tag{2.1.1}$$

over all $u \in L_2(0, \infty; U)$, where $y(t) = y(t; y_0)$ is the corresponding solution of (1.1.1) given explicitly by (1.1.7). The advantages of assumption (1.1.6) of the present Part I are reaped at the very outset of the analysis, as a result of the following properties [L-T.1, Theorem in section 3 and Remark 3.3], a consequence of (H.3):

$$(Lu)(t) = \int_0^t e^{A(t-\tau)} Bu(\tau) d\tau \tag{2.1.2}$$

$$continuous: \; L_2(0, \infty; U) \longrightarrow L_2(0, \infty; Y) \cap C_b([0, \infty]; Y) \tag{2.1.3}$$

$$(Ww)(t) = \int_0^t e^{A(t-\tau)} Gw(\tau) d\tau \tag{2.1.4}$$

$$continuous: \; L_2(0, \infty; Y) \longrightarrow L_2(0, \infty; Y) \cap C_b([0, \infty]; Y) \tag{2.1.5}$$

where C_b is the space of Y-valued continuous functions which are uniformly bounded on $[0, \infty]$.

The dual versions of (2.1.2)-(2.1.5) are

$$(L^* f)(t) = B^* \int_t^\infty e^{A^*(\tau - t)} f(\tau) d\tau \tag{2.1.6}$$

$$continuous: \; L_2(0, \infty; Y) \cap L_1(0, \infty; Y) \longrightarrow L_2(0, \infty; U) \tag{2.1.7}$$

$$(W^* v)(t) = G^* \int_t^\infty e^{A^*(\tau - t)} v(\tau) d\tau \tag{2.1.8}$$

$$continuous: \; L_2(0, \infty; Y) \cap L_1(0, \infty; Y) \longrightarrow L_2(0, \infty; Y) \tag{2.1.9}$$

These properties will be profitably used in Theorem 2.1.1 (ii), (iii).

Theorem 2.1.1 *(i) With reference to the minimization problem (2.1.1), there exists a unique optimal pair denoted by* $\{u_w^0(\cdot\,;y_0), y_w^0(\cdot\,;y_0)\}$, *with corresponding optimal cost denoted by*

$$\begin{aligned} J_w^0(y_0) &= J(u_w^0(\cdot\,;y_0), y_w^0(\cdot\,;y_0)) \\ &= \int_0^\infty [\|Ry_w^0(t;y_0)\|_Y^2 + \|u_w^0(t;y_0)\|_U^2 - \gamma^2\|w(t)\|_Y^2]dt \end{aligned} \tag{2.1.10}$$

This statement does not require the stability hypothesis (1.1.6).

(ii) The optimal pair is related by

$$u_w^0(\cdot\,;y_0) = -L^*R^*Ry_w^0(\cdot\,;y_0) \tag{2.1.11}$$

and is explicitly given in terms of the problem data by the following formulas:

$$\begin{aligned} -u_w^0(\cdot\,;y_0) &= [I + L^*R^*RL]^{-1}L^*R^*R[e^{A\cdot}\,y_0 + Ww] \in L_2(0,\infty;U) \\ &= -u_{w=0}^0(\cdot\,;y_0) - u_w^0(\cdot\,;y_0 = 0) \end{aligned} \tag{2.1.12}$$

$$\begin{aligned} y_w^0(\cdot\,;y_0) &= [I + LL^*R^*R]^{-1}[e^{A\cdot}\,y_0 + Ww] \in L_2(0,\infty;Y) \\ &= y_{w=0}^0(\cdot\,;y_0) + y_w^0(\cdot\,;y_0 = 0) \end{aligned} \tag{2.1.13}$$

where both inverse operators in (2.1.12) and (2.1.13) are well defined as bounded operators on all of $L_2(0,\infty;U)$ *and* $L_2(0,\infty;Y)$ *respectively (for* $[I + LL^*R^*R]^{-1}$ *see [L-T.2 p.891, below Eq. (2.8e)]). Moreover, the optimal dynamics is, of course,*

$$y_w^0(t;y_0) = e^{At}y_0 + \{Lu_w^0(\cdot\,;y_0)\}(t) + \{Ww(\cdot)\}(t) \in C_b([0,\infty];Y) \cap L_2(0,\infty;Y) \tag{2.1.14}$$

(iii) The optimal cost $J_w^0(y_0)$ *in (2.1.10) is given explicitly in terms of the data by the following formulas:*

$$\begin{aligned} J_w^0(y_0) &= (e^{A\cdot}\,y_0 + Ww, R^*R[I + LL^*R^*R]^{-1}[e^{A\cdot}\,y_0 + Ww])_{L_2(0,\infty;Y)} \\ &\qquad -\gamma^2(w,w)_{L_2(0,\infty;Y)} \end{aligned} \tag{2.1.15a}$$

$$= J_{w=0}^0(y_0) + J_w^0(y_0 = 0) + \chi_{y_0,w} \tag{2.1.15b}$$

$$J^0_{w=0}(y_0) = (e^{A\cdot}\, y_0, R^*R[I + LL^*R^*R]^{-1}(e^{A\cdot}\, y_0))_{L_2(0,\infty;Y)} \tag{2.1.16a}$$

$$= \|[I + (R^*R)^{1/2}LL^*(R^*R)^{1/2}]^{-1/2}(R^*R)^{1/2}(e^{A\cdot}\, y_0)\|^2_{L_2(0,\infty;Y)} \tag{2.1.16b}$$

$$J^0_w(y_0 = 0) = (Ww, R^*R[I + LL^*R^*R]^{-1}Ww)_{L_2(0,\infty;Y)} - \gamma^2(w,w)_{L_2(0,\infty;Y)} \tag{2.1.17a}$$

$$= -(w, E_\gamma w)_{L_2(0,\infty;Y)} \tag{2.1.17b}$$

$$= \|[I + (R^*R)^{1/2}LL^*(R^*R)^{1/2}]^{-1/2}(R^*R)^{1/2}Ww\|^2_{L_2(0,\infty;Y)} - \gamma^2\|w\|^2_{L_2(0,\infty;Y)} \tag{2.1.17c}$$

where

$$\begin{aligned} E_\gamma &= \gamma^2 I - W^*R^*R[I + LL^*R^*R]^{-1}W \\ &= \gamma^2 I - W^*(R^*R)^{1/2}[I + (R^*R)^{1/2}LL^*(R^*R)^{1/2}]^{-1}(R^*R)^{1/2}W \\ &= \gamma^2 I - S;\ S \text{ nonnegative self-adjoint operator in } \mathcal{L}(L_2(0,\infty;Y)) \end{aligned} \tag{2.1.18}$$

The cross terms in (2.1.15b) are linear in w:

$$\chi_{y_0,w} = (e^{A\cdot}\, y_0, R^*R[I + LL^*R^*R]^{-1}Ww)_{L_2(0,\infty;Y)} + (Ww, R^*R[I + LL^*R^*R]^{-1}e^{A\cdot}\, y_0)_{L_2(0,\infty;Y)} \tag{2.1.19a}$$

$$= 2(e^{A\cdot}\, y_0, R^*R[I + LL^*R^*R]^{-1}Ww)_{L_2(0,\infty;Y)} \tag{2.1.19b}$$

$$= 2(e^{A\cdot}\, y_0, (R^*R)^{1/2}[I + (R^*R)^{1/2}LL^*(R^*R)^{1/2}]^{-1}(R^*R)^{1/2}Ww)_{L_2(0,\infty;Y)} \tag{2.1.19c}$$

In going from (2.1.16a) to (2.1.16b), as well as from (2.1.17a) to (2.1.17b), and from (2.1.19b) to (2.1.19c) we have used the identity

$$(R^*R)^{1/2}[I + LL^*R^*R]^{-1} = [I + (R^*R)^{1/2}LL^*(R^*R)^{1/2}]^{-1}(R^*R)^{1/2} \in \mathcal{L}(L_2(0,\infty;Y)) \tag{2.1.20}$$

so that

$$R^*R[I + LL^*R^*R]^{-1} = (R^*R)^{1/2}[I + (R^*R)^{1/2}LL^*(R^*R)^{1/2}]^{-1}(R^*R)^{1/2}$$
$$= \text{self-adjoint operator in } \mathcal{L}(L_2(0,\infty;Y)) \tag{2.1.21}$$

Proof: (i) For fixed $w \in L_2(0,\infty;Y)$, the optimal problem is a standard quadratic (strictly convex) problem in u, which has a unique optimal solution.

(ii) Under the stability assumption (1.1.6), whereby properties (2.1.2)-(2.1.5) hold, we consider the Lagrangean

$$\mathrm{L}(u,y,\lambda) = 1/2[(R^*Ry,y)_{L_2(0,\infty;Y)} + (u,u)_{L_2(0,\infty;U)}]$$
$$+(\lambda, y - e^{A\cdot}\, y_0 - Lu - Ww)_{L_2(0,\infty;Y)} \tag{2.1.22}$$

to which we apply Liusternik's general Lagrange Multiplier Theorem [L.1, Theorem 1, p243] (note that for any $g \in L_2(0,\infty;Y)$, we can take $u = w = 0$ and $y = e^{A\cdot}\, y_0 + g$): there exist $u_w^0 \in L_2(0,\infty;U)$, y_w^0, $\lambda_w^0 \in L_2(0,\infty;Y)$ such that $\mathrm{L}_u = \mathrm{L}_y = \mathrm{L}_\lambda = 0$ at $(u_w^0, y_w^0, \lambda_w^0)$. From (2.1.22) we obtain in the appropriate inner products:

$$\mathrm{L}_y = 0: \qquad (R^*Ry, \delta y) + (\lambda, \delta y) = 0 \quad \forall \delta y \in L_2(0,\infty;Y) \tag{2.1.23}$$

$$\lambda_w^0 = -R^*Ry_w^0 \tag{2.1.24}$$

$$\mathrm{L}_u = 0: \qquad (u, \delta u) - (L^*\lambda, \delta u) = 0 \quad \forall \delta u \in L_2(0,\infty;U) \tag{2.1.25}$$

$$u_w^0 = L^*\lambda_w^0 \tag{2.1.26}$$

Then (2.1.24) inserted into (2.1.26) yields the basic relationship (2.1.11) between the optimal y_w^0 and the optimal u_w^0, as desired, where we have explicitly indicated the dependence on y_0.

Inserting the optimal dynamics (2.1.14) into (2.1.11) yields readily u_w^0 in (2.1.12). Moreover, applying L to (2.1.11) and inserting the resulting Lu_w^0 in the optimal dynamics (2.1.14) yields readily y_w^0 in (2.1.13). [We recall from [L-T.2, p891, below Eq. (2.8e)] that $[I + LL^*R^*R]^{-1} \in \mathcal{L}(L_2(0,\infty;Y))$].

(iii) By (2.1.11), inserting $(u_w^0, u_w^0) = -(u_w^0, L^*R^*Ry_w^0) = -(Lu_w^0, R^*Ry_w^0)$ into (2.1.10) yields

$$
\begin{aligned}
J_w^0(y_0) &= (y_w^0, R^*Ry_w^0) + (u_w^0, u_w^0) - \gamma^2(w,w) = (y_w^0 - Lu_w^0, R^*Ry_w^0) - \gamma^2(w,w) \\
\text{(by (2.1.14))} \quad &= (e^{A\cdot} y_0 + Ww, R^*Ry_w^0) - \gamma^2(w,w)
\end{aligned}
\tag{2.1.27}
$$

Inserting, now, y_w^0 from (2.1.13) into (2.1.27) yields (2.1.15a) as desired. To show the remaining of part (iii), we need to verify identity (2.1.20):

$$
\begin{aligned}
(R^*R)^{1/2}[I + LL^*R^*R]^{-1} &= \{[I + LL^*R^*R](R^*R)^{-1/2}\}^{-1} \\
&= \{(R^*R)^{-1/2}[I + (R^*R)^{1/2}LL^*(R^*R)^{1/2}]\}^{-1} \\
&= [I + (R^*R)^{1/2}LL^*(R^*R)^{1/2}]^{-1}(R^*R)^{1/2} \quad \square
\end{aligned}
\tag{2.1.28}
$$

2.2 Strict positive-definiteness of the operator E_γ for $\gamma > \gamma_c$

We return to the self-adjoint operator E_γ in $\mathcal{L}(L_2(0,\infty;Y))$ defined by (2.1.18). Define the critical value γ_c of γ by:

$$
\begin{aligned}
\gamma_c{}^2 &\equiv \|W^*R^*R[I + LL^*R^*R]^{-1}W\|_{L_2(0,\infty;Y)} \\
&= \|W^*(R^*R)^{1/2}[I + (R^*R)^{1/2}LL^*(R^*R)^{1/2}]^{-1}(R^*R)^{1/2}W\|_{L_2(0,\infty;Y)} \\
&= \sup_{\|w\|=1}(W^*R^*R[I + LL^*R^*R]^{-1}Ww, w)_{L_2(0,\infty;Y)}
\end{aligned}
\tag{2.2.1}
$$

since, by (2.1.21)

$$
\begin{aligned}
W^*R^*R[I + LL^*R^*R]^{-1}W &= W^*(R^*R)^{1/2}[I + (R^*R)^{1/2}LL^*(R^*R)^{1/2}]^{-1}(R^*R)^{1/2}W \\
&= \text{nonnegative self} - \text{adjoint operator in } \mathcal{L}(L_2(0,\infty;Y))
\end{aligned}
\tag{2.2.2}
$$

As a consequence of (2.2.1), we obtain:

Corollary 2.2.1 *The self-adjoint operator $E_\gamma \in \mathcal{L}(L_2(0,\infty;Y))$ in (2.1.18) is (strictly) positive if and only if $\gamma > \gamma_c$, (defined by (2.2.1)):*

$$
(E_\gamma w, w)_{L_2(0,\infty;Y)} \geq (\gamma^2 - \gamma_c{}^2)\|w\|^2_{L_2(0,\infty;Y)}
\tag{2.2.3}
$$

in which case $E_\gamma{}^{-1} \in \mathcal{L}(L_2(0,\infty;Y))$. $\square$

3 Maximization of $J^0_w(y_0)$ over w: existence of a unique optimal w^*.

In this section we return to the optimal $J^0_w(y_0)$ in (2.1.10) for $w \in L_2(0,\infty;Y)$ fixed and consider the problem:

$$\text{maximize } J^0_w(y_0), \text{ equivalently, minimize } -J^0_w(y_0), \text{ over all } w \in L_2(0,\infty;Y). \tag{3.1}$$

Theorem 3.1 *(i) For $\gamma > \gamma_c$ (defined in (2.2.1)), the following estimate holds true for any $\epsilon > 0$ and every $w \in L_2(0,\infty;Y)$:*

$$-J^0_w(y_0) \geq [\ \gamma^2 - ({\gamma_c}^2 + \epsilon)]\|w\|^2_{L_2(0,\infty;Y)} - J^0_{w=0}(y_0) - C_\epsilon\|y_0\|^2_Y \tag{3.2}$$

(ii) For $\gamma > \gamma_c$ (defined in (2.2.1)), there exists a unique optimal solution $w^*(\cdot\ ; y_0) \in L_2(0,\infty;Y)$ for the optimal problem (3.1):

$$\max_{w\in L_2(0,\infty;Y)} J^0_w(y_0) \equiv J^0_{w=w^*}(y_0) \equiv J^*(y_0). \tag{3.3}$$

(iii) If $0 < \gamma < \gamma_c$, then $\sup_w J^0_w(y_0) = +\infty$. □

<u>Proof:</u> (i) We return to (2.1.15b) which gives $-J^0_w(y_0)$ as the sum of three contributions: a quadratic term in w, given by $-J^0_w(y_0 = 0) = (w, E_\gamma w)$ in (2.1.17b), which satisfies (2.2.3); a linear term in w, given by $-\chi_{y_0,w}$ in (2.1.19b), and satisfying with $V = W^*[I + R^*RLL^*]^{-1}R^*Re^{A\cdot}$

$$\begin{aligned} |\chi_{y_0,w}| &\leq 2\|w\|_{L_2(0,\infty;Y)}\|Vy_0\|_{L_2(0,\infty;Y)} \\ &\leq \epsilon\|w\|^2_{L_2(0,\infty;Y)} + \epsilon^{-1}\|\ V\|^2\|y_0\|^2_Y \end{aligned} \tag{3.4}$$

where the norm of V is in $\mathcal{L}(Y; L_2(0,\infty;Y))$; finally a constant term in w given by $-J^0_{w=0}(y_0)$. Thus, (3.2) follows with $C_\epsilon = \epsilon^{-1}\|V\|^2$.

(ii) The expression of $-J^0_w(y_0)$ given by (2.1.15b) as a quadratic functional, bounded below by part (i), guarantees that there exists a unique optimal solution w^* in $L_2(0,\infty;Y)$.

(iii) If $0 < \gamma < \gamma_c$, then

$$\inf_{\|w\|=1} (E_\gamma w, w)_{L_2(0,\infty;Y)} = \rho < 0$$

Here, for $\epsilon > 0$ sufficiently small, there exists w_ϵ, $\|w_\epsilon\| = 1$ such that $(E_\gamma w_\epsilon, w_\epsilon)_{L_2(0,\infty;Y)} < \rho + \epsilon < 0$. Then, define $w_k = k w_\epsilon \in L_2(0,\infty;Y)$, for a real constant k. From (2.1.15b), (2.1.16a), (2.1.17b), we have

$$\begin{aligned} -J^0_{w_k}(y_0) &= k^2 (E_\gamma w_\epsilon, w_\epsilon)_{L_2(0,\infty;Y)} - 2k(w_\epsilon, V y_0) - J^0_{w=0}(y_0) \\ &\longrightarrow -\infty \quad as \quad k \longrightarrow \infty \end{aligned}$$

as desired, since $(E_\gamma w_\epsilon, w_\epsilon) < 0$. □

With the optimal w^* provided by Theorem 3.1(ii), we return to the optimal pair $\{u^0_w, y^0_w\}$ over u of Theorem 2.1.1. and set, along with (3.3):

$$u^*(\cdot\,; y_0) \equiv u^0_{w=w^*}(\cdot\,; y_0) \in L_2(0,\infty;U); \qquad y^*(\cdot\,; y_0) \equiv y^0_{w=w^*}(\cdot\,; y_0) \in L_2(0,\infty;Y); \tag{3.5}$$

Theorem 3.2 *(i) The unique optimal $w^*(\cdot\,; y_0)$ provided by Theorem 3.1(ii) is given explicitly in terms of the problem data by (see (2.1.8) and (3.5)):*

$$\gamma^2 w^*(\cdot\,; y_0) = W^* R^* R y^*(\cdot\,; y_0) \in L_2(0,\infty;Y), \quad \gamma > \gamma_c \tag{3.6}$$

(ii) Thus, for $\gamma > \gamma_c$ (defined by (2.2.1)), the original minimax problem (1.2.2) has a unique solution $\{u^*(\cdot\,; y_0), y^*(\cdot\,; y_0), w^*(\cdot\,; y_0)\}$ satisfying (3.6) and given by

$$-u^*(\cdot\,; y_0) = [I + L^* R^* R L]^{-1} L^* R^* R\, [e^{A\cdot} y_0 + W w^*(\cdot\,; y_0)] \in L_2(0,\infty;U) \tag{3.7}$$

$$y^*(\cdot\,; y_0) = [I + L L^* R^* R]^{-1} [e^{A\cdot} y_0 + W w^*(\cdot\,; y_0)] \in L_2(0,\infty;Y) \tag{3.8}$$

$$u^*(\cdot\,; y_0) = -L^* R^* R y^*(\cdot\,; y_0) \tag{3.9}$$

with optimal dynamics

$$y^*(t;y_0) = e^{At}y_0 + \{Lu^*(\cdot\ ;y_0)\}(t) + \{Ww^*(\cdot\ ;y_0)\}(t) \tag{3.10}$$

which therefore satisfies

$$\{[I + LL^*R^*R - \gamma^{-2}WW^*R^*R]y^*(\cdot\ ;y_0)\}(t) = e^{At}y_0 \tag{3.11}$$

Proof: (i) and (ii) Either we use the Lagrange multiplier approach as in the proof of Theorem 2.1.1(i), this time with free parameters $\{u_w^0, y_w^0, \lambda, w\} \in L_2(0,\infty;U) \times [L_2(0,\infty;Y)]^3$:

$$\mathrm{L}(u_w^0, y_w^0, \lambda, w) = 1/2[(R^*Ry_w^0, y_w^0) + (u_w^0, u_w^0)] + (\lambda, y_w^0 - Lu_w^0 - e^{A\cdot}\ y_0 - Ww) \tag{3.12}$$

whereby for the unique optimal $\{u^*, y^*, \lambda^*, w^*\}$ we get:

$$\mathrm{L}_{u_w^0} = 0: \qquad (u_w^0, \delta u_w^0) - (L^*\lambda, \delta u_w^0) = 0, \quad \forall \delta u_w^0 \in L_2(0,\infty;U) \tag{3.13}$$

$$u^* \equiv u^0_{w=w^*} = L^*\lambda^* \tag{3.14}$$

$$\mathrm{L}_{y_w^0} = 0: \qquad (R^*Ry_w^0, \delta y_w^0) + (\lambda, \delta y_w^0) = 0, \quad \forall \delta y_w^0 \in L_2(0,\infty;Y) \tag{3.15}$$

$$\lambda^* = -R^*Ry^0_{w=w^*} \equiv -R^*Ry^* \tag{3.16}$$

$$u^* = -L^*R^*Ry^* \tag{3.17}$$

$$\mathrm{L}_w = 0: \qquad -\ \gamma^2(w, \delta w) - (W^*\lambda, \delta w) = 0 \quad \forall \delta w \in L_2(0,\infty;Y) \tag{3.18}$$

$$-\ \gamma^2 w^* = W^*\lambda^* = -W^*R^*Ry^*. \tag{3.19}$$

Thus, (3.6) - (3.9) are proved. Or else, we set to zero the variation with respect to w of

$$J_w^0(y_0) = (R^*Ry_w^0, y_w^0) + (u_w^0, u_w^0) -\ \gamma^2(w, w) \tag{3.20}$$

i.e., we set

$$(R^*Ry_w^0, \delta y_w^0) + (u_w^0, \delta u_w^0) -\ \gamma^2(w, \delta w) = 0, \tag{3.21}$$

where from (1.1.7), we insert $\delta y_w^0 = L\delta u_w^0 + W\delta w$ into (3.21) and recall (2.1.11) for $w = w^*$, to re-obtain (3.17) and (3.19), from (3.21). Finally, inserting (3.6) and (3.9) into (3.10) yields (3.11).

□

Remark 3.1 We are not authorized to boundedly invert on $L_2(0,\infty;Y)$ the operator $[I + LL^*R^*R - \gamma^{-2}WW^*R^*R]$ for all $\gamma > \gamma_c$; only for γ sufficiently large, in fact $\gamma^2 > \gamma_1^2 = \|I + LL^*R^*R\|/\|WW^*R^*R\|$, in $L_2(0,\infty;Y)$-norms. Then, for $\gamma > \gamma_1$, proceeding as in [L-T.2, Lemma 5.1], by use of this inversion, we can obtain that $y^*(t;y_0) \equiv \Phi(t)y_0 \in C_b([0,\infty];Y)$, via (2.1.14), satisfies the semigroup property $\Phi(t+\tau) = \Phi(t)\Phi(\tau) = \Phi(\tau)\Phi(t)$, (refer to the subsequent Lemma 4.2.1 for details) so that $\Phi(t)$ is a strongly continuous semigroup on Y, as desired. Below, in section 4.3, we shall show this conclusion holds, in fact, for all $\gamma > \gamma_c$. This will require a modification of our approach. □

4 Explicit expressions of $\{u^*, y^*, w^*\}$ for $\gamma > \gamma_c$ in terms of the data via E_γ^{-1}.

4.1 Explicit expression of $w^*(\cdot\,;y_0)$ for $\gamma > \gamma_c$ in terms of the data via E_γ^{-1}.

Proposition 4.1.1 *For $\gamma > \gamma_c$ (defined by (2.2.1)), we have*

$$w^*(\cdot\,;y_0) = E_\gamma^{-1}W^*R^*R[I + LL^*R^*R]^{-1}(e^{A\cdot}y_0) \in L_2(0,\infty;Y) \tag{4.1.1}$$

Proof: We insert (3.8) into (3.6) thereby obtaining

$$[\gamma^2 I - W^*R^*R[I + LL^*R^*R]^{-1}W]w^*(\cdot\,;y_0) = W^*R^*R[I + LL^*R^*R]^{-1}(e^{A\cdot}y_0) \tag{4.1.2}$$

Recalling the definition (2.1.18) of E_γ, we rewrite (4.1.2) as

$$E_\gamma w^*(\cdot\,;y_0) = W^*R^*R[I + LL^*R^*R]^{-1}(e^{A\cdot}y_0) \tag{4.1.3}$$

from which (4.1.1) follows for $\gamma > \gamma_c$ by Corollary 2.2.1. □

Remark 4.1.1 If we apply $W^*R^*R[I+LL^*R^*R]^{-1}$ to Eq. (3.11), we get

$$W^*R^*R[I+LL^*R^*R]^{-1}\{[I+LL^*R^*R]y^*(\cdot\,;y_0)\ -\gamma^{-2}WW^*R^*Ry^*(\cdot\,;y_0)\} = W^*R^*R[I+LL^*R^*R]^{-1}(e^{A\cdot}\,y_0) \tag{4.1.4}$$

We then use (3.6) to obtain Eq. (4.1.2). But, Eq. (4.1.2) is solvable for w^* for $\gamma>\gamma_c$, while Eq. (3.11) is solvable for y^* for γ sufficiently large, see Remark 3.1. On the other hand, inserting w^* given by (4.1.1) into the right hand sides of (3.7) and (3.8), produces explicit expressions for u^* and y^* for all $\gamma>\gamma_c$ in terms of the problem data, which we omit writing explicitly. □

4.2 A transition property for w^* for $\gamma > \gamma_c$

We have the following important property:

Theorem 4.2.1 *For $\gamma>\gamma_c$ (defined in (2.2.1), we have:*

$$w^*(t+\sigma;y_0)=w^*(\sigma;y^*(t;y_0)) \quad \text{a.e. in} \quad t,\sigma>0 \tag{4.2.1}$$

for t fixed, the equality being intended in $L_2(0,\infty;Y)$ □.

Proof: Step 1 We begin with the following:

Lemma 4.2.1 *For $\gamma>\gamma_c$ (defined in (2.2.1), we have:*

$$[I+LL^*R^*R-\gamma^{-2}WW^*R^*R][y^*(t+\cdot\,;y_0)-y^*(\cdot\,;y^*(t;y_0))]=0 \tag{4.2.2}$$

Remark 4.2.1 Recall Remark 3.1: for γ sufficiently large, not necessarily for all $\gamma>\gamma_c$, the operator in (4.2.2) is boundedly invertible on $L_2(0,\infty;Y)$, so that we obtain from (4.2.2): $y^*(t+\cdot\,;y_0)=y^*(\cdot\,;y^*(t;y_0))$ first in $L_2(0,\infty;Y)$, then in $C_b([0,\infty];Y)$, by (2.1.14) which is the sought after semigroup property of the operator $\Phi(t)$ in Remark 3.1, at least for γ large. □

Proof of Lemma 4.2.1: (in the style of [L-T.2, Lemma 5.1]). For $\gamma>\gamma_c$, we return to (3.11), which we rewrite explicitly using (2.1.2)-(2.1.9) as:

$$y^*(t;y_0)+\{LL^*R^*Ry^*(\cdot\,;y_0)\}(t)-\gamma^{-2}\{WW^*R^*Ry^*(\cdot\,;y_0)\}(t)=e^{At}y_0 \tag{4.2.3}$$

$$\{LL^*R^*Ry^*(\cdot\,;y_0)\}(t) = \int_0^t e^{A(t-\tau)}BB^* \int_\tau^\infty e^{A^*(\alpha-\tau)}R^*Ry^*(\alpha;y_0)d\alpha d\tau \tag{4.2.4}$$

$$\{WW^*R^*Ry^*(\cdot\,;y_0)\}(t) = \int_0^t e^{A(t-\tau)}GG^* \int_\tau^\infty e^{A^*(\alpha-\tau)}R^*Ry^*(\alpha;y_0)d\alpha d\tau \tag{4.2.5}$$

Thus, (4.2.3) written for t replaced by $t+\sigma$ becomes explicitly

$$\begin{aligned} y^*(t+\sigma;y_0) \quad &+\int_0^{t+\sigma} e^{A(t+\sigma-\tau)}BB^* \int_\tau^\infty e^{A^*(\alpha-\tau)}R^*Ry^*(\alpha;y_0)d\alpha d\tau \\ &-\gamma^{-2}\int_0^{t+\sigma} e^{A(t+\sigma-\tau)}GG^* \int_\tau^\infty e^{A^*(\alpha-\tau)}R^*Ry^*(\alpha;y_0)d\alpha d\tau = e^{A(t+\sigma)}y_0 \end{aligned} \tag{4.2.6}$$

On the other hand, by (4.2.3) with t replaced by σ and y_0 replaced by $y^*(t;y_0)$, we obtain

$$\begin{aligned} y^*(\sigma;y^*(t;y_0)) \quad &+\int_0^\sigma e^{A(\sigma-\beta)}BB^* \int_\beta^\infty e^{A^*(r-\beta)}R^*Ry^*(r;y^*(t;y_0))drd\beta \\ &-\gamma^{-2}\int_0^\sigma e^{A(\sigma-\beta)}GG^* \int_\beta^\infty e^{A^*(r-\beta)}R^*Ry^*(r;y^*(t;y_0))drd\beta = e^{A\sigma}y^*(t;y_0) \end{aligned} \tag{4.2.7}$$

where, by (4.2.3)-(4.2.5)

$$\begin{aligned} e^{A\sigma}y^*(t;y_0) = e^{A(\sigma+t)}y_0 \quad &-\int_0^t e^{A(t+\sigma-\tau)}BB^* \int_\tau^\infty e^{A^*(\alpha-\tau)}R^*Ry^*(\alpha;y_0)d\alpha d\tau \\ &-\int_0^t e^{A(t+\sigma-\tau)}GG^* \int_\tau^\infty e^{A^*(\alpha-\tau)}R^*Ry^*(\alpha;y_0)d\alpha d\tau \end{aligned} \tag{4.2.8}$$

Next, subtracting and adding on the right of (4.2.8)

$$\begin{aligned} &\int_t^{t+\sigma} e^{A(t+\sigma-\tau)}BB^* \int_\tau^\infty e^{A^*(\alpha-\tau)} \quad R^*Ry^*(\alpha;y_0)d\alpha d\tau \\ &\qquad = \int_0^\sigma e^{A(\sigma-\beta)}BB^* \int_{t+\beta}^\infty e^{A^*(\alpha-(t+\beta))}R^*Ry^*(\alpha;y_0)d\alpha d\beta \qquad (4.2.9) \\ (\alpha - t = r) &\qquad = \int_0^\sigma e^{A(\sigma-\beta)}BB^* \int_\beta^\infty e^{A^*(r-\beta))}R^*Ry^*(t+r;y_0)drd\beta \end{aligned}$$

and a similar term with B replaced by $-G/\gamma$, we finally rewrite (4.2.7) as

$$\begin{aligned} y^*(\sigma; y^*(t; y_0)) &+ \int_0^\sigma e^{A(\sigma-\beta)} BB^* \int_\beta^\infty e^{A^*(r-\beta)} R^* R y^*(r; y^*(t; y_0)) dr d\beta \\ &-\gamma^{-2} \int_0^\sigma e^{A(\sigma-\beta)} GG^* \int_\beta^\infty e^{A^*(r-\beta)} R^* R y^*(r; y^*(t; y_0)) dr d\beta \\ &- \int_0^\sigma e^{A(\sigma-\beta)} BB^* \int_\beta^\infty e^{A^*(r-\beta)} R^* R y^*(r+t; y_0) dr d\beta \\ &+\gamma^{-2} \int_0^\sigma e^{A(\sigma-\beta)} GG^* \int_\beta^\infty e^{A^*(r-\beta)} R^* R y^*(r+t; y_0) dr d\beta \end{aligned} \tag{4.2.10}$$

$$\begin{aligned} = &- \int_0^{t+\sigma} e^{A(t+\sigma-\tau)} BB^* \int_\tau^\infty e^{A^*(\alpha-\tau)} R^* R y^*(\alpha; y_0) d\alpha d\tau \\ &+\gamma^{-2} \int_0^{t+\sigma} e^{A(t+\sigma-\tau)} GG^* \int_\tau^\infty e^{A^*(\alpha-\tau)} R^* R y^*(\alpha; y_0) d\alpha d\tau + e^{A(t+\sigma)} y_0 \end{aligned} \tag{4.2.11}$$

We finally subtract (4.2.11) from (4.2.6), and after a cancellation of six terms we obtain

$$\begin{aligned} [y^*(t+\sigma; y_0) &- y^*(\sigma; y^*(t; y_0))] \\ &+ \int_0^\sigma e^{A(\sigma-\beta)} BB^* \int_\beta^\infty e^{A^*(r-\beta)} R^* R[y^*(t+r; y_0) - y^*(r; y^*(t; y_0))] dr d\beta \\ &-\gamma^{-2} \int_0^\sigma e^{A(\sigma-\beta)} GG^* \int_\beta^\infty e^{A^*(r-\beta)} R^* R[y^*(t+r; y_0) - y^*(r; y^*(t; y_0))] dr d\beta = 0 \end{aligned} \tag{4.2.12}$$

Recalling (2.1.2)-(2.1.9), we rewrite (4.2.12) precisely as in (4.2.2) and Lemma 4.2.1 is proved. □

Step 2 Starting from (4.2.2) and applying to it the operator $W^*R^*R[I + LL^*R^*R]^{-1}$ as in Remark 4.1.1, we obtain

$$\begin{aligned} W^*R^*R \quad &[y^*(t + \cdot\,; y_0) - y^*(\cdot\,; y^*(t; y_0))] \\ &-\gamma^{-2} W^*R^*R[I + LL^*R^*R]^{-1} WW^*R^*R[y^*(t + \cdot\,; y_0) - y^*(\cdot\,; y^*(t; y_0)] = 0 \end{aligned} \tag{4.2.13}$$

Step 3

Lemma 4.2.2 *For $\gamma > \gamma_c$, (defined in (2.2.1)) and with reference to (3.6) we have:*

$$\gamma^2 w^*(t+\sigma; y_0) = \{W^*R^*Ry^*(t + \cdot\,; y_0)\}(\sigma) \tag{4.2.14}$$

Proof of Lemma 4.2.2: By direct verification. By (3.6), rewritten explicitly via (2.1.8) we have after a change of variable $\beta - t = \tau$:

$$\begin{aligned}\gamma^2 w^*(t+\sigma; y_0) &= G^* \int_{t+\sigma}^{\infty} e^{A^*(\beta-(t+\sigma))} R^* R y^*(\beta; y_0) d\beta \\ &= G^* \int_{\sigma}^{\infty} e^{A^*(\tau-\sigma)} R^* R y^*(t+\tau; y_0) d\tau\end{aligned} \tag{4.2.15}$$

which is precisely (4.2.14). □

Step 4 Using (4.2.14) and (3.6) in (4.2.13), we rewrite (4.2.13) as

$$\begin{aligned}\gamma^2[w^*(t+\sigma; y_0) &- w^*(\sigma; y^*(t; y_0))] \\ &- W^* R^* R[I + LL^* R^* R]^{-1} W[w^*(t+\cdot\ ; y_0) - w^*(\cdot\ ; y^*(t; y_0))] = 0\end{aligned} \tag{4.2.16}$$

or, recalling the definition of E_γ in (2.1.18):

$$E_\gamma[w^*(t+\cdot\ ; y_0) - w^*(\cdot\ ; y^*(t; y_0))] = 0 \tag{4.2.17}$$

Thus, by Corollary 2.2.1, if $\gamma > \gamma_c$, then $E_\gamma{}^{-1} \in L_2(0, \infty; Y)$, and so by (4.2.17) we obtain

$$w^*(t+\cdot\ ; y_0) - w^*(\cdot\ ; y^*(t; y_0)) = 0 \text{ in } L_2(0, \infty; Y) \tag{4.2.18}$$

as desired. Theorem 4.2.1 is proved. □

The implication on Ww^* of the property (4.2.1) of w^* is examined next.

Corollary 4.2.1 *For all $\gamma > \gamma_c$ (defined in (2.2.1)), we have for all $t, \sigma > 0$*

$$\{Ww^*(\cdot\ ; y_0)\}(t+\sigma) - \{Ww^*(\cdot\ ; y_0)\}(\sigma) - e^{A\sigma}\{Ww^*(\cdot\ ; y_0)\}(t) \equiv 0 \tag{4.2.19}$$

Proof: By (2.1.4) we compute

$$\begin{aligned}&\{Ww^*(\cdot\ ; y_0)\}(t+\sigma) - e^{A\sigma}\{Ww^*(\cdot\ ; y_0)\}(t) - \{Ww^*(\cdot\ ; y_0)\}(\sigma) \\ &= \int_0^{t+\sigma} e^{A(t+\sigma-\tau)} Gw^*(\tau; y_0) d\tau - \int_0^t e^{A(t+\sigma-\tau)} Gw^*(\tau; y_0) d\tau - \int_0^{\sigma} e^{A(\sigma-\beta)} Gw^*(\beta; y_0) d\beta\end{aligned} \tag{4.2.20}$$

We now add and subtract (use $\tau - t = \beta$):

$$\int_t^{t+\sigma} e^{A(t+\sigma-\tau)} G w^*(\tau; y_0) d\tau = \int_0^{\sigma} e^{A(\sigma-\beta)} G w^*(t+\beta; y_0) d\beta \tag{4.2.21}$$

to the right hand side of (4.2.20) to obtain, after a cancellation

$$\begin{aligned} \{Ww^*(\cdot\,; y_0)\}(t+\sigma) \quad & -\{Ww^*(\cdot\,; y_0)\}(\sigma) - e^{A\sigma}\{Ww^*(\cdot\,; y_0)\}(t) \\ &= \int_0^{\sigma} e^{A(\sigma-\beta)} G[w^*(t+\beta; y_0) - w^*(\beta; y_0)] d\beta \equiv 0 \end{aligned} \tag{4.2.22}$$

where, in the last step, we have used (4.2.1). □

4.3 The semigroup property for y^* for $\gamma > \gamma_c$ and its stability

Defining, as in Remark 3.1, the operator $\Phi(t)$ (which depends on γ) by

$$y^*(t; x) = \Phi(t)x \ \in C_b([0, \infty]; Y), \quad \forall\, x \in Y \tag{4.3.1}$$

we obtain the semigroup property:

Theorem 4.3.1 *For $\gamma > \gamma_c$ (see (2.2.1)), $y_0 \in Y$, and t, $\sigma > 0$ we have:*

$$y^*(t+\sigma; y_0) = y^*(\sigma; y^*(t; y_0)) \ \in C_b([0, \infty]; Y) \tag{4.3.2}$$

so that $\Phi(t)$ is a s.c. semigroup on Y. □

Proof: Step 1

Lemma 4.3.1 *For $\gamma > \gamma_c$, $y_0 \in Y$, and t, $\sigma > 0$ we have*

$$\begin{aligned} &[y^*(t+\sigma; y_0) - y^*(\sigma; y^*(t; y_0)] + \{LL^*R^*R[y^*(t+\cdot\,; y_0) - y^*(\cdot\,; y^*(t; y_0)]\}(\sigma) \\ &= \{Ww^*(\cdot\,; y_0)\}(t+\sigma) - \{Ww^*(\cdot\,; y_0)\}(\sigma) - e^{A\sigma}\{Ww^*(\cdot\,; y_0)\}(t) \end{aligned} \tag{4.3.3}$$

Proof: We return to the optimal dynamics (3.10) rewritten via (3.9) as

$$y^*(t; y_0) + \{LL^*R^*Ry^*(\cdot\,; y_0)\}(t) = e^{At}y_0 + \{Ww^*(\cdot\,; y_0)\}(t) \tag{4.3.4}$$

From here on, the proof proceeds as the proof of Lemma (4.2.1) below (4.2.3). Details may be omitted. □

Step 2 We now apply Corollary 4.2.1, Eq. (4.2.19) to (4.3.3) and obtain

$$[I + LL^*R^*R][y^*(t+\cdot\,;y_0) - y^*(\cdot\,;y^*(t;y_0))] = 0 \tag{4.3.5}$$

Since $[I + LL^*R^*R]^{-1} \in \mathcal{L}(L_2(0,\infty;Y))$, see [L-T.2, F-L-T.1] , we obtain from (4.3.5)

$$y^*(t+\sigma;y_0) - y^*(\sigma;y^*(t;y_0)) = 0, \tag{4.3.6}$$

first in $L_2(0,\infty;Y)$, and then, by the regularity of y^*, in $C_b([0,\infty];Y)$. □

Corollary 4.3.1 *The s.c. semigroup $\Phi(t)$ in (4.3.1), guaranteed by Theorem 4.3.1, is exponentially stable: there exists $C \geq 1$, $\rho > 0$ such that*

$$||\Phi(t)||_{\mathcal{L}(Y)} \leq Ce^{-\rho t}, \quad t \geq 0 \tag{4.3.7}$$

Proof: The s.c. semigroup $\Phi(t)$ satisfies $y^*(t;x) = \Phi(t)x \in L_2(0,\infty;Y)$ for all $x \in Y$ by optimality (3.5) and then conclusion (4.3.7) follows via a known theorem in [D.1]. □

5 The Riccati operator, P, for $\gamma > \gamma_c$

The property of $\Phi(t)$ in (4.3.1) as a s.c. uniformly stable semigroup on Y – as guaranteed by section 4.3 – allows us to readily fall into the abstract treatment of [L-T.2, section 5], [F-L-T.1]. Accordingly, we shall first define (in terms of the problem data) an operator $P \in \mathcal{L}(Y)$ for $\gamma > \gamma_c$, and we shall next show that P is, in fact, a solution of an algebraic Riccati (operator) equation and indeed the unique such solution (within a certain class). As in this section we shall very closely follow [F-L-T.1], under the simplifying stability assumption (1.1.6), details will be either sketched or directed to this reference where appropriate.

5.1 Definition of P and its preliminary properties

For $\gamma > \gamma_c$, and recalling (1.1.6), we define the operator $P \in \mathcal{L}(Y)$ by (see (4.3.1)):

$$Px = \int_0^\infty e^{A^*\sigma} R^* R y^*(\sigma; x) d\sigma = \int_0^\infty e^{A^*\sigma} R^* R \Phi(\sigma) x d\sigma \tag{5.1.1a}$$

$$= \int_t^\infty e^{A^*(\tau - t)} R^* R y^*(\tau - t; x) d\tau = \int_t^\infty e^{A^*(\tau - t)} R^* R \Phi(\tau - t) x d\tau \tag{5.1.1b}$$

Theorem 5.1.1 *With reference to (5.1.1), we have for $\gamma > \gamma_c$:*

(i)

$$u^*(t; y_0) = -B^* P y^*(t; y_0) = -B^* P \Phi(t) y_0 \quad \text{a.e. in } t; \ y_0 \in Y \tag{5.1.2a}$$

$$B^* P \Phi(t): \ \text{continuous } Y \longrightarrow L_2(0, \infty; U) \tag{5.1.2b}$$

(ii)

$$\gamma^2 w^*(t; y_0) = G^* P y^*(t; y_0) = G^* P \Phi(t) y_0 \tag{5.1.3a}$$

$$G^* P \Phi(t): \ \text{continuous } Y \longrightarrow L_2(0, \infty; Y) \tag{5.1.3b}$$

(iii) the operator $P \in \mathcal{L}(Y)$ satisfies the symmetric relation for $x_1, x_2 \in Y$:

$$(Px_1, x_2) = \int_0^\infty [(Ry^*(t; x_1), Ry^*(t; x_2))_Y + (u^*(t; x_1), u^*(t; x_2))_U - \gamma^2 (w^*(t; x_1), w^*(t; x_2))_Y] dt \tag{5.1.4}$$

from which it follows that P is a nonnegative self-adjoint operator: $P = P^* \geq 0$ on Y and that the optimal cost of problem (2.1.1) is

$$\begin{aligned} (Py_0, y_0)_Y &= J^*(y_0) \quad [\text{optimal cost in } (3.3)] \\ &= J(u^*(\cdot\,; y_0), y^*(\cdot\,; y_0), w^*(\cdot\,; y_0)) \end{aligned} \tag{5.1.5}$$

<u>Proof:</u> (i) One applies B^* to (5.1.1b) with x replaced by $y^*(t; y_0)$ and obtains (5.1.2) via the semigroup property $\Phi(\tau - t)\Phi(t) = \Phi(\tau)$ as well as (2.1.6) and (3.9). The proof for (ii) is similar, this time by use of (2.1.8) and (3.6).

(ii) For $x_1, x_2 \in Y$ we write from (5.1.1) using $e^{A\cdot} x_2$ from (3.10) :

$$(Px_1, x_2)_Y = (R^* R y^*(\cdot\,; x_1), e^{A\cdot} x_2)_{L_2(0, \infty; Y)}$$

$$= (Ry^*(\cdot\,;x_1), Ry^*(\cdot\,;x_2)) - (L^*R^*Ry^*(\cdot\,;x_1), u^*(\cdot\,;x_2))_{L_2(0,\infty;U)}$$

$$-(W^*R^*Ry^*(\cdot\,;x_1), w^*(\cdot\,;x_2))_{L_2(0,\infty;Y)} \tag{5.1.6}$$

and (5.1.4) follows from (5.1.6) recalling (3.9) and (3.6). □

5.2 P satisfies the Algebraic Riccati Equation

Next, we call A_F the infinitesimal generator of the s.c. uniformly stable semigroup, $\Phi(t)$ in (4.3.1):

$$\Phi(t)x = e^{A_Ft}x,\ x \in Y;\quad \frac{d\Phi(t)x}{dt} = A_F\Phi(t)x = \Phi(t)A_Fx,\ x \in D(A_F) \tag{5.2.1}$$

Lemma 5.2.1 *For P and A_F defined by (5.1.1) and (5.2.1) we have*

$$Y \supset D(B^*P) \supset D(A_F); \qquad Y \supset D(G^*P) \supset D(A_F) \tag{5.2.2}$$

so that $D(B^*P)$ is dense in Y.

Proof: The proof is a (simplified) version of the proof of [F-L-T.1, Lemma 4.5 pp 338-9] using (H.3), respectively (H.4). □

Next, we identify A_F.

Lemma 5.2.2 *(i) For $x \in Y$ and $t > 0$ we have that*

$$\frac{d\Phi(t)}{dt} = [A - BB^*P - \gamma^{-2}GG^*P]\Phi(t)x \quad \in [D(A^*)]' \tag{5.2.3}$$

(ii) Moreover,

$$[A - BB^*P - \gamma^{-2}GG^*P]x = A_Fx,\quad x \in D(A_F). \tag{5.2.4}$$

Proof: See proof of [F-L-T.1, Lemma 4.6]. Here we use (4.3.1), (5.2.1), (5.1.2), and (5.1.3). □

Lemma 5.2.3 *With P and A_F defined by (5.1.1) and (5.2.1) we have*

(a) $$A^*P \in \mathcal{L}(D(A_F); Y) \tag{5.2.5}$$

(b) $$A_F^* P \in \mathcal{L}(D(A); Y) \tag{5.2.6}$$

(c) $$-A^* P = R^* Rx + PA_F x \in Y, \quad x \in D(A_F) \tag{5.2.7}$$

(d) $$-A_F^* P = R^* Rx + PAx \in Y, \quad x \in D(A) \tag{5.2.8}$$

Proof: It is a (simplified) version of the proof of [F-L-T.1, Lemma 4.7] for (a) and (b), and of [F-L-T.1, Lemma 4.8] for (c) and (d). In the latter case, we differentiate (Px_1, x_2) as given by (5.1.1b), with $x_1 \in D(A_F)$ and $x_2 \in D(A)$. □

Corollary 5.2.1 *We have*

$$\left.\begin{array}{l}(B^* Px_1, B^* Px_2)_U \\ \\ (G^* Px_1, G^* Px_2)_U\end{array}\right\} = \text{well}-\text{defined for } x_1, x_2 \in D(A); \text{ or else } x_1, x_2 \in D(A_F) \tag{5.2.9}$$

$$B^* P,\ G^* P \in \mathcal{L}(D(A); U) \cap \mathcal{L}(D(A_F); U) \tag{5.2.10}$$

Proof: See [F-L-T.1, Corollary 4.9], or [M-T.2; Lemma 15.1]. □

We finally obtain the ultimate goal of our analysis.

Theorem 5.2.1 *For $\gamma > \gamma_c$, the operator P defined by (5.1.1) satisfies the Algebriac Riccati Equation, ARE_γ*

$$(Px, Az)_Y + (Ax, Pz)_Y + (Rx, Rz)_Y = (B^* Px, B^* Pz)_Y - \gamma^{-2}(G^* Px, G^* Pz)_Y \tag{5.2.11}$$

for all $x, z \in D(A)$, or else for all $x, y \in D(A_F)$.

Proof: We combine Lemma 5.2.2 with Corollary 5.2.1. □

6 Final remarks on the general case

The above direct, explicit treatment of the min-max problem (1.2.2) under the additional stability assumption (1.1.6) should be contrasted with the general case, as given in [M-T.1] (under the additional Finite Cost Condition and Detectability Condition which are automatically satisfied if (1.2.2) is assumed). To summarize the general case in [M-T.1]:

(i) Now, the minimization of J over u is first carried out over a fixed $[0,T]$, $T<\infty$, as in section 2 of the present paper, and it is then followed by a limit process as $T\uparrow\infty$, as in [F-L-T.1]. However, a known idea of a "decoupling" procedure is employed, whereby the limit process eventually leads to formulas such as

$$u^0_{w,\infty}(\cdot\,;y_0) = -B^*p_{w,\infty}(\cdot\,;y_0) \tag{6.1}$$

$$p_{w,\infty}(t;y_0) = P_{0,\infty}y^0_{w,\infty}(\cdot\,;y_0) + r_{w,\infty}(t), \quad \forall t>0, \quad in\ Y \tag{6.2}$$

where $\{u^0_{w,\infty}(\cdot\,;y_0), y^0_{w,\infty}(\cdot\,;y_0)\}$ is the optimal pair of the minimization of J over the time interval $[0,\infty]$ holding $w\in L_2(0,\infty;Y)$ fixed. Moreover, $P_{0,\infty}$ is the Algebraic Riccati operator *corresponding to the case* $w\equiv 0$ and provided by [F-L-T.1] (reflected in the first subscript of P). It is an important feature of the "decoupling" formula (6.2) that the function $r_{w,\infty}(t)$ satisfies the differential equation

$$\dot r_{w,\infty}(t) = -A^*_{P_{0,\infty}}r_{w,\infty}(t) - P_{0,\infty}Gw(t) \tag{6.3}$$

with *stable* generator $A_{P_{0,\infty}} = A - BB^*P_{0,\infty}$ [F-L-T.1]. Similarly, the equations for $p_{w,\infty}$ and $y^0_{w,\infty}$ can be written as

$$\dot p_{w,\infty}(t;y_0) = -A^*_{P_{0,\infty}}p_{w,\infty}(t;y_0) + P_{0,\infty}Bu^0_{w,\infty}(t;y_0) - R^*Ry^0_{w,\infty}(t;y_0) \tag{6.4}$$

$$\dot y^0_{w,\infty}(t;y_0) = A_{P_{0,\infty}}y^0_{w,\infty}(t;y_0) - BB^*r_{w,\infty}(t) + Gw(t) \tag{6.5}$$

both with *stable* generators as well, where $B^*r_{w,\infty}(t)\in L_2(0,\infty;U)$.

(ii) The presence of *stable* generators for $r_{w,\infty}, p_{w,\infty}$, and $y^0_{w,\infty}$ then allows one to carry out (with greater technical difficulties) the maximization of $J^0_w(y_0)$ over all $w\in L_2(0,\infty;Y)$ *directly* over the infinite time interval $[0,\infty]$. We refer to [M-T.1] for a complete treatment.

7 References

[B.1] V. Barbu, H^∞-Boundary Control with State Feedback; the Hyperbolic Case, preprint 1992.

[B-B.1] T. Başar and P. Bernhard. *H^∞-Optimal Control and Related Minimax Design Problems. A Dynamic Game Approach*, Birkhaüser Boston (1991).

[D.1] R. Datko. Extending a Theorem of Liapunov to Hilbert Space, *J. Math. Anal. Appl.*, **32** (1970), pp. 610-616.

[F-L-T.1] F. Flandoli, I. Lasiecka, and R. Triggiani. Algebraic Riccati Equations with Non-Smoothing Observation Arising in Hyperbolic and Euler-Bernoulli Equations, *Annali di Matematica Pura et Applicata*, **IV** (1989), Vol CLIII, pp. 307-382.

[L.1] D.G. Luenberger, *Optimization by Vector Space Methods*, John Wiley, New York.

[L-T.1] I. Lasiecka and R. Triggiani. A Lifting Theorem for the Time Regularity of Solutions to Abstract Equations with Unbounded Operators and Applications to Hyperbolic Equations, *Proc. Amer. Math. Soc.* 102, 4 (1988), pp. 745-755.

[L-T.2] I. Lasiecka and R. Triggiani. Riccati Equations for Hyperbolic Partial Differential Equations with $L_2(0,T;L_2(\Gamma))$-Dirichlet Boundary Terms, *SIAM J. Control Optimiz.*, **24** (1986), pp. 884-924.

[L-T.3] I. Lasiecka and R. Triggiani. *Differential and Algebraic Riccati Equations with Application to Boundary/Point Control Problems: Continuous Theory and Approximation Theory*, Volume # 164 in the Springer-Verlag Lectures Notes LNCIS series (1991), pp. 160.

[M-T.1] C. McMillan and R. Triggiani. Min-Max Game Theory and Algebraic Riccati Equations for Boundary Control Problems with Continuous Input-Solution Map. Part II: the General Case (to appear). Presented at: (i) International Conference on Differential Equations and Mathematical Physics held at Georgia Institute of Technology, Atlanta GA, March 22-28, 1992; (ii) IFIP Conference "Boundary Control and Boundary Variation" held at Sophia Antipoles, France, June 3-5, 1992.

Hyperbolic Singular Perturbation of a Quasilinear Parabolic Equation

B. NAJMAN University of Zagreb, Zagreb, Croatia

The quasi–linear initial value problems

$$\varepsilon u'' + B(t,u)u' + A(t,u)u = f(t,u)\ ,\ u(0) = u_{0\varepsilon}\ ,\ u'(0) = u_{1\varepsilon}\ (\varepsilon>0) \quad (1)$$

and

$$B(t,u)u' + A(t,u)u = f(t,u)\ ,\ u(0) = u_{00} \quad (2)$$

in a separable Hilbert space X are considered; the convergence rate of the mild solutions u_ε of (1) to the mild solution u_0 of (2) on a common existence interval is given in Theorems 4 and 5. The precise assumptions are stated there; the main assumption is that $A(t,u)$ and $B(t,u)$ are self–adjoint and

$$\mathrm{Re}(A(t,u)v|B(t,u)v) \geqslant 0 \quad (v \in D(A(t,u) \cap D(B(t,u))). \quad (3)$$

This assumption appears in Clément, Prüss (1989)in the linear case. The convergence in linear case is studied in Section 1. Except for its independent interest, the linear results are the basis for the treatment of the general case in Section 2.

We are primarily interested in the case when B is large ($A^{1/2}$ is $B^{1/2}$ –bounded), since the case of "small" B (e.g. $B = \beta I$) can be treated by different methods.

We estimate the convergence rate under the assumption that the solution exist; the existence problem is not considered as it is an independent problem (which has received considerable attention, comp. e.g. Esham, Weinacht (1989), Kato (1975) and their references). We refer to Esham, Weinacht (1989) also for a result similar to ours in a special situation.

1. THE LINEAR CASE

The equations (1) and (2) become

$$\varepsilon u'' + B(t)u' + A(t)u = f(t) \quad (t \in J)$$

$$u(0) = u_{0\varepsilon}\ ,\ u'(0) = u_{1\varepsilon}\ . \quad (4)$$

and

$$B(t)u' + A(t)u = f(t) \quad (t \in J)$$
$$u(0) = u_{00} \, . \tag{5}$$

which are considered on $J = [0,T]$; in (4) $0 < \varepsilon \leqslant 1$. Our standing assumptions are:

(I) For every $t \in J$ the operators $A(t)$ and $B(t)$ are self–adjoint positive definite operators in X. The space $D(A(t)) \cap D(B(t))$ is dense in X .

Moreover

$$\mathrm{Re}(A(t)u|B(t)u) \geqslant 0 \qquad (u \in D(A(t)) \cap D(B(t))),$$
$$\sup_{t\in J} \|B(t)^{-1}\| < \infty \tag{6}$$

(II) $t \mapsto A(t)u, B(t)u$ are continuous and a.e. differentiable for all $u \in D(A(t)) \cap D(B(t))$.

Further we assume

$$u_{0\varepsilon} \in D(A(0)^{1/2}) \cap D(B(0)^{1/2}) \, , \, u_{1\varepsilon} \in X \tag{7}$$

$$f \in L^2(J,X) \, , \, B(\cdot)^{-1/2}f \in L^2(J,X) \tag{8}$$

We say that u is a solution of (4) (or (5) if $\varepsilon = 0$) if $u \in C(J,X)$, $u(t) \in D(A(t)) \cap D(B(t))$ for almost all $t \in J$ and the distributional derivative u' satisfies $u'(t) \in D(B(t)^{1/2})$, $B(\cdot)^{1/2} u' \in C(J,X)$ and

$$\varepsilon(u''|v) + (u'|B(t)v) + (u|A(t)v) = (f(t)|v)$$
$$(u(0)|v) = (u_{0\varepsilon}|v) \, , \, (B(0)^{1/2} u'(0)|B(0)^{-1/2}v) = (u_{1\varepsilon}|v)$$

for all $v \in D(A(t) \cap D(B(t))$; if $\varepsilon = 0$ the second initial condition is absent.

We assume that such solutions exist for all $\varepsilon \leqslant 1$ and therefore all the formal manipulations below are justified. The a priori estimates proved in Proposition 1 can be used (cf. Lions (1969)) to prove their existence under appropriate conditions. Where necessary, we assume that the solutions additionally satisfy $u'(t) \in D(B(t))$ (a.e.) .

1.1. A priori estimates

In this section we shall need following conditions:

$$\int_0^T \|A(t)^{-1/2}A'(t)A(t)^{-1/2}\|dt < \infty \tag{9}$$

$$\int_0^T \|B(t)^{-1/2}B'(t)B(t)^{-1/2}\|dt < \infty \tag{10}$$

$$\int_0^T [\|B'(t)B(t)^{-1}\| + \|B(t)^{-1/2}B'(t)B(t)^{-1}\|^2] < \infty . \tag{11}$$

PROPOSITION 1. a) Assume (9). There exists $C > 0$ such that for $t \in J$, $0 < \varepsilon \leqslant 1$

$$\varepsilon\|u_\varepsilon'(t)\|^2 + \|A(t)^{1/2}u_\varepsilon(t)\|^2 + \int_0^t \|B(s)^{1/2}u_\varepsilon'(s)\|^2 ds$$

$$\leqslant C[\|A(0)^{1/2}u_{0\varepsilon}\|^2 + \varepsilon\|u_{1\varepsilon}\|^2 + \int_0^t \|B(s)^{-1/2}f(s)\|^2 ds] . \tag{12}$$

b) Assume (9) and (10). There exists $C > 0$ such that for $t \in J$, $0 \leqslant \varepsilon \leqslant 1$

$$\|B(t)^{1/2}u_\varepsilon(t)\|^2 \leqslant C[\|B(0)^{1/2}u_{0\varepsilon}\|^2$$

$$+ \|A(0)^{1/2}u_{0\varepsilon}\|^2 + \varepsilon\|u_{1\varepsilon}\|^2 + \int_0^t \|B(s)^{-1/2}f(s)\|^2 ds] , \tag{13}$$

and for $\varepsilon = 0$ the condition (9) is not needed and

$$\|B(t)^{1/2}u_0(t)\|^2 \leqslant C[\|B(0)^{1/2}u_{00}\|^2 + \int_0^t \|B(s)^{-1/2}f(s)\|^2 ds]. \tag{14}$$

c) Assume (9), (11) and $u_{0\varepsilon} \in D(B)$. There exists $C > 0$ such that for $t \in J$, $0 < \varepsilon \leqslant 1$

$$\|B(t)u_\varepsilon(t)\|^2 \leqslant C[\|B(0)u_{0\varepsilon}\|^2 + \varepsilon\|A(0)^{1/2}u_{0\varepsilon}\|^2$$

$$+ \varepsilon^2\|u_{1\varepsilon}\|^2 + \int_0^t [\|f(s)\|^2 + \varepsilon\|B(s)^{-1/2}f(s)\|^2]ds . \tag{15}$$

For $\varepsilon = 0$ the condition (9) is not needed and the second term in (11) can be omitted. In this case

$$\|Bu_0(t)\|^2 \leqslant C[\|Bu_{00}\|^2 + \int_0^t \|f(s)\|^2 ds] . \tag{16}$$

Proof. a) Multiplying (4) by u_ε' we find

$$\frac{d}{dt}[\varepsilon\|u_\varepsilon'(t)\|^2 + \|A(t)^{1/2}u_\varepsilon(t)\|^2] + 2\|B(t)^{1/2}u_\varepsilon'(t)\|^2 =$$

$$= (A'(t)u_\varepsilon(t)|u_\varepsilon(t)) + 2\mathrm{Re}(f(t)|u_\varepsilon'(t)) .$$

Integrating, it follows that

$$\varepsilon\|u_\varepsilon'(t)\|^2 + \|A(t)^{1/2}u_\varepsilon(t)\|^2 + 2\int_0^t \|B(s)^{1/2}u_\varepsilon'(s)\|^2 ds \leqslant \varepsilon\|u_{1\varepsilon}\|^2 + \|A(0)^{1/2}u_{0\varepsilon}\|^2$$

$$+ \int_0^t [\|A(s)^{-1/2}A'(s)A(s)^{-1/2}\| \, \|A(s)^{1/2}u_\varepsilon(s)\|^2 + \|B(s)^{1/2}u_\varepsilon'(s)\|^2 + \|B(s)^{-1/2}f(s)\|^2]ds .$$

Gronwall's inequality implies (12) . If $\varepsilon = 0$ the proof of (13) is similar.

b) Multiplying (4) by u_ε we find

$$\frac{d}{dt}[\|B(t)^{1/2}u_\varepsilon(t)\|^2 + 2\varepsilon\mathrm{Re}(u_\varepsilon'(t)|u_\varepsilon(t)] + 2\|A(t)^{1/2}u_\varepsilon(t)\|^2$$

$$= (B'(t)u_\varepsilon(t)|u_\varepsilon(t)) + 2\mathrm{Re}(f(t)|u_\varepsilon(t))+2\varepsilon\|u_\varepsilon'(t)\|^2 .$$

Integrating, it follows that

$$\|B(t)^{1/2}u_\varepsilon(t)\|^2+2\int_0^t\|A(s)^{1/2}u_\varepsilon(s)\|^2ds \leqslant \|B(0)^{1/2}u_{0\varepsilon}\|^2+2\varepsilon|(u_{1\varepsilon}|u_{0\varepsilon})|+2\varepsilon|(u_\varepsilon'(t)|u_\varepsilon(t))|$$
$$+\int_0^t[\|B(s)^{-1/2}B'(s)B(s)^{-1/2}\| \, \|B(s)^{1/2}u_\varepsilon(s)\|^2$$
$$+2\varepsilon\|u_\varepsilon'(s)\|^2+\|B(s)^{1/2}u_\varepsilon(s)\|^2+\|B(s)^{-1/2}f(s)\|^2)ds .$$

From (10),(12), $2|(u_{1\varepsilon}|u_{0\varepsilon})| \leqslant \|u_{1\varepsilon}\|^2+\|u_{0\varepsilon}\|^2$,$2\varepsilon|(u_\varepsilon'(t)|u_\varepsilon(t))|$
$\leqslant \frac{1}{2}\|B(t)^{1/2}u_\varepsilon(t)\|^2 + 2\varepsilon^2\|B(t)^{-1/2}u_\varepsilon'(t)\|^2$ and Gronwall's inequality follows (13).

For $\varepsilon = 0$ the proof is similar.

c) Multiplying (4) by Bu_ε we find

$$\frac{d}{dt}[\|B(t)u_\varepsilon(t)\|^2+2\varepsilon\mathrm{Re}(u_\varepsilon'(t)|B(t)u_\varepsilon(t))]+2\mathrm{Re}(A(t)u_\varepsilon(t)|B(t)u_\varepsilon(t))$$
$$= 2\mathrm{Re}(f(t)|B(t)u_\varepsilon(t))+2\mathrm{Re}(B'(t)u_\varepsilon(t)|B(t)u_\varepsilon(t))+$$
$$2\varepsilon\|B(t)^{1/2}u_\varepsilon'(t)\|^2+2\varepsilon\mathrm{Re}(u_\varepsilon'(t)|B'(t)u_\varepsilon(t)).$$

Integrating, estimating and using the hypothesis (I) it follows that

$$\|B(t)u_\varepsilon(t)\|^2 \leqslant \|B(0)u_{0\varepsilon}\|^2+2\varepsilon|(u_\varepsilon'(t)|B(t)u_\varepsilon(t))|+2\varepsilon|(u_{1\varepsilon}|B(0)u_{0\varepsilon})|$$
$$+\int_0^t[2\varepsilon\|B(s)^{1/2}u_\varepsilon'(s)\|^2+2\varepsilon|(u_\varepsilon'(s)|B'(s)u_\varepsilon(s))|+|(B'(s)u_\varepsilon(s)|B(s)u_\varepsilon(s))$$
$$+ \|B(s)u_\varepsilon(s)\|^2 + \|f(s)\|^2]ds .$$

Using (12), it follows that

$$\|B(t)u_\varepsilon(t)\|^2 \leqslant C\{\|B(0)u_{0\varepsilon}\|^2+\varepsilon\|A(0)^{1/2}u_{0\varepsilon}\|^2+\varepsilon^2\|u_{1\varepsilon}\|^2$$
$$+ \int_0^t[(1+\|B'(s)B(s)^{-1}\|+\varepsilon\|B(s)^{-1/2}B'(s)B(s)^{-1}\|^2)\|B(s)u_\varepsilon(s)\|^2$$
$$+ \|f(s)\|^2+\varepsilon\|B(s)^{-1/2}f(s)\|^2]ds\}.$$

From (11) and Gronwall's inequality follows (15).

For $\varepsilon = 0$ the proof is similar.

REMARK. In a similar way, from (9) it follows that

$$\varepsilon\|A(t)^{-1/2}u_\varepsilon'(t)\| + \|u_\varepsilon(t)\|^2 \leqslant$$

$$\leqslant C[\|u_{0\varepsilon}\|^2 + \varepsilon\|A(0)^{-1/2}u_{1\varepsilon}\|^2 + \frac{1}{\varepsilon}\int_0^t \|A(s)^{-1/2}f(s)\|^2 ds] .$$

We shall not need this result.

1.2. The perturbation of the limiting equation

Consider two equations of the form (5), that is

$$B_i(t)y' + A_i(t)y = f , \; y_i(0) = x . \tag{16$_i$}$$

Let y_i be the solutions and $w = y_1 - y_2$. Then w satisfies

$$B_1(t)w' + A_1(t)w = g(t) , \; w(0) = 0 \tag{17}$$

with

$$g(t) = [\hat{A}(t) - \hat{B}(t)]A_2(t)y_2(t) + \hat{B}(t)f(t) \tag{18}$$

and also

$$B_1(t)^{-1/2}g(t) = [\tilde{A}(t) - \tilde{B}(t)B_2(t)^{-1/2}A_2(t)B_2(t)^{-1/2}]B_2(t)^{1/2}y_2(t) + \tilde{B}(t)B_2((t)^{-1/2}f(t) \tag{19}$$

where

$$\hat{B}(t) = [B_2(t) - B_1(t)]B_2(t)^{-1} , \; \hat{A}(t) = [A_2(t) - A_1(t)]A_2(t)^{-1} ,$$

$$\tilde{B}(t) = B_1(t)^{-1/2}\,[B_2(t) - B_1(t)]B_2(t)^{-1/2} , \; \tilde{A}(t) = B_1(t)^{-1/2}[A_2(t) - A_1(t)]B_2(t)^{-1/2}.$$

We shall use following assumptions:

$$\int_0^T(\|\tilde{A}(t)\|^2 + \|\tilde{B}(t)\|^2)dt < \infty \tag{20}$$

$$\sup_{t\in[0,T]}\|B_2(t)^{-1/2}A_2(t)B_2(t)^{-1/2}\| < \infty \tag{21}$$

$$\int_0^T(\|B_1'(t)B_1(t)^{-1}\| + \|B_2'(t)B_2(t)^{-1}\|^2)dt < \infty \tag{22}$$

$$\int_0^T(\|\hat{A}(t)\|^2 + \|\hat{B}(t)\|^2)dt < \infty \tag{23}$$

$$\sup_{t\in[0,T]} \|A_2(t)B_2(t)^{-1}\| < \infty . \tag{24}$$

PROPOSITION 2. a) Assume $x \in D(B_2(0)^{1/2})$, B_1 and B_2 satisfy (10), f satisfies (8), and (20), (21) hold. There exists $C > 0$ such that for all $t \in J$

$$\|B_1(t)^{1/2}w(t)\|^2 \leqslant \int_0^t(\|\widetilde{A}(s)\|^2+\|\widetilde{B}(s)\|^2)ds(\|B_2(0)^{1/2}x\|^2$$
$$+\|B_2(s)^{-1/2}f(s)\|^2+\int_0^t\|B_2(r)^{-1/2}f(r)\|^2dr)ds\ . \tag{25}$$

b) Assume $x \in D(B_2(0)$, f satisfies (8), and (22), (23), (24) hold. There exists $C > 0$ such that for all $t \in J$

$$\|B_1(t)w(t)\|^2 \leqslant C\int_0^t(\|\hat{A}(s)\|^2+\|\hat{B}(s)\|^2)(\|B_2(0)x\|^2+\|f(s)\|^2+\int_0^t\|f(r)\|^2dr)ds\ . \tag{26}$$

Proof. a) From Proposition 1b) we find

$$\|B_1(t)^{1/2}w(t)\|^2 \leqslant C\ \int_0^t\|B_1(s)^{-1/2}g(s)\|^2ds\ .$$
$$\|B_2(t)^{1/2}y_2(t)\|^2 \leqslant C(\|B_2(0)^{1/2}x\|^2+\int_0^t\|B_2(s)^{-1/2}f(s)\|^2ds)\ .$$

and therefore (25) follows from (19).

b) From Proposition 1c) we find

$$\|B_1(t)w(t)\|^2 \leqslant C\int_0^t\|g(s)\|^2ds\ .$$

$$\|B_2(t)y_2(t)\|^2 \leqslant C[\|B_2(0)x\|^2 + \int_0^t\|f(s)\|^2ds]\ .$$

The estimate (26) follows from (24) and (18).

1.3. The estimate of $u_\varepsilon - u_0$

Set $v_\varepsilon = u_\varepsilon - u_0$, $v_{0\varepsilon} = u_{0\varepsilon} - u_{00}$. Then v_ε is the solution of

$$B(t)v' + A(t)v = -\varepsilon u_\varepsilon''$$
$$v(0) = v_{0\varepsilon}\ . \tag{27}$$

PROPOSITION 3. In addition to the standing assumptions assume (9).

a) If (10) holds then

$$\|B(t)^{1/2}v_\varepsilon(t)\|^2 \leqslant C[\|B(0)^{1/2}v_{0\varepsilon}\|^2 + \varepsilon\|A(0)^{1/2}u_{0\varepsilon}\|^2 +$$
$$+ \varepsilon^2\|u_{1\varepsilon}\|^2 + \varepsilon + \varepsilon\int_0^t\|B(s)^{-1/2}f(s)\|^2ds]\ . \tag{28}$$

b) If (11) holds, then

$$\|B(t)v_\varepsilon(t)\|^2 \leqslant C[\|B(0)v_{0\varepsilon}\|^2 + \varepsilon\|A(0)^{1/2}u_{0\varepsilon}\|^2 + \varepsilon^2\|u_{1\varepsilon}\|^2 +$$
$$+ \varepsilon + \varepsilon\int_0^t\|B(s)^{-1/2}f(s)\|^2ds]\ . \tag{29}$$

Proof. a) Multiplying (27) by v_ε we find

$$\frac{d}{dt}[\|B(t)^{1/2}v_\varepsilon(t)\|^2 + 2\varepsilon\, \mathrm{Re}(u_\varepsilon'(t)|v_\varepsilon(t))] + 2\|A(t)^{1/2}v_\varepsilon(t)\|^2 =$$

$$= 2\varepsilon\, \mathrm{Re}(u_\varepsilon'(t)|v_\varepsilon'(t)) + (B'(t)v_\varepsilon(t)|v_\varepsilon(t)) .$$

Integrating and then succesively applying the Cauchy–Schwarz inequality, (6), (10) and Gronwall's lemma, we obtain

$$\|B(t)^{1/2}v_\varepsilon(t)\|^2 + \int_0^t\|A(s)^{1/2}v_\varepsilon(s)\|^2 ds \leqslant C[\|B(0)^{1/2}v_{0\varepsilon}\|^2$$

$$+ \varepsilon^2\|B(0)^{-1/2}u_{1\varepsilon}\|^2 + \varepsilon^2\|u_\varepsilon'(t)\|^2 + \int_0^t[\|u_\varepsilon'(s)\|^2 + \|u_0'(s)\|^2]ds$$

Since $u_0' \in L^2(J,X)$ by (6), the RHS is less or equal to

$$C[\|B(0)^{1/2}v_{0\varepsilon}\|^2+\varepsilon^2\|B(0)^{-1/2}u_{1\varepsilon}\|^2+\varepsilon+\varepsilon^2\|u_\varepsilon'(t)\|^2+\varepsilon\int_0^t\|B(s)^{1/2}u_\varepsilon'(s)\|^2ds] ,$$

hence (12) implies (28).

(b) Multiplying (27) by Bv_ε we find

$$\frac{d}{dt}[\|B(t)v_\varepsilon(t)\|^2 + 2\varepsilon\, \mathrm{Re}(u_\varepsilon'(t)|B(t)v_\varepsilon(t))] \leqslant 2\, \mathrm{Re}(B'(t)v_\varepsilon(t)|B(t)v_\varepsilon(t))$$

$$+ 2\varepsilon\, \mathrm{Re}(u_\varepsilon'(t)|B'(t)v_\varepsilon(t)) + 2\varepsilon\, \mathrm{Re}(u_\varepsilon'(t)|B(t)v_\varepsilon(t)) .$$

Integrating we obtain

$$\|B(t)v_\varepsilon(t)\|^2 \leqslant 2\varepsilon|(u_\varepsilon'(t)|B(t)v_\varepsilon(t))|+\|B(0)v_{0\varepsilon}\|^2+2\varepsilon(u_{1\varepsilon}|B(0)v_{0\varepsilon})$$

$$+ 2\int_0^t[\varepsilon|(u_\varepsilon'(s)|B'(s)v_\varepsilon(s))|+\varepsilon|(u_\varepsilon'(s)|B(s)v_\varepsilon'(s))|+|(B'(s)v_\varepsilon(s)|B(s)v_\varepsilon(s)|]ds$$

Applying the Cauchy–Schwarz inequality, (11) and Gronwall's lemma, the RHS can be estimated by $C[\varepsilon^2\|u_\varepsilon(t)\|^2 + \|B(0)v_{0\varepsilon}\|^2 + \varepsilon^2\|u_{1\varepsilon}\|^2 + \varepsilon\int_0^t|(u_\varepsilon'(s)|B'(s)v_\varepsilon(s) + B(s)v_\varepsilon'(s))|ds]$.

From (12) and $B(\cdot)^{1/2}u_0' \in L^2(J,X)$ we deduce (29).

REMARK. The solution u_ε of the equation

$$\varepsilon u'' + B(t)u' + A(t)u = f , \quad u(s) = u_0 , \quad u'(s) = u_1$$

can be written as

$$u_\varepsilon(t) = C_\varepsilon(t,s)u_0 + S_\varepsilon(t,s)u_1 + \frac{1}{\varepsilon} S_\varepsilon(t , \cdot)*f \tag{30}$$

where $S_\varepsilon(t , \cdot)*f = \int_0^t S_\varepsilon(t,r)f(r)dr$.

We write shortly $C_\varepsilon(t) := C_\varepsilon(t,0)$, $S_\varepsilon(t) := S_\varepsilon(t,0)$.

2. THE QUASILINEAR CASE

Consider the full equations (1) and (2). Throughout the section we assume that there exist a Banach space Y and a neighbourhood V of u_{00} in Y such that

$$Y \text{ is densely embedded in } D(B(0,u_{00})^{1/2}). \tag{31}$$

The functions A,B and f are defined on $J \times V$; we state the conditions on these functions below. For a function $v : J \to V$ denote by

$$u(t) = C_\varepsilon(t,s;v)u_0 + S_\varepsilon(t,s;v)u_1 + \frac{1}{\varepsilon} S_\varepsilon(t, \cdot ;v)*f$$

(cf. (30)) the weak solution of the initial value problem

$$\varepsilon u'' + B(t,v(t))u' + A(t,v(t))u = f(t) , \quad u(s) = u_0 , \quad u'(s) = u_1 ,$$

and similarly by

$$u(t) = C_0(t,s;v)u_0 + C_0(t, \cdot ;v)*f$$

the weak solution of the initial value problem

$$B(t,v(t))u' + A(t,v(t))u = f(t) , \quad u(s) = u_0 .$$

Shorter set $C_\varepsilon(t;v) := C_\varepsilon(t,0;v)$ and analogously $S_\varepsilon(t;v)$, $C_0(t;v)$.

The <u>mild solution of (1)</u> on J is the function $u_\varepsilon \in C(J,Y)$ such that $u_\varepsilon(J) \subset V$ and which satisfies

$$u_\varepsilon(t) = C_\varepsilon(t;u_\varepsilon)u_{0\varepsilon} + S_\varepsilon(t;u_\varepsilon)u_{1\varepsilon} + \frac{1}{\varepsilon} S_\varepsilon(t, \cdot ;u_\varepsilon)*f(t,u_\varepsilon(\cdot)) . \tag{32}$$

Analogously, the mild solution of (2) on J is the function $u_0 \in C(J,Y)$ such that $u_0(J) \subset V$ and which satisfies

$$u_0(t) = C_0(t;u_0)u_{00} + C_0(t, \cdot ;u_0)*f(\cdot,u_0(\cdot)) . \tag{33}$$

Throughout the section we assume:

(C) For every $(t,u) \in J \times V$ the operators $A(t,u)$ and $B(t,u)$ are self–adjoint positive definite operators in X . The space Y is contained in $D(A(t,u)) \cap D(B(t,u))$. Moreover,

$$\operatorname{Re}(A(t,u)v|B(t,u)v) \geqslant 0 \qquad (v \in D(A(t,u)) \cap D(B(t,u)))$$

and

$$\sup \{\|B(t,u)^{-1}\| : t \in J,\ u \in V\} < \infty . \tag{34}$$

For every $v \in Y$ the maps $(t,u) \mapsto A(t,u)v$, $B(t,u)v$ from $J \times Y$ in X are C^1.

It follows that for every $u \in C^1(J,Y)$ with $u(J) \subset V$ the functions $A(\cdot,u(\cdot))$, $B(\cdot,u(\cdot))$ satisfy (I) and (II) from Sec. 1. Set

$$A := A(0,u_{00}), \quad B := B(0,u_{00}).$$

We state our first set of assumptions.

Assume that there exists $C > 0$ such that for all $s,t \in J$, $u,v \in V$ the following estimates hold:

$$D(B(t,u)^{1/2}) = D(B^{1/2}), \quad D(A(t,u)^{1/2}) = D(A^{1/2}),$$

$$\|B(t,u)^{1/2}B(s,v)^{-1/2}\| \leqslant C, \tag{35}$$

$$\|A(t,u)^{1/2}A(s,v)^{-1/2}\| \leqslant C, \tag{36}$$

$$\|A(t,u)^{1/2}B(s,v)^{-1/2}\| \leqslant C. \tag{37}$$

Further it is assumed that for every $u \in C^1(J,Y)$ with $u(J) \subset V$ following inclusions hold:

$$B^{-1/2}(\frac{d}{dt} B(\cdot,u(\cdot)))B^{-1/2} \in L^2(J,X) \tag{38}$$

$$A^{-1/2}(\frac{d}{dt} A(\cdot,u(\cdot)))A^{-1/2} \in L^2(J,X). \tag{39}$$

As a consequence of (37), (38) and (39) we conclude

$$B^{-1/2}B(\cdot,u(\cdot))B^{-1/2} \in L^2(J,X), \quad B^{-1/2}A(\cdot,u(\cdot))B^{-1/2} \in L^2(J,X). \tag{40}$$

Further we assume the existence of $g \in L^2(J)$ such that for all $t \in J$, $u,v \in V$

$$\|B^{-1/2}[B(t,u)-B(t,v)]B^{-1/2}\|+\|B^{-1/2}[A(t,u)- A(t,v)]B^{-1/2}\| \leqslant g(t)\|B^{1/2}(u-v)\|, \tag{41}$$

$$\|B^{-1/2}[f(t,u) - f(t,v)]\| \leqslant g(t)\|B^{1/2}(u-v)\|. \tag{42}$$

We assume that (2) has a mild solution u_0 such that

$$B^{-1/2}f(\cdot,u_0(\cdot)) \in C(J,X). \tag{43}$$

Moreover

$$u_{0\varepsilon} \in D(B^{1/2}) \cap D(A^{1/2}) \cap V \ (\varepsilon \geqslant 0), \quad u_{1\varepsilon} \in X \ (\varepsilon > 0). \tag{44}$$

THEOREM 4. Assume that the equations (1) and (2) have mild solutions u_ε and u_0 which are $C^1(J,Y)$. Further assume that (31), (C) and (35)–(44) hold. Then there exists $C > 0$ and $T' \in (0,T]$ such that for all $t \in [0,T']$:

$$\|B^{1/2}[u_\varepsilon(t)-u_0(t)]\| \leqslant C[\|B^{1/2}(u_{0\varepsilon}-u_{00})\| + \|A^{1/2}(u_{0\varepsilon}-u_{00})\| + \varepsilon^{1/2}(1 + \|A^{1/2}u_{0\varepsilon}\| + \|u_{1\varepsilon}\|)] \; . \tag{45}$$

Proof. Set $v_\varepsilon = u_\varepsilon - u_0$. Then it follows from (32) and (33) that

$$v_\varepsilon = I_{1,\varepsilon} + \dots + I_{7,\varepsilon}$$

with

$$I_{1,\varepsilon}(t) = C_\varepsilon(t;u_\varepsilon)(u_{0\varepsilon}-u_{00})$$

$$I_{2,\varepsilon}(t) = [C_\varepsilon(t;u_\varepsilon) - C_0(t;u_\varepsilon)]u_{00}$$

$$I_{3,\varepsilon}(t) = [C_0(t;u_\varepsilon) - C_0(t;u_0)]u_{00}$$

$$I_{4,\varepsilon}(t) = S_\varepsilon(t;u_\varepsilon)u_{1\varepsilon}$$

$$I_{5,\varepsilon}(t) = \frac{1}{\varepsilon} S_\varepsilon(t, \cdot \,;u_\varepsilon)*[f(\cdot \,;u_\varepsilon) - f(\cdot \,;u_0)]$$

$$I_{6,\varepsilon}(t) = [\frac{1}{\varepsilon} S_\varepsilon(t, \cdot \,;u_\varepsilon) - C_0(t, \cdot \,;u_\varepsilon)]*f(\cdot \,;u_0)$$

$$I_{7,\varepsilon}(t) = [C_0(t, \cdot \,;u_\varepsilon) - C_0(t, \cdot \,;u_0)]*f(\cdot \,;u_0) \; .$$

Denoting $A_\varepsilon(t) = A(t,u_\varepsilon(t))$, $B_\varepsilon(t) = B(t,u_\varepsilon(t))$, we find from (13), (25) and (28) that

$$\|B^{1/2}I_{1,\varepsilon}(t)\| \leqslant C(\|B^{1/2}(u_{0\varepsilon}-u_{00})\| + \|A^{1/2}(u_{0\varepsilon}-u_{00})\|)$$

$$\|B^{1/2}I_{2,\varepsilon}(t)\| \leqslant C\, \varepsilon^{1/2}(1 + \|A^{1/2}u_{0\varepsilon}\|)$$

$$\|B^{1/2}I_{3,\varepsilon}(t)\| \leqslant C\|B^{1/2}u_{00}\|^{1/2}(\textstyle\int_0^t[\|B^{-1/2}(A_\varepsilon(s) - A_0(s))B^{-1/2}\|^2 + \|B^{-1/2}(B_\varepsilon(s) - B_0(s))B^{-1/2}\|]ds)^{1/2}$$

$$\|B^{1/2}I_{4,\varepsilon}(t)\| \leqslant C\, \varepsilon^{1/2}\|u_{1\varepsilon}\|$$

$$\|B^{1/2}I_{5,\varepsilon}(t)\| \leqslant C(\textstyle\int_0^t\|B^{-1/2}[f(s,u_\varepsilon(s)) - f(s,u_0(s))]\|^2ds)^{1/2}$$

$$\|B^{1/2}I_{6,\varepsilon}(t)\| \leqslant C\, \varepsilon^{1/2}[1 + (\textstyle\int_0^t\|B^{-1/2}f(s,u_0(s))\|^2ds]^{1/2}$$

$$\|B^{1/2}I_{7,\varepsilon}(t)\| \leqslant C\{(\int_0^t[\|B^{-1/2}(A_\varepsilon(s) - A_0(s))B^{-1/2}\|^2 + \|B^{-1/2}(B_\varepsilon(s) - B_0(s))B^{-1/2}\|^2]ds)^{1/2} (\|B^{-1/2}f(s,u_0(s))\|^2 + \int_0^t\|B^{-1/2}f(r,u_0(r))\|^2dr\}^{1/2}$$

Setting $\varphi_\varepsilon(t) = \max\{\|B^{1/2}v_\varepsilon(s)\|,\ 0 \leqslant s \leqslant t\}$,$h(t) = \|g\|_{L^2(0,t)}$, we find from (35) – (43) :

$$\varphi_\varepsilon(t) \leqslant C[\|B^{1/2}(u_{0\varepsilon}-u_{00})\|+\|A^{1/2}(u_{0\varepsilon}-u_{00})\|+\varepsilon^{1/2}(1+\|A^{1/2}u_{0\varepsilon}\|+\|u_{1\varepsilon}\|)+h(t)\varphi_\varepsilon(t)] .$$

Since $\lim_{t\to 0} h(t) = 0$, picking T' sufficiently small we find (45).

REMARKS. 1. As mentioned in the introduction, we consider only the convergence of solutions; we do not state sufficient conditions for the existence of solutions of (1) and (2).

2. The assumptions of Theorem 4 are not minimal. For example, we do not actually need $u_\varepsilon \in C^1(J,Y)$; it is sufficient that $u = u_\varepsilon$ satisfies (38) and (39). Also, the assumptions (38) and (39) are implicit; we have not written explicit conditions on A and B which would imply them.

At the end we estimate $\|Bv_\varepsilon(t)\|$. We need stronger assumptions:

$$Y \text{ is embedded in } D(B) , \tag{46}$$

$$D(B) \subset D(A) , \tag{47}$$

and $D(B(t,u)) = D(B)$, $D(A(t,u)) = D(A)$ for all $t \in J$, $u \in V$. Moreover there exists $C > 0$ such that for all $s,t \in J$, $u,v \in V$ the following estimates hold

$$\|B(t,u)\, B(s,v)^{-1}\| \leqslant C , \tag{48}$$

$$\|A(t,u)A(s,v)^{-1}\| \leqslant C . \tag{49}$$

Further for every $u \in C^1(J,Y)$ with $u(J) \subset V$

$$\frac{d}{dt}(B(\cdot,u(\cdot))B^{-1} \in C(J,X) . \tag{50}$$

We also assume the existence of $g \in L^2(J)$ such that for all $t \in J$, $u,v \in V$

$$\|[A(t,u) - A(t,v)]A^{-1}\|+\|[B(t,u) - B(t,v)]B^{-1}\| \leqslant g(t)\|B(u-v)\| \tag{51}$$

$$\|f(t,u) - f(t,v)\| \leqslant g(t)\|B(u-v)\| . \tag{52}$$

We assume that (2) has a mild solution u_0 such that

$$f(\cdot, u_0(\cdot)) \in L^2(J,X) . \tag{53}$$

Moreover

$$u_{0\varepsilon} \in V \quad (\varepsilon \geqslant 0) , \quad u_{1\varepsilon} \in X \ (\varepsilon > 0) . \tag{54}$$

Note that (47), (48) and (49) imply $\|A(t,u)B(t,u)^{-1}\| \leqslant C$ for all $t \in J$, $u \in V$, and (54) implies $u_{0\varepsilon} \in D(A) \subset D(A^{1/2})$.

THEOREM 5. Assume that (46) – (54) and (C) hold and that the equation (1) and (2) have mild solutions u_ε and u_0 which are $C^1(J,Y)$. Then there exists $C > 0$ and $T' \in (0,T]$ such that for all $t \in [0,T']$

$$\|B[u_\varepsilon(t)-u_0(t)]\| \leqslant C[\|B(u_{0\varepsilon}-u_{00})\|+\varepsilon^{1/2}(1+\|A^{1/2}(u_{0\varepsilon}-u_{00})\|+\varepsilon\|u_{1\varepsilon}\|)]. \tag{55}$$

We omit the proof since it is similar to the proof of Theorem 4.

Applications. The crucial assumption is (3). We refer to (Clément,Ph., Prüss,J., 1989) for its verification in the linear case. It seems difficult to satisfy it in the nonlinear case unless $B(t,u) = B(t)$; moreover the existence problem is much more difficult in the nonlinear case. The assumptions we made, in particular the condition (C), are too strong to cover the simplest case $B(t) = \beta I$ which can be treated by different methods. we refer to Milani (1987) and to Esham, Weinacht (1989) for models with quasilinear A and $B(t,u) = \beta I$.

REFERENCES:

Clément,Ph., Prüss,J., (1989), On Second Order Differential Equations in Hilbert Space, Boll. U.M.I. (7) 3–B, 623–638.

Esham,B.F., Weinacht,R.J., (1989),Hyperbolic–parabolic singular perturbations for quasilinear equations, SIAM J.Math.Anal. 20, 1344–1365.

Kato,T., (1975), Quasi–linear Equations of Evolution, with Applications to Partial Differential Equations, Lect. Notes Math. 448, Springer Verlag, Berlin, 25–70.

Lions,J.L, (1969), Quelques méthodes de résolution des problèmes aux limites non linéaires, Dunod, Gauthier–Villars, Paris.

Milani,A., (1987), Long time existence and singular perturbation results for quasi–linear hyperbolic equations with small parameter and dissipation term II, Non Linear Anal. TMA 11, 1371–1381.

The Laplace–Stieltjes Transform in Banach Spaces and Abstract Cauchy Problems

FRANK NEUBRANDER Louisiana State University, Baton Rouge, Louisiana

1. INTRODUCTION

This paper is an introduction to some of the fundamental results of vector-valued Laplace-Stieltjes transform theory and their application to the abstract Cauchy problem $u'(t) = Au(t), u(0) = x$, where A is a closed linear operator on an infinite dimensional Banach space E.

Vector-valued Laplace transform theory dates back to Einar Hille's monograph on "Functional Analysis and Semi-Groups" (E. Hille, (1948)). It was clear to E. Hille that the Laplace transform formula $R(\lambda, A) = \int_0^\infty e^{-\lambda t}T(t)\,dt$ which connects the resolvent $R(\lambda, A)$ of A with the semigroup $T(t)$ generated by A is a crucial tool in the study of the abstract Cauchy problem. The Laplace transform approach to any linear evolution equation always leads to a Laplace formula like the one above, i.e. the Laplace transform of a linear evolution equation with a solution family $P(t)$ leads to a resolvent type equation whose solution (or resolvent) family $R(\lambda)$ is given by $R(\lambda) = \int_0^\infty e^{-\lambda t}P(t)\,dt$. Thus, vector-valued Laplace transform theory is a critical tool for the study of linear evolution equations. In trying to extend the classical numerical theory of the Laplace transform

$$r(\lambda) = \int_0^\infty e^{-\lambda t} f(t)\,dt$$

to functions with values in a Banach space E, E. Hille remarks on several occasions that the key results of the classical theory can be lifted to the abstract setting if the Banach space E is reflexive, but not in general. In fact it was shown by S. Zaidman (1960) (see also Theorem 2.4 below) that classical Laplace transform theory extends to a Banach space E if and only if E has the Radon-Nikodym property (e.g. reflexive spaces, l^1, and all spaces which do not contain c_0). However, the solution families $\lambda \to R(\lambda)$ and $t \to P(t)$ associated with an evolution equation take values in $\mathcal{L}(E)$,the Banach space of bounded operators on E, which does not have the Radon-Nikodym property (at least no infinite dimensional example is known).

From a historical point of view, this "negative" result was the end of the transform approach to the abstract Cauchy problem and other evolution equations. Instead Laplace transform methods were replaced by equally powerful methods which relied heavily on the algebraic properties of the functions $\lambda \to R(\lambda)$ and $t \to P(t)$. For example, in dealing with the abstract Cauchy problem, the essential algebraic properties of the solution families used in the theory are the resolvent equation $R(\mu, A) - R(\lambda, A) = (\lambda - \mu)R(\lambda, A)R(\mu, A)$ and the semigroup property $T(t+s) = T(t)T(s)$. In the drastically revised version of E. Hille's monograph from 1957, E. Hille and R.S. Phillips comment on the changes in methods as follows:

"Thus in keeping in spirit of the times the algebraic tools now play a major role and are introduced early in the book; they lead to a more satisfactory operational calculus and spectral theory... On the other hand, the Laplace-Stieltjes transform methods, used by Hille for such purposes, have not been replaced but rather supplemented by the new tools."

In more recent monographs on semigroup theory the role of the Laplace transform is reduced to a footnote at best, and almost every proof is based on the semigroup property $T(t+s) = T(t)T(s)$. The major disadvantage of the "algebraic approach" to the abstract Cauchy problem and other linear evolution equations becomes obvious if one compares the mathematical theories associated with them (semigroup theories, cosine families, the theory of integro-differential equations, etc.). It is striking how similar the results and techniques are. Still, every type of linear evolution equation requires its own theory because the algebraic properties of the solution and resolvent families involved change. Hence it is tempting to search for a general analytic principle behind all these theories. In recent years the vector-valued Laplace transform has been reconsidered for that reason. It has turned out that many key results of the classical theory, if modified properly, may be fully extended to the abstract setting. Moreover, many basic aspects in the various theories of linear evolution equations can be deduced, improved, and generalized by using vector-valued transform techniques.

The Laplace transform part of this paper builds on results from W. Arendt (1987) and B. Hennig, F. Neubrander (1990). In applying the Laplace transform methods to the abstract Cauchy problem, we use methods similar to those in a paper by M. Hieber, A. Holderrieth, and F. Neubrander (1992).

2. THE LAPLACE-STIELTJES TRANSFORM

In this paper, the space of vector-valued functions on which we will consider the Laplace-Stieltjes transform will be the Banach space $Lip_0([0,\infty);X)$ which consists of all Lipschitz continuous functions $\alpha : [0,\infty) \mapsto E$ with $\alpha(0) = 0$ and with norm

$$\|\alpha\|_{Lip} := \inf\{M > 0 \ : \ \|\alpha(t) - \alpha(s)\| \leq M|t-s| \text{ for all } t, s \geq 0\}.$$

For $\alpha \in Lip_0([0,\infty);X)$ and $\lambda \in \mathbb{C}$ with $Re\lambda > 0$ the *Laplace-Stieltjes integral* $r(\lambda)$ is given by the improper Stieltjes or Bochner integral

$$r(\lambda) := \int_0^\infty e^{-\lambda s}\, d\alpha(s) = \lambda \int_0^\infty e^{-\lambda t}\alpha(t)\, dt. \tag{2.1}$$

The function r is called the Laplace-Stieltjes transform L_S of α; i.e., $r = L_S(\alpha)$.

In order to see how the Laplace-Stieltjes transform relates to the Laplace transform, let $f \in L^\infty([0,\infty);X)$. Then $\alpha(t) := \int_0^t f(s)\, ds$ is in $Lip_0([0,\infty);X)$ and the Laplace-Stieltjes transform reduces to the Laplace transform; i.e., $\int_0^\infty e^{-\lambda t}\, d\alpha(t) = \int_0^\infty e^{-\lambda t} f(t)\, dt$. Recall that a space E has the Radon-Nikodym property if and only if any $\alpha \in Lip_0([0,\infty);X)$ is the antiderivative of an L^∞-function (see W. Arendt (1987)). Hence, if a space E has the Radon-Nikodym property, then $Lip_0([0,\infty);X)$ is isometrically isomorphic to $L^\infty([0,\infty);X)$ and the Laplace-Stieltjes transform reduces to the Laplace transform. If E does not have the Radon-Nikodym property, then the Laplace-Stieltjes transform is a true generalization of the Laplace transform. As we shall see, this is precisely the generalization needed in order to handle transformable functions with values in an arbitrary Banach space E.

The first result is the main technical tool in vector-valued Laplace-Stieltjes transform theory. It seems to be one of those "folk theorems" which is difficult to trace back to its origin. By $e_{-\lambda}$ we denote the exponential function $t \mapsto e^{-\lambda t}$, and by $\chi_{[0,t]}$ the characteristic function of the interval $[0,t]$.

THEOREM 2.1 *The spaces $Lip_0([0,\infty);X)$ and $\mathcal{L}(L^1(0,\infty);X)$ are isometrically isomorphic. The isomorphism assigns to every operator $T \in \mathcal{L}(L^1(0,\infty);X)$ a function $\alpha \in Lip_0([0,\infty);X)$ with $\|\alpha\|_{Lip} = \|T\|$ such that*

$$Tg = \int_0^\infty g(t)\,d\alpha(t)$$

for all continuous $g \in L^1(0,\infty)$, $T\chi_{[0,t]} = \alpha(t)$ for all $t \geq 0$, and $Te_{-\lambda} = r(\lambda)$ for all $\lambda \in \mathbb{C}$ with $Re\lambda > 0$.

Proof: Let $\alpha \in Lip_0([0,\infty);X)$. For a function g with support in $[a,b]$ which is either continuous or of bounded variation define $T_\alpha g := \int_a^b g(t)\,d\alpha(t)$. Let $M \geq \|\alpha\|_{Lip}$. Then

$$\begin{aligned}\|T_\alpha g\| &= \|\int_a^b g(t)d\alpha(t)\| = \|\lim_{|\pi|\to 0}\sum_{i=1}^n g(\xi_i)(\alpha(t_i)-\alpha(t_{i-1}))\| \\ &\leq M\lim_{|\pi|\to 0}\sum_{i=1}^n |g(\xi_i)|(t_i - t_{i-1}) = M\int_a^b |g(t)|\,dt = M\|g\|_1.\end{aligned}$$

It follows that $\|T_\alpha g\| \leq \|\alpha\|_{Lip}\|g\|_1$ for all functions g with compact support which are either continuous or of bounded variation. Since the union of these functions is total in $L^1(0,\infty)$, there exists a unique extension of the operator T_α to $\mathcal{L}(L^1(0,\infty);X)$ with $\|T_\alpha\| \leq \|\alpha\|_{Lip}$. If $T_\alpha = 0$, then $T_\alpha\chi_{[0,b]} = \int_0^b d\alpha(t) = \alpha(b) = 0$ for all $b \geq 0$. Moreover, if g is a continuous function in $L^1(0,\infty)$, then $g = \lim_{n\to\infty} g\chi_{[0,n]}$ in $L^1(0,\infty)$. Thus,

$$T_\alpha g = \lim_{n\to\infty} T_\alpha(g\chi_{[0,n]}) = \lim_{n\to\infty}\int_0^n g(t)\,d\alpha(t) = \int_0^\infty g(t)\,d\alpha(t).$$

To show that the mapping $\alpha \to T_\alpha$ is onto, let $T \in \mathcal{L}(L^1(0,\infty);X)$. Define $\alpha(t) := T\chi_{[0,t]}$. Then

$$\|\alpha(t)-\alpha(s)\| = \|T\chi_{[s,t]}\| \leq \|T\|\|\chi_{[s,t]}\|_1 = \|T\||t-s|.$$

So $\alpha \in Lip_0([0,\infty);X)$, $\|\alpha\|_{Lip} \leq \|T\|$, and $T\chi_{[0,t]} = \alpha(t) = \int_0^t d\alpha(s) = T_\alpha\chi_{[0,t]}$. Since the characteristic functions $\{\chi_{[0,t]}, t \geq 0\}$ are total in $L^1(0,\infty)$, we conclude that $T = T_\alpha$ and $\|T_\alpha\| = \|\alpha\|_{Lip}$ for all $\alpha \in Lip_0([0,\infty);X)$. ◇

The main task of Laplace-Stieltjes transform theory is to study how properties of the function α and $r = L_S(\alpha)$ affect each other. Thus Theorem 2.1 will be crucial for the following reason. Both the set of characteristic functions $\{\chi_{[0,t]} : t > 0\}$ and the set of exponential functions $\{e_{-\lambda} : \lambda > 0\}$ are total subsets of $L^1(0,\infty)$. Hence, the operator T_α as defined above is completely determined by its action on either of the subsets. Properties of the transformed function $r(\lambda) = T_\alpha e_{-\lambda}$ will be reflected in the operator T_α, and, in particular, in the determining function $T_\alpha \chi_{[0,t]} = \alpha(t)$ and vice versa. The first illustration of the power of this theorem is in the following proof of the *Uniqueness Theorem* for Laplace-Stieltjes transform theory.

THEOREM 2.2 *Let $a, b > 0$, $\lambda_n := a + nb$ for $n \in \mathbb{N}_0$, and $\alpha \in Lip_0([0,\infty); X)$. If $r(\lambda_n) = \int_0^\infty e^{-\lambda_n t}\, d\alpha(t) = 0$ for all $n \in \mathbb{N}_0$, then $\alpha(t) = 0$ for all $t \geq 0$.*

Proof: Define T_α as in Theorem 2.1. If $T_\alpha e_{-\lambda_n} = r(\lambda_n) = 0$ for all $n \in \mathbb{N}_0$, then $T_\alpha = 0$, by the totality of the set $\{e_{-\lambda_n}\}$ in $L^1(0,\infty)$. In particular, $\alpha(t) = T_\alpha \chi_{[0,t]} = 0$.
$\diamond$

So far we have seen that $Lip_0([0,\infty); X)$ is isometrically isomorphic to the space $\mathcal{L}(L^1(0,\infty); X)$, and if E has the Radon-Nikodym property, to $L^\infty([0,\infty); X)$. We will now come to the Laplace-Stieltjes transform representation of $Lip_0([0,\infty); X)$. Let $\alpha \in Lip_0([0,\infty); X)$ and $r = L_S(\alpha)$. Then $\lambda \mapsto r(\lambda)$ is a holomorphic function from the open right halfplane $\{\lambda \in \mathbb{C} : Re\lambda > 0\}$ into E and

$$r^{(n)}(\lambda) = \int_0^\infty e^{-\lambda t}(-t)^n d\alpha(t) \tag{2.2}$$

for all $n \in \mathbb{N}_0$ and $\lambda \in \mathbb{C}$ with $Re\lambda > 0$. Let $x^* \in E^*$ and define $\beta(t) := \langle \alpha(t), x^* \rangle$. Then $\beta \in Lip_0([0,\infty))$ and $\|\beta\|_{Lip} \leq \|\alpha\|_{Lip}\|x^*\|$. It follows from the fundamental theorem of calculus that β is differentiable a.e., and that $\beta(t) = \int_0^t \beta'(s)\, ds$ for all $t \geq 0$. Clearly, $\|\beta'(t)\| \leq \|\alpha\|_{Lip}\|x^*\|$ for almost all $t \geq 0$. It follows that

$$\langle \frac{1}{k!}\lambda^{k+1} r^{(k)}(\lambda), x^* \rangle = \frac{1}{k!}\lambda^{k+1} \int_0^\infty e^{-\lambda t}(-t)^k \beta'(t)\, dt.$$

Hence, by $|\langle \frac{1}{k!}\lambda^{k+1} r^{(k)}(\lambda), x^* \rangle| \leq \|\alpha\|_{Lip}\|x^*\|$ for all $x^* \in E^*$ and $\lambda > 0$, we obtain

$$\|r\|_W := \sup_{k \in \mathbb{N}_0} \sup_{\lambda > w} \|\frac{1}{k!}\lambda^{k+1} r^{(k)}(\lambda)\| \leq \|\alpha\|_{Lip}. \tag{2.3}$$

Let $C_W^\infty((0,\infty), E) := \{r \in C^\infty((0,\infty), E) : \|r\|_W < \infty\}$. Then the Widder space $C_W^\infty((0,\infty); X)$ is a Banach space and the *Widder-Arendt Representation Theorem* of Laplace-Stieltjes transform theory (see W. Arendt (1987) or B. Hennig, F. Neubrander (1990)) can be stated as follows.

THEOREM 2.3 *The Laplace-Stieltjes transform L_S is an isometric isomorphism from $Lip_0([0,\infty); X)$ onto $C_W^\infty((0,\infty); X)$.*

Proof: It follows from (2.3) that L_S maps $Lip_0([0,\infty);X)$ into $C_W^\infty((0,\infty);X)$ and that $\|L_S(\alpha)\|_W = \|r\|_W \leq \|\alpha\|_{Lip}$. By the Uniqueness Theorem 2.2, L_S is one-to-one. It remains to be shown that L_S is onto. Let $r \in C_W^\infty((0,\infty);X)$. Define $T_k \in \mathcal{L}(L^1(0,\infty);X)$ by

$$T_k f := \int_0^\infty f(t)(-1)^k \frac{1}{k!}\left(\frac{k}{t}\right)^{k+1} r^{(k)}\left(\frac{k}{t}\right) dt.$$

Then $\|T_k\| \leq \|r\|_W$ for all $k \in \mathbb{N}_0$ and

$$T_k e_{-\lambda} = (-1)^k \frac{1}{k!}\int_0^\infty e^{-\lambda t}(\frac{k}{t})^{k+1} r^{(k)}(\frac{k}{t})\, dt.$$

Using change of variables and integration by parts one obtains (with some work; for details, see D.V. Widder (1948) or B. Hennig, F. Neubrander (1990))

$$T_k e_{-\lambda} = u\int_0^\infty \frac{1}{k!}\left(\frac{k}{u}\right)^{k+1} e^{\frac{-kt}{u}} t^k h(t)\, dt$$

where $h(t) := \frac{1}{t}r(\frac{1}{t})$ and $1/u := \lambda$. The functions

$$\tau_k(t) := \frac{1}{k!}\left(\frac{k}{u}\right)^{k+1} e^{\frac{-kt}{u}} t^k$$

are "approximate identities"; i.e., they are positive, their L^1-norm is one and for all $\epsilon > 0$ there exists an interval I containing u such that $\int_{t\notin I} \tau_k(t)\, dt < \epsilon$ for all k large enough. Hence,

$$T_k e_{-\lambda} \longrightarrow uh(u) = r(\frac{1}{u}) = r(\lambda)$$

as $k \to \infty$. The set of exponential functions $\{e_{-\lambda} : \lambda > 0\}$ is total in $L^1(0,\infty)$. By the Theorem of Banach-Steinhaus, there exists a bounded operator $T \in \mathcal{L}(L^1(0,\infty);X)$ with $\|T\| \leq \|r\|_W$ such that $T_k f \to Tf$ for all $f \in L^1(0,\infty)$. In particular,

$$r(\lambda) = \lim_{k\to\infty} T_k e_{-\lambda} = Te_{-\lambda}.$$

It follows from the Representation Theorem 2.1 that there exists $\alpha \in Lip_0([0,\infty);X)$ with $\|\alpha\|_{Lip} = \|T\| \leq \|r\|_W$ such that $Tg = \int_0^\infty g(t)\, d\alpha(t)$ for all piecewise continuous $g \in L^1(0,\infty)$. Hence, for all $\lambda > 0$,

$$r(\lambda) = Te_{-\lambda} = \int_0^\infty e^{-\lambda t}\, d\alpha(t).$$

$\diamond$

In trying to extend Widder's original proof of the above theorem (see Widder (1948)) to Laplace transforms of Banach space valued functions, E. Hille (1948) remarks that Widder's

"*... argument seems to hold for any abstract space in which bounded sets are weakly compact, but it is not clear in the present writing whether or not his conditions are always sufficient for the existence of the representation.*"

This statement was made precise by S. Zaidman (1960) and W. Arendt (1987). The result is as follows:

THEOREM 2.4 *The Laplace transform is an isometric isomorphism between the spaces $L^\infty([0,\infty);X)$ and $C_W^\infty((0,\infty);X)$ if and only if E has the Radon-Nikodym property.*

Proof: A Banach space E has the Radon-Nikodym property if and only if the antiderivative operator $(If)(t) = \int_0^t f(s)ds$ is an isometric isomorphism between the spaces $L^\infty([0,\infty);X)$ and $Lip_0([0,\infty);X)$. By Theorem 2.3, the Laplace-Stieltjes transform L_S is an isometric isomorphism from $Lip_0([0,\infty);X)$ onto $C_W^\infty((0,\infty);X)$. Now the statement follows from the fact that the Laplace transform is the composition of L_S with I. ◊

Closely related to the Widder-Arendt Representation Theorem 2.3 is the following *Post-Widder Inversion* of the Laplace-Stieltjes transform.

THEOREM 2.5 *Let $r \in C_W^\infty((0,\infty);X)$. Then, for all $t > 0$,*

$$L_S^{-1}(r)(t) = \alpha(t) = \lim_{k\to\infty}(-1)^k \frac{1}{k!}\left(\frac{k}{t}\right)^{k+1}\left(\frac{r(\lambda)}{\lambda}\right)^{(k)}_{\lambda=k/t}. \tag{2.4}$$

Proof: Let $r \in C_W^\infty((0,\infty);X)$. By Theorem 2.3, there exists $\alpha \in Lip_0([0,\infty);X)$ with $r(\lambda) = \int_0^\infty e^{-\lambda t}d\alpha(t)$ for all $\lambda > 0$. Using the approximate identities $\tau_k(t)$ defined in the proof of Theorem 2.3, it follows that

$$\alpha(t) = \lim_{k\to\infty}\frac{1}{k!}\left(\frac{k}{t}\right)^{k+1}\int_0^\infty e^{\frac{-ks}{t}} s^k \alpha(s)ds$$

for all $t > 0$. Combining (2.2) with (2.1), one obtains (2.4). ◊

Another method frequently used to invert the Laplace-Stieltjes transform is the *Complex Inversion* formula. We note that it follows from the remarks preceding (2.2) and Theorem 2.3 that any $r \in C_W^\infty((0,\infty);X)$ has an analytic extension on the half-plane $\{\lambda \in \mathbb{C} : Re\lambda > 0\}$.

THEOREM 2.6 *Let $r \in C_W^\infty((0,\infty);X)$. Then, for all $t > 0$,*

$$L_S^{-1}(r)(t) = \alpha(t) := \frac{1}{2\pi i} \lim_{n\to\infty} \int_{1-in}^{1+in} e^{\lambda t} \frac{r(\lambda)}{\lambda} \, d\lambda. \tag{2.5}$$

Proof: Let $r \in C_W^\infty((0,\infty);X)$. By Theorem 2.1 and 2.3 there exist $\alpha \in Lip_0([0,\infty);X)$ and $T \in \mathcal{L}(L^1(0,\infty);X)$ such that $r(\lambda) = \int_0^\infty e^{-\lambda t} d\alpha(t) = Te_{-\lambda}$ for all $\lambda > 0$ and $\alpha(t) = T\chi_{[0,t]}$ for all $t > 0$. Define

$$\alpha_n(t) := \frac{1}{2\pi i} \int_{1-in}^{1+in} e^{\lambda t} \frac{r(\lambda)}{\lambda} \, d\lambda = T\left(\frac{1}{2\pi i} \int_{1-in}^{1+in} e^{\lambda t} \frac{e_{-\lambda}}{\lambda} \, d\lambda \right).$$

Then

$$\|\alpha(t) - \alpha_n(t)\| \leq \|T\| \|\chi_{[0,t]} - \frac{1}{2\pi i} \int_{1-in}^{1+in} e^{\lambda t} \frac{e_{-\lambda}}{\lambda} \, d\lambda \|_1.$$

Now the statement follows from the fact that the functions $\frac{1}{2\pi i} \int_{1-in}^{1+in} e^{\lambda t} \frac{e_{-\lambda}}{\lambda} \, d\lambda$ converge towards the characteristic function $\chi_{[0,t]}$ in $L^1(0,\infty)$ as $n \to \infty$. ◇

Another consequence of Theorem 2.1 is the Laplace-Stieltjes transform version of the Trotter-Kato approximation theorem (see also C. Lizama (1989)).

THEOREM 2.7 *Let $\alpha_n \in Lip_0([0,\infty);X)$ with $\|\alpha_n\|_{Lip} \leq M$ for all $n \in \mathbb{N}$ and $r_n = L_S(\alpha_n)$. Then the following statements are equivalent.*

(i) There exist $a, b > 0$ such that $\lim_{n\to\infty} r_n(\lambda_k)$ exists for all $k \in \mathbb{N}_0$, where $\lambda_k := a + kb$.

(ii) There exists $\alpha \in Lip_0([0,\infty);X)$ with $\|\alpha\|_{Lip} \leq M$ such that $\lim_{n\to\infty} r_n(\lambda) = \int_0^\infty e^{-\lambda t} d\alpha(t)$ uniformly in λ from compact subsets of $(0,\infty)$.

(iii) $\lim_{n\to\infty} \alpha_n(t)$ exists for all $t \geq 0$.

(iv) There exists $\alpha \in Lip_0([0,\infty);X)$ with $\|\alpha\|_{Lip} \leq M$ such that $\alpha_n(t) \to \alpha(t)$ uniformly in t from compact subintervals of $[0,\infty)$.

Proof: By Theorem 2.1 there exist $T_n \in \mathcal{L}(L^1(0,\infty); X)$ with $\|T_n\| \leq M$ such that $T_n e_{-\lambda} = r_n(\lambda)$ and $T_n \chi_{[0,t]} = \alpha_n(t)$ for all $n \in \mathbb{N}$, all $t \geq 0$ and all $\lambda > 0$. Each of the statements imply that the operators T_n converge on a total subset of $L^1(0,\infty)$. By the theorem of Banach-Steinhaus there exists $T \in \mathcal{L}(L^1(0,\infty); X)$ such that $T_n f \to Tf$ as $n \to \infty$ for all $f \in L^1(0,\infty)$. Let $b > 0$. Then the sets $\chi_b := \{\chi_{[0,t]} : 0 \leq t \leq b\}$ and $E_b := \{e_{-\lambda} : \frac{1}{b} \leq \lambda \leq b\}$ are compact in $L^1(0,\infty)$. Hence, the uniformly bounded sequence T_n converges uniformly on χ_b and E_b (H.H. Schaefer (1980), Satz III.4.5). Now the statement follows from Theorem 2.1. ◇

THEOREM 2.8 *For $n \in \mathbb{N}_0$ let E_n be Banach spaces, $\alpha_n \in Lip([0,\infty), E_n)$, $r_n = L_S(\alpha_n)$ and $P_n \in \mathcal{L}(E_0, E_n)$. Assume that there exist constants $M, K > 0$ such that $\|P_n\| \leq K$ and $\|\alpha_n\|_{Lip} \leq M$ for all $n \in \mathbb{N}_0$. Then the following statements are equivalent.*

(i) There exist $a, b > 0$ such that $\lim_{n\to\infty} \|r_n(\lambda_k) - P_n r_0(\lambda_k)\|_n = 0$ for all $k \in \mathbb{N}_0$, where $\lambda_k := a + kb$.
(ii) $\lim_{n\to\infty} \|r_n(\cdot) - P_n r_0(\cdot)\|_n = 0$ uniformly on compact subsets of $(0,\infty)$.
(iii) $\lim_{n\to\infty} \|\alpha_n(t) - P_n \alpha_0(t)\|_n = 0$ for all $t \geq 0$.
(iv) $\lim_{n\to\infty} \|\alpha_n(\cdot) - P_n \alpha_0(\cdot)\|_n = 0$ uniformly on compact subsets of $[0,\infty)$.

Proof: By Theorem 2.1 there exist $T_n \in \mathcal{L}(L^1(0,\infty), E_n)$ with $\|T_n\| \leq M$ such that $T_n e_{-\lambda} = r_n(\lambda)$ and $T_n \chi_{[0,t]} = \alpha_n(t)$ for all $n \in \mathbb{N}_0$, all $t \geq 0$ and all $\lambda > 0$. Define operators $U_n \in \mathcal{L}(L^1(0,\infty), E_n)$ by $U_n := T_n - P_n T$. Then $\|U_n\| \leq M + KM =: C$ for all $n \in \mathbb{N}_0$. To be able to apply the theorem of Banach-Steinhaus, define operators $\tilde{U}_n \in \mathcal{L}(L^1(0,\infty), l^\infty(E_i))$ by

$$\tilde{U}_n f := (0, 0, 0, ..., U_n f, 0, 0, 0, ...).$$

Then $\|\tilde{U}_n f\|_{l^\infty(E_i)} = \|U_n f\|_n$ and $\|\tilde{U}_n f\|_{l^\infty(E_i)} \leq C$ for all $n \in \mathbb{N}_0$. Each of the statements imply that the operators $\tilde{U}_n$ converge on a total subset of $L^1(0,\infty)$ towards 0. By the theorem of Banach-Steinhaus the operators $\tilde{U}_n$ and hence the operators U_n converge to 0 for all $f \in L^1(0,\infty)$. The uniform convergence follows as in the proof of Theorem 2.7. ◇

Because of its importance in applications, a survey on vector-valued Laplace-Stieltjes transform theory should include M. Sova's characterization of the Laplace transform of functions $\alpha \in Lip_0([0,\infty); X)$ which admit an analytic extension into a sector $\Sigma(\beta) := \{0 \neq \lambda \in \mathbb{C} : |arg(\lambda)| < \beta\}$ for some $\beta > 0$.

Let $w \in \mathbb{R}$ and $0 < \beta \leq \pi$. Then the closure of the sector $\Sigma(w, \beta) := \{\lambda \in \mathbb{C}\backslash\{w\} : |arg(\lambda - w)| < \beta\}$ will be denoted by $\Sigma'(w, \beta)$, and the closure of $\Sigma(w, \beta)$

without the point w by $\Sigma"(w,\beta)$. For a proof of the following theorem we refer to M. Sova (1979) and F. Neubrander (1989).

THEOREM 2.9 *Let $w \geq 0$ and let r be a function from (w,∞) into a Banach space E. The following statements are equivalent.*

(i) There exist $M,\beta > 0$ and $\alpha \in C(\Sigma'(\beta),E)$ which is analytic in $\Sigma(\beta)$, $\alpha(0) = 0$ and $\|\alpha(t+h)-\alpha(t)\| \leq M|h|e^{wRe(t+h)}$ for $t,h \in \Sigma'(\beta)$ such that $r(\lambda) = \int_0^\infty e^{-\lambda t}\,d\alpha(t)$ for all $\lambda > w$.

(ii) There exists $M,\gamma > 0$ such that r is continuous in $\Sigma"(w,\pi/2+\gamma)$, analytic in $\Sigma(w,\pi/2+\gamma)$ and satisfies $\|(\lambda - w)r(\lambda)\| \leq M$ for all $\lambda \in \Sigma"(w,\pi/2+\gamma)$.

(iii) r is analytic in the halfplane $H_w := \{\lambda : Re\lambda > w\}$ and there exist $M,C > 0$ such that $\|Re(\lambda - w)r(\lambda)\| \leq M$ and $\|\frac{1}{k!}(Im\lambda)^{k+1}r^{(k)}(\lambda)\| \leq MC^{k+1}$ for all $\lambda \in H_w$ and all $n \in \mathbb{N}_0$.

Applying vector-valued Laplace-Stieltjes transform theory to evolution equations, it is desirable to allow for functions α with arbitrary exponential growth. This can be done by using the following "shifting" procedure.

For $w \geq 0$ let $Lip_w([0,\infty),E)$ be the space of all functions $\beta : [0,\infty) \to E$ with $\beta(0) = 0$ and $\|\beta(t+h) - \beta(t)\| \leq hMe^{w(t+h)}$ for all $t,h \geq 0$ and some constant M. Then $Lip_w([0,\infty),E)$ is a Banach space with norm

$$\|\beta\|_{Lip(w)} := \inf\{M : \|\beta(t+h) - \beta(t)\| \leq hMe^{w(t+h)} \text{ for all } t,h \geq 0\},$$

and I_w defined by $I_w\beta(t) := \int_0^t e^{-ws}\,d\beta(s)$ is an isometric isomorphism between the spaces $Lip_w([0,\infty),E)$ and $Lip_0([0,\infty);X)$.

For $w \geq 0$ let $C_W^\infty((w,\infty),E)$ denote the Banach space of all functions $r \in C^\infty((w,\infty),E)$ with norm

$$\|r\|_{W,w} := \sup_{k\in\mathbb{N}_0}\sup_{\lambda>w}\|\frac{1}{k!}(\lambda - w)^{k+1}r^{(k)}(\lambda)\| < \infty.$$

The shift S_w defined by $S_w r(\lambda) := r(\lambda - w)$ is an isometric isomorphism between $C_W^\infty((0,\infty),E)$ and $C_W^\infty((w,\infty),E)$. Hence, the Laplace-Stieltjes transform $L_S = S_w L_S I_w$ is an isometric isomorphism between $Lip_w([0,\infty),E)$ and $C_W^\infty((w,\infty),E)$ and all theorems mentioned in this section can be rephrased for these spaces.

3. THE ABSTRACT CAUCHY PROBLEM

In order to apply Laplace-Stieltjes transform techniques to the abstract Cauchy problem

$$u'(t) = Au(t), u(0) = x, \qquad (ACP)$$

where A is a closed operator on a Banach space E (or to any other linear evolution equation), it is convenient to consider not only mild, but also integrated solutions (see also G. Lumer (1990)). By a mild solution of (ACP) we mean an exponentially bounded, Bochner integrable function u which satisfies (ACP_0) $u(t) = A\int_0^t u(s)\,ds + x$ for almost all $t \geq 0$. For a mild solution u we set $v(t) := \int_0^t u(s)\,ds$. This gives a function $v \in Lip_w([0,\infty),E)$ for some $w \geq 0$ which satisfies the integral equation (ACP_1) $v(t) = A\int_0^t v(s)\,ds + tx$ for all $t \geq 0$. If (ACP) or (ACP_0) describe the states $u(t)$ of an evolutionary system, then (ACP_1) describes the evolution of the cumulative states $v(t)$ of the system with initial value x.

Repeating this procedure yields the n-times integrated Cauchy problem

$$v(t) = A\int_0^t v(s)\,ds + \frac{t^n}{n!}x. \qquad (ACP_n)$$

A function $v \in Lip_w([0,\infty),E)$ which satisfies (ACP_n) for all $t \geq 0$ is called a strong integrated solution of (ACP_n). An exponentially bounded, Bochner integrable function v which satisfies (ACP_n) for almost all $t \geq 0$ is called a mild integrated solution of (ACP_n). Clearly, the normalized antiderivative a mild integrated solution of (ACP_n) yields a strong integrated solution of (ACP_{n+1}). In general, the derivative of a strong integrated solution does not exist. However, if E has the Radon-Nikodym property, then v is a strong integrated solution of (ACP_n) if and only if v' is a mild integrated solution of (ACP_{n-1}).

THEOREM 3.1 *Let A be a closed operator on a Banach space E, let $x \in E$ be fixed, $1 \leq n \in \mathbb{N}$ and $M, w \geq 0$. Then the following statements are equivalent.*

(i) There exists a strong integrated solution $v \in Lip_w([0,\infty),E)$ of (ACP_n) such that $\|v\|_{Lip(w)} \leq M$.

(ii) There exists $y \in C^\infty((\omega,\infty),E)$ such that $(\lambda I - A)y(\lambda) = x$ and $\|\frac{y(\lambda)}{\lambda^{n-1}}\|_{W,w} \leq M$.

(iii) There exists $y : (0,\infty) \to E$ and $v \in Lip_w([0,\infty),E)$ with $\|v\|_{Lip(w)} \leq M$ such that $(\lambda I - A)y(\lambda) = x$ and $y(\lambda) = \lambda^{n-1}\int_0^\infty e^{-\lambda t}\,dv(t)$ for all $\lambda > w$.

Proof: The equivalence of (ii) and (iii) follows immediately from Theorem 2.3 and the "shifting procedure" described at the end of the previous section.

Assume that (i) holds. Then

$$\begin{aligned} y(\lambda) &:= \lambda^{n-1} \int_0^\infty e^{-\lambda t}\, dv(t) = \lambda^n \int_0^\infty e^{-\lambda t} v(t)\, dt \\ &= \lambda^{n+1} \int_0^\infty e^{-\lambda t} \int_0^t v(s)\, ds\, dt \end{aligned}$$

exists for $\lambda > w$, is contained in the domain $D(A)$ of the operator A (by the closedness of A), and

$$\int_0^\infty e^{-\lambda t} \frac{t^n}{n!} x\, dt = \lambda y(\lambda) - x.$$

This proves that (i) implies (iii).

Conversely, assume (iii). Define $w(t) := \int_0^t v(s)ds$. Then, by Theorem 2.6, $v(t) = \frac{1}{2\pi i} \int_{w+1+i\mathbb{R}} e^{\lambda t} \frac{y(\lambda)}{\lambda^n}\, d\lambda$, and $w(t) = \frac{1}{2\pi i} \int_{w+1+i\mathbb{R}} e^{\lambda t} \frac{y(\lambda)}{\lambda^{n+1}}\, d\lambda$. It follows from the closedness of A that $w(t) \in D(A)$ for all $t \geq 0$ and that

$$Aw(t) = \frac{1}{2\pi i} \int_{w+1+i\mathbb{R}} e^{\lambda t} \left(\frac{y(\lambda)}{\lambda^n} - \frac{x}{\lambda^{n+1}} \right) d\lambda = v(t) - \frac{t^n}{n!} x.$$

This proves that (iii) implies (i). ◇

One should notice that except for closedness, no assumption is made on A in the theorem above. As a consequence, only integrated solutions are obtained. It can be shown that the time regularity of the solutions improves by the factor one, if an algebraic structure is added to the problem, i.e. if one assumes that the resolvent (or a regularized resolvent) of A exists on a right halfplane (see M. Hieber, A. Holderrieth, F. Neubrander (1992)). To obtain mild solutions of (ACP) without assuming the existence of a regularized resolvent of A, it is necessary to make additional assumptions on the Banach space E. The following *local Hille-Yosida Theorem* is an immediate consequence of Theorem 3.1.

COROLLARY 3.2 *Let A be a closed operator on a Banach space E with the Radon-Nikodym property, let $x \in E$ be fixed, and $M, w \geq 0$. Then the following statements are equivalent.*

(i) There exists a mild solution u of (ACP) with $\|u(t)\| \leq Me^{wt}$ for almost all $t \geq 0$.
(ii) There exists $y \in C^\infty((\omega, \infty), E)$ such that $(\lambda I - A)y(\lambda) = x$ and $\|\frac{1}{k!}y^{(k)}(\lambda)\| \leq \frac{M}{(\lambda - w)^{k+1}}$ for all $\lambda > w$ and $k \in \mathbb{N}_0$.

Assume that the pointwise resolvent equation $\lambda y - Ay = x$ has a solution $y = y(\lambda)$ which is analytic in a halfplane $H_w = \{\lambda \in \mathbb{C} : Re\lambda > w \geq 0\}$. In order to apply statement (ii) of Theorem 3.1 one has to show that there exist $n \in \mathbb{N}$ and $M > 0$ such that $\|\frac{y(\lambda)}{\lambda^{n-1}}\|_{W,w} \leq M$. Next we will indicate the proof of the fact that if $\|y(\lambda)\| \leq C|\lambda|^k$ for all $\lambda \in H_w$ and some constants $C > 0$, $k \in \mathbb{N}_0 \cup \{-1\}$, then $\|\frac{y(\lambda)}{\lambda^{k+2}}\|_{W,w} \leq M$, where the constant M depends only on C and w.

THEOREM 3.3 *Let A be a closed operator on a Banach space E and let $x \in E$ be fixed. Assume that there exists $w \geq 0$ such that the equation $\lambda y - Ay = x$ has a solution $y = y(\lambda)$ for all $\lambda > w$. If the function y has an analytic continuation into the halfplane $\{\lambda \in \mathbb{C} : Re\lambda > w\}$ where it satisfies $\|y(\lambda)\| \leq C|\lambda|^k$ for some $C > 0$ and some $k \in \mathbb{N}_0 \cup \{-1\}$, then there exists a locally Hölder continuous mild integrated solution v of (ACP_{k+2}) for all $t \geq 0$ which is $O(e^{w't})$ for all $w' > w$.*

Proof: It follows from the analyticity of y and the growth assumptions made that

$$v(t) := \frac{1}{2\pi i}\int_{w'+i\mathbb{R}} e^{\lambda t}\frac{y(\lambda)}{\lambda^{k+2}}\,d\lambda$$

is well-defined for all $t > 0$ and $O(e^{w't})$. It can be shown (see W. Arendt, H. Kellermann (1989), Prop. 3.1) that for all $0 < \epsilon < 1$ and $T > 0$ there exists $C_T > 0$ such that $\|v(t) - v(s)\| \leq C_T|t-s|^\epsilon$ for all $t, s \in [0,T]$, that $v(0) = 0$, and that $y(\lambda) = \lambda^{k+2}\int_0^\infty e^{-\lambda t}\,d\alpha(t)$, where $\alpha(t) := \int_0^t v(s)\,ds$. It follows from Theorem 3.1 that α is a strong integrated solution of (ACP_{k+3}). Because of the differentiability of α, the function $v = \alpha'$ is a mild integrated solution of (ACP_{k+2}) for all $t \geq 0$. ◇

THEOREM 3.4 *Let A be a closed operator on a Banach space E and let $x \in E$ be fixed. Assume that there exists $w \geq 0$ such that the equation $\lambda y - Ay = x$ has a solution $y = y(\lambda)$ for all $\lambda > w$. If the function y has an analytic continuation into a sector $\Sigma(w, \pi/2 + \gamma)$ for some $\gamma > 0$ such that $\|(\lambda - w)y(\lambda)\| \leq M|\lambda|^k$ for some constants $M > 0$, $k \in \mathbb{N}_0$ and all $\lambda \in \Sigma''(w, \pi/2 + \gamma)$, then there exists a strong integrated solution $v \in Lip_w([0,\infty), E)$ of (ACP_{k+1}) which is analytic in a sector $\Sigma(\beta)$ for some $\beta > 0$. Moreover, the function $v^{(k+1)}$ solves $u'(t) = Au(t)$ for all $t > 0$.*

Proof: The statement follows immediately from the Theorems 2.9 and 3.1. ◇

THEOREM 3.5 *Let A be a closed operator on a Banach space E, let $x \in E$ be fixed, $n \geq 1$, $M > 0$, and $a > w \geq 0, b > 0$. For all $m \in \mathbb{N}$ let A_m be operators, $x_m \in E$, and $y_m = y_m(\lambda)$ with $\|z_m\|_{W,w} \leq M$ for all $m \in \mathbb{N}$ where $z_m(\lambda) := \frac{1}{\lambda^{n-1}} y_m(\lambda)$. If $y_m(\lambda) \to y(\lambda)$ and $x \leftarrow x_m = \lambda y_m(\lambda) - A_m y_m(\lambda) \to \lambda y(\lambda) - Ay(\lambda)$ for all $\lambda = a + kb, k \in \mathbb{N}$, then $t \to \alpha(t) := \lim_{m\to\infty} L_S^{-1}(z_m)(t) \in Lip_w([0,\infty), E)$ is a strong integrated solution of (ACP_n).*

Proof: It follows from Theorem 2.3 and the remarks concerning the "shifting procedure" at the end of Section 2, that there are functions $\alpha_m \in Lip_w([0,\infty), E)$ with $\|\alpha_m\|_{Lip(w)} \leq M$ such that $z_m = L_S(\alpha_m)$ for all $m \in \mathbb{N}$. Now the statement follows with Theorem 2.7 and Theorem 3.1. ◇

So far we have seen that for the existence of an exponentially bounded integrated solution of (ACP_n) for some $n \in \mathbb{N}$, it is necessary and sufficient to assume that there exists $w \geq 0$ such that

(a) $x \in \bigcap_{\lambda > w} Range(\lambda I - A)$,
(b) there exists a function y with $y(\lambda) \in (\lambda I - A)^{-1}x$ for all $\lambda > w$ which has an analytic, polynomially bounded extension into the right halfplane H_w.

If an operator A and $x \in E$ satisfy the range condition (a) and the growth condition (b) above, then the integrated solution v is given by the inverse Laplace-Stieltjes transform L_S^{-1} (as described in Theorem 2.5 and Theorem 2.6) of the function $\lambda \to y(\lambda)/\lambda^n$ for some $n \in \mathbb{N}$ large enough. To obtain uniqueness of exponentially bounded integrated solutions a condition on the point spectrum $p\sigma(A)$ of A is needed. Using the uniqueness theorem for analytic functions, it was shown by E. Hille (see E. Hille, R.S. Phillips (1957), Theorem 23.7.1) that there exists at most one exponentially bounded integrated solution if $p\sigma(A)$ is not dense in any right halfplane. The Uniqueness Theorem 2.2 for Laplace-Stieltjes transforms leads to the following result.

THEOREM 3.6 *Let A be a closed operator on a Banach space E and $n \in \mathbb{N}_0$. If there exist $a, b > 0$ such that the equidistant points $\lambda_k = a + kb$, $k \in \mathbb{N}_0$ do not belong to the point spectrum of A, then (ACP_n) has at most one exponentially bounded Bochner integrable solution.*

Proof: Let v be a mild integrated solution of (ACP_n) for $x = 0$. Then $\alpha(t) := \int_0^t v(s)\, ds$ is a strong integrated solution of (ACP_m) for all $m \geq 0$. Let $y := L_S(\alpha)$. By Theorem 3.1, there exists $w \geq 0$ such that $\lambda y(\lambda) - Ay(\lambda) = 0$ for all $\lambda > w$. Because the points λ_k do not belong to the point spectrum of A, it follows that $y(\lambda_k) = 0$ for all $\lambda_k > w$. By Theorem 2.2, $\alpha(t) = 0$ for all $t \geq 0$. ◇

REFERENCES

1. Arendt, W. (1987). Vector valued Laplace transforms and Cauchy problems, *Israel J. Math.*, 59: pp. 327-352.
2. Arendt, W. and Kellermann, H. (1989). Integrated solutions of Volterra integro-differential equations and applications, *Volterra Integrodifferential Equations in Banach Spaces and Applications* (G. Da Prato and M. Iannelli, eds.). Pitman.
3. Hennig, B. and Neubrander, F (1990). On representations, inversions and approximations of Laplace transforms in Banach spaces, *Applicable Analysis*, to appear.
4. Hieber, M., Holderrieth, A. and Neubrander, F (1992). Regularized semigroups and systems of linear partial differential equations, *Ann. Scuola Norm. Pisa*, 19: pp. 363-379.
5. Hille, E. (1948). *Functional Analysis and Semi-Groups*, Amer. Math. Soc. Coll. Publ. 31, New York.
6. Hille, E. and Phillips, R.S. (1957). *Functional Analysis and Semi-Groups*, Amer. Math. Soc. Coll. Publ. 31, Providence, Rhode Island.
7. Lizama, C. (1989). On the convergence and approximation of integrated semigroups, *Preprint.*
8. Lumer, G. (1990). Solutions généralisées et semi-groupes intégrés, *C.R. Acad. Sci. Paris*, 310, série I: pp. 557-582.
9. Neubrander, F. (1989). Abstract elliptic operators, analytic interpolation semigroups, and Laplace transforms of analytic functions. *Preprint. Semesterbericht Funktionalanalysis. Tübingen*, 1989.
10. Schaefer, H.H. (1980). *Topological Vector Spaces*, Springer, New York- Heidelberg-Berlin.
11. Sova, M. (1979). The Laplace transform of analytic vector-valued functions (complex conditions). *Casopis pro pestovani mat.*, 104: pp. 267-280.
12. Widder, D.V. (1948). *The Laplace Transform*, Princeton University Press, Princeton, New Jersey.
13. Zaidman, S. (1960). Sur un théorème de I. Miyadera concernant la reprèsentation des fonctions vectorielles par des intégrales de Laplace, *Tôhoku Math. J.*, 12: pp. 47-51.

Stable Asymptotics for Differential Equations in a Hilbert Space and Applications to Boundary Value Problems in Domains with Conical Points

SERGE NICAISE University of Valenciennes, Valenciennes, France

0. Introduction

In this paper, we firstly give sufficient conditions for a family of operators A_α defined in a fixed Hilbert space X, depending on a real parameter α in a neighbourhood of 1, in order to insure that A_α tends to A as α goes to 1 in the generalized sense of Kato [6]. As usual, this leads to the "continuity" of the spectrum of A_α with respect to α, which we need later on.

In a second step, we consider the following differential equation associated with A_α :

$$\frac{du_\alpha}{dt} - A_\alpha u_\alpha(t) = f_\alpha(t) \quad \text{in} \quad \mathbf{R}. \tag{0.1}$$

According to [12], we give a comparison result in different weighted Sobolev spaces and we study the polynomial resolution corresponding to (0.1). Here due to the continuous dependence of the A_α's with respect to α, we are looking for stable results i.e. results which are continuous with respect to α. This is done in §2 using the continuity of the spectrum and of the resolvent and following the ideas of [2].

As an application, we consider the Dirichlet problem for elliptic operators of order $2m$ with smooth coefficients in plane domains with a conical point. We actually obtain an asymptotic representation of a weak solution near this point, which is stable under small variations of the angle. For operators with constant coefficients, this is proved by reducing the boundary value problem

$$(0.2) \qquad \begin{cases} L_0 u = f & \text{in} \quad C_\omega, \\ u = \dfrac{\partial u}{\partial \nu} = \cdots = \dfrac{\partial^{m-1} u}{\partial \nu^{m-1}} = 0 & \text{on} \quad \partial C_\omega, \end{cases}$$

where C_ω is a plane cone of opening ω, into a differential equation (0.1) in a fixed Hilbert space. Of course, the operators A_α we get fulfil the conditions of §1 and 2, since we have displayed them in that way. For non constant coefficients, as in [11] (who treated second order operators), we use an iterative procedure starting from low regularity up to the desired regularity. As explained in [2] for second order operators, these results will be useful in order to prove general edge asymptotics in dimension 3.

1. A criterion for generalized convergence in the sense of Kato.

Let X be a Hilbert space with a norm denoted by $\| \cdot \|$. Let further A_α be a family of closed operators on X, for $\alpha \in I$, where I is a fixed neighbourhood of 1. For convenience, we write $A_1 = A$. We make the following assumptions :

$$(1.1) \qquad D(A_\alpha) = D(A), \quad \forall \alpha \in I.$$

$$(1.2) \qquad A_\alpha \quad \text{has a compact resolvent}, \quad \forall \alpha \in I.$$

There exists a continuous function c on I such that $c(1) = 0$ and

$$(1.3) \qquad \|A_\alpha u - Au\| \leq c(\alpha)\|u\|_{D(A)}, \quad \forall u \in D(A), \quad \alpha \in I\,,$$

where we recall that $\|u\|_{D(A)} = \|u\| + \|Au\|$, for all $u \in D(A)$.

Theorem 1.1. *Under the previous assumptions, A_α tends to A in the generalized sense of Kato (cf. §IV.2.6 of [6]) or equivalently*

$$(1.4) \qquad R(\lambda, A_\alpha) \to R(\lambda, A) \quad \text{in norm, as} \quad \alpha \to 1, \forall \lambda \in \rho(A),$$

where $R(\lambda, A) = (\lambda - A)^{-1}$ is the resolvent operator of A.

Consequently, if λ is an eigenvalue of A with algebraic multiplicity k (in the sequel, we only speak about algebraic multiplicity and write simply multiplicity), then for all sufficiently small $\varepsilon > 0$, there exists a neighbourhood I_ε of 1 such that

A_α has exactly k eigenvalues (repeated according to their multiplicity) in the open ball $B(\lambda, \varepsilon)$. Finally, for a fixed closed curve γ containing a finite number k of eigenvalues of A, then there exists a neighbourhood J of 1 such that for all $\alpha \in J$, A_α has exactly k eigenvalues inside γ; moreover we can number the eigenvalues of A_α $\{\lambda_{j\alpha}\}_{j=1}^k$ so that

$$\lambda_{j\alpha} \to \lambda_{j1}, \quad \text{as} \quad \alpha \to 1, \ \forall j = 1, \cdots, k \,. \tag{1.5}$$

Proof : For a fixed $\lambda \in \rho(A)$, we may write

$$\lambda - A_\alpha = \{I + (A - A_\alpha)R(\lambda, A)\}(\lambda - A) \,.$$

Therefore, λ will belong to $\rho(A_\alpha)$, if

$$\|(A - A_\alpha)R(\lambda, A)\| < 1/2.$$

This estimate holds for α close enough to 1, using (1.3) and the fact that $R(\lambda, A)$ is a bounded operator from $(X, \|\cdot\|)$ into $(D(A), \|\cdot\|_{D(A)})$. Moreover, for such α's $R(\lambda, A_\alpha)$ will be uniformly bounded with respect to α. Now, (1.4) follows directly form (1.3) and the following easily checked identity :

$$R(\lambda, A_\alpha) - R(\lambda, A) = R(\lambda, A_\alpha)(A_\alpha - A)R(\lambda, A), \quad \forall \lambda \in \rho(A) \,.$$

The remainder follows from Theorems IV.2.25 and IV.3.16 of [6]. ∎

2. Stable asymptotics for differential equations in a Hilbert space.

Let us consider the following differential equation associated with A_α : for a given f_α (from $\mathbf{R}$ into X), we look for a solution u_α of

$$\frac{du_\alpha}{dt} - A_\alpha u_\alpha = f_\alpha \quad \text{on} \quad \mathbf{R}. \tag{2.1}$$

As in [12], we want to give a comparison result in different weighted Sobolev spaces and to consider the polynomial resolution (which correspond to particular right-hand sides f_α). But here, in both cases, we want to give a stable asymptotics; this means that we shall give an asymptotic of the solution which is continuous with respect to α. To solve these stabilization procedure, we follow the ideas of [2]. We firstly need to recall the notion of divided difference (see §8 of [2]) : Let $\mu_1, \cdots, \mu_K$ be arbitrary complex numbers, then the divided difference of a holomorphic function w at $\mu_1, \cdots, \mu_K$ is defined by

$$w[\mu_1, \cdots, \mu_K] = \frac{1}{2i\pi} \int_\gamma \frac{w(\lambda)}{\pi_{j=1}^K (\lambda - \mu_j)} d\lambda\,, \tag{2.2}$$

where γ is a simple curve surrounding all the μ_j's. In the same way, we set

$$\mathcal{S}[\mu_1, \cdots, \mu_K; t] = w_t[\mu_1, \cdots, \mu_K]\,, \tag{2.3}$$

when w_t is the holomorphic function $w_t : z \to e^{zt}$.

We are now able to solve (2.1) with right-hand sides defined by divided differences.

Theorem 2.1. *Let $\mu_{1\alpha}, \cdots, \mu_{K\alpha} \in \mathbf{C}$ and $f_\alpha \in X$, for all $\alpha \in I$, satisfying*

$$\begin{cases} \mu_{j\alpha} \to u_{j1} & \text{in} \quad \mathbf{C}, \quad \forall j = 1, \cdots, K\,, \\ f_\alpha \to f_1 & \text{in} \quad X, \quad \text{as} \quad \alpha \to 1\,. \end{cases} \tag{2.4}$$

Denote by γ a fixed closed curve of $\mathbf{C}$ surrounding $\mu_{11}, \cdots, \mu_{K1}$. By Theorem 1.1, there exists a neighbourhood J of 1 such that

i) int γ contains $\mu_{j\alpha}$, for all $\alpha \in J$, $j = 1, \cdots, K$.

ii) A_α has exactly k eigenvalues $\{\lambda_{j\alpha}\}_{j=1}^k$ inside γ fulfilling (1.5).

Then for all $\alpha \in J$, there exists a solution u_α (from $\mathbf{R}$ into $D(A)$) of

$$\frac{du_\alpha}{dt} - A_\alpha u_\alpha(t) = \mathcal{S}[\mu_{1\alpha}, \cdots, \mu_{K\alpha}; t] f_\alpha \quad \text{on} \quad \mathbf{R}\,, \tag{2.5}$$

which admits the following expansion :

$$\begin{aligned} u_\alpha(t) = & \sum_{j=1}^K c_{j\alpha} \mathcal{S}[\mu_{1\alpha}, \cdots, \mu_{j\alpha}; t] \\ & + \sum_{j'=1}^k d_{j'\alpha} \mathcal{S}[\mu_{1\alpha}, \cdots, \mu_{K\alpha}, \lambda_{1\alpha}, \cdots, \lambda_{j'\alpha}; t], \end{aligned} \tag{2.6}$$

where $c_{j\alpha}, d_{j'\alpha}$ are contiuous in J with values in $D(A)$, and fulfil

$$\begin{aligned} A_\alpha c_{j\alpha} \to A c_{j1} \quad & \text{in} \quad X, \quad \forall j = 1, \cdots, K, \\ A_\alpha d_{j'\alpha} \to A d_{j'1} \quad & \text{in} \quad X, \quad \forall j' = 1, \cdots, k, \quad \text{as} \quad \alpha \to 1. \end{aligned}$$

Proof : Using Lemma 1.3.4 of [5], we deduce that u_α given by

$$u_\alpha(t) = \frac{1}{2i\pi}\int_\gamma \frac{e^{zt}R(z,A_\alpha)f_\alpha}{\Pi_{j=1}^{K}(z-\mu_{j\alpha})}dz, \tag{2.7}$$

is a solution of (2.5).

Since the only poles of $R(z,A_\alpha)$ inside γ are the $\lambda_{j\alpha}$'s, for $j \in \{1,\cdots,k\}$, we see that

$$\chi(z,\alpha) = R(z,A_\alpha)\prod_{j'=1}^{k}(z-\lambda_{j'\alpha}) \tag{2.8}$$

is holomorphic in a neighbourhood of $\overline{\text{int}\gamma}$. Moreover, owing to (1.4) and (1.5), we have

$$\chi(z,\alpha) \to \chi(z,1) \quad \text{in norm, as} \quad \alpha \to 1,\ \forall z \in \gamma. \tag{2.9}$$

We now set

$$v_\alpha : z \to \chi(z,\alpha)f_\alpha \,.$$

Then (2.7) becomes

$$u_\alpha(t) = (w_t \cdot v_\alpha)[\mu_{1\alpha},\cdots,\mu_{K\alpha},\lambda_{1\alpha},\cdots,\lambda_{k\alpha}] \,.$$

By the Leibniz formula for divided difference (see Lemma 8.1 of [2]), we get (2.6), with

$$c_{j\alpha} = v_\alpha[\mu_{j\alpha},\cdots,\mu_{K\alpha},\lambda_{1\alpha},\cdots\lambda_{k\alpha}], \quad \forall j = 1,\cdots,K,$$
$$d_{j'\alpha} = v_\alpha[\lambda_{j'\alpha},\cdots,\lambda_{k\alpha}], \quad \forall j' = 1,\cdots,k.$$

The continuity results on the $c_{j\alpha}$'s and the $d_{j'\alpha}$'s can be easily deduced from (2.9), (2.4) and (1.5). ∎

This theorem allows us to give a stable polynomial resolution since in that case we only have $K = 1$ and $\mu_{1\alpha} = \mu$ independent of α (see §4 of [12] and §3 hereafter).

Let us now pass to the comparison result between two different weighted Sobolev spaces (see Definition 2.2 of [12]). According to [12], we need more assumptions on the A_α's : we suppose that there exist two positive real numbers δ and N and a neighbourhood $\mathcal{U}$ of 1 such that

$$\rho(A_\alpha) \supset \Sigma_{\delta,N} = \{\lambda \in \mathbf{C} : |\arg\lambda \pm \pi/2| \leq \delta \text{ and } |\lambda| \geq N\}, \ \forall \alpha \in \mathcal{U} . \tag{2.10}$$

Moreover, we suppose that the assumption $(H1)$ of [12] is satisfied by A_α uniformly in $\mathcal{U}$ i.e. there exists a closed subspace S of X and a constant $C > 0$ (independent of α) such that

$$\|\lambda R(\lambda, A_\alpha)f\|_X + \|R(\lambda, A_\alpha)f\|_{D(A)} \leq C \cdot \|f\|_X, \ \forall f \in S, \lambda \in \Sigma_{\delta,N}, \alpha \in \mathcal{U}. \tag{2.11}$$

Theorem 2.2. *Let $\alpha(1), \alpha(2)$ be two real numbers such that $\alpha(1) < \alpha(2)$. For $j = 1$ and 2, assume that the line $Re\lambda = -\alpha(j)$ contains no eigenvalue of A and let $f_\alpha \in L^2_{\alpha(j)}(\mathbf{R}, S)$, for all $\alpha \in \mathcal{U}$, fulfil*

$$f_\alpha \to f_1 \quad \text{in} \quad L^2_{\alpha(j)}(\mathbf{R}, S), \quad \text{as} \quad \alpha \to 1 . \tag{2.12}$$

Then there exists a neighbourhood $\mathcal{U}'$ of 1 such that for all $\alpha \in \mathcal{U}'$, (2.1) has a solution $u^{(j)}_\alpha \in H^1_{\alpha(j)}(\mathbf{R}, X) \cap L^2_{\alpha(j)}(\mathbf{R}, D(A))$; moreover their difference is given by

$$u^{(1)}_\alpha(t) - u^{(2)}_\alpha(t) = \sum_{\lambda \in Sp(A), -\alpha(2) < Re\lambda < -\alpha(1)} R_{\lambda\alpha}, \tag{2.13}$$

where for all eigenvalue λ of A of multiplicity k, we set

$$R_{\lambda\alpha} = \sum_{j=1}^{k} c_{j\alpha} \mathcal{S}[\lambda_{1\alpha}, \cdots, \lambda_{j\alpha}; t] , \tag{2.14}$$

when $\{\lambda_{j\alpha}\}_{j=1}^k$ denotes the k eigenvalues of A_α in a neighbourhood of λ satisfying (1.5). The $c_{j\alpha}$'s are continuous from $\mathcal{U}'$ with values in $D(A)$ and

$$A_\alpha c_{j\alpha} \to A c_{1\alpha} \quad \text{in} \quad X, \text{ as } \quad \alpha \to 1 .$$

Finally, there exists a constant $C_1 > 0$ (independent of α) such that

$$\|u^{(j)}_\alpha\|_{1,\alpha(j),X} + \|u^{(j)}_\alpha\|_{0,\alpha(j),D(A)} \leq C_1 \|f\|_{0,\alpha(j),S}, \quad \forall j = 1, 2. \tag{2.15}$$

Proof : From (2.10) and Theorem 1.1, we deduce that there exists a neighbourhood $\mathcal{U}'$ of 1 such that the line $Re\lambda = -\alpha(j)$ contains no eigenvalue of A_α, for all $\alpha \in \mathcal{U}'$, $j = 1$ and 2. Therefore, for $\alpha \in \mathcal{U}'$, the existence of $u_\alpha^{(j)}$ and the estimate (2.15) follow from Theorem 2.7 of [12] and from the uniform estimate (2.11). Theorem 2.7 of [12] also shows (2.13) with $R_{\lambda\alpha}$ given by (see (2.25) of [12])

$$R_{\lambda\alpha} = \frac{\sqrt{2\pi}}{2i\pi} \int_\gamma e^{tz} R(z, A_\alpha)(\mathfrak{F} f_\alpha)(-iz)dz, \tag{2.16}$$

where γ is a fixed curve surrounding $\lambda \in Sp(A)$ ((2.16) actually holds for all α in a sufficiently small neighbourhood of 1).

In order to obtain (2.14), we use analogous arguments than in Theorem 2.1; indeed using $\chi(z, \alpha)$ given by (2.8), we remark that

$$R_{\lambda\alpha} = (w_t V_\alpha)[\lambda_{1\alpha}, \cdots, \lambda_{k\alpha}] ,$$

where the function V_α is defined by

$$V_\alpha : z \to \sqrt{2\pi}\chi(z, \alpha)(\mathfrak{F} f_\alpha)(-iz) ,$$

which is a holomorphic function in a neighbourhood of $\overline{\text{int}\gamma}$ (since $(\mathfrak{F} f)(-iz)$ is analytical in the strip $-\alpha(2) < Re z < -\alpha(1)$, as shown in Theorem 2.7 of [12]). We conclude by the Leibniz formula for divided differences, the continuity of the $c_{j\alpha}$'s being a consequence of (1.4) and (2.12). ∎

Remark 2.3 : For convenience, we have supposed that the assumption (H1) of [12] is fulfilled uniformly in α; Theorem 2.2 still holds if we replace (H1) by the assumption (H2) of [12].

3. The Dirichlet problem for elliptic operators of order $2m$ with constant coefficients in infinite cones of the plane.

Let L_0 be a properly elliptic operator, homogeneous with constant coefficients in $\mathbf{R}^2$. We consider the Dirichlet problem in the cone $C_\omega = \{re^{i\theta} : r > 0, \quad 0 < \theta < \omega\}$, $\omega \in]0, 2\pi]$:

$$L_0 u = f \quad \text{in} \quad C_\omega , \tag{3.1}$$

$$u = \frac{\partial u}{\partial \nu} = \cdots = \frac{\partial^{m-1} u}{\partial \nu^{m-1}} = 0 \quad \text{on} \quad \partial C_\omega. \tag{3.2}$$

We shall reduce (3.1)-(3.2) into a differential equation in a Hilbert space in order to apply the results of §2. This is made in the two usual following steps (see [1] and [12], for instance):

a) using polar coordinates (r, θ) and the Euler change of variable $r = e^t$, (3.1)-(3.2) is equivalent to

$$\sum_{\ell=0}^{2m} A_\ell(\theta, D_\theta) D_t^\ell v = g \quad \text{in} \quad B_\omega, \tag{3.3}$$

$$v = \frac{\partial v}{\partial \nu} \cdots = \frac{\partial^{m-1} v}{\partial \nu^{m-1}} = 0 \quad \text{on} \quad \partial B_\omega, \tag{3.4}$$

where $B_\omega = \{(t, \theta) : t \in \mathbf{R}, 0 < \theta < \omega\}$ and

$$A_\ell(\theta, D_\theta) = \sum_{k=0}^{2m-\ell} a_{\ell,k}(\theta) \frac{\partial^k}{\partial \theta^k}, \tag{3.5}$$

when $a_{\ell,k}$ are infinitely differentiable functions [7]. Due to the ellipticity assumption, $a_{2m,0}(\theta)$ is different from 0 for every θ, therefore without loss of generality we may suppose that $a_{2n,0}(\theta) \equiv 1$.

b) We use the argument of reduction of order in D_t, and introduce the vectors

$$V = (v, D_t v, \cdots, D_t^{2m-1} v), \tag{3.6}$$

$$F = (0, 0, \cdots, 0, g)\ , \tag{3.7}$$

then (3.3) can be written as

$$D_t V - \mathcal{A} V = F \quad \text{in} \quad \mathbf{R}\ , \tag{3.8}$$

where

$$\mathcal{A}(u_0, u_1, \cdots, u_{2m-1}) = (u_1, u_2, \cdots, u_{2m-1}, -\sum_{\ell=0}^{2m-1} A_\ell u_\ell). \tag{3.9}$$

Since (3.3) is set on B_ω and in order to take into account the boundary conditions (3.4), we introduce

$$X_\omega = \prod_{\ell=0}^{2m-1} H^{2m-1-\ell}(]0, \omega[)\ ,$$

$$D(\mathcal{A}) = \{(v_\ell)_{\ell=0,\cdots,2m-1} \in X_\omega : v_\ell \in H^{2m-\ell}(]0,\omega[)\ , \forall \ell = 0,\cdots,2m-1 \text{ and } v_0 \in \mathring{H}^m(]0,\omega[)\}\ .$$

We have just proven that $v \in H^{2m}_{loc}(B_\omega)$ is a solution of (3.3)-(3.4) iff V given by (3.6) belongs to $D(\mathcal{A})$ and fulfils (3.8) (see Lemma 3.1 of [12]). In the same way, λ_0 is an eigenvalue of the operator $\mathcal{A}$ in X_ω iff the operator $\mathcal{L}_\omega(\lambda_0)$ defined hereafter is not invertible (as usual, we say that λ_0 is an eigenvalue of $\mathcal{L}_\omega(\lambda)$). Analogous equivalence holds for the associated Jordan chains as explained in §3 of [12].

$$\mathcal{L}_\omega(\lambda) : H^{2m}(]0,\omega[) \cap \mathring{H}^m(]0,\omega[) \to L^2(]0,\omega[: \ u \mapsto \sum_{\ell=0}^{2m} \lambda^\ell A_\ell(\theta, D_\theta)u\ .$$

The most important problem in (3.8) is that it is set in a Hilbert space X_ω depending on ω. In order to reduce it to a fixed one, we make the change of variable

$$\Psi_\alpha :]0,\omega_0[\to]0,\omega[: \theta \to \alpha\theta\ ,$$

when $\alpha = \omega/\omega_0$, ω_0 being supposed to be a critical angle. This change of variable induces an isomorphism between X_ω and X_{ω_0} ; moreover the differential equation (3.8) set in X_ω becomes

$$D_t V_\alpha - \mathcal{A}_\alpha V_\alpha = F_\alpha \quad \text{in} \quad X_{\omega_0}\ , \tag{3.10}$$

where $\mathcal{A}_\alpha$ is given by (3.9) replacing θ by $\alpha\theta$ (in the operators A_ℓ) and $F_\alpha = F \circ \Psi_\alpha$.

Let us check that the operators $\mathcal{A}_\alpha$ satisfy the assumptions of §2: (1.1) and (1.2) are clearly satisfied, (1.3) is considered in the

Lemma 3.1. *There exists a neighbourhood I of 1 and a constant $C > 0$ such that*

$$\|(\mathcal{A}_\alpha - \mathcal{A})u\|_{X_{\omega_0}} \le C|1-\alpha|\|V\|_{D(\mathcal{A})}, \quad \forall V \in D(\mathcal{A})\ .$$

Proof : Fix $V \in D(\mathcal{A})$, from the definition of the $\mathcal{A}_\alpha$'s, we have

$$\begin{aligned}\|(\mathcal{A}_\alpha - \mathcal{A})V\|_{X_{\omega_0}} &= \|\sum_{\ell=0}^{2m-1} (A_{\ell\alpha} - A_\ell)v_\ell\|_{L^2(]0,\omega_0[)} \\ &\le \sum_{\ell=0}^{2m-1} c_\ell(\alpha)\|v_\ell\|_{H^{2m-\ell}(]0,\omega_0[)}\ ,\end{aligned}$$

where we have set

$$c_\ell(\alpha) = \sum_{k=0}^{2m-\ell} \sup_{0\le\theta\le\omega_0} |a_{\ell k}(\alpha\theta)/\alpha^k - a_{\ell k}(\theta)|.$$

Due to the smooth properties of the $a_{\ell k}$'s, we have

$$c_\ell(\alpha) \le C \cdot |1-\alpha| .$$

Therefore, it remains to show that

$$\sum_{\ell=0}^{2m-1} \|v_\ell\|_{H^{2m-\ell}(]0,\omega_0[)} \le C \cdot \|V\|_{D(\mathcal{A})} .$$

This last one follows from the ellipticity of A_0, since Agmon-Douglis-Nirenberg a priori estimates imply that

$$\|v_0\|_{H^{2m}(]0,\omega_0[)} \le C \cdot \{\|A_0 v_0\|_{L^2(]0,\omega_0[)} + \|v_0\|_{L^2(]0,\omega_0[)}\}.$$

■

In the same way, using the estimate (3.14) of [12] fulfilled by $\mathcal{A}$ in X_{ω_0} and a perturbation argument, we can show that the $\mathcal{A}_\alpha$'s satisfy (2.11) uniformly in a neighbourhood of 1, with $S = \{F \in X_{\omega_0}$ in the form (3.7)$\}$.

We are now ready to give a stable decomposition for a weak solution of (3.1)-(3.2). Applying Theorems 2.1 and 2.2 to problems (3.10) and going back to the original problem, we have the (see Theorem 5.11 and Lemma 10.4 of [3])

Theorem 3.2. *Let $\omega_0 \in]0, 2\pi]$ and $\ell \in \mathbb{N} \cup \{0\}$ be fixed. Assume that the line $Re\lambda = \ell + 2m - 1$ contains no eigenvalue of $\mathcal{L}_{\omega_0}(\lambda)$. Suppose given f_ω in $H^\ell(C_\omega)$, for all ω in a neighbourhood $\mathcal{U}$ of ω_0, satisfying*

$$\|f_\omega \circ \Psi_{\omega/\omega_0} - f_{\omega_0}\|_{H^\ell(C_{\omega_0})} \to 0, \quad \text{as} \quad \omega \to \omega_0 .$$

Then there exists a neighbourhood $\mathcal{U}'$ of ω_0 such that for all $\omega \in \mathcal{U}'$ we have the following results : if $u_\omega \in \mathring{H}^m(C_\omega)$ is a solution of (3.1)-(3.2) with data f_ω, then

$$u_\omega = u_{0\omega} + \Sigma_\lambda R_{\lambda\omega} + \Sigma_n S_{n\omega} , \tag{3.11}$$

where $u_{0\omega} \in H^{2m+\ell}(C_\omega)$, the first sum extends to all eigenvalues λ of $\mathcal{L}_{\omega_0}(\lambda)$ in the strip $Re\lambda \in]m-1, \ell+2m-1[$, if λ is such an eigenvalue of multiplicity k, *according to Theorem 1.1, there exist k eigenvalues $\lambda_{1\omega}, \cdots, \lambda_{k\omega}$ of $\mathcal{L}_\omega(\lambda)$ such that $\lambda_{j\omega} \to \lambda$, as $\omega \to \omega_0$, for all $j = 1, \cdots, k$; and*

$$R_{\lambda\omega} = \sum_{j=1}^{k} c_{j\omega} \mathcal{S}[\lambda_{1\omega}, \cdots, \lambda_{j\omega}; \ell nr]. \tag{3.12}$$

The second sum extends to all nonnegative integer $n \in [0, k-1]$ and

$$S_{n\omega} = k_{n\omega} \mathcal{S}[n + 2m; \ell nr], \tag{3.13}$$

if $n + 2m$ is not an eigenvalue of $\mathcal{L}_{\omega_0}(\lambda)$, otherwise if $n + 2m$ is an eingenvalue of $\mathcal{L}_{\omega_0}(\lambda)$ of multiplicity k', then denoting by $\mu_{1\omega}, \cdots, \mu_{k'\omega}$ the k' eigenvalues of $\mathcal{L}_\omega(\lambda)$ satisfying $\mu_{j'\omega} \to n + 2m$ as $\omega \to \omega_0$, for all $j' = 1, \cdots, k'$, we have

$$S_{n\omega} = k_{n\omega} \mathcal{S}[n + 2m; \ell nr] + \sum_{j'=1}^{k'} k_{nj'\omega} \mathcal{S}[n + 2m, \mu_{1\omega}, \cdots, \mu_{j'\omega}; \ell nr]. \tag{3.14}$$

In the expressions (3.12), (3.13) and (3.14), the coefficients $c_{j\omega}$, $k_{n\omega}$ and $k_{nj'\omega}$ belong to $H^{2m}(]0, \omega[)$ and depend continously on ω in the following sense :

$$\|c_{j\omega} \circ \Psi_{\omega/\omega_0} - c_{j\omega_0}\|_{H^{2m}(]0,\omega_0[)} \to 0, \quad \text{as} \quad \omega \to \omega_0$$

and analogously for $k_{n\omega}$ and $k_{nj'\omega}$.

Remark 3.3 : For convenience, we have treated here the Dirichlet problem for an elliptic operator in dimension 2. Nevertheless, it is possible to consider other boundary conditions and higher dimensions (for instance, for rotationally symmetric cones of $\mathbf{R}^3$ as considered in [13] for the Lamé system).

In our example, if the functions $a_{\ell,k}$ would be analytic functions (which is the case for the usual example as the Laplace operator or the biharmonic one), the family $\mathcal{A}_\alpha$ would be a holomorphic family of type (A) in the sense of Kato (see §VII. 2 of [6]); in that case, Lemma 3.1 and Theorem 1.1 are in accordance with Theorem VII.1.8 of [6]. Let us notice that we cannot hope an analytical dependence if we apply a non-analytic change of variable Ψ_α (as considered by [9] in non-regular cone of $\mathbf{R}^n$ or for transmission problems, where two (or more) parameters appear).

4. The Dirichlet problem for elliptic differential equations of order $2m$ in domains with conical points.

The aim of this paragraph is to extend the results of Maz'ya and Rossmann [11] concerning second order operators to operators of order $2m$. We follow the method of §3 of [11] with the necessarry adaptations, using firstly the results of §3 and secondly an interative procedure.

Let L be a properly elliptic operator of order $2m$ ($m \in \mathbf{N}$) with smooth coefficients in $\mathbf{R}^2$. As usual, we denote by L_0 its principal part frozen at 0.

We recall the weighted Sobolev spaces we need : for $\ell \in \mathbf{N} \cup \{0\}$ and $\beta \in \mathbf{R}$, we define the Hilbert spaces :

$$V_\beta^\ell(C_\omega) = \{u \in \mathcal{D}'(C_\omega) : r^{\beta-\ell+|\alpha|}D^\alpha u \in L^2(C_\omega), \quad \forall |\alpha| \leq \ell\},$$
$$W_\beta^\ell(C_\omega) = \{u \in \mathcal{D}'(C_\omega) : r^{\beta}D^\alpha u \in L^2(C_\omega), \quad \forall |\alpha| \leq \ell\}.$$

If C'_ω denotes $C_\omega \cap B(0,1)$, we define analogously $V_\beta^\ell(C'_\omega)$ and $W_\beta^\ell(C'_\omega)$.

Lemma 4.1. *Let $u_\omega \in \mathring{H}^m(C_\omega)$ be a solution of*

$$Lu_\omega = f_\omega \quad \text{in} \quad C_\omega, \tag{4.1}$$

where $f_\omega \in W_{-\varepsilon}^\ell(C_\omega)$, satisfying

$$\|u_\omega\|_{H^m(C_\omega)} + \|f_\omega\|_{W_{-\varepsilon}^\ell(C_\omega)} \leq M, \tag{4.2}$$

for some $M > 0$ (independent of ω), for all ω in a neighbourhood of a fixed $\omega_0 \in]0, 2\pi]$.

If $\varepsilon \in]0, 1/2[$, then

$$u_\omega \in V_{m-\varepsilon}^{2m}(C'_\omega), \tag{4.3}$$

and there exists a constant C and a neighbourhood $\mathcal{U}'$ of ω_0 such that

$$\|u_\omega\|_{V_{m-\varepsilon}^{2m}(C'_\omega)} \leq C, \quad \forall \omega \in \mathcal{U}'\,. \tag{4.4}$$

Proof : Since $W_{-\varepsilon}^\ell(C'_\omega) \hookrightarrow V_{m-\varepsilon}^0(C'_\omega)$, u_ω may be seen as a solution of (4.1) with f_ω in $V_{m-\varepsilon}^0(C'_\omega)$. But in [8], it is proved that the strip $Re\lambda \in [m-3/2, m-1/2]$ has no eigenvalue of $\mathcal{L}_\omega(\lambda)$, for all ω ; therefore if $\varepsilon \in]0, 1/2[$, the strip $Re\lambda \in [m-1, m-1+\varepsilon]$ has no eigenvalue of $\mathcal{L}_\omega(\lambda)$ too. Owing to Theorem 4.19 of [4], we deduce (4.3). Finally, the estimate (4.4) follows from the smoothness of the coefficients of L. ∎

In the sequel, for a fixed $\ell \in \mathbf{N} \cup \{0\}$, we shall say that an opening $\omega_0 \in]0, 2\pi]$ is critical if one of the following holds :

i) There exists an eigenvalue $\lambda_0 \in [m-1, \ell+2m-1]$ of $\mathcal{L}_{\omega_0}(\lambda)$ of multiplicity ≥ 2 such that for all neighbourhood $\mathcal{V}$ of λ_0, there exists a neighbourhood of ω_0 such that $\mathcal{L}_\omega(\lambda)$ has more than two different eigenvalues in $\mathcal{V}$ (called branching point in [2]).

ii) There exist two different eigenvalues λ_1, λ_2 in the strip $Re\lambda \in [m-1, \ell+2m-1]$ of $\mathcal{L}_{\omega_0}(\lambda)$ and two non-negative intergers k_1, k_2 such that $\lambda_1 + k_1 = \lambda_2 + k_2$ and $Re\lambda_1 + k_1 \in [m-1, \ell+2m-1]$ (called crossing point in [2]).

iii) $\mathcal{L}_{\omega_0}(\lambda)$ has an eigenvalue in $[m-1, \ell+2m-1]$ which is an integer (also called crossing point in [2]).

Let us now fix a critical opening ω_0. From the properties of the spectrum of $\mathcal{L}_{\omega_0}(\lambda)$, it is clear that we can fix $\varepsilon_1 > 0$ such that

a) The strip $Re\lambda \in]j, j+\varepsilon_1]$ has no eigenvalue of $\mathcal{L}_{\omega_0}(\lambda)$, for all $j = m-1, \cdots, \ell+2m-1$.

b) Due to the point a above and Theorem 1.1, for all $\varepsilon' \in]0, \varepsilon_1[$, there exists a neighbourhood $\mathcal{U}_{\varepsilon'}$ of ω_0 such that the line $Re\lambda = j+\varepsilon'$ has no eigenvalue of $\mathcal{L}_\omega(\lambda)$, for all $\omega \in \mathcal{U}_{\varepsilon'}$ and all $j = m-1, \cdots, \ell+2m-1$. Moreover, we may suppose that the number of eigenvalues of $\mathcal{L}_\omega(\lambda)$ in the strip $Re\lambda \in]j+\varepsilon', j+1+\varepsilon'[$ is constant for all ω in $\mathcal{U}_{\varepsilon'}$, $j = m-1, \cdots, \ell+2m-2$. Accordingly, for $\omega \in \mathcal{U}_{\varepsilon'}$, we denote by $\{\lambda_{i\omega}\}_{i=1}^{I}$ the set of eigenvalues of $\mathcal{L}_\omega(\lambda)$ in the strip $Re\lambda \in]m-1, \ell+2m-1+\varepsilon'[$ fulfilling

$$\lambda_{i\omega} \to \lambda_{i\omega_0} \quad \text{as} \quad \omega \to \omega_0, \quad \forall i = 1, \cdots, I.$$

c) For all $\varepsilon' \in]0, \varepsilon_1[$ and a, b two integers satisfying $m-1 \leq a \leq b \leq \ell+2m-1$, we set

$$\Lambda_{a,b} = \{(i,j) \in \{0, 1, \cdots, I\} \times (\mathbf{N} \cup \{0\}) \text{ such that } a+\varepsilon' < Re\lambda_{i\omega_0} + j < b+\varepsilon'\},$$

with the agreement $\lambda_{0\omega} = 0$. Finally, for $\beta \in \mathcal{P}(\Lambda_{a,b})$ (the set of nonempty subset of $\Lambda_{a,b}$), if

$$\beta = \{(i_1, j_1), (i_2, j_2), \cdots, (i_k, j_k)\},$$

then we define

$$S_{\beta\omega}(r) = \mathcal{S}[\lambda_{i_1\omega} + j_1, \cdots, \lambda_{i_k\omega} + j_k; \ell n r];$$

for an arbitrary integer ℓ, we also set

$$\beta - \ell = \{(i_1, j_1 - \ell), \cdots, (i_k, j_k - \ell)\} .$$

We are now ready to state the (extension of Lemma 9 of [11])

Lemma 4.2. *For some $\varepsilon > 0$, let $u_\omega \in V^{j+2m}_{k-\varepsilon+1}(C'_\omega) \cap \mathring{H}^m(C_\omega)$ be a solution of (4.1) with $f_\omega \in V^j_{k-\varepsilon}(C'_\omega)$ depending continuously on ω, where j, k are two integers such that $j \geq 0$ and $-m \leq j - k \leq \ell$. We suppose that*

$$\|u_\omega\|_{V^{2m}_{k-\varepsilon+1-j}(C'_\omega)} + \|f_\omega\|_{V^j_{k-\varepsilon}(C'_\omega)} \leq M , \tag{4.5}$$

for some $M > 0$, and all ω in a neighbourhood of ω_0. Then there exists $0 < \varepsilon' < \min(\varepsilon, \varepsilon_1)$ such that for all $\omega \in \mathcal{U}_{\varepsilon'}$, we have

$$u_\omega = u_{0\omega} + \sum_{\beta \in \Lambda_{\ell+2m-2-k, \ell+2m-1-k}} c_{\beta\omega}(\theta) S_{\beta\omega}(r) , \tag{4.6}$$

where $u_{0\omega} \in V^{j+2m}_{k-\varepsilon'}(C'_\omega)$, and $c_{\beta\omega} \in H^{2m}(]0, \omega[)$ depends continuously on ω. Moreover, there exists $M_1 > 0$ such that

$$\|u_{0\omega}\|_{V^{2m}_{k-j-\varepsilon'}(C'_\omega)} \leq M_1 . \tag{4.7}$$

Proof : We may write

$$L_0 u_\omega = f_\omega + (L_0 - L) u_\omega .$$

From the Taylor theorem with an integral remainder, we know that

$$L_0 - L : V^{j+2m}_{k-\varepsilon+1}(C'_\omega) \to V^j_{k-\varepsilon}(C'_\omega)$$

continuously (with a norm uniformly bounded with respect to ω). From the assumptions, we deduce that $L_0 u_\omega \in V^j_{k-\varepsilon}(C'_\omega)$. Owing to Theorem 1.2 of [7] and the results of §2, u_ω admits the decomposition (4.6) (in fact, the singular part contains only the β's of the form $\{(i_1, 0), (i_2, 0), \cdots, (i_k, 0)\}$ so that $L_0(u_\omega - u_{0\omega}) = 0$).

The estimate (4.7) follows from (4.5) and the uniform estimate of $\mathcal{L}_\omega(\lambda)^{-1}$, for all ω in $\mathcal{U}_{\varepsilon'}$ (see (2.15)). ■

If $m = 1$, the previous lemma suffices to prove the general result; for $m \geq 2$, we need another technical lemma.

Lemma 4.3. *For some $\varepsilon > 0$, let $u_\omega \in V^{p+2m}_{m-\varepsilon}(C'_\omega) \cap \mathring{H}^m(C_\omega)$ be a solution of (4.1) with $f_\omega \in V^p_{-\varepsilon}(C_\omega)$ depending continuously on ω, where $p \in \{0, 1, \cdots, \ell\}$.*

Suppose that there exists $M > 0$ such that

$$\|u_\omega\|_{V^{2m}_{m-\varepsilon-p}(C'_\omega)} + \|f_\omega\|_{V^p_{-\varepsilon}(C_\omega)} \leq M ,$$

for all ω in a neighbourhood of ω_0. Then there exists $0 < \varepsilon' < \max(\varepsilon, \varepsilon_1)$ such that for all $q = 1, \cdots, m$, we have

$$u_\omega = u_{\omega q} + \Sigma_{\omega q}, \tag{4.8}$$

where $u_{\omega q} \in V^{p+2m}_{m-q-\varepsilon'}(C'_\omega)$ satisfying

$$\|u_{\omega q}\|_{V^{2m}_{m-q-\varepsilon'-p}(C'_\omega)} \leq M_1 , \tag{4.9}$$

$$\Sigma_{\omega q} = \sum_{\beta \in \mathcal{P}(\Lambda_{p+m-1,p+m+q-1})} c_{\beta\omega}(\theta) S_{\beta\omega}(r) , \tag{4.10}$$

$$L\Sigma_{\omega q} = \sum_{i=0}^{1} \sum_{\beta \in \mathcal{P}(\Lambda_{p+m+q-2+i,p+m+q-1+i})} \sum_{|\gamma| \leq 2m-1+i} a^i_{\omega\beta\gamma}(x) S_{\beta-|\gamma|,\omega}(r) , \tag{4.11}$$

where $c_{\beta\omega} \in H^{2m}(]0, \omega[)$, $a^i_{\omega\beta\gamma} \in C^\infty(\bar{C}'_\omega)$ depend continuously on ω.

Proof : We prove (4.8)-(4.11) by recurrence on q. For $q = 1$, this is a direct consequence of Lemma 4.2 taking $j = p$ and $k = m - 1$ since

$$V^p_{-\varepsilon}(C'_\omega) \hookrightarrow V^p_{m-1-\varepsilon}(C'_\omega) .$$

We now show that if (4.8)-(4.11) hold for q, then they hold for $q + 1 \leq m$. We define $T_{\omega q}$ as in (4.11) replacing $a^i_{\omega\beta\gamma}(x)$ by $a^i_{\omega\beta\gamma}(0)$ and summing only on $|\gamma| = 2m - 1 + i$. The Taylor theorem allows us to say that

$$T'_{\omega q} = L\Sigma_{\omega q} - T_{\omega q} \in V^p_{m-\varepsilon-q-1}(C'_\omega) . \tag{4.12}$$

By Theorem 2.1, we can compute explicitly the solution $w_{\omega q}$ of

$$L_0 w_{\omega q} = -T_{\omega q} ,$$

which admits the following expansion :

$$w_{\omega q} = \sum_{\beta\in\mathcal{P}(\Lambda_{p+m+q-1,p+m+q})} c_{\omega\beta}(\theta)S_{\omega\beta}(r) , \tag{4.13}$$

with $c_{\omega\beta} \in H^{2m}(]0,\omega[)$ depends continuously on ω.

Due to (4.13) and the recurrence hypotheses, we see that $u_{\omega q} - w_{\omega q} \in V^{p+2m}_{m-q-\varepsilon'}(C'_\omega)$ fulfils

$$L(u_{\omega q} - w_{\omega q}) = f_\omega - T'_{\omega q} - (L - L_0)w_{\omega q} \in V^{p}_{m-q-1-\varepsilon'}(C'_\omega) .$$

Applying Lemma 4.2 to $u_{\omega q} - w_{\omega q}$ (with $j = p$ and $k = m - q - 1$), we deduce that

$$u_{\omega q} - w_{\omega q} = u_{\omega q+1} + \sum_{\beta\in\mathcal{P}(\Lambda_{p+m+q-1,p+m+q})} d_{\beta\omega}(\theta)S_{\beta\omega}(r) , \tag{4.14}$$

with $u_{\omega q+1} \in V^{p+2m}_{m-q-1-\varepsilon'}(C'_\omega)$ and $d_{\beta\omega} \in H^{2m}(]0,\omega[)$ depends continuously on ω. Replacing $u_{\omega q}$ by the expansion (4.13) in the decomposition (4.8) of u_ω, we obtain the desired expansion of u_ω for $q + 1$. The technical assumption (4.11) for $\Sigma_{\omega q+1}$ is long but easy to check.

This proves the iterative procedure. ∎

We are now able to give the stable asymptotics for the weak solution of (4.1).

Theorem 4.4. *For some $\varepsilon > 0$, let $f_\omega \in W^{\ell}_{-\varepsilon}(C_\omega)$ depending continuously on ω, let $u_\omega \in \mathring{H}^m(C_\omega)$ be a solution of (4.1) fulfilling (4.2).*

Then there exist $0 < \varepsilon' < \max(\varepsilon, \varepsilon_1)$, $M' > 0$ and a neighbourhood $\mathcal{V}$ of ω_0 such that for all $\omega \in \mathcal{V}$, $j \in \{0, 1, \cdots, \ell\}$, u_ω admits the following decomposition :

$$u_\omega = u_{\omega j} + \Sigma_{\omega j} \tag{4.16}$$

with $u_{\omega j} \in V^{j+2m}_{-\varepsilon'}(C'_\omega)$ fulfils

$$\|u_{\omega j}\|_{V^{2m}_{-j-\varepsilon'}(C'_\omega)} \leq M' , \tag{4.17}$$

$$\Sigma_{\omega j} = \sum_{\beta\in\mathcal{P}(\Lambda_{m-1,j+2m-1})} c_{\beta\omega}(\theta)S_{\beta\omega}(r) , \tag{4.18}$$

with $c_{\beta\omega} \in H^{2m}(]0,\omega[)$ *depends continuously on* ω.

Proof : As Lemma 4.3, we prove this result by recurrence on j. For $j = 0$, since $f \in V^0_{-\varepsilon}(C_\omega)$, applying Lemma 4.1, we deduce that $u_\omega \in V^{2m}_{m-\varepsilon}(C'_\omega)$. So Lemma 4.3 for $p = 0$ gives the result.

The proof of the iterative procedure is analogous to the later one. The only difference is that one has to consider $L\Sigma_{\omega j} - P_j(f)$ instead of $L\Sigma_{\omega j}$, where $P_j(f)$ is the Taylor expansion of order $j-1$ of f at 0. ■

REFERENCES

[1] S. AGMON and L. NIRENBERG, Properties of solutions of ordinary differential equations in Banach space, *Comm. Pure Appl. Math., 16 (1963), 121-239.*

[2] M. COSTABEL and M. DAUGE, General edge asymptotics of solutions of second order elliptic boundary value problems I and II, *Preprint, Publications du Laboratoire d'Analyse Numérique, Université de Paris VI (1991).*

[3] M. DAUGE, Elliptic boundary value problem on corner domains - Smoothness and asymptotics of solutions, *Lecture Notes in Math., Vol. 1341, Springer-Verlag, New-York - Berlin (1988).*

[4] M. DAUGE, S. NICAISE, M. BOURLARD and M.S. LUBUMA, Coefficients des singularités pour des problèmes aux limites elliptiques sur un domaine à singularités coniques I, *RAIRO Modél. Math. Anal. Num., 24 (1990), 27-52.*

[5] H.O. FATTORINI, The Cauchy problem, *Encyclopedia of Math. and its appl., 18, Addison-Wesley Publ. Comp., 1983.*

[6] T. KATO, Perturbation theory for linear operators, *Springer-Verlag, New-York, 1966.*

[7] V.A. KONDRATIEV, Boundary value problems for elliptic equations in domains with conical or angular points, *Trans. Moscow Math. Soc., 16 (1967), 227-313.*

[8] V.A. KOZLOV and V.G. MAZ'YA, Spectral properties of the operators bundles generated by elliptic boundary value problems in a cone, *Functional Analysis and its applications, 22 (1988), 114-121.*

[9] V.A. KOZLOV and J. ROSSMANN, On the behaviour of the spectrum of parameter-depending operators under small variation of the domain and application to operator pencils generated by elliptic boundary value problems in a cone, *Math. Nachr., 153 (1991), 123-129.*

[10] V.G. MAZ'YA and B.A. PLAMENEVSKII, Estimates in L_p and in Hölder classes and the Miranda-Agmon maximum principle for solutions of elliptic boundary value problems in domains with singular points on the boundary, *Amer. Math. Soc. Transl (2), 123 (1984), 1-56.*

[11] V.G. MAZ'YA and J. ROSSMANN, On a problem of Babuska (Stable asymptotics of the solution of the Dirichlet problem for elliptic equations of second order in domains with angular points), *Preprint, University of Linköping (1990).*

[12] S. NICAISE, Differential equations in Hilbert spaces and applications to boundary value problems in nonsmooth domains, *Journal of Functional Analysis, 95 (1991), 195-218.*

[13] A.M. SÄNDIG, Coefficient formulae for asymptotic expansions of solutions of elliptic boundary value problems near conical boundary points, *Z. für Analysis und ihne Anwendungen, 1992, to appear.*

Skeel's Condition Number for Operators in $C^0(\Omega)$ and Application to the Regularized Cogenerator Equation of a Submarkovian Resolvent Family

LUC PAQUET University of Mons, Mons, Belgium

1. INTRODUCTION

Around 1979, a new condition number for linear systems was introduced by Skeel [12] in Numerical Analysis. Let us consider a real or complex invertible matrix **A** with N rows and N columns and a right member **b**, a non-zero vector with N real or complex components. Let x denotes the exact solution of the linear system **Ax=b**. The purpose of condition numbers is to measure how fast vary the solution when we perturb the datas A and b. The importance in Numerical Analysis of condition numbers comes from the fact that by backward error analysis the computed solution in finite precision arithmetic is showned to be the exact solution of a nearby problem, nearby being linked to the machine precision. It is then important to know how much the exact solution of that perturbed equation may differ from the exact solution of the original problem. Roughly speaking, the condition number is the limit of the ratio of the relative variation of the solution to the relative variation of the datas when the size of the perturbation goes to zero. The novelity of Skeel's theory is in the way to measure the relative variation of the datas: instead of using norms he uses modulus. The meaning of $|\delta A|\le\varepsilon|A|$ and $|\delta b|\le\varepsilon|b|$ for some small positive number ε is of course completely different than $\|\delta A\|\le\varepsilon\|A\|$ and $\|\delta b\|\le\varepsilon\|b\|$ where $\|.\|$ unless otherwise stated denotes the supremum norm. It means that for every $1\le i,j\le N$ we must have $|\delta a_{i,j}|\le\varepsilon|a_{i,j}|$ and for every $1\le i\le N$: $|\delta b_i|\le\varepsilon|b_i|$ i.e. that every component of δA and δb must be small in modulus with respect to the corresponding component of A or respectively b if $\varepsilon>0$ is small. If norms were used instead, it would simply means that $|\delta a_{i,j}|$ must be roughly speaking small with respect to the largest elements in modulus of A and that $|\delta b_i|$ must be small with respect to the biggest components of b. Thus the perturbations satisfying the inequalities $|\delta A| \leq \varepsilon |A|$ and $|\delta b| \leq \varepsilon |b|$ are quite different than those satisfying $\|\delta A\| \leq \varepsilon \|A\|$ and $\|\delta b\| \leq \varepsilon \|b\|$. Important also is that Skeel's condition number is really the condition number of the whole linear system and not only of the matrix A; its value depends on the second member b.

In the first part of the present work we generalize Skeel's theory to equations $Au=f\neq 0$ in $C_0(\Omega)$, the vector space of all continuous functions on the locally compact metrisable and separable space Ω, vanishing at infinity. In this equation, A is a linear bounded invertible opertor in $C_0(\Omega)$ such that $|A|$ sends $C_0(\Omega)$ into $C_0(\Omega)$. The particular case of matrix equations corresponds to the compact space $\Omega = \{1,2,3,...,N\}$ endowed with the discrete topology. We also extend to our general setting, the "simplifided Skeel's condition number" $K_{S,S}(A,f)$ which is essentially the same thing as the Skeel's condition number except that in this case, only the operator A is perturbed and not the second member f. The simplified Skeel's condition number is always smaller than the Skeel's condition number, this last number being itself smaller or equal to two times the simplified Skeel's condition number. The supremum over all f in $C_0(\Omega)\backslash\{0\}$ of $K_{S,S}(A,f)$ is called the Bauer-Skeel's condition number $K_{B,S}(A)$ [13] and is equal to $\| |A^{-1}||A| \|$. Of course, this last number depends only of A.

In the second part of this paper, we apply our previous extension of Skeel's theory to the "regularized" equation:

$$(V+\varepsilon I)\, u_\varepsilon = f, \qquad (1,1)$$

where V is the cogenerator of a Feller semi-group $(P_t)_{t\geq 0}$ and is supposed to be a bounded operator. Let $\{R_\lambda\}_{\lambda\geq 0}$ $(R_0=V)$ be the resolvent family associated to the semi-group $(P_t)_{t\geq 0}$. We will suppose that the points of Ω are of measure zero with respect to the measures $(\sigma_{\lambda,x})_{x\in\Omega}$ associated to each R_λ, for all $\lambda\geq 0$. Under this hypothesis, we study the behaviour of $K_{B,S}(V+\varepsilon I)$ as $\varepsilon \to 0^+$ and of the simplified Skeel condition number $K_{S,S}(V+\varepsilon I,f)$ as $\varepsilon \to 0^+$ as well as the dependance of this latest with respect to the second member $f\neq 0$. For $K_{B,S}(V+\varepsilon I)$ we get the simple formula : $K_{B,S}(V+\varepsilon I)= 1+2\varepsilon^{-1}\|V\|$ which shows that $K_{B,S}(V+\varepsilon I)$ blows up to infinity like $2\varepsilon^{-1}\|V\|$ as $\varepsilon \to 0^+$. This implies that it can not be worse for $K_{S,S}(V+\varepsilon I,f)$, but we would like to get a more precise idea of the behaviour of $K_{S,S}(V+\varepsilon I,f)$ as $\varepsilon \to 0^+$ exhibiting how it depends of the second member f. Restricting us to a second member f in the domain of the generator $\mathbf{G}$ of the Feller semi-group $(P_t)_{t\geq 0}$ (this restriction is natural as $D(\mathbf{G})=R(V)$), we get that $K_{S,S}(V+\varepsilon I,f) \geq (2/\varepsilon)(1+\alpha(\varepsilon))(\|f\|/\|\mathbf{G}f\|)$ where $\alpha(\varepsilon) \longrightarrow 0$ as $\varepsilon \longrightarrow 0^+$. If we suppose moreover that f is a surmedian function i.e. $\lambda R_\lambda f \leq f$ for $\lambda > 0$ or equivalentely $\mathbf{G}f \leq 0$, then we can prove a kind of reciprocal inequality : $K_{S,S}(V+\varepsilon I,f) \leq 1+(2/\varepsilon)(1+\beta(\varepsilon))(\|f\|/\|\mathbf{G}f\|)$ where $\beta(\varepsilon) \longrightarrow 0$ as $\varepsilon \longrightarrow 0^+$. Thus the coefficient of the term in $1/\varepsilon$ is given by $2(\|f\|/\|\mathbf{G}f\|)$ which will be a damping coefficient if $\|\mathbf{G}f\|$ is large with respect to $\|f\|$.

Finally, we prove a result which implies in particular that our hypothesis, that the points of Ω are of measure zero with respect to the measures $(\sigma_{\lambda,x})_{x\in\Omega}$ associated to R_λ, for all $\lambda \geq 0$, is satisfied, if $(P_t)_{t\geq 0}$ is the semi-group generated in $C_0(\Omega)$ by a second order real elliptic operator L whose coefficients are only continuous on $\mathbf{R}^N$, $L1 \leq 0$, and Ω is a bounded open subset of $\mathbf{R}^N$ which is quasi-regular at infinity with respect to $\mathbf{1-L}$ where $\mathbf{L}$ is the local operator associated to L [7],[11].

We will give no proofs for the first part of this work which should appear elsewhere concentrating on the second part which I feel better to be in the scope of these proceedings.

2. EXTENSION OF SKEEL'S THEORY.

Let us first precise some notations. By Ω, unless ortherwise specified, we will denote an arbitrary locally compact space, metrisable and separable. $C_0(\Omega)$ will denote the Banach space of all continuous complex-valued functions on Ω which vanish at infinity endowed with the supremum norm. That the continuous function f on Ω vanish at infinity means that for every $\varepsilon > 0$, there exists a compact K_ε of Ω such that out of K_ε, $|f|$ is smaller than ε. $\|\cdot\|$ unless ortherwise specified, will always denote the supremum norm and when applied to an operator the corresponding subordinated norm. $B_b(\Omega)$ will denote the Banach space of all complex-valued bounded Borel measurable functions on Ω endowed with the supremum norm. $B(C_0(\Omega))$ will denote the Banach space of all linear bounded operators on $C_0(\Omega)$, endowed with the usual norm operator subordinated to the supremum norm. Also to be concise, when we speak of a *measure* on Ω, *it will always mean a Borel regular measure on* Ω *of totally finite variation.*

DEFINITION 2.1. Let A be a bounded linear operator on $C_0(\Omega)$. By $|A|$ we mean the linear bounded operator in $B_b(\Omega)$ defined by $(|A|f)(x) = \int f(y)d|\mu_x|(y)$, $\forall x \in \Omega$, $\forall f \in B_b(\Omega)$, where μ_x denotes the measure associated by the Riesz representation theorem to the linear continuous form on $C_0(\Omega)$ which sends f onto Af(x).

DEFINITION 2.2. Let A be a linear bounded invertible operator in $C_0(\Omega)$ and f a non- zero function in $C_0(\Omega)$. Let u be the solution of Au=f. We will call, Skeel condition number of the linear equation Au=f, the number, if it exists, $K_S(A,f)$=

$$\lim_{\varepsilon\to 0^+} \sup\left\{ \frac{\|\delta u\|/\|u\|}{\varepsilon} ;\ \delta A\in B(C_0(\Omega)),\ |\delta A|\le\varepsilon\ |A|, \delta f\in C_0(\Omega),\ |\delta f|\le\varepsilon\,|f|,\ (A+\delta A)(u+\delta u)=f+\delta f \right\}.$$

(2,2,1)

Let us observe that the condition $|\delta A|\leq\varepsilon|A|$ is equivalent to $|\delta\mu_x|\leq\varepsilon|\mu_x|$, where $(\delta\mu_x)_{x\in\Omega}$ (resp. $(\mu_x)_{x\in\Omega}$) denotes the family of measures associated by the Riesz representation theorem to δA (resp.A).

A priori, it is not clear, whether or not the limit exists in (2,2,1), but under the hypothesis that |A| sends $C_0(\Omega)$ into $C_0(\Omega)$, we get the following extension of Skeel's theory [12]:

THEOREM 2.3. Let A be a linear bounded invertible operator in $C_0(\Omega)$ such that |A| sends $C_0(\Omega)$ into $C_0(\Omega)$. Let f be a non-zero element of $C_0(\Omega)$ and u the solution of Au = f. Then the limit in (2,2,1) exists and moreover:

$$\mathbf{K_S(A,f)} = \frac{\left\| |\mathbf{A}^{-1}|\ |\mathbf{A}|\ |\mathbf{u}| + |\mathbf{A}^{-1}|\ |\mathbf{f}| \right\|}{\|\mathbf{u}\|}.$$

(2,3,1)

A variant of the previous theory consists in only perturbing the operator A and not the second member:

DEFINITION 2.4. Let A be a linear bounded invertible operator in $C_0(\Omega)$ and f a non- zero function in $C_0(\Omega)$. Let u be the solution of Au=f. We will call, simplified Skeel condition number of the linear equation Au=f, the number, if it exists, $K_{S,S}(A,f)$=

$$\lim_{\varepsilon\to 0^+} \sup\left\{ \frac{\|\delta u\|/\|u\|}{\varepsilon} ;\ \delta A\in B(C_0(\Omega)),\ |\delta A|\le\varepsilon\ |A|,\ (A+\delta A)(u+\delta u)=f \right\}.$$

(2,4,1)

PROPOSITION 2.5. Let A be a linear bounded invertible operator in $C_0(\Omega)$ such that $|A|$ sends $C_0(\Omega)$ into $C_0(\Omega)$. Let f be a non-zero element of $C_0(\Omega)$ and u the solution of $Au = f$. Then the limit in (2,4,1) exists and moreover:

$$\mathbf{K_{S,S}(A,f)} = \frac{\left\| \,|\mathbf{A}^{-1}|\; |\mathbf{A}|\; |\mathbf{u}|\, \right\|}{\|\mathbf{u}\|} \quad . \tag{2,5,1}$$

From (2,3,1) and (2,5,1) follows immediately :

PROPOSITION 2.6. $K_{S,S}(A,f) \le K_S(A,f) \le 2\, K_{S,S}(A,f)$. (2,6,1)

(2,6,1) shows that if we are only interested by the order of magnitude of $K_S(A,f)$ then the knowledge of $K_{S,S}(A,f)$ is sufficient.

DEFINITION 2.7. Let A be a linear bounded invertible operator in $C_0(\Omega)$ such that $|A|$ sends $C_0(\Omega)$ into $C_0(\Omega)$. Generalizing the case of matrices [13], we will call the supremum of all $K_{S,S}(A,f)$ for f running over $C_0(\Omega)\backslash\{0\}$ the Bauer-Skeel condition number of A and we will denote it $K_{B,S}(A)$.

PROPOSITION 2.8. $K_{B,S}(A) = \|\, |A^{-1}|\, |A|\, \|$

For the sake of completeness, let us recall what is the classical condition number:

DEFINITION 2.9. Let A be a linear bounded invertible operator on a Banach space X. Generalizing the case of matrices , we will call, $\|A^{-1}\|\; \|A\|$ the classical condition number of A and we will denote it $K(A)$.

By 2.7, $K_{B,S}(A) \le K(A)$. A nice geometric interpretation was given for it by W. Kahan [6] in 1966 for matrices. Its proof which uses Hahn-Banach theorem extends straightforwardly to linear bounded invertible operator on a Banach space X, so we have:

THEOREM 2.10. Let A be a linear bounded invertible operator on a Banach space X. Then K(A) is equal to the least upper bound of ||A-

$B\|/\|A\|$ when B is running over all bounded linear non-invertible operators on X.

It shows that K(A) is equal to the "relative distance" of A to the non-invertible operators (linear and bounded) on X.

3. APPLICATION TO SEMI-GROUP THEORY.

We will now apply the previous theory to the regularized cogenerator equation: $(V+\varepsilon I)\, u_\varepsilon = f$, $\varepsilon>0$, where V is the cogenerator of a Feller semi-group $(P_t)_{t\geq 0}$ on $C_0(\Omega)$. As usual Ω will denote an arbitrary locally compact space, metrizable and separable. We suppose also that V is a bounded operator and we denote by $(R_\lambda)_{\lambda>0}$ the cogenerated submarkovian resolvent family. G will denote the generator of the semi-group $(P_t)_{t\geq 0}$ or equivalently of the resolvent $(R_\lambda)_{\lambda>0}$. As V is a positive operator, $|V+\varepsilon I|=V+\varepsilon I$ on $C_0(\Omega)$ and thus $|V+\varepsilon I|$ sends $C_0(\Omega)$ into $C_0(\Omega)$. To be able to compute $K_{S,S}(V+\varepsilon I,f)$, we will need the hypothesis that the points are of measure zero with respect to the measures $(\sigma_{\lambda,x})_{x\in\Omega}$ associated to R_λ, for all $\lambda\geq 0$, ($R_0 = V$). Before giving a more general result in that direction, we shall need some definitions:

DEFINITION 3.1. G will be said to be semi-compact iff given some sequence $(g_n)_{n\in\mathbb{N}}$ of functions contained in the domain of G, D(G), with $\sup_{n\in\mathbb{N}} \|g_n\|+\|Gg_n\|$ finite, then $(g_n)_{n\in\mathbb{N}}$ possesses a convergent subsequence $(g_{n_k})_{k\in\mathbb{N}}$ in $C_0(\Omega)$. This is equivalent to the compactness of R_λ for some or every λ in the resolvent set of G.

DEFINITION 3.2. Let υ be a positive measure on Ω. G will be said to be υ-closed iff given some sequence $(g_n)_{n\in\mathbb{N}}$ of functions contained in the domain of G, with $\sup_{n\in\mathbb{N}} \|g_n\|+\|Gg_n\|$ finite and $g,f\in C_0(\Omega)$ such that $\|g_n-g\| \longrightarrow 0$ as $n \longrightarrow +\infty$ and $\langle\varphi\upsilon,Gg_n\rangle \longrightarrow \langle\varphi\upsilon,f\rangle$, for every φ belonging to $C_0(\Omega)$, then $g\in D(G)$ and $Gg=f$.

THEOREM 3.3. Let Ω be endowed with a positive finite measure υ. Let us suppose that G is the generator of a Feller semi-group $(P_t)_{t\geq 0}$ on

$C_0(\Omega)$ that is semi-compact, υ-closed and such that 0 is in the resolvent set of G. Then for any $x \in \Omega$ and any $\lambda \geq 0$, the measure $\sigma_{\lambda,x}$ associated by the Riesz representation theorem to the continuous linear form on $C_0(\Omega)$ which sends $f \longmapsto R_\lambda f(x)$ is absolutely continuous with respect to the measure υ.

Proof: (i) Let K be a Borel set such that $\upsilon(K) = 0$. As υ is a positive measure, this implies $\upsilon|_K = 0$. We have to show that $\sigma_{\lambda,x}(K) = 0$. Due to the regularity of the measure $\sigma_{\lambda,x}$ we may suppose that K is a compact set. By Urysohn's lemma , we can construct a decreasing sequence $(f_n)_{n \in \mathbb{N}}$ of continuous functions with compact supports in Ω with values in [0,1] such that $\lim_{n \to +\infty} f_n(y) = \chi_K(y)$, $\forall y \in \Omega$, where χ_K denotes the characteristic function of K.
(ii) Let $\lambda \geq 0$ and let us set for $n \in \mathbf{N}$, $g_n = R_\lambda f_n$. Thus $(g_n)_{n \in \mathbb{N}}$ is a sequence contained in D(G). Let us show that $(g_n)_{n \in \mathbb{N}}$ possesses a convergent subsequence $(g_{n_k})_{k \in \mathbb{N}}$ in $C_0(\Omega)$. Due to the semi-compactness of G it suffices to show that $\sup_{n \in \mathbb{N}} \|g_n\| + \|Gg_n\| < +\infty$. First, the sequence $(g_n)_{n \in \mathbb{N}}$ is obviously bounded in $C_0(\Omega)$ due to $\|g_n\| = \|R_\lambda f_n\| \leq \|R_\lambda\|$. Now, the sequence $(Gg_n)_{n \in \mathbb{N}}$ is also bounded in $C_0(\Omega)$. This follows from $Gg_n = GR_\lambda f_n = \lambda R_\lambda f_n - f_n$ which implies that $\|Gg_n\| = \|\lambda R_\lambda f_n - f_n\| \leq 2\|f_n\| \leq 2$. This shows that $\sup_{n \in \mathbb{N}} \|g_n\| + \|Gg_n\| < +\infty$.
(iii) Let us call g the limit of the subsequence $(g_{n_k})_{k \in \mathbb{N}}$. g is in $C_0(\Omega)$. Let us show that g is in D(G) and that $(\lambda - G)g = 0$. This will imply that g=0 because λ is in the resolvent set of G. To show this we will use the hypothesis that G is υ-closed. Let φ be a function belonging to $C_0(\Omega)$. We have that: $\int G\, g_{n_k} \varphi\, d\nu = \int (G R_\lambda f_{n_k}) \varphi\, d\nu = \int [(\lambda R_\lambda - I) f_{n_k}] \varphi\, d\nu = \lambda \int g_{n_k} \varphi\, d\nu - \int f_{n_k} \varphi\, d\nu$. Now, $\int g_{n_k} \varphi\, d\nu \longrightarrow \int g \varphi\, d\nu$ because $g_{n_k} \longrightarrow g$ in $C_0(\Omega)$. $f_{n_k}(x)$ tends to $\chi_K(x)$, $\forall x \in \Omega$, and $|f_{n_k}| \leq 1$. Thus by the Lebesgue dominated convergence theorem $\int f_{n_k} \varphi\, d\nu \longrightarrow \int \chi_K \varphi\, d\nu = 0$ because $\upsilon|_K = 0$. It results from all this, that $\int G\, g_{n_k} \varphi\, d\nu \longrightarrow \lambda \int g \varphi\, d\nu$, $\forall \varphi \in C_0(\Omega)$. From the ν-closedness of G, it then follows that $g \in D(G)$ and that $Gg = \lambda g$. This implies g=0.
(iv) We have proved that $g_{n_k} = R_\lambda f_{n_k} \longrightarrow 0$ in $C_0(\Omega)$. This implies that for every $x \in \Omega$, $\int f_{n_k}\, d\sigma_{\lambda,x} \longrightarrow 0$. But $f_{n_k}(y)$ tends to $\chi_K(y)$, $\forall y \in \Omega$, and $|f_{n_k}| \leq 1$. Thus by the Lebesgue dominated convergence theorem $\int f_{n_k}\, d\sigma_{\lambda,x} \longrightarrow \int \chi_K\, d\sigma_{\lambda,x}$. This implies $\int \chi_K\, d\sigma_{\lambda,x} = 0$. Thus $\sigma_{\lambda,x}(K) = 0$ for every Borel set K which is of υ-measure 0. By the Radon-Nikodym theorem this is equivalent to the absolute continuity of the $\sigma_{\lambda,x}$ with respect to υ.

Q.E.D.

COROLLARY 3.4. Under the same hypothesis of theorem 3.3 and if moreover the points of Ω are of υ-measure zero, then they are also of $\sigma_{\lambda,x}$-measure zero for any $x \in \Omega$ and any $\lambda \geq 0$.

For the sake of easy reference, we will call **(H)** the **hypothesis**: "*the points of* Ω *are of* $\sigma_{\lambda,x}$ *-measure zero for any* $x \in \Omega$ *and any* $\lambda \geq 0$" and at the end of this paper, we will show that the hypothesis of corollary 3.4 are true for a general second order elliptic operator L with real continuous coefficients on $\mathbf{R}^N$ such that $L1 \leq 0$, Ω a bounded open subset of $\mathbf{R}^N$ which is quasi-regular at infinity with respect to $\mathbf{1-L}$ where $\mathbf{L}$ is the local operator associated to L [7],[11] and υ the Lebesgue measure restricted to Ω.

THEOREM 3.5. Let V be the bounded cogenerator of a Feller semi-group $(P_t)_{t \geq 0}$ on $C_0(\Omega)$ whose resolvent satisfies **(H)** and G its generator. Let $f \in C_0(\Omega)$, $\varepsilon > 0$, and u_ε the solution in $C_0(\Omega)$ of $(V+\varepsilon I)u_\varepsilon = f$. Then:

(i) $$K_{S,S}(V+\varepsilon I, f) = \frac{\| \, |u_\varepsilon| + 2\,\varepsilon^{-1} V |u_\varepsilon| \, \|}{\|u_\varepsilon\|} ; \qquad (3,5,1)$$

(ii) $$K_{B,S}(V+\varepsilon I) = 1 + 2\,\varepsilon^{-1} \|V\| ; \qquad (3,5,2)$$

(iii) $$K(V+\varepsilon I) = \| V+\varepsilon I \| \, \|(V+\varepsilon I)^{-1}\| = (1 + \varepsilon^{-1}\|V\|)(1 + \varepsilon^{-1}\|R_{\varepsilon^{-1}}\|). \qquad (3,5,3)$$

<u>Proof</u>: (i) By (2,5,1): $$K_{S,S}(V+\varepsilon I, f) = \frac{\| \, |(V+\varepsilon I)^{-1}| \, |(V+\varepsilon I)| \, |u_\varepsilon| \, \|}{\|u_\varepsilon\|}. \qquad (3,5,4)$$

Now: $(V+\varepsilon I)^{-1} = (\varepsilon I - G^{-1})^{-1} = ((\varepsilon G - I)G^{-1})^{-1} = -G(I-\varepsilon G)^{-1} = -G\,\varepsilon^{-1}(\varepsilon^{-1} - G)^{-1} = -\varepsilon^{-1} G R_{\varepsilon^{-1}} = -\varepsilon^{-1}(\varepsilon^{-1} R_{\varepsilon^{-1}} - I) = \varepsilon^{-1}(I - \varepsilon^{-1} R_{\varepsilon^{-1}}) =: -G_{\varepsilon^{-1}}$ (3,5,5) where $G_{\varepsilon^{-1}}$ denotes the Yosida approximation [8] (p. 20) of the generator G: $G = s\text{-}\lim_{\varepsilon \to 0+} G_{\varepsilon^{-1}}$. (3,5,5) shows that the family of measures associated to $(V+\varepsilon I)^{-1}$ by the Riesz representation theorem is equal to $(\varepsilon^{-1}(\delta_x - \varepsilon^{-1} \sigma_{\varepsilon^{-1},x}))_{x \in \Omega}$. Due to the hypothesis (H) and the

fact that the total variation of a sum of mutually singular measures is equal to the sum of the total variations of each measure, we have that: $|\varepsilon^{-1}(\delta_x - \varepsilon^{-1}\sigma_{\varepsilon-1,x})| = \varepsilon^{-1}(\delta_x + \varepsilon^{-1}\sigma_{\varepsilon-1,x})$. This implies that $|(V+\varepsilon I)^{-1}| = \varepsilon^{-1}(I+\varepsilon^{-1}R_{\varepsilon-1})$. For $V+\varepsilon I$, as it is a positive operator, we have that $|(V+\varepsilon I)| = V+\varepsilon I$. Thus: $|(V+\varepsilon I)^{-1}| \, |(V+\varepsilon I)| = \varepsilon^{-1}(I+\varepsilon^{-1}R_{\varepsilon-1})(V+\varepsilon I) = \varepsilon^{-1}V + I + \varepsilon^{-2}R_{\varepsilon-1}V + \varepsilon^{-1}R_{\varepsilon-1}$. But for $\lambda > 0$, $R_\lambda V = (\lambda I - G)^{-1}V = [(\lambda G^{-1}-I)G]^{-1}V = [(I+\lambda(-G)^{-1})(-G)]^{-1}V = V(I+\lambda V)^{-1}V$ because $V = (-G)^{-1}$. Thus $R_\lambda V = (1/\lambda) V(I+\lambda V)^{-1} [(I+\lambda V)-I] = (1/\lambda)V - (1/\lambda) R_\lambda$ from which we deduce that $\varepsilon^{-2} R_{\varepsilon-1}V = \varepsilon^{-1}V - \varepsilon^{-1}R_{\varepsilon-1}$. It then follows that:
$$|(V+\varepsilon I)^{-1}| \, |(V+\varepsilon I)| = I + 2\varepsilon^{-1}V . \qquad (3,5,6)$$
Replacing by $(3,5,6)$ in $(3,5,4)$, we get $(3,5,1)$.

(ii) By 2.8, $K_{B,S}(V+\varepsilon I) = || \, |(V+\varepsilon I)^{-1}| \, |(V+\varepsilon I)| \, ||$. Thus, by $(3,5,6)$:

$K_{B,S}(V+\varepsilon I) = || \, I + 2\varepsilon^{-1}V \, || = \sup_{x\in\Omega} || \, \delta_x + 2\varepsilon^{-1}\sigma_{0,x} \, ||$

$\qquad = \sup_{x\in\Omega} (|| \, \delta_x || + ||2\varepsilon^{-1}\sigma_{0,x} ||) \text{ by (H)}$

$\qquad = \sup_{x\in\Omega} (1 + 2\varepsilon^{-1}||\sigma_{0,x} ||)$

$\qquad = 1 + 2\varepsilon^{-1}\sup_{x\in\Omega} ||\sigma_{0,x} || = 1 + 2\varepsilon^{-1}||V||,$

which proves $(3,5,2)$.

(iii) $|| \, V+\varepsilon I \, || \, ||(V+\varepsilon I)^{-1}|| = || \, V+\varepsilon I \, || \, || \, |(V+\varepsilon I)^{-1}| \, ||$. But, we have shown above that $|(V+\varepsilon I)^{-1}| = \varepsilon^{-1}(I+\varepsilon^{-1}R_{\varepsilon-1})$ and thus: $|| \, |(V+\varepsilon I)^{-1}| \, ||$
$= ||\varepsilon^{-1}(I+\varepsilon^{-1}R_{\varepsilon-1})|| = \varepsilon^{-1} \sup_{x\in\Omega} || \, \delta_x + \varepsilon^{-1}\sigma_{\varepsilon-1,x} ||$
$= \varepsilon^{-1}(1 + \varepsilon^{-1}\sup_{x\in\Omega} ||\sigma_{\varepsilon-1,x} ||) = \varepsilon^{-1}(1+\varepsilon^{-1}||R_{\varepsilon-1}||).$

$|| \, V+\varepsilon I \, || = \sup_{x\in\Omega} ||\sigma_{0,x} + \varepsilon \delta_x|| = \sup_{x\in\Omega} (\varepsilon+||\sigma_{0,x}||) = \varepsilon + ||V||.$
Putting together these equalities, we get:
$|| \, V+\varepsilon I \, || \, ||(V+\varepsilon I)^{-1}|| = (\varepsilon + ||V||) \varepsilon^{-1}(1+\varepsilon^{-1}||R_{\varepsilon-1}||)$
$\qquad = (1 + \varepsilon^{-1}||V||)(1+\varepsilon^{-1}||R_{\varepsilon-1}||),$ which proves $(3,5,3)$.

Q.E.D.

By $(3,5,2)$ and $(3,5,3)$ $K_{B,S}(V+\varepsilon I)$ and $K(V+\varepsilon I)$ can not differ a lot in magnitude. More precisely:

COROLLARY 3.6. $0.5\, K(V+\varepsilon I) \leq K_{B,S}(V+\varepsilon I) \leq K(V+\varepsilon I).$

Proof: By $(3,5,3)$: $0.5\, K(V+\varepsilon I) = (1 + \varepsilon^{-1}||V||)\, 0.5\, (1+ \varepsilon^{-1}||R_{\varepsilon-1}||) \leq (1 + \varepsilon^{-1}||V||) \leq (1 + 2\varepsilon^{-1}||V||) = K_{B,S}(V+\varepsilon I)$ by $(3,5,2)$. The second inequality in the statement is general.

Q.E.D.

COROLLARY 3.7. $2\,\varepsilon^{-1}\dfrac{||V|f-\varepsilon^{-1}R_{\varepsilon-1}f|||}{||f-\varepsilon^{-1}R_{\varepsilon-1}f||}\le K_{S,S}(V+\varepsilon I,f)$

$$\le 1+2\,\varepsilon^{-1}\dfrac{||V|f-\varepsilon^{-1}R_{\varepsilon-1}f|||}{||f-\varepsilon^{-1}R_{\varepsilon-1}f||}.\qquad (3,7,1)$$

Proof: By (3,5,1):

$$2\,\varepsilon^{-1}\dfrac{||V|u_\varepsilon|||}{||u_\varepsilon||}\le K_{S,S}(V+\varepsilon I,f)\le 1+2\,\varepsilon^{-1}\dfrac{||V|u_\varepsilon|||}{||u_\varepsilon||}.\qquad (3,7,2)$$

From $u_\varepsilon=(V+\varepsilon I)^{-1}f$, we get from (3,5,5): $u_\varepsilon=\varepsilon^{-1}(f-\varepsilon^{-1}R_{\varepsilon-1}f)$, and replacing u_ε by this expression in (3,7,2) we get (3,7,1).

Q.E.D.

COROLLARY 3.8. $-\ K_{S,S}(V+\varepsilon I,f)\le 1+2\,\varepsilon^{-1}||V||,$ $\qquad(3,8,1)$

$-$ for $f\in D(G)$: $K_{S,S}(V+\varepsilon I,f)\ge 2\,\varepsilon^{-1}(1+\alpha(\varepsilon))\dfrac{||f||}{||Gf||}$, as $\varepsilon\longrightarrow 0^+$, $\qquad(3,8,2)$

where $\alpha(\varepsilon)\longrightarrow 0$ as $\varepsilon\longrightarrow 0^+$.

Proof: (i) From $K_{S,S}(V+\varepsilon I,f)\le K_{B,S}(V+\varepsilon I)$ and from (3,5,2) follows (3,8,1).

(ii) From (3,7,1): $K_{S,S}(V+\varepsilon I,f)\ge 2\,\varepsilon^{-1}\dfrac{||V|f-\varepsilon^{-1}R_{\varepsilon-1}f|||}{||f-\varepsilon^{-1}R_{\varepsilon-1}f||}.$ $\qquad(3,8,3)$

$|V(f-\varepsilon^{-1}R_{\varepsilon-1}f)|\le V|f-\varepsilon^{-1}R_{\varepsilon-1}f|$ because V is a positive operator which implies that the family of measures associated to V by the Riesz representation theorem are positive. This implies that:

$||\ V|f-\varepsilon^{-1}R_{\varepsilon-1}f|\ ||\ge||\ V(f-\varepsilon^{-1}R_{\varepsilon-1}f)\ ||=\varepsilon||\ V(V+\varepsilon I)^{-1}f||$ from (3,5,5)

$=||\ V(I+\varepsilon^{-1}V)^{-1}f\ ||=||\ R_{\varepsilon-1}f||$ because $R_\lambda=V(I+\lambda V)^{-1}.$ $\qquad(3,8,4)$

Also $f-\varepsilon^{-1}R_{\varepsilon-1}f=-R_{\varepsilon-1}Gf.$ $\qquad(3,8,5)$

By (3,8,4) and (3,8,5), we get from (3,8,3):

$$K_{S,S}(V+\varepsilon I,f)\ge 2\,\varepsilon^{-1}\dfrac{||\ R_{\varepsilon-1}f\ ||}{||\ R_{\varepsilon-1}Gf\ ||}=2\,\dfrac{||\ \varepsilon^{-1}\ R_{\varepsilon-1}f\ ||}{||\ R_{\varepsilon-1}Gf\ ||}$$

$$\ge 2\,\dfrac{||f||-||\ \varepsilon^{-1}\ R_{\varepsilon-1}f\ -f\ ||}{\varepsilon[\ ||Gf||+||\ \varepsilon^{-1}R_{\varepsilon-1}Gf\ -\ Gf||\ ]}$$

$$= \frac{2}{\varepsilon} \frac{\|f\|}{\|Gf\|} \cdot \frac{1 - \dfrac{\| \varepsilon^{-1} R_{\varepsilon-1}f - f\|}{\|f\|}}{1 + \dfrac{\| \varepsilon^{-1}R_{\varepsilon-1}Gf - Gf\|}{\|Gf\|}} =: \frac{2}{\varepsilon} (1+\alpha(\varepsilon)) \frac{\|f\|}{\|Gf\|},$$

where $\alpha(\varepsilon) \longrightarrow 0$ as $\varepsilon \longrightarrow 0^+$ because the resolvent associated to a contraction semi-group is an L_∞ - resolvent family [9]. This proves (3,8,2).

Q.E.D.

3.8 show us that $K_{S,S}(V+\varepsilon I,f)$ can not blow up more rapidly than $\dfrac{2\|V\|}{\varepsilon}$ but blows at least as quick as $\dfrac{2\|f\|}{\|Gf\|}\dfrac{1}{\varepsilon}$ modulo less rapidly increasing terms. In the case, f is surmedian with respect to $(R_\lambda)_{\lambda>0}$ and belongs to $D(G)$, we will prove a reciprocal inequality to (3,8,2) showing that in this case, the behaviour of $K_{S,S}(V+\varepsilon I,f)$ is given by $\dfrac{2\|f\|}{\|Gf\|}\dfrac{1}{\varepsilon}$, modulo less rapidly increasing terms. First let us recall, what is a surmedian function with respect to a resolvent family $(R_\lambda)_{\lambda>0}$.

DEFINITION 3.9. [10] Let f be a real valued function in $C_0(\Omega)$. f is said to be surmedian with respect to the submarkovian resolvent family $(R_\lambda)_{\lambda>0}$ iff $\lambda R_\lambda f \leq f$, for every $\lambda>0$. If $f \in D(G)$, this is equivalent to $Gf \leq 0$, due to $Gf = \lim_{\lambda \longrightarrow +\infty} \lambda(\lambda R_\lambda f - f)$.

PROPOSITION 3.10. Let $f \in D(G)$ be a surmedian function. Then:
$$K_{S,S}(V+\varepsilon I,f) \leq 1 + 2\varepsilon^{-1}(1+\beta(\varepsilon))\frac{\|f\|}{\|Gf\|}, \quad \text{as } \varepsilon \longrightarrow 0^+, \qquad (3,10,1)$$
where $\beta(\varepsilon) \longrightarrow 0$ as $\varepsilon \longrightarrow 0^+$.

Proof: f being surmedian: $V|f-\varepsilon^{-1}R_{\varepsilon-1}f| = V(f-\varepsilon^{-1}R_{\varepsilon-1}f) = -VGR_{\varepsilon-1}f = R_{\varepsilon-1}f$. From (3,7,1), we then get:

$$K_{S,S}(V+\varepsilon I,f) \leq 1+ 2\varepsilon^{-1}\frac{\|R_{\varepsilon-1}f\|}{\|f-\varepsilon^{-1}R_{\varepsilon-1}f\|} = 1+2\frac{\|\varepsilon^{-1}R_{\varepsilon-1}f\|}{\|R_{\varepsilon-1}Gf\|}$$

$$= 1+ 2\varepsilon^{-1}\frac{\|\varepsilon^{-1}R_{\varepsilon-1}f\|}{\|\varepsilon^{-1}R_{\varepsilon-1}Gf\|}$$

$$= 1+2\,\varepsilon^{-1}\frac{\|f\|}{\|Gf\|}\left(\frac{\|\varepsilon^{-1}R_{\varepsilon^{-1}}f\|/\|f\|}{\|\varepsilon^{-1}R_{\varepsilon^{-1}}Gf\|/\|Gf\|}\right)$$

$$= 1+2\,\varepsilon^{-1}\frac{\|f\|}{\|Gf\|}\left(1+\beta(\varepsilon)\right);\ \text{where } \beta(\varepsilon)\longrightarrow 0 \text{ as } \varepsilon\longrightarrow 0^+,$$

because $\dfrac{\|\varepsilon^{-1}R_{\varepsilon^{-1}}f\|/\|f\|}{\|\varepsilon^{-1}R_{\varepsilon^{-1}}Gf\|/\|Gf\|} \longrightarrow 1$ as $\varepsilon\longrightarrow 0^+$.

Q.E.D.

All this has been proved under the hypothesis **(H)**, which is true if corollary 3.4 is true. We now show that the hypothesis of corollary 3.4 are true for a very general class of second order real elliptic operators L on $\mathbf{R}^N$ and for very general bounded open subsets Ω of $\mathbf{R}^N$. Let us precise the setting:

$\mathbf{L} = \sum_{i,j=1}^{N} a_{i,j}\,\partial^2_{i,j} + \sum_{i=1}^{N} b_i\,\partial_i + c$, **where** $a_{i,j}$, b_i , c **are real-valued continuous**

functions on $\mathbf{R}^N$, with $c\le 0$, and $\sum_{i,j=1}^{N} a_{i,j}\,\xi_i\,\xi_j > 0$, $\forall\,\xi\in \mathbf{R}^N\setminus\{0\}$. Let

$\mathbf{L}$, the associated local operator on $\mathbf{R}^N$ in the sense of [7][11] i.e. for V an arbitrary open set of $\mathbf{R}^N$, we set: $D(\mathbf{L},V)=\{f\in W^{2,p}_{loc}(V)$ for some $p>N$, and $Lf\in C(V)\}$ and $\mathbf{L}^V$: $D(\mathbf{L},V)\longrightarrow C(V)$: $f\longmapsto Lf$. The local operator on $\mathbf{R}^N$ associated to L is the family $\mathbf{L}=\{\mathbf{L}^V$; V non-void open set of $\mathbf{R}^N\}$. For each non-void open set V in $\mathbf{R}^N$, we consider the largest restriction $\mathbf{L}_V$ of the operator $\mathbf{L}^V$ which operates in $C_0(V)$ i.e. $D(\mathbf{L}_V)=\{f\in C_0(V);\ f\in D(\mathbf{L},V)$ and $\mathbf{L}^V f\in C_0(V)\}$. In [11], I have shown that the hypothesis of Lumer's basic theorem 5.4 in [7] are verified. From this theorem, it then follows that $\mathbf{L}_V$ generates a Feller semi-group iff V is quasi-regular at infinity with respect to $1-\mathbf{L}$ i.e. iff there exists a compact set K contained in V and a function $h\in D(\mathbf{L},V\setminus K)$, $h>0$, $(1-L)h\ge 0$ in $V\setminus K$ such that $\forall\varepsilon>0$, $\exists\,K_\varepsilon$ compact satisfying $V\supset K_\varepsilon\supset K$ with $h<\varepsilon$ in $V\setminus K_\varepsilon$. This class of open sets is very general and contains in particular the class of bounded open sets which are regular for the Dirichlet problem associated to L. In the following, Ω will denote a bounded open subset of $\mathbf{R}^N$ which is quasi-regular at infinity with respect to $1-\mathbf{L}$ and υ the Lebesgue measure restricted to Ω. If $\mathbf{L}_V$ generates a Feller semi-group, its resolvent will be denoted $(R^V_\lambda)_{\lambda>0}$.

PROPOSITION 3.11. $\mathbf{L}_\Omega$ is semi-compact and 0 is in the resolvent set of $\mathbf{L}_\Omega$. Moreover $\mathbf{L}_\Omega$ is also υ- closed.

Proof: (i) First, let us show that $R_\lambda^\Omega = (\lambda I - L_\Omega)^{-1}$ is a compact operator for some $\lambda>0$. Let us consider $(f_n)_{n\in\mathbb{N}}$ a bounded sequence of functions in $C_0(\Omega)$. Let B be an open ball containing $\overline{\Omega}$. Let us denote by $\tilde{f}_n$ the function in $C_0(B)$, which extends f_n by 0 in $B\setminus\overline{\Omega}$. Let us set $g_n = R_\lambda^\Omega f_n$ and $\hat{g}_n = R_\lambda^B \tilde{f}_n$. First, let us show that we can extract a convergent subsequence in $C_0(B)$ from $(\hat{g}_n)_{n\in\mathbb{N}}$. B being very regular, by [2] for p>N: $D(L_B) = \{f\in C_0(B)\cap W^{2,p}(B);\ Lf\in C_0(B)\}$. As $L^{p_1}(B) \supset L^{p_2}(B)$ if $p_1\le p_2$, by maximality of the generators, $D(L_B)$ is independant of p, for p larger than N. Thus, also $D(L_B) = \{f\in C_0(B)\cap W^{2,p}(B)$ for some $p>N$; $Lf\in C_0(B)\}$. The sequence $(\hat{g}_n)_{n\in\mathbb{N}}$ is contained in $D(L_B)$ and bounded in the supremum norm due to $\|\hat{g}_n\|\le\|R_\lambda^B\|\ \|\tilde{f}_n\|\le\|R_\lambda^B\|\ \|f_n\|$, $(f_n)_{n\in\mathbb{N}}$ being itself bounded. Moreover $\|L_B\hat{g}_n\|=\|L_B R_\lambda^B \tilde{f}_n\|\le\|(\lambda R_\lambda^B - I)\tilde{f}_n\|\le\|\lambda R_\lambda^B \tilde{f}_n\| + \|\tilde{f}_n\|\le 2\|\tilde{f}_n\|\le 2\|f_n\|$, and thus is also bounded. By [2] the mapping from $C_0(B)$ in $W^{2,p}(B)\cap C_0(B)$ which sends $f\in C_0(B)$ on $u\in W^{2,p}(B)\cap C_0(B)$ such that $Lu=f$ is a bounded mapping from $C_0(B)$ in $W^{2,p}(B)\cap C_0(B)$. There exists thus a constant C>0, such that: $\|u\|_{W^{2,p}(B)} \le C\,\|Lu\|_{C_0(B)}$.

From this inequality follows that the sequence $(\hat{g}_n)_{n\in\mathbb{N}}$ is bounded in $W^{2,p}(B)$ for the $W^{2,p}$-norm. By the Rellich-Kondrashov theorem, the canonical injection from $W^{2,p}(B)$ into $C(\overline{B})$ is a compact operator. These two facts together imply that $(\hat{g}_n)_{n\in\mathbb{N}}$ possesses a convergent subsequence $(\hat{g}_{n_k})_{k\in\mathbb{N}}$ in $C_0(B)$. In particular the subsequence $(\hat{g}_{n_k})_{k\in\mathbb{N}}$ is a Cauchy sequence in $C_0(B)$. Let us show now by comparison with the subsequence $(\hat{g}_{n_k})_{k\in\mathbb{N}}$, that the subsequence $(g_{n_k})_{k\in\mathbb{N}}$ is a Cauchy sequence in $C_0(\Omega)$, where g_n denotes $R_\lambda^\Omega f_n$. The comparison between these two subsequences rely on the fact that $(\lambda-L)g_n=f_n$ in Ω and $(\lambda-L)\hat{g}_n=\tilde{f}_n$ in $B\supset\Omega$, which implies that $g_n - \hat{g}_n|_\Omega$ is $L-\lambda$ harmonic in Ω. Moreover, on the boundary of Ω, g_n vanish and thus $(g_n-\hat{g}_n)|_{\partial\Omega} = -\hat{g}_n|_{\partial\Omega}$, which implies that: $((g_n-g_m) - (\hat{g}_n - \hat{g}_m))|_{\partial\Omega} = -(\hat{g}_n - \hat{g}_m)|_{\partial\Omega}$, $\forall\ m,n\in\mathbb{N}$. Applying then the maximum principle,we get that:

$$\|(g_n-g_m) - (\hat{g}_n - \hat{g}_m)\|_{\infty,\Omega} = \|(\hat{g}_n - \hat{g}_m)\|_{\infty,\partial\Omega}$$ because

$(g_n-g_m)-(\hat{g}_n-\hat{g}_m) = (g_n-\hat{g}_n) - (g_m-\hat{g}_m)$ and is thus $L-\lambda$ harmonic in Ω. Thus: $\|g_n-g_m\|\le 2\|\hat{g}_n-\hat{g}_m\|$. This last inequality and the fact that $(\hat{g}_{n_k})_{k\in\mathbb{N}}$ is a Cauchy sequence in $C_0(B)$ implies that $(g_{n_k})_{k\in\mathbb{N}}$ is a Cauchy sequence in $C_0(\Omega)$. Thus, given a bounded sequence $(f_n)_{n\in\mathbb{N}}$ of functions in $C_0(\Omega)$, we have shown that $(g_n = R_\lambda^\Omega f_n)_{n\in\mathbb{N}}$ possesses a convergent subsequence $(g_{n_k})_{k\in\mathbb{N}}$. This proves that R_λ^Ω is a compact operator.

Now it is straightforward to prove that $\mathbf{L}_\Omega$ is semi-compact. Let $(g_n)_{n\in N}$ be a sequence of functions contained in $D(\mathbf{L}_\Omega)$, such that $\sup_{n\in N} \|g_n\|+\|\mathbf{L}_\Omega g_n\|< +\infty$. Then $g_n = R_\lambda^\Omega(\lambda I-\mathbf{L}_\Omega)g_n$. By $\sup_{n\in N} \|g_n\|+\|\mathbf{L}_\Omega g_n\| < +\infty$, it follows that $((\lambda I-\mathbf{L}_\Omega)g_n)_{n\in N}$ is a bounded sequence in $C_0(\Omega)$. R_λ^Ω being a compact operator, it then follows that $(g_n)_{n\in N}$ possesses a convergent subsequence. This proves that $\mathbf{L}_\Omega$ is semi-compact.
(ii) Let us prove now that $\mathbf{L}_\Omega$ is υ- closed. Let us consider some sequence $(g_n)_{n\in N}$ of functions contained in the domain of $\mathbf{L}_\Omega$, with $\sup_{n\in N} \|g_n\|+\|\mathbf{L}_\Omega g_n\|< +\infty$ and $g,f\in C_0(\Omega)$ such that $\|g_n-g\| \to 0$ as $n \to +\infty$ and $\langle\varphi\upsilon,\mathbf{L}_\Omega g_n\rangle \to \langle\varphi\upsilon,f\rangle$, for every φ belonging to $C_0(\Omega)$. We must prove that this implies that $g\in D(\mathbf{L}_\Omega)$ and $\mathbf{L}_\Omega g=f$. From $\sup_{n\in N} \|g_n\|+\|\mathbf{L}_\Omega g_n\|< +\infty$, it follows by the interior estimates of Agmon, Douglis and Nirenberg, that for every open relatively compact subset $\mathbf{D}$ of Ω, the sequence $(g_{n|\mathbf{D}})_{n\in N}$ is bounded in $W^{2,p}(\mathbf{D})$, for $p>N$ [1]. But $W^{2,p}(\mathbf{D})$ [15] p.210 is a reflexive Banach space and the closed unit ball in a reflexive Banach space is weakly sequentially compact [3] p.69. By using then Cantor diagonalization process, we can extract from the sequence $(g_n)_{n\in N}$ a subsequence $(g_{n_k})_{k\in N}$ weakly convergent in $W^{2,p}(\mathbf{D})$, for every relatively compact open subset $\mathbf{D}$ of Ω. This convergence implies the convergence in the sense of distributions. As the sequence $(g_n)_{n\in N}$ converges also to g in $C_0(\Omega)$, it follows that the weak-limit of $(g_{n_k})_{k\in N}$ in every $W^{2,p}(\mathbf{D})$ must coincide with g. In particular, $\forall\varphi \in C_{00}(\Omega)$, the space of all continuous functions with compact support, and for every α multi-index with $|\alpha|\le 2$: $\langle\varphi\upsilon,D^\alpha g_{n_k}\rangle \to \langle\varphi\upsilon,D^\alpha g\rangle$. This implies that for every $h \in C(\mathbf{R}^N)$: $\langle\varphi\upsilon,hD^\alpha g_{n_k}\rangle = \langle h\varphi\upsilon,D^\alpha g_{n_k}\rangle \to \langle h\varphi\upsilon,D^\alpha g\rangle=\langle\varphi\upsilon,hD^\alpha g\rangle$. Applying $\langle\varphi\upsilon,hD^\alpha g_{n_k}\rangle \to \langle\varphi\upsilon,hD^\alpha g\rangle$ with $h=a_{i,j},b_i,c$, and linearity we get that $\langle\varphi\upsilon, L g_{n_k}\rangle \to \langle\varphi\upsilon, Lg\rangle$, $\forall\varphi \in C_{00}(\Omega)$. As it converges also to $\langle\varphi\upsilon,f\rangle$, we get $Lg=f$. Thus $g \in W^{2,p}_{loc}(\Omega)\cap C_0(\Omega)$ and $Lg=f \in C_0(\Omega)$. This implies that $g \in D(\mathbf{L}_\Omega)$ and $\mathbf{L}_\Omega g=f$. This proves that $\mathbf{L}_\Omega$ is υ-closed.

Q.E.D.

By the previous result and 3.4, the measures $\sigma_{\lambda,x}$ are absolutely continuous with respect to the Lebesgue measure restricted to Ω for every $x\in\Omega$ and every $\lambda\ge 0$ where the $\sigma_{\lambda,x}$ are the measures associated to R_λ^Ω, the resolvent of $\mathbf{L}_\Omega$. This is known [2] if Ω has a C^2-boundary, but our result is far more general as it applies to the most general class possible of bounded open subsets; for example, it holds for the Laplace operator in the complex plane and Ω the interior of the Julia set of a holomorphic polynomial because such a set is even regular for the

Dirichlet problem [14]. Moreover it shows that the R_λ^Ω are compact operators for every λ in the resolvent set of $\mathbf{L}_\Omega$ provided Ω is a bounded open subset of $\mathbf{R}^N$ which is quasi-regular at infinity with respect to $1-\mathbf{L}$.

Going back to the general setting described at the beginning of this section, let us conclude by mentionning some possible application. If we replace the equation Vu=f, for $f \in D(G)\backslash\{0\}$ by the equation $(V+\varepsilon I)u_\varepsilon = f$, we can bound the error as follows:

PROPOSITION 3.12. If $f \in D(G)\backslash\{0\}$, then $\|u-u_\varepsilon\| \leq \|(\varepsilon^{-1}R_{\varepsilon^{-1}}-I)u\|$, and thus tends to 0 as $\varepsilon \to 0^+$. If moreover, $f \in D(G^2)$, then:

$$\frac{\|u-u_\varepsilon\|}{\|u\|} \leq \varepsilon \frac{\|Gu\|}{\|u\|} = \varepsilon \frac{\|G^2f\|}{\|Gf\|}$$

Proof: $u=V^{-1}f= -Gf$ and $u_\varepsilon= -\varepsilon^{-1}(\varepsilon^{-1}R_{\varepsilon^{-1}}- I)f= -\varepsilon^{-1}R_{\varepsilon^{-1}}Gf= \varepsilon^{-1}R_{\varepsilon^{-1}}u$. Thus: $\|u-u_\varepsilon\| \leq \|(\varepsilon^{-1}R_{\varepsilon^{-1}}-I)u\|$. If moreover, $f \in D(G^2)$, then $u \in D(G)$ and thus: $\|u-u_\varepsilon\| \leq \|(\varepsilon^{-1}R_{\varepsilon^{-1}}-I)u\| \leq \|R_{\varepsilon^{-1}}Gu\| \leq \varepsilon\|Gu\|$.

Q.E.D.

If moreover f is surmedian and $\|Gf\|$ is large with respect to $\|f\|$, then by (3,10,1) the regularized equation $(V+\varepsilon I)u_\varepsilon = f$ will be "well conditioned". In that case, it could be a good strategy to solve numerically $(V+\varepsilon I)u_\varepsilon = f$, rather than Vu=f.

REFERENCES

[1] L. Bers, F John, M. Schechter, "*Partial Differential Equations*", Lectures in Applied Mathematics (1957), Volume 3A, American Mathematical Society, Fourth Printing (1974), Providence, Rhode Island.

[2] J-M Bony, "*Principe du maximum dans les espaces de Sobolev*", C.R. Acad. Sc. Paris, t. 265 (18 septembre 1967), série A, pages 333 à 336.

[3] M-M Day, "*Normed Linear Spaces*", Third Edition, Ergebnisse Der Mathematik Und Ihrer Grenzgebiete, Band 21, Springer-Verlag (1973).

[4] J. Dieudonné, "*Eléments d'Analyse*", tome 2, Gauthier-Villars (1969).

[5] P.R. Halmos, "*Measure Theory*", Van Nostrand Reinhold Company (1950).

[6] W. Kahan, *"Numerical Linear Algebra"*, Canadian Math. Bull. 9 (1966), p.757-801.

[7] G. Lumer, *"Problème de Cauchy pour Opérateurs Locaux et "Changement de Temps" "*, Annales de l'Institut Fourier, tome XXV-Fascicule 3 et 4, (1975), p. 409-446.

[8] G. Lumer, *"Cours sur les semi-groupes"*, Publication Interne n° 151 de l'U.E.R. de Mathématiques Pures et Appliquées de l'Université de Lille 1, (Mars 1979), 59655 Villeneuve d'Ascq Cedex (France).

[9] G. Lumer, *"Semigroups, Potential Theory, Evolution Equations"*, Unpublished Manuscript, Institut de Mathématique et d'Informatique, Université de Mons-Hainaut, 20 Place du Parc, 7000 Mons (Belgique).

[10] P-A. Meyer, *"Probabilités et Potentiel"*, Hermann, Paris (1966).

[11] L. Paquet, *"Equations d'évolution pour opérateurs locaux et équations aux dérivées partielles"*, C.R. Acad. Sc. Paris, t. 286 (30 Janvier 1978), série A, pages 215 à 218.

[12] R.D. Skeel, *"Scaling for Numerical Stability in Gaussian Elimination"*, Journal of the Association for Computing Machinery, Vol. 26, No.3, July 1979, p. 494-526.

[13] G.W. Stewart, J-g Sun, *"Matrix Perturbation Theory"*, Academic Press, Inc. (1990).

[14] P. Tortrat, *"Aspects Potentialistes de l'Itération des Polynômes"*, Séminaire de Théorie du Potentiel Paris, No.8, Lecture notes in Mathematics 1235, Springer-Verlag (1987), pages 195-209.

[15] F. Trèves, *"Basic Linear Partial Differential Equations"*, Academic Press (1975).

Rational Approximations of Analytic Semigroups

S. PISKAREV Luleå University of Technology, Luleå, Sweden

This lecture is a summary of joint work with M. Crouzeix, S. Larsson and V. Thomée 1991.

1. Let A be a densely defined, closed linear operator which generates in a Banach space E a bounded strongly continuous semigroup $\exp(tA)$:

$$\| \exp(tA) \| \leq C \qquad \text{for} \quad t \geq 0.$$

It is well known that the C_0-semigroup $\exp(tA)$ gives the solution of the well-posed initial-value problem

$$u'(t) = Au(t) \;\text{ for }\; t \geq 0, \;\; u(0) = u^0, \tag{1}$$

by the formula $u(t) = \exp(tA)\, u^0$, $t \geq 0$. Let us consider the semidiscrete approximation of the problem (1) in the Banach spaces E_n :

$$u_n'(t) = A_n u_n(t) \qquad \text{for} \;\; t \geq 0, \quad u_n(0) = u_n^0 ,$$

with operators A_n , which generate C_0-semigroups and are consistent with the operator A. The general approximation scheme can be described in the following way. Let E_n and E be Banach spaces and $\{ p_n \}$ is a system of linear bounded operators $p_n : E \to E_n$ with the property :

$$\| p_n x \|_{E_n} \to \| x \|_E \qquad \text{as } n \to \infty \qquad \text{for any } x \in E.$$

The sequence $\{ x_n \}$, $x_n \in E_n$, is said to be convergent to the element $x \in E$ iff $\| x_n - p_n x \| \to 0$ as $n \to \infty$; we write $x_n \to x$. The sequence of linear bounded operators $\{ B_n \}$, $B_n : E_n \to E_n$, is said to be convergent to the bounded linear operator B iff for every $x \in E$ and every sequence $\{x_n\}$, $x_n \in E_n$, the following implication holds :

$$x_n \to x \quad \Longrightarrow \quad B_n x_n \to Bx \;;$$

we write $B_n \to B$.

Usially E_n is finitedimensional, although, in general, $\dim E_n \to \infty$ and $\| A_n \| \to \infty$ as $n \to \infty$.

THEOREM ABC . The following conditions (A) and (B) are equivalent to condition (C).

(A) there exist $\lambda \in \rho(A) \cap \bigcap_n \rho(A_n)$ such that the resolvents convergence

$$R(\lambda, A_n) \to R(\lambda, A) ;$$

(B) there are some constants $M_1 \geq 1$ and ω such that

$$\| \exp(tA_n) \| \leq M_1 \exp(\omega t) \qquad \text{for } t \geq 0;$$

(C) for every finite $T > 0$ we have

$$\max_{t \in [0,T]} \| \exp(tA_n) u_n^0 - p_n \exp(tA) u^0 \| \to 0$$

as $n \to \infty$ whenever $u_n^0 \to u^0$.

It is possible to consider completely discrete schemes based on rational $A(\theta)$-acceptable approximation of $\exp(tA_n)$. We say that $r(z)$ is $A(\theta)$-acceptable if

$$| r(z) | \leq 1 \quad \text{for } z \in C \backslash \Sigma_\theta ,$$

and accurate of order q, where $q \geq 1$, if

$$r(z) = \exp(z) + O(|z|^{q+1}) \quad \text{for } z \to 0 \text{ and } z \in C \backslash \Sigma_\theta ,$$

where $\Sigma_\theta = \{ z = \rho \exp(i\mu): \rho > 0, |\mu| < \pi - \theta \}$. The approximation to $\exp(tA)$ can be constructed by $r(\tau A_n)^k$, where $t = k\tau$, $\tau > 0$. Of course, we have consider the situation when $\sigma(A_n) \subseteq C \backslash \Sigma_\theta$. If $\theta = \pi/2$ then we say that $r(z)$ is A-acceptable.

THEOREM 1. [Brenner, Thomée 1979]. Assume that condition (B) is satisfied with $\omega = 0$ and r is A-acceptable . Then there is a constant C depending on $r(z)$ such that

$$\| r(\tau A_n)^k \| \leq C M_1 \sqrt{k} \qquad \text{for } k \geq 0, \ \tau > 0.$$

This theorem is sharp, the $\sqrt{k}$ cannot be removed, in general.

THEOREM 2. [Brenner,Thomée 1979]. Assume that the conditions of Theorem 1 are fulfilled and $r(z)$ is accurate of order q. Then

$$\| r(\tau A_n)^k x_n - \exp(tA_n) x_n \| \leq C M_1 t \tau^q \| A_n^{q+1} x_n \| \quad \text{for } t = k\tau, \tau > 0.$$

In order to consider the approximation of analytic semigroups we have

to change conditions (B) and (C) in Theorem ABC into the conditions :
(B') there are some constants $M \geq 1$ and $\alpha \in (0, \pi/2)$ such that

$$\| R(z;A_n) \| \leq M/|z| \qquad \text{for} \quad z \in \Sigma_\alpha ;$$

(C') for every finite $T > 0$ and some β we have for every μ

$$\max_{z \in \Sigma(\beta,\mu)} \| \exp(zA_n)\, u_n^0 - p_n \exp(zA) u^0 \| \rightarrow 0$$

as $n \rightarrow \infty$ and whenever $u_n^0 \rightarrow u^0$. Here $\Sigma(\beta, \mu) = \{z \in \Sigma_\beta : |z| \leq \mu \}$.

THEOREM 3.[Sobolevskii, Hoang 1977; Larsson, Thomée, Wahlbin 1991]. Assume that the condition (B') is satisfied and $r(z)$ is A-acceptable and accurate of order q. Then

$$\| r(\tau A_n)^k u_n - \exp(tA_n) u_n \| \leq C M \tau^q \|A_n^q u_n\| \quad \text{for} \ t = k\tau, \ \tau > 0.$$

THEOREM 4. [Larsson, Thomée, Wahlbin 1991; Piskarev 1979]. Assume that the conditions of Theorem 3 are fulfilled and either $| r(\infty) | < 1$ or $\tau_n \|A_n\| = O(1)$. Then

$$\| r(\tau_n A_n)^k u_n - \exp(tA_n) u_n \| \leq C(1+M) \tau_n^q t^{-q} \|u_n\| \quad \text{for} \ t = k\tau_n \ , u_n \in E_n .$$

2. The main result of the paper is
THEOREM 5. Assume that condition (B') is satisfied with some α and r is accurate of order q and A(θ)-acceptable for some $\theta \in (\alpha, \pi/2]$. Then

$$\| r(\tau A_n)^k \| \leq C(1+M) \quad \text{for} \quad k \geq 0, \tau > 0,$$

Moreover,

$$\| r(\tau A_n)^k - \exp(tA_n) - \gamma^k \exp(-ka\tau^{-b}(-A_n)^{-b}) \| \leq C M (k^{-q} + k^{-1/b}), \qquad (2)$$

where $\gamma = r(\infty)$ and $a > 0$, b are some constants.
REMARK. For example, the Crank-Nicolson scheme is given by the fraction $r(z) = (1 + z/2)/(1 - z/2)$. In this case we have $\gamma = -1$, $a = 4$, $b = 1$.
REMARK. The result of Theorem 5 can be very useful applied to numerical treatment of the Cauchy problem for the integrodifferential equation

$$u'(t) = A\,u(t) + \int_0^t B(t,s)u(s)\, ds + f(t), \quad u(0) = u^0,$$

where the operator-function $B(t,s)$ is smooth, $D(B) \subseteq D(A)$, and the operator A generates an analytic semigroup. Using integration by part and the Gronwall inequality we can easily show the stability of an approximation scheme of the form

$$U_n^k = r(\tau A_n)^k U_n^0 + Q_n \sum_{k=1}^{n} r(\tau A_n)^{n-k}(\sigma_n{}^k(U) + f^k), \; U_n^0 = u_n^0 ,$$

where Q_n and σ_n are suitable discrete operators.

Proof of Theorem 5. First of all we have to note that there exist a positive number a and an integer $b \in [1, \pi/(2\theta)]$ such that

$$r(z) = \gamma \exp(-a(-w)^b) + O(|w|^{b+1}) \quad \text{as } w = z^{-1} \to 0,$$

and because of theorem 4 we are interested only in the case $|\gamma| = 1$. By deLaubenfels1987,1988, it follows that the semigroup $\exp(-t(-A_n)^{-b})$ exists and is bounded uniformly in t. It is easy to see that the function $f_k(\tau A_n) = r(\tau A_n)^k - \exp(k\tau A_n) - \gamma^k \exp(-ka\tau^{-b}(-A)^{-b})$ can be defined by the formula

$$f_k(\tau A_n) = 1/2\pi i \int_\Gamma (r(\tau z)^k - \exp(k\tau z) - \gamma^k \exp(-ka(-\tau z)^{-b})) R(z;A_n)\, dz$$

with a suitable contour Γ (see condition (B') and $A(\theta)$-acceptability).To see this we can write

$$f_k(\tau A_n) = 1/2\pi i \int_\Gamma (r(\tau z)^k - \gamma^k) R(z;A_n)\, dz - 1/2\pi i \int_\Gamma \exp(k\tau z) R(z;A_n) dz$$
$$- \gamma^k/2\pi i \int_\Gamma (\exp(-ka\tau^{-b} (-z)^{-b}) - I) R(z;A_n)\, dz.$$

Here the first term on the right is equal to $r(\tau A_n)^k - \gamma^k I$, since the rational function $r(\tau z)^k - \gamma^k$ is equal to 0 at $z = \infty$. The second term is $-\exp(tA_n)$. It is possible to show that the third term is equal to $\gamma^k(\exp(-t(-A_n)^{-b}) - I)$. To get the estimation (2) for $\| f_k(\tau A_n) \|$ we consider the following inequalities

$$| r(z)^k - \exp(kz) | \le C k |z|^{q+1} \exp(- c k |z|) \quad \text{for } |z| \le 1, z \in \Sigma_\alpha ,$$

$$| r(z)^k - \gamma^k \exp(-ak(-z)^{-b}) | \le C k |z|^{-b-1} \exp(- c |z|^{-b}) \quad \text{for } |z| \ge 1, z \in \Sigma_\alpha ,$$

(we use the substitutions $z= - p \exp(i\varphi)$, $s = k p$, $s = k p^{-b}$ respectively)

$$\int_0^1 |f_k(z)|dp/p \le \int_0^1 | r(z)^k - \exp(kz) | \, dp/p + \int_0^1 |\exp(-ka(-z)^{-b}| \, dp/p$$

$$\le C\,k \int_0^1 p^q \exp(-ckp)dp + \int_0^1 \exp(-ckp^{-b}) \, dp/p$$

$$\le C\,k^{-q} \int_0^k s^q \exp(-cs) \, ds + 1/b \int_k^\infty \exp(-cs) \, ds/s$$

$$\le C\,k^{-q} + C\exp(-ck) \le C\,k^{-q},$$

$$\int_1^\infty |f_k(z)| \, dp/p \le \int_1^\infty | r(z)^k - \gamma^k \exp(-ka(-z)^{-b}) | \, dp/p + \int_1^\infty | \exp(kz)| \, dp/p$$

$$\le C\,k \int_1^\infty p^{-b-1} \exp(-ckp^{-b}) \, dp/p + \int_1^\infty \exp(-ckp) \, dp/p$$

$$= C\,k^{-1/b} \, 1/b \int_0^k s^{1/b} \exp(-cs) \, ds + \int_k^\infty \exp(-cs) \, ds/s$$

$$\le C\,k^{-1/b} \, 1/b \int_0^\infty s^{1/b} \exp(-cs) \, ds + \int_k^\infty \exp(-cs) \, ds$$

$$= C\,k^{-1/b} + C\exp(-ck) \le C\,k^{-1/b}.$$

REFERENCES

Brenner P. and Thomée V.(1979). On rational approximation of semigroups, SIAM J.Numer.Anal., **16** : 683.

Crouzeix M., Larsson S., Piskarev S. and Thomée V.(1991). The stability of rational approximations of analytic semigroups. Preprint, **28**, Department of Mathematics, Chalmers University of Technology.

de Laubenfels R.(1987). Powers of generators of holomorphic semigroups, Proc.Amer.Math.Soc., **99** : 105.

de Laubenfels R.(1988).Inverses of operators,Proc.Amer.Math.Soc, **104** : 443.

Larsson S., Thomée V. and Wahlbin L.B.(1991). Finite element methods for a strongly damped wave equation,IMA J. Numer.Anal., **11** : 115.

Piskarev S.(1979). Error estimates for approximation of semigroups of operators by Padé fractions,Soviet Math.(Iz.VUZ), **23** : 31.

Sobolevskii P.E. , Hoang Van Lai.(1977)."Difference schemes of optimal type for approximation of solutions of parabolic equations", Differ.and Integro-differ.Equations, pp.37-43.(Russian).

On the Cauchy–Dirichlet Problem for an Equation of Third Order

J. POPIOŁEK Warsaw University (Bialystok), Bialystok, Poland

1 INTRODUCTION

In the domain

$$\mathcal{R}_T^+ = \{ (x,t) \in \mathbf{R}^2 : x > 0,\ 0 < t < T \} \tag{1}$$

$(T = const > 0)$, we consider the following partial differential equation

$$\mathbf{L}u \equiv D_x^3 u - D_t^2 u = 0, \tag{2}$$

where $D_x^3 = \frac{\partial^3}{\partial x^3}$, $D_t^2 = \frac{\partial^2}{\partial t^2}$.

It was proved in DZHURAEV and POPIOLEK [6], [7], that equation (2) is one of the canonical forms of third-order partial differential equations and it is called the equation with characteristics multiple. The problem consisting in finding solutions of equations with characteristics multiple by the method of the Laplace's transformation has been intensively examined in papers of BLOCK [2], [3] and DEL VECCHIO [10]. The relevant fundamental solutions of (2) were found by CATTABRIGA [4]. The papers of CATTABRIGA [4], DZHURAEV and ABDINAZAROV [5], ABDINAZAROV [1], were devoted to solve some boundary-value problems for equation (2). In this paper we deal with the Cauchy-Dirichlet problem for this equation.

2 FUNDAMENTAL SOLUTIONS

The function

$$\mathrm{U}(x,t;y,\tau) = (t-\tau)^{-\frac{2}{3}}\, \Phi\left(\frac{x-y}{(t-\tau)^{\frac{2}{3}}} \right), \tag{3}$$

where

$$\Phi(z) = \int_0^{+\infty} \exp\left[-\tfrac{1}{\sqrt{2}}\xi^{\frac{3}{2}}\right]\cos\left(\tfrac{1}{\sqrt{2}}\xi^{\frac{3}{2}} + \xi z\right)d\xi \tag{4}$$

will be called the *fundamental solution* of the equation (2).

Let us note that the function Φ given by formula (4) satisfies the differential equation

$$\Phi''(z) - \tfrac{4}{9}z^2\Phi'(z) - \tfrac{10}{9}z\Phi(z) = 0$$

and has the following properties (CATTABRIGA [4])

$$\int_0^{+\infty}\Phi(z)\,dz = \frac{\pi}{3}, \quad \int_{-\infty}^{0}\Phi(z)\,dz = \frac{2\pi}{3}. \tag{5}$$

By CATTABRIGA [4], for the function $\mathbf{U}$ and its derivatives the following inequalities hold

$$|\, D_x^i D_t^j \mathbf{U}(x,t;y,\tau)\,| \le C(t-\tau)^{-\frac{2i+3j+2}{3}}\left|\frac{x-y}{(t-\tau)^{\frac{2}{3}}}\right|^{-\frac{2i+3j+2+3\kappa}{2}}, \tag{6}$$

when $\frac{x-y}{(t-\tau)^{\frac{2}{3}}} \longrightarrow +\infty$,

$$|\, D_x^i D_t^j \mathbf{U}(x,t;y,\tau)\,| \le C(t-\tau)^{-\frac{2i+3j+2}{3}}\exp\left[-c\left|\frac{x-y}{(t-\tau)^{\frac{2}{3}}}\right|^3\right], \tag{7}$$

when $\frac{x-y}{(t-\tau)^{\frac{2}{3}}} \longrightarrow -\infty$, where $C, c = const > 0$, $(x,t), (y,\tau) \in \mathcal{R}_T$, $\tau < t$, $\kappa = \frac{1-(-1)^j}{2}$, $i,j = 0,1,\ldots$.

Let us introduce the function

$$\mathbf{V}(x,t;y,\tau) = (t-\tau)^{\frac{1}{3}}\,\Psi\left(\frac{x-y}{(t-\tau)^{\frac{2}{3}}}\right), \tag{8}$$

where Ψ such that the following relation

$$D_t\mathbf{V} = -D_\tau\mathbf{V} = \mathbf{U} \tag{9}$$

holds good.

Since the function Ψ satisfies the equation

$$-\tfrac{2}{3}z\Psi'(z) + \tfrac{1}{3}\Psi(z) = \Phi(z),$$

then the function $\mathbf{V}$ fulfils equation (2).

It can be verified directly that the function Ψ is of the form

$$\Psi(z) = \begin{cases} z^{\frac{1}{2}}\left(A_+ + \frac{3}{2}\int_z^{+\infty}\xi^{-\frac{3}{2}}\Phi(\xi)\,d\xi\right) & \text{if } z > 0 \\ |\,z\,|^{\frac{1}{2}}\left(A_- + \frac{3}{2}\int_{-\infty}^{z}\xi^{-\frac{3}{2}}\Phi(\xi)\,d\xi\right) & \text{if } z < 0, \end{cases} \tag{10}$$

where A_+, A_- are certain positive real numbers.

3 PROPERTIES OF SOME INTEGRALS

At present, we introduce the following integrals

$$\mathbf{I}(x,t;\nu) = \frac{3}{\pi}\int_0^{+\infty} \mathbf{V}(x,t;y,0)\,\nu(y)\,dy\,, \tag{11}$$

where **V** is given by formula (8) and ν is continuous in $(0, +\infty)$,

$$\mathbf{J}(x,t;\mu) = \frac{3}{\pi}\int_0^{+\infty} \mathbf{U}(x,t;y,0)\,\mu(y)\,dy\,, \tag{12}$$

where **U** is given by formula (3) and μ is continuous in $(0, +\infty)$,

$$\mathbf{Z}(x,t;\varrho) = \frac{3}{\pi}\int\int_{\mathcal{R}_t^+} \mathbf{V}(x,t;y,\tau)\,\varrho(y,\tau)\,dy\,d\tau\,, \tag{13}$$

where

$$\mathcal{R}_t^+ = \{(y,\tau)\in \mathbf{R}^2 : 0 < y < +\infty,\ 0 < \tau < t\}$$

and the function ϱ is continuous in $\mathcal{R}_T^+$

and

$$\mathbf{K}(x,t;\varphi) = \frac{3}{\pi}\int_0^{t} D_y^2\,\mathbf{V}(x,t;0,\tau)\,\varphi(\tau)\,d\tau\,, \tag{14}$$

where φ is continuous in $[0,T]$.

We give some lemmas

LEMMA 1 *The integral (11) satisfies equation (2) for $(x,t)\in\mathcal{R}_T^+$. Moreover, if the function ν is continuous at point $x_0 \in (0,+\infty)$, then the relations are valid*

$$\lim_{P\longrightarrow P_0} \mathbf{I}(x,t;\nu) = 0\,,$$

$$\lim_{P\longrightarrow P_0} D_t\,\mathbf{I}(x,t;\nu) = \nu(x_0)\,,$$

where $P = (x,t)\in\mathcal{R}_T^+$, $P_0 = (x_0,0)$.

LEMMA 2 *The integral (12) satisfies equation (2) for $(x,t)\in\mathcal{R}_T^+$. Furthermore, if the function μ is continuous at point $x_0\in(0,+\infty)$, then*

$$\lim_{P\longrightarrow P_0} \mathbf{J}(x,t;\mu) = \mu(x_0)\,.$$

If, now, the function μ possesses continuous derivatives to the third order in $(0,+\infty)$ and the following inequality is valid

$$|\,\mu^{(k)}(x)\,| \le M_\mu\,,$$

where $M_\mu = const > 0$, $k = 0,1,2,3$, $x \in (0, +\infty)$, *then*

$$\lim_{P \longrightarrow P_0} D_t \mathbf{J}(x,t;\mu) = 0.$$

LEMMA 3 *If the function* ϱ *is continuous in* $\mathcal{R}_T^+$ *and satisfies the Hölder condition*

$$| \varrho(y,\tau) - \varrho(\overline{y},\tau) | \leq M_\varrho | y - \overline{y} |^\alpha,$$

where $M_\varrho = const > 0$. (y,τ), $(\overline{y},\tau) \in \mathcal{R}_T^+$, $0 < \alpha \leq 1$, *then the equality*

$$\mathbf{L}\,\mathbf{Z}(x,t;\varrho) = -\varrho(x,t)$$

holds true.

The proofs of Lemmas 1 - 3 are similar to those in Popiołek [9].
Now, we prove the following

LEMMA 4 *The integral (14) satisfies equation (2) for* $(x,t) \in \mathcal{R}_T^+$. *Moreover, if function* φ *is continuous at point* $t_0 \in (0,T]$, *then the relation is valid*

$$\lim_{P \longrightarrow P_0} \mathbf{K}(x,t;\varphi) = -\varphi(t_0), \tag{15}$$

where $P = (x,t) \in \mathcal{R}_T^+$, $P_0 = (0,t_0)$.

Proof: The first part of the lemma may be proved analogously to that of Lemma 1 in Popiołek [9], basing on the relations

$$\lim_{\tau \to t} D_t^k D_y^2 \mathbf{V}(x,t;0,\tau) = 0,$$

where $k = 0,1$.
In order to prove the second part we decompose the integral (14) as follows

$$\mathbf{K}(x,t;\varphi) = \frac{3}{\pi}\varphi(t_0) \int_0^t D_y^2 \mathbf{V}(x,t;0,\tau)\,d\tau + \tag{16}$$

$$+ \frac{3}{\pi} \int_0^t D_y^2 \mathbf{V}(x,t;0,\tau)[\varphi(\tau) - \varphi(t_0)]\,d\tau.$$

Now, we investigate the behaviour of the integral

$$K_1(x,t) = \frac{3}{\pi} \int_0^t D_y^2 \mathbf{V}(x,t;0,\tau)\,d\tau.$$

Making use of the definition (8) of the function **V** and the relation (12') in CATTABRIGA [4], we obtain

$$K_1(x,t) = -\frac{2}{\pi}\int_0^t \frac{x}{(t-\tau)^{\frac{5}{3}}}\,\Phi\left(\frac{x}{(t-\tau)^{\frac{2}{3}}}\right)\,d\tau\,.$$

Introducing the new integration variable $z = x\,(t-\tau)^{-\frac{2}{3}}$ we find that

$$K_1(x,t) = -\frac{3}{\pi}\int_{xt^{-\frac{2}{3}}}^{+\infty} \Phi(z)\,dz\,.$$

In view of the relation (5), we get

$$\lim_{P\to P_0} K_1(x,t) = -1\,.$$

Basing on the continuity the function φ at point t_0 and proceeding as in KRZYŻAŃSKI [8], one can prove that the second integral of the sum on the right-hand side of (16) tends to zero as $P \to P_0$.

As a result of the aforegoing considerations we get relation (15), which completes the proof of Lemma 4.

4 THE LINEAR PROBLEM

We pose the linear Cauchy-Dirichlet problem $(\mathcal{C}-\mathcal{D}-\mathcal{L})$:

Find a solution u of the partial differential equation

$$\mathbf{L}\,u(x,t) = f(x,t) \tag{17}$$

in $\mathcal{R}_T^+$ satisfying the boundary condition

$$u(0,t) = g(t)\,, \tag{18}$$

where $t \in [0,T]$, and the initial conditions

$$D_t^k u(x,0) = h_k(x)\,, \tag{19}$$

where $x \in [0,+\infty)$, $k = 0,1$ and f, g, h_0, h_1 are given functions such that the following compatibility conditions

$$g(0) = h_0(0)\,,\quad g'(0) = h_1(0)$$

hold good.

We make the following assumptions

(A.1) The function f is defined and continuous for $(x,t) \in \mathcal{R}_T^+$ and satisfies the conditions

$$| f(x,t) | \leq M_f,$$
$$| f(x,t) - f(y,t) | \leq M_f \, | x - y |^{\alpha},$$

where $M_f = const > 0$, $(x,t), (y,t) \in \mathcal{R}_T^+$, $0 < \alpha \leq 1$.

(A.2) The function g is continuous in $[0, T]$ and satisfies the inequality

$$| g(t) | \leq M_g,$$

where $M_g = const > 0$, $t \in [0, T]$.

(A.3) The function h_0 possesses continuous derivatives to the third order in $[0, +\infty)$ and satisfies the inequality

$$| h_0^{(k)}(x) | \leq M_h^0,$$

where $M_h^0 = const > 0$, $k = 0, 1, 2, 3$, $x \in [0, +\infty)$.

(A.4) The function h_1 is continuous in $[0, +\infty)$ and satisfies the inequality

$$| h_1(x) | \leq M_h^1,$$

where $M_h^1 = const > 0$, $x \in [0, +\infty)$.

We now prove

THEOREM 1 *If assumptions (A.1) - (A.4) are satisfied, then there exists a solution u of the problem $(\mathcal{C} - \mathcal{D} - \mathcal{L})$; moreover, the following inequality holds*

$$\| u \|_0 \leq \mathcal{M} (\| f \|_0 + \langle g \rangle_0 + | h_0 |_0 + | h_1 |_0), \tag{20}$$

where $\| v \|_0 \equiv \sup_{x,t} | v(x,t) |$, $\langle w \rangle_0 \equiv \sup_t | w(t) |$, $| z |_0 \equiv \sup_x | z(x) |$, $\mathcal{M} = 2 \max(1, T, \frac{1}{2} T^2)$.

Proof: We shall seek a solution of the problem $(\mathcal{C} - \mathcal{D} - \mathcal{L})$ in the following form

$$u(x,t) = \mathbf{J}(x,t;h_0) + \mathbf{I}(x,t;h_1) - \mathbf{K}(x,t;\varphi) - \mathbf{Z}(x,t;f), \tag{21}$$

where φ is unknown function.

By Lemmas 1 - 4 and Remark 1 in POPIOŁEK [9], we can assert that the function u, given by formula (21), satisfies the equation (17) and the initial conditions (19). Imposing on the said function the boundary condition (18), we get the following equality

$$\varphi(t) = g(t) - \mathbf{I}(0,t;h_1) - \mathbf{J}(0,t;h_0) + \mathbf{Z}(0,t;f). \tag{22}$$

Thus, if we substitute the function φ, given by (22), into the formula (21), we obtain a solution of the $(\mathcal{C} - \mathcal{D} - \mathcal{L})$ problem.

It remains to prove the estimation (20).

Let us observe that by Theorem 1 in POPIOŁEK [9], it is sufficient to estimate the integral $\mathbf{K}(x, t; \varphi)$.

By virtue of formulae (29) - (31) in POPIOŁEK [9], assumptions (A.1) - (A.4) and (22), the following inequality is valid

$$\langle \varphi \rangle_0 \leq \mathcal{M} \left(\| f \|_0 + | h_0 |_0 + | h_1 |_0 \right) + \langle g \rangle_0 .$$

Repeating the argument used in the proof of Lemma 4 in KRZYŻAŃSKI [8], we get

$$| \mathbf{K}(x, t; \varphi) | \leq \langle \varphi \rangle_0 . \tag{23}$$

Thus, by the above-obtained results and Theorem 1 in POPIOŁEK [9], we can conclude that estimate (20) holds true, as required.

5 THE NONLINEAR PROBLEM

We shall examine the nonlinear Cauchy-Dirichlet problem $(\mathcal{C} - \mathcal{D} - \mathcal{N})$:

Find a solution u of the partial differential equation

$$\mathbf{L}\, u(x, t) = \mathbf{F}[\, x, t; u(x, t)\,] \tag{24}$$

in $\mathcal{R}_T^+$ satisfying the boundary condition

$$u(0, t) = g(t) \, , \tag{25}$$

where $t \in [\,0, T\,]$, and the initial conditions

$$D_t^k u(x, 0) = h_k(x) \, , \tag{26}$$

where $x \in [\,0, +\infty\,)$, $k = 0, 1$ and $\mathbf{F}$, g, h_0, h_1 are given functions such that the following compatibility conditions

$$g(0) = h_0(0) \, , \quad g'(0) = h_1(0)$$

hold good.

We need one more assumption

(A.5) The function $\mathbf{F}$ is defined and continuous for $(x,t) \in \mathcal{R}_T^+$, $|u| < \infty$ and satisfies the conditions

$$|\mathbf{F}[x,t;u]| \leq M_F(1 + |u|^{k_F}),$$

$$|\mathbf{F}[x,t;u] - \mathbf{F}[\overline{x},t;\overline{u}]| \leq m_F(|x - \overline{x}|^{h_F} + |u - \overline{u}|),$$

where M_F, $m_F = const > 0$, $k_F \in (0,1)$, $h_F \in (0,1]$.

THEOREM 2 *If assumptions (A.2) - (A.5) are satisfied, and the following condition*

$$m_F T^2 < 1 \tag{27}$$

holds good, then there exists a solution u of the $(\mathcal{C} - \mathcal{D} - \mathcal{N})$ problem. The said solution is unique.

Proof: We seek a solution of the $(\mathcal{C} - \mathcal{D} - \mathcal{N})$ problem in the following form

$$u(x,t) = \mathbf{J}(x,t;h_0) + \mathbf{I}(x,t;h_1) - \mathbf{K}(x,t;\varphi) - \mathbf{Z}(x,t;F[u]), \tag{28}$$

where

$$\varphi(t) = g(t) - \mathbf{I}(0,t;h_1) - \mathbf{J}(0,t;h_0) + \mathbf{Z}(0,t;\mathbf{F}[u]) \tag{29}$$

and

$$\mathbf{Z}(x,t;\mathbf{F}[u]) = \frac{3}{\pi}\int\int_{\mathcal{R}_t^+} \mathbf{V}(x,t;y,\tau)\,\mathbf{F}[y,\tau;u(y,\tau)]\,dy\,d\tau\,.$$

Proceeding by the usual method we shall define a sequence of successive approximations for equation (28)

$$u_{n+1}(x,t) = \mathbf{J}(x,t;h_0) + \mathbf{I}(x,t;h_1) - \mathbf{K}(x,t;\varphi_n) - \mathbf{Z}(x,t;F[u_n]) \tag{30}$$

$(n = 0, 1, \ldots)$, where

$$\varphi_n(t) = g(t) - \mathbf{I}(0,t;h_1) - \mathbf{J}(0,t;h_0) + \mathbf{Z}(0,t;\mathbf{F}[u_n]) \tag{31}$$

and u_0 is an arbitrarily fixed function such that

$$\| u_0 \|_0 \leq \lambda, \tag{32}$$

with λ - a positive parameter.

It can be checked that if the parameter λ satisfies the conditions

$$\mathcal{M}(M_h^0 + M_h^1 + M_g + M_F + M_F\lambda^{k_F}) \leq \lambda,$$

then on the basis (20) , (30) , (31) , (32) and assumption (A.5) we obtain

$$\| u_1 \|_0 \leq \lambda .$$

Using mathematical induction, one can prove that the sequence of functions $\{ u_n \}$ defined by formula (30) is a sequence of equibounded functions. We shall prove that the said sequence is uniformly convergent.

On the basis of (30) we obtain

$$u_{n+1} - u_n = -\mathbf{K}(x,t;\varphi_n) + \mathbf{K}(x,t;\varphi_{n-1}) - \tag{33}$$

$$-\mathbf{Z}(x,t;\mathbf{F}[u_n]) + \mathbf{Z}(x,t;\mathbf{F}[u_{n-1}]),$$

where φ_n and φ_{n-1} are understood as in (31).

In view of (14) we can write

$$\mathbf{K}(x,t;\varphi_n) - \mathbf{K}(x,t;\varphi_{n-1}) = \mathbf{K}(x,t;\varphi_n - \varphi_{n-1}),$$

whence and by formula (23) we can conclude that

$$| \mathbf{K}(x,t;\varphi_n) - \mathbf{K}(x,t;\varphi_{n-1}) | \leq | \varphi_n - \varphi_{n-1} | . \tag{34}$$

As a result of (31) and (34), we obtain

$$| \mathbf{K}(x,t;\varphi_n) - \mathbf{K}(x,t;\varphi_{n-1}) | \leq | \mathbf{Z}(0,t;\mathbf{F}[u_n]) - \mathbf{Z}(0,t;\mathbf{F}[u_{n-1}]) | . \tag{35}$$

It follows from (13) that

$$\mathbf{Z}(x,t;\mathbf{F}[u_n]) - \mathbf{Z}(x,t;\mathbf{F}[u_{n-1}]) = \mathbf{Z}(x,t;\mathbf{F}[u_n] - \mathbf{F}[u_{n-1}]),$$

hence by Lemma 3 in Popiołek [9], we have

$$| \mathbf{Z}(x,t;\mathbf{F}[u_n]) - \mathbf{Z}(x,t;\mathbf{F}[u_{n-1}]) | \leq \tfrac{1}{2}T^2 \, | \mathbf{F}[u_n] - \mathbf{F}[u_{n-1}] |,$$

whence and by assumption (A.5), we get

$$| \mathbf{Z}(x,t;\mathbf{F}[u_n]) - \mathbf{Z}(x,t;\mathbf{F}[u_{n-1}]) | \leq \tfrac{1}{2} m_F T^2 \, | u_n - u_{n-1} | . \tag{36}$$

On joining (33), (35) and (36) we obtain

$$| u_{n+1} - u_n | \leq m_F T^2 \, | u_n - u_{n-1} | . \tag{37}$$

Consequently, as follows from the condition (27) we have right to state that the sequence $\{ u_n \}$, defined by formula (30), is uniformly convergent and, putting

$$\tilde{u}(x,t) = \lim_{n\to\infty} u_n(x,t)\,, \quad (x,t) \in \overline{\mathcal{R}}_T^+\,,$$

we can conclude that the function $\tilde{u}$ is the required solution of the $(\mathcal{C} - \mathcal{D} - \mathcal{N})$ problem. We still have to prove the uniqueness of the above-obtained solution.
If $\overline{u}$ and $\underline{u}$ are two solutions of the $(\mathcal{C} - \mathcal{D} - \mathcal{N})$ problem, then putting $u = \overline{u} - \underline{u}$ and taking into consideration the estimate (20) we obtain $u \equiv 0$, by which we end the proof of Theorem 2.

References

[1] ABDINAZAROV, S., (1989), *On an equation of third order*, Izv. Akad. Nauk Uz. S. S. R., 6, 3-7 (in Russian).

[2] BLOCK, H., (1912), *Sur les équations linéaires aux dérivées partielles à caractéristiques multiples*, Arkiv för Mat. Astr. och Fys., I, 7:13, 1-34, II, 7:21, 1-30.

[3] BLOCK, H., (1913), *Sur les équations linéaires aux dérivées partielles à caractéristiques multiples*, Arkiv för Mat. Astr. och Fys., III, 8:23, 1-51.

[4] CATTABRIGA, L., (1961), *Potenziali di linea e di dominio per equazioni non paraboliche in due variabili a caratteristiche multiple*, Rend. Sem. Math. Univ. Padova, 31, 1-45.

[5] DZHURAEV, T. D., ABDINAZAROV, S., (1989), *On the theory of equations with characteristics multiple*, Dokl. Akad. Nauk Az. S. S. R., 4:1, 42-45 (in Russian).

[6] DZHURAEV, T. D., POPIOŁEK, J., (1989), *Canonical forms of third-order partial differential equations*, Uspekhi Mat. Nauk, 44:4(268), 237-238 = Russian Math. Surveys, 44:4, 203-204.

[7] DZHURAEV, T. D., POPIOŁEK, J., (1991), *On classification and reduction to canonical forms of partial differential equations of third order*, Differencial'nye Uravnenija, 27:10, 1734-1745 = Differential Equations, 27:10.

[8] KRZYŻAŃSKI, M., (1971), *Partial differential equations of second order*, vol. I, PWN, Warsaw.

[9] POPIOŁEK, J., *On Cauchy's problem for an equation of third order* (to appear).

[10] DEL VECCHIO, E., (1916-17), *Sur deux problèmes d'intégration pour les équations paraboliques* $\partial^3 z/\partial\xi^3 - \partial z/\partial\eta = 0$, $\partial^3 z/\partial\xi^3 - \partial^2 z/\partial\eta^2 = 0$, BLOCK, H., *Remarque à la note précédente*, Arkiv för Mat. Astr. och Fys., 11, 3-20.

Linear Evolutionary Integral Equations on the Line

J. PRÜSS University of Paderborn, Paderborn, Germany

0. Introduction

Let X, Y be Banach spaces such that $Y \overset{d}{\hookrightarrow} X$, $\{A(t)\}_{t\geq 0} \subset \mathcal{B}(Y, X)$ measurable, and consider the Volterra equation on the line

$$u(t) = g(t) + \int_0^\infty A(\tau)u(t-\tau)d\tau\ , \quad t \in \mathbb{R}, \tag{0.1}$$

where $g \in L^1_{loc}(\mathbb{R} : X)$. This paper is devoted to a study of the solvability behavior of (0.1) in various spaces of functions on the line.

In case $g(t) \equiv 0$ and $u(t) \equiv 0$ for $t < 0$, (0.1) reduces to the Volterra equation on the halfline

$$u(t) = g(t) + \int_0^t A(\tau)u(t-\tau)d\tau\ , \quad t \in \mathbb{R}_+, \tag{0.2}$$

which together with its differentiated version has been the subject of many papers in recent years. It is impossible to give here an account of the full literature concerning (0.2); instead we refer only to the forthcoming monograph Prüss [16], which contains the theory of (0.2) as well as of (0.1) and a fairly complete list of references.

Todays theory for (0.2) centers around the concept of the *resolvent* $S(t)$, which whenever it exists is unique and represents the solution by the *variation of parameters formula*

$$u(t) = \frac{d}{dt}\int_0^t S(t-s)f(s)ds\ , \quad t \in \mathbb{R}_+. \tag{0.3}$$

The resolvent $S(t)$ for (0.2) is a strongly continuous family of bounded linear operators in X leaving invariant Y, and also strongly continuous in Y, such that the *resolvent equations* hold.

$$S(t)y = y + \int_0^t A(t-s)S(s)yds = y + \int_0^t S(t-s)A(s)yds\ , \quad y \in Y\ ,\ t \in \mathbb{R}_+. \tag{0.4}$$

To measure the time regularity of forcing terms required to obtain strong solutions of (0.2), the notion of a-regularity of resolvents is useful. Let $a \in L^1_{loc}(\mathbb{R}_+)$; a resolvent $S(t)$ for (0.2) is called *a-regular* if the convolution of a with S over the halfline, i.e.

$$(a * S)(t) = \int_0^t a(t-s)S(s)ds \quad t > 0,$$

is strongly continuous from X to Y. For example, a C_0-semigroup is always 1-regular, while a cosine family is t-regular.

There are two main differences between (0.1) and (0.2). An important property of (0.2) not shared by (0.1) is its *locality*. When studying (0.2) on the halfline we may actually restrict attention to compact intervals $J = [0, T]$, while when working with (0.1) we are always forced to consider unbounded intervals. On the other hand, (0.1) enjoys the property of *translation invariance* which is not present in (0.2). Thus one cannot expect e.g. *periodic* or *almost periodic solutions* of (0.2) when the forcing function g has such a property, for this one must consider (0.1). The proper question concerning periodic or almost periodic behaviour of (0.2) would be to find solutions which are *asymptotically periodic* or *asymptotically almost periodic*. Like in the finite dimensional case, under reasonable assumptions it then turns out that the solutions of (0.2) and (0.1) are asymptotic to each other, in particular having solved the periodic problem for (0.1), we obtain also a solution of the asymptotically periodic problem for (0.2).

These considerations should give enough motivation to study the solvability behaviour of (0.1). In addition, problems from viscoelasticity or heat conduction in materials with memory or electrodynamics with memory in general lead to problems of the form (0.1) rather than (0.2); the latter is only obtained after restricting attention to the positive halfline, by assuming $u(t) = g(t) = 0$ for $t < 0$ as above, or by considering the history of $u(t)$ for $t < 0$ as known and incorporating it into the forcing function $g(t)$ for $t > 0$.

(0.1) in a somewhat more special form was studied for the first time in the author's 'Habilitationschrift' [12]; slightly improved versions of the results obtained there were published in Prüss [14]. This paper can be considered as a continuation of that work. Da Pato and Lunardi [6] study the differentiated version of (0.1) with main part A_0 a generator of an analytic semigroup, $Y = \mathcal{D}(A_0)$ equipped with the graph norm of A_0, plus a perturbation part in spaces of uniformly continuous bounded functions on the line, or uniformly Hölder continuous functions. Their results are fairly complete as regards to their setting, however, is restricted to the parabolic case and require perturbations which admit analytic extension to a sector symmetric to the positive halfline, i.e. fairly nice perturbations. In Clément and da Prato [4] and in Clément and Prüss [5] the sum approach based on the da Prato-Grivard theorem resp. the Dore-Venni theorem is applied to an equation motivated by the theory of heat conduction in material with memory, which formally can be rewritten as (0.1) with $A(t) = a(t)A_0$, A_0 generating an analytic semigroup and $a(t)$ a real-valued kernel with special properties.

The principal idea in the papers [12], [14] follows the classical approach in finite dimensions. It is concerned with the *admissibility* of certain translation invariant spaces of functions $g : \mathbb{R} \to X$ and characterizations of the latter in terms of the *symbol* $H(\lambda)$ of (0.1), defined formally by

$$H(\lambda) = (I - \hat{A}(\lambda))^{-1}/\lambda\,, \quad \lambda \in \mathbb{C}\,, \tag{0.5}$$

which is just the Laplace transform of the resolvent $S(t)$ of (0.2). Observe that formally the Fourier transform $\tilde{u}$ of the solution u of (0.1) is given by

$$\tilde{u}(\rho) = H(i\rho)\tilde{g}(\rho), \quad \rho \in \mathbb{R}. \tag{0.6}$$

Defining the *spectrum* Λ_0 of (0.1) as

$$\Lambda_0 = \{\rho \in \mathbb{R} : i\rho - i\rho\hat{A}(i\rho) \in \mathcal{B}(Y, X) \text{ not boundedly invertible}\}, \tag{0.7}$$

it is natural to expect that the solvability behaviour of (0.1) should be reasonable, provided we restrict the forcing functions in such a way that

$$\Lambda_0 \cap \operatorname{supp} \tilde{g} = \emptyset.$$

Therefore it is very useful to have a concept which allows for localization of (0.1) in the frequency domain. This is achieved by the notion of the spectrum $\sigma(f)$ of a function f on the line which goes back to Carleman; one of its several equivalent definitions is

$$\sigma(f) = \operatorname{supp} \widetilde{D_f},$$

where D_f denotes the distribution associated with f and $\widetilde{D_f}$ means its Fourier transform in the sense of distributions.

In this paper we present a fairly general framework for the description of the solvability properties of (0.1) in terms of admissability, the spectrum of the forcing functions involved and the spectrum Λ_0 of (0.1). Section 1 contains the definition and necessary background on $\sigma(f)$; proofs have been included for the sake of completeness and since there seems to be no reference which covers the needed results, in infinite dimensions, at least. In Section 2 the class of homogeneous spaces is introduced and the above mentioned localization in terms of such spaces is explained. The following three sections form the backbones of our approach. Admissability is introduced, necessary conditions in terms of the symbol for admissability are derived and consequences in particular for periodic and almost periodic behaviour of (0.1) are derived. The crucial result for this part of the theory is the unique solvability of (0.1), whenever $\sigma(f)$ is compact and does not intersect Λ_0.

Sections 6 and 7 are devoted to two results containing sufficient conditions for admissibility in the hyperbolic case; see the papers cited above and Prüss [16] for the much better behaved parabolic case. Theorem 4 is a perturbation result which is based on the Paley-Wiener lemma in infinite dimensions; see Prüss [12] or Gripenberg [10]. In the case of Hilbert spaces X and Y, Parseval's theorem leads to a full characterization of admissability in terms of the symbol $H(\lambda)$ if only mild regularity restrictions are imposed. The corresponding result, Theorem 5, is a considerable extension of Theorem 3 of Prüss [14] which contains as a special case Gearhart's characterization of the spectrum of a C_0-semigroup of contractions in a Hilbert space; this is worked out in detail in Example 2. Finally, in Section 8 we apply the forgoing results to the viscoelastic Timoshenko beam model; cp. Grimmer et al [9].

1. The Spectrum of Functions

Let $f \in L^1_{loc}(\mathbb{R} : X)$ be of *subexponential growth*, where X denotes a complex Banach space; this means

$$\int_{-\infty}^{\infty} |f(t)|e^{-\epsilon|t|}dt < \infty, \quad \text{for each } \epsilon > 0.$$

The *Fourier-Carleman transform* $\hat{f}$ of f is then defined by

$$\hat{f}(\lambda) = \begin{cases} \int_0^\infty e^{-\lambda t} f(t)dt , & \text{Re } \lambda > 0 \\ -\int_{-\infty}^0 e^{-\lambda t} f(t)dt , & \text{Re } \lambda < 0. \end{cases}$$

Obviously, $\hat{f}(\lambda)$ is holomorphic in $\mathbb{C} \setminus i\mathbb{R}$. Define

$$\rho(f) = \{\rho \in \mathbb{R} : \hat{f}(\lambda) \text{ admits analytic continuation to some ball } B_\varepsilon(i\rho)\}; \tag{1.1}$$

then $\sigma(f) = \mathbb{R} \setminus \rho(f)$ is called the *spectrum* of f. The Fourier-Carleman transform and this approach of defining the spectrum of functions in the scalar case are due to Carleman [3]; see also Katznelson [11] and Prüss [14].

Example 1 (i) Suppose $f \in L^1_{loc}(\mathbb{R}; X)$ is τ-periodic. Then an easy computation yields

$$\hat{f}(\lambda) = (1 - e^{-\lambda\tau})^{-1} \int_0^\tau e^{-\lambda t} f(t)dt , \quad \text{Re } \lambda \neq 0, \tag{1.2}$$

i.e. $\hat{f}(\lambda)$ extends to a meromorphic function with simple poles at most at $\lambda_n = 2\pi i n/\tau$, $n \in \mathbf{Z}$. The residuum at $\lambda = \lambda_n$ is given by

$$\text{Res}\hat{f}(\lambda)|_{\lambda=\lambda_n} = f_n = \frac{1}{\tau}\int_0^\tau e^{-2\pi i n t/\tau} f(t)dt , \quad n \in \mathbf{Z} ,$$

the nth Fourier coefficient of f. Therefore we have

$$\sigma(f) = \{2\pi n/\tau : n \in \mathbf{Z} , \ f_n \neq 0\} \subset (2\pi/\tau)\mathbf{Z} . \tag{1.3}$$

(ii) Let $f \in L^1(\mathbb{R}; X)$ and let $\tilde{f}(\rho)$ denote its Fourier transform

$$\tilde{f}(\rho) = \int_{-\infty}^\infty f(t) e^{-i\rho t} dt , \quad \rho \in \mathbb{R}. \tag{1.4}$$

Then

$$\lim_{\sigma\to 0+} (\hat{f}(i\rho + \sigma) - \hat{f}(i\rho - \sigma)) = \lim_{\sigma\to 0+} \int_{-\infty}^\infty f(t) e^{-i\rho t} e^{-\sigma|t|} dt = \tilde{f}(\rho),$$

for every $\rho \in \mathbb{R}$; therefore $\rho_0 \notin \sigma(f)$ implies $\tilde{f}(\rho) \equiv 0$ in a neighborhood of ρ_0, and so $\rho_0 \notin \text{supp}\, \tilde{f}$. Conversely, if $\rho_0 \notin \text{supp}\, \tilde{f}$ then $\hat{f}(\lambda)$ is continuous on a ball $B_\varepsilon(i\rho_0)$, hence by Morera's theorem (cp. Conway [Con78]). $\hat{f}(\lambda)$ is holomorphic there, i.e. $\rho_0 \notin \sigma(f)$. This implies the relation

$$\sigma(f) = \text{supp}\, \tilde{f} . \quad \text{for all} \quad f \in L^1(\mathbb{R}; X). \qquad \square \tag{1.5}$$

Below we will show that (1.5) holds for all $f \in L^1_{loc}(\mathbb{R}; X)$ of polynomial growth provided the Fourier transform of f is understood in the sense of vector-valued distributions.

Next we summarize some of the properties of the spectrum of functions of subexponential growth.

Proposition 1 *Let $f, g \in L^1_{loc}(\mathbb{R}; X)$ be of subexponential growth, $\alpha \in \mathbb{C} \setminus \{0\}$, $K \in BV(\mathbb{R}; \mathcal{B}(X; Z))$ with* supp dK *compact. Then*
(i) $\sigma(f)$ is closed ; *(ii) $\sigma(\alpha f) = \sigma(f)$;*

(iii) $\sigma(f(\cdot+h)) = \sigma(f)$; *(iv)* $\sigma(\dot{f}) \subset \sigma(f)$
(v) $\sigma(f+g) \subset \sigma(f) \cup \sigma(g)$ *(vi)* $\sigma(dK * f) \subset \sigma(f)$;
(vii) $\sigma(\widetilde{dK} f) \subset \sigma(f) - \operatorname{supp} dK$

Proof: (i) and (ii) are trivial by the definition of $\sigma(f)$. (iii) follows from the identity

$$\hat{f}(\cdot+h)(\lambda) = e^{\lambda h}\hat{f}(\lambda) - \int_0^h e^{\lambda(h-t)} f(t)dt \,, \qquad \operatorname{Re}\lambda \neq 0, \tag{1.6}$$

since the second term on the right is an entire function.
(iv) If $\dot{f} \in L^1_{loc}(\mathbb{R} : X)$ is of subexponential growth, then

$$\widehat{\dot{f}}(\lambda) = \lambda\hat{f}(\lambda) - f(0)\,, \qquad \operatorname{Re}\ \lambda \neq 0;$$

this implies (iv).
(v) If $\rho \notin \sigma(f) \cup \sigma(g)$ then $\hat{f}(\lambda)$ and $\hat{g}(\lambda)$ admit analytic extension to a ball $B_\varepsilon(i\rho_0)$, hence $\rho \notin \sigma(f+g)$.
(vi) Since supp dK is compact, i.e. $K(t)$ is constant for $t \geq N$, $t \leq -N$, say, $dK * f$ is welldefined and

$$\int_{-\infty}^{\infty} e^{-\varepsilon|t|}|dK * f|dt \leq \int_{-N}^{N} |dK(\tau)|(\int_{-\infty}^{\infty} e^{-\varepsilon|t+\tau|}|f(t)|dt)$$
$$\leq (\int_{-N}^{N} |dK(\tau)|e^{\varepsilon|\tau|}) \int_{-\infty}^{\infty} e^{-\varepsilon|t|}|f(t)|dt,$$

hence $dK * f$ is again of subexponential growth. A simple computation yields for Re $\lambda \neq 0$

$$(dK * f)\hat{}(\lambda) = (\int_{-\infty}^{\infty} dK(\tau)e^{-\lambda\tau})\hat{f}(\lambda) + \int_{-\infty}^{\infty} dK(\tau)\int_0^{\tau} f(t-\tau)e^{-\lambda t}dt \tag{1.7}$$

and so (vi) follows since the coefficient of $\hat{f}(\lambda)$ and the second term are entire.
(vii) We have for Re $\lambda > 0$

$$\begin{aligned}(\widetilde{dK}\cdot f)\hat{}(\lambda) &= \int_{-\infty}^{\infty} \widetilde{dK}(t)f(t)e^{-\lambda t}dt = \int_{-\infty}^{\infty} dK(\rho)\int_{-\infty}^{\infty} f(t)e^{-(\lambda+i\rho)t}dtds \\ &= \int_{-\infty}^{\infty} dK(\rho)\hat{f}(\lambda+i\rho)d\rho.\end{aligned} \tag{1.8}$$

and similarly for Re $\lambda < 0$. This formula defines an analytic extension to all $\lambda = i\rho_0$, such that $\rho_0 + \rho \notin \sigma(f)$, i.e. (vii) holds. □

The spectrum of a function also enjoys a certain continuity property which turns out to be very useful.

Theorem 1 *Let $f_n, f \in L^1_{loc}(\mathbb{R}; X)$ be of subexponential growth and such that*
(i) $\int_{-\infty}^{\infty} e^{-\varepsilon|t|}|f_n(t) - f(t)|dt \to 0$ *as* $n \to \infty$, *for each* $\varepsilon > 0$;
(ii) there are $c > 0$ *and* $k \in \mathbb{N}_0$ *such that*

$$|\hat{f}_n(\lambda)| \leq c|Re\ \lambda|^{-k}\,, \quad 0 < |Re\ \lambda| < \delta\,,\ n \in \mathbb{N}.$$

Then $\sigma(f_n) \subset \Lambda$ *for all* $n \in \mathbb{N}$ *implies* $\sigma(f) \subset \Lambda$*: equivalently*

$$\sigma(f) \subset \cap_{m\geq 1}\overline{\cup_{n\geq m}\sigma(f_n)}. \tag{1.9}$$

Proof: (i) implies $\widehat{f_n}(\lambda) \to \hat{f}(\lambda)$ uniformly on compact subsets of $\mathbb{C} \setminus i\mathbb{R}$. Let $\Lambda \subset \mathbb{R}$ be closed and such that $\sigma(f_n) \subset \Lambda$ for all $n \in \mathbb{N}$. Choose $\rho_0 \notin \Lambda$, and let $r \leq \text{dist}(\rho_0, \Lambda)/4$, $\delta/2$. We want to prove that $\hat{f}(\lambda)$ admits analytic extension to $B_\varepsilon(i\rho_0)$; for this it sufficient to prove

$$|\widehat{f_n}(\lambda)| \leq M \quad \text{for all } n \in \mathbb{N}\ ,\ |\lambda - i\rho_0| \leq r, \tag{1.10}$$

by Montel's theorem (cp. Conway [Con78]. For this purpose observe that $\widehat{f_n}(\lambda)$ is holomorpic on $B_\varepsilon(i\rho_0)$, hence Cauchy's theorem yields with

$$h(\lambda) = [(\lambda - i\rho_0)(1 + \frac{(\lambda - i\rho_0)^2}{4r^2})]^k$$

the relation

$$h(\lambda)\widehat{f_n}(\lambda) = \frac{1}{2\pi i}\int_{|z-i\rho_0|=2r} h(z)\widehat{f_n}(z)\frac{dz}{z-\lambda}, \quad \lambda \in B_r(i\rho_0)\ ,\ n \in \mathbb{N}.$$

Since for $|z - i\rho_0| = 2r$ we have $|h(z)| = 2^k|\text{ Re } z|^k$, (ii) yields

$$|h(\lambda)\widehat{f_n}(\lambda)| \leq \frac{c}{2\pi}\int_{|z-i\rho_0|=2r} 2^k|\text{ Re } z|^k \cdot |\text{ Re } z|^{-k}\frac{|dz|}{|z-\lambda|} \leq 2^{k+1}c, \quad \lambda \in B_r(i\rho_0)\ ,\ n \in \mathbb{N},$$

and so

$$\sup_{\lambda \in B_r(i\rho_0)} |\widehat{f_n}(\lambda)| = \sup_{|\lambda - i\rho_0|=r} |\widehat{f_n}(\lambda)| \leq 2^{k+1}c \cdot \sup_{|\lambda - i\rho_0|=r} |\frac{1}{h(\lambda)}| \leq 2^{k+1}c(\frac{4}{3})^k r^{-k} = M,$$

independent of $n \in \mathbb{N}$, i.e. (1.10) follows and the proof is complete. □

Observe that (i) of Theorem 1 is implied by

$$|f_n| \leq g \quad \text{a.e.}\ , \quad f_n \to f \text{ a.e.} \tag{1.11}$$

where $g \in L^1_{loc}(\mathbb{R})$ is of subexponential growth, while (ii) follows if g is *growing polynomially*, i.e. $\int_{-\infty}^{\infty} |g(t)|(1+|t|)^{-m}dt < \infty$ for some $m \in \mathbb{N}$. In fact, then e.g. for Re $\lambda > 0$

$$\begin{aligned}|\widehat{f_n}(\lambda)| &\leq \int_0^\infty g(t)e^{-Re\lambda t}dt \leq (\int_0^\infty g(t)(1+t)^{-m}dt) \cdot \sup_{t>0}[(1+t)^m e^{-Re\lambda t}] \\ &\leq (\int_0^\infty g(t)(1+t)^{-m}dt)(\frac{m}{\text{Re }\lambda})^m e^{1-m}, \quad 0 < \text{Re }\lambda < 1.\end{aligned}$$

We have seen in Example 1, (ii) that $\sigma(f) = \text{supp}\tilde{f}$ holds for $f \in L^1(\mathbb{R})$. This characterization of the spectrum allows for a considerable extension if the Fourier transform is understood in the sense of distributions. For this purpose let $f \in L^1_{loc}(\mathbb{R}; X)$ be of polynomial growth; the tempered distribution D_f corresponding to f is then defined according to

$$[D_f, \varphi] = \int_{-\infty}^{\infty} f(t)\varphi(t)dt\ , \quad \varphi \in \mathcal{S}, \tag{1.12}$$

where $\mathcal{S}$ denotes the Schwartz space of all C^∞-functions on $\mathbb{R}$ with each of its derivatives decaying faster than any polynomial. The Fourier transform of D_f is defined by

$$[\widetilde{D_f}, \varphi] = [D_f, \tilde{\varphi}]\ , \quad \varphi \in \mathcal{S}. \tag{1.13}$$

Recall also the definition of supp D_f. A number $\rho \in \mathbb{R}$ belongs to supp D_f if for every $\varepsilon > 0$ there is $\varphi \in \mathcal{S}$ such that supp $\varphi \subset B_\varepsilon(\rho)$ and $[D_f, \varphi] \neq 0$.

Proposition 2 *Let $f \in L^1_{loc}(\mathbb{R}; X)$ be of polynomial growth. Then*
(i) $\sigma(f) = \operatorname{supp}\widetilde{D_f}$;
(ii) $\sigma(f) = \emptyset \iff f \equiv 0$;
(iii) $\sigma(f) = \{\rho_1, \ldots, \rho_n\} \iff f(t) = \sum_{k=1}^n p_k(t)e^{i\rho_k t}$, for some polynomials $p_k(t)$.
(iv) $\sigma(f)$ compact $\iff f$ admits extension to an entire function of exponential growth.

Proof: (i) If $\rho_0 \notin \sigma(f)$ then $\hat{f}(\lambda)$ admits holomorphic extension to some ball $B_\varepsilon(i\rho_0)$. Let $\varphi \in \mathcal{S}$ be such that supp $\varphi \subset B_\varepsilon(\rho_0)$; then

$$\begin{aligned}\int_{-\infty}^{\infty} (\hat{f}(i\rho + \sigma) - \hat{f}(i\rho - \sigma))\varphi(\rho)d\rho &= \int_{-\infty}^{\infty} (\int_{-\infty}^{\infty} f(t)e^{-i\rho t}e^{-\sigma|t|}dt)\varphi(\rho)d\rho \\ &= \int_{-\infty}^{\infty} f(t)e^{-\sigma|t|}\tilde{\varphi}(t)dt\end{aligned}$$

implies with $\sigma \to 0$

$$\begin{aligned}[\widetilde{D_f}, \varphi] &= \int_{-\infty}^{\infty} f(t)\tilde{\varphi}(t)dt = \lim_{\sigma\to 0+}\int_{-\infty}^{\infty} f(t)e^{-\sigma|t|}\tilde{\varphi}(t)dt \\ &= \lim_{\sigma\to 0+}\int_{-\infty}^{\infty} (\hat{f}(i\rho + \sigma) - \hat{f}(i\rho - \sigma))\varphi(\rho)d\rho = 0,\end{aligned}$$

i.e. $\rho_0 \notin \operatorname{supp}\widetilde{D_f}$. This shows the inclusion $\operatorname{supp}\widetilde{D_f} \subset \sigma(f)$.

To prove the converse inclusion we use Theorem 1. For this purpose let $\psi \in C_0^\infty(\mathbb{R})$ be such that supp $\psi \subset (-1, 1)$, $\psi \geq 0$, and $\int_{-\infty}^{\infty} \psi(\rho)d\rho = 1$; define $\psi_n(\rho) = n\psi(n\rho)$. Then $\widetilde{\psi_n} \to 1$ as $n \to \infty$, uniformly for bounded t, and $\widetilde{\psi_n} \in \mathcal{S}$. Since f is at most polynomially growing, $f_n = \widetilde{\psi_n} f$ belongs to $L^1(\mathbb{R}; X)$, $f_n \to f$ uniformly for bounded t, and $|f_n| \leq |f| = g$. Thus (1.11) holds, which implies (i) and (ii) of Theorem 1 since f is growing polynomially. Example 1. (ii) yields

$$\sigma(f_n) = \operatorname{supp}\widetilde{f_n} \subset B_\varepsilon(\operatorname{supp}\widetilde{D_f}) \quad \text{for} \quad n \geq n(\varepsilon).$$

In fact, if $\varphi \in \mathcal{S}$ is such that supp $\varphi \subset B_{\varepsilon/2}(\rho_0)$, then

$$\begin{aligned}\int_{-\infty}^{\infty} \widetilde{f_n}(\rho)\varphi(\rho)d\rho &= \int_{-\infty}^{\infty} f_n(t)\tilde{\varphi}(t)dt = \int_{-\infty}^{\infty} f(t)\widetilde{\psi_n}(t)\widetilde{\varphi_n}(t)dt \\ &= \int_{-\infty}^{\infty} f(t)\widetilde{\psi_n * \varphi}(t)dt = [\widetilde{D_f}, \psi_n * \varphi];\end{aligned}$$

with supp $(\psi_n * \varphi) \subset$ supp φ + supp $\psi_n \subset B_{\varepsilon/2}(\rho_0) + (-1/n, 1/n)$ we conclude $[\widetilde{D_f}, \psi_n * \varphi] = 0$ if dist $(\rho_0, \operatorname{supp}\widetilde{D_f}) \geq \varepsilon$ and $\varepsilon/2 > 1/n$. Therefore by Theorem 1 we obtain $\sigma(f) \subset B_\varepsilon(\operatorname{supp}\widetilde{D_f})$ for every $\varepsilon > 0$, i.e. $\sigma(f) \subset \operatorname{supp}\widetilde{D_f}$.
(ii) $\sigma(f) = \emptyset$ implies $\widetilde{D_f} = 0$ by (i), hence $D_f = 0$, and so $f = 0$.
(iii) Consider first the special case $\sigma(f) = \{0\}$; we then have to show that f is a polynomial. For this purpose consider the Laurent expansion of $\hat{f}(\lambda)$

$$\hat{f}(\lambda) = \sum_{-\infty}^{\infty} a_n\lambda^n, \quad \lambda \in \mathbb{C} \setminus \{0\}.$$

The coefficients a_n are given by

$$a_n = \frac{1}{2\pi i}\int_{|z|=r} \hat{f}(z)z^{-n-1}dz\ , \quad n \in \mathbf{Z}\ ,$$

where $r > 0$ is arbitrary. Since f is of polynomial growth by assumption, there follows the estimate

$$|\hat{f}(\lambda)| \leq c|\text{Re }\lambda|^{-k}\ , \quad 0 < |\text{Re }\lambda| < 1, \tag{1.14}$$

for some $k \in \mathbb{N}_0$. This implies

$$\left|\frac{1}{2\pi i}\int_{|z|=r} z^{n+k-1}(1+\frac{z^2}{r^2})^k\hat{f}(z)dz\right| \leq c2^k r^n, \quad \text{for all} \quad n \in \mathbf{Z}\ , r > 0,$$

hence

$$c2^k r^n \geq \left|\sum_{l=0}^{k}\binom{k}{l} r^{-2l}\cdot\frac{1}{2\pi i}\int_{|z|=r} z^{n+k-1+2l}\hat{f}(z)dz\right| = \left|\sum_{l=0}^{k}\binom{k}{l} r^{-2l}a_{-n-k-2l}\right|$$

Multiplying with r^{2k} and $r \to 0$ then implies $a_{-n-k} = 0$ for all $n \geq 1$, i.e. $\lambda = 0$ is a pole of order at most k of $\hat{f}(\lambda)$. Let $f_0(t) = \sum_{l=0}^{k} a_{-l-1}t^l/l!$; obviously $\widehat{f_0}(\lambda) = \sum_{-k}^{-1} a_l\lambda^{-l}$ and so $\hat{f}(\lambda) - \widehat{f_0}(\lambda)$ is entire, i.e. $\sigma(f - f_0) = \emptyset$, hence $f = f_0$. The general case can now easily be reduced to this special case.

(iv) Suppose $\sigma(f)$ is compact; we then define

$$f_0(z) = \frac{-1}{2\pi i}\int_\Gamma \hat{f}(\lambda)e^{\lambda z}d\lambda\ , \quad z \in \mathbb{C}\ ,$$

where Γ denotes any closed rectifiable simple contour surrounding $\sigma(f)$ clockwise. It is easy to verify that for any $|\text{Re }\lambda| > \max_{\mu\in\Gamma}|\text{Re }\mu|$ the identity

$$\widehat{f_0}(\lambda) = \frac{1}{2\pi i}\int_\Gamma \hat{f}(\mu)\frac{d\mu}{\mu-\lambda}$$

holds. Thus it remains to show that the right hand side of this equation equals $\hat{f}(\lambda)$. For this purpose let $R > 0$ be sufficiently large, i.e. $R > |\lambda|$, $\max\{|\rho| : \rho \in \sigma(f)\}$. Then Cauchy's theorem yields the identity

$$(1+\frac{\lambda^2}{R^2})^k\hat{f}(\lambda) = \frac{1}{2\pi i}\int_{\Gamma\cup\Gamma_R}(1+\frac{\mu^2}{R^2})^k\hat{f}(\mu)\frac{d\mu}{\mu-\lambda} \tag{1.15}$$

where $\Gamma = \Gamma^- \cup \Gamma^+$, $\Gamma_R = \Gamma_R^- \cup \Gamma_R^+$ and k from (1.14). Similarly to the estimate in step (iii) of this proof, we obtain

$$\left|\frac{1}{2\pi i}\int_{\Gamma_R}(1+\frac{\mu^2}{R^2})^k\hat{f}(\mu)\frac{d\mu}{\mu-\lambda}\right| \leq c2^kR^{-k}\sup_{|\lambda|=R}\frac{1}{|\mu-\lambda|} \to 0 \quad \text{as} \quad R \to \infty,$$

hence with $R \to \infty$, (1.15) implies $\widehat{f_0}(\lambda) = \hat{f}(\lambda)$ for all Re $\lambda \neq 0$. By uniqueness of the Laplace-transform this yields $f_0 \equiv f$. The converse follows from the Taylor series of entire functions. □

2. Homogeneous Spaces

In general, (0.1) is more difficult to handle than its local version (0.2), however, in contrast to (0.2) it enjoys the property of *translation invariance*, i.e. if u is a solution of (0.1) in the sense of Definition 11.2 below, then $(T_h u)(t) = u(t+h)$, $t \in \mathbb{R}$, is again a solution of (0.1), with g replaced by $T_h g$, for each $h \in \mathbb{R}$. To be able to exploit this property, we consider the solvability behavior of (0.1) only in spaces of locally integrable functions on $\mathbb{R}$ which are translation invariant, more precisely homogeneous in the sense of

Definition 1 *A Banach space $\mathcal{H}(X)$ of locally integrable X-valued functions is said to be homogeneous if*
(i) $\mathcal{H}(X) \hookrightarrow L^1_{loc}(\mathbb{R}; X)$, i.e. for each $T > 0$ there is a constant $C(T) > 0$ such that

$$\int_{-T}^{T} |f(t)|dt \leq C(T)|f|_{\mathcal{H}(X)};$$

(ii) $\{T_h : h \in \mathbb{R}\} \subset \mathcal{B}(\mathcal{H}(X))$ is strongly continuous and uniformly bounded;
(iii) $\mathcal{B}(X) \hookrightarrow \mathcal{B}(\mathcal{H}(X))$, i.e. $f \in \mathcal{H}(X)$, $K \in \mathcal{B}(X)$ imply $Kf \in \mathcal{H}(X)$;
are satisfied.

Examples of homogeneous spaces are $L^p(X) := L^p(\mathbb{R}; X)$, $1 \leq p < \infty$, $C_{ub}(X) := C_{ub}(\mathbb{R}; X)$, the space of bounded uniformly continuous X-valued functions, the space of almost periodic functions $AP(X)$, the space of continuous ω-periodic functions $P_\omega(X)$, etc.. $L^\infty(X) := L^\infty(\mathbb{R}; X)$ and $C_b(X) := C_b(\mathbb{R}; X)$ are not homogeneous since the group of translations is not strongly continuous in these spaces.

The largest space we will consider is $BM(X)$, the space of functions *bounded in mean*, defined by

$$BM(X) = \{f \in L^1_{loc}(\mathbb{R}; X) : |f|_1^b < \infty\},$$

where the norm $|\cdot|_1^b$ in $BM(X)$ is given by

$$|f|_1^b = \sup\{\int_t^{t+1} |f(s)|ds \ : \ t \in \mathbb{R}\}.$$

$BM(X)$ is not homogeneous, however, its closed subspace $BM^0(X)$, defined by

$$BM^0(X) = \{f \in BM(X) : |T_h f - f|_1^b \to 0 \text{ as } h \to 0\},$$

has this property; in fact, it is the largest homogeneous space, as (a) of the next proposition shows.

If $\mathcal{H}(X)$ is homogeneous, the generator of the translation group will be denoted by $D_{\mathcal{H}}$, or simply by $D = d/dt$ if there is no danger of confusion. We then may define

$$\mathcal{H}^m(X) = \mathcal{D}(D^m_{\mathcal{H}}), \quad |f|_{\mathcal{H}^m(X)} := \sum_{l=0}^{m} |D^l f|_{\mathcal{H}(X)}, \quad m \in \mathbb{N},$$

and

$$\mathcal{H}^\infty(X) = \cap_{m \geq 1} \mathcal{H}^m(X);$$

obviously $\mathcal{H}^m(X)$ are again Banach spaces, and $\mathcal{H}^\infty(X)$ is a Frechét space with the seminorms $p_m(f) = |D^m f|_{\mathcal{H}(X)}$, $m \in \mathbb{N}_0$. Of interest is also the space $\mathcal{H}^a(X)$ of entire vectors w.r.t. the translation group, i.e.

$$\mathcal{H}^a(X) = \{f \in \mathcal{H}^\infty(X) : \overline{\lim}_{n\to\infty} |D^n f|_{\mathcal{H}(X)}^{1/n} < \infty\}.$$

This is precisely the space of all functions $f \in \mathcal{H}(X)$ such that $T_h f$ extends to an entire function on $\mathbb{C}$ of exponential growth. Since $\{T_h\}$ is a C_0-group on $\mathcal{H}(X)$ it is wellknown that $\mathcal{H}^a(X)$ is dense in $\mathcal{H}(X)$; this is the second assertion of the next proposition.

Proposition 3 *Let $\mathcal{H}(X)$ be homogeneous. Then*
(i) $\mathcal{H}(X) \hookrightarrow BM^0(X)$;
(ii) $\mathcal{H}^a(X)$ is dense in $\mathcal{H}(X)$;
(iii) each $K \in BV(\mathbb{R}: \mathcal{B}(X))$ induces via $Tf := dK * f$ an operator $T \in \mathcal{B}(\mathcal{H}(X))$; there is a constant $C > 0$ such that $|T| \le C \text{Var}\, K|_{-\infty}^{\infty}$.

Proof: (i) For $f \in \mathcal{H}(X)$ we have by (i) and (ii) of Definition 1

$$\int_t^{t+1} |f(s)|ds = \int_0^1 |(T_t f)(s)|ds \le C(1)|T_t f|_{\mathcal{H}(X)} \le C(1)M|f|_{\mathcal{H}(X)}, \quad t \in \mathbb{R},$$

where $M = \sup\{|T_t|_{\mathcal{B}(\mathcal{H}(X))} : t \in \mathbb{R}\}$.
(ii) This is standard in semigroup theory; see e.g. Davies [7], p.30.
(iii) This follows from the representation

$$\begin{aligned} Tf &= dK * f = \int_{-\infty}^{\infty} dK(\tau) f(\cdot - \tau) = \int_{-\infty}^{\infty} dK(\tau) T_{-\tau} f \\ &= \lim_{\delta \to 0} \sum_{i=1}^{N} (K(\tau_i) - K(\tau_{i-1})) T_{-\tau_i} f, \end{aligned}$$

where $\delta = \sup\{\tau_i - \tau_{i-1} : i = 1, \ldots, N\}$ denotes the size of the decomposition $-\infty < \tau_0 < \ldots < \tau_N < \infty$; observe that the function $\tau \mapsto T_\tau f \in \mathcal{H}(X)$ is uniformly continuous on $\mathbb{R}$. □

Observe that in particular any function $F : \mathbb{R} \to \mathcal{B}(X)$ such that $F(\cdot)x$ is measurable for each $x \in X$, and $|F(t)| \le \varphi$ a.e., $\varphi \in L^1(\mathbb{R})$, induces by Proposition 3 (iii) via $Tf := F * f$ an operator $T \in \mathcal{B}(\mathcal{H}(X))$, for every homogeneous space $\mathcal{H}(X)$.

Functions $f \in BM^0(X)$ are of polynomial growth, since

$$\int_{-\infty}^{\infty} |f(t)| \frac{dt}{1+t^2} \le \sum_{j=0}^{\infty} \frac{1}{1+j^2} \left(\int_0^1 |T_j f(t)|dt + \int_{-1}^0 |T_{-j} f(t)|dt \right) \le c|f|_1^b,$$

for some constant $c > 0$, independent of f. Therefore, $\sigma(f)$, the spectrum of f, introduced in Section 1 via the Fourier-Carleman transform is welldefined. This gives rise to the family $\mathcal{H}_\Lambda(X)$ of subspaces of a homogeneous space $\mathcal{H}(X)$, defined by

$$\mathcal{H}_\Lambda(X) = \{f \in \mathcal{H}(X) : \sigma(f) \subset \Lambda\}.$$

Theorem 1 shows that $\mathcal{H}_\Lambda(X)$ is closed, whenever $\Lambda \subset \mathbb{R}$ is closed. For the spaces $C_{ub\Lambda}(X)$ we use the abbreviation $\Lambda(X)$, i.e.

$$\Lambda(X) = \{f \in C_{ub}(X) : \sigma(f) \subset \Lambda\}.$$

If $\mathcal{H}(X)$ is homogeneous and $\Lambda \subset \mathbb{R}$ is closed, then Proposition 1 implies that $\mathcal{H}_\Lambda(X)$ is again homogeneous. This gives a fine structure of subspaces suitable to describe the solvability behavior of (0.1) in terms of the spectrum of the inhomogeneity f.

If $\mathcal{H}(X)$ is homogeneous, it is not difficult to verify $\mathcal{H}^a(X) \subset \mathcal{H}^\sigma(X)$, where

$$\mathcal{H}^\sigma(X) = \{f \in \mathcal{H}(X) : \sigma(f) \text{ compact }\};$$

therefore the space $\mathcal{H}^\sigma(X)$ is also dense in $\mathcal{H}(X)$.

Next given $b \in \mathcal{W}(\mathbb{R}) = \{b \in L^1(\mathbb{R}) \ : \ \hat{b}(\rho) \neq 0 \text{ for all } \rho \in \mathbb{R}\} \cup \{\delta_0\}$, the *Wiener class*, we define

$$\mathcal{H}^b(X) = \{f \in \mathcal{H}(X) \ : \ f = b * g \text{ for some } g \in \mathcal{H}(X)\} ,$$

i.e. $\mathcal{H}^b(X)$ is the range of the convolution operator on $\mathcal{H}(X)$ induced by b. By a classical result (cp. Widder [17], p.207), there follows $\mathcal{H}^\sigma(X) \subset \mathcal{H}^b(X)$, for each $b \in \mathcal{W}(\mathbb{R})$, hence these spaces are also dense in $\mathcal{H}(X)$. It should be noted that for the kernels $b_\beta(t) = e^{-t}t^{\beta-1}/\Gamma(\beta)$, $t > 0$, $b_\beta(t) = 0$ for $t < 0$, one obtains $\mathcal{H}^{b_\beta}(X) = \mathcal{H}^\beta(X) = \mathcal{D}(D^\beta)$, for all $\beta > 0$; in fact, it is easy to verify the relation $(I + D)^{-\beta} f = b_\beta * f$.

3. Admissibility

Lets turn attention to (0.1) now. If $A \in L^1(\mathbb{R}_+; \mathcal{B}(Y, X))$ and $u \in BM^0(Y)$ then $A * u$ is welldefined, belongs to $BM^0(X)$, and $|A * u|_1^b \leq |A|_1 |u|_1^b$. For many applications, however, integrability of $A(t)$ is too stringent, but then the convolution $A * u$ over $\mathbb{R}$ is in general not defined. For this reason we restrict attention to kernels $A(t)$ of the form

$$A(t) = A_0(t) + A_1(t) , \quad t \geq 0 , \text{ where}$$
$$A_0 \in L^1(\mathbb{R}_+; \mathcal{B}(Y, X)) \text{ and } A_1 \in BV(\mathbb{R}_+; \mathcal{B}(Y, X)), \tag{3.1}$$

and consider the differentiated version of (0.1).

$$\dot{u}(t) = \int_0^\infty A_0(\tau)\dot{u}(t - \tau)d\tau + \int_0^\infty dA_1(\tau)u(t - \tau) + f(t) , \quad t \in \mathbb{R}. \tag{3.2}$$

In the sequel we always assume (3.1) and confine our study to (3.2). The cases (0.1) with $A \in L^1(\mathbb{R}_+; \mathcal{B}(Y, X))$ and on the second order equation where (0.1) is differentiated twice can be treated similarly, or sometimes even reduced to (3.2).

Definition 2 *Let $f \in BM^0(X)$. Then*
(a) $u \in BM^0(X)$ is called strong solution of (3.2) if $u \in BM^0(Y)$, $u - A_0 * u \in BM^1(X)$, and $(u - A_0 * u)^\cdot = dA_1 * u + f$ a.e. on $\mathbb{R}$.
(b) $u \in BM^0(X)$ is called mild solution of (3.2) if there are $f_n \in BM^0(X)$, $f_n \to f$ in $BM^0(X)$, and strong solutions $u_n \in BM^0(Y)$ of (3.2) with f replaced by f_n, such that $u_n \to u$ in $BM^0(X)$.

The solvability behavior of (3.2) is described in terms of admissibility of homogeneous spaces which we introduce next.

Definition 3 *Let $b \in \mathcal{W}(\mathbb{R})$. A homogeneous space $\mathcal{H}(X)$ is called <u>b-admissible</u> for (3.2), if for each $f \in \mathcal{H}^b(X)$ there is a unique strong solution $u \in \mathcal{H}(X) \cap BM^0(Y)$ of (3.2), and $(f_n) \subset \mathcal{H}^b(X)$, $f_n \to 0$ in $\mathcal{H}(X)$, implies $u_n \to 0$ in $\mathcal{H}(X)$. A homogeneous space $\mathcal{H}(X)$ is called 0-admissible if it is δ_0-admissible.*

Thus the kernel $b \in \mathcal{W}(\mathbb{R})$ measures the amount of regularity in time of an inhomogeneity f needed to allow for a strong solution. Suppose the homogeneous space $\mathcal{H}(X)$ is b-admissible. Then we may define the *solution operator* $G : \mathcal{H}^b(X) \to \mathcal{H}(X) \cap BM^0(Y)$ by means of $Gf = u$, where $u \in \mathcal{H}(X) \cap BM^0(Y)$ denotes the unique solution of (3.2) with $f \in \mathcal{H}^b(X)$. Then by definition of b-admissibility, G admits a unique bounded linear extension to $G \in \mathcal{B}(\mathcal{H}(X))$ since $\mathcal{H}^b(X)$ is dense in $\mathcal{H}(X)$; see Section 2. Moreover, by the closed graph theorem $b * G \in \mathcal{B}(\mathcal{H}(X), BM^0(Y))$ as well, where we deliberately used the notation $b * G$ for the operator $G(b * f)$. Therefore, Gf is a mild solution of (3.2), for each $f \in \mathcal{H}(X)$.

The translation invariance and uniqueness of strong solutions for $f \in \mathcal{H}^b(X)$ then imply that G commutes with the group of translations, hence also with its generator D, in particular $G \in \mathcal{B}(\mathcal{H}^m(X))$ for each $m \geq 0$, as well as $b * G \in \mathcal{B}(\mathcal{H}^m(X), BM^m(Y))$. Observe also the relation

$$G(\varphi * f) = \varphi * Gf , \quad f \in \mathcal{H}(X) , \varphi \in L^1(\mathbb{R}), \tag{3.3}$$

which follows again from translation invariance and from Proposition 3. These observations are collected in

Proposition 4 *Suppose the homogeneous space $\mathcal{H}(X)$ is b-admissible, and let the solution operator G be defined as above. Then*
(i) $G \in \mathcal{B}(\mathcal{H}^m(X))$, and $b * G \in \mathcal{B}(\mathcal{H}^m(X), BM^m(Y))$ for each $m \geq 0$;
(ii) G commutes with the translation group, i.e. $T_h G = G T_h$ for all $h \in \mathbb{R}$;
(iii) G commutes with D;
(iv) $G(\varphi * f) = \varphi * Gf$ for all $f \in \mathcal{H}(X)$, $\varphi \in L^1(\mathbb{R})$.

To obtain conditions necessary for admissibility, suppose $\Lambda(X)$ is b-admissible. For $\rho \in \Lambda$ the function $f(t) = e^{i\rho t}x$, $x \in X$, then belongs to $\Lambda^a(X)$, hence there is a unique strong solution $u \in \Lambda(X) \cap BM^0(Y)$ of (3.2). Translation invariance, i.e. Proposition 4 then yields

$$\dot{u} = DGf = G\dot{f} = i\rho Gf = i\rho u,$$

hence $u(t) = e^{i\rho t}y$ for some unique $y \in Y$. (3.2) now implies

$$(i\rho - i\rho\hat{A}_0(i\rho) - \widehat{dA_1}(i\rho))y = x,$$

hence the operators $i\rho - \hat{A}_0(i\rho) - \widehat{dA_1}(i\rho) = i\rho(I - \hat{A}(i\rho)) \in \mathcal{B}(Y, X)$ are invertible for each $\rho \in \Lambda$. Moreover, boundedness of G in $\Lambda(X)$, and of $b * G$ from $\Lambda(X)$ to $BM^0(Y)$ yield the estimates

$$|y|_X = |u|_{\Lambda(X)} \leq |G|_{\mathcal{B}(\Lambda(X))}|f|_{\Lambda(X)} \leq c|x|_X,$$
$$|\tilde{b}(\rho)||y|_Y = |b * u|_{BM^0(Y)} \leq |b * G|_{\mathcal{B}(\Lambda(X), BM^0(Y))}|f|_{\Lambda(X)} \leq c|x|_X.$$

With the notation

$$H(\lambda) = \frac{1}{\lambda}(I - \hat{A}(\lambda))^{-1} = (\lambda - \lambda\hat{A}_0(\lambda) - \widehat{dA_1}(\lambda))^{-1}, \tag{3.4}$$

which was used before, this implies

$$|H(i\rho)|_{\mathcal{B}(X)} + |\check{b}(\rho)||H(i\rho)|_{\mathcal{B}(X,Y)} \leq M \,, \quad \rho \in \Lambda.$$

Defining the *real spectrum* of (3.2) by

$$\Lambda_0 = \{\rho \in \mathbb{R} : i\rho - i\rho\hat{A}_0(i\rho) - \widehat{dA_1}(i\rho) \in \mathcal{B}(Y,X) \text{ is not invertible}\}, \tag{3.5}$$

we have proved

Proposition 5 *Suppose $\Lambda(X)$ is b-admissible, $\Lambda \subset \mathbb{R}$ closed. Then $\Lambda \cap \Lambda_0 = \emptyset$, and there is a constant $M > 0$ such that*

$$|H(i\rho)|_{\mathcal{B}(X)} + |\check{b}(\rho)||H(i\rho)|_{\mathcal{B}(X,Y)} \leq M \,, \quad \textit{for all } \rho \in \Lambda, \tag{3.6}$$

where $H(\lambda)$ is defined by (3.4).

The converse of Proposition 5 is in general false, even in the semigroup case $A(t) \equiv A$. However, for large classes of equations or closed sets $\Lambda \subset \mathbb{R}$ characterization of admissibility in terms of the spectral condition $\Lambda \cap \Lambda_0 = \emptyset$ and uniform bounds on $H(i\rho)$ like (3.6) is possible. A large part of the subsequent sections deals with this question.

4. Λ-Kernels for $\Lambda \subset \mathbb{R}$ Compact

A main tool in the sequel will be the following

Lemma 1 *Suppose $\varphi \in L^1(\mathbb{R})$ is such that $\tilde{\varphi} \in C_0^\infty(\mathbb{R})$ and supp $\tilde{\varphi} \cap \Lambda_0 = \emptyset$. Then there is $G_\varphi \in W^{\infty,1}(\mathbb{R}; \mathcal{B}(X,Y))$ such that*

$$\dot{G}_\varphi = A_0 * \dot{G}_\varphi + dA_1 * G_\varphi + \varphi = \dot{G}_\varphi * A_0 + G_\varphi * dA_1 + \varphi. \tag{4.1}$$

Moreover, G_φ admits extension to an entire function $G_\varphi(z)$ with values in $\mathcal{B}(X,Y)$ of exponential growth.

Proof: Since $\tilde{\varphi}$ has compact support and supp $\tilde{\varphi} \cap \Lambda_0 = \emptyset$, $H(i\rho)$ is bounded in $\mathcal{B}(X,Y)$ on supp $\tilde{\varphi}$, and therefore $G_\varphi(z)$ defined by

$$G_\varphi(z) = \frac{1}{2\pi}\int_{-\infty}^{\infty} H(i\rho)\tilde{\varphi}(\rho)e^{i\rho z}d\rho \,, \quad z \in \mathbb{C}\,, \tag{4.2}$$

is an entire function with values in $\mathcal{B}(X,Y)$. Clearly G_φ and all its derivatives are bounded on $\mathbb{R}$ and of exponential growth on $\mathbb{C}$, and it is easily verified that (4.1) holds. Thus it remains to prove $G_\varphi^{(n)} \in L^1(\mathbb{R}; \mathcal{B}(X,Y))$ for each $n \in \mathbb{N}_0$. Replacing $\tilde{\varphi}$ by $\tilde{\varphi} \cdot (i\rho)^n$, i.e. φ by $\varphi^{(n)}$, it is enough to show $G_\varphi \in L^1(\mathbb{R}; \mathcal{B}(X,Y))$ since $G_\varphi^{(n)} = G_{\varphi^{(n)}}$.

(i) Suppose first that in addition to (3.1) we have

$$\int_0^\infty |A_0(t)|_{\mathcal{B}(Y,X)}(1+t^2)dt < \infty \,, \quad \text{and} \quad \int_0^\infty |dA_1(t)|_{\mathcal{B}(Y,X)}(1+t^2)dt < \infty.$$

Then $i\rho\hat{A}(i\rho)$ is twice continuously differentiable on $\mathbb{R}$, hence $\Phi(\rho) = H(i\rho)\tilde{\varphi}(\rho)$ is so as well. Integrating by parts twice, (4.2) becomes

$$G_\varphi(t) = \frac{-1}{2\pi t^2}\int_{-\infty}^{\infty} \Phi''(\rho)e^{i\rho t}d\rho,$$

and so we obtain an estimate $|G_\varphi(t)|_{\mathcal{B}(X,Y)} \le C(1+t^2)^{-1}$, i.e. $G_\varphi \in L^1(\mathbb{R};\mathcal{B}(X,Y))$.
(ii) To remove the moment conditions, consider $A_m(t)$ given by

$$A_m(t) = A_{0m}(t) + A_{1m}(t)\ , \quad t > 0\ , \text{ where}$$

$$A_{0m}(t) = A_0(t) \text{ for } t \le m\ ,\ A_{0m}(t) = 0 \text{ for } t > m,$$
$$A_{1m}(t) = A_1(t) \text{ for } t \le m\ ,\ A_{1m}(t) = A_1(m) \text{ for } t > m.$$

Then $\lambda\hat{A}_m(\lambda) \to \lambda\hat{A}(\lambda)$ in $\mathcal{B}(Y,X)$ as $m \to \infty$, uniformly on compact subsets of $\overline{\mathbb{C}}_+$, and A_{0m}, dA_{1m} have moments of any order. The relation

$$i\rho(I-\hat{A}_m(i\rho)) = i\rho(I-\hat{A}(i\rho))(I - H(i\rho)i\rho(\hat{A}_m(i\rho)-\hat{A}(i\rho))) \tag{4.3}$$

therefore shows that $H_m(i\rho) = (i\rho - i\rho\hat{A}_m(i\rho))^{-1}$ exists on a neighborhood U of supp $\tilde{\varphi}$ and $|H_m(i\rho)|_{\mathcal{B}(X,Y)} \le M < \infty$ on supp $\tilde{\varphi}$, provided m is large enough. (4.3) then yields

$$H(i\rho)\tilde{\varphi}(\rho) = H_m(i\rho)\tilde{\varphi}(\rho) + H(i\rho)\tilde{\varphi}(\rho)i\rho[\hat{A}(i\rho)-\hat{A}_m(i\rho)]H_m(i\rho)\tilde{\varphi}_1(\rho), \tag{4.4}$$

where $\tilde{\varphi}_1 \in C_0^\infty$ is such that $\tilde{\varphi}_1 = 1$ on supp $\tilde{\varphi}$, $0 \le \tilde{\varphi}_1 \le 1$, and supp $\tilde{\varphi}_1 \subset U$. By step (i) of this proof there are $G_m, G_{jm} \in L^1(\mathbb{R};\mathcal{B}(X,Y))$ such that $\tilde{G}_m = \tilde{\varphi}H_m$, $\tilde{G}_{jm} = (i\rho)^{1-j}\tilde{\varphi}_1 H_m$. Let

$$K = (A_0 - A_{0m}) * G_{0m} + (dA_1 - dA_{1m}) * G_{1m},$$

which belongs to $L^1(\mathbb{R};\mathcal{B}(X))$; then (4.4) becomes the convolution equation

$$G_\varphi = G_m + G_\varphi * K. \tag{4.5}$$

Since
$$I - \tilde{K} = I - i\rho(\hat{A}-\hat{A}_m)H_m\tilde{\varphi}_1 = (i\rho - i\rho\hat{A}_m(1-\tilde{\varphi}_1) - i\rho\hat{A}\tilde{\varphi}_1)H_m$$

is invertible on $\mathbb{R}$ by the choice of $\tilde{\varphi}_1$, if m is large enough, the vector-valued Paley-Wiener lemma (cp. Prüss [12] or Gripenberg [10]) yields a resolvent kernel $R \in L^1(\mathbb{R};\mathcal{B}(X))$ for (4.5), and so the representation

$$G_\varphi = G_m + G_m * R$$

shows $G_\varphi \in L^1(\mathbb{R};\mathcal{B}(X,Y))$. The proof is complete. □

By means of Lemma 1 we can now prove the converse of Proposition 5 for compact Λ.

Theorem 2 *Suppose $\Lambda \subset \mathbb{R}$ is compact, $\Lambda \cap \Lambda_0 = \emptyset$, and let $\mathcal{H}(X)$ be homogeneous. Then $\mathcal{H}_\Lambda(X)$ is 0-admissible and there is $G_\Lambda \in W^{\infty,1}(\mathbb{R};\mathcal{B}(X,Y))$ such that the solution operator G for $\mathcal{H}_\Lambda(X)$ is represented by*

$$(Gf)(t) = \int_{-\infty}^{\infty} G_\Lambda(\tau)f(t-\tau)d\tau\ , \quad t \in \mathbb{R}\ ,\ f \in \mathcal{H}_\Lambda(X). \tag{4.6}$$

Proof: Since $\Lambda \subset \mathbb{R}$ is compact and $\Lambda \cap \Lambda_0 = \emptyset$, there is a neighborhood $\Lambda_{3\varepsilon} = \Lambda + \overline{B_{3\varepsilon}}(0)$ of Λ such that $\Lambda_{3\varepsilon} \cap \Lambda_0 = \emptyset$; observe that Λ_0 is closed. Choose $\varphi_0 \in L^1(\mathbb{R})$ such that $\tilde{\varphi}_0 \in$

$C_0^\infty(\mathbb{R})$, $\check{\varphi}_0 \equiv 1$ on Λ_ε, $\check{\varphi}_0 \equiv 0$ on $\mathbb{R} \setminus \Lambda_{2\varepsilon}$, $0 \le \check{\varphi}_0 \le 1$, and let $G_\Lambda \in W^{\infty,1}(\mathbb{R};\mathcal{B}(X,Y))$ denote the function from Lemma 1 with $\varphi = \varphi_0$. If $f \in \mathcal{H}_\Lambda(X)$, then $\sigma(f) \subset \Lambda$ hence $\varphi_0 * f = f$, and so $u = G_\Lambda * f$ is a strong solution of (3.2), by (4.1). On the other hand, if $u \in \mathcal{H}_\Lambda(X)$ is a strong solution of (3.2) then (4.1) yields $G_\Lambda * f = \varphi_0 * u = u$, hence uniqueness. Since $G_\Lambda \in L^1(\mathbb{R};\mathcal{B}(X,Y))$, we finally obtain boundedness of the solution operator G in $\mathcal{H}_\Lambda(X)$, as well as from $\mathcal{H}_\Lambda(X)$ to $BM^0(Y)$, and so $\mathcal{H}_\Lambda(X)$ is 0-admissible.
□

Observe that 0-admissible implies b-admissible for any $b \in \mathcal{W}(\mathbb{R})$. Lemma 1 also yields dense solvability and uniqueness in $\mathcal{H}_\Lambda(X)$ whenever $\Lambda \cap \Lambda_0 = \emptyset$, as we show now.

Proposition 6 *Let $\mathcal{H}(X)$ be homogeneous and suppose $\Lambda \cap \Lambda_0 = \emptyset$. Then*
(i) $\sigma(u) \subset \Lambda_0$ for each strong solution of (3.2) with $f = 0$;
(ii) for each $f \in \mathcal{H}_\Lambda(X)$ there is at most one strong solution $u \in \mathcal{H}_\Lambda(X)$;
(iii) for each $f \in \mathcal{H}_\Lambda^\sigma(X)$ there is a strong solution $u \in \mathcal{H}_\Lambda(X)$;
(iv) if $f \in \mathcal{H}_\Lambda(X)$, and $u \in \mathcal{H}_\Lambda(X)$ is a strong solution, then $\sigma(u) = \sigma(f)$.

Proof: (i) Since $\sigma(u) = \operatorname{supp} \check{D}_u$, where D_u denotes the distribution generated by $u \in BM^0(Y) \subset BM^0(X)$ we have to show

$$[\check{D}_u, \check{\varphi}] = [D_u, \check{\check{\varphi}}] = (\varphi * u)(0) = 0$$

for each $\check{\varphi} \in C_0^\infty(\mathbb{R})$ such that $\operatorname{supp} \check{\varphi} \cap \Lambda_0 = \emptyset$. But given such φ, let G_φ denote the kernel from Lemma 1; (4.1) and (3.2) then imply $\varphi * u = 0$, in particular $(\varphi * u)(0) = 0$.
(ii) follows from (i), since $\sigma(u) \subset \Lambda_0 \cap \Lambda = \emptyset$ implies $u = 0$.
(iii) If $f \in \mathcal{H}_\Lambda^\sigma(X)$ then $\sigma(f) \subset \Lambda$ is compact and does not intersect Λ_0. Applying Theorem 2 to the set $\sigma(f)$ we obtain a strong solution $u \in \mathcal{H}_{\sigma(f)}(X) \subset \mathcal{H}_\Lambda(X)$ of (3.2).
(iv) Suppose $u \in \mathcal{H}_\Lambda(X)$ is a strong solution of (3.2), and let $\check{\varphi} \in C_0^\infty(\mathbb{R})$ be such that $\operatorname{supp} \check{\varphi} \cap \sigma(u) = \emptyset$. Then $\sigma(\varphi * u) = \emptyset$, hence $\varphi * u = 0$ by Proposition 2, and so (3.2) implies also $\varphi * f = 0$. But this implies as above $[\check{D}_f, \check{\varphi}] = 0$, i.e. $\sigma(f) \subset \sigma(u)$.

Conversely, let $\check{\varphi} \in C_0^\infty(\mathbb{R})$ satisfy $\operatorname{supp} \check{\varphi} \cap \sigma(f) = \emptyset$. Then $\varphi * f = 0$, hence $\varphi * u$ is a strong solution of (3.2) with $f = 0$. Since $\sigma(\varphi * u) \subset \operatorname{supp} \check{\varphi}$ but also $\sigma(\varphi * u) \subset \Lambda_0$ by (i) of this Proposition, we obtain $\sigma(\varphi * u) = \emptyset$, hence $\varphi * u = 0$ by Proposition 2 again; this implies $\sigma(u) = \sigma(f)$. □

A reasoning similar to that of step (iv) of the proof of Proposition 6 yields

Corollary 1 *Suppose $\mathcal{H}(X)$ is a homogeneous space which is b-admissible, and let G denote the solution operator. Then*

$$\sigma(Gf) = \sigma(f) \quad \textit{for each } f \in \mathcal{H}(X).$$

Proof: Consider first $f \in \mathcal{H}^b(X)$ and let $u = Gf$. Then as in (iv) of the proof of Proposition 6 we have $\sigma(f) \subset \sigma(u)$. Conversely, let $\check{\varphi} \in C_0^\infty(\mathbb{R})$ satisfy $\operatorname{supp} \check{\varphi} \cap \sigma(f) = \emptyset$. Then $\varphi * u$ belongs to $\mathcal{H}(X)$ and is a strong solution of (3.2) with $f = 0$, hence by admissibility $\varphi * u = 0$. This then implies $\sigma(u) = \sigma(f)$, i.e. we have $\sigma(Gf) = \sigma(f)$ for each $f \in \mathcal{H}^b(X)$.

For the general case consider $\tilde{\varphi} \in C_0^\infty(\mathbb{R})$ such that $0 \le \tilde{\varphi} \le 1$, $\tilde{\varphi}(\rho) = 1$ for $|\rho| \le 1$, $\tilde{\varphi}(\rho) = 0$ for $|\rho| \ge 2$, and let $\varphi_n(t) = n\varphi(nt)$. φ_n is then an approximation of the identity, i.e.

$$\varphi_n * f = \int_{-\infty}^{\infty} \varphi_n(\tau) T_{-\tau} f d\tau = \int_{-\infty}^{\infty} \varphi(\tau) T_{-\tau/n} f d\tau \to f \quad \text{in } \mathcal{H}(X)$$

since the translation group is strongly continuous in $\mathcal{H}(X)$. But $\sigma(\varphi_n * f) \subset \sigma(f) \cap$ supp $\tilde{\varphi}_n \subset \sigma(f) \cap \bar{B}_{2n}(0)$ is compact, i.e. $\varphi_n * f \in \mathcal{H}^\sigma(X)$. From Proposition 4 we obtain $\varphi_n * Gf = G\varphi_n * f$, hence these are strong solutions, and so $\sigma(\varphi_n * Gf) = \sigma(\varphi_n * f)$. But with Theorem 1 we obtain

$$\sigma(Gf) = \cap_{m\ge 1}\overline{\cup_{n\ge m}\sigma(\varphi_n * Gf)} = \cap_{m\ge 1}\overline{\cup_{n\ge m}\sigma(\varphi_n * f)} = \sigma(f). \quad \square$$

From Proposition 6 it is evident, that the main part in proving admissability of $\mathcal{H}_\Lambda(X)$, where $\Lambda \cap \Lambda_0 = \emptyset$, consists in establishing a bound M such that

$$|u|_{\mathcal{H}(X)} \le M|f|_{\mathcal{H}(X)} \quad \text{and} \quad |b * u|_{BM^0(Y)} \le M|f|_{\mathcal{H}(X)}$$

holds for each $f \in \mathcal{H}_\Lambda(X)$. Before we study this question for unbounded sets $\Lambda \subset \mathbb{R}$ let us discuss further consequences of admissibility.

5 Almost Periodic Solutions

Recall that a function $f \in C_b(X)$ is called *almost periodic* (a.p.) if $\{T_\tau f : \tau \in \mathbb{R}\}$ is relatively compact in $C_b(X)$. Since almost periodic functions are uniformly continuous, the space $AP(X)$ of all a.p. functions is a closed translations invariant subspace of $C_{ub}(X)$, which is also homogeneous. Note that periodic functions are characterized among a.p. functions by the stronger property that $\{T_\tau f : \tau \in \mathbb{R}\}$ is compact in $C_b(X)$.

The *Bohr transform* given by

$$\alpha(\rho, f) := \lim_{N\to\infty} N^{-1} \int_0^N e^{-i\rho t} f(t) dt\,, \quad \rho \in \mathbb{R}\,,\ f \in AP(X),$$

is welldefined and continuous from $AP(X)$ to X. For $f \in AP(X)$ its exponent set

$$\exp(f) = \{\rho \in \mathbb{R} : \alpha(\rho, f) \ne 0\}$$

is at most countable, and $\sigma(f) = \overline{\exp(f)}$. This can be proved by means of Bochner's approximation theorem, which states that, given a fixed countable subset $\{\rho_j\}_1^\infty$, there are convergence factors $\gamma_{nj} \in \mathbb{R}$ with $\gamma_{nj} \to 1$ as $n \to \infty$ such that the trigonometric polynomials

$$f_n = B_n f = \sum_{j=1}^{n} \gamma_{nj} \alpha(\rho_j, f) e^{i\rho_j t}\,, \quad n \in \mathbb{N},$$

converge to f uniformly on $\mathbb{R}$, provided $\exp(f) \subset \{\rho_j\}_1^\infty$. In virtue of this result, it is also clear that $f \in AP(X)$ is uniquely determined by its Bohr transform.

All of these results and many other can be found in the monograph of Amerio and Prouse [1].

Recall also that a function $f \in C_{ub}(X)$ is called *asymptotically almost periodic* (a.a.p.) (to the right) if $\{T_\tau f|_{[0,\infty)}\}_{\tau\ge 0} \subset C_b(\mathbb{R}_+; X)$ is relatively compact. It can be shown that then there is an a.p. function f_a such that $f_0(t) = f(t) - f_a(t) \to 0$ as $t \to \infty$, i.e. $f_0 \in C_0^+(X)$, where

$$C_0^+(X) = \{f \in C_{ub}(X) : f(t) \to 0 \quad \text{as } t \to \infty\}$$

is another closed subspace of $C_{ub}(X)$. The a.p. function f_a is necessarily unique, since

$$\alpha(\rho, f_a) = \alpha(\rho, f)\,, \quad \rho \in \mathbb{R},$$

and the Bohr transform is unique. Moreover, $|f_a|_\infty \le |f|_\infty$, and therefore we have the direct sum decomposition of the space $AAP(X)$ of a.a.p. functions

$$AAP(X) = AP(X) \oplus C_0^+(X).$$

Note that

$$C_l^+(X) = \{f \in C_{ub}(X) : \lim_{t \to \infty} f(t) = f(\infty) \text{ exists }\}$$

is a closed subspace of $AAP^+(X)$. It is easy to see that $C_0^+(X)$, $C_l^+(X)$, and $AAP^+(X)$ are homogeneuous.

After these preparations we are ready to prove the following result on a.p. and a.a.p. solutions of (3.2).

Theorem 3 *Suppose $\Lambda(X)$ is b-admissible, where $\Lambda \subset \mathbb{R}$ is closed, and let G denote the corresponding solution operator. Then*
(i) $f \in AP(X)$, $\exp(f) \subset \Lambda$, imply $Gf \in AP(X)$, $\exp(Gf) = \exp(f)$, and

$$\alpha(\rho; Gf) = H(i\rho)\alpha(\rho, f)\,, \quad \rho \in \exp(f). \tag{5.1}$$

(ii) $f \in AAP(X)$, $\sigma(f) \subset \Lambda$, imply $Gf \in AP(X)$, $(Gf)_0 = Gf_0$, $(Gf)_a = Gf_a$ and also (5.1) holds;
(iii) If $0 \in \Lambda$, $f \in C_l^+(X)$, $\sigma(f) \subset \Lambda$, imply $Gf \in C_l^+(X)$ and

$$(Gf)(\infty) = H(0)f(\infty). \tag{5.2}$$

Proof: (i) Suppose $\Lambda(X)$ is b-admissible. Then if $\rho \in \Lambda$, as in the discussion before Proposition 5 we obtain $G(ae^{i\rho t}) = H(i\rho)ae^{i\rho t}$; note that $\Lambda \cap \Lambda_0 = \emptyset$. Hence if $f(t) = \sum_{n=1}^N a_n e^{i\rho_n t}$ is a trigonometric polynomial with $\exp(f) = \{\rho_n\}_1^N \subset \Lambda$ we obtain by linearity $(Gf)(t) = \sum_{n=1}^N H(i\rho_n)e^{i\rho_n t}$, i.e. (i) of Theorem 3 holds for trigonometric polynomials. The general case then follows by Bochner's approximation theorem mentioned above since G is bounded and the Bohr transform is continuous.
(ii) If $f = f_a + f_0 \in AAP^+(X) \cap \Lambda(X)$, then $T_{\tau_n} f \to f_a$ uniformly on compact subsets of $\mathbb{R}$, for some sequence $\tau_n \to \infty$. Hence by Theorem 1 we obtain $\sigma(f_a) \subset \sigma(f) \subset \Lambda$, and so $\sigma(f_0) = \sigma(f - f_a) \subset \sigma(f) \cup \sigma(f_a) \subset \sigma(f)$ as well. Since $\Lambda(X)$ is b-admissable we therefore have $Gf = Gf_a + Gf_0$, where $Gf_a \in AP(X)$ by step (i) of this proof, and (5.1) holds again. Thus it remains to show $Gf_0 \in C_0^+(X)$, i.e. $(Gf_0)(t) \to 0$ as $t \to \infty$. By Proposition 6 it is enough to assume $\sigma(f_0)$ compact, i.e. w.l.o.g. Λ compact. Let $G_\Lambda \in W^{\infty,1}(\mathbb{R}; \mathcal{B}(X, Y))$ denote a Λ-kernel for (3.2) which exists by Theorem 2. Then $Gf_0 = G_\Lambda * f_0$ and since $G_\Lambda \in L^1(\mathbb{R}; \mathcal{B}(X, Y))$, $(Gf_0)(t) \to 0$ as $t \to \infty$ follows from $f_0(t) \to 0$ as $t \to \infty$.

(iii) This is a consequence of (ii) with $f_a(t) \equiv f(\infty)$. □

Observe that in case $f = b * g$, with $g \in \Lambda(X) \cap AP(X)$ and $\Lambda(X)$ is b-admissible then $u \in BM^0(Y)$ is a.p. in the sense of Stepanov in Y, i.e. $\{T_\tau u : t \in \mathbb{R}\} \subset BM^0(Y)$ is relatively compact.

If $f \in C_{ub}(X)$ is ω-periodic, then $\exp(f) \subset (2\pi/\omega)\mathbf{Z}$ and

$$\alpha(2\pi n/\omega, f) = f_n = \frac{1}{\omega}\int_0^\omega e^{-2\pi int/\omega} f(t)dt\ , \quad n \in \mathbf{Z}\ ,$$

are the Fourier-coefficients of f. Then with $\Lambda = (2\pi/\omega)\mathbf{Z}$ we have $\Lambda(X) = P_\omega(X)$, and Theorem 3 yields the following simple

Corollary 2 *Suppose $\Lambda = (2\pi/\omega)Z$ and $\Lambda(X)$ is b-admissible. Then for every $f \in P_\omega(X)$ there is a unique mild solution $u \in P_\omega(X)$ of (3.2) and for their Fourier coefficients we have*

$$u_n = H(2\pi in/\omega)f_n\ , \quad n \in Z\ . \tag{5.3}$$

In passing we note the following result for a.p. solutions, without assuming any admissibility.

Proposition 7 *Let $f \in AP(X)$ and suppose $u \in BM^0(Y)$ is a strong solution of (3.2) which is a.p. in Y in the sense of Stepanov. Then*

$$(i\rho - i\rho\hat{A}_0(i\rho) - \widehat{dA_1}(i\rho))\alpha(\rho, u) = \alpha(\rho, f) \quad \text{for all}\ \ \rho \in \mathbb{R}. \tag{5.4}$$

In particular, the condition

$$\alpha(\rho, f) \in \mathcal{R}(i\rho - i\rho\hat{A}_0(i\rho) - \widehat{dA_1}(i\rho))\ , \quad \rho \in \mathbb{R},$$

is necessary for existence of u.

This follows from direct computation on the Fourier coefficients of u by using (3.2).

6. Admissibility for Hyperbolic Equations

For $\Lambda \subset \mathbb{R}$ unbounded, admissibility of $\mathcal{H}_\Lambda(X)$ in the hyperbolic case is in general quite delicate to prove. There are not many tools available for this at present, the Paley-Wiener lemma, the subordination principle, and Parseval's theorem in the Hilbert space case are the almost only ones. The main difficulty is the lack of decay properties of $H(i\rho)$ as $|\rho| \to \infty$ which is in contrast to the parabolic case, and for this reason also multiplier theory does not work well, except for $\mathcal{H}(X) = L^2(\mathbb{R}; X)$, when X and Y are Hilbert spaces.

We begin the discussion here with an application of the Paley-Wiener lemma to the perturbed equation

$$\dot{u} = A_0 * \dot{u} + dA_1 * u + B_0 * u + b * B_1 * u + f, \tag{6.1}$$

where the unperturbed equation (3.2) has some admissibility properties at infinity.

Theorem 4 *Suppose A_0, A_1 satisfy (3.1), let $B_0 \in L^1(I\!R_+; \mathcal{B}(X))$, $B_1 \in L^1(I\!R_+; \mathcal{B}(Y, X))$, and $b \in \mathcal{W}(I\!R)$. Let $\Lambda \subset I\!R$ be closed, $\mathcal{H}(X)$ homogeneous, and assume*
(i) there is $R > 0$ such that $\mathcal{H}_{\Lambda_R}(X)$ is b-admissible for (3.2), where $\Lambda_R = \{\rho \in \Lambda : |\rho| \geq R\}$;
(ii) $\Lambda \cap \Lambda_1 = \emptyset$, for the spectrum Λ_1 of (6.1).
Then $BM^0_\Lambda(X)$ is b-admissible for (6.1). Moreover, if (3.2) admits a Λ_R-kernel G_R, integrable in $\mathcal{B}(X)$, $b * G_R$ integrable in $\mathcal{B}(X, Y)$, then there is a Λ-kernel G_Λ for (6.1) such that $G_\Lambda - G_R \in L^1(I\!R; \mathcal{B}(X))$, $b * (G_\Lambda - G_R) \in L^1(I\!R; \mathcal{B}(X, Y))$. In this case $\mathcal{H}_\Lambda(X)$ is b-admissible, for any homogeneous space $\mathcal{H}(X)$.

<u>*Proof:*</u> Choose $N \geq R+1$ and let $G_N \in \cap_{m\geq 0} W^{m,1}(\mathrm{I\!R}; \mathcal{B}(X,Y))$ be a kernel for $\Lambda_N = \Lambda \cap \bar{B}_{N+2}$ for (6.1), according to Theorem 2; observe that this theorem applies by assumption (ii). Choose a cut-off function $\hat{\varphi} \in C_0^\infty(\mathrm{I\!R})$, $0 \leq \hat{\varphi} \leq 1$, such that $\hat{\varphi} = 1$ for $|\rho| \leq N+1$, $\hat{\varphi} = 0$ for $|\rho| \geq N+2$ and decompose any $f \in \mathcal{H}_\Lambda(X)$ as

$$f = \varphi * f + (f - \varphi * f) = f_1 + f_2.$$

Then $\sigma(f_1) \subset \Lambda_N$ and $\sigma(f_2) \subset \Lambda \cap \{\rho \in \mathrm{I\!R} : |\rho| \geq N+1\} = \Lambda_\infty$; the admissibility of $\mathcal{H}_\Lambda(X)$ is this way reduced to admissibility of $\mathcal{H}_{\Lambda_\infty}(X)$ we may therefore assume $\Lambda_N = \emptyset$, where $N \geq R+1$ is sufficiently large, specified later.

Let G_0 denote the solution operator for (3.2) in $\mathcal{H}_\Lambda(X)$ which exists and belongs to $\mathcal{B}(\mathcal{H}_\Lambda(X))$, with $R_0 = b * G_0 \in \mathcal{B}(BM^0_\Lambda(X), BM^0_\Lambda(Y))$, by assumption (i). Then (6.1) may be rewritten as the fixed point equation

$$u = G_0 f + G_0 B_0 * u + G_0 B_1 * b * u \tag{6.2}$$

or equivalently with $u = G_0 v$

$$v = f + B_0 * G_0 v + B_1 * R_0 v. \tag{6.3}$$

Thus the solution operator G and $b * G$ for (6.1) is given by

$$G = G_0(I - B_0 * G_0 - B_1 * R_0)^{-1} \quad , \quad R = R_0(I - B_0 * G_0 - B_1 * R_0)^{-1},$$

and to prove b-admissibility of $\mathcal{H}_\Lambda(X)$ it is therefore sufficient to show

$$|B_0 * G_0 + B_1 * R_0| < 1.$$

For this purpose consider the Fejer approximations

$$B_j^N(t) = \frac{1}{2\pi} \int_{-N}^{N} (1 - |\rho|/N) \hat{B}_j(i\rho) e^{i\rho t} d\rho\,, \quad t \in \mathrm{I\!R}\,,\ j = 0, 1;$$

since $B_0 \in L^1(\mathrm{I\!R}; \mathcal{B}(X))$ we have $B_0^N \to B_0$ in $L^1(\mathrm{I\!R}; \mathcal{B}(X))$, and similarly $B_1^N \to B_1$ in $L^1(\mathrm{I\!R}; \mathcal{B}(Y, X))$. Now for $f \in BM^0_\Lambda(X)$ we have

$$\sigma(B_0^N * f) \subset \sigma(f) \cap \operatorname{supp} \hat{B}_0^N \subset \sigma(f) \cap \bar{B}_N(0) = \emptyset,$$

hence $B_0^N * f = 0$, and similarly $B_1^N * g = 0$ for $g \in BM_\Lambda^0(Y)$. This implies

$$|B_0 * G_0 + B_1 * R_0| \leq |B_0 - B_0^N|_1 |G_0| + |B_1 - B_1^N|_1 |R_0| < 1,$$

provided N is sufficiently large. Note that the norms of G_0 and R_0 do not depend on $N \geq R+1$.

If there exists a Λ_R-kernel G_R which is integrable in $\mathcal{B}(X)$, $b*G_R$ integrable in $\mathcal{B}(X,Y)$, then $K = B_0 * G_R + B_1 * b * G_R$ belongs to $L^1(\mathbb{R};\mathcal{B}(X))$. By the infinite dimensional Paley-Wiener lemma (see Gripenberg [10], Prüss [12], [14]) there is a resolvent kernel $R \in L^1(\mathbb{R};\mathcal{B}(X))$ for K. It follows readily from (6.2) that then

$$G_\Lambda = G_R + G_R * R,$$

is a Λ-kernel for (6.1). □

Next we draw a connection between integrability properties of the resolvent $S(t)$ for the local version (0.2) of (0.1) and admissibility.

Proposition 8 *Suppose (0.2) admits a resolvent $S(t)$, which is integrable in $\mathcal{B}(X)$ and such that $b * S$ is integrable in $\mathcal{B}(X,Y)$, for some $b \in \mathcal{W}(\mathbb{R}_+)$. Then each homogeneous space $\mathcal{H}(X)$ is b-admissible.*

Proof: We will show the representation

$$(Gf)(t) = \int_0^\infty S(\tau)f(t-\tau)d\tau\ , \quad t \in \mathbb{R}. \tag{6.4}$$

Since $S(t)$ is integrable in $\mathcal{B}(X)$ by assumption, Proposition 3 shows $G \in \mathcal{B}(\mathcal{H}(X))$ for each homogeneous space. On the other hand, since $b*S$ is integrable in $\mathcal{B}(X,Y)$ we obtain $b * G \in \mathcal{B}(\mathcal{H}(X); BM^0(Y))$. From the first resolvent equation we obtain with $R = b * S$

$$(R - A_0 * R)^{\cdot} = b + dA_1 * R\ , \quad t \in \mathbb{R},$$

hence $u = b*Gg = Gb*g = R*g$ with $g \in \mathcal{H}(X)$ is a strong solution of (3.2) for $f = b*g$. Conversely, if u is a strong solution of (3.2), by the second resolvent equation,

$$(R - R * A_0)^{\cdot} = b + R * dA_1,$$

we obtain $b * (u - S * f) = 0$, hence $u = S * f$, by the uniqueness property of the class $\mathcal{W}(\mathbb{R}_+)$. □

7. Admissibility in Hilbert Spaces

If X and Y are both Hilbert spaces one may use Parseval's theorem to obtain admissibility of homogeneous spaces. The simplest result in this direction is certainly the following which concerns admissibility of $L^2_\Lambda(X)$.

Proposition 9 *Suppose X, Y are Hilbert spaces, $\Lambda \subset \mathbb{R}$ is closed, and $b \in W(\mathbb{R})$. Then $L^2_\Lambda(X)$ is b-admissible if $\Lambda \cap \Lambda_0 = \emptyset$ and there is some constant $M \geq 1$ such that*

$$|H(i\rho)|_{\mathcal{B}(X)} + |\tilde{b}(\rho)||H(i\rho)|_{\mathcal{B}(X,Y)} \leq M\ , \quad \rho \in \Lambda. \tag{7.1}$$

This is an easy consequence of the representation

$$\check{u}(\rho) = H(i\rho)\check{f}(\rho)\,, \quad \rho \in \Lambda,$$

of the solution of (3.2) for $f \in L^2_\Lambda(X)$ with $\sigma(f)$ compact, and Parseval's theorem; cp. Proposition 6. The details are left to the reader.

If one imposes mild regularity restrictions on $A(t)$ and the function $b(t)$, and assumes existence of an exponentially bounded b-regular resolvent $S(t)$, the full converse of Proposition 5 can be obtained. The resulting theorem contains as a special case Gearhart's characterization of the spectrum of a C_0-semigroup in a Hilbert space; see Gearhart [8] and Prüss [13]. It is an extension of Theorem 3 in Prüss [14].

Before we state it, a property of scalar-valued has to be introduced.

Definition 4 *Let $b \in L^1_{loc}(I\!R_+)$ be a subexponential growth. The kernel $b(t)$ is called 1-regular, if*

$$sup_{Re\lambda>0}|\lambda\hat{b}'(\lambda)/\hat{b}(\lambda)| < \infty.$$

For example, every nonnegative, nonincreasing, convex function is 1-regular; cp. Prüss [15]. This notion has turned out to be quite useful in connection with Volterra equations and has been employed for the study of equations of parabolic type in Prüss [15]; see Prüss [16] for a systematic study.

Theorem 5 *Suppose X, Y are Hilbert spaces, $\Lambda \subset I\!R$ is closed, and let $b \in L^1(I\!R_+)$ be 1-regular and such that $\hat{b}(0) \neq 0$. Assume that (0.2) admits a b-regular resolvent, exponentially bounded, and in addition*

$$|\lambda\hat{A}'(\lambda)|_{\mathcal{B}(Y,X)} \leq C|\hat{b}(\lambda)|\,, \qquad Re\,\lambda > 0\,,\ |\lambda| \geq R, \tag{7.2}$$

for some constants $R, C > 0$. Then $\Lambda(X)$ is b-admissible if and only if $\Lambda \cap \Lambda_0 = \emptyset$ and (7.1) holds. In this case for each $x \in X$ there is $h_x \in L^1(X) \cap L^2(X)$ such that $b * h_x \in L^1(Y) \cap L^2(Y)$ and $\widetilde{h_x}(\rho) = H(i\rho)x$ for all $\rho \in \Lambda$. Moreover, if Λ_0 is compact, then there is a Λ-kernel $G_\Lambda : I\!R \to \mathcal{B}(X)$, strongly continuous for $t \neq 0$, such that $G_\Lambda(\cdot)x \in L^1(I\!R) \cap L^\infty(I\!R)$, $G_\Lambda(0+)x - G_\Lambda(0-)x = x$, $\lim_{|t|\to\infty} G_\Lambda(t)x = 0$, and $\check{G}_\Lambda(\rho)x = H(i\rho)x$ on Λ, for each $x \in X$.

Proof: The necessity part follows from Proposition 5. To prove the sufficiency of (7.1) we may assume, as in the proof of Theorem 4, $\Lambda \cap \bar{B}_N(0) = \emptyset$, where $N > 0$ can be arbitrarily large.

(a) First we show that (7.1) holds also on $\Lambda_\varepsilon = \Lambda + \bar{B}_\varepsilon(0)$ with M replaced by 4M, provided $\varepsilon > 0$ is chosen small enough. For this purpose consider the identity

$$\begin{aligned} I - \hat{A}(i\rho) &= [I + i\rho_0(\hat{A}(i\rho_0) - \hat{A}(i\rho))H(i\rho_0)](I - \hat{A}(i\rho_0)) \\ &= [I + K(\rho, \rho_0)](I - \hat{A}(i\rho_0))\ , \quad \rho_0 \in \Lambda\,,\ |\rho - \rho_0| \leq \varepsilon; \end{aligned}$$

we show $|K(\rho, \rho_0)| \leq 1/2$ whenever $\rho_0 \in \Lambda$, $|\rho - \rho_0| \leq \varepsilon$, provided $\varepsilon > 0$ is sufficiently small. This then implies $\Lambda_0 \cap \Lambda_\varepsilon = \emptyset$ and (7.1) with M replaced by 4M.

From 1-regularity of $b(t)$ we obtain the estimate

$$|\hat{b}(i\rho)| \leq |\hat{b}(i\rho_0)| + \int_{\rho_0}^{\rho} |\hat{b}'(i\tau)|d\tau \leq |\hat{b}(i\rho_0)| + \frac{c}{\rho_0 - \varepsilon} \int_{\rho_0-\varepsilon}^{\rho_0+\varepsilon} |\hat{b}(i\tau)|d\tau,$$

hence

$$\int_{\rho_0-\varepsilon}^{\rho_0+\varepsilon} |\hat{b}(is)|ds \leq 2\varepsilon(1 - \frac{2c\varepsilon}{|\rho_0| - \varepsilon})^{-1}|\hat{b}(i\rho_0)| \leq 4\varepsilon|\hat{b}(i\rho_0)|,$$

if $\rho_0 \in \Lambda$ and $\varepsilon \leq N/(1+4c)$. This yields with (7.2)

$$\begin{aligned} |K(\rho,\rho_0)|_{\mathcal{B}(X)} &\leq |\rho_0| \int_{\rho_0}^{\rho} |d\hat{A}(is)|_{\mathcal{B}(Y,X)} \cdot |H(i\rho_0)|_{\mathcal{B}(X,Y)} \\ &\leq \frac{CM|\rho_0|}{|\rho_0| - \varepsilon} \int_{\rho_0-\varepsilon}^{\rho_0+\varepsilon} |\hat{b}(is)|ds \cdot |\hat{b}(i\rho_0)|^{-1} \leq 8CM\varepsilon \end{aligned}$$

for each $\rho_0 \in \Lambda$, $|\rho - \rho_0| \leq \varepsilon$, provided $\varepsilon > 0$ is small enough; recall $|\rho_0| \geq N$.
(b) By assumption, (0.2) admits a resolvent $S(t)$ such that

$$|S(t)|_{\mathcal{B}(X)} + |(b*S)(t)|_{\mathcal{B}(X,Y)} \leq M_0 e^{(\omega-1)t}, \quad t \geq 0,$$

for some constants $M_0, \omega > 0$. But then the function $u(t) = e^{-\omega t}S(t)xe_0(t)$ belongs to $L^2(X)$ for each $x \in X$, and similarly $v(t) = e^{-\omega t}(b*S)(t)xe_0(t)$ belongs to $L^2(Y)$ for each $x \in X$. By Parseval's theorem therefore $\tilde{u}(\rho) = \hat{S}(\omega + i\rho)x = H(\omega + i\rho)x$ belongs to $L^2(X)$ as well, and similarly $\tilde{v}(\rho) = \hat{b}(\omega + i\rho)H(\omega + i\rho)x$ is in $L^2(Y)$ for every $x \in X$.
(c) Choose a function $\varphi \in C^\infty(\mathbb{R})$ such that $0 \leq \varphi \leq 1$, $|\varphi'|_\infty < \infty$, $\varphi \equiv 1$ on $\Lambda_{\varepsilon/3}$, $\varphi \equiv 0$ on $\mathbb{R} \setminus \Lambda_{2\varepsilon/3}$, and consider the identity

$$H(i\rho)x = (1 + \frac{\omega}{i\rho})H(\omega + i\rho)x + H(i\rho)(\omega + i\rho)[\hat{A}(i\rho) - \hat{A}(\omega + i\rho)]H(\omega + i\rho)x,$$

which after multiplication with $\varphi(\rho)$ yields

$$\varphi(\rho)H(i\rho)x = (1 + \frac{\omega}{i\rho})\varphi(\rho)\tilde{u}(\rho) + \varphi(\rho)H(i\rho)\frac{(\omega + i\rho)}{\hat{b}(\omega + i\rho)}[\hat{A}(i\rho) - \hat{A}(\omega + i\rho)]\tilde{v}(\rho). \quad (7.3)$$

Since $\tilde{u} \in L^2(X)$, $\tilde{v} \in L^2(Y)$ and $0 \notin \operatorname{supp} \varphi$ there follows $\varphi(\rho)H(i\rho)x \in L^2(X)$ and similarly $\varphi(\rho)\hat{b}(i\rho)H(i\rho)x \in L^2(Y)$, provided $\hat{b}(i\rho)/\hat{b}(\omega + i\rho)$ and $(\omega + i\rho)[\hat{A}(i\rho) - \hat{A}(\omega + i\rho)]/\hat{b}(\omega + i\rho)$ are bounded for $\rho \in \Lambda_\varepsilon$. To prove this we employ again 1-regularity of $b(t)$ and (7.2). The identity

$$\hat{b}(i\rho + s) = \hat{b}(i\rho + \omega) - \int_s^\omega \hat{b}'(i\rho + \tau)d\tau$$

yields with 1-regularity

$$|\hat{b}(i\rho + s)| \leq |\hat{b}(i\rho + \omega)| + \frac{c}{|\rho|}\int_0^\omega |\hat{b}(i\rho + \tau)|d\tau, \quad 0 \leq s \leq \omega,$$

hence

$$\int_0^\omega |\hat{b}(i\rho + \tau)|d\tau \leq \frac{\omega}{1 - \frac{c\omega}{N}}|\hat{b}(i\rho + \omega)|, \quad |\rho| > N, \quad (7.4)$$

provided $N > c\omega$: in particular $\hat{b}(i\rho)/\hat{b}(\omega + i\rho)$ is bounded on Λ_ε. On the other hand, (7.2) and (7.4) yield

$$\begin{aligned}|\hat{A}(i\rho) - \hat{A}(\omega + i\rho)|_{\mathcal{B}(Y,X)} &\leq \int_0^\omega |\hat{A}'(s + i\rho)|ds \leq \frac{C}{|\rho|}\int_0^\omega |\hat{b}(s + i\rho)|ds \\ &\leq \frac{C}{|\rho|}\cdot\frac{N\omega}{N - c\omega}|\hat{b}(\omega + i\rho)|\,, \quad \rho \in \Lambda_\varepsilon,\end{aligned}$$

hence $(\omega + i\rho)[\hat{A}(i\rho) - \hat{A}(\omega + i\rho)]/\hat{b}(\omega + i\rho)$ is bounded on Λ_ε as well. Thus, by Parseval's theorem, for each $x \in X$ there is $h_x \in L^2(X)$ such that $b * h_x \in L^2(Y)$ and $\widetilde{h_x}(\rho) = \varphi(\rho)H(i\rho)x$, in particular $\widetilde{h_x}(\rho) = H(i\rho)x$ for $\rho \in \Lambda_\varepsilon$.
(d) Next we show $h_x \in L^1(X)$ and $b * h_x \in L^1(Y)$ for each $x \in X$. For this observe first that in virtue of (7.2), $\hat{A}(i\rho)$ is not only Lipschitz on Λ_ε but also strongly differentiable a.e., since X and Y are Hilbert spaces, which enjoy the Radon-Nikodym property. Differentation of $\widetilde{h_x}$ therefore yields

$$\begin{aligned}\widetilde{h_x}'(\rho) &= \varphi'(\rho)H(i\rho)x + i\varphi(\rho)H'(i\rho)x \\ &= \varphi'(\rho)H(i\rho)x - \varphi(\rho)H(i\rho)x/\rho + i\varphi(\rho)H(i\rho)i\rho\hat{A}'(i\rho)H(i\rho)x;\end{aligned}$$

since $0 \notin \operatorname{supp}\varphi$, φ' is bounded and $\operatorname{supp}\varphi' \subset \Lambda_\varepsilon$, by (7.2) we see that $\widetilde{h_x}' \in L^2(X)$, hence $th_x \in L^2(X)$ by Parseval's theorem. Similarly, employing 1-regularity of $b(t)$ once more we obtain also $t \cdot b * h_x \in L^2(Y)$. But then

$$|h_x|_1 = \int_{-1}^1 |h_x(t)|dt + \int_{|t|\geq 1} |h_x(t)t|dt/|t| \leq \sqrt{2}(|h_x|_2 + |th_x|_2) < \infty,$$

i.e. $h_x \in L^1(X)$; similarly $b * h_x \in L^1(Y)$ for each $x \in X$.
(e) Next we apply a duality argument to obtain $h_x^* \in L^1(X) \cap L^2(X)$ such that $\widetilde{h_x^*}(-\rho) = \varphi(\rho)H(i\rho)^*x$ for each $x \in X$. For this purpose observe that $S^*(t)x$ is weakly continuous and strongly measurable on $\mathbb{R}_+$; the latter follows again from the Radon-Nikodym property. Therefore $H(\omega + i\rho)^*x$ belongs to $L^2(X)$ for each $x \in X$, and by boundedness of $(\omega + i\rho)(\hat{A}(i\rho) - \hat{A}(\omega + i\rho))/\hat{b}(i\rho)$, as above we obtain also $\varphi(\rho)H(i\rho)^*x \in L^2(X)$, for each $x \in X$. Also by duality we get $(\varphi(\rho)H(i\rho)^*x)' \in L^2(X)$, hence as in (c) and (d) there is $h_x^* \in L^1(X) \cap L^2(X)$ such that $\widetilde{h_x^*}(-\rho) = \varphi(\rho)H(i\rho)^*x$, $\rho \in \mathbb{R}$, $x \in X$.

By similar arguments, for each $y \in Y$ there is $g_y^* \in L^1(X) \cap L^2(X)$ such that $g_y^*(-\rho) = \varphi(\rho)(\hat{b}(i\rho)H(i\rho))^*y$.
(f) The mappings $x \mapsto h_x$, $x \mapsto h_x^*$ from X to $L^1(X)$ are linear and closed; hence by the closed graph theorem, there is a constant $C > 0$ such that

$$|h_x|_1 + |h_x^*|_1 \leq C|x|_X \quad , \quad x \in X.$$

Similarly,

$$|g_y^*|_1 \leq C|y|_Y \ , \ y \in Y \ , \ \text{and} \ \ |b * h_x|_{L^1(Y)} \leq C|x|_X \ , \ x \in X.$$

To prove boundedness of the solution operator G from $\Lambda(X)$ to $C_{ub}(X)$ and of $b * G$ from $\Lambda(X)$ to $C_{ub}(Y)$ it is enough to verify the relations

$$(x, Gf(t))_X = \int_{-\infty}^{\infty} (h_x^*(t - \tau), f(\tau))_X d\tau\,, \quad t \in \mathbb{R}, \tag{7.5}$$

and

$$(y, b * Gf(t))_Y = \int_{-\infty}^{\infty} (g_y^*(t-\tau), f(\tau))_X d\tau\ , \quad t \in \mathbb{R}, \tag{7.6}$$

for each $f \in \Lambda(X)$ such that $\sigma(f)$ is compact. In fact, (7.5) and (7.6) yield the estimates

$$|Gf|_{L^\infty(X)}| \leq C|f|_{L^\infty(X)} \text{ and } |b * Gf|_{L^\infty(Y)}| \leq C|f|_{L^\infty(X)},$$

for each $f \in \Lambda(X)$ with compact spectrum, which by Proposition 6 is sufficient for b-admissibility of $\Lambda(X)$.

So let such f be given, choose another cutoff function $\tilde{\varphi}_0 \in C_0^\infty(\mathbb{R})$, real and symmetric, such that $\tilde{\varphi}_0 \equiv 1$ on a neighborhood of $\sigma(f)$, and let $G_0 \in W^{\infty,1}(\mathbb{R}; \mathcal{B}(X,Y))$ denote the kernel from Lemma 1 such that $\tilde{G}_0(\rho) = \tilde{\varphi}(\rho)H(i\rho)\tilde{\varphi}_0(\rho)$, $\rho \in \mathbb{R}$. Then with $\varphi_0 * f = f$ and $\widetilde{G_0^*}(\rho)x = \tilde{G}_0(-\rho)^* x = \tilde{\varphi}_0(\rho)\widetilde{h_x^*}(\rho)$, i.e. $G_0^* x = \varphi_0 * h_x^*$, we obtain

$$\begin{aligned}(x, Gf(t))_X &= (x, (G_0 * f)(t))_X = \int_{-\infty}^{\infty} (G_0^*(t-\tau)x, f(\tau))_X d\tau \\ &= \int_{-\infty}^{\infty} (h_x^*(t-\tau), (\varphi_0 * f)(\tau))_X d\tau = \int_{-\infty}^{\infty} (h_x^*(t-\tau), f(\tau))_X d\tau\ , \quad t \in \mathbb{R},\end{aligned}$$

i.e. (7.5) holds. Similarly, one also obtains (7.6).

(g) Finally, suppose Λ_0 is compact; w.o.l.g. we then may choose $\varphi(\rho) \equiv 1$ for $|\rho|$ sufficiently large, say $|\rho| \geq N_0$. Then $1 - \varphi \in C_0^\infty(\mathbb{R})$, $\varphi(0) = 0$, hence $1 - \varphi = \tilde{\psi}$ for some $\psi \in W^{\infty,1}(\mathbb{R})$ with $\int_{-\infty}^{\infty} \psi(t)dt = 1$. Then the decomposition of the first term in (7.3),

$$\varphi(\rho)(1 + \frac{\omega}{i\rho})H(\omega + i\rho)x = H(\omega + i\rho)x - \tilde{\psi}(\rho)H(\omega + i\rho)x + \omega\frac{\varphi(\rho)}{i\rho}H(\omega + i\rho)x,$$

shows that this term is the Fourier transform of a bounded function $w_1(t)$ which is continuous for $t \neq 0$, satisfies the jump relation $\lim_{t\to 0+} w_1(t) - \lim_{t\to 0-} w_1(t) = w_1(0+) - w_1(0-) = x$ as well as $\lim_{|t|\to\infty} w_1(t) = 0$. On the other hand, the second term in (7.3) can be written in the form

$$H(\omega + i\rho)\frac{(\omega + i\rho)}{\hat{b}(i\rho)}[\hat{A}(i\rho) - \hat{A}(\omega + i\rho)]\varphi(\rho)\hat{b}(i\rho)H(i\rho)x = H(\omega + i\rho)\tilde{w}_2(\rho),$$

where $w_2 \in L^2(X)$, hence is the Fourier transform of $Se^{-\omega \cdot} * w_2$, which belongs to $C_0(\mathbb{R}; X)$. The poof is complete. □

From the proof of Theorem 5 it follows that in case $A(t) = B(t) + C(t)$, where $B \in L^1_{loc}(\mathbb{R}_+; \mathcal{B}(Y, X))$ and $C \in L^1_{loc}(\mathbb{R}_+; \mathcal{B}(X))$, only $B(t)$ needs to be subject to (7.2). For $C(t)$ the weaker condition

$$|\lambda \hat{c}'(\lambda)| \leq M\ ; \quad \operatorname{Re} \lambda > 0\ ,\ |\lambda| \geq R \tag{7.7}$$

is sufficient. This remark is quite useful, as we shall see in Section 8.

By an interpolation argument the admissibility assertion in Theorem 5 can be extended to $L^p_\Lambda(X)$, for all $1 \leq p < \infty$.

Corollary 3 *Let the assumptions of Theorem 5 be satisfied and let (7.1) be fullfilled. Then $L^p_\Lambda(X)$ is b-admissible,* $1 \le p < \infty$.

Proof: For a simple function $f = \sum_{i=1}^n x_i \chi_i$, where $x_i \in X$ and χ_i are characteristic functions of disjoint measurable sets of finite measure, we define

$$Gf = \sum_{i=1}^n h_{x_i} * \chi_i.$$

We then have obviously the estimate $|Gf|_1 \le C|f|_1$; cp. (f) of the proof of Theorem 5. On the other hand, as there we also have $|Gf|_\infty \le C|f|_\infty$, hence by the vector-valued Riesz-Thorin interpolation theorem (see e.g. Bergh and Löfström [2]) $G \in \mathcal{B}(L^p(X))$ for each $1 \le p < \infty$. Similarly, $b * G \in \mathcal{B}(L^p(X), L^p(Y))$, and therefore $L^p_\Lambda(X)$ is b-admissible for every $p \in [1, \infty)$. □

Since for hyperbolic problems Λ_0 need not be compact in general, it remains open whether in the situation of Theorem 5 a Λ-kernel G_Λ always exists. Even if so, it seems to be very difficult to obtain pointwise estimates of $G_\Lambda(t)$, in particular to prove the integrability of G_Λ. For this reason it also seems to be quite unclear whether Corollary 3 can be extended to homogeneous spaces other than $L^p_\Lambda(X)$, e.g. it is not known whether $BM^0_\Lambda(X)$ is b-admissible.

Let us conclude this section with an illustrative application of Theorem 5 to obtain the characterization of the spectrum of a C_0-semigroup in a Hilbert space, i.e. Gearhart's theorem mentioned above.

Example 2 Let A generate a C_0-semigroup e^{At} in the Hilbert space X. Choose $Y = X_A$, $b(t) = e^{-t}e_0(t)$, $\Lambda = (2\pi/\tau)\mathbf{Z}$. With $A(t) \equiv A$ the assumptions of Theorem 5 are easily seen to be true: the resolvent $S(t)$ is of course the C_0-semigroup e^{At}. A moment of reflection shows also that $\Lambda(X)$ is b-admissible if and only if the evolution equation

$$\dot{u} = Au + f \tag{7.8}$$

admits a mild τ-periodic solution u, for each $f \in C_{ub}(X)$ τ-periodic. Via the variation of parameters formula the latter is equivalent to $1 \in \rho(e^{A\tau})$. Theorem 5 characterizes this property by

$$\{2\pi ni/\tau : n \in \mathbf{Z}\} \subset \rho(A), \ \sup\{|(2\pi ni/\tau - A)^{-1}| : n \in \mathbf{Z}\} < \infty, \tag{7.9}$$

which is precisely Gearhart's condition; cp. Gearhart [8], Prüss [13].

8. Viscoelastic Timoshenko Beams

Consider the viscoelastic Timoshenko beam model

$$\begin{aligned} \ddot{w} &= da_1 * (w_{xx} + \phi_x) + f_s, \\ \ddot{\phi} &= de_1 * \phi_{xx} - \gamma da_1 * (w_x + \phi) + f_b, \end{aligned} \tag{8.1}$$

where we have assumed for simplicity that the density $\rho_0 = 1$, the shear correction coefficient $\kappa = 1$, and the length of the beam $l = 1$. We want to consider (8.1) on the line, for the case when one end is clamped and the other is free, i.e. w.r.t. the boundary conditions

$$w(t,0) = \phi(t,0) = 0 \quad , \quad \phi_x(t,1) = w_x(t,1) + \phi(t,1) = 0. \tag{8.2}$$

Shear and tensile moduli $a_1(t)$ and $e_1(t)$ will be assumed to be those of a rigid solid, which means

$$\begin{aligned} &a_1(t)\ ,\ e_1(t) \text{ are nonnegative, nonincreasing, convex;} \\ &a_1(0+)\ ,\ e_1(0+) < \infty \quad ; \quad e_\infty = e_1(\infty)\ ,\ a_\infty = a_1(\infty) > 0. \end{aligned} \tag{8.3}$$

Thus (8.1) is of hyperbolic type, it is of 'perturbation type' since

$$\frac{\widehat{de_1}(\lambda)}{\widehat{da_1}(\lambda)} = \frac{e_1(0+) + \widehat{\dot{e}_1}(\lambda)}{a_1(0+) + \widehat{\dot{a}_1}(\lambda)} = \widehat{dk}(\lambda)\ , \quad \lambda > 0$$

for some $k \in BV_{loc}(\mathbb{R}_+)$ with $k(0+) = \lim_{\lambda\to\infty} \widehat{dk}(\lambda) = e_1(0+)/a_1(0+) > 0$. To obtain an abstract formulation of (8.1), let $X = L^2(0,1) \times L^2(0,1)$ and

$$Y = \{(w,\phi) \in W^{2,2}([0,1]) \times W^{2,2}([0,1]) : w(0) = \phi(0) = 0\ ,\ \phi'(1) = w'(1) + \phi(1) = 0\}$$

be equipped with their natural norms, and define $A_1(t) \in \mathcal{B}(Y,X)$ by means of

$$A_1(t) = \begin{pmatrix} a_1(t)\partial_x^2 & a_1(t)\partial_x \\ -\gamma a_1(t)\partial_x & e_1(t)\partial_x^2 - \gamma a_1(t) \end{pmatrix}\ , \ t \in \mathbb{R}_+.$$

Then with $u = (w,\phi)^T$, (8.1) can be rewritten as

$$\ddot{u} = dA_1 * u + f\ , \quad t \in \mathbb{R}, \tag{8.4}$$

a second order problem. Apparently, (8.4) is not of the form (3.2), however, convolving (8.4) with $c(t) = e^{-t}h_0(t)$, h_0 the Heaviside function, it becomes

$$\dot{u} = c * \dot{u} + c * dA_1 * u + g\ , \quad t \in \mathbb{R}, \tag{8.5}$$

with $g = c * f$ which is of the form (3.2). Since $c \in \mathcal{W}(\mathbb{R}_+)$ it is easily seen that (8.4) and (8.5) are equivalent.

We want to apply Theorem 5 to obtain admissability of $\Lambda(X)$, as well as Corollary 3 for $L^p_\Lambda(X)$, $1 \le p < \infty$, where $\Lambda \subset \mathbb{R}$ closed is such that $\Lambda \cap \Lambda_0 = \emptyset$. For this purpose note first that there is a b-regular exponentially bounded resolvent $S(t)$ for the local versions of (8.4) and (8.5), where e.g. $b(t) = tc(t)$ is a possible choice. In the case $a_1 \equiv e_1$ this is a consequence of the special form $A_1(t) = a_1(t)A_0$, where A_0 is negative definite and the fact that $a_1(t)$ is of positive type; the general case then follows by perturbation techniques; cp. Prüss [16]. Obviously $b \in L^1(\mathbb{R}_+)$ is 1-regular and such that $\hat{b}(0) \neq 0$.

To verify (7.2) for (8.5), observe that we have to show

$$|\lambda[\hat{c}\hat{A}_1]'(\lambda)|_{\mathcal{B}(Y,X)} \le C|\hat{b}(\lambda)|\ , \quad \text{Re}\,\lambda > 0\ ,\ |\lambda| \ge R,$$

and

$$|\lambda\hat{c}'(\lambda)| \le M\ , \quad \text{Re}\,\lambda > 0\ ,\ |\lambda| \ge R,$$

by the remark following the proof of Theorem 5. The second of these inequalities is obvious, while the first follows from

$$|\lambda^2 \hat{a}_1'(\lambda)| + |\lambda^2 \hat{e}_1'(\lambda)| \le M_0 \,,$$

as an easy calculation shows; but this is a consequence of (8.3), since e.g.

$$|\lambda^2 \hat{a}_1'(\lambda)| = |\widehat{td\dot{a}_1}(\lambda) + 2\hat{\dot{a}}_1(\lambda)| \le \int_0^\infty t d\dot{a}_1(t) - 2\int_0^\infty \dot{a}_1(t)dt \le 3a_1(0+),$$

by convexity of $a_1(t)$.

Thus the assumptions of Theorem 5 are satisfied and it remains to compute Λ_0 and to show boundedness of $H(i\rho)$ and $\hat{b}(i\rho)H(i\rho)$ in $\mathcal{B}(X)$ resp. in $\mathcal{B}(X,Y)$. This means that we have to study the invertibility properties of $\lambda^2 - \widehat{dA}(\lambda)$ on the imaginary axis, i.e. for $\lambda = i\rho$, $\rho \in \mathbb{R}$. So let $u = (w, \phi)$, $f = (g, \psi)$ and consider the equation $(\rho^2 + \widehat{dA}(i\rho))u = f$, i.e.

$$\begin{aligned} \rho^2 w &+ \widehat{da_1}(i\rho)w'' + \widehat{da_1}(i\rho)\phi' = g \\ \rho^2 \phi &+ \widehat{de_1}(i\rho)\phi'' - \gamma\widehat{da_1}(i\rho)(w' + \phi) = \psi \\ w(0) &= \phi(0) = 0 \,,\ \phi'(1) = w'(1) + \phi(1) = 0. \end{aligned} \tag{8.6}$$

Taking the inner product in $L^2(0,1)$ of the first equation with w, integrating by parts once, with the help of the boundary conditions we obtain the identity

$$\rho^2 |w|^2 = \widehat{da_1}(i\rho)[|w'|^2 + (\phi, w')] + (g, w); \tag{8.7}$$

similarly, the second equation yields after multiplication with ϕ

$$\rho^2 |\phi|^2 = \widehat{de_1}(i\rho)|\phi'|^2 + \gamma\widehat{da_1}(i\rho)[(w', \phi) + |\phi|^2] + (\psi, \phi). \tag{8.8}$$

Here we used the notation $(\cdot,\cdot)$ for the inner product and $|\cdot|$ for the norm in $L^2(0,1)$. Multiplying (8.7) with γ and adding the result to (8.8) leads to

$$\rho^2(\gamma|w|^2 + |\phi|^2) = \widehat{de_1}(i\rho)|\phi'|^2 + \gamma\widehat{da_1}(i\rho)|w' + \phi|^2 + \gamma(g, w) + (\psi, \phi), \tag{8.9}$$

and taking real and imaginary parts in this identity we finally obtain

$$\rho^2(\gamma|w|^2 + |\phi|^2) = \text{ Re } \widehat{de_1}(i\rho)|\phi'|^2 + \gamma \text{ Re } \widehat{da_1}(i\rho)|w' + \phi|^2 + \gamma \text{ Re } (g, w) + \text{ Re } (\psi, \phi), \tag{8.10}$$

and

$$0 = \text{ Im } \widehat{de_1}(i\rho)|\phi'|^2 + \gamma \text{ Im } \widehat{da_1}(i\rho)|w' + \phi|^2 + \gamma \text{ Im } (g, w) + \text{ Im } (\psi, \phi). \tag{8.11}$$

The properties (8.3) of e_1 and a_1 imply

$$\begin{aligned} \text{Re } \widehat{de_1}(i\rho) \ge e_\infty > 0 \quad &, \quad \text{Re } \widehat{da_1}(i\rho) \ge a_\infty > 0. \\ \rho \text{ Im } \widehat{de_1}(i\rho) \ge 0 \quad &, \quad \rho \text{ Im } \widehat{da_1}(i\rho) \ge 0, \\ \lim_{|\rho|\to\infty} \widehat{de_1}(i\rho) = e_1(0+) \quad &, \quad \lim_{|\rho|\to\infty} \widehat{da_1}(i\rho) = a_1(0+), \end{aligned}$$

as it is easily verified. Moreover, for $\rho > 0$, Im $\widehat{de_1}(i\rho) = 0$ if and only if $e_1(t)$ is pieceweise linear, with nodes only at $t \in 2\pi\mathbb{N}/\rho$. If $e_1(t) \not\equiv e_\infty$, then there is a smallest such ρ, say $\rho_e > 0$, and then Im $\widehat{de_1}(i\rho) = 0$ if and only if $\rho \in \mathbf{Z}\ \rho_e$. The same is true of course for a_1, with a smallest $\rho_a > 0$, in case $a_1(t) \not\equiv a_\infty$.

To compute Λ_0 observe that either $\rho^2 + \widehat{dA}(i\rho)$ is invertible in $\mathcal{B}(X, Y)$ or 0 is an eigenvalue of $\rho^2 + \widehat{dA}(i\rho)$; the latter is equivalent to $\rho \in \Lambda_0$. Let $g = \psi = 0$ and assume Im $\widehat{de_1}(i\rho) \neq 0$: then (8.11) implies $\phi' \equiv 0$, hence $\phi \equiv 0$ by (8.6), and then $\widehat{da_1}(i\rho)(w' + \phi) = 0$, hence $\rho^2 w = 0$, by (8.6) again. Therefore Im $\widehat{de_1}(i\rho) \neq 0$ implies $\rho \notin \Lambda_0$. Similarly, if $g = \psi = 0$ and Im $\widehat{da_1}(i\rho) \neq 0$, (8.11) yields $w' + \phi \equiv 0$, and then by (8.6) again $\rho^2 w \equiv 0$, i.e. $w \equiv 0$ and $\phi \equiv 0$, i.e. $\rho \notin \Lambda_0$.

Thus $\rho \in \Lambda_0$ implies Im $\widehat{de_1}(i\rho)$ =Im $\widehat{da_1}(i\rho) = 0$, i.e. since $0 \notin \Lambda$, $e_1(t)$ and $a_1(t)$ must be of the special form explained above. In particular, if e_1 or a_1 is log-convex, or $-\dot{e}_1$ or $-\dot{a}_1$ is convex, then $\Lambda_0 = \emptyset$.

Now let $\Lambda \subset \mathbb{R}$ be closed such that Im $\widehat{de_1}(i\rho) \neq 0$, Im $\widehat{da_1}(i\rho) \neq 0$ for all $\rho \in \Lambda$ and

$$\sup\{\frac{\text{Re } \widehat{de_1}(i\rho)}{\rho \text{ Im } \widehat{de_1}(i\rho)} \cdot \frac{\text{Re } \widehat{da_1}(i\rho)}{\rho \text{ Im } \widehat{da_1}(i\rho)} : \rho \in \Lambda\} = N < \infty. \tag{8.12}$$

Then (8.10) and (8.11) yield for $|\rho| \geq 1$

$$\begin{aligned} \rho^2(\gamma|w|^2 + |\phi|^2) &\leq \gamma \text{ Re } (g, w) + \text{ Re } (\psi, \phi) - N\rho(\gamma \text{ Im } (g, w) + \text{ Im } (\psi, \phi)) \\ &\leq C_1|\rho|(\gamma|g|\,|w| + |\psi|\,|\phi|), \end{aligned}$$

hence

$$\rho^2(\gamma|w|^2 + |\phi|^2) \leq C_1^2(\gamma|g|^2 + |\psi|^2)\,, \quad \rho \in \Lambda\,,\ |\rho| \geq 1. \tag{8.13}$$

with (8.10) this implies by Re $\widehat{de_1}(i\rho) \geq e_\infty > 0$, Re $\widehat{da_1}(i\rho) \geq a_\infty > 0$,

$$e_\infty|\phi'|^2 + \gamma a_\infty|w' + \phi|^2 \leq C_2^2(\gamma|g|^2 + |\psi|^2)\,, \tag{8.14}$$

and (8.6) gives

$$|w''|^2 + |\phi''|^2 \leq C_3^2(\gamma|g|^2 + |\psi|^2)\rho^2\,, \quad \rho \in \Lambda\,,\ |\rho| \geq 1. \tag{8.15}$$

Estimates (8.13) and (8.15) yield

$$\rho|(\rho^2 + \widehat{dA_1}(i\rho))^{-1}|_{\mathcal{B}(X)} + \frac{1}{|\rho|}|(\rho^2 + \widehat{dA_1}(i\rho))^{-1}|_{\mathcal{B}(X,Y)} \leq C_4\,, \quad \rho \in \Lambda\,,\ |\rho| \geq 1,$$

which implies

$$|H(i\rho)|_{\mathcal{B}(X)} + |\hat{b}(i\rho)|\,|H(i\rho)|_{\mathcal{B}(X,Y)} \leq C_5\,, \quad \rho \in \Lambda, \tag{8.16}$$

for (8.5) . Invoking Theorem 5 then yields b-admissibility of $\Lambda(X)$, and Corollary 3 that of $L^p_\Lambda(X)$. We may now apply Theorem 3 or Corollary 2 to deduce results on almost periodic or periodic solutions of (8.1); the details of this, however, are left to the reader.

References

[1] L. Amerio and G. Prouse. *Almost-Periodic Functions and Functional Equations.* van Nostrand Reinbold Co., New York, 1971.

[2] J. Bergh and J. Löfström. *Interpolation Spaces. An Introduction*, volume 223 of *Grundl. Math. Wiss.* Springer Verlag, Berlin, Heidelberg, New York, 1976.

[3] T. Carleman. *L'integrale de Fourier et questions qui s'y rattachent.* ALmqvist and Wiksell, Uppsala, 1944.

[4] Ph. Clément and G. Da Prato. Existence and regularity results for an integral equation with infinite delay in a Banach space. *Integral Eqns. Operator Theory*, 11:480–500, 1988.

[5] Ph. Clément and J. Prüss. Completely positive measures and Feller semigroups. *Math. Ann.*, 287:73–105, 1990.

[6] G. Da Prato and A. Lunardi. Solvability on the real line of a class of linear Volterra integrodifferential equations of parabolic type. *Ann. Mat. Pura Appl.*, 55:67–118, 1988.

[7] E. B. Davies. *One Parameter Semigroups*, volume 15 of *London Math. Soc. Monographs*. Academic Press, New York, 1980.

[8] L. Gearhart. Spectral theory for contraction semigroups on Hilbert spaces. *Trans. Amer. Math. Soc.*, 236:385–394, 1978.

[9] R. Grimmer, R. Lenczewski, and W. Schappacher. Wellposedness of hyperbolic equations with delay in the boundary conditions. In: Ph. Clément, S. Invernizzi, E. Mitidieri, and I. Vrabie, editors, *Trends in Semigroup Theory and Applications*, pages 215–228. Marcel Dekker, 1989.

[10] G. Gripenberg. Asymptotic behaviour of resolvents of abstract Volterra equations. *J. Math. Anal. Appl.*, 122:427–438, 1987.

[11] Y. Katznelson. *An Introduction to Harmonic Analysis.* Dover Publ. Inc., 2 edition, 1976.

[12] J. Prüss. Lineare Volterra Gleichungen in Banach Räumen. Habilitationsschrift, Universität-GH Paderborn, 1984.

[13] J. Prüss. On the spectrum of C_0-semigroups. *Trans. Amer. Math. Soc.*, 284:847–857, 1984.

[14] J. Prüss. Bounded solutions of Volterra equations. *SIAM J. Math. Anal.*, 19:133–149, 1988.

[15] J. Prüss. Maximal regularity of linear vector valued parabolic Volterra equations. *J. Integral Eq. Appl.*, 3:63–83, 1991.

[16] J. Prüss. *Linear Evolutionary Integral Equations and Applications.* Birkhäuser, Basel, 1993. to appear.

[17] D.V. Widder. *The Laplace Transform.* Princeton University Press, Princeton, 1941.

The Model Nonlocal Nonlinear Equation

I. A. SHISHMAREV and P. I. NAUMKIN Moscow State University, Moscow, Russia

1 INTRODUCTION

We consider the following model nonlinear nonlocal equation

$$u_t + uu_x + \mathbb{K}(u) = 0, \tag{1}$$

where

$$\mathbb{K}(u) = \frac{1}{2\pi}\int_{-\infty}^{\infty} e^{ipx} K(p)\hat{u}(p,t)dp$$

$\hat{u}(p,t)$ is the Fourier transform of the function $u(x,t)$. The dispersion relation for equation (1) is equal to the symbol $K(p)$ of the operator $\mathbb{K}$ and under appropriate choice of $K(p)$ equation (1) reduces to various well-known equations. For example, if $K(p) = p^2$, we obtain the Burger equation

$$u_t + uu_x - u_{xx} = 0.$$

If $K(p) = ip^3$, we have the Korteweg–de Vries equation

$$u_t + uu_x - u_{xxx} = 0.$$

If $K(p) = ip|p|$, we find the Benjamin–Ono equation (Benjamin, 1967; Ono, 1975)

$$u_t + uu_x + \frac{1}{\pi}\int_{-\infty}^{\infty} \frac{u_{ss}(s,t)ds}{x-s} = 0.$$

If $K(p) = -p^2 + p^4$, we get the Kuramoto–Sivashinsky equation (Kuramoto, 1982)

$$u_t + uu_x + u_{xx} + u_{xxxx} = 0.$$

If $K(p) = ip(\frac{1}{p}\, thp)^{1/2}$, we obtain the Whitham equation (Whitham, 1974)

$$u_t + uu_x + \int_{-\infty}^{\infty} k_g(x-s)u_s(s,t)ds = 0,$$

where $k_g(x) = \frac{1}{2\pi}\int_{-\infty}^{\infty} e^{ipx}(\frac{1}{p}\, thp)^{1/2}\, dp$. If $K(p) = -ip^3 K_0(B|p|)$, where K_0 is the modified Bessel function, we find the Leibovich equation (Leibovich, 1970)

$$u_t + uu_x + \frac{\partial^2}{\partial x^2}\int_{-\infty}^{\infty} \frac{u_s(s,t)}{\sqrt{(x-s)^2 + B^2}}\, ds = 0.$$

If $K(p) = \frac{ip}{\sqrt{1+p^2}}$, we have the Klimontovich equation (Klimontovich, 1964)

$$u_t + uu_x + \int_{-\infty}^{\infty} K_0(|x-s|)u_s(s,t)ds = 0.$$

And if $K(p) = \beta p^2 + \gamma p^2 \frac{1}{1+p^2}$, we get the Rudenko–Soluyan equation (Rudenko and Soluyan, 1975)

$$u_t + uu_x - \beta u_{xx} - \gamma \frac{\partial}{\partial x}\int_{-\infty}^{\infty} e^{-|x-s|}u_s(s,t)ds = 0.$$

Thus equation (1) is very general and can describe different wave processes.

2 CONSERVATION LAWS

Let $u(x,t)$, the solution of equation (1), vanish at infinity sufficiently fast. Equation (1) is invariant under the Galilean transformation $t' = t, x' = x + ct, u'(x',t') = u(x,t) + c$. If the symbol $K(p)$ of the operator $\mathbb{K}$ in equation (1) is conservative, i.e., $\operatorname{Re} K(p) = 0$, then the following conservation laws holds

$$I_1 = \int_{-\infty}^{\infty} u(x,t)dx = \text{const},$$

$$I_2 = \int_{-\infty}^{\infty} u^2(x,t)dx = \text{const},$$

$$I_3 = \int_{-\infty}^{\infty} \left(\frac{u^3}{3} + u\tilde{\mathbb{K}}(u)\right) dx = \text{const},$$

where $\tilde{\mathbb{K}}(u)$ has the symbol $\tilde{K}(p) = \frac{K(p)}{ip}$. Using the integral of energy I_3, we can write equation (1) in the Hamiltonian form

$$u_t = -\frac{\partial}{\partial x}\frac{\delta \mathcal{H}}{\delta u}, \mathcal{H} = \frac{1}{2}I_3.$$

The center of mass $R = \int_{-\infty}^{\infty} xu\, dx$ moves with the constant speed

$$\frac{dR}{dt} = I_2 + \tilde{K}(o)I_1.$$

3 THE CRESTING OF WAVES

Below we show that if the operator $\mathbb{K}(u)$ is strongly dissipative, then discontinuous initial data are smoothing with time. This happens, for example, with the solutions of Burgers, Korteweg-de Vries–Burgers and Kuramoto–Sivashinsky equations. Now we consider the opposite situation: The crested (piecewise continuous) initial data remain to be crested all the time of existence of solution of the Cauchy problem for equation (1). The function $u(x,t)$ which is continuous in the strip $\Pi = [0,T] \times R_1$ and has the piecewise continuous first and second derivatives with respect to x is called the generalized solution of the Cauchy problem for equation (1) if $u(x,t)$ tends to initial data $\bar{u}(x)$ as $t \to 0$ and satisfies equation (1) in the generalized sense, i.e., for any function $\psi(x,t) \in C_0^\infty(\Pi)$ the integral identity

$$\int_0^T \int_{-\infty}^{\infty} \left[u\left(\psi_t + \frac{1}{2}\psi_x u\right) + \psi_x \tilde{\mathbb{K}}(u) \right] dx\, dt = 0$$

is valid. This solution is classical out of the lines of discontinuity and these lines $x(t)$ satisfy the Gugonio condition: $\dot{x}(t) = u(x,t)$.

THEOREM 1 Let $\tilde{K}(p) = \frac{K(p)}{ip} \in \mathcal{L}_1(R_1) \cap C^1(R_1 - 0)$ and the initial data $\bar{u}(x)$ be piecewise smooth, besides $\bar{u}'(+0) - \bar{u}'(-0) \neq 0$.

Then the first derivative of the generalized solution $u(x,t)$ will be discontinuous during all the time of existence of solution.

4 BREAKING OF WAVES

Now we give a positive answer to the problem of Whitham (Whitham, 1974) on the breaking of waves described by equation (1) with a symbol $K(p) = ip(\frac{thp}{p})^{1/2}$, corresponding to the potential theory of waves on water. The breaking of wave means that $\max_{x \in R_1} |u_x(x,t)| \to \infty$ as $t \to T_0 < \infty$ is the so-called breaking time.

THEOREM 2 Let (1) the kernel

$$k(x) = \frac{1}{2\pi} \int_{-\infty}^{\infty} e^{ipx} \tilde{K}(p) dp$$

satisfy the following conditions

$$k(x) \in C^1(R_1 - 0) \cap \mathcal{L}_1(R_1), \quad |k(x)| \leq C|x|^{-\alpha},$$

$$|k'(x)| \leq C|x|^{-1-\alpha}, \quad x \in [-a, a] - 0,$$

$$\int_{|x| \geq a} |k'(x)| dx \leq C,$$

$$\alpha = \frac{3}{5} - \gamma, \quad \gamma \in \left(0, \frac{1}{10}\right], \quad a \in (0,1], \quad c > 0,$$

(2) the symbol $K(p)$ be conservative or dissipative, i.e.

$$K_1(p) \equiv \operatorname{Re} K(p) \geq 0,$$

(3) initial data be sufficiently steep

$$m_0^2 > \frac{7c}{\gamma a}\left(u_1 + \sqrt{J}\right) + \left(\frac{2b}{\gamma}\right)^2,$$

where

$$u_1 = \max_{x \in R_1} |\bar{u}'(x)|, \quad b = \sup_{|p| \le p_0} (0, -K_1(p)), \quad J = \int_{-\infty}^{\infty} (\bar{u}'''(x))^2 \, dx.$$

Then a solution $u(x,t)$ of the Cauchy problem for equation (1) belongs to the class $C^\infty([0,T_0); H^\infty(R_1))$ is unique and breaks at time $T_0 \le \frac{1}{m_0(1-\gamma)}$.

5 EXISTENCE OF GLOBAL SOLUTION

Equation (1) has traveling wave solutions. Thus, under certain conditions, the solution of the Cauchy problem for equation (1) exists in the large.

THEOREM 3 Let (1) operator $\mathbb{K}$ be strictly dissipative

$$K_1(p) \ge 0, \qquad \text{for } |p| \le p_0, \quad K_1(p) \ge \mathcal{E} > 0, \qquad \text{for } |p| \ge p_0 > 0,$$

(2) $\bar{u}(x) \in H^\infty(R_1)$,
(3) $\|\bar{u}\|_{H^2(R_1)} \le C$,
where $C = C(\mathcal{E}, h) > 0$ is sufficiently small.

Then there exists a unique solution $u(x,t)$ for the Cauchy problem for the equation (1) from the class $C^\infty([0,\infty); H^\infty(R_1))$. The characteristic condition of this theorem is the smallness of the steepness of the initial data. From the comparison of theorems 2 and 3 we conclude that the requirement of a large steepness of initial perturbations is not only sufficient but also a necessary condition for the breaking of a wave in finite time.

Theorem 3 shows that even small dissipation ensures the existence of the classical solution $u(x,t)$ in the large. Indeed, in the conservative case, when $K_1(p) = 0$, a wave breaks down (Theorem 2) but if $K_1(p) \ge \mathcal{E} > 0$ then the solution exists for all time. If dissipation is stronger, we will see that breaking is impossible and, moreover, the nonsmooth initial perturbations become smooth for all subsequent times.

6 SMOOTHING OF SOLUTIONS

THEOREM 4 Let (1) the operator $\mathbb{K}$ in equation (1) be strongly dissipative

$$K_1(p) \ge \mathcal{E}|p|^\alpha, \text{ for } |p| \ge p_0 > 0, \tag{2}$$

where $\mathcal{E} > 0, \alpha > 1$.

(2) initial data $\bar{u}(x) \in H^\rho(R_1), \rho \ge 0, \rho > \frac{3}{2} - \alpha$.

Then there exists a unique solution for the Cauchy problem for equation (1) from the class $C^\infty((0,T], H^\infty(R_1)) \cap C^0([0,T], H^\rho(R_1))$ for some $T > 0$.

In the following theorem, we state that the solution is smoothing and exists for all $t \ge 0$.

THEOREM 5 Let (1) the condition (2) be fulfilled with

$$\alpha > 3/2, \quad (2) \bar{u}(x) \in \mathcal{L}_2(R_1).$$

Then the unique solution $u(x,t)$ to the Cauchy problem for equation (1) exists and belongs to the class

$$C^\infty((0,\infty); H^\infty(R_1)) \cap C^0([0,\infty); \mathcal{L}_2(R_1))$$

EXAMPLES The conditions of Theorem 5 are valid for the Burgers, Korteweg–de Vries–Burgers and Kuramoto–Sivashinsky equations.

7 PERIODIC PROBLEM

In the case where the wave medium has periodic structure, it is natural to replace the definition of pseudodifferential operator $\mathbb{K}$ in equation (1) by the following

$$\mathbb{K}(u) = \frac{1}{2\pi} \sum_{p=-\infty}^{\infty} e^{ipx} K_p \hat{u}_p(t)$$

where $u(x,t) \in C_\pi^\infty(R_1) = \{u(x,t) \in C^\infty, u(x,t) = u(x+2\pi, t), x \in R_1\}$, $\hat{u}_p(t)$ are the Fourier coefficients of function $u(x,t)$, $\{K_p\}$ is the symbol of operator $\mathbb{K}(u)$.

The results, analogous to Theorems 1–5, are valid for the solutions of the periodic problem for equation (1) (see Naumkin and Shishmarev, 1990). Now let us consider a periodic problem with a small interaction. To account for the effect of smallness of a nonlocal interaction it is natural to consider equation (1) with a small parameter μ in front of the operator $\mathbb{K}$. We thus consider the problem

$$u_t + uu_x + \mu\mathbb{K}(u) = 0, \quad u|_{t=0} = \bar{u}(x). \tag{3}$$

The purpose of this section is to study the asymptotics of a solution of problem (3) as $\mu \to 0$.

THEOREM 6 Suppose the following conditions are satisfied

(1) $k(x) = \frac{1}{2\pi} \sum_{p=-\infty}^{\infty} \frac{K_p}{ip} e^{ipx} \in C^1(\Omega - 0)$, $|k(x)| \le c|x|^{-\alpha}$, $|k'(x)| \le c|x|^{-1-\alpha}$, $x \in \Omega - 0$, $\alpha \le \frac{3}{5} - \gamma$, $\gamma \in (0, \frac{1}{10}]$, $c > 0$, $\Omega = [-\pi, \pi]$, and $\operatorname{Re} K_p \ge 0$, for $|p| \ge p_0 > 0$, $p \in Z$.

(2) $\bar{u}(x) \in C_\pi^\infty(R_1)$.

(3) The steepness $m_0 = \min_{x \in \Omega} \bar{u}'(x)$ of the initial data is sufficiently large

$$m_0^2 > \frac{b^2}{\gamma^2} + \frac{40c}{\gamma}(u_1 + J^{1/2}),$$

where

$$u_1 = \max_{x \in \Omega} |\bar{u}'(x)|, \quad J = \int_\Omega |\bar{u}'''(x)|^2 \, dx, \quad b = 2 \max_{|p| \le p_0} |\operatorname{Re} K_p|$$

Then the following assertions are true:

(1) For any $\mu \in [0,1]$ there exists a unique 2π-periodic solution $u_\mu(x,t)$ of the problem (3) with $u_\mu(x,t) \in\in C^\infty([0, T_\mu); C_\pi^\infty(R_1))$, where T_μ is the breaking time.

(2) For $u_\mu(x,t)$ there is the asymptotic expansion $u_\mu(x,t) = \sum_{k=0}^{\infty} \mu^k u_k(x,t)$, where the remainder term δ_n has the estimate $\max_{t \in [0, T_0 - \beta]} \| \frac{\partial^m \delta_n}{\partial t^m} \|_{(l)} \le C_0 \mu^{n+1}$, $\beta > 0$ is arbitrary, $m, l \ge 0$ are any integers, $\| \cdot \|_{(l)}$ is the norm of the space $C^l([0, 2\pi])$, and $C_0 = C_0(l, m, n, \beta)$.

8 ASYMPTOTICS FOR $t \to \infty$

We consider the case where the potential $v(x,t) = \int_{-\infty}^x u(x,t)dx$ vanish at infinity sufficiently fast. Then from equation (1) we obtain

$$v_t + v_x^2 + \mathbb{K}(v) = 0, \quad v|_{t=0} = \bar{v}(x), \quad x \in R_1, \quad t \ge 0. \tag{4}$$

The solution $v(x,t)$ of the Cauchy problem (4) is real valued, so the symbol $K(p)$ of the operator $\mathbb{K}$ must satisfy the condition $K(-p) = K^*(p)$. Thus the conservative part $K_2(p) = \operatorname{Im} K(p)$ is an odd function of $p \in R_1$ and nonconservative part $K_1(p) = \operatorname{Re} K(p)$ is an even one. The following statement is valid.

THEOREM 7 Let (1) the symbol $K(p)$ satisfy the conditions $K_1(p) = \lambda + w|p|^\delta + O(|p|^{\sigma+\delta})$, $|K_2(p)| \le C_1|p|^\beta$, $|p| \le 1$, $\lambda \ge 0$, $0 < \delta < 3$, $\delta < \beta$, $w > 0$; $K_1(p) \ge \lambda + c_2 m^\delta(p) M(p)$, $p \in R_1$; $|K(p_1) - K(p_2)| \le c_3|\Delta p|^{\sigma_2} M^\alpha(p_1)$, $|\Delta p| = |p_1 - p_2| \le 1, p_1, p_2 \in R_1, \sigma_1, \sigma_2 \in (0,1]$, $\alpha \ge 0$; $m(p) \equiv \min(1, |p|)$, $M(p) \equiv \max(1, |p|)$.

(2) The Fourier transform $\hat{\bar{v}}(p)$ of the initial data $\bar{v}(x) \in \mathcal{L}_1(R_1)$ be such as $|\hat{\bar{v}}(p)| \le \mathcal{E} M^{-3-\alpha}(p)$, $|\hat{\bar{v}}(p_1) - \hat{\bar{v}}(p_2)| \le c_4|\Delta p|^{\sigma_3} M^{-3}(p_1)$, $|\Delta p| \le 1, p_1, p_2 \in R_1$, $\sigma_3 \in (0,1]$, $C_i > 0$, $i = 1, \ldots, 4$, and $0 < \mathcal{E} < C$, where C is a constant that depends on $K(p)$. Then the asymptotic formula for the solution $v(x,t)$ of the problem (4)

$$v(x,t) = Ae^{-\lambda t} t^{-1/\delta} \int_0^\infty \cos p\xi e^{-wp^\delta}\, dp + O(e^{-\lambda t} t^{-1/\delta - \beta}) \tag{5}$$

is valid for $t \to \infty$ uniformly with respect to $\xi = |x| t^{-1/\delta} > 0$, where $\beta > 0$ is a constant, the constant A is explicitly expressed in terms of the symbol $K(p)$, and initial function $\bar{v}(x)$, namely,

$$A = 2 \sum_{n=0}^{\infty} \phi_n(0) \mathcal{E}^{n+1}, \quad \phi_0(0) = \frac{1}{\mathcal{E}} \int_{-\infty}^{\infty} \bar{v}(y)\, dy,$$

$$\phi_1(0) = h - \frac{1}{\mathcal{E}^2} \int_{-\infty}^{\infty} \frac{\hat{\bar{v}}(y)\hat{\bar{v}}(-y)\, dy}{K(0) - K(y) - K(-y)}, \ldots$$

EXAMPLES The symbol $K(p) = C_1|p|^\delta + iC_2|p|^\delta \operatorname{sgn} p$, $\delta \in (0,3)$, $\nu > 1$, $C_1 > 0$, C_2 is arbitrary, satisfies all the conditions of Theorem 7. In particular, if $\delta = 2$ and $\nu = 3/2$, then $K(p)$ corresponds to the Korteweg–de Vries–Burgers equation. The symbol $K(p) = ap^2 + bp^4, a, b > 0$, of the Kuramoto–Sivashinsky equation also satisfies the conditions of Theorem 7, and for the solutions $v(x,t)$ the asymptotic formula (5) is true with $\delta = 2$ and $\lambda = 0$.

In conclusion, we note that if the potential $v(x,t)$ does not vanish at infinity (i.e., $\int_{-\infty}^{\infty} \bar{u}(x)\, dx \ne 0$) we can consider the original equation (1) for the function $u(x,t)$. The asymptotic formula analogous to (5) is valid in that case if the parameter δ in the conditions of Theorem 7 satisfies the requirement $0 < \delta < 2$.

REFERENCES

Benjamin T.B. (1967). Internal waves of permanent form in fluids of great depth. J. Fluid Mech., 29:559.

Klimontovich Y.L. (1964). Statistical theory of nonequilibrium processes in plasma. Moscow,

Kuramoto Y. (1982). Chemical oscillations, waves and turbulence. Berlin: Springer.

Leibovich S.J. (1970). Weakly nonlinear waves in rotating fluids. J. Fluid Mech., 42:802.

Naumkin P.I. and Shishmarev I.A. (1990). A periodic problem for Whitham's equation. Math. USSR, Sbornic, 7:403.

Ono H. (1975). Algebraic solitary waves in stratified fluids. J. Phys. Soc. Jpn., 39:1082.

Rudenko O.V. and Soluyan S.I. (1975). Theoretical foundations of nonlinear acoustics. Moscow.

Whitham G. (1970). Linear and nonlinear waves. New York: Wiley.

Quasilinear Parabolic Equations and Semiflows

GIERI SIMONETT University of Zürich, Zürich, Switzerland

1. Introduction

In this note we consider the following abstract **quasilinear Cauchy problem**

$$(QCP) \quad \begin{cases} \dot{u} + A(u)u = f(u) , & t > 0 , \\ u(0) = x . \end{cases} \tag{1.1}$$

We assume that there are given two Banach spaces X_0 and X_1 with $X_1 \overset{d}{\hookrightarrow} X_0$, i.e. X_1 is densely embedded in X_0. For a nonempty subset $U \subset X_0$ let

$$(A, f) : U \longrightarrow \mathcal{L}(X_1, X_0) \times X_0$$

where $\mathcal{L}(X_1, X_0)$ denotes the Banach space of bounded linear operators from X_1 into X_0. We assume that $-A(u)$ generates a strongly continuous analytic semigroup on X_0 and moreover that the domain of definition of $A(u)$ - equipped with the graph norm - coincides with X_1 (up to equivalent norms) for each $u \in U$. We shall abbreviate this by writing

$$A(u) \in \mathcal{H}(X_1, X_0) , \quad u \in U . \tag{1.2}$$

By a solution of the quasilinear Cauchy problem (QCP) we mean a function $u : J_u \to X_0$ on a nontrivial interval J_u with

$$u \in C(J_u, U) \in C^1(J_u \setminus \{0\}, X_0) \cap C(J_u \setminus \{0\}, X_1) \tag{1.3}$$

such that

$$\dot{u}(t) + A(u(t))u(t) = f(u(t)) , \quad t \in J_u \setminus \{0\} , \quad u(0) = x .$$

Problem (1.1) is also called a **parabolic quasilinear evolution equation.** Problems of this type have been investigated by several authors. We would like to mention the following references: [**1, 2, 3**], [4], [8], [**12**]. In fact, there are quite different approaches to deal with the nonlinear evolution equation (1.1).

On the one hand one can take an approach based on the concept of parabolic fundamental solutions and try to obtain a solution as a fixed point in appropriate function spaces. This has been carried out in [**1**]. By a detailed analysis based on interpolation theory, Amann was able to prove that quasilinear parabolic equations generate (local) semiflows in appropriate phase spaces (which are interpolation spaces between X_1 and X_0). With this result quasilinear equations can be considered more ore less as ordinary differential equations.

On the other hand, the nonlinear equation (1.1) can also be treated by the method of maximal regularity developed in [**6**]. In this case one has to work in the very special spaces of 'maximal regularity' (which are the little Hölder spaces or the little Nikolskii spaces in some applications). However, this technique does not take into consideration the special structure of quasilinear equations.

Angenent gave in [4] an improvement of the result of maximal regularity which renders a possibility to treat quasilinear equations but now taking advantage of the smoothing property of solutions (i.e. that the solutions become more regular than the initial values). Moreover Angenent proved with a very elegant idea the smooth dependence of solutions on the data. In fact, he shows the smooth dependence in the topology of X_1 for initial values lying in X_1.

We would like to give a refinement and improvement of his results. We study the continuous dependence of solutions on the initial values in the topology of X_α, where X_α is a continuous interpolation space between X_1 and X_0. In fact, we show that the map $(t, x) \mapsto u(t, x)$ defines a (Lipschitz) continuous semiflow on X_α, if A and f are locally Lipschitz continuous. Then we give further regularity results: If A and f are C^k, or C^∞, or analytic, then the map $(t, x) \mapsto u(t, x)$ has the same property, as a function with values in X_α. A detailed analysis shows, that this statement can even be improved, cf. Remark 3.8. We obtain results on smooth dependence in any Banach space lying between X_1 and X_α. Thus, we use maximal regularity and then are able to state results in spaces, which do not necessarily have the property of maximal regularity. An additional result shows, that the derivative $\partial_2 u(\cdot, u)$ solves the linearized problem associated to (1.1). Our new ingredients and results are contained in Proposition 2.1, Theorem 3.1, Theorem 3.2, Proposition 3.5, Theorem 3.7, and Remark 3.8 *a*).

2. A Result on Maximal Regularity

Let (X_0, X_1) be a pair of Banach spaces with $X_1 \overset{d}{\hookrightarrow} X_0$. For $\alpha \in (0,1)$ we denote by

$$X_\alpha := (X_0, X_1)_\alpha$$

the interpolation spaces obtained by the **continuos interpolation method** of Da Prato and Grisvard, cf. [**6**]. The spaces X_α, $0 < \alpha < 1$, can be characterized in the following way. For fixed $T > 0$ let $V_{\alpha,T}(X_0, X_1)$ be the function space

$$\begin{aligned} V_{\alpha,T}(X_0,X_1) :=& \{u \in C^1((0,T], X_0) \cap C((0,T], X_1); \\ & \lim_{t\to 0} t^{1-\alpha}(\|u'(t)\|_{X_0} + \|u(t)\|_{X_1}) = 0\} \,, \end{aligned} \tag{2.1}$$

equipped with the norm

$$\|u\|_{V_{\alpha,T}(X_0,X_1)} := \sup_{t\in(0,T]} t^{1-\alpha}(\|u'(t)\|_{X_0} + \|u(t)\|_{X_1}) \,. \tag{2.2}$$

We may extend this definition to $\alpha = 1$ by setting

$$V_{1,T}(X_0, X_1)) := C^1([0,T], X_0) \cap C([0,T], X_1) \,.$$

Given any $u \in V_{\alpha,T}(X_0, X_1)$ and $0 < s < t \leq T$ we obtain

$$\|u(t) - u(s)\|_{X_0} \leq \|u\|_{V_{\alpha,T}(X_0,X_1)} \int_s^t \frac{d\tau}{\tau^{1-\alpha}} \leq \|u\|_{V_{\alpha,T}(X_0,X_1)} \int_0^{t-s} \frac{d\tau}{\tau^{1-\alpha}} \,,$$

which shows that $V_{\alpha,T}(X_0, X_1)$ is continuously embedded in $UC^\alpha((0,T], X_0)$, the space of uniformly α-Hölder continuos functions on the interval $(0,T]$ with values in X_0. Thus, each function $u \in V_{\alpha,T}(X_0, X_1)$ can be extended to $[0,T]$. It is not difficult to see that the map

$$R_\alpha : \quad V_{\alpha,T}(X_0, X_1) \to X_0 \,, \quad u \mapsto R_\alpha u := u(0) \tag{2.3}$$

defines a bounded linear operator. Now, the elements of X_α are exactly the traces of functions belonging to $V_{\alpha,T}(X_0, X_1)$. More precisely we obtain

$$X_\alpha = \operatorname{im} R_\alpha \,, \quad 0 < \alpha < 1 \,, \tag{2.4}$$

and

$$\|x\|_\alpha := \inf\{\|u\|_{V_{\alpha,T}(X_0,X_1)} \,;\; u \in V_{\alpha,T}(X_0,X_1),\; x = R_\alpha u\} \tag{2.5}$$

defines a norm on X_α (cf. [**6,5**] and [**9**, Appendix]). It is immediately clear that

$$R_\alpha \in \mathcal{L}(V_{\alpha,T}(X_0, X_1), X_\alpha) \,. \tag{2.6}$$

The elements of $V_{\alpha,T}(X_0, X_1)$ are continuous in the topology of X_α. More precisely we obtain the following

2.1 PROPOSITION

$V_{\alpha,T}(X_0, X_1) \hookrightarrow C([0,T], X_\alpha) \cap C^{\alpha-\beta}([0,T], X_\beta)$ for $0 \leq \beta < \alpha \leq 1$, where $C^{\alpha-\beta}([0,T], X_\beta)$ denotes the space of Hölder continuous functions of exponent $\alpha - \beta$.

PROOF The proof is given in [**11**, Satz 5.1]. □

For $A \in \mathcal{H}(X_1, X_0)$, the elements of X_α can be characterized by the following property

$$x \in X_\alpha \iff (t \mapsto e^{-tA}x) \in V_{\alpha,T}(X_0, X_1)\ , \quad 0 < \alpha < 1\ . \tag{2.7}$$

We first consider the linear Cauchy problem

$$(CP)_{(A,f,x)} \qquad \begin{cases} \dot{u} + Au = f(t)\ , & 0 < t \leq T\ , \\ u(0) = x\ . \end{cases}$$

Setting

$$\begin{aligned} C_\alpha(X_0) := C_{\alpha,T}(X_0) &:= \{f \in C((0,T], X_0);\ \lim_{t \to 0} t^{1-\alpha}\|f(t)\|_{X_0} = 0\}\ , \\ \|f\|_{C_{\alpha,T}(X_0)} &:= \sup_{t \in (0,T]} t^{1-\alpha}\|f(t)\|_{X_0} \end{aligned}$$

for $0 < \alpha < 1$ and $C_{1,T}(X_0) := C([0,T], X_0)$ we easily deduce, thanks to (2.6), that

$$(\partial_t + A, R_\alpha) \in \mathcal{L}\big(V_{\alpha,T}(X_0, X_1),\ C_{\alpha,T}(X_0) \times X_\alpha\big)\ , \tag{2.8}$$

holds, where ∂_t denotes the derivative. Following [4] we define

$$\begin{aligned} \mathcal{M}_\alpha(X_1, X_0) := \{&A \in \mathcal{H}(X_1, X_0)\ ; \\ &(\partial_t + A, R_\alpha) \in Isom\big(V_{\alpha,T}(X_0, X_1),\ C_{\alpha,T}(X_0) \times X_\alpha\big)\}\ . \end{aligned} \tag{2.9}$$

Observe that $A \in \mathcal{M}_\alpha(X_1, X_0)$ just means that the Cauchy problem $(CP)_{(A,f,x)}$ has, for each $(f, x) \in C_{\alpha,T}(X_0) \times X_\alpha$, a unique solution

$$u := (\partial_t + A, R_\alpha)^{-1}(f, x) \in V_{\alpha,T}(X_0, X_1)\ .$$

Moreover, u depends continuously on the data (f, x). Note that $\dot{u}$ and Au belong to the same function space as f. The pair (X_0, A) is then said to have the property of **maximal regularity**. Of course, we have to show that the set $\mathcal{M}_\alpha(X_1, X_0)$ is nonvoid.

2.2 REMARKS

a) $\mathcal{M}_\alpha(X_1, X_0)$, $0 < \alpha \le 1$, was introduced by Angenent in [**4**]. This definition generalizes the class $\mathcal{M}_1(X_1, X_0)$ of Da Prato and Grisvard in [**6**].

b) Let $A \in \mathcal{M}_\alpha(X_1, X_0)$ and $(f, x) \in C_\alpha(X_0) \times X_\alpha$ be given and set

$$J_A f := J_{A,T} f := (\partial_t + A, R_\alpha)^{-1}(f, 0) ,$$
$$x(\cdot) := (\partial_t + A, R_\alpha)^{-1}(0, x) .$$

We then have

$$\begin{aligned} (J_A f)(t) &= \int_0^t e^{-(t-\tau)A} f(\tau)\, d\tau , \\ x(t) &= e^{-tA} x , \qquad\qquad t \in (0, T] . \end{aligned} \tag{2.10}$$

(Each solution of the Cauchy problem $(CP)_{(A,f,x)}$ necessarily satisfies the variation of constants formula). Together with (2.7) we obtain the following characterization

$$A \in \mathcal{M}_\alpha(X_1, X_0) \iff J_A\big(C_\alpha(X_0)\big) \subset C_\alpha(X_1) . \tag{2.11}$$

Indeed, take $f \in C_\alpha(X_0)$. This certainly implies $f \in L_1((0,T), X_0) \cap C((0,T], X_0)$, where $L_1((0,T), X_0)$ denotes the Banach space of (Bochner) integrable functions on $(0,T)$. Now, [**10**, Theorem 4.2.4] shows that $J_A f$ belongs to $C^1((0,T], X_0)$ and that $(J_A f)' = f - AJ_A f$ holds. Together with the assumption $J_A f \in C_\alpha(X_0)$ this implies $J_A f \in V_\alpha(X_0, X_1)$. For any $(f, x) \in C_\alpha(X_0) \times X_\alpha$, the function $J_A f + x(\cdot)$ - which belongs to $V_\alpha(X_0, X_1)$ by the given argument and by (2.7) - is the unique solution of the Cauchy problem. Now the open mapping theorem gives that the bounded linear operator in (2.8) is indeed an isomorphism.

c) Let $X_\gamma := (X_0, X_1)_\gamma$, $\gamma \in (0,1)$, be an interpolation space, given by an arbitrary interpolation method $(\cdot,\cdot)_\gamma$ of exponent γ. Suppose $A \in \mathcal{M}_\alpha(X_1, X_0)$ and $B \in \mathcal{L}(X_\gamma, X_0)$ for $0 \le \gamma < 1$, where $X_\gamma := X_0$ for $\gamma = 0$. Then we have the following perturbation result

$$A + B \in \mathcal{M}_\alpha(X_1, X_0) \tag{2.12}$$

cf. [**4**, Lemma 2.5].

d) The definitions in (2.4),(2.5) and (2.9) are independent of T.

2.3 THEOREM

Suppose we have two Banach spaces E_1, E_0 *with* $E_1 \overset{d}{\hookrightarrow} E_0$ *and an* $A \in \mathcal{H}(E_1, E_0)$. *Let* $X_0 := (E_0, E_1)_\theta$ *be a continuous interpolation space for an arbitrary* $\theta \in (0,1)$. *Let* X_1 *be the domain of definition of* A_{X_0} *, the part of* A *in* X_0.

Then $A_{X_0} \in \mathcal{M}_\alpha(X_1, X_0)$ *for each* $\alpha \in (0,1]$.

PROOF Cf. [**4**, Theorem 2.14]. Another proof, working with the semigroup e^{-tA} rather than with the resolvent of $-A$, is given in [**11**, Theorem 5.4]. □

3. Maximal Regularity and Quasilinear Equations

Under the assumption of maximal regularity, we show that the autonomous quasilinear Cauchy problem (QCP) generates a semiflow on X_α. In addition we prove the smooth dependence on the initial values.

Let X_0 and X_1 be Banach spaces with $X_1 \overset{d}{\hookrightarrow} X_0$. Fix two reals α and β with

$$0 < \beta < \alpha \le 1 . \tag{3.1}$$

Let $U := U_\alpha$ be an open subset of the continuous interpolation space X_α (with $X_\alpha := X_1$ for $\alpha = 1$). We denote by U_β the same set, but equipped with the norm of X_β. (Note that $X_\alpha \hookrightarrow X_\beta$ so that U is indeed a subset of X_β). Let us assume

$$\begin{aligned} &(A, f) \in C^{1-}(U_\beta\,,\, \mathcal{M}_\alpha(X_1, X_0) \times X_0)\,, \quad 0 < \beta < \alpha < 1\,, \\ &(A, f) \in C^{1}\;\,(U_\beta\,,\, \mathcal{M}_1(X_1, X_0) \times X_0)\,, \quad 0 < \beta < \alpha = 1\,. \end{aligned} \tag{3.2}$$

3.1 THEOREM *For each* $x \in U_\alpha$ *the quasilinear Cauchy problem*

$$\dot{u} + A(u)u = f(u)\,, \quad t > 0\,, \quad u(0) = x \tag{3.3}$$

has a unique maximal solution $u(\cdot, x)$, *defined on the maximal interval of existence* $[0, t^+(x))$ *and*

$$u(\cdot, x) \in C([0, t^+(x)), U_\alpha) \cap C^\alpha([0, t^+(x)), X_0)\,.$$

Moreover $u(\cdot, x) \in V_{\alpha,T}(X_0, X_1)$ *for each* $0 < T < t^+(x)$.

$$\mathcal{D} := \bigcup_{x \in U_\alpha} [0, t^+(x)) \times \{x\}$$

is open in $\mathbb{R}^+ \times U_\alpha$ *and* $(t, x) \mapsto u(t, x) \in C^{0,1-}(\mathcal{D}, U_\alpha)$, *i.e. the map defines a* C^{1-}*-semiflow on* U_α .

PROOF The proof is given in [**11**]. □

With the additional assumption

$$(A, f) \in C^k(U_\beta\,, \mathcal{L}(X_1, X_0) \times X_0)\,, \quad k \in \mathbb{N}^* \cup \{\infty, \omega\} \tag{3.4}$$

we prove the following

THEOREM 3.2 *The equation* (3.3) *defines a* C^k *smooth semiflow in* U_α *i.e.*

$$(t, x) \mapsto u(t, x) \in C^{0,k}(\mathcal{D}, U_\alpha)\,, \quad k \in \mathbb{N}^* \cup \{\infty, \omega\}\,.$$

Set $v := \partial_2 u(\cdot, x)h$ *with* $h \in X_\alpha$. *Then we have* $v \in V_{\alpha,T}(X_0, X_1)$ *for* $0 < T < t^+(x)$ *and* v *is the unique solution of the linearized Cauchy problem*

$$\begin{aligned} \dot{v} + A(u(t,x))v &= -\partial A(u(t,x))[v, u(t,x)] + \partial f(u(t,x))v\,, \quad 0 < t < t^+(x)\,,\\ v(0) &= h\,. \end{aligned} \tag{3.5}$$

We shall prove the assertion shortly by using the implicit function theorem. But first we need some technical tools. Let $\mathcal{N}_A$ and $\mathcal{N}_f$ denote the Nemytskii operators induced by A resp. f. Thus, for a map $u : [0, T] \to U_\alpha$ we define

$$(\mathcal{N}_A(u)\,, \mathcal{N}_f(u))(t) := (A(u(t))\,, f(u(t)))\,, \quad 0 \le t \le T\,.$$

3.3 LEMMA *Let* $U_\alpha \subset X_\alpha$ *be open and fix* $T > 0$. *Then*

a) $C([0, T], U_\alpha)$ *is open in* $C([0, T], X_\alpha)$,

b) $(\mathcal{N}_A\,, \mathcal{N}_f) \in C^k(C([0, T], U_\alpha), C([0, T], \mathcal{L}(X_1, X_0)) \times C([0, T], X_0))$,

$(\partial\mathcal{N}_A(u)v\,, \partial\mathcal{N}_f(u)v)(t) = (\partial A(u(t))v(t)\,, \partial f(u(t))v(t))\,, \ 0 \le t \le T$,

for $u \in C([0, T], U_\alpha)$ *and* $v \in C([0, T], X_\alpha)$.

c) *The map* $m : C([0, T], \mathcal{L}(X_1, X_0)) \times V_{\alpha,T}(X_0, X_1) \longrightarrow C_{\alpha,T}(X_0)$

$$(B, v) \longmapsto Bv\,, \quad (Bv)(t) := B(t)v(t)$$

is bilinear and continuous.

PROOF We just prove c). Let $B = B(\cdot)$ be an element in $C([0,T], \mathcal{L}(X_1, X_0))$ and v a map in $V_{\alpha,T}(X_0, X_1)$. It easily follows that $(\ t \mapsto B(t)v(t)\ ) \in C((0,T], X_0)$. Moreover, we have

$$t^{1-\alpha}\|B(t)v(t)\|_{X_0} \leq \|B(t)\|_{\mathcal{L}(X_1,X_0)} t^{1-\alpha}\|v(t)\|_{X_1} \leq \sup_{0\leq t\leq T} \|B(t)\|_{\mathcal{L}(X_1,X_0)}\|v\|_{V_{\alpha,T}}$$

for $0 < t \leq T$ and, since the map is obviously bilinear, the assertion follows. □

Let i be the inclusion of Proposition 2.1, i.e.

$$i\ :\ V_{\alpha,T}(X_0, X_1) \longrightarrow C([0,T], X_\alpha)$$

and let $V \subset V_{\alpha,T}(X_0, X_1)$ be the open subset, given by

$$V\ := i^{-1}(C([0,T], U_\alpha))\ . \tag{3.6}$$

We define the nonlinear map

$$\begin{array}{llll} \phi\ : V \times U_\alpha \subset V_{\alpha,T}(X_0, X_1) \times X_\alpha & \longrightarrow & C_{\alpha,T}(X_0) \times X_\alpha \\ \qquad (u,x) & \longmapsto & (\partial_t u + A(u)u - f(u)\ ,\ R_\alpha u - x)\ . \end{array} \tag{3.7}$$

Here and in the following, we do not always distinguish between the maps and the induced substitution operators. Observe that

$$(u,x) \mapsto (\partial_t u\ ,\ R_\alpha u - x) \in C^\omega(V \times U_\alpha, C_{\alpha,T} \times X_\alpha) \tag{3.8}$$

(i.e. the map is analytic) since it is the restriction of a linear and continuous map to an open subset.

3.4 PROPOSITION

$$\phi \in C^k(V \times U_\alpha\,,\, C_{\alpha,T} \times X_\alpha)\ , \qquad k \in \mathbb{N}^* \cup \{\infty, \omega\}\ ,$$

and

$$\partial_1\phi(u,x)v = (\partial_t v + A(u)v + \partial A(u)[v,u] - \partial f(u)v\ ,\ R_\alpha v)\ .$$

PROOF Thanks to (3.8), it suffices to show $u \mapsto \mathcal{N}_A(u)u - \mathcal{N}_f(u) \in C^k(\,V,\ C_{\alpha,T}(X_0))$. By considering the following diagrams

$$V \xrightarrow{(i,id)} C([0,T], U_\alpha) \times V_{\alpha,T} \xrightarrow{(\mathcal{N}_A,id)} C([0,T], \mathcal{L}(X_1, X_0)) \times V_{\alpha,T} \xrightarrow{m} C_{\alpha,T}(X_0)\ ,$$

$$V \xrightarrow{i} C([0,T], U_\alpha) \xrightarrow{\mathcal{N}_f} C([0,T], X_0) \hookrightarrow C_{\alpha,T}(X_0)$$

the assertion is a consequence of Lemma 3.3 and of (3.8). □

We show that $\partial_1\phi(u_0,x_0)$ is an isomorphism for any given $(u_0,x_0) \in V \times U_\alpha$. First observe that the equation $\partial_1\phi(u_0,x_0)v = (k,x)$ is equivalent to

$$\begin{cases} \dot{v}(t) + A(u_0(t))v(t) + \partial A(u_0(t))[v(t),u_0(t)] - \partial f(u_0(t))v(t) = k(t), \ 0 < t \le T \ , \\ v(0) = x \ . \end{cases} \tag{3.9}$$

3.5 PROPOSITION *We have*

$$\partial_1\phi(u_0,x_0) \in Isom(V_{\alpha,T}\,, C_{\alpha,T}(X_0) \times X_\alpha) \text{ for } (u_0,x_0) \in V_{\alpha,T} \times X_\alpha \text{ and}$$

$v := -[\partial_1\phi(u_0,x_0)]^{-1}\partial_2\phi(u_0,x_0)h$, $h \in X_\alpha$, *is the solution of*

$$\begin{aligned} \dot{v}(t) + A(u_0(t)) &= -\partial A(u_0(t))[v(t),u_0(t)] + \partial f(u_0(t))v(t) \ , \ 0 < t \le T \ , \\ v(0) &= h \ . \end{aligned} \tag{3.10}$$

PROOF Assume first that $\alpha < 1$. We have to prove that the Cauchy problem in (3.9) has a unique solution for any given $(k,x) \in C_{\alpha,T}(X_0) \times X_\alpha$. Due to the assumption (3.4) we have

$$\mathcal{A}(t) := A(u_0(t)) - \partial f(u_0(t)) \in \mathcal{M}_\alpha(X_1,X_0) + \mathcal{L}(X_\beta,X_0)\ , \quad 0 \le t \le T\ .$$

With Proposition 2.1 and the perturbation result in Remark 2.2c) we obtain

$$\mathcal{A} \in C([0,T],\mathcal{M}_\alpha(X_0,X_1))\ . \tag{3.11}$$

Defining

$$C(t) := -\partial A(u_0(t))[\ \cdot\ ,u_0(t)]\ , \quad 0 < t \le T\ . \tag{3.12}$$

we can rewrite equation (3.9), obtaining

$$\dot{v}(t) + \mathcal{A}(t)v(t) - C(t)v(t) = k(t)\ , \quad 0 < t \le T\ , \quad v(0) = x \tag{3.13}$$

or equivalently

$$\dot{v} + \mathcal{A}(0)v = (\mathcal{A}(0) - \mathcal{A}(\cdot))v + C(\cdot)v + k\ , \quad 0 < t \le T\ , \quad v(0) = x\ . \tag{3.14}$$

Due to the property of maximal regularity we can solve this equation by a fixed point argument. Indeed, set

$$\mathcal{W}_\tau := \{v \in V_{\alpha,\tau}(X_0, X_1)\ ;\ v(0) = x\}\ ,$$

for a fixed $\tau \in (0, T]$ and define

$$G:\ \mathcal{W}_\tau \longrightarrow W_\tau\ ,\quad G(v) := x(\cdot) + J_{\mathcal{A}(0)}((\mathcal{A}(0) - \mathcal{A}(\cdot))v + C(\cdot)v + k)\ .$$

G is a contraction for a sufficiently small τ. In fact, we have

$$\|G(u) - G(v)\|_{V_{\alpha,\tau}(X_0,X_1)} \le c\|(\mathcal{A}(0) - \mathcal{A}(\cdot))(u - v)\|_{C_{\alpha,\tau}(X_0)} + c\|C(\cdot)(u - v)\|_{C_{\alpha,\tau}(X_0)}$$

with a constant c which is independent of τ. The first term can be estimated by

$$\|(\mathcal{A}(0) - \mathcal{A}(\cdot))(u - v)\|_{C_{\alpha,\tau}} \le \|\mathcal{A}(0) - \mathcal{A}(\cdot)\|_{C([0,\tau],\mathcal{L}(X_1,X_0))}\|(u - v)\|_{V_{\alpha,\tau}} \qquad (3.15)$$

and the second by

$$\begin{aligned} t^{1-\alpha}\|C(t)(u(t)-v(t))\|_{X_0} &= t^{1-\alpha}\|\partial A(u_0(t))[u(t) - v(t), u_0(t)]\|_{X_0} \\ &\le \|\partial A(u_0(t))\|_{\mathcal{L}(X_\beta,X_1;X_0)} t^{1-\alpha}\|u_0(t)\|_{X_1}\|u(t) - v(t)\|_\beta \\ &\le c \sup_{0\le t\le\tau} \|\partial A(u_0(t))\|_{\mathcal{L}(X_\beta,X_1;X_0)}\|u_0\|_{V_{\alpha,\tau}}\tau^{\alpha-\beta}\|u - v\|_{V_{\alpha,\tau}} \end{aligned}$$

due to (3.4) and Proposition 2.1. (Hereby, $\mathcal{L}(X_\beta, X_1; X_0)$ denotes the space of all continuous bilinear forms on $X_\beta \times X_1$ with values in X_0). The last estimate together with (3.11) and (3.15) implies the existence of a $T_1 \in (0, T]$ such that

$$\|G(u) - G(v)\|_{V_{\alpha,T_1}(X_1,X_0)} \le (1/2)\,\|u - v\|_{V_{\alpha,T_1}(X_0,X_1)}\ .$$

Let $v_1 \in V_{\alpha,T_1}(X_0, X_1)$ be the (uniquely determined) fixed point of G. Then, v_1 solves the equation (3.9) on the interval $(0, T_1]$. In a next step we can, if necessary, solve the same equation with initial value $v_1(T_1)$ on some interval $[T_1, T_2]$ and so on. The case $\alpha = 1$ has been treated by [6, p.351], cf. also [4, p.100].

Observe that the derivative of ϕ with respect to the second variable is given by

$$\partial_2\phi(u_0, x_0)h = (0, -h)$$

for $h \in X_\alpha$. Now

$$v := -[\partial_1\phi(u_0, x_0)]^{-1}\partial_2\phi(u_0, x_0)h = [\partial_1\phi(u_0, x_0)]^{-1}(0, h)$$

is the solution of (3.10) since $\partial_1\phi(u_0, x_0)v = (0, h)$. □

PROOF OF THEOREM 3.2

We fix $x_0 \in U_\alpha$ and $T \in (0, t^+(x_0))$. Let $u(\cdot, x_0)$ be the solution of the quasilinear Cauchy problem (3.3) with initial value x_0 and define

$$u_0 := u(\cdot, x_0)|(0, T] \ . \tag{3.16}$$

For $(u, x) \in V_{\alpha,T}(X_0, X_1) \times X_\alpha$ we have, due to the definition of ϕ, that

$$u \quad \text{solves} \quad \dot{u} + A(u)u = f(u) \ , \ 0 < t \le T \ , \ u(0) = x \iff \phi(u, x) = 0 \ . \tag{3.17}$$

Therefore we conclude (with Proposition 3.5)

$$\phi(u_0, x_0) = 0 \quad \textit{and} \quad \partial_1 \phi(u_0, x_0) \in Isom(V_{\alpha,T}(X_0, X_1), \, C_{\alpha,T}(X_0) \times X_\alpha) \ . \tag{3.18}$$

Now, the implicit function theorem gives the existence of an open neighborhood W of (u_0, x_0) in $V \times U_\alpha$, an open neighborhood U_0 of x_0 in U_α and a mapping $g \in C^k(U_0, V_{\alpha,T}(X_0, X_1))$ such that

$$(u, x) \in W \ , \quad \phi(u, x) = 0 \iff x \in U_0 \ , \quad u = g(x) \ . \tag{3.19}$$

Moreover

$$\partial g(x_0) = -[\partial_1 \phi(g(x_0), x_0)]^{-1} \partial_2 \phi(g(x_0), x_0) \ . \tag{3.20}$$

The characterization in (3.17) gives

$$g(x) = u(\cdot, x)|(0, T] \ . \tag{3.21}$$

Hence, we obtain

$$[\ x \mapsto u(\cdot, x)\ ] \in C^k(U_0, V_{\alpha,T}(X_0, X_1)) \ . \tag{3.22}$$

We consider now the diagram

$$\begin{array}{ccccccc} U_0 & \xrightarrow{C^k} & V_{\alpha,T}(X_0, X_1) & \xrightarrow{i} & C([0,T], X_\alpha) & \xrightarrow{\mathrm{ev}_t} & X_\alpha \\ x & \longmapsto & u(\cdot, x) & \longmapsto & u(\cdot, x) & \longmapsto & \mathrm{ev}_t u(\cdot, x) = u(t, x) \end{array}$$

and conclude that

$$x \longmapsto u(t, x) \in C^k(U_0, X_\alpha) \quad \text{for} \quad t \in [0, T] \ . \tag{3.23}$$

(3.21) implies

$$\partial_2 u(\cdot, x)h = \partial g(x)h \in V_{\alpha,T}(X_0, X_1)$$

for $h \in X_\alpha$. Setting $v := \partial_2 u(\cdot, x_0)h$ we get, thanks to (3.20),

$$v = \partial g(x_0)h = -[\partial_1 \phi(u_0, x_0)]^{-1} \partial_2 \phi(u_0, x_0)h \ . \tag{3.24}$$

Now Proposition 3.5 gives all the assertions. □

In addition, the solution $u(t,x)$ of the quasilinear problem (3.3) is also k- times differentiable with respect to the first variable.

3.6 PROPOSITION *Assume that* (3.4) *holds. Then*

$$u(\cdot,x) \in C^k((0,t^+(x)), X_1) \cap C^{k+1}((0,t^+(x)), X_0) . \tag{3.25}$$

Moreover we have, for $m = 1,2, \ldots, k$,

$$[\, t \longmapsto t^m \partial_1^m u(t,x)\,] \in C_{\alpha,T}(X_1) \cap BC((0,T], X_\alpha) , \quad 0 < t \leq T < t^+(x) . \tag{3.26}$$

PROOF Here we use a result of Angenent [4, Corollary 2.10 and Corollary 2.13]. The technical details can be verified by similar arguments as given above. □

By putting together Theorem 3.2 and Proposition 3.6 we obtain

3.7 THEOREM *Set*

$$\overset{\circ}{\mathcal{D}} := \{\, (t,x) \in \mathcal{D} \; ; \; t > 0 \,\} .$$

Then it follows

$$(t,x) \mapsto u(t,x) \in C^k(\overset{\circ}{\mathcal{D}}, U_\alpha) . \tag{3.27}$$

Moreover, $v := \partial_2 u(\cdot,x)h$, $h \in X_\alpha$, *is the unique solution of the linearized problem*

$$\begin{aligned} \dot{v} + A(u(t,x))v &= -\partial A(u(t,x))[v, u(t,x)] + \partial f(u(t,x))v , \quad 0 < t < t^+(x) , \\ v(0) &= h . \end{aligned}$$

3.8 REMARKS

a) An inspection of the proofs given above shows that the semiflow is also smooth with respect to the topology of X_1. With the assumption (3.4) we obtain

$$(t,x) \mapsto u(t,x) \in C^k(\overset{\circ}{\mathcal{D}}, U_1) , \quad k \in \mathbb{N}^* \cup \{\infty, \omega\} , \tag{3.28}$$

where $U_1 := U_\alpha \cap X_1$, *equipped with the topology of* X_1 .

In fact, by following the proof of Theorem 3.2 until (3.22) and then considering the diagram

$$\begin{array}{ccccc} U_0 & \xrightarrow{C^k} & V_{\alpha,T}(X_0,X_1) & \xrightarrow{\text{ev}_t} & X_1 \\ x & \longmapsto & u(\cdot,x) & \longmapsto & \text{ev}_t u(\cdot,x) = u(t,x) \end{array}$$

we obtain that

$$x \longmapsto u(t,x) \in C^k(U_0, X_1) \quad \text{for} \quad t \in (0,T] \; ,$$

since $\text{ev}_t \in \mathcal{L}(V_{\alpha,T}(X_0,X_1),X_1)$ for $t \in (0,T]$. Now, the assertion follows by the same arguments as above. □

This observation renders a possibility to prove the smooth dependence of $(t,x) \mapsto u(t,x)$ in each Banach space X satisfying $X_1 \hookrightarrow X \hookrightarrow X_\alpha$ for some $\alpha \in (0,1)$. In fact, we then have

$$(t,x) \mapsto u(t,x) \in C^k(\mathring{\mathcal{D}}_X \, , U_X) \, , \quad k \in \mathbb{N}^* \cup \{\infty, \omega\} \; ,$$

where $\mathring{\mathcal{D}}_X := \mathring{\mathcal{D}} \cap (\mathbb{R}^+ \times X)$.

b) The idea of using the implicit function theorem goes back to Angenent. However, Proposition 2.1, Theorem 3.1, Theorem 3.2, Proposition 3.5 and Theorem 3.7 as well as Remark 3.8 *a*) seem to be new (in the context of quasilinear equations and maximal regularity).

4. References

[1] H. Amann, *Dynamic theory of quasilinear parabolic equations -I. Abstract evolution equations,* Nonlinear Anal., Theory, Methods & Appl. **12** (1988), 895–919.

[2] H. Amann, *Dynamic theory of quasilinear parabolic equations II. Reaction diffusion systems,* Differential and Integral Equations **3** (1990), 13–75.

[3] H. Amann, Parabolic Evolution Equations, book in preparation.

[4] S.B. Angenent, *Nonlinear analytic semiflows,* Proc. Roy. Soc. Edinburgh, **115A** (1990), 91-107.

[5] Ph. Clément, H.J.A.M. Heijmans et al. One-Parameter Semigroups, North Holland, CWI Monograph 5, Amsterdam, 1987.

[6] G. Da Prato and P. Grisvard, *Equations d'évolution abstraites nonlinéaires de type parabolique,* Ann. Mat. Pura Appl. (4), **120** (1979), 329–396.

[7] A. Lunardi, *Analyticity of the maximal solution of an abstract nonlinear parabolic equation,* Nonlinear Anal., Theory, Methods & Appl. **6** (1982), 503–521.

[8] A. Lunardi, *Abstract quasilinear parabolic equations,* Math. Ann. **267** (1984), 395-415

[9] A. Lunardi, *Interpolation spaces between domains of elliptic operators and spaces of continuous functions with applications to nonlinear parabolic equations,* Math. Nachr. **121** (1985), 295–318.

[10] A. Pazy, Semigroups of Linear Operators and Applications to Partial Differential Equations, Springer, New York, 1983.

[11] G. Simonett, Zentrumsmannigfaltigkeiten für quasilineare parabolische Gleichungen, Institut für angewandte Analysis und Stochastik, Report Nr.2, Berlin, 1992.

[12] P.E. Sobolevskii, *Equations of parabolic type in a Banach space,* Am. Math. Soc.Transl., Ser2 **49** (1966), 1-62

On the Hille–Yosida Operators

EUGENIO SINESTRARI University of Roma "La Sapienza," Rome, Italy

1 INTRODUCTION

Let $(X, \|\cdot\|)$ be a Banachspace and $L(X)$ the algebra of bounded linear operators of X into itself.

A linear operator $A : D(A) \subset X \to X$ is called a Hille-Yosida operator if there exist $\omega \in R$ such that if $\lambda > \omega$ then $(\lambda - A)^{-1} \in L(X)$ and

$$M := \sup\left\{\|(\lambda-\omega)^n(\lambda-A)^n\|_{L(X)} \; ; \; \lambda > \omega \, , \; n \in N\right\} < \infty \tag{1}$$

A HY-operator is the generator of a semigroup if and only if $\overline{D(A)} = X$: in this case it is known that the abstract Cauchy problem

$$u'(t) = Au(t) + f(t) \, , \; t \in [0,T] \, , \; u(0) = u_0 \tag{2}$$

can be investigated with the aid of the variation of constants formula which is no more available when $D(A)$ is not dense on X.

For this reason in Da Prato, Sinestrari (1985) a new method has been proposed to study problem (2) when A is only a HY-operator: the results obtained extend the classical ones: for instance the Phillips theorem has the following generalization.

Theorem 1.1
Let A be a HY-operator satisfying (1). If $f \in W^{1,1}(]0,T[\,;X)$, $u_0 \in D(A)$ and $u_1 := Au_0 + f(0) \in \overline{D(A)}$ then problem (2) has a unique (strict) solution $u \in C^1([0,T]\,;X)$ and for each $t \in [0,T]$ we have

$$\|u(t)\| \le Me^{\omega t}\left(\|u_0\| + \int_0^t \|e^{-\omega s}f(s)\|ds\right) \tag{3}$$

$$\|u'(t)\| \leq Me^{\omega t}\left(\|u_1\| + \int_0^t \|e^{-\omega s}f'(s)\|ds\right) \tag{4}$$

For the proof see theorem 8.1 of Da Prato, Sinestrari (1987) where also different kind of solutions are considered and other existence results are proved: they will be summarized in the next section where also alternative approaches to prove some of those results are mentioned.

Recently (Da Prato, Sinestrari 1992) the theory has been extended to consider problem (2) when A depends on t: the main results will be listed in section 3. Finally applications to partial differential equations of hyperbolic type will be given in the last section.

2 THE AUTONOMOUS CASE

In addition to the strict solution of (2) considered in theorem 1.1 we consider also less regular solutions of problem (2).

Definition 2.1
Let $f \in L^1(]0,T[\,;X)$ and $u_0 \in \overline{D(A)}$: a function $u \in C([0,T]\,;X)$ is an integral solution of problem (2) if for each $t \in [0,T]$ $\int_0^t u(s)ds \in D(A)$ and $u(t) = u_0 + A\int_0^t u(s)ds + \int_0^t f(s)ds$.

Definition 2.2
Let $f \in L^p(]0,T[\,;X)$, $1 \leq p \leq \infty$ and $u_0 \in \overline{D(A)}$: a function $u \in L^p(]0,T[\,;X)$ is an L^p-solution of problem (2) if there exist $\{f_k\} \subset L^p(]0,T[\,;X)$ and $\{u_{0k}\} \subset X$ such that problem

$$u_k'(t) = Au_k(t) + f_k(t)\ ,\ t \in [0,T]\ a.e.\ ,\ u_k(0) = u_{0k} \tag{5}$$

has a solution $u_k \in W^{1,p}(]0,T[\,;X)$ and
$\lim_{k\to\infty}\left(\|f_k - f\|_{L^p(]0,T[\,;X)} + \|u_k - u\|_{L^p(]0,T[\,;X)} + \|u_{0k} - u_0\|\right) = 0.$

Definition 2.3
Let $f \in C([0,T]\,;X)$ and $u_0 \in \overline{D(A)}$: a function $u \in C([0,T]\,;X)$ is a C-solution of problem (2) if there exist $\{f_k\} \subset C([0,T]\,;X)$ and $\{u_{0k}\} \subset X$ such that problem

$$u_k'(t) = Au_k(t) + f_k(t)\ ,\ t \in [0,T]\ ,\ u_k(0) = u_{0k} \tag{6}$$

has a solution $u_k \in C^1([0,T]\,;X)$ and
$\lim_{k\to\infty}\left(\|f_k - f\|_{C([0,T]\,;X)} + \|u_k - u\|_{C([0,T]\,;X)} + \|u_{0k} - u_0\|\right) = 0.$

It can be proved (see Da Prato, Sinestrari (1987) theorem 5.1) that any solution just defined satisfies the a-priori estimate (3) and hence is unique. In addition to theorem 1.1 we have also the following existence results (see theorems 7.2, 8.3 and 9.1 of Da Prato, Sinestrari (1987))

Theorem 2.4
For any $f \in L^p(]0,T[\,;X)$ and $u_0 \in \overline{D(A)}$ there exist a unique L^p-solution of problem (2) and coincides with the integral solution. The same result holds by replacing L^p with C.

Theorem 2.5
Given $f \in L^p(]0,T[\,;X)$ such that $Af \in L^p(]0,T[\,;X)$ and $u_0 \in D(A)$ such that $Au_0 \in \overline{D(A)}$ there exists a unique $u \in W^{1,p}(]0,T[\,;X)$ such that $Au \in C([0,T]\,;X)$ and verifying

$$u'(t) = Au(t) + f(t)\ ,\ t \in [0,T]\ a.e.\ ,\ u(0) = u_0 \tag{7}$$

If in addition $f \in C([0,T]\,;X)$ we have also $u \in C^1([0,T]\,;X)$ and (7) holds for each $t \in [0,T]$.

The method used to study problem (2) when A is only a HY-operator is to consider the time-derivative as an operator in the space $L^p(]0,T[\,;X)$. More precisely by defining $B : D(B) \subset L^p(]0,T[\,;X) \longrightarrow L^p(]0,T[\,;X)$ as

$$D(B) = \left\{u \in W^{1,p}(]0,T[\,;X)\ ;\ u(0) = 0\right\}\ ,\ Bu = -u'$$

we obtain a HY-operator with dense domain (because $p \neq \infty$) and we can write problem (2) as

$$B(u - u_0) + Au + f = 0$$

By substituting B with its Yosida approximation we are able to give a representation formula for the solution of the approximating problem; from this it is possible to deduce the a-priori estimate (3) for the solutions of problem (2); the existence results are obtained by taking first $f \in C_0^\infty(]0,T[\,;X)$, $u_0 = 0$ and by getting estimates also for the higher order derivatives of the solutions of the approximating problem: the general case is proved by density argument.

Let us mention now some alternative approaches to problem (2): they give some of the preceding results and can be classified in three types: the first is a inhomogeneous version of the Yosida's approximation method, the second is based on the theory of non linear semigroups and the last one derives from the theory of integrated semigroups.

The first approach is due to Kato (1986): the basic tool is the Yosida approximation A_n of A and the first step is to solve the approximating problem

$$u_n'(t) = A_n u_n(t) + f(t)\ ,\ t \in [0,T]\ ,\ u_n(0) = \left(I - n^{-1}A\right)u_0$$

when $f \in W^{2,1}(]0,T[\,;X)$, $u_0 \in D(A)$ and $Au_0 + f(0) \in \overline{D(A)}$: some a-priori estimates similar to (3) and (4) let consider the case $f \in W^{1,1}(]0,T[\,;X)$ by an approximation in $W^{1,1}$-norm. The same method has been used in Yagi (1991) to extend this result to a multi-valued operator A.

In Bénilan, Egberts (1989) a similar extension is given of theorem 1.1 and the last part of theorem 2.5: they use some results of the theory of non linear semigroups announced in Bénilan, Crandall, Pazy (1988).

The third approach is based on the theory of integrated semigroups: in fact a HY-operator is the generator of a locally Lipschitz continuous integrated semigroup: this enables to prove theorem 1.1 as shown in Kellerman, Hieber (1989); in Thieme (1990) the following representation formula for any solution of problem (2) is proved

$$u(t) = T(t)u_0 + \lim_{\lambda \to +\infty} \int_0^t T(t-s)\lambda(\lambda - A)^{-1} f(s)ds \ , \ t \in [0,T]$$

where T is the semigroup generated by the part of A in $\overline{D(A)}$: in this paper also semilinear problems are considered.

3 THE NONAUTONOMOUS CASE

The alternative approaches just mentioned do not seem to generalize to the time-dependent case as done in Da Prato, Sinestrari (1992): in this paper the same methods of Da Prato, Sinestrari (1987) are applied to the Cauchy problem

$$u'(t) = A(t)u(t) + f(t) \ , \ t \in [0,T] \ ; \ u(0) = u_0 \tag{8}$$

under the following assumptions:
(i) for all $t \in [0,T]$, $A(t) : D \to X$ is a linear operator between the Banach spaces D and X
(ii) $D \subset X$ and there exists $c > 0$ such that for all $t \in [0,T]$ and $x \in D$ we have

$$c^{-1}\|x\|_D \le \|x\| + \|A(t)x\| \le c\|x\|_D$$

(iii) there exist $\omega \in R$ such that if $\lambda > \omega$ then $(\lambda - A(t))^{-1} \in L(X)$ for all $t \in [0,T]$ and

$$M := \sup\left\{\|(\lambda-\omega)^n(\lambda - A(t_1))^{-1}...(\lambda - A(t_n))^{-1}\| \ ; \ \lambda > \omega \ , \ 0 \le t_n \le ... \le t_1 \le T\right\} < \infty$$

After defining the strict solutions and the L^p or C-solutions of (8) as in the autonomous case, the following existence theorems are proved in Da Prato, Sinestrari (1992) (see theorems 3.5, 4.2 and 5.1)

Theorem 3.1
Suppose that $A \in Lip([0,T]; L(D,X))$ and that there exists a sequence $\{A_k\}$ in $C^\infty([0,T]; L(D,X))$ verifying (i)–(iii) with c, ω, M independent of k and such that

$$\lim_{k\to\infty} \|A - A_k\|_{C([0,T];L(D,X))} = 0 \ \ and \ \ \sup_k \|A_k'\|_{C([0,T];L(D,X))} < \infty \tag{9}$$

Then, for any $f \in L^p(]0,T[\ ; X)$ and $u_0 \in \overline{D}$ there exists a unique L^p-solution of problem (8) and for each $t \in [0,T]$ we have

$$\|u(t)\| \le Me^{\omega t}\left(\|u_0\| + \int_0^t e^{-\omega s}\|f(s)\|ds\right) \tag{10}$$

Theorem 3.2
Suppose that $A \in C^1([0,T]; L(D,X))$ and that there exists a sequence $\{A_k\}$ in $C^\infty([0,T]; L(D,X))$ verifying (i)–(iii) with c, ω, M independent of k and such that

$$\lim_{k\to\infty} \|A - A_k\|_{C^1([0,T];L(D,X))} = 0 \tag{11}$$

Then, for any $f \in C([0,T]\ ;X)$ and $u_0 \in \overline{D}$, there exists a unique C-solution of problem (8) verifying estimate (10). If in addition $f \in W^{1,1}(]0,T[\ ;X)$, $u_0 \in D$ and $A(0)u_0 + f(0) \in \overline{D}$ then the solution is strict.

4 APPLICATIONS

In Da Prato, Sinestrari (1987) there are several examples of differential operators with non dense domain which are HY-operators in suitable function spaces (whish must be non reflexive according to a known result by Kato (1959)) and with different boundary conditions (Cauchy, Dirichlet, periodic etc.); a combination with an analogous theory (see Sinestrari (1985)) for analytic semigroups non continuous at the origin gives rise to applications to ultraparabolic partial differential problems as

$$u_t(t,\tau,x) + u_\tau(t,\tau,x) = \Delta u(t,\tau,x) + f(t,\tau,x)$$

for $t,\tau \in [0,T]$ and $x \in \overline{\Omega}$ (where Ω is a bounded and open set of R^n) or to a Laplace equation in infinite dimensional spaces arising in the theory of stochastic control: see sections 16 and 17 of Da Prato, Sinestrari (1987) for more details.

Another application of the autonomous theory has been done by Grimmer, Sinestrari (1992) where one-dimensional hyperbolic problems are studied in spaces of continuous functions (the restriction to one dimension is forced by a negative result of Brenner (1966)): more precisely we consider the Cauchy-Dirichlet problem in a bounded interval for a symmetric hyperbolic system in canonical form:

$$\begin{aligned} D_t w_i(t,x) &= \lambda_i D_x w_i(t,x) + f_i(t,x) \\ w_i(0,x) &= \alpha_i(x) \\ w_h(t,0) &= \sum_{k=1}^{n-p} \beta_{hk} w_{p+k}(t,0) \\ w_{p+k}(t,\ell) &= \sum_{h=1}^{p} \gamma_{kh} w_h(t,\ell) \end{aligned} \tag{12}$$

where $t \in [0,T]$, $x \in [0,\ell]$, $1 \le p < n$, $i = 1,...,n$; $h = 1,...,p$; $k = 1,...,n-p$. We assume that $\lambda_1 < ... < \lambda_p < 0 < \lambda_{p+1} < ... < \lambda_n$ and that β_{hk}, γ_{kh} are constants.

Supposing that $f_i \in W^{1,1}(]0,T[\ ;\ C[0,\ell])$, $\alpha_i \in C^1[0,\ell]$ and some (necessary) compatibility conditions are satisfied there exists a unique solution $(w_1...w_n) \in C^1([0,T] \times [0,\ell])$ of problem (12): see section 3 of Grimmer, Sinestrari (1992) for the proof.

A self-suggesting application of this result is given (see theorem 2.3 of the same paper) to a very classical problem for the one-dimensional wave equation:

$$\begin{aligned} w_{tt}(t,x) &= w_{xx}(t,x) + f(t,x) \\ w(0,x) &= u_0(x) \\ w_t(0,x) &= u_1(x) \\ w(t,0) &= w(t,\ell) = 0 \end{aligned} \tag{13}$$

where $t \in [0,T]$ and $x \in [0,\ell]$. The result (which seems to be strangely new) gives not only a strict solution but also a pointwise estimate of w and its derivatives:

Theorem 4.1
Given $f \in W^{1,1}(]0,T[;C[0,\ell])$, $u_0 \in C^2[0,\ell]$ and $u_1 \in C^1[0,\ell]$ such that $u_0(0) = u_0(\ell) = u_1(0) = u_1(\ell) = u_0''(0) + f(0,0) = u_0''(\ell) + f(0,\ell) = 0$ there exists a unique $w \in C^2([0,T] \times [0,\ell])$, solution of problem (13).

In addition we have for each $(t,x) \in [0,T] \times [0,\ell]$

$$|w(t,x)| \le \|u_0\| + 2(\|u_0'\| + \|u_1\|)t + \int_0^t (t-s)\|f(s,\cdot)\|ds$$

$$|w_t(t,x)| + |w_x(t,x)| \le 2\left(\|u_0'\| + \|u_1\| + \int_0^t \|f(s,\cdot)\|ds\right)$$

$$|w_{tt}(t,x)| + |w_{xt}(t,x)| \le 2\left(\|u_0'' + f(0,\cdot)\| + \|u_1'\| + \int_0^t \|f(s,\cdot)\|ds\right)$$

where $\|\cdot\|$ denotes the sup-norm.
Remark that the conditions on f, u_0, u_1 (except the time regularity of f) are also necessary to have a solution in $C^2([0,T] \times [0,\ell])$.

As mentioned before, in Thieme (1990) are studied lipschitz perturbations of HY-operators with applications to non linear functional differential equations of first order arising from problems of age-dependent population dynamics.

Finally we mention a simple application of the non autonomous theory. Let us consider the following first order partial differential problem

$$\begin{aligned} u_t(t,x) &+ a(t,x)u_x(t,x) = f(t,x) \\ u(0,x) &= u_0(x) \\ u(t,0) &= u(t,\ell) \end{aligned} \tag{14}$$

where $t \in [0,T]$ and $x \in [0,\ell]$. We will suppose that $x \to a(t,x)$ is discontinuous so that the method of characteristics cannot be used. By choosing $X = L^\infty(0,\ell)$ and $D = Lip_\#[0,\ell] := \{u \in Lip[0,\ell] \ ; \ u(0) = u(\ell)\}$ we deduce from theorem 3.2 the following result

Theorem 4.2
Let $a \in C^1([0,T] \ ; \ L^\infty(0,\ell))$ and let $\gamma > 0$ be such that $\gamma^{-1} < a(t,x) < \gamma$ for $t \in [0,T]$ and $x \in [0,\ell]$a.e.. For each $f \in W^{1,1}(]0,T[;L^\infty(0,\ell))$ and $u_0 \in Lip_\#[0,\ell]$ such that $a(0,\cdot)u_0' + f(0,\cdot) \in C_\#[0,\ell]$ there exists a unique solution u of problem (14) such that $u, u_t \in C([0,T] \times [0,\ell])$ and $u_x(t,x)$ exists for $t \in [0,\ell]$ and $x \in [0,\ell]$ a.e..
In addition we have for each $t \in [0,T]$

$$\sup_{x\in[0,\ell]} |u(t,x)| \le \sup_{x\in[0,\ell]} |u_0(x)| + \int_{t_0}^t ess.\ \sup_{x\in[0,\ell]} |f(s,x)|ds$$

For the proof see section 6 of Da Prato, Sinestrari (1992) where the possibility of choosing also $X = Lip_\#[0,\ell]$ and $D = \{u \in X \ ; \ u' \in X\}$ is proved (this gives rise to a family of non dissipative operators $A(t)$ which are shown to satisfy the conditions of theorem 3.2).

REFERENCES

Bénilan,P., Crandall,M., and Pazy,A. (1988). "Bonnes solutions" d'un problème d'évolution semi-linéaire. *C.R. Acad. Sci. Paris*, 306 : 527–530.

Bénilan,P. and Egberts,P. (1989). Mild solutions, in Operator Semigroups and Evolution Equations. *Semesterbericht Funktionalanalysis Tübingen* : 25–35.

Brenner,P. (1966). The Cauchy problem for symmetric hyperbolic systems in L_p. *Math. Scand.*, 19 : 27–37.

Da Prato,G. and Sinestrari,E. (1985). On the Phillips and Tanabe regularity theorems. *Semesterbericht Funktionalanalysis Tübingen* : 117–124.

Da Prato,G. and Sinestrari,E. (1987). Differential operators with non dense domain. *Ann. Sc. Norm. Sup. Pisa*, 14 : 285–344.

Da Prato,G. and Sinestrari,E. (1992). Non autonomous evolution operators of hyperbolic type. *Semigroup Forum* (to appear).

Grimmer,R. and Sinestrari, E. (1992). Maximum norm in one-dimensional hyperbolic problems. *Diff. Integral Eq.*, 5 : 421–432.

Kato,T.(1959). Remarks on pseudo-resolvents and infinitesimal generators of semigroups. *Proc. Japan Acad.*, 35 : 467–468.

Kato,T. (1986). The Cauchy problem for operators with nondense domain. (*preprint*).

Kellerman,H. and Hieber,M. (1989), Integrated semigroups. *J. Funct. Anal.*, 84 : 160–180.

Sinestrari,E. (1985). On the abstract Cauchy problem of parabolic type in spaces of continuous functions. *J. Math. Anal. Appl.*, 107 : 16–66.

Thieme,H. (1990). Semiflows generated by lipschitz perturbations of non-densely defined operators. *Diff. Integral Eq.*, 3 : 1035–1066.

Yagi,A. (1991). A generator theorem of semigroup for multivalued linear operators. *Osaka J. Math.*, 28 : 385–410.

Boundedness of Functions of Positive Operators

P. E. SOBOLEVSKII Hebrew University of Jerusalem, Jerusalem, Israel

1. The linear operator A in Banach space E with the dense domain $D(A)$ is called positive (see [1]) if its spectrum lies inside of the angle $L = L(2\varphi)$ of the size $2\varphi \in (0, 2\pi)$ with the vertex at the origin, which is symmetric relatively to the non-negative part of the real axis. On the sides of the angle and outside of it the estimate of the resolvent

$$\|(z - A)^{-1}\|_{E \to E} \leq M(\varphi)(1 + |z|)^{-1}(\varphi \leq |argz| \leq \pi) \tag{1}$$

takes place.

Let the function $\psi(z)$ be analytic inside of the angle and continuous on L except the finite number of points with summable singularities. Let $\psi(z)$ permit the estimate $|\psi(z)| \leq c|z|^{-\varepsilon}$ when $|z| \to +\infty$ for any $\varepsilon > 0$. Then

$$\psi(A) = \frac{1}{2\pi i} \int_L \psi(z)(z - A)^{-1} dz \tag{2}$$

defines the bounded operator in E.

This formula defines the function of the operator A generated by the scalar function $\psi(z)$. It is supposed that in formula (2) the contour L is passed in the positive direction (counterclockwise).

EXAMPLE: The formula

$$A^{-(\alpha+i\beta)} = \frac{1}{2\pi i} \int_L z^{-(\alpha+i\beta)}(z - A)^{-1} dz \tag{3}$$

defines the negative fractional powers $A^{-(\alpha+i\beta)}$ of the positive operator A for an arbitrary $0 < \alpha < 1, -\infty < \beta < +\infty$. The operators $A^{\alpha+i\beta}$ can be defined as the inverse to the operators $A^{-(\alpha+i\beta)}$. The operators $A^{\alpha+i\beta}$ are already unbounded (since the operator A is unbounded) and they have domains dense in E. The function $\varphi(z) = z^{i\beta}$ is analytic inside of L and bounded on L, but $|\varphi(z)|$ does not tend to zero when $|z| \to +\infty$. Therefore, generally speaking, we cannot use formula (3) when we define the operator $A^{i\beta}$.

Let $\Omega \subset R^n$ be an open bounded set with a smooth boundary $\partial\Omega$ and $\bar{\Omega} = \Omega \cup \partial\Omega$. Let the differential expression

$$au(x) = (-1)^m \Sigma_{|\alpha| \leq 2m} a_\alpha(x) D_x^\alpha u(x)$$

be given on $\bar{\Omega}$. Let the system of boundary operators

$$B_j u(x) = \Sigma_{|\beta| \leq m_j} b_{j\beta}(x) D_x^\beta u(x) (j = 1, \cdots, m)$$

be given on $\partial\Omega$, and let the positive elliptic condition be satisfied (see [1]). Then the formula $Au(x) = au(x)$ defines the positive operator $A = A_p$, which acts in $L_p(\Omega)$ (with arbitrary $1 < p < +\infty$), and which is defined on the functions $u(x) \in W_p^{2m}(\Omega)$ satisfying the boundary conditions $B_j u(x) = 0 (x \in \partial\Omega, j = 1, \cdots, m)$. Therefore the operators $A^{-(\alpha+i\beta)}$ are defined for every $0 < \alpha < 1, -\infty < \beta < +\infty$ and are bounded in $L_p(\Omega)$.

Seeley [2] proved that the norms $\|A^{(\alpha+i\beta)}\|_{L_p \to L_p}$ are bounded uniformly on α. It means that the imaginary powers $A^{i\beta}$ are bounded in $L_p(\Omega)$. Hence an exact description of the domains $D(A^\alpha)$ in terms of smoothness and satisfaction of boundary conditions can be given. The inclusion theorems, exact in the sense that the inclusion operator $D(A^\alpha) = D(A_p^\alpha)$ to some $E_1 = E_1(\alpha, p)$ is continuous but not compact, were obtained earlier [6, 3]. This fact by virtue of the identity $A^{-(\alpha+i\beta)} = A^{-\alpha} \cdot A^{-i\beta}$ yields the boundedness of the operator $A^{-(\alpha+i\beta)}$ from L_p to $E_1(\alpha, p)$. It follows also from theorem 2 given below without using Seeley theorem.

The positive operator A is called strongly positive if $\varphi < \frac{\pi}{2}$ in estimate (1). For example, the elliptic operators in L_p are strongly positive. The Cauchy problem

$$v'(t) + Av(t) = 0 (0 \leq t \leq 1), v(0) = v^0$$

is correctly solvable in $C([0,1], E)$ for the strongly positive operator A.

Let us consider the corresponding Crank-Nikolson difference scheme

$$(u_k - u_{k-1}) \cdot \tau^{-1} + A \cdot (u_k + u_{k-1}) \cdot 2^{-1} = 0 (k = 1, \cdots, N; \tau = N^{-1}), u_0 = v^0.$$

The stability of the scheme means that the operators $[(1-\tau/2A)\cdot(1+\tau/2A)^{-1}]^n (n = 1,\cdots,N)$ are bounded uniformly on N.

The estimate

$$\|[(I-\tau/2A)\cdot(I+\tau/2A)^{-1}]^n \cdot A^{-\alpha}\|_{L_P \to E_1(\alpha,p)} \leq M(\alpha,p)(n=1,\cdots,N)$$

follows from theorem 2 for the elliptic operator A.

An analogous fact takes place for any Padé scheme.

2. We shall say that $D(A^\alpha)$ with some $\alpha \in (0,1)$ is continuously included in Banach space E_1 if $D(A^\alpha) \subset E_1$ and the inequality

$$\text{(4)} \qquad \|u\|_{E_1} \leq M\|A^\alpha u\|_E [u \in D(A^\alpha)]$$

takes place for some M which does not depend on u.

We shall say that the linear operator F is (E_1, E, α)-bounded if $D(A^\alpha) \subset D(F), Fu \in E_1$ and the inequality

$$\text{(5)} \qquad \|Fu\|_{E_1} \leq M_1\|A^\alpha u\|_E [u \in D(A^\alpha)]$$

takes place.

The inequality

$$\text{(6)} \qquad \|u\|_{E_1} \leq \widetilde{M}\|Au\|_E^{\alpha}\|u\|_E^{1-\alpha} [u \in D(A)]$$

follows from (4) and from the moments inequality for the fractional powers of positive operators [1].

We shall say that $D(A^\alpha)$ is weakly continuously included in E_1, if $D(A) \subset E_1$ and inequality (6) takes place. Generally speaking, inequality (4) does not follow from (6).

The weaker inequality

$$\text{(7)} \quad \|u\|_{E_1} \quad \bar{M}\cdot\frac{\varepsilon_1+\varepsilon_2}{\varepsilon_1\varepsilon_2}\cdot M(A)\cdot\|A^{\alpha+\varepsilon_2}u\|_E^{\frac{\varepsilon_1}{\varepsilon_1+\varepsilon_2}}\cdot\|A^{\alpha-\varepsilon_1}u\|_E^{\frac{\varepsilon_2}{\varepsilon_1+\varepsilon_2}}[u \in D(A^{\alpha+\varepsilon_2})]$$

takes place [3] for every $0 < \varepsilon_1 \quad \alpha, 0 < \varepsilon_2 \quad 1-\alpha$.

Inequality (5) due to the moments inequality gives

$$\|Fu\|_{\varepsilon_1} \leq \widetilde{M}_1 \|Au\|_E^{\alpha} \cdot \|u\|_E^{1-\alpha} [u \in D(A)]. \tag{8}$$

We shall say that the linear operator F is (E_1, E, α)-weakly bounded if $D(A) \subset D(F), Fu \in E_1$ and inequality (8) takes place. If F is a closed operator from E to E_1, then the inequality

$$\|Fu\|_{E_1} \leq \bar{M}_1 \cdot \frac{\varepsilon_1 + \varepsilon_2}{\varepsilon_1 \cdot \varepsilon_2} \cdot M(A) \cdot \|A^{\alpha+\varepsilon_2} u\|_E^{\frac{\varepsilon_1}{\varepsilon_1+\varepsilon_2}} \cdot \|A^{\alpha-\varepsilon_1} u\|_E^{\frac{\varepsilon_2}{\varepsilon_1+\varepsilon_2}} [u \in D(A^{\alpha+\varepsilon_2})] \tag{9}$$

follows from (8) [3].

THEOREM I. *Let $\psi(z)$ be analytic inside of L, continuous and bounded on L with*

$$\sup_L |\psi(z)| = R. \tag{10}$$

Let $D(A^\alpha)$ be weakly continuously included in E_1. Then $\psi(A)$ is (E, E_1, α)-weakly bounded and the inequality

$$\|\psi(A)u\|_{E_1} \leq R \cdot \check{M} \|Au\|_E^{\alpha} \cdot \|u\|_E^{1-\alpha} [u \in D(A)] \tag{11}$$

takes place.

THEOREM 2.. *Let $\psi(z)$ satisfy the conditions of theorem I and let A be an elliptic operator. Then $\psi(A)$ is (E_1, E, α)-bounded and the inequality*

$$\|\psi(A)u\|_{E_1} \leq R \cdot \bar{M} \cdot \|A^\alpha u\|_E \, [u \in D(A^\alpha)] \tag{12}$$

takes place for $E = L_p(\Omega), E_1 = W_2^k(\Omega)$ with $1 < p < +\infty, 0 \leq \frac{k}{2m} < \alpha < \frac{k}{2m} + \frac{n}{2m} \cdot \frac{1}{p}, \frac{1}{q} = \frac{1}{p} - \frac{2m}{n}(\alpha - \frac{k}{2m})$ and $\bar{M} = \bar{M}(\alpha, p, k)$.

3. We denote by $E_{\alpha,p}(0 < \alpha < 1, 1 \leq p \leq +\infty)$ (see [5] the set of $u \in E$ with the finite norm

$$|u|_{\alpha,p} = [\int_0^\infty \|A(\lambda + A)^{-1} u\|_E^p \frac{d\lambda}{\lambda}]^{1/p}. \tag{13}$$

LEMMA I.. *Let $\psi(z)$ satisfy the conditions of theorem I. Then the operator $\psi(A)$ is bounded in $E_{\alpha,p}$ and the inequality*

$$|\psi(A)u|_{\alpha,p} \leq R \cdot \alpha^{-1}(1-\alpha)^{-1} \cdot C_1 |u|_{\alpha,p} (u \in E_{\alpha,p}) \tag{14}$$

takes place.

PROOF: For every $\lambda > 0$ the function $(\lambda + z)^{-1}\psi(z)$ is analytic with respect to z on L and it tends to zero as $|z|^{-1}$ then $|z| \to +\infty$. Then, according to formula (2) the representation

$$\lambda^\alpha A(\lambda + A)^{-1}\psi(A)u =$$

$$= \frac{1}{2\pi i}\int_L \frac{\lambda^\alpha}{\lambda + z} \cdot \psi(z) \cdot A(z - A)^{-1}u\,dz \tag{15}$$

takes place for $u \in D(A)$.

Since $z = \rho \cdot \exp(\pm i\varphi), \rho \geq 0$ for $z \in L$, then according to (1) the estimates

$$|\lambda + z|^{-1} \leq (\cos\varphi/2)^{-1} \cdot (\lambda + \rho)^{-1},$$

$$\|A(z - A)^{-1}u\|_E \leq (2\cos\varphi/2 + 1) \cdot M(\varphi) \cdot \|A(\rho + A)^{-1}u\|_E$$

take place and formula (15) yields the inequality

$$\|\lambda^\alpha A(\lambda + A)^{-1}\psi(A)u\|_E$$

$$R \cdot \widetilde{M}(\varphi) \cdot \int_0^\infty \frac{\lambda^\alpha}{\lambda + \rho} \cdot \|A(\rho + A)^{-1}u\|_E d\rho.$$

Having carried out the substitution $\rho = \lambda s$ in the right-side integral, we obtain the inequality

$$\|\lambda^\alpha A(\lambda + A)^{-1}\psi(A)u\|_E$$

$$R \cdot \widetilde{M}(\varphi) \cdot \int_0^\infty \frac{1}{1 + s} \cdot \|\lambda^\alpha A(\lambda s + A)^{-1}u\|_E ds.$$

Using the Minkovsky integral inequality, we obtain the estimate

$$|\psi(A)u|_{\alpha,p}$$

$$R \cdot \widetilde{M}(\varphi) \cdot \int_0^\infty \frac{1}{1 + s}\left[\int_0^\infty \|\lambda^\alpha A(\lambda s + A)^{-1}u\|_E^p \frac{d\lambda}{\lambda}\right]^{\frac{1}{p}} ds.$$

At last by the substitution $\lambda s = \tau$ in the inner integral, we obtain the inequality

$$|\psi(A)u|_{\alpha,p} \leq R \cdot \widetilde{M}(\varphi) \cdot \frac{\pi}{\sin \pi\alpha} \cdot |u|_{\alpha,p}.$$

Since $D(A)$ is dense in $E_{\alpha,p}$, this fact yields inequality (14).

LEMMA 2.. *$E_{\alpha,1} \subset E_1$ and the inequality*

$$|u|_{E_1} \leq E_2|u|_{\alpha,1}(u \in E_{\alpha,1}) \tag{16}$$

takes place.

PROOF: The inequality

$$|u|_{E_!} \leq \widetilde{M} \cdot M_1^\alpha \cdot M_0^{1+\alpha} \cdot |u|_{\alpha,1}[u \in D(A)],$$

$$M_0 = \sup_{\lambda>0} \|\lambda(\lambda + A)^{-1}\|_{D\to E}, M_1 = \sup_{\lambda>0} \|A(\lambda + A)^{-1}\|_{E\to E},$$

follows the identity

$$u = \int_0^\infty (\lambda + A)^{-1} \cdot A \cdot (\lambda + A)^{-1} u d\lambda \; [u \in D(A)]$$

and from (6) and (1).

Since $D(A)$ is dense in $E_{\alpha,1}$, then inequality (16) is true.

LEMMA 3.. *The inequality*

$$|u|_{\alpha,1} \leq \alpha^{-1}(1-\alpha)^{-1} \cdot C_3\|Au\|_E^\alpha \cdot \|u\|_E^{1-\alpha} \tag{17}$$

takes place.

PROOF: The inequality

$$|u|_{\alpha,1} \leq$$

$$\int_0^N \lambda^{\alpha-1} \cdot M_1 \cdot \|u\|_E d\lambda + \int_N^\infty \lambda^{\alpha-2} \cdot M_0 \cdot \|Au\|_E d\lambda$$

follows the definition (13) and estimate (1) for any $N > O$.

Having calculated the minimum on $N > 0$ of the right side of the inequality, we obtain inequality (17) for $C_3 = M_0^\alpha M_1^{1-\alpha}$.

Let us begin proving theorem I. The inequality

$$\|\psi(A)u\|_{E_1} \leq C_2|\psi(A)u|_{\alpha,1}$$

follows from inequality (16).

Then using inequality (14) for $p = 1$ we obtain the inequality

$$\|\psi(A)u\|_{E_1} \leq C_2 \cdot R \cdot \alpha^{-1} \cdot (1-\alpha)^{-1} \cdot C_1 \cdot |u|_{\alpha,1}.$$

Finally, using (17), we obtain the estimate

$$\|\psi(A)u\|_{E_1} \leq C_2 \cdot R \cdot \alpha^{-1} \cdot (1-\alpha)^{-1} \cdot C_1 \cdot \alpha^{-1} \cdot (1-\alpha)^{-1} \cdot C_3 \cdot \|Au\|_E^{\alpha} \cdot \|u\|_E^{1-\alpha}.$$

Hence inequality (11) with $\check{M} = C_1 \cdot C_2 \cdot C_3 \cdot \alpha^{-2}(1-\alpha)^{-2}$ is now obtained.

3. Let us begin proving theorem 2.

Let $0 < \varepsilon < \alpha - k/2m$. Then from result [3] it follows that $A^{-(\alpha-\varepsilon)}$ is a bounded operator from $L_p(\Omega)$ to $W^k_{q_\varepsilon}(\Omega)$ for $\frac{1}{q_\varepsilon} = \frac{1}{p} - \frac{2m}{n}(\alpha - \varepsilon - \frac{k}{2m}$. Using the moments inequality, we come to the following inequality:

$$\|u\|_{W^k_{q_\varepsilon}} \quad M(p,\varepsilon) \cdot \|Au\|_{L_p}^{\alpha-\varepsilon} \cdot \|u\|_{L_p}^{1-\alpha+\varepsilon}[u \in D(A) = D(A_p)].$$

This is an inequality of kind (6). Thus, thanks to theorem I the inequality

$$\|\psi(A)u\|_{W^k_{q_\varepsilon}} \leq R \cdot \bar{M}(p,\varepsilon) \cdot \|Au\|_{L_p}^{\alpha-\varepsilon} \cdot \|u\|_{L_p}^{1-\alpha+\varepsilon}[u \in D(A)]$$

takes place.

Using an inequality of kind (9), we obtain the inequality

$$\|\psi(A)u\|_{W^k_{q_\varepsilon}} \leq R \cdot C(p,\varepsilon) \cdot \|A^{\alpha}u\|_{L_p}^{1/2} \|A^{\alpha-2\varepsilon}u\|_{L_p}^{1/2}[u \in D(A^{\alpha})]$$

for $0 < \varepsilon < \alpha/2$.

Having carried out the replacement $A^{\alpha}u = v$, we obtain the inequality

$$\|\psi(A)(A)^{-\alpha}v\|_{W^k_{q_\varepsilon}}$$

$$R \cdot C(p,\varepsilon) \cdot \|v\|_{L_p}^{1/2} \cdot \|A^{-2\varepsilon}v\|_{L_p}^{1/2}[v \in E = L_p(\Omega)] \tag{18 .}$$

Let $0 < \varepsilon < \frac{n}{4m}(1 - \frac{1}{p})$. Then from results [6] it follows, that $A^{-2\varepsilon}$ is a bounded operator from L_{π_ε} to L_p for $\frac{1}{\pi_\varepsilon} = \frac{1}{p} + \frac{4m\varepsilon}{n}$.

Hence, inequality (18) yields the inequality

$$\|\psi(A) \cdot A^{-\alpha} v\|_{W^k_{q_\varepsilon}}$$

(19 .) $$R \cdot \tilde{C}(p,\varepsilon) \cdot \|v\|^{1/2}_{L_p} \|v\|^{1/2}_{L\pi_\varepsilon} [v \in L_p(\Omega)]$$

Applying inequality (19) to the characteristic function $\chi_e(x)$ of the measurable set $e \in \Omega$, we obtain the estimate

$$\|\psi(A) \cdot A^{-\alpha} \chi_e\|_{W^k_{q_\varepsilon}} \leq R \cdot \tilde{C}(p,\varepsilon) \cdot (mease)^{\frac{1}{p}+\frac{2m\varepsilon}{n}}.$$

It means that on the characteristic functions the operator $\psi(A)A^{-\alpha}$ acts as a bounded operator from L_{p_ε} with $\frac{1}{p_\varepsilon} = \frac{1}{p} + \frac{2m\varepsilon}{n}$ to $W^k_{q_\varepsilon}$ and the inequality

$$\|\psi(A) \cdot A^{-\alpha} \chi_e\|_{W^k_{q_\varepsilon}} \leq R \cdot C(p,\varepsilon) \cdot \|\chi_e\|_{L_{p_\varepsilon}}$$

takes place.

Since the last estimate takes place for any $p \in (1, +\infty)$ (and sufficiently small $\varepsilon > 0$), the statement of theorem 2 follows from it and from the interpolation theorem [5].

REFERENCES

1. Krasnoselskii, M.A., Zabreiko, P.P.Pustylnik, E.J. and Sobolevskii, P.E. (1966). Integral operators in the spaces of summable functions, Nauk, Moskow, p. 500.
2. Seeley, R. (1970). "Fraction powers of boundary problems", Actes Cong. Int. Mathematisenen (1971), Paris, N2, p. 795.
3. Sobolevskii, P.E. (1966). On fraction powers of weakly positive operators, J. RAC USSR, 166, N6, p. 12.
4. Stein, E.M. and Weiss, G. (1959). An extension of a theorem of Marei Kiewiez and of its applications, J. of Math. and Mech., 8, N2, p. 17.
5. Tribel, H. (1980). Interpolation theory, Functional spaces, Differential Operators, Mir, Moskow, p. 664.
6. Zabreiko, P.P., Krasnoselskii, M.A. and Pustylnick, E.J. (1965), On fraction powers of elliptic operators, J. RAC USSR, 165, N5, p.153.

On (Semi)Groups Having Empty Resolvent Set

FRANCISZEK HUGON SZAFRANIEC Jagiellonian University, Krakow, Poland

The example

This note[1] serves more as an invitation to a problem connected with subnormal operators, than that it has pretention of building (a piece of) theory. Let us start with an interesting example.

Suppose that $\mathcal{H}$ is a reproducing kernel Hilbert space of the kernel

$$K(z,u) = e^{z\bar{u}}, \quad z,u \in \mathbb{C}.$$

It is composed, cf. (Bargmann, 1961) of the entire functions f such that

$$\int_{\mathbb{C}} |f(z)|^2 e^{-\frac{1}{2}|z|^2} \mathrm{d}x\mathrm{d}y < \infty, \quad z = x + \mathrm{i}y.$$

This is the so called Bargmann or Bargmann–Fock space. For any $z \in \mathbb{C}$ consider an unbounded densely defined operator $e(z)$ in $\mathcal{H}$ defined as

$$(e(z)f)(\zeta) = e^{z\zeta} f(\zeta), \quad \zeta \in \mathbb{C}$$

where f's are in $\mathcal{D} = \mathrm{lin}\{e_z p\colon\ p \in \mathbb{C}[Z],\ e_z(u) = e^{zu},\ z,u \in \mathbb{C}\}$. $\mathcal{D}$ is dense in $\mathcal{H}$ and invariant under any $e(z)$. Then

$$e(0) = I, \quad e(z_1 + z_2)f = e(z_1)e(z_2)f, \quad f \in \mathcal{D}.$$

It is a matter of direct verification that for $f \in \mathcal{D}$

$$\|e(z+u)f - e(z)f\| \to 0 \text{ as } u \to 0$$

and

$$\|(e(z+u)f - e(z)f)/z - Mf\| \to 0 \text{ as } u \to 0$$

where $(Mf)(\zeta) = \zeta f(\zeta)$, $\zeta \in \mathbb{C}$. This means that M (with domain $\mathcal{D}(M) = \mathcal{D}$) is a generator of $\{e(z)\}_{z \in \mathbb{C}}$. This is the prototype of our (semi)group.

M is known (Bargmann, 1961) to be unitarily equivalent to the operator

$$2^{-\frac{1}{2}}\left(x - \frac{\mathrm{d}}{\mathrm{d}x}\right)$$

with a suitable domain in $\mathcal{L}_2(\mathbb{R})$. This is the famous creation operator of quantum mechanics. This operator, or equivalently M, has *empty resolvent set*,

---

The research resulting in this paper has been supported by the grant of KBN (Komitet Badań Naukowych, Warsaw).

so the habitual resolvent methods do not work here; one has to look for another approach. Fortunately, the point spectrum $\sigma_p(M^*)$ of M^* (which is the differentiation operator) is equal to $\mathbb{C}$ and the eigenvectors are $e^{z\bar{u}}$, $u \in \mathbb{C}$; this may be of some help.

Though $\mathcal{H}$ has its own inner product (expressed in terms of K), sometimes it may be useful to realize that $\mathcal{H}$ is a closed subspace of $\mathcal{L}_2(\mathbb{C}, \exp(-\frac{1}{2}|z|^2 dxdy))$. This means that M as well as $e(z)$'s extend from $\mathcal{H}$ to normal operators (still being the multiplication operators) in this $\mathcal{L}_2$-space and these normal extensions have empty resolvent set as well; this is another feature of our example.

Semigroups of unbounded operators were considered in (Hueghes, 1977) and (Szafraniec, 1990b) for instance. However, since,

$$\|e(z)1\| = e^{\frac{1}{2}|z|^2}, \; z \in \mathbb{C},$$

our considerations do not fit in with that context.

The abstract set-up

Now we make an attempt at putting the aforesaid in some abstact framework. Let $\mathcal{H}$ be a complex Hilbert space and S a densely defined operator in $\mathcal{H}$; $\mathcal{D}(S)$ stands for domain of S.

THEOREM 1. *Let S be a operator in $\mathcal{H}$ with invariant domain. Suppose that S^* has a total set of linearily independent eigenvectors. Then there is a dense linear subspace $\mathcal{D}$ of $\mathcal{H}$ containing $\mathcal{D}(S)$ and a family $\{e(z)\}_{z \in \mathbb{C}}$ of closable operators having $\mathcal{D}$ as a common invariant domain and such that*

$$Se(z)f = e(z)Sf, \; f \in \mathcal{D}(S) \tag{1}$$

$$e(0) = I, \; e(z_1 + z_2)f = e(z_1)e(z_2)f, \; f \in \mathcal{D} \tag{2}$$

$$\|e(z+u)f - e(z)f\| \to 0 \text{ as } u \to 0, \; f \in \mathcal{D} \tag{3}$$

and

$$\|(e(z+u)f - e(z)f)/z - Sf\| \to 0 \text{ as } u \to 0, \; f \in \mathcal{D}(S). \tag{4}$$

Proof. Let x_λ be an eigenvector of S^* corresponding to the eigenvalue λ. For $z \in \mathbb{C}$ set

$$e_*(z)x_\lambda = e^{\bar{z}\lambda}x_\lambda$$

and extend it linearily to $\mathcal{D}_* = \text{lin}\{x_\lambda \colon \lambda \in \sigma_p(S^*)\}$. Then (2) holds for $f \in \mathcal{D}_*$. Set

$$e_0(z) = e_*(z)^*, \; z \in \mathbb{C}.$$

Take $f \in \mathcal{D}(S)$. We want to show that f belongs to each $\mathcal{D}(e_0(z))$. For any $x \in \mathcal{D}_*$ $<f, e_*(z)x>$ is an analytic function in z and $\sum_{n=0}^{\infty} z^n/n! <S^n f, x>$ is its Taylor series at 0. Thus

$$< \sum_{n=0}^{\infty} (z^n/n!)S^n f, x> = \sum_{n=0}^{\infty} z^n/n! <S^n f, x> = <f, e_*(z)x>.$$

Consequently, $f\in\mathcal{D}(e_0(z))$ and

$$e_0(z)f=\sum_{n=0}^{\infty}(z^n/n!)S^nf \text{ for } f\in\mathcal{D}(S). \tag{5}$$

Set $\mathcal{D}=\text{lin}\{e_0(z)f\colon f\in\mathcal{D}(S),\ z\in\mathbb{C}\}$; $\mathcal{D}$ is invariant for each $e_0(z)$. Denote by $e(z)$ the restriction of $e_0(z)$ to $\mathcal{D}$.

Notice that (2) can be obtained from the same equality satisfied by $e_*(z)$ as mentioned earlier. Let x_λ be an eigenvector of S^*. Then, for $e(u)f\in\mathcal{D}$, we have

$$<e(z)Se(u)f,x_\lambda> = <e(u)f,S^*e_*(z)x_\lambda> = <e(u)f,S^*e^{\bar{z}\lambda}x_\lambda>$$

$$= <e(u)f,e^{\bar{z}\lambda}S^*x_\lambda> = <e(u)f,\lambda e^{\bar{z}\lambda}x_\lambda> = <e(u)f,e_*(z)\lambda x_\lambda>$$

$$= <f,e_*(u)e_*(z)S^*x_\lambda> = <f,e_*(z)e_*(u)S^*x_\lambda> = <Se(z)e(u)f,x_\lambda>$$

and this gives (1). (3) and (4) can get straightforwardly using (5). This completes the proof. ■

Call S *subnormal* if there is a Hilbert space $\mathcal{K}\supset\mathcal{H}$ and a normal operator N in $\mathcal{K}$ such that

$$\mathcal{D}(S)\subset\mathcal{D}(N)\cap\mathcal{H} \text{ and } Sf=Nf \text{ for } f\in\mathcal{D}(S).$$

S is said to *cyclic* if there is a vector $f\in\mathcal{D}(S)$ such that $\mathcal{D}(S)=\text{lin}\{S^nf,\ n\geq 0\}$. It is known (Stochel, Szafraniec 1989a) that for a cyclic operator S subnormality is equivalent to existence of a positive Borel measure μ, say, on $\mathbb{C}$ such that

$$<S^mf,S^nf> = \int_{\mathbb{C}}\lambda^m\bar{\lambda}^n\mu(d\lambda),\ f\in\mathcal{D}(S),\ m,n\in\mathbb{N}. \tag{6}$$

Call such a measure a *representing measure of* S. Representing measures are not uniquely determined; cf (Stochel, Szafraniec, 1989b) for a general approach and (Szafraniec, 1989a) for a concrete example.

THEOREM 2. *Let S be a cyclic subnormal operator in $\mathcal{H}$. Suppose that S^* has a total set of eigenvectors and there exists a measure μ representing S such that*

$$\sigma_p(S^*)^*\cap\text{supp}\mu\neq\emptyset.$$

Then the conclusion of Theorem 1 *holds with $e(z)$'s being subnormal.*

Proof. Combining Theorem 6, Corollary 11 and Proposition 9 of (Stochel, Szafraniec, 1989b) we get that $e(z)$ is unitarily equivalent to multiplication by the function e_z, $e_z(\zeta)=e^{z\zeta}$, $\zeta\in\sigma_p(S^*)^*\cap\text{supp}\mu$, which is subnormal. ■

Remark. In this section we made a one-way journey from subnormal generators to groups of (subnormal) unbounded operators. The creation operator and the entire group it generates was used as a pattern. The return, describing groups with subnormal generators, seems to be more difficult because of a lack of good characterization of unbounded subnormal operators.

References

Bargmann V. (1961), On a Hilbert space of analytic functions and an associated integral transform. I, *Comm. Pure Appl. Math.*, **19** 187–214.

Hughes R.J (1977), Semigroups of unbounded linear operators in Banach space,

Trans. Amer. Math. Soc., **230**, 113–145.
Stochel J., Szafraniec F.H. (1989a), On normal extensions of unbounded operators. II, *Acta Sci. Math. (Szeged)*, **53**, 153 – 177. Stochel J., Szafraniec F.H. (1989b), On normal extensions of unbounded operators. III. Spectral properties, *Publ. RIMS, Kyoto Univ.*, **25**, 105 – 139.
Szafraniec F.H. (1990a), A RKHS of entire functions and its multiplication operator. An explicit example, in "*Linear Operators in Function Spaces*", Proceedings, Timişoara (Romania), June 6 – 16, 1988, eds. H. Helson, B. Sz. – Nagy and F. – H. Vasilescu, *Operator Theory: Advances and Applications*, vol. **43**, pp. 309 – 312, Birkhäuser, Basel.
Szafraniec F.H. (1990b), Kato – Protter type inequalities, bounded vectors and the exponential function, *Ann. Polon. Math.*, **51**, 303 – 312.

Generalized Spectral Decomposition for the r-adic Renyi Map

S. TASAKI International Solvay Institute for Physics and Chemistry, Brussels, Belgium

I. ANTONIOU University of Brussels, Brussels, Belgium

Abstract Spectral properties of the Perron–Frobenius operator of r-adic map are studied under various settings of its domain. By increasing the smoothness of functions in the domain, the essential spectral radius is shown to decrease according to the manner discussed by Tangerman, Pollicott and Ruelle. And the explicit expressions of eigenfunctions corresponding to these spectra are obtained. Finally, in the framework of the rigged Hilbert space with a polynomial test function space, a generalized spectral decomposition of U of Gelfand–Maurin type is formulated.

1. Introduction

One of the main problems in the non-equilibrium statistical physics is to derive quantities such as lifetimes and diffusion coefficients, which characterize the irreversible phenomena, from the underlying dynamics. Prigogine and his co-workers (Hasegawa et al., 1991, Petrosky, Prigogine and Tasaki, 1991, Petrosky and Prigogine, 1991, Prigogine et al., 1991, Antoniou and Prigogine, 1992) have developed a "complex spectral theory", by which the characteristic quantities of irreversible phenomena are obtained as complex eigenvalues of the generator of motion. Since the generator of motion of a conservative dynamical system is Hermitian on a certain Hilbert space, the inclusion of complex eigenvalues requires the extension of the generator to a wider functional space. As was shown by Antoniou and Prigogine (1992) for a simple quantum system, this extension is formulated in a framework of the rigged Hilbert spaces (Gelfand and Vilenkin, 1964, see also e.g., Bohm and Gadella, 1989, Parravicini, Gorini and Sudarshan, 1980) and the generator acquires a generalized spectral decomposition containing complex eigenvalues. These results are the extension of the works of Gelfand (Gelfand and Vilenkin,1964, Gelfand and Shilov,1967) and Maurin (1968) for complex eigenvalues. Recently, Hasegawa and Saphir (1992a,b) have applied the complex spectral theory to the Bernoulli map, the simplest exact dynamical system (see e.g., Lasota and Mackey,1985) on the unit interval [0,1], and derive a generalized spectral decomposition of its Perron–Frobenius operator. The same result is obtained by Gaspard (1992) for the general r-adic Renyi map by using the Euler summation formula. One of the interesting features of their generalized spectral decomposition is that it contains Ruelle's resonances (Ruelle, 1986,1987a,b,1989) as eigenvalues. In this paper, we will study the spectral properties of the Perron–Frobenius operator of the r-adic Renyi map with respect to different domains and formulate a spectral decomposition associated with Ruelle's resonances in terms of an appropriate rigged Hilbert space. The purpose of this study is three-fold: 1) to illustrate the mathematical structure of the "complex spectral theory", 2) to show the relation between the generalized spectral decomposition and Ruelle's resonances, and 3) to clarify the mechanism through which the generalized spectral decomposition emerges.

The paper will be arranged as follows. After summarizing the results on polynomial eigenfunctions of the Perron–Frobenius operator U of the r-adic map (section 2), square integrable eigenfunctions of U are constructed and their analytic properties will be studied (section 3). The differentiability of eigenfunctions is found to depend on the magnitude of eigenvalues, i.e., smoother eigenfunctions correspond to eigenvalues with smaller magnitudes. It is this property that causes the decrease of essential spectral radius of U by increasing the smoothness of functions in the domain. This fact is studied in detail in section 4. For expanding maps, such a change of the spectrum of the Perron–Frobenius operators was studied by Tangerman (1986), Pollicott (1990) and Ruelle (1989) and the estimate of the essential spectral radius was derived. However, in the present example, the spectrum of U can be completely determined and all the corresponding eigenfunctions can be obtained explicitly (cf. Theorems 2 and 3). In section 5, the generalized spectral decomposition of U will be rigorously formulated in terms of an appropriate rigged Hilbert space as an extension of Gelfand–Maurin construction. This is a typical situation in the "complex spectral theory", which is formulated on the framework of rigged Hilbert spaces. In the following, the sets of integers, real numbers and complex numbers are denoted respectively by **Z**, **R** and **C**, and the lth derivative of a function $\rho(x)$ is abbreviated as $\rho^{(l)}(x)$ without notice.

2. r-adic Renyi Map and Polynomial Eigenfunctions

The r-adic Renyi map is defined on the unit interval [0,1] as

$$x_{n+1} = S(x_n) = r x_n \qquad (\text{mod } 1) , \tag{2.1}$$

where $r \geq 2$ is a positive integer. This map is exact and thus mixing and ergodic (Lasota and Mackey, 1985, p.70). The Perron–Frobenius operator U governs the evolution of densities $\rho(x)$ (Lasota and Mackey,1985,p.37) and is given by

$$U\rho(x) \equiv \sum_{y,S(y)=x} \frac{1}{|S'(y)|}\rho(y) = \frac{1}{r}\sum_{j=0}^{r-1} \rho(\frac{x+j}{r}) . \tag{2.2}$$

The Perron–Frobenius operator U is defined on the space of absolutely Lebesgue integrable functions on [0,1] as ρ is a probability density. Here we restrict U to the Hilbert space L^2 of square integrable densities, because this is the starting point for the construction of rigged Hilbert spaces.

From the definition (2.2), it is obvious that the Perron–Frobenius operator U maps polynomials of degree n to polynomials of degree n. Thus, it admits polynomial eigenfunctions. Irrespective to the value of r, these eigenpolynomials are the Bernoulli polynomials $B_n(x)$ (Hasegawa and Saphir, 1992a,b, Gaspard, 1992) defined by the following generating function

$$\frac{te^{tx}}{e^t - 1} = \sum_{n=0}^{\infty} \frac{B_n(x)}{n!} t^n \qquad (|t| < 2\pi) , \tag{2.3}$$

and the corresponding eigenvalues are $1/r^n$. This fact is a restatement of a well-known multiplication theorem of the Bernoulli polynomials (e.g., Olver, 1974, p.284). Moreover, the Bernoulli polynomials are the only eigenpolynomials as any polynomial can be uniquely expressed as a linear combination of them. Here, we shall summarize the results as a theorem.

THEOREM 1 The nth order Bernoulli polynomial $B_n(x)$ is an eigenfunction of the Perron–Frobenius operator U with eigenvalue $1/r^n$

$$U\ B_n(x) = \frac{1}{r} \sum_{j=0}^{r-1} B_n\left(\frac{x+j}{r}\right) = \frac{1}{r^n}\ B_n(x) \qquad (n = 0, 1, \cdots) . \tag{2.4}$$

The Bernoulli polynomials are the only polynomial eigenfunctions.

As was pointed out by Hasegawa and Saphir (1992a,b) and by Gaspard (1992), the eigenvalues corresponding to these eigenpolynomials coincide with Ruelle's resonances determined as the zeros of the Fredholm determinant $d(z) \equiv \det(1 - z^{-1}U)$ (Ruelle, 1986) and provide a generalized spectral decomposition of U. The precise definition of this spectral decomposition will be formulated in section 5.

In the forthcoming sections, we will use the following properties of the Bernoulli polynomials (see e.g., Olver, 1974, Section 8-1).

PROPOSITION 1

$$B_n(1) - B_n(0) = \begin{cases} 1 , & (n = 1) \\ 0 , & (\text{otherwise}) \end{cases} \tag{2.5}$$

$$\left(\frac{d}{dx}\right)^l \frac{B_n(x)}{n!} = \frac{B_{n-l}(x)}{(n-l)!} \qquad (n \geq 1,\ \ 0 \leq l \leq n-1) . \tag{2.6}$$

3. Non-polynomial Eigenfunctions

In the Hilbert space L^2, the trigonometric functions $\{\exp(2\pi nix)\}_{n\in\mathbf{Z}}$ form a complete orthonormal basis. These functions are transformed by U as

$$Ue^{2\pi nix} = \frac{1}{r}\sum_{j=0}^{r-1}\exp\left(2\pi ni\frac{x+j}{r}\right) = \begin{cases} \exp\left(2\pi i\frac{n}{r}x\right), & (\frac{n}{r}\in\mathbf{Z}) \\ 0. & (\frac{n}{r}\notin\mathbf{Z}) \end{cases} \tag{3.1}$$

One of them, namely $\rho_0(x)\equiv 1$ is the unique stationary density of U. It is convenient to parameterize the other trigonometric functions as

$$g_n^s(x) \equiv \exp(2\pi i s r^n x), \qquad (n=0,1,2,\cdots) \tag{3.2}$$

where s is an integer which cannot be divided by r:

$$s\in\mathbf{Z_r}\equiv\{s\in\mathbf{Z} : s/r\notin\mathbf{Z}\}. \tag{3.3}$$

Then the relation (3.1) becomes

$$Ug_n^s(x) = \begin{cases} g_{n-1}^s(x), & (n\geq 1) \\ 0. & (n=0) \end{cases} \tag{3.4}$$

Eq.(3.4) means that the operator U is the adjoint unilateral shift in the orthogonal complement of constant functions and that a set $\{g_0^s(x)\}_{s\in\mathbf{Z_r}}$ is a basis of the wandering generating space (Sz-Nagy and Foias, 1970) of the shift. As the dimension of the wandering generating subspace is the multiplicity of the shift, the operator U has infinite 'semi-Lebesgue' spectrum (Parry, 1981, p.52). Due to the shift property (3.4), it is easy to construct the eigenfunctions in terms of $g_n^s(x)$.

THEOREM 2

i) The spectral radius of U in the Hilbert space L^2 is one, i.e., its spectrum lies in the unit disk in the complex plane $\{\lambda\in\mathbf{C} : |\lambda|\leq 1\}$. For $\lambda=1$, there exists a unique eigenfunction $\rho_0(x)\equiv 1$ and, for $|\lambda|<1$, there exist countably many eigenfunctions of the following form

$$\psi_\lambda^s(x)\equiv\sum_{n=0}^{\infty}\lambda^n g_n^s(x). \qquad (s\in\mathbf{Z_r}) \tag{3.5}$$

Any eigenfunction of U corresponding to an eigenvalue $|\lambda|<1$ is their superposition.

ii) Let l be a non-negative integer satisfying

$$\frac{-\ln|\lambda|}{\ln r} > l \geq \frac{-\ln|\lambda|}{\ln r} - 1. \tag{3.6}$$

Then $\psi_\lambda^s(x)$ is l-times continuously differentiable on [0,1] (For $x = 0, 1$, derivatives are taken from one side). Moreover, its lth derivative $\psi_\lambda^{s(l)}(x)$ is Hölder continuous of exponent α: $0 < \alpha < -\ln(|\lambda r^l|)/\ln r$.

Proof: i) It is a well-known result that the spectrum of U lies in the unit disk (e.g., Reed and Simon, 1980, p.192).

It is obvious that $\rho_0(x) \equiv 1$ is the eigenfunction of U with eigenvalue 1. Its uniqueness is guaranteed by the ergodicity of the r-adic map (Lasota and Mackey, 1985, Theorem 4.2.2).

Next, we shall show that, for $|\lambda| < 1$, $\{\psi_\lambda^s(x)\}_{s\in\mathbf{Z}_r}$ are well-defined and form a basis of the space of eigenfunctions corresponding to the eigenvalue λ. Since $|g_n^s(x)| = 1$ on [0,1], if $|\lambda| < 1$, the series (3.5) converges absolutely and uniformly. Thus, $\psi_\lambda^s(x)$ is continuous and belongs to L^2. Let $\psi_\lambda(x) = \sum_{s\in\mathbf{Z}_r}\sum_{n=0}^{\infty} c_n^s\ g_n^s(x) \in L^2$ be an eigenfunction of U with eigenvalue λ,

$$U\ \psi_\lambda(x) = \lambda\ \psi_\lambda(x)\ , \tag{3.7}$$

then (3.4) gives

$$c_{n+1}^s = \lambda\ c_n^s = \cdots = \lambda^n\ c_0^s\ , \tag{3.8}$$

and thus

$$\psi_\lambda(x) = \sum_{s\in\mathbf{Z}_r}\sum_{n=0}^{\infty}\lambda^n\ c_0^s\ g_n^s(x) = \sum_{s\in\mathbf{Z}_r} c_0^s\psi_\lambda^s(x)\ . \tag{3.9}$$

This completes the proof of i).

ii) As we have

$$\left(\frac{d}{dx}\right)^t g_n^s(x) = (2\pi i s r^n)^t\ g_n^s(x)\ , \tag{3.10}$$

the term-by-term tth derivative of (3.5) is

$$\sum_{n=0}^{\infty}\lambda^n\left(\frac{d}{dx}\right)^t g_n^s(x) = (2\pi i s)^t\sum_{n=0}^{\infty}(\lambda r^t)^n\ g_n^s(x)\ . \tag{3.11}$$

Let $0 \le t \le l$, then because of $|g_n^s(x)| = 1$ and $|\lambda| r^t \le |\lambda| r^l < 1$ (see (3.6)), the series (3.11) converges absolutely and uniformly on [0,1]. Thus, $\psi_\lambda^s(x)$ is l-times continuously differentiable and its tth derivative is given by (3.11). The Hölder continuity of $\psi_\lambda^{s(l)}(x)$ is proved as follows. Because of an inequality

$$|\exp(ix) - 1| \le 2^{1-\alpha}|x|^\alpha\ , \qquad (x \in \mathbf{R},\ 0 < \alpha < 1)$$

we have

$$|g_n^s(x) - g_n^s(y)| = |\exp(2\pi i s r^n(x-y)) - 1| \le 2(\pi|s|)^\alpha r^{\alpha n}|x-y|^\alpha\ . \tag{3.12}$$

Therefore, for $x \neq y$ and $|\lambda| r^{l+\alpha} < 1$,

$$\frac{|\psi_\lambda^{s(l)}(x) - \psi_\lambda^{s(l)}(y)|}{|x-y|^\alpha} \leq (2\pi|s|)^l \sum_{n=0}^{\infty} (|\lambda| r^l)^n \frac{|g_n^s(x) - g_n^s(y)|}{|x-y|^\alpha}$$
$$\leq 2^{l+1}(\pi|s|)^{l+\alpha} \sum_{n=0}^{\infty} [|\lambda| r^{l+\alpha}]^n < +\infty . \tag{3.13}$$

QED

Since the eigenpolynomial $B_n(x)$ corresponds to an eigenvalue $1/r^n < 1$, it should be a superposition of $\psi_\lambda^s(x)$. Indeed, by rearranging the Fourier expansions of the Bernoulli polynomials (see Olver, 1974, p.284), we obtain

PROPOSITION 2

$$B_0(x) = \rho_0(x) , \qquad \frac{B_n(x)}{n!} = \sum_{s \in \mathbf{Z}_r} \frac{-1}{(2\pi i s)^n} \psi_{1/r^n}^s(x) . \quad (n = 1, 2, \cdots) \tag{3.14}$$

By Theorem 2. ii), when $1 > |\lambda| \geq 1/r$, the eigenfunctions $\psi_\lambda^s(x)$ become nowhere differentiable. As was pointed out by Hata (1985), the function $\psi_{1/r}^s(x)$ is expressed by well-known Weierstrass' functions

$$\psi_{1/r}^s(x) = \alpha_r(sx) + i\beta_r(sx) , \tag{3.15}$$

where $\alpha_r(x)$ and $\beta_r(x)$ are Weierstrass' functions

$$\alpha_r(x) = \sum_{n=0}^{\infty} \frac{1}{r^n} \cos(2\pi r^n x) , \qquad \beta_r(x) = \sum_{n=0}^{\infty} \frac{1}{r^n} \sin(2\pi r^n x) . \tag{3.16}$$

4. Change of the Spectrum of U

Due to Theorem 2, when the domain of U is restricted to the space of l-times continuously differentiable functions, the eigenfunctions $\psi_\lambda^s(x)$ corresponding to eigenvalues $1/r^l \leq |\lambda| < 1$ will be excluded and the radius of essential spectrum of U (i.e., the remainder of the spectrum after removing isolated eigenvalues) may decrease. For expanding maps, the change of the spectrum of their Perron–Frobenius operators was studied by Tangerman (1986), Pollicott (1990) and Ruelle (1989), and the estimate of the essential spectral radius is obtained (Pollicott, 1990, Proposition 4 and Ruelle, 1989, Theorem 3.2). For the r-adic maps, more precise results can be obtained.

First we specify domains of U. Let us denote the space of Hölder continuous functions of exponent α $(0 < \alpha \leq 1)$ as H^α and the space of l-times continuously differentiable functions on [0,1] as C^l (For $x = 0, 1$, the derivatives are taken from one side). We then introduce the following spaces

$$C^{(l,\alpha)} \equiv \{\rho \in C^l : \rho^{(l)} \in H^\alpha\} , \tag{4.1}$$

$$C_p^{(l,\alpha)} \equiv \{\rho \in C^{(l,\alpha)} : \rho^{(s)}(1) = \rho^{(s)}(0) \ \ (s = 0, 1, \cdots l-1)\} . \tag{4.2}$$

Both spaces are Banach spaces with respect to the norm

$$\|\rho\|_{l,\alpha} \equiv \sum_{j=0}^{l} \|\rho^{(j)}\|_\infty + \|\rho^{(l)}\|_\alpha , \tag{4.3}$$

where $\|\cdot\|_\infty$ is a usual maximum norm $\|\rho\|_\infty = \max_{x\in[0,1]} |\rho(x)|$ and $\|\cdot\|_\alpha$ is a semi-norm defined on H^α

$$\|g\|_\alpha \equiv \sup\{\frac{|g(x)-g(y)|}{|x-y|^\alpha} : x, y \in [0,1], x \neq y\} . \quad (g(x) \in H^\alpha) \tag{4.4}$$

Note that $C^{(0,\alpha)} = C_p^{(0,\alpha)}$. The Perron–Frobenius operator U maps both $C^{(l,\alpha)}$ and $C_p^{(l,\alpha)}$ onto themselves. The spectrum of U depends on the choice of its domain as stated by the following theorem.

THEOREM 3 Let l be a non-negative integer and α be $0 < \alpha \leq 1$.

i) On the functional space $C_p^{(l,\alpha)}$, the Perron–Frobenius operator U admits a decomposition

$$U\rho(x) = \int_0^1 dx' \rho(x') + \hat{U}\rho(x) , \quad (\rho \in C_p^{(l,\alpha)}) \tag{4.5}$$

where the first term corresponds to an isolated point spectrum at $\lambda = 1$ and the operator $\hat{U}$ has its spectrum inside the whole disk $\{\lambda \in \mathbf{C} : |\lambda| \leq 1/r^{l+\alpha}\}$.

ii) On the functional space $C^{(l,\alpha)}$ $(l \geq 1)$, the Perron–Frobenius operator U admits a decomposition

$$U\rho(x) = \int_0^1 dx' \rho(x') + \sum_{j=1}^{l} \frac{\rho^{(j-1)}(1) - \rho^{(j-1)}(0)}{j!\ r^j} B_j(x) + \bar{U}\rho(x) , \quad (\rho \in C^{(l,\alpha)}) \tag{4.6}$$

where the first two terms correspond to isolated point spectra at $\lambda = 1, 1/r, 1/r^2 \cdots 1/r^l$ and the operator $\bar{U}$ has its spectrum inside the whole disk $\{\lambda \in \mathbf{C} : |\lambda| \leq 1/r^{l+\alpha}\}$.

In both cases, eigenfunctions of the residual operators are given by $\psi_\lambda^s(x)$ of (3.5) with eigenvalue $\lambda : \ |\lambda| < 1/r^{l+\alpha}$.

For its proof, we need the following lemma on Hölder continuous functions.

LEMMA 1 Define an operator $\hat{U}$ by $\hat{U}\rho(x) \equiv U\ \rho(x) - \int_0^1 dx'\rho(x')$. Let $g(x) \in H^\alpha$, then $\hat{U}g(x) \in H^\alpha$ and

$$|\int_0^1 dx e^{-2\pi i n x} g(x)| \le \frac{\|g\|_\alpha}{2^{\alpha+1}|n|^\alpha} , \tag{4.7}$$

$$\|\hat{U}g\|_\alpha \le \frac{\|g\|_\alpha}{r^\alpha} , \tag{4.8}$$

$$\|\hat{U}g\|_\infty \le \frac{\|g\|_\alpha}{r^\alpha} . \tag{4.9}$$

Proof: i) Since $|g(x) - g(y)| \le \|g\|_\alpha |x-y|^\alpha$ and

$$\int_0^1 dx e^{-2\pi i n x} g(x) = \frac{1}{2}\int_0^1 dx e^{-2\pi i n x}\{g(x) - g(x + \frac{1}{2n})\} , \tag{4.10}$$

(e.g., Hardy an Rogosinski,1968,p.24), we obtain (4.7)

$$|\int_0^1 dx e^{-2\pi i n x} g(x)| \le \frac{1}{2}\int_0^1 dx |g(x) - g(x + \frac{1}{2n})| \le \frac{\|g\|_\alpha}{2^{\alpha+1}|n|^\alpha} . \tag{4.11}$$

ii) Inequality (4.8) and $\hat{U}g(x) \in H^\alpha$ follow from

$$|\hat{U}g(x) - \hat{U}g(y)| \le \frac{1}{r}\sum_{j=0}^{r-1} |g(\frac{x+j}{r}) - g(\frac{y+j}{r})| \le \frac{\|g\|_\alpha}{r^\alpha}|x-y|^\alpha . \tag{4.12}$$

iii) Since $\int_0^1 dx' g(x') = 1/r \sum_{j=0}^{r-1} \int_0^1 dy g(y/r + j/r)$, we have

$$\begin{aligned} |\hat{U}g(x)| &\le \frac{1}{r}\sum_{j=0}^{r-1}\int_0^1 dy |g(\frac{x+j}{r}) - g(\frac{y+j}{r})| \\ &\le \frac{1}{r}\sum_{j=0}^{r-1}\frac{\|g\|_\alpha}{r^\alpha}\int_0^1 dy |x-y|^\alpha \le \frac{\|g\|_\alpha}{r^\alpha} , \end{aligned} \tag{4.13}$$

which implies $\|\hat{U}g\|_\infty \le \|g\|_\alpha / r^\alpha$. *QED*

Proof of Theorem 3: i) From Theorem 2, for any $\lambda : |\lambda| < 1/r^{l+\alpha}$, the function $\psi_\lambda^s(x)$ is l-times continuously differentiable, $\psi_\lambda^{s(l)}(x)$ is Hölder continuous of exponent α and its derivatives $\psi_\lambda^{s(t)}(x)$ $(0 \le t \le l)$ are periodic (see (3.11)). Moreover, as it has zero average, it is an eigenfunction of $\hat{U}$ with eigenvalue λ. Therefore, the whole disk $\{\lambda : |\lambda| \le 1/r^{l+\alpha}\}$ is inside the spectrum of $\hat{U}$. We will show that the spectral radius

$R(\hat{U})$ of $\hat{U}$ is bounded by $1/r^{l+\alpha}$ with the aid of a well-known formula (Reed and Simon, 1980, Theorem VI.6)

$$R(\hat{U}) = \lim_{m\to\infty} \|\hat{U}^m\|^{1/m} . \tag{4.14}$$

i – a) Estimate of $\|(\hat{U}^m\rho)^{(l)}\|_\alpha$

Let $\rho(x) \in C_p^{(l,\alpha)}$, then $\int_0^1 dx\rho'(x) = \rho(1) - \rho(0) = 0$ and we have

$$\frac{d}{dx}\hat{U}\rho(x) = \frac{1}{r}U\frac{d\rho}{dx}(x) = \frac{1}{r}\hat{U}\frac{d\rho}{dx}(x) .$$

Similarly, as $\rho^{(t)}(1) = \rho^{(t)}(0)$ $(0 \le t \le l-1)$,

$$(\frac{d}{dx})^l\hat{U}^m\rho(x) = \frac{1}{r^{lm}}\hat{U}^m\rho^{(l)}(x) . \tag{4.15}$$

Since $\rho^{(l)}(x) \in H^\alpha$, Lemma 1 (4.8) gives

$$\begin{aligned}\|(\hat{U}^m\rho)^{(l)}\|_\alpha &= \frac{1}{r^{lm}} \|\hat{U}^m\rho^{(l)}\|_\alpha \le \frac{1}{r^{lm}}\frac{1}{r^\alpha} \|\hat{U}^{m-1}\rho^{(l)}\|_\alpha \\ &\le \cdots \le \left(\frac{1}{r^{l+\alpha}}\right)^m \|\rho^{(l)}\|_\alpha .\end{aligned} \tag{4.16}$$

i – b) Estimate of $\|(\hat{U}^m\rho)^{(l)}\|_\infty$

Inequalities (4.8), (4.9) of Lemma 1 and eq.(4.15) give

$$\begin{aligned}\|(\hat{U}^m\rho)^{(l)}\|_\infty &= \frac{1}{r^{lm}} \|\hat{U}^m\rho^{(l)}\|_\infty \le \frac{1}{r^{lm}}\frac{1}{r^\alpha} \|\hat{U}^{m-1}\rho^{(l)}\|_\alpha \\ &\le \cdots \le \left(\frac{1}{r^{l+\alpha}}\right)^m \|\rho^{(l)}\|_\alpha .\end{aligned} \tag{4.17}$$

i – c) Estimate of $\|(\hat{U}^m\rho)^{(t)}\|_\infty$ $(t = 0, 1, \cdots l-1)$

The function $\rho(x)$ can be expended into a Fourier series

$$\rho(x) = \int_0^1 dx'\rho(x') + \sum_{s\in \mathbf{Z}_r}\sum_{n=0}^\infty c_n^s\, g_n^s(x) , \tag{4.18}$$

where c_n^s is the Fourier coefficient

$$c_n^s = \int_0^1 dx' g_n^{s*}(x')\rho(x') = \int_0^1 dx' \exp(-2\pi i s r^n x')\rho(x') . \tag{4.19}$$

Thus, from (3.4), we obtain

$$\hat{U}^m\rho(x) = \sum_{s\in \mathbf{Z}_r}\sum_{n=m}^\infty c_n^s\, g_{n-m}^s(x) = \sum_{s\in \mathbf{Z}_r}\sum_{n=0}^\infty c_{n+m}^s\, g_n^s(x) , \tag{4.20}$$

which leads to (cf.(3.10))

$$|(\hat{U}^m \rho)^{(t)}(x)| \leq \sum_{s\in \mathbf{Z}_r} \sum_{n=0}^{\infty} (2\pi |s| r^n)^t |c^s_{n+m}| \,. \tag{4.21}$$

As $\rho(x)$ is l-times continuously differentiable and $\rho^{(t)}(1) = \rho^{(t)}(0)$ $(0 \leq t \leq l-1)$, we obtain

$$\begin{aligned} |c^s_{n+m}| &= (\frac{1}{|2\pi i s r^{n+m}|})^l \Big| \int_0^1 dx' \exp(-2\pi i s r^{n+m} x') \rho^{(l)}(x') \Big| \\ &\leq \frac{\|\rho^{(l)}\|_\alpha}{2^{\alpha+1}(2\pi)^l} (\frac{1}{|s| r^n})^{l+\alpha} (\frac{1}{r^{l+\alpha}})^m \,, \end{aligned} \tag{4.22}$$

where $\rho^{(l)}(x) \in H^\alpha$ and inequality (4.7) of Lemma 1 have been used. Thus, since $l + \alpha - t > 1$ for $0 \leq t \leq l-1$, we obtain

$$\begin{aligned} \|(\hat{U}^m \rho)^{(t)}\|_\infty &\leq \frac{\|\rho^{(l)}\|_\alpha}{2^{\alpha+1}(2\pi)^{l-t}} (\frac{1}{r^{l+\alpha}})^m \sum_{s\in \mathbf{Z}_r} \frac{1}{|s|^{l+\alpha-t}} \sum_{n=0}^{\infty} (\frac{1}{r^{l+\alpha-t}})^n \\ &\leq \frac{\|\rho^{(l)}\|_\alpha}{2^\alpha (2\pi)^{l-t}} \frac{\zeta(l+\alpha-t)}{1 - r^{t-\alpha-l}} (\frac{1}{r^{l+\alpha}})^m \,, \end{aligned} \tag{4.23}$$

where $\zeta(\beta) \equiv \sum_{n=1}^\infty 1/n^\beta$ is Riemann's zeta function.

By combining (4.16), (4.17) and (4.23), we obtain

$$\|\hat{U}^m \rho\|_{l,\alpha} = \sum_{t=0}^{l} \ \|(\hat{U}^m \rho)^{(t)}\|_\infty + \ \|(\hat{U}^m \rho)^{(l)}\|_\alpha \leq M \frac{\|\rho^{(l)}\|_\alpha}{r^{m(l+\alpha)}} \leq M \frac{\|\rho\|_{l,\alpha}}{r^{m(l+\alpha)}} \,, \tag{4.24}$$

where M is a positive constant depending only on l and α

$$M = 2 + \sum_{t=0}^{l-1} \frac{1}{2^\alpha (2\pi)^{l-t}} \frac{\zeta(l+\alpha-t)}{1 - r^{t-\alpha-l}} \,. \tag{4.25}$$

Thus,

$$R(\hat{U}) = \lim_{m\to\infty} \ \|\hat{U}^m\|^{1/m} \leq \lim_{m\to\infty} M^{1/m} \frac{1}{r^{l+\alpha}} = \frac{1}{r^{l+\alpha}} \,. \tag{4.26}$$

The decomposition (4.5) is obvious from the definition of $\hat{U}$.

ii) By using the Bernoulli polynomials $B_j(x)$, we will introduce an operator $\hat{B}$

$$\hat{B}\rho(x) \equiv \rho(x) - \sum_{j=1}^{l} \{\rho^{(j-1)}(1) - \rho^{(j-1)}(0)\} \frac{B_j(x)}{j!} \,, \tag{4.27}$$

which is a projection operator from $C^{(l,\alpha)}$ to $C_p^{(l,\alpha)}$. Indeed, from the properties (2.5) and (2.6) (see Proposition 1) of the Bernoulli polynomials, we have for $0 \le t \le l-1$

$$\begin{aligned}(\hat{B}\rho)^{(t)}(1) - (\hat{B}\rho)^{(t)}(0) =& \rho^{(t)}(1) - \rho^{(t)}(0) \\ &- \sum_{j=1}^{l} \{\rho^{(j-1)}(1) - \rho^{(j-1)}(0)\} \frac{B_{j-t}(1) - B_{j-t}(0)}{(j-t)!} = 0 ,\end{aligned}$$

or $\hat{B}\rho(x) \in C_p^{(l,\alpha)}$. And due to the boundary condition for $C_p^{(l,\alpha)}$, we have

$$\hat{B} f(x) = f(x) \quad (\forall\ f(x) \in C_p^{(l,\alpha)}) . \tag{4.28}$$

Also it is bounded as

$$\|\hat{B}\rho\|_{l,\alpha} \le \ \|\rho\|_{l,\alpha} + \sum_{j=1}^{l} \frac{|\rho^{(j-1)}(1)| + |\rho^{(j-1)}(0)|}{j!} \ \|B_j\|_{l,\alpha} \le K \ \|\rho\|_{l,\alpha} , \tag{4.29}$$

with $K = 1 + 2\sum_{j=1}^{l} \ \|B_j\|_{l,\alpha}/j! > 0$.

Now we shall define $\bar{U} \equiv \hat{U}\hat{B}$, then the spectrum of $\bar{U}$ lies in the whole disk $\{\lambda \in \mathbf{C} : |\lambda| \le 1/r^{l+\alpha}\}$. Indeed, for each value λ in this disk, the functions $\psi_\lambda^s(x) \in C^{(l,\alpha)}$ are eigenfunctions of $\bar{U}$ (cf. the proof of i)). Also the spectral radius $R(\bar{U})$ is bounded by $1/r^{l+\alpha}$. Indeed, as $\hat{B}\hat{U}\hat{B} = \hat{U}\hat{B}$ (which follows from (4.28)), we have

$$\begin{aligned}\|\bar{U}^m \rho\|_{l,\alpha} =& \ \|(\hat{U}\hat{B})^m \rho\|_{l,\alpha} = \ \|\hat{U}^m \hat{B}\rho\|_{l,\alpha} \\ \le& \frac{M}{r^{m(l+\alpha)}} \ \|\hat{B}\rho\|_{l,\alpha} \le \frac{MK}{r^{m(l+\alpha)}} \ \|\rho\|_{l,\alpha} ,\end{aligned} \tag{4.30}$$

which leads to $R(\bar{U}) \le 1/r^{l+\alpha}$.

In addition, since the Bernoulli polynomials are eigenfunctions of U (Theorem 1), we have

$$\begin{aligned}\bar{U}\rho(x) =& \hat{U}\hat{B}\rho(x) = U\hat{B}\rho(x) - \int_0^1 dx'\rho(x') \\ =& U\rho(x) - \sum_{j=1}^{l} \frac{\rho^{(j-1)}(1) - \rho^{(j-1)}(0)}{j!\ r^j} B_j(x) - \int_0^1 dx'\rho(x') ,\end{aligned} \tag{4.31}$$

which is the decomposition (4.6). *QED*

Remark.1 In the proof of i), the periodic boundary condition on $C_p^{(l,\alpha)}$ is used to show (4.15) and to estimate $\|(\hat{U}^m\rho)^{(t)}\|_\infty$ $(0 \le t \le l-1)$ through the Fourier expansion.

Remark.2 To make the estimate of $\|(\hat{U}^m\rho)^{(l-1)}\|_\infty$ (cf.(4.23)), the condition $\alpha > 0$ is used. However, the above results can be easily extended to C^l. In this case,

$\|(\bar{U}^m\rho)^{(t)}\|_\infty$ is estimated for $0 \le t \le l-2$ through the Fourier expansion and for $t = l, l-1$ like i-a) and i-b).

Remark.3 This theorem is consistent with the results of Ruelle (1989). In his work, resonances are associated with the point spectra in the decomposition like (4.6). Thus, the eigenvalues for Bernoulli's polynomials precisely correspond to Ruelle's resonances. Through the present analysis, the appearance of Ruelle's resonances can be explained qualitatively as follows: Smoother eigenfunctions correspond to eigenvalues smaller in magnitude. Thus, by increasing the smoothness of functions in the domain, large eigenvalues corresponding to rough eigenfunctions are excluded. However, special linear combinations of such rough eigenfunctions may be smooth and remain as eigenfunctions corresponding to isolated eigenvalues, which are nothing but Ruelle's resonances.

Remark.4 By identifying 0 and 1, the interval [0,1] can be regarded as a circle, where r-adic transformation (2.1) is C^∞. In this view, it is natural to consider the problem on the function space $C_p^{(l,\alpha)}$, where we have no Ruelle's resonances. To have them, we should use functions with discontinuity at $x = 0$. Thus the appearance of Ruelle's resonances depends not only on the smoothness of functions, but also on their boundary properties.

5. Generalized Spectral Decomposition of U

According to Theorem 3, by increasing the smoothness of functions in the domain, we may characterize U only through its isolated point spectra and have an expansion in terms of the Bernoulli polynomials (cf.(4.6)). At first sight, we can choose infinitely differentiable functions or analytic functions as an appropriate domain. But, it is not the case. For example, the image $U\exp(4\pi i x) = \exp(2\pi i x)$ of an analytic function $\exp(4\pi i x)$ cannot be expanded in terms of the Bernoulli polynomials. Indeed, from the generating function (2.3), we should have

$$Ue^{4\pi i x} = e^{2\pi i x} = \frac{e^{2\pi i}-1}{2\pi i}\sum_{n=0}^{\infty}\frac{B_n(x)}{n!}(2\pi i)^n = 0 \times \sum_{n=0}^{\infty}\frac{B_n(x)}{n!}(2\pi i)^n\,. \tag{5.1}$$

Thus, the sum is ill-defined. Here we shall choose a space of all polynomials as a domain and introduce a rigged Hilbert space where U admits a generalized spectral decomposition in the sense of Gelfand (Gelfand and Vilenkin,1964, Gelfand and Shilov,1967) and Maurin (1968).

Let $\mathcal{D}$ be a space of all polynomials equipped with the inductive topology of LF-space (Treves, 1967, p.130). In this topology, the convergence of a sequence $\{p_n(x)\}_{n=0}^{\infty}$ is defined as

$$p_n(x) \to p(x) \quad (n \to \infty)\,,\ \text{in } \mathcal{D} \iff \begin{cases} \exists M > 0, \text{s.t.} \forall n,\ (\text{degree of } p_n(x)) \le M \\ \lim_{n\to\infty} \|p_n - p\|_\infty = 0\,. \end{cases}$$

The space $\mathcal{D}$ has the following properties.

THEOREM 4

i) $\mathcal{D}$ is a nuclear LF-space and thus, complete and barreled.

ii) $\mathcal{D}$ is dense in L^2.

iii) $\mathcal{D}$ is invariant under U and U is continuous in $\mathcal{D}$.

Proof: i) in Treves (1967), p.130 and p.526.
ii) e.g., in Hardy and Rogosinski (1968), p.21.
iii) It follows from the facts that U maps polynomials of degree n to polynomials degree n and that U is bounded. *QED*

By considering the topological dual $\mathcal{D}^\dagger$ of $\mathcal{D}$, we have the Gelfand triplet or rigged Hilbert space: $\mathcal{D} \subset L^2 \subset \mathcal{D}^\dagger$. From Theorem 4. iii), the adjoint $U^\dagger$ can be continuously extended to $\mathcal{D}^\dagger$. This is therefore an appropriate rigged Hilbert space to give meaning to the spectral decomposition of U.

THEOREM 5

i) Define $\tilde{B}_n \in \mathcal{D}^\dagger$ as

$$\langle \tilde{B}_n, \rho \rangle \equiv \rho^{(n-1)}(1) - \rho^{(n-1)}(0) \qquad (\forall \rho(x) \in \mathcal{D}) . \tag{5.2}$$

Then, $\tilde{B}_n$ is a generalized eigenfunction of $U^\dagger$ (i.e., the left eigenfunction of U, see Parravicini, Gorini and Sudarshan, 1980) with eigenvalue $1/r^n$:

$$U^\dagger \tilde{B}_n = \frac{1}{r^n} \tilde{B}_n . \tag{5.3}$$

ii) The Bernoulli polynomials $\{B_n(x)/n!\} \subset \mathcal{D}$ and $\{\tilde{B}_n\} \subset \mathcal{D}^\dagger$ form a biorthonormal system

$$\langle \tilde{B}_n, \frac{B_m}{m!} \rangle = \delta_{nm} . \tag{5.4}$$

iii) For any $\rho(x) \in \mathcal{D}$, the completeness relation

$$\rho(x) = \int_0^1 dx' \rho(x') + \sum_{n=1}^{\infty} \frac{B_n(x)}{n!} \langle \tilde{B}_n, \rho \rangle , \tag{5.5}$$

is fulfilled.

iv) For any $\rho(x) \in \mathcal{D}$, U admits a generalized spectral decomposition

$$U\rho(x) = \int_0^1 dx' \rho(x') + \sum_{n=1}^{\infty} \frac{B_n(x)}{n!} \frac{1}{r^n} \langle \tilde{B}_n, \rho \rangle , \tag{5.6}$$

Proof: i) As the differentiation of any order is a continuous operation on $\mathcal{D}$, $\tilde{B}_n$ is well-defined as an element of $\mathcal{D}^\dagger$. Then, since

$$\begin{aligned}(U\rho)^{(t)}(x) &= \left(\frac{d}{dx}\right)^t \frac{1}{r}\sum_{j=0}^{r-1} \rho\left(\frac{x+j}{r}\right) \\ &= \frac{1}{r^{t+1}} \sum_{j=0}^{r-1} \rho^{(t)}\left(\frac{x+j}{r}\right) = \frac{1}{r^t}\, U\rho^{(t)}(x) \ ,\end{aligned} \tag{5.7}$$

we have for any $\rho(x) \in \mathcal{D}$

$$\begin{aligned}\langle U^\dagger \tilde{B}_n, \rho\rangle &= \langle \tilde{B}_n, U\rho\rangle = (U\rho)^{(n-1)}(1) - (U\rho)^{(n-1)}(0) \\ &= \frac{1}{r^{n-1}}\{U\rho^{(n-1)}(1) - U\rho^{(n-1)}(0)\} \\ &= \frac{1}{r^n}\{\rho^{(n-1)}(1) - \rho^{(n-1)}(0)\} = \frac{1}{r^n}\langle \tilde{B}_n, \rho\rangle \ .\end{aligned} \tag{5.8}$$

ii) It follows immediately from Proposition 1.
iii) Let $\rho(x) \in \mathcal{D}$ be a polynomial of degree m. Then it can be uniquely expressed as a superposition of the Bernoulli polynomials with degree not greater than m

$$\rho(x) = \sum_{n=0}^{m} a_n \frac{B_n(x)}{n!} \ . \quad (a_n \in \mathbf{C}) \tag{5.9}$$

Then from Proposition 1, we obtain

$$\begin{aligned}&\int_0^1 dx'\rho(x') = \sum_{n=0}^{m} \frac{a_n}{(n+1)!}\{B_{n+1}(1) - B_{n+1}(0)\} = a_0 \ , \\ &\langle \tilde{B}_n, \rho\rangle = \sum_{j=n}^{m} \frac{a_j}{(j-n+1)!}\{B_{j-n+1}(1) - B_{j-n+1}(0)\} = a_n \ , \ (1 \le n \le m) \\ &\langle \tilde{B}_n, \rho\rangle = 0 \ , \quad (n > m)\end{aligned} \tag{5.10}$$

and thus (5.5).
iv) The decomposition (5.6) follows from (5.5) and Theorem 1. *QED*

Remark. 1 Hasegawa and Saphir (1992b) use polynomials as test functions, but their argument is incomplete as they do not consider the topological aspects.

Remark. 2 Gaspard (1992) proposes the use of a Fréchet subspace of analytic functions as a test function space. The above results can be easily extended to this case.

Remark. 3 For any complex valued function $f(z)$ analytic in a domain including the unit disk $|z| \le 1$, an operator function $f(U)$ is well-defined and admits a generalized spectral decomposition

$$f(U)\rho(x) = f(1)\int_0^1 dx'\rho(x') + \sum_{n=1}^{\infty} \frac{B_n(x)}{n!} f\left(\frac{1}{r^n}\right)\langle \tilde{B}_n, \rho\rangle \ . \tag{5.11}$$

Remark.4 As shown in section 3, the adjoint $U^\dagger$ acts on a subspace of L^2 as a unilateral shift. It is well known that the unilateral shift does not admit non-trivial invariant subspaces (Dowson, 1978, Example 1.38), i.e., it cannot have eigenfunctions. As shown in Theorem 5. i), the adjoint $U^\dagger$ does have eigenfunctions in the extended space $\mathcal{D}^\dagger$. Therefore, in rigged Hilbert spaces, operators may have completely different properties from those in the original Hilbert spaces. It is this feature of rigged Hilbert spaces which allows generators of motion to have complex eigenvalues in the "complex spectral theory" of Prigogine et al..

Acknowledgement

We thank Prof. I. Prigogine for his encouragements and fruitful discussions during this work. One of us (ST) thanks Prof. G. Lumer and organizers of "3rd International Workshop-Conference on Evolution Equations, Control Theory and Biomathematics" for their invitation and hospitality. We are grateful to Prof. L.P. Horwitz, Dr. P. Gaspard, Dr. H. Hasegawa and Mr. W. Saphir for discussions and to Dr. Z. Suchanecki for discussions and useful comments. We also acknowledge the Belgian Government (under the contract "Pole d'attraction interuniversitaire"), the European Communities Commission (contract n° PSS*0143/B), the U.S. Department of Energy, Grant N° FG05-88ER13897, and the Robert A. Welch Foundation for financial support of this work.

REFERENCES

1. Antoniou, I. and Prigogine, I. (1992). Dynamics and Intrinsic Irreversibility, *Nuovo Cimento*, in press.
2. Bohm, A. and Gadella, M. (1989). Dirac kets, Gamow vectors and Gelfand triplets, *Springer Lect. Notes on Physics, 348*, Springer, Berlin.
3. Dowson, H.R. (1978). *Spectral Theory of Linear Operators*, Academic Press, London.
4. Gaspard, P. (1992). *r*-adic one-dimensional maps and Euler summation formula, to appear in *J. of Physics A, Mathematical and General.*
5. Gelfand, I. and Vilenkin, N. (1964). *Generalized Functions, Vol.4, Applications to Harmonic Analysis*, Academic Press, New York.
6. Gelfand, I. and Shilov, G. (1967). *Generalized Functions, Vol.3, Theory of Differential Equations*, Academic Press, New York.
7. Hardy, G.H. and Rogosinski, W.W. (1968). *Fourier Series*, Cambridge University Press, Cambridge.
8. Hasegawa, H., Petrosky, T., Prigogine, I. and Tasaki, S. (1991). Quantum Mechanics and the Direction of Time, *Found. Phys. 21*, 263.

9. Hasegawa, H. and Saphir, W.C. (1992a). Decaying Eigenstates for Simple Chaotic Systems, *Physics Letters A161*, 471.
10. Hasegawa, H. and Saphir, W.C. (1992b). Unitarity and Irreversibility in Chaotic Systems, preprint, University of Texas.
11. Hata, M. (1985). On the functional equation $\frac{1}{r}\{f(\frac{x}{p}) + \cdots + f(\frac{x+p-1}{p})\} = \lambda f(\mu x)$, *J. Math. Kyoto Univ. 25-2*, 357.
12. Lasota, A. and Mackey, M.C. (1985). *Probabilistic properties of deterministic systems*, Cambridge University Press, Cambridge.
13. Maurin, K. (1968). *General Eigenfunction Expansions and Unitary Representations of Topological Groups*, Polish Sci. Publ., Warsaw.
14. Olver, F.W.J. (1974). *Asymptotics and Special Functions*, Academic Press, New York.
15. Parravicini, G., Gorini, V. and Sudarshan, E.C.G. (1980). Resonances, scattering theory and rigged Hilbert spaces, *J. Math. Phys. 21*, 2208.
16. Parry, W. (1981). *Topics in Ergodic Theory*, Cambridge University Press, Cambridge.
17. Petrosky, T., Prigogine, I. and Tasaki, S. (1991). Quantum Theory of Non-integrable Systems, *Physica, A173*, 175.
18. Petrosky, T. and Prigogine, I. (1991). Alternative Formulation of Classical and Quantum Dynamics for Non-integrable Systems, *Physica, A175*, 146.
19. Pollicott, M. (1990). The differential zeta function for Axiom A attractors, *Ann. Math. 131*, 331.
20. Prigogine, I., Petrosky, T., Hasegawa, H., and Tasaki, S. (1991). Integration of Non-integrable Systems, *Solitons and Chaos* (I. Antoniou and F. Lambert, eds.), Springer, Berlin, p.3.
21. Reed, M. and Simon, B (1980). *Functional Analysis*, Academic Press, New York.
22. Ruelle, D. (1986). Resonances of Chaotic Dynamical Systems, *Phys. Rev. Lett. 56*, 405.
23. Ruelle, D. (1987a). Resonances for Axiom A Flows, *J. Diff. Geom. 25*, 99.
24. Ruelle, D. (1987b). One-Dimensional Gibbs States and Axiom A Diffeomorphisms, *J. Diff. Geom. 25*, 117.
25. Ruelle, D. (1989). The Thermodynamic Formalism for Expanding Maps, *Commun. Math. Phys. 125*, 239.
26. Sz-Nagy, B. and Foias, C. (1970). *Harmonic Analysis of Operators in Hilbert Space*, North-Holland, Amsterdam.
27. Tangerman, F. (1986). Meromorphic continuation of Ruelle zeta functions, Boston University thesis (unpublished).
28. Treves, F. (1967). *Topological Vector Spaces, Distributions and Kernels*, Academic Press, New York.

Can We Understand What Life Is?

RENÉ THOM Institut des Hautes Etudes Scientifiques, Bures-sur-Yvette, France

Under this form, in a general sense, the question obviously calls for an affirmative answer. Some processes of living matter appear immediately intelligible to our naïve (prescientific, non-instrumental) observation. (Predation is one example : "The cat eats the mouse"). Certain processes of inanimate matter, on the other hand, may seem mysterious to us (the odd shape of a rock sculptured by erosion, for instance). The usual scientific attitude consists in rejecting immediate intelligibility as something of no import and substituting for it a mediate, or constructed, intelligibility. The now prevalent doctrine of reductionism proposes the following approach : we use analytical instruments that perform better than our senses, and, with the finer elements so observed, reconstruct by means of a mechanical or physical model the biological facts studied. But when we talk about "mechanism", a certain ambiguity is perceived. It seems reasonable to say that some mechanical processes are immediately intelligible at our level, as, for example, the collision of two bowls during a game. Now it would be difficult to deny that this immediate intelligibility comes from the sensori-motor sensitivity regulating the way we act on surrounding bodies. Thus even this mechanical evidence is of biological origin, so that any biological theorization based thereon will incur the "philosophical" reproach of circularity. In fact the reference to mechanism appeared in the history of Biology with Descartes' theory of the machine-animal. The theory itself grew from anatomical data allowing an organ to be assimilated to an instrument : the heart to a pump driving liquid blood through its vessels, the crystalline part of the eye to a lens, the movement of a joint to that of a rope on a pulley, the lungs to a pair of bellows, and so on. Today one likes to think of the central nervous system as a gigantic computer. These explanations work, they satisfy our intuition (except perhaps for the last one). But for someone with a philosophical turn of mind, they appear more like *explananda* than true explanations. For when we consider the temporal cycle of ontogenesis, we raise the problem of the embryology of these structures with their finalist vocation. Towards 1880, the German embryologist, W. Roux, had tried in his *Entwicklungs-Mechanik* to account for tissue formation by mechanical interactions of this type (thus, according to him, the circumvolutions of the encephalus were due to compression of growing tissue in the brain-pan). However if we accept the validity of this type of explanation, we are clearly just shifting the problem. Whilst explanations of local processes are possible, their spatio-temporal aggregation raises an enormous quandary in Embryology.

A specific correspondence,a "coding", would also be required between the location of genes and the corresponding localization in the organism, something which has not been debated up to now. The same local event in the body would affect differently the nuclei of the different cells according to their position, giving rise to a *pot pourri* of all kinds of physiological reactions.

Modern Biology, of course, believes it has the answer : "It's all in the genes". But this reply is only acceptable if we think of a local physico-chemical determinism that prescribes the mechanical, physical and biochemical behaviour of each living cell at each moment. Now our contemporaries do not think much of determinism. "Prigoginian fluctuation" and the "chaos" rediscovered by dynamicians, throw some doubt on the foreseeable nature of phenomena within the cell. Yet in many cases an organism's survival depends on rigorous determinism (particularly in higher organisms). And so we are faced with a dilemma : either we postulate determinism in spite of theoretical considerations tending to point the other way, or we make do with a "qualitative" determinism concerned with what is "most likely" to happen (defined by Aristotle as *natural* movements (*physikè matabolè*) as opposed to *accidental* movements to be thrown out of the *épistèmé*.)

We have to be convinced of one thing : the fact that a regulation mechanism "nearly always" succeeds in no way implies that the mechanism is subject to strict quantitative determinism as found in the laws of classical physics. From this point of view, sufficient attention has not been drawn to the importance of theorems establishing the structural stability of gradient or gradient-like dynamical systems. A structure of this type is sufficient to define global behaviour that will "almost always" occur, the bifuractions of which can be controlled by manipulating a finite number of parameters. This is the framework of "elementary" catastrophe theory, which I persist in believing to be essential for the modelling of the processes of "homeostasis" and embryology. There are difficulties, of course, due to the fact that the presence of one or of a few *perverse* molecules can trigger such bifurcations, linked with the often sharp variations of the enzymatic activity of certain enzymes. (An organism can be killed by the introduction of a single viral molecule.) I think that we should attempt to define the biochemical values having a space-time signification (morphogenetic gradients) whose existence seems *a priori* necessary for the building of organic structures. The *perversity* of a molecule will then be defined by its capacity of perturbing the regulating mechanisms at work in this parameter space. Another important theoretical problem concerns the role of non-local physical agents in biophysical and biochemical behaviour. (In man, speech and sight play a considerable part in behaviour determination.) The existence of these agents underlines the shortcomings of thermodynamic vision, which talks about a state of equilibrium without *ever* being able to say how long it takes to get there. We still know very little about the morphology of phase changing during breaks in the metastability of a state (even in the inanimate world) ; and the translocal effects related to changes in the structure of macromolecular systems can scarcely be described by any of the formalisms known to science today. All this goes to show that a complete formalization of metabolism is likely to remain in the shape of a project for many years to come, and that there is at least something to be said for an approximate description of a qualitative nature, in the sense of Henri Poincaré's Qualitative Dynamics, accounting for general phenomena (regulation and embryology) in normal conditions, but *taking the effects of metastability into account.*

In this respect one may wonder whether any intellection of life is not bound to be forever incomplete. The representation of life which we might have achieved, as a mental act (C), would necessarily be part (connected with our brain) of a global metabolism. If this representation were true to life, that would imply that the set (E) of the states of living matter which I constitute possesses the property of having a representation R of E on the subset $C = R(E)$. This would mean that E would have to be infinite, thus excluding any complete finite modelling of life. In spite of its highly theoretical character, this reasoning nonetheless shows the essential difficulty that there is in seeking to confer on Biology the theoretical aim of exhaustive description. This may explain, though it cannot excuse, why Biologists are incapable of equipping themselves with a theoretical Biology. For in limiting oneself to the description of metabolical processes that are *true in general*, and in using models *of a generic character* (stable and of minimal complexity), one could go quite far in the description and understanding of organogenesis phenomena, which, for the moment, seem to an open mind to belong to the world of miracles.

Reference

THOM René, *Esquisse d'une Sémiophysique*, InterÉditions, Paris 1989
Traduction anglaise : *Semiophysics: a Sketch,* Addison-Wesley, Redwood City, CA, 1990

On a Problem from Thermal Convection

WOLF von WAHL University of Bayreuth, Bayreuth, Germany

Let us consider an infinite layer $\mathbb{R}^2 \times (-\frac{1}{2}, \frac{1}{2})$ heated from below together with the Boussinesq-equations

$$(1) \qquad \begin{cases} \underline{u}' - Pr\Delta\underline{u} + \underline{u}\cdot\nabla\underline{u} - Pr\vartheta\underline{k} + Pr\nabla\pi &= 0, \\ \nabla\cdot\underline{u} &= 0, \\ \vartheta' - \Delta\vartheta - Ru_z + \underline{u}\cdot\nabla\vartheta &= 0. \end{cases}$$

x, y refer to the variables in $\mathbb{R}^2$, z varies in $[-\frac{1}{2}, \frac{1}{2}]$. $\underline{k}$ is the third unit-vector, f_x, f_y, f_z refer to the first, second, and third component of a vector-field f. Pr is the Prandtl-number, R is the Rayleigh-number. $\underline{u}, \pi$ are the velocity field and the pressure of the fluid, ϑ is the fluctuation of the temperature from a static state. The temperature is kept fixed at $z = -\frac{1}{2}$ and $z = +\frac{1}{2}$ with values T_0, T_1, $T_0 > T_1$. Beside initial values for $\underline{u}, \vartheta$ at time $t = 0$ we prescribe boundary conditions at $z = \pm\frac{1}{2}$. These are the usual ones, namely

Stress-free boundaries: $u_z = \partial_z u_x = \partial_z u_y = 0$, $\vartheta = 0$ at $z = \pm\frac{1}{2}$.

Rigid boundaries: $\underline{u} = 0$, $\vartheta = 0$ at $z = \pm\frac{1}{2}$. We are interested in solutions which are periodic in the x, y plane with respect to a rectangle $\mathcal{P} = (-\frac{\pi}{\alpha}, \frac{\pi}{\alpha}) \times (-\frac{\pi}{\beta}, \frac{\pi}{\beta})$ with wave-numbers α, β in x- and y-direction. Moreover we are concerned here with the case $Pr = +\infty$.

Any solenoidal field $\underline{u}$ in the layer $\mathbb{R}^2 \times (-\frac{1}{2}, \frac{1}{2})$, which has the periodicity prescribed above, can be decomposed in the following way:

$$(2) \qquad \underline{u} = \operatorname{curl}\operatorname{curl}\varphi\underline{k} + \operatorname{curl}\psi\underline{k} + F$$

with periodic functions φ, ψ and a vector field F which depends on z only and has constant third component. $P = \operatorname{curl}\operatorname{curl}\varphi\underline{k}$ is called the poloidal part of $\underline{u}$, $\operatorname{curl}\psi\underline{k}$ is the toroidal part and F is the so called mean flow. Then (1) can be transformed into an equivalent form in which the pressure is eliminated and the nonlinearity is almost local. Cf. for this [5]. Let us take a strong solution of the new system which exists locally in time (Pr is still finite). $|\vartheta|$ can be bounded independently of time t and Pr as we will prove later. Following the proofs in [7,8] for the existence of a local (in time) strong solution of this new system it is not too difficult to verify the following assertions: If $Pr \to +\infty$, then the length of the maximal interval of existence tends to $+\infty$. On any finite time interval we obtain the system which formally arises from the poloidal-toroidal-mean-flow system corresponding to (1) if we divide by Pr and then set $Pr = +\infty$. The main step consists in proving that for Pr large

$$\Phi_{Pr} = (\varphi_{Pr}, \psi_{Pr}, F_{Pr}, \vartheta_{Pr})$$

stays bounded on any time interval in appropriate norms, as well as $(\frac{1}{\sqrt{Pr}}\Phi_{Pr})' = \frac{1}{\sqrt{Pr}}\Phi'_{Pr} = \frac{1}{\sqrt{Pr}}(\varphi'_{Pr}, \psi'_{Pr}, F'_{Pr}, \vartheta'_{Pr})$ whence it follows that $\frac{1}{\sqrt{Pr}}\Phi'_{Pr}$ tends to 0 as $Pr \to \infty$. The regularity properties of the strong solution are preserved. In the case $Pr = +\infty$ there is moreover neither a toroidal field nor a mean flow and the system for φ, ϑ reads

$$(3) \qquad \begin{cases} \Delta^2(-\Delta_2)\varphi &= (-\Delta_2)\vartheta, \text{ [1]} \\ \vartheta' - \Delta\vartheta + \underline{\delta}\varphi \cdot \nabla\vartheta &= R(-\Delta_2)\varphi, \\ \vartheta(0) &= \vartheta_0. \end{cases}$$

The boundary conditions for φ, ϑ are as follows:

$$\varphi = \partial^2_{zz}\varphi = 0 = \vartheta \text{ at } z = \pm\frac{1}{2} \text{ for stress-free boundaries,}$$

$$\varphi = \partial_z\varphi = 0 = \vartheta \text{ at } z = \pm\frac{1}{2} \text{ for rigid boundaries.}$$

$\underline{\delta}.$ is simply $\operatorname{curl}\operatorname{curl}.\underline{k}$, i.e. the vectorial operator $\begin{pmatrix} \partial^2_{xz} \\ \partial^2_{yz} \\ -\Delta_2 \end{pmatrix}$ with $\Delta_2 = \partial^2_{xx} + \partial^2_{yy}$.

Due to [6, p. 132] the temperature T in (1) is bounded by T_0. Pr is finite now. From the point of view of Physics this is more or less self-understanding. For ϑ in (1) we have

$$\vartheta = \frac{R}{T_0 - T_1}[T - (\frac{T_1 + T_0}{2} - (T_0 - T_1)z)]$$

in dimensionless writing. Set $\Omega = \mathcal{P} \times (-\frac{1}{2}, \frac{1}{2})$. Thus we obtain an $L^\infty(\Omega)$-bound for $|\vartheta|$. The bound is of the form $c \cdot R = c(R)$ and does neither depend on Pr nor on t. Thus it is kept if $Pr \to \infty$. The dimensionless system (1) in its poloidal-, toroidal-, mean-flow form is particularly suitable if one lets Pr tend to $+\infty$. Let us mention that there is another dimensionless form of (1) where Pr only appears in the row for ϑ. This particular system can be used to let Pr tend to 0, which is not our concern here. Due to their definition Pr and R are always coupled. It is thus a mathematical idealization

[1]The solution is $\varphi = (\Delta^2)^{-1}[\vartheta - (1/|\mathcal{P}|)\int_{\mathcal{P}} \vartheta\, dx\, dy]$. Thereby we correct line 5 in [7, p. 273]. This correction has however no influence on the remaining part of the proof in [7, p. 273] since $\underline{\delta}\varphi = \underline{\delta}(\Delta^2)^{-1}\vartheta$.

to consider Pr and R as independent parameters. In particular we have shown that the limit-process $Pr \to +\infty$ mathematically leads to (3). This problem makes sense, its unique global strong solution can be estimated uniformly with respect to time as we will show. As for (3) our bound on $\|\vartheta(t)\|_{L^\infty(\Omega)}$ implies

$$\|\triangle^2\varphi(t)\| \le c$$

$$\begin{aligned}
&\frac{d}{dt}\|\nabla\vartheta\|^2(t) + 2\|\triangle\vartheta(t)\|^2 \le \\
&\le c\|\nabla\vartheta(t)\|\|\triangle\vartheta(t)\| + c \\
&\le \varepsilon\|\triangle\vartheta(t)\|^2 + c(\varepsilon)\|\vartheta(t)\|\|\triangle\vartheta(t)\| + c \\
&\le 2\varepsilon\|\triangle\vartheta(t)\|^2 + c(\varepsilon),
\end{aligned}$$

$$\|\nabla\vartheta(t)\|^2 \le c, \ c = c(R),$$

where the norm $\|.\|$ refers to the $L^2(\Omega)$-norm. Differentiating the system (3) with respect to t we obtain

$$\|\triangle^2\varphi'(t)\|^2 \le 2\|\vartheta'(t)\|^2,$$

(4)
$$\begin{aligned}
&\frac{d}{dt}\|\vartheta'\|^2(t) + 2\|\nabla\vartheta'(t)\|^2 \le \\
&\le c\|\underline{\delta}\varphi'(t)\|_{L^\infty(\Omega)} \cdot \|\nabla\vartheta(t)\|\|\vartheta'(t)\| + c\|(-\triangle_2)\varphi'(t)\|\|\vartheta'(t)\|.
\end{aligned}$$

Also from (3) we obtain

$$\begin{aligned}
(-\triangle)^{-\frac{1}{2}}\vartheta' &= -(-\triangle)^{\frac{1}{2}}\vartheta - (-\triangle)^{-\frac{1}{2}}(\underline{\delta}\varphi\cdot\nabla\vartheta) + R(-\triangle)^{-\frac{1}{2}}(-\triangle_2)\varphi \\
\|(-\triangle)^{-\frac{1}{2}}\vartheta'\| &\le c\|\nabla\vartheta\| + c.
\end{aligned}$$

Since $\|\vartheta'\|^2 \le \|\nabla\vartheta'\|\|(-\triangle)^{-\frac{1}{2}}\vartheta'\|$, the estimate (4) implies

$$\begin{aligned}
\frac{d}{dt}\|\vartheta'\|^2(t) + 2\|\nabla\vartheta'(t)\|^2 &\le c\|\nabla\vartheta'(t)\|^{\frac{1}{2}} \cdot \|\vartheta'(t)\| \\
&\le c\|\nabla\vartheta'(t)\|^{\frac{3}{2}}.
\end{aligned}$$

This gives a bound $\|\vartheta'(t)\|^2 \le c$ and consequently $\|\triangle\vartheta(t)\| \le c$ with $c = c(R)$. Observe that (3) has a global (in time) regular solution with the property $\int_{\mathcal{P}} \varphi(t,x,y,z)\,dz = 0$, $z \in (-\frac{1}{2},\frac{1}{2})$, $t \ge 0$. This was shown in [7, ch. V].

What was not shown in [7] is that (3) admits uniform (in t) bounds on higher order norms of φ and ϑ. It's proved here. We thus have the possibility to study possible attractors and their dimension. The problem is now to estimate the dimension of the attractor from below and above in terms of R. In particular it would be interesting to know if non regular oscillations occurr.

The question for an attractor containing non regular oscillations has to be understood as the question for a chaotic attractor. This matter has been discussed numerically for two dimensional solutions in [4,3].

In a somewhat different form the present problem with $Pr = +\infty$ appears when convection in the mantle of the earth is considered. Then the layer is spherical; cf. [2]. It's known that any solenoidal field in a spherical layer, which is free of productivity, admits a decomposition into a poloidal and a toroidal field (cf. [1]). Thus we arrive at a system similar to (3).

References

[1] Backus, G.: Poloidal and Toroidal Fields in Geomagnetic Field Modelling. Rev. Geophys. 24, 75-109(1986).

[2] Houseman, G.: The dependence of convection planform on mode of heating. Nature 332, 346-349(1988).

[3] Stewart, C. A.: The Route to Chaos in Thermal Convection at Infinite Prandtl Number II: Formation and Breakup of the Torus. Preprint. Institut Non Lineaire de Nice. UMR CNRS 129. INLN 91.20(1991).

[4] Stewart, C. A., Turcotte, D. L.: The Route to Chaos in Thermal Convection at Infinite Prandtl Number 1. Some Trajectories and Bifurcations. Journal of Geophysical Research 94, B10, 13707-13717(1989).

[5] Schmitt, B., Wahl, Wolf von: Decomposition of Solenoidal Fields into Poloidal Fields, Toroidal Fields and the Mean Flow. Applications to the Boussinesq-Equations. The Navier-Stokes Equations II - Theory and Numerical Methods. Proceedings, Oberwolfach 1991. J. G. Heywood, K. Masuda, R. Rautmann, S. A. Solonnikov (eds.). Lecture Notes in Mathematics 1530, 291-305. Springer: Berlin, Heidelberg, New York (1992).

[6] Temam, R.: Infinite-Dimensional Dynamical Systems in Mechanics and Physics. Applied Mathematical Sciences 68. Springer: Berlin, Heidelberg, New York (1988).

[7] Wahl, W. von: The Boussinesq-Equations in Terms of Poloidal and Toroidal Fields and the Mean Flow. Lecture Notes. Bayreuther Math. Schriften 40, 203-290(1992).

[8] Wahl, W. von: The Boussinesq-Equations in Terms of Poloidal and Toroidal Fields and the Mean Flow. Lecture Notes. Improved and Corrected Version. Preprint, Univ. of Bayreuth (1992).